MITSUBISHI

ECLIPSE
1990-93 REPAIR MANUAL

President — Dean F. Morgantini, S.A.E.
Vice President–Finance — Barry L. Beck
Vice President–Sales — Glenn D. Potere

Executive Editor — Kevin M. G. Maher
Production Manager — Ben Greisler, S.A.E.
Project Managers — Michael Abraham, George B. Heinrich III, S.A.E.
Will Kessler, A.S.E., S.A.E., Richard Schwartz

Schematics Editor — Christopher G. Ritchie

CHILTON™ Automotive Books
PUBLISHED BY **W. G. NICHOLS, INC.**

Manufactured in USA
© 1993 Chilton Book Company
1020 Andrew Drive
West Chester, PA 19380
ISBN 0-8019-8415-7
Library of Congress Catalog Card No. 92-05494
6789012345 6543210987

Contents

Contents

SAFETY NOTICE

Proper service and repair procedures are vital to the safe, reliable operation of all motor vehicles, as well as the personal safety of those performing repairs. This manual outlines procedures for servicing and repairing vehicles using safe, effective methods. The procedures contain many NOTES, CAUTIONS and WARNINGS which should be followed along with standard procedures to eliminate the possibility of personal injury or improper service which could damage the vehicle or compromise its safety.

It is important to note that the repair procedures and techniques, tools and parts for servicing motor vehicles, as well as the skill and experience of the individual performing the work vary widely. It is not possible to anticipate all of the conceivable ways or conditions under which vehicles may be serviced, or to provide cautions as to all of the possible hazards that may result. Standard and accepted safety precautions and equipment should be used when handling toxic or flammable fluids, and safety goggles or other protection should be used during cutting, grinding, chiseling, prying, or any other process that can cause material removal or projectiles.

Some procedures require the use of tools specially designed for a specific purpose. Before substituting another tool or procedure, you must be completely satisfied that neither your personal safety, nor the performance of the vehicle will be endangered.

Although information in this manual is based on industry sources and is complete as possible at the time of publication, the possibility exists that some vehicle manufacturers made later changes which could not be included here. While striving for total accuracy, W. G. Nichols, Inc. cannot assume responsibility for any errors, changes or omissions that may occur in the compilation of this data.

PART NUMBERS

Part numbers listed in this reference are not recommendations by Chilton for any product by brand name. They are references that can be used with interchange manuals and aftermarket supplier catalogs to locate each brand supplier's discrete part number.

SPECIAL TOOLS

Special tools are recommended by the vehicle manufacturer to perform their specific job. Use has been kept to a minimum, but where absolutely necessary, they are referred to in the text by the part number of the tool manufacturer. These tools can be purchased, under the appropriate part number, from your local dealer or regional distributor, or an equivalent tool can be purchased locally from a tool supplier or parts outlet. Before substituting any tool for the one recommended, read the SAFETY NOTICE at the top of this page.

ACKNOWLEDGMENTS

W. G. Nichols, Inc. expresses appreciation to Mitsubishi Motor Sales of America for their generous assistance.

1

GENERAL INFORMATION AND MAINTENANCE

HOW TO USE THIS BOOK

Chilton's Total Car Care Manual for the 1990-93 Mitsubishi Eclipse, Plymouth Laser and Eagle Talon is intended to help you learn more about the inner workings of your vehicle and save you money on its maintenance and repairs.

The first two sections will be the most used, since they contain maintenance and tune-up information and procedures. Studies have shown that a properly tuned and maintained car can get at least 10% better gas mileage than an out-of-tune car. The other sections deal with the more complex systems of your car. Operating systems from engine through brakes are covered to the extent that the average do-it-yourselfer becomes mechanically involved. This book will not explain such things as rebuilding the differential for the simple reason that the expertise required and the investment in special tools make this task uneconomical. It will give you detailed instructions to help you change your own brake pads and shoes, replace spark plugs, and do many more jobs that will save you money, give you personal satisfaction, and help you avoid expensive problems.

A secondary purpose of this book is a reference for owners who want to understand their car and/or their mechanics better. In this case, no tools at all are required.

Before removing any bolts, read through the entire procedure. This will give you the overall view of what tools and supplies will be required. There is nothing more frustrating than having to walk to the bus stop on Monday morning because you were short one bolt on Sunday afternoon. So read ahead and plan ahead. Each operation should be approached logically and all procedures thoroughly understood before attempting any work.

All sections contain adjustments, maintenance, removal and installation procedures, and repair or overhaul procedures. When repair is not considered practical, we tell you how to remove the part and then how to install the new or rebuilt replacement. In this way, you at least save the labor costs. Backyard repair of such components as the alternator is just not practical.

Two basic mechanic's rules should be mentioned here. One, whenever the left side of the car or engine is referred to, it is meant to specify the driver's side of the car. Conversely, the right side of the car means the passenger's side. Secondly, most screws and bolt are removed by turning counterclockwise, and tightened by turning clockwise.

Safety is always the most important rule. Constantly be aware of the dangers involved in working on an automobile and take the proper precautions. (See the section in this section Servicing Your Vehicle Safely and the SAFETY NOTICE on the acknowledgment page.)

Pay attention to the instructions provided. There are 3 common mistakes in mechanical work:

1. Incorrect order of assembly, disassembly or adjustment. When taking something apart or putting it together, doing things in the wrong order usually just costs you extra time; however, it CAN break something. Read the entire procedure before beginning disassembly. Do everything in the order in which the instructions say you should do it, even if you can't immediately see a reason for it. When you're taking apart something that is very intricate (for example, a carburetor), you might want to draw a picture of how it looks when assembled at one point in order to make sure you get everything back in its proper position. (We will supply exploded views whenever possible). When making adjustments, especially tune-up adjustments, do them in order; often, one adjustment affects another, and you cannot expect even satisfactory results unless each adjustment is made only when it cannot be changed by any order.

2. Overtorquing (or undertorquing). While it is more common for over-torquing to cause damage, undertorquing can cause a fastener to vibrate loose causing serious damage. Especially when dealing with aluminum parts, pay attention to torque specifications and utilize a torque wrench in assembly. If a torque figure is not available, remember that if you are using the right tool to do the job, you will probably not have to strain yourself to get a fastener tight enough. The pitch of most threads is so slight that the tension you put on the wrench will be multiplied many times in actual force on what you are tightening.

A good example of how critical torque is can be seen in the case of spark plug installation, especially where you are putting the plug into an aluminum cylinder head. Too little torque can fail to crush the gasket, causing leakage of combustion gases and consequent overheating of the plug and engine parts. Too much torque can damage the threads, or distort the plug which changes the spark gap.

➡**There are many commercial products available for ensuring that fasteners won't come loose, even if they are not torqued just right (a very common brand is Loctite®. If you're worried about getting something together tight enough to hold, but loose enough to avoid mechanical damage during assembly, one of these products might offer substantial insurance. Read the label on the package and make sure the products is compatible with the materials, fluids, etc. involved before choosing one.**

3. Cross threading. This occurs when a part such as a bolt is screwed into a nut or casting at the wrong angle and forced. Cross threading is more likely to occur if access is difficult. It helps to clean and lubricate fasteners, and to start threading with the part to be installed going straight in. Then, start the bolt, spark plug, etc. with your fingers. If you encounter resistance, unscrew the part and start over again at a different angle until it can be inserted and turned several turns without much effort. Keep in mind that many parts, especially spark plugs, used tapered threads so that gentle turning will automatically bring the part you're treading to the proper angle if you don't force it or resist a change in angle. Don't put a wrench on the part until its's been threaded a couple of turns by hand. If you suddenly encounter resistance, and the part has not seated fully, don't force it. Pull it back out and make sure it's clean and threading properly.

Always take your time and be patient; once you have some experience, working on your car will become an enjoyable hobby.

TOOLS AND EQUIPMENT

The service procedures in this book presuppose a familiarity with hand tools and their proper use. However, it is possible that you may have a limited amount of experience with the sort of equipment needed to work on an automobile. This section is designed to help you assemble a basic set of tools that will handle most of the jobs you may undertake.

In addition to the normal assortment of screwdrivers and pliers, automotive service work requires an investment in wrenches, sockets and the handles needed to drive them, plus various measuring tools such as torque wrenches and feeler gauges.

You will find that virtually every nut and bolt on your vehicle is metric. Therefore, despite a few close size similarities, standard inch-size tools will not fit and must not be used. You will need a set of metric wrenches as your most basic tool kit, ranging from about 6-17mm in size. High quality forged wrenches are available in three styles: open end, box end and combination open/box end. The combination tools are generally the most desirable as a starter set; the wrenches shown in the accompanying illustration are of the combination type.

The other set of tools inevitably required is a ratchet handle and socket set. This set should have the same size range as your wrench set. The ratchet, extensions and flex drives for the sockets are available in many sizes; it is advisable to choose a ³⁄₈″ drive set initially. One break in the inch/metric sizing war is that metric sized sockets sold in the U.S. have inch-sized drive (¹⁄₄″, ³⁄₈″, ¹⁄₂″ and etc.). Thus, if you already have an inch-sized socket set, you need only buy new metric sockets in the sizes needed. Sockets are available in 6-point and 12-point versions; six point types are stronger and are a good choice for a first set. The choice of a drive handle for the sockets should be made with some care. If this is your first set, invest the extra money in a flex-head ratchet; it will get into many places otherwise accessible only through a long chain of universal joints, extensions and adapters . An alternative is a flex handle, which lacks the ratcheting feature but has a head which pivots 180°; such a tool is shown below the ratchet handle in the illustration. In addition to the range of sockets mentioned, a rubber lined spark plug socket should be purchased.

The most important thing to consider when purchasing hand tools is quality. Don't be misled by the low cost of bargain tools. Forged wrenches, tempered screwdriver blades and fine tooth ratchets are much better investments than their less expensive counterparts. The skinned knuckles and frustration inflicted by poor quality tools make any job an unhappy chore. Another consideration is that quality tools come with an unbeatable replacement guarantee; if the tool breaks, you get a new one, no questions asked.

Most jobs can be accomplished using the tools on the accompanying lists. There will be an occasional need for a special tool, such as snap ring pliers; that need will be mentioned in the text. It would not be wise to buy a large assortment of tools on the premise that someday they will be needed. Instead, the tools should be acquired one at a time, each for a specific job, both to avoid unnecessary expense and to be certain that you have the right tool.

The tools needed for basic maintenance jobs, in addition to the wrenches and sockets mentioned, include:

1. Jackstands, for support.
2. Oil filter wrench.
3. Oil filter spout or funnel.
4. Grease gun.
5. Battery post and clamp cleaner.
6. Container for draining oil.
7. Many rags for the inevitable spills.

In addition to these items there are several others which are not absolutely necessary but handy to have around. These include a transmission funnel and filler tube, a drop (trouble) light on a long cord, an adjustable (crescent) wrench and slip joint pliers.

A more advanced list of tools, suitable for tune-up work, can be drawn up easily. While the tools are slightly more sophisticated, they need not be outrageously expensive. The key to these purchases is to make them with an eye towards adaptability and wide range. A basic list of tune-up tools could include:

8. Tachometer/dwell meter.
9. Spark plug gauge and gapping tool.
10. Feeler gauges for valve adjustment.
11. Timing light.

Note that if your vehicle has electronic ignition, you will have no need for a dwell meter. On most vehicles a tachometer is provided on the instrument panel of the vehicle. You will need both the wire type (spark plugs) and the flat type (valves) feeler gauges. The choice of a timing light should be made carefully. A light which works on the DC current supplied by the vehicle battery is the best choice; it should have a xenon tube for brightness. Since most of the vehicles have electronic ignition or will have it in the future, the light should have an inductive pickup which clamps around the No. 1 spark plug cable.

In addition to these basic tools, there are several other tools and gauges which you may find useful. These include:

12. A compression gauge. The screw-in type is slower to use but eliminates the possibility of faulty reading due to escaping pressure.
13. A manifold vacuum gauge.
14. A test light.
15. A combination volt/ohmmeter.
16. An induction meter, used to determine whether or not there is current flowing in a wire, an extremely helpful tool for electrical troubleshooting.

Finally, you will find a torque wrench necessary for all but the most basic of work. The beam type models are perfectly adequate. The newer click type (breakaway) torque wrenches are more accurate but are much more expensive and must be periodically recalibrated.

Special Tools

Special tools are available from:
Kent-Moore Corporation
29784 Little Mack
Roseville, Michigan 48066
In Canada:
Kent-Moore of Canada, Ltd.,
2395 Cawthra
Mississauga, Ontario
Canada L5A 3P2

SERVICING YOUR VEHICLE SAFELY

It is virtually impossible to anticipate all of the hazards involved with automotive maintenance and service, but care and common sense will prevent most accidents.

The rules of safety for mechanics range from 'don't smoke around gasoline", to 'use the proper tool(s) for the job". The trick to avoiding injuries is to develop safe work habits and take every possible precaution.

Do's

• Do keep a fire extinguisher and first aid kit within easy reach.

• Do wear safety glasses or goggles when cutting, drilling, grinding or prying, even if you have 20-20 vision. If you wear glasses for the sake of vision, they should be made of hardened glass that can serve also as safety glasses or wear safety goggles over your regular glasses.

• Do shield your eyes whenever you work around the battery. Batteries contain sulfuric acid. In case of contact with the eyes or skin, flush the area with water or a mixture of water/baking soda and get medical attention immediately.

• Do use safety stands for any under vehicle service procedures. Jacks are for raising and lowering vehicles; safety stands are for making sure the vehicle stays raised until you want it to come down. Whenever the car is raised, block the wheels remaining on the ground and set the parking brake.

• Do use adequate ventilation when working with any chemicals or hazardous materials. Like carbon monoxide, the asbestos dust resulting from brake lining wear can be poisonous in sufficient quantities.

• Do disconnect the negative battery cable when working on the electrical system. The secondary ignition system can contain up to 40,000 volts.

• Do follow manufacturer's directions whenever working with potentially hazardous materials. Both brake fluid and antifreeze are poisonous if taken internally.

• Do properly maintain your tools. Loose hammerheads, mushroomed punches and chisels, frayed or poorly grounded electrical cords, excessively worn screwdrivers, spread wrenches (open end), cracked sockets, slipping ratchets or faulty droplight sockets can cause accidents.

• Do use the proper size and type of tool for the job being done.

• Do, when possible, pull on a wrench handle rather than push on it and adjust your stance to prevent a fall.

• Do be sure the adjustable wrenches are tightly closed on the nut or bolt and pulled so that the face is on the side of the fixed jaw.

• Do select a wrench or socket that fits the nut or bolt. The wrench or socket should sit straight, not cocked.

• Do strike squarely with a hammer; avoid glancing blows.

• Do set the parking brake and block the drive wheels if the work requires the engine running.

Don'ts

• Don't run an engine in a garage or anywhere else without proper ventilation-EVER! Carbon monoxide is poisonous; it takes a long time to leave the human body and you can build up a deadly supply of it in your system by simply breathing in a little every day. You may not realize you are slowly poisoning

yourself. Always use power vents, windows, fans or open the garage doors.

• Don't work around moving parts while wearing a necktie or other loose clothing. Short sleeves are much safer than long, loose sleeves; hard-toed shoes with neoprene soles protect your toes and give a better grip on slippery surfaces. Jewelry such as watches, fancy belt buckles, beads or body adornment of any kind is not safe working around a car. Long hair should be hidden under a hat or cap.

• Don't use pockets for toolboxes. A fall or bump can drive a screwdriver deep into your body. Even a wiping cloth hanging from the back pocket can wrap around a spinning shaft or fan.

• Don't smoke when working around gasoline, cleaning solvent or other flammable material.

• Don't smoke when working around the battery. When the battery is being charged, it gives off explosive hydrogen gas.

• Don't use gasoline to wash your hands; there are excellent soaps available. Gasoline may contain lead, which can enter the body through a cut, accumulating in the body until you are very ill. Gasoline also removes all the natural oils from the skin so that bone dry hands will suck up oil and grease.

• Don't service the air conditioning system unless you are equipped with the necessary equipment and training. The refrigerant (R-12) is under pressure and when released, will instantly freeze any surface it contacts, including your eyes. Although the refrigerant is normally non-toxic, R-12 becomes a deadly poisonous gas in the presence of an open flame. One good whiff of the vapors from burning refrigerant can be fatal.

SERIAL NUMBER IDENTIFICATION

Vehicle Identification Number

▶ See Figures 1, 2, 3, 4 and 5

The Vehicle Identification Number (VIN) is located on a plate which is attached to the left top side of the instrument panel. These numbers are visible from the outside of the vehicle.

All Vehicle Identification Numbers contain 17 digits. The vehicle number is a code which tells country, make, vehicle type, engine, body and many other important characteristics of that specific vehicle.

There is also a vehicle information code plate which is riveted to the bulkhead in the engine compartment. The plate shows the VIN, model code, engine model, transaxle model and body

color codes. The engine code used on this plate differs from the code letter used in the 8th position of the Vehicle Identification Number (VIN). Either code can be used to identify the particular engine in the vehicle. Since the vehicle owners card is usually carried, it may be easier to to use the code letter in the VIN for engine reference. A second reason for referring to the VIN for engine

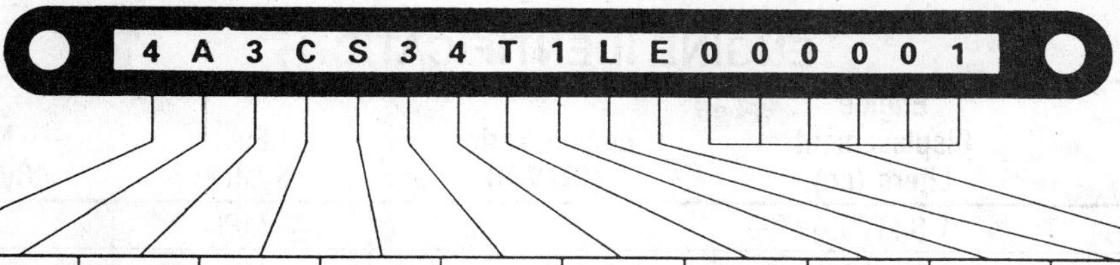

1st Digit	2nd Digit	3rd Digit	4th Digit	5th Digit	6th Digit	7th Digit	8th Digit	9th Digit	10th Digit	11th Digit	12th to 17th Digits
Country	Make	Vehicle type	Others	Line	Price class	Body	Engine	*Check digits	Model year	Plant	Serial number
4-USA	A-Mitsubishi	3-Passenger Car	C-Automatic Seat Belt	S-Eclipse T-Eclipse 4WD	3-Medium 4-High 5-Premium 6-Special	4-3 door Hatchback	T-1.8 liter (107 cu.in.) [SOHC-MPI] R-2.0 liters (122 cu.in.) [DOHC-MPI] U-2.0 liters (122 cu.in.) [DOHC-MPI-Turbo]	1 2 3 . . . 9 x	L-1990 Year	E-DSM	000001 to 999999

NOTE * "Check digit" means a single number or letter x used to verify the accuracy of transcription of vehicle identification number.

Fig. 2 Vehicle identification number code chart

identification is that code 4G63, located on the vehicle information code plate, does identify the engine as a 2.0L DOHC engine, but does not tell you if the engine is equipped with a turbocharger. If the 8th VIN number is a U, there is no doubt that the engine in question is a 2.0L DOHC engine equipped with a turbocharger.

The engine codes found on the vehicle information code plate are as follows:

4G37 — 1.8L SOHC engine

4G63 — 2.0L DOHC engine

A vehicle safety certification label is attached to the face of the left door pillar post. This label indicates the month and year of manufacture, Gross Vehicle Weight Rating (G.R.V.W.) front and rear, and Vehicle Identification Number (VIN).

Fig. 3 Vehicle safety certification label

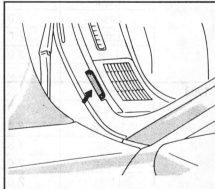

Fig. 1 Vehicle identification number location.

ENGINE IDENTIFICATION

Year	Model	Engine Displacement Liters (cc)	Engine Series (ID/VIN)	Fuel System	No. of Cylinders	Engine Type
1990	Eclipse	1.8 (1755)	T	MPI	4	SOHC
	Eclipse	2.0 (1997)	R	MPI	4	DOHC
	Eclipse	2.0 (1997)	U	Turbo	4	DOHC
	Laser	1.8 (1755)	T	MPI	4	SOHC
	Laser	2.0 (1997)	R	MPI	4	DOHC
	Laser	2.0 (1997)	U	Turbo	4	DOHC
	Talon	2.0 (1997)	R	MPI	4	DOHC
	Talon	2.0 (1997)	U	Turbo	4	DOHC
1991	Eclipse	1.8 (1755)	T	MPI	4	SOHC
	Eclipse	2.0 (1997)	R	MPI	4	DOHC
	Eclipse	2.0 (1997)	U	Turbo	4	DOHC
	Laser	1.8 (1755)	T	MPI	4	SOHC
	Laser	2.0 (1997)	R	MPI	4	DOHC
	Laser	2.0 (1997)	U	Turbo	4	DOHC
	Talon	2.0 (1997)	R	MPI	4	DOHC
	Talon	2.0 (1997)	U	Turbo	4	DOHC
1992	Eclipse	1.8 (1755)	T	MPI	4	SOHC
	Eclipse	2.0 (1997)	R	MPI	4	DOHC
	Eclipse	2.0 (1997)	U	Turbo	4	DOHC
	Laser	1.8 (1755)	T	MPI	4	SOHC
	Laser	2.0 (1997)	R	MPI	4	DOHC
	Laser	2.0 (1997)	U	Turbo	4	DOHC
	Talon	2.0 (1997)	R	MPI	4	DOHC
	Talon	2.0 (1997)	U	Turbo	4	DOHC
1993	Eclipse	1.8 (1755)	B	MPI	4	SOHC
	Eclipse	2.0 (1997)	E	MPI	4	DOHC
	Eclipse	2.0 (1997)	F	Turbo	4	DOHC
	Laser	1.8 (1755)	B	MPI	4	SOHC
	Laser	2.0 (1997)	E	MPI	4	DOHC
	Laser	2.0 (1997)	F	Turbo	4	DOHC
	Talon	1.8 (1755)	B	MPI	4	SOHC
	Talon	2.0 (1997)	E	MPI	4	DOHC
	Talon	2.0 (1997)	F	Turbo	4	DOHC

MPI—Multi-Point Fuel Injection
Turbo—Turbocharged
SOHC—Single Overhead Camshaft
DOHC—Double Overhead Camshaft

Engine Identification Number

▶ See Figures 6 and 7

The engine model number is stamped at the front side on the top edge of the cylinder block. The same 4 character code as on the vehicle information code plate is used. The engine serial number is also stamped near the engine model number. As mentioned above, the engine can also be identified by the 8th digit in the VIN number.

Transaxle Identification

▶ See Figures 4, 6, 7 and 8

The transaxle model code is located on the vehicle information code plate. The transaxle identification number is etched on a boss located on the front upper portion of the case.

ROUTINE MAINTENANCE

Air Cleaner

▶ See Figure 9

REMOVAL & INSTALLATION

All vehicles covered in this manual are equipped with a disposable paper cartridge air cleaner element. At every tune-up or sooner, if the car is operated in a dusty area, remove the air filter element and inspect its condition. Check the element by holding a light up to the filter. If light can be seen through the filter, then it should be OK. Replace the filter if light can not be seen through the filter or if it appears extremely dirty. Loose dust can sometimes be removed by striking the filter against a hard surface several times or by blowing through it with compressed air from the inside out. The filter should be inspected at every oil changed and replaced as required. Before installing either the original or a replacement filter, wipe out the inside of the air cleaner housing with a clean rag or paper towel.

Except 2.0L Turbocharged Engine
▶ See Figures 10, 11, 12, 13 and 14

1. Disconnect the negative battery cable.
2. Disconnect the air-flow sensor connector located on the top of the air cleaner cover.
3. Remove the air intake hose.
4. Unclamp the air cleaner cover.
5. Push the air intake hose backward and remove the air cleaner cover from the housing.

➡Care must be taken when removing the air cleaner cover from the housing. The air-flow sensor is attached to the cover and could be damaged during cover removal.

6. Remove the air cleaner element. Thoroughly clean the air cleaner housing prior to replacing the air filter.
   To install:
7. Install the new air filter element into the housing.
8. Install the air cleaner cover into position and secure in place.
9. Install the air intake hose. Reconnect the air-flow sensor harness connector.
10. Connect the negative battery cable.

2.0L Turbocharged Engine

1. Disconnect the negative battery cable.
2. Disconnect the air-flow sensor connector.
3. Disconnect the boost hose.
4. Disconnect the solenoid valve with hoses.
5. Disconnect the air intake hose.
6. Remove the air cleaner retainer bolts and the air cleaner assembly.
7. Unclamp the cover and remove from the housing.

➡Care must be taken when removing the air cleaner cover. The air flow sensor is attached and could be damaged during cover removal.

8. Remove the air cleaner element. Thoroughly clean the air cleaner housing prior to replacing the air filter.
   To install:
9. Install the new air cleaner element into the housing. Install and secure the cover in place.
10. Install the air cleaner assembly and the retainer bolts.
11. Connect the air intake hose.
12. Connect the solenoid valve.
13. Connect the boost hose.
14. Connect the air-flow sensor connector.
15. Connect the negative battery cable.

Fuel Filter

▶ See Figure 15

REMOVAL & INSTALLATION

✳✳CAUTION

Do not use conventional fuel filters, hoses or clamps when servicing fuel injection systems. They are not compatible with the injection system and could fail, causing personal injury or damage to the vehicle. Use only hoses and clamps specifically designed for fuel injection systems.

1. Relieve the fuel system pressure as follows:
   a. Loosen the fuel filler cap to release fuel tank pressure.
   b. Disconnect the fuel pump harness connector located at the rear of the fuel tank.
   c. Start the vehicle and allow it to run until it stalls from lack of fuel. Turn the key to the **OFF** position.
   d. Disconnect the negative battery cable, then reconnect the fuel pump connector and reinstall the fuel filler cap.
   e. Wrap shop towels around the fitting that is being disconnected to absorb residual fuel in the lines.
2. The filter is located in the engine compartment. Hold the fuel filter nut securely with a backup wrench.
3. Cover the hose connection with shop towels to prevent any splash of fuel that could be caused by residual pressure in the fuel pipe line. Remove the eye bolt. Discard the gaskets.
4. Loosen the main pipe flare nut while holding the fuel filter nut securely. Separate the flare nut connection at the filter.

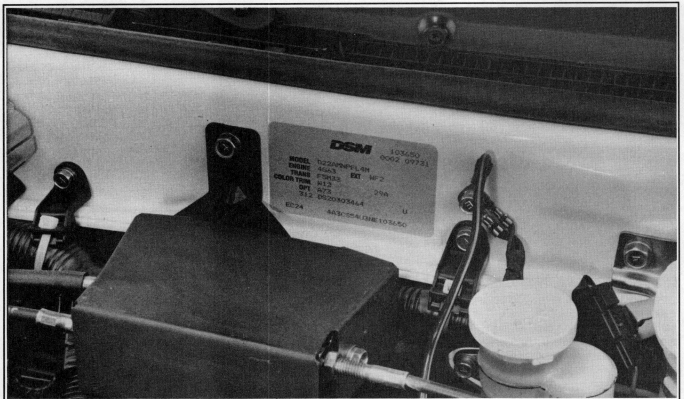

Fig. 5 The vehicle model, engine model, transaxle model, and body color code are all noted on the vehicle information code plate

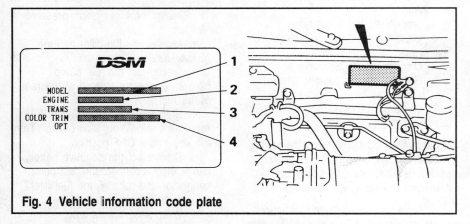

Fig. 4 Vehicle information code plate

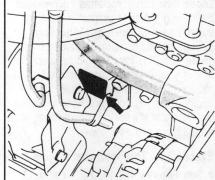

Fig. 6 Engine model stamping located on the front side of the cylinder block

5. Remove the mounting bolts and remove the fuel filter.

To install:

6. Install the filter to its bracket only finger-tight. Movement of the filter will ease attachment of the fuel lines.

7. Insert the main pipe at the connector part of the filter and manually screw in the main pipe's flare nut. Make sure new O-rings are installed prior to installation.

8. While holding the fuel filter nut, tighten the eye bolts to 22 ft. lbs. (30 Nm). Tighten the flare nut to 25 ft. lbs. (35 Nm).

9. Tighten the filter mounting bolts to 10 ft. lbs. (14 Nm).

10. Connect the negative battery cable. Turn the key to the **ON** position to pressurize the fuel system and check for leaks.

11. If repairs of a leak are required, remember to release the fuel pressure before opening the fuel system.

PCV Valve

♦ **See Figures 16, 17, 18, 19 and 20**

The Positive Crankcase Ventilation (PCV) valve is part of a system which is designed to protect the atmosphere from harmful vapors. Blow-by gas from the crankcase, as well as fumes from crankcase oil are diverted into the combustion chamber where they are burned during engine operation. Proper

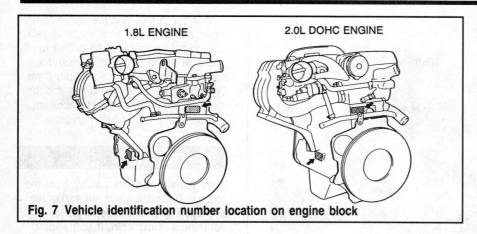

Fig. 7 Vehicle identification number location on engine block

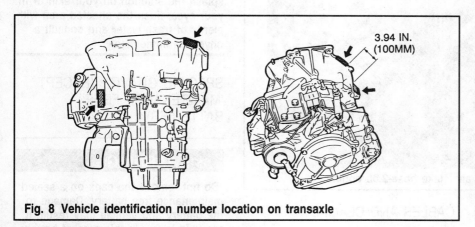

Fig. 8 Vehicle identification number location on transaxle

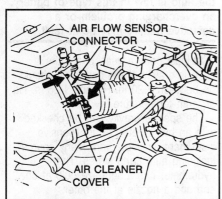

Fig. 9 Air cleaner housing assembly and air-flow sensor — 1.8L engine

operation of this system will improve engine performance as well as decrease the amount of harmful vapors released into the atmosphere.

REMOVAL & INSTALLATION

1. Disconnect the negative battery cable.

2. Remove the air intake hose and air cleaner assembly.

3. Disconnect the ventilation hose from the Positive Crankcase Ventilation (PCV) valve.

4. Unscrew the PCV valve from the rocker cover.

5. Inspect the valve as follows:
   a. Insert a thin stick into the positive ventilation valve from the threaded side to check that the plunger moves.
   b. If the plunger does not move, the PCV valve is clogged.

6. Clean or replace the valve.

7. Install the PCV valve into the rocker cover and tighten the valve to 6 ft. lbs. (8 Nm) torque.

8. Reconnect the ventilation hose to the valve.

9. Install the air intake hose and the air cleaner assembly.

10. Reconnect the negative battery cable.

Evaporative Canister

The charcoal canister is part of the Evaporative Emission Control System.

This system prevents the escape of raw gasoline vapors from the vehicle's fuel system to escape into the atmosphere.

The canister is designed to absorb fuel vapors under specific conditions. The canister is about the size of a coffee can and is located in the engine compartment.

Prior to removing the canister assembly from the vehicle, label the vacuum hose connections at the top of the canister prior to removal. For proper functioning of the system, it is essential that the hoses be returned to their original locations.

SERVICING

The canister itself does not require replacement at a specific mileage however, the Evaporative Emission Control System does required a careful operations check at 50,000 miles or every 5 years, whichever comes first. Testing procedures are listed in Section 4 of this manual.

Battery

▶ See Figures 21, 22, 23, 24, 25 and 26

FLUID LEVEL (EXCEPT 'MAINTENANCE FREE" BATTERIES)

✱✱WARNING

Do not remove the caps on a sealed maintenance free battery. Damage to the battery and vehicle may result. If the fluid is low in this type of battery, an overcharging problem or a defective battery may be at fault.

Check the battery electrolyte level at least once a month, or more often during hot weather during periods of extended vehicle operation. The level can be checked through the case on translucent polypropylene batteries; the cell caps must be removed on other models. The electrolyte level in each cell should be at the bottom of the split ring inside, or the line mark on the outside of the case.

If the level is found to be low, add only clean water to the battery until the solution is at the desired level. Each cell

Fig. 10 Removing the clamp from the air intake hose-2.0L engine

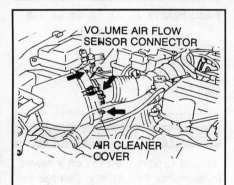

VOLUME AIR FLOW
SENSOR CONNECTOR

AIR CLEANER
COVER

Fig. 11 Air cleaner housing assembly and related components — 2.0L engine

is separated from the other and must be checked and filled individually.

If water is being added to a battery that is subject to freezing temperatures, it is advisable to run the engine to allow the water electrolyte solution to combine thus preventing icing.

CABLES AND CLAMPS

Once a year the battery terminals and the cable clamps should be cleaned. Loosen the clamps and remove the cables, the negative cable first. On top post batteries, a special puller is used to remove the cable clamps; these are inexpensive and are available from most auto parts stores. The side terminal battery cables are secured with a bolt.

Clean the cable clamps and the battery terminal with wire brush, until corrosion, dirt and grease are removed and the metal is shiny. It is especially important to clean the inside of the clamp thoroughly, since a small deposit of foreign material or oxidation will prevent electrical flow. Special tools are available for cleaning these parts, one type for conventional batteries and another type for side terminal batteries.

Before installing the battery cables, loosen the battery hold-down clamp or strap, remove the battery and check the battery tray for soundness. Replace the battery and tighten the hold-down clamp or strap securely. Be careful not to over tighten the hold-down, which could crack the battery case.

After the clamps and the terminals are clean, reinstall the cables, negative cable last; do not hammer on the cables to install on the batter posts. Tighten the clamps securely but do not distort them. Once installed, coat the clamps and the terminals with a thin coat of petroleum jelly or equivalent, to help prevent corrosion.

✳✳CAUTION

Keep flames or sparks away from the battery. Batteries gives off explosive hydrogen gas. The electrolyte contains sulfuric acid. If you should splash the solution on your skin or in your eyes, flush the affected area with plenty of fresh water and consult a physician immediately.

SPECIFIC GRAVITY (EXCEPT 'MAINTENANCE FREE" BATTERIES)

✳✳WARNING

Do not remove the caps on a sealed maintenance free battery. Damage to the battery and vehicle may result. If the fluid is low in this type of battery, an overcharging problem or a defective battery may be at fault.

At least once a year, check the specific gravity of the battery. It should be between 1.20 in. Hg and 1.26 in. Hg at room temperature.

The specific gravity can be checked with an hydrometer, an inexpensive instrument available from many sources, including auto parts stores. The hydrometer has a squeeze bulb at one end and a nozzle at the other. Electrolyte from the battery is drawn into the hydrometer until the float is lifted from its seat. The specific gravity is then read by noting the position of the float. During testing, make sure the float is not pined against the top of the hydrometer. If so, the test readings will be inaccurate. Generally, if after charging, the specific gravity between any 2 cells varies more than 50 points (0.050), the battery is bad and should be replaced.

It is not possible to check the specific gravity (in this manner) on sealed, maintenance free batteries. Instead, the indicator built into the top of the case must be relied on to display any signs of

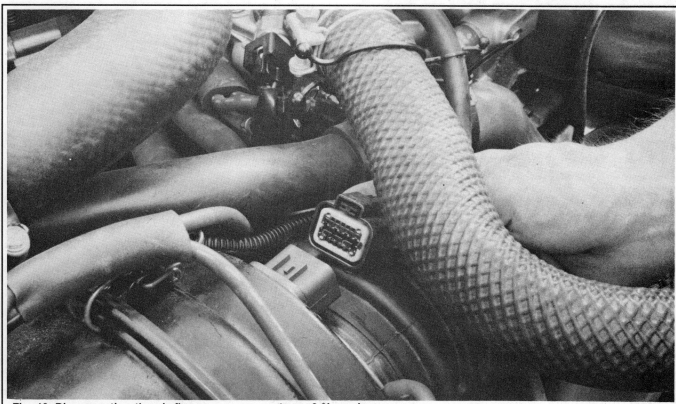

Fig. 12 Disconnecting the air flow sensor connector — 2.0L engine

battery deterioration. If the indicator is dark, the battery is assumed to be OK. If the indicator is light, the specific gravity is low and the battery should be charged or replaced. Unfortunately, the indicator at the top of the battery is not always as reliable as one would hope. Periodic testing is recommended for an accurate assessment of a batteries condition.

REPLACEMENT

When it becomes necessary to replace the battery, select one with a rating equal to or greater than the original. Deterioration of system components makes the batteries job harder as time goes on. The slow increase in electrical resistance over time makes it prudent to install a new battery with a greater capacity than the old.

1. Make sure all accessories and the ignition key are in the OFF position. Raise the hood and support.

2. Remove the negative battery cable first and then the positive terminal.

3. Lubricate the hold-down with penetrating oil before loosening. Remove the battery clamp or hold-down.

4. Using an approved battery carrier, lift the battery out of the tray and place away from the vehicle.

5. Clean and inspect the battery tray.

6. Prepare a solution of baking soda and water, mixing to a thick paste. Paint the solution onto any corroded areas and allow to set for about one minute. Reapply the mixture a second time.

7. Rinse the area with warm water to remove baking soda mixture.

8. Apply a coat of rust converter to the affected areas and allow to dry. Apply a second application of rust converter once the first coat has dried.

9. Apply two coats of high quality metal primer to the affected areas and allow to completely dry before going any further.

10. Apply a coat of undercoating to the battery tray area and allow to dry overnight, if possible.

11. Clean the battery with soap and water.

12. Install the battery into the vehicle and tighten the hold-down bolt or clamp.

13. Apply a light coat of petroleum jelly to the terminal posts.

14. Tighten the battery cable terminals to 10 ft. lbs. (15 Nm).

15. Clean the cable ends that connect to the body and starter motor, as required.

16. If a slow to no start problem still exists, it is possible that a battery cable or the starter motor may be defective.

Belts

INSPECTION

▶ **See Figures 27 and 28**

Check the drive belts for cracks, fraying, wear and proper tension every 6,000 miles. It is recommended that the belts be replaced every 24 months or 24,000 miles.

1. Place a straight-edge along the top edge of the belt and across 2 pulleys. Allow both ends of the straight-edge to rest on top of each pulley for support.

2. Measure the deflection of the belt from the straight-edge with a force of about 22 lbs. applied midway between

Fig. 13 Removing the air cleaner cover from the housing — 2.0L engine

Fig. 14 Removing the air filter from the housing — 2.0L engine

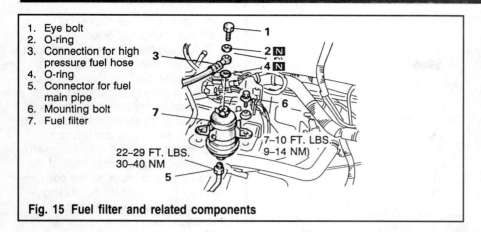

1. Eye bolt
2. O-ring
3. Connection for high pressure fuel hose
4. O-ring
5. Connector for fuel main pipe
6. Mounting bolt
7. Fuel filter

22–29 FT. LBS.
30–40 NM

7–10 FT. LBS.
9–14 NM

Fig. 15 Fuel filter and related components

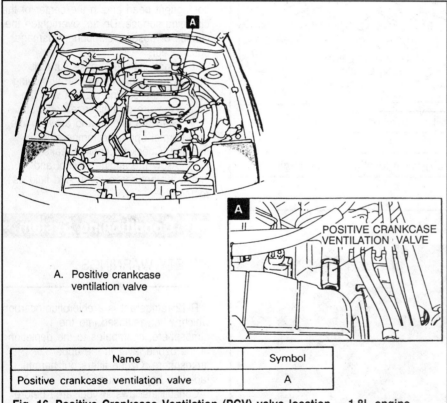

A. Positive crankcase ventilation valve

POSITIVE CRANKCASE VENTILATION VALVE

Name	Symbol
Positive crankcase ventilation valve	A

Fig. 16 Positive Crankcase Ventilation (PCV) valve location — 1.8L engine

the 2 pulleys. Deflection should measure as follows:

Air Conditioning Compressor Drive Belt
1.8L engine — 0.160-0.200 in. (4.0-5.0mm)
2.0L engine — 0.180-0.200 in. (4.5-5.0mm)
Alternator/Water Pump Drive Belt
1.8L engine — 0.315-0.433 in. (8.0-11.0mm)
2.0L engine — 0.354-0.453 in. (9.0-11.5mm)

3. Belt tension can also be checked with a tension gauge. Measure the belt

tension between any 2 pulleys. The desired value should be 55-110 lbs. (250-500 N).

ADJUSTMENT

♦ See Figures 29, 30, 31, 32 and 33

Excessive belt tension will cause damage to the alternator and water pump pulley bearings, while, on the other hand, loose belt tension will produce slip and premature wear on the

belt. Therefore, be sure to adjust the belt tension to the proper level.

To adjust the tension on a drive belt, loosen the adjusting bolt or fixing bolt locknut on the alternator, alternator bracket or tension pulley. Then move the alternator or turn the adjusting bolt to adjust belt tension. Once the desired value is reached, secure the bolt or locknut and recheck tension.

REMOVAL & INSTALLATION

1. Loosen the adjusting bolt or fixing bolt locknut on the alternator, alternator bracket or tension pulley.
2. Move the alternator or turn the adjusting bolt to release the belt tension.
3. Take note of the exact routing of the belt prior to removal. Lift the drive belt from the pulleys and remove from the engine compartment.

➡**If the belt being removed is located behind another belt, removal of the first belt is required.**

To install:
4. Position the replacement belt around the pulleys making sure belt routing is correct.
5. Adjust the belt until the correct tension is reached.
6. Once the desired tension is reached, secure the bolt or locknut and the pivot bolts. Recheck the belt tension and correct as required.

Hoses

REMOVAL & INSTALLATION

♦ See Figure 34

When draining the coolant, keep in mind that animals are attracted by the ethylene glycol in antifreeze, and are likely to drink any coolant that is left in an uncovered container or in puddles on the ground. This may prove to be fatal if a sufficient quantity is ingested. Always drain coolant into an sealable container. Coolant is also a hazard to the environment and should be disposed of accordingly. Coolant should be reused unless contaminated or several years old. If coolant can't be reused, it should be disposed of in an appropriate manner, as not to cause injury to animals or the environment. Many auto

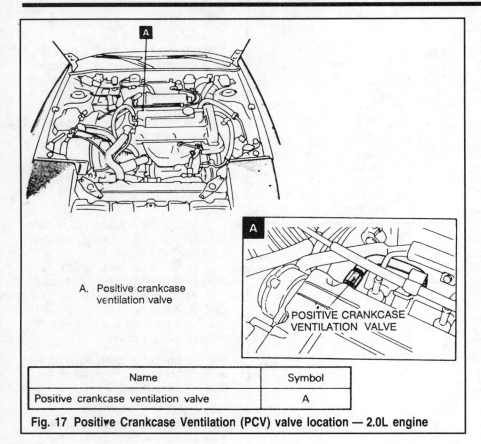

A. Positive crankcase ventilation valve

POSITIVE CRANKCASE VENTILATION VALVE

Name	Symbol
Positive crankcase ventilation valve	A

Fig. 17 Positive Crankcase Ventilation (PCV) valve location — 2.0L engine

Fig. 18 Removing the hose clamp and ventilation hose on the Positive Crankcase Ventilation (PCV) valve — 2.0 engine

parts stores offer free disposal of antifreeze to their customers.

1. Loosen the radiator cap and open the petcock. Drain the cooling system.

2. Loosen the hose clamps at each end of the hose to be removed.

3. Working the hose back and forth, slide it off of it's connection. Remove the hose from the engine compartment.

To install:

4. Install a replacement hose of the same length and shape as the one removed to the connections, making sure new clamps are in place prior to installation.

5. Tighten the clamps making sure they are positioned beyond the component bead and in the center of the clamping surface. Do not overtighten the clamps or hose or component damage could occur.

6. Once installed, make sure all hoses are out of the way of all moving parts. Fill the cooling system with the proper mixture of antifreeze and water.

7. Start the engine and allow to idle until normal operating temperature is reached. Check for coolant leaks.

8. Allow the engine to cool and check the antifreeze level. Add fluid as required.

Air Conditioning System

SAFETY WARNINGS

R-12 refrigerant is a chlorofluorocarbon which, when released into the atmosphere, contributes to the depletion of the ozone layer in the upper atmosphere. Ozone filters out harmful radiation from the sun.

Consult the laws in your area before servicing the air conditioning system in your vehicle. In some states, it is illegal to perform repairs involving refrigerant unless the work is done by a certified technician.

• Avoid contact with a charged refrigeration system, even when working on another part of the air conditioning system or vehicle. If a heavy tool comes into contact with a section of copper tubing or a heat exchanger, it can easily cause the relatively soft material to rupture.

• When it is necessary to apply force to a fitting which contains refrigerant, as when checking that all system couplings are securely tightened, use a wrench on

Fig. 19 Removing the Positive Crankcase Ventilation (PCV) valve from the rocker cover — 2.0 engine

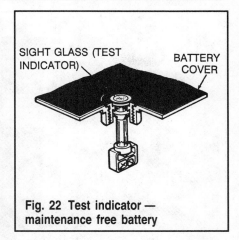

SIGHT GLASS (TEST INDICATOR)   BATTERY COVER

Fig. 22 Test indicator — maintenance free battery

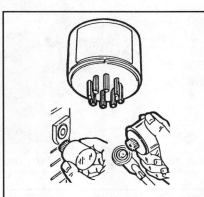

Fig. 23 Clean the battery post with a wire brush or the special tool shown

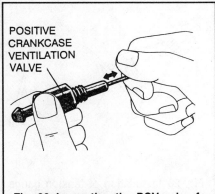

POSITIVE CRANKCASE VENTILATION VALVE

Fig. 20 Inspecting the PCV valve for plunger movement

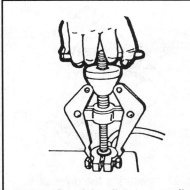

Fig. 21 Special pullers are available to remove the cable clamps

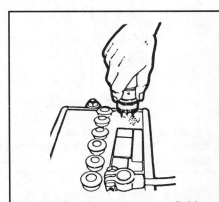

Fig. 24 Special tools are available for cleaning the posts and clamps on side terminal batteries

both parts of the fitting involved, if possible. This will avoid putting torque on refrigerant tubing. (It is advisable, when possible, to use tube or line wrenches when tightening these flare nut fittings.)

• A toxic gas is formed when R-12 contacts any flame. Avoid refrigerant exposure to open flames when working on any air conditioning system.

• Never start a system without first verifying that both service valves are backseated, if equipped, and that all fittings are throughout the system are snugly connected.

• Avoid applying heat to any refrigerant line or storage vessel. Charging may be aided by using water heated to less than +125°F (+51°C) to warm the refrigerant container. Never allow a refrigerant storage container to sit out in the sun, or near any other source of heat, such as a radiator.

• Always wear goggles when working on a system to protect the eyes. If refrigerant contacts the eye, it is advisable in all cases to see a physician as soon as possible.

• Frostbite from liquid refrigerant should be treated by first gradually warming the area with cool water, and

Fig. 26 Proper positioning of the battery, battery hold-down and battery cables in the engine compartment

Fig. 25 Clean the inside of the clamps with a wire brush or the special tool

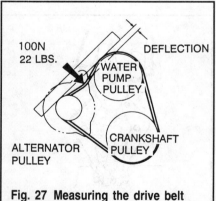

Fig. 27 Measuring the drive belt deflection using the straight edge method

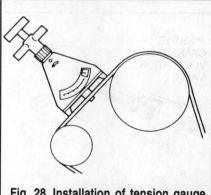

Fig. 28 Installation of tension gauge to check for correct belt tension

then gently applying petroleum jelly. A physician should be consulted.

• Always keep refrigerant container fittings capped when not in use. Avoid sudden shock to the container which

might occur from dropping it, or from banging a heavy tool against it. Never carry a container in the passenger compartment of a car.

• Always completely discharge the system using the proper equipment, before painting the vehicle (if the paint is to be baked on), or before welding anywhere near refrigerant lines.

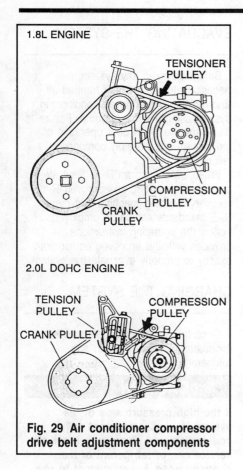

Fig. 29 Air conditioner compressor drive belt adjustment components

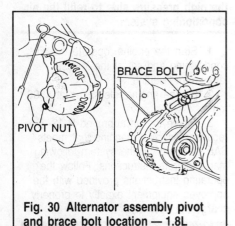

Fig. 30 Alternator assembly pivot and brace bolt location — 1.8L engine

SYSTEM INSPECTION

✳✳CAUTION

The compressed refrigerant used in the air conditioning system expands into the atmosphere at a temperature of -2°F or lower. This will freeze any surface that it contacts, including your eyes. In addition, the refrigerant

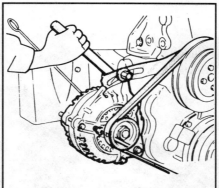

Fig. 31 Increasing the belt tension using prybar to apply pressure at stator part of alternator

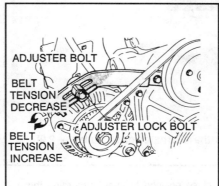

Fig. 32 Alternator mounting and adjustment fasteners — 2.0L engine

decomposes into a poisonous gas in the presence of a flame. Do not open or disconnect any part of the air conditioning system.

Checking For Oil Leaks

Refrigerant leaks can sometimes appear as oily areas on various components due to the compressor oil that is transported along with refrigerant, throughout the air conditioning system. During your visual inspection of system components, look for oily spots on all air conditioning hoses and lines, especially in areas of hose and tube connections and crimpings. If oily deposits are present, the system may have a freon leak at that point.

Keep The Condenser Clear

Periodically inspect the front of the condenser for bent fins or foreign material (dirt, leaves, twigs, paper, etc.). If any cooling fins are bent, straighten the fins carefully using needle nose

pliers. Remove any debris that may restrict air flow over the condenser. Reduced air flow over the condenser will decrease the efficiency of the air conditioning system.

REFRIGERANT LEVEL CHECKS

Sight Glass Check

The sight glass is a refrigerant level indicator. To check the refrigerant level, clean the sight glass and start the engine. Push the air conditioner button to operate the compressor, place the blower switch to the high position and move the temperature control lever to the extreme left. After operating for a few minutes in this manner, check the sight glass.

• If the sight glass is clear, the magnetic clutch is engaged, the compressor discharge line is warm and the compressor inlet is cool; the system has a full charge.

• If the sight glass is clear, the magnetic clutch is engaged, and their is no significant temperature difference between compressor inlet and discharge lines; the system has lost some refrigerant.

• If the sight glass is clear and the magnetic clutch is disengaged; the clutch is faulty or the system is out of refrigerant. An aid in finding the problem might be to check the low pressure switch and the clutch coil for continuity. Also check the pressure in the system using a set of air conditioning gages.

• If the sight glass shows foam or bubbles, the system could be low on charge. Occasional foam or bubbles are normal when the ambient temperature is above 110°F (43°C) or below 70°F (21°C). Adjust the engine speed to 1500 rpm and block the air flow through the condenser to increase the compressor discharge pressure to 206-220 psi. If the sight glass still shows bubbles or foam, system charge level is low.

The refrigerant system will not be low on charge unless there is a leak. Find and repair the leak prior to adding refrigerant to the system.

GAUGE SETS

Most of the service work performed in air conditioning requires the use of a set of 2 gauges, one for the high (head)

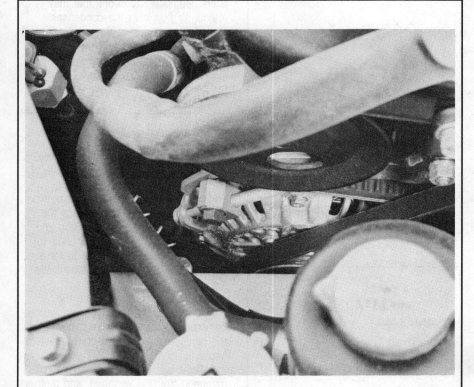

Fig. 33 Correct positioning of the belt on the alternator drive pulley — 2.0L engine

pressure side of the system, the other for the low (suction) side.

The low side gauge records both pressure and vacuum. Vacuum readings are calibrated from 0 to 30 inches Hg and the pressure graduations read from 0 to no less than 60 psi.

The high side gauge measures pressure from 0 to at last 600 psi.

Both gauges are threaded into a manifold that contains two hand shut-off valves. Proper manipulation of these valves and the use of an approved freon recovery/recycling machine allow the user to perform the following services:

1. Test high and low side pressures.
2. Remove air, moisture, and/or contaminated refrigerant.
3. Purge the system of refrigerant.
4. Charge the system with refrigerant.

The manifold valves are designed so that they have no direct effect on gauge readings but serve only to provide for or cut off the flow of refrigerant through the manifold. During all testing and hook-up operations, the valves are kept in a

close position to avoid disturbing the refrigeration system. The valves are opened only to purge the system or refrigerant or to charge it.

DISCHARGING THE SYSTEM

R-12 refrigerant is a chlorofluorocarbon which, when released into the atmosphere, can contribute to the depletion of the ozone layer in the upper atmosphere. Ozone filters out harmful radiation from the sun. When discharging an R-12 filled air conditioning system, an approved R-12 recovery/recycling machine that meets SAE standards should be employed. Follow the operating instructions provided with the approved equipment exactly to properly discharge the system.

To discharge the air conditioning system, use the refrigerant recovery unit to remove the refrigerant gas from the system. Refer to your specific refrigerant recovery and recycling unit instruction manual for proper operation of the unit.

EVACUATING THE SYSTEM

Before charging any system, it is necessary to remove any trapped air and moisture from the air conditioning system using a vacuum pump. Failure to do so may result in poor operation of the system and possibly component failure.

When evacuating an R-12 filled air conditioning system, an approved R-12 recovery/recycling machine that meets SAE standards should be employed. Follow the operating instructions provided with the approved equipment exactly to properly evacuate the system.

CHARGING THE SYSTEM

Once the system has been properly evacuated, we are ready to add refrigerant.

✳✳CAUTION

If the high pressure side of the system is used, refrigerant will counterflow and sometimes break the service can of refrigerant of the charging hose. Never attempt to use the high pressure side to refill the air conditioning system.

1. Start the engine, operate the air conditioning and set at the lowest temperature setting.
2. Fix the engine idle speed at 1500 rpm.
3. Attach the an approved R-12 recovery/recycling machine that meets SAE standards to the system following manufacturers instructions. Follow the operating instructions provided with the approved equipment exactly to properly charge the system.

✳✳CAUTION

Never exceed the recommended maximum charge for the system. The maximum amount of R-12 refrigerant used in the system is 33 oz. (935g).

LEAK TESTING THE SYSTEM

Checking the air conditioning system for leaks can be a difficult procedure

Fig. 34 Radiator drain cock assembly, normally located on the lower left side of the radiator

because of the engine compartment congestion. The recommended method is the use of an approved electronic refrigerant leak detector.

1. An R-12 charge of at least a half pound is needed for leak testing. Check all fitting connections and the front of the compressor for wetness and oil. A slight amount of oil in front of the compressor is normal.

2. Apply some soapy water to the fittings in the engine compartment. Check for bubbles from the fittings.

3. If white smoke comes out of the A/C vents, this may indicate a leak in the evaporator or fittings inside the evaporator housing.

4. Coat the compressor with a soap solution and check for bubbles around the compressor housing halves and fittings.

5. Check all rubber hoses for cracks and deterioration. A hose that looks OK, may be deteriorated and leaking slowly.

6. Check for a damaged condenser.

Windshield Wipers

For maximum effectiveness and longest element life, the windshield and wiper blades should be kept clean. Dirt, tree sap, road tar and so on will cause streaking, smearing and blade deterioration if left on the windshield. It is advisable to wash the windshield carefully with a commercial glass cleaner at least once a month. Wipe off the rubber blades with a wet rag afterwards. Do not attempt to move the wipers back and forth by hand; damage to the motor and drive mechanism may result.

If the blades are found to be cracked, broken or torn they should be replaced immediately. Replacement intervals will vary with usage. If the wiper pattern is smeared or streaked, or if the blade chatters across the glass, the blades should be replaced. It is easiest and most sensible to replace them in pairs.

REPLACING WIPER BLADES

▶ See Figure 35

1. Lift the wiper arm away from the glass.

2. Depress the pinch release lever on the upper surface of the arm and separate the end link from the arm assembly.

3. Slide the end link off of the wiper blade taking note of it's exact position on the blade.

4. Slide the wiper blade from the wiper arm assembly and remove the remaining end link.

5. Installation is the reverse of the removal procedure.

If the arms have been replaced in the past, it is possible that a different style blade be on the vehicle. One possible type of refill has 2 metal tabs which are unlocked by squeezing them together. The rubber blade can then be withdrawn from the frame jaws. A new one is installed by inserting it into the front frame jaws and sliding it rearward to engage the remaining frame jaws. There are usually 4 jaws; be certain when installing that the refill is engaged in all of them. At the end of its travel, the tabs will lock into place on the front jaws of the wiper blade frame.

Regardless of the type of refill used, make sure that all of the frame jaws are engaged as the refill is pushed into place and locked. The metal blade holder and frame will scratch the glass if allowed to touch it.

Tire and Wheels

TIRE ROTATION

▶ **See Figure 36**

Tires should be rotated periodically to get the maximum tread lift available. Normally this is done once a year or 7,000 miles. If front end problems are suspected have them corrected before rotating the tires. On installation of the

TIRE DIAGNOSTIC CHART

Symptom		Probable cause		Remedy
Rapid wear at shoulders		Under-inflation or lack of rotation		Adjust the tire pressure
Rapid wear at center		Over-inflation or lack of rotation		
Cracked treads		Under-inflation		

TIRE DIAGNOSTIC CHART

Symptom	Probable cause	Remedy
Wear on one side	Excessive camber	Inspect the camber
Feathered edge	Incorrect toe-in	Adjust the toe-in
Bald spots	Unbalanced wheel	Adjust the imbalanced wheels
Scalloped wear	Lack of rotation of tires or worn or out-of-alignment suspension	Rotate the tires Inspect the front suspension alignment

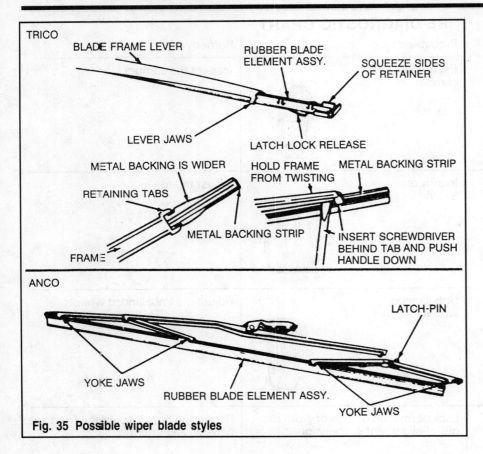

Fig. 35 Possible wiper blade styles

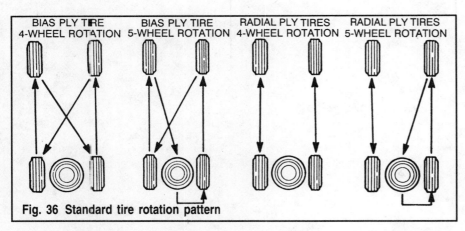

Fig. 36 Standard tire rotation pattern

wheel assemblies, torque the lug nuts to 87-101 ft. lbs. (120-140 Nm).

➡**Mark the wheel position and direction of rotation on radial tires before removing them.**

If equipped with aluminum wheels, install the lug nuts and tighten finger-tight. Once tighten finger-tight, tighten to specifications in a crisscross sequence. Do not use an impact wrench or push the wrench by foot to tighten the wheel nuts. Do not apply oil to the threaded portions of the stud or nuts.

On high speed tire (205/55R16 88V) vehicles, the left and the right tires are respectively specified. Attach the wheel marked 'ROTATION LEFT SIDE" on the left side of the vehicle, and the wheel marked 'ROTATION RIGHT SIDE" on the right side of the vehicle.

TIRE DESIGN

All 4 tires should be of the same construction type. Radial, bias, or bias-belted tires should not be mixed. The wheels must be the correct width for the tire. Tire dealers have charts of tire and rim compatibility. A mismatch can cause sloppy handling and rapid tire wear. The tread width should match the rim width (inside bead to inside bead) within an inch. For radial tires, the rim width should be 80% or less of the tire (not tread) width. The height (mounted diameter) of the new tires can greatly change speedometer accuracy, engine speed at a given road speed, fuel mileage, acceleration, and ground clearance. Tire manufacturers furnish full measurement specifications.

TIRE INFLATION

The tires should be checked frequently for proper air pressure. Make sure that the tires are cool, as you will get a false reading when the tires are heated because air pressure increases with temperature. A chart in the glove compartment or on the driver's door pillar gives the recommended inflation pressure. Maximum fuel economy and tire life will result if pressure is maintained at the highest figure given on chart. When checking pressures, do not neglect the spare tire. The tires should be checked before driving since pressure can increase as much as 6 pounds per square inch (psi) due to heat buildup.

➡**Some spare tires require pressures considerably higher than those used in other tires.**

While you are checking the tire pressure, take a look at the tread. The tread should be wearing evenly across the tire. Excessive wear in the center of the tread could indicate overinflation. Excessive wear on the outer edges could indicate underinflation. An irregular wear pattern is usually a sign of incorrect front wheel alignment or wheel balance. A front end that is out of alignment will usually pull the car to one side of a flat road when the steering wheel is released. Incorrect wheel balance will produce vibration in the steering wheel, while unbalanced rear wheels will result in floor, seat or trunk vibration.

It is a good idea to have your own accurate tire gauge, and to check pressures weekly. Not all gauges on service station air pumps can be trusted for accuracy.

Tires should be replaced when a tread wear indicator appears as a solid band across the tread.

CARE OF SPECIAL WHEELS

Aluminum is vulnerable to abrasive cleaners, detergents and alkalies.

Maintenance of aluminum wheels includes frequent washing and waxing. Failure to heed this warning will cause the protective coating to be damaged. If a vehicle washing detergent has been used, or salt from sea water or road chemicals adhere, wash the vehicle as soon as possible. After washing the

vehicle, apply body or wheel wax to the aluminum type wheels to prevent corrosion.

When cleaning the vehicle with steam, do not direct steam onto the aluminum type wheels. The finish on the wheels may be damaged.

FLUIDS AND LUBRICANTS

Fluid Disposal

Used fluids such as engine oil, transaxle fluid, antifreeze and brake fluid are hazardous wastes and must be disposed of properly. Before draining any fluids, consult with the local authorities; in many areas, waste oils, etc. is being accepted as a part of recycling programs. A number of service stations and auto parts stores are also accepting waste fluids for recycling.

Be sure of the recycling center's policies before draining any fluids, as many will not accept different fluids that have been mixed together, such as oil and antifreeze.

Fuel Recommendations

All vehicles except those equipped with DOHC engines have been designed to run on unleaded fuel having a minimum octane rating of 87 or 91 RON (Research Octane Number).

Vehicles equipped with DOHC engines have been designed to run on unleaded fuel having a minimum octane rating of 91 or 95 RON (Research Octane Number).

The use of a fuel too low in octane (a measurement of anti-knock quality) will result in spark knock. Since many factors such as altitude, terrain, air temperature and humidity affect the operating efficiency, knocking may result even though the recommended fuel is being used. If persistent knocking occurs, it may be necessary to switch to a higher grade of fuel. Continuous or heavy knocking may result in engine damage and should be diagnosed, if continued.

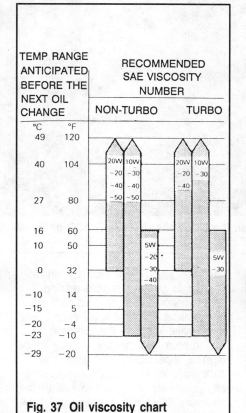

Fig. 37 Oil viscosity chart

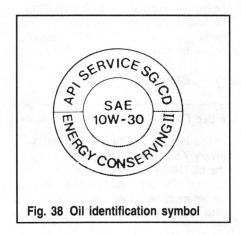

Fig. 38 Oil identification symbol

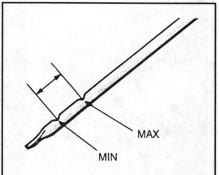

Fig. 39 Inspection of oil level using dipstick. Note the positioning of the maximum (MAX) and the minimum (MIN) marks on the oil dipstick.

Engine Oil Recommendations

▶ See Figures 37 and 38

Engine oil should be used which conforms to the requirements of the API classification 'For Service SG' or 'For Service SG/CD', and have the proper SAE grade number for the expected temperature range of vehicle operation.

❊❊CAUTION

Non-detergent or straight mineral oil must never be used.

In order to improve fuel economy and conserve energy new, lower friction engine oils have been developed. These oils are readily available and can be identified by such labels as 'Energy Conserving', 'Energy Saving', 'Improved Economy', etc.

A standard symbol appears on the top of oil containers and has 3 distinct areas for identifying various aspects of the oil. The top portion will indicate quality of the oil. The center portion will show SAE viscosity grade, such as SAE 10W-30.

Fig. 40 The oil level should be between the 2 notches (MAX AND MIN) on the oil dipstick. Always use a clean rag when wiping off dipstick.

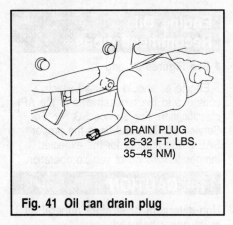

DRAIN PLUG
26–32 FT. LBS.
35–45 NM)

Fig. 41 Oil pan drain plug

'Energy Conserving" shown in the lower portion, indicates that the oil has fuel-saving capabilities.

Engine

OIL LEVEL CHECK

▶ See Figures 39 and 40

The best time to check the engine oil is before operating the engine or after it has been sitting for at least 10 minutes in order to gain an accurate reading. This will allow the oil to drain back in the crankcase. To check the engine oil level, make sure that the vehicle is resting on a level surface, remove the oil dipstick, wipe it clean and reinsert the stick firmly for an accurate reading. The oil dipstick has two marks to indicate the minimum (MIN.) and maximum (MAX) oil level. If the oil is at or below the 'MIN" mark on the dipstick, oil should be added as necessary. The oil level should be maintained in the safety margin, neither going above the 'MAX" mark or below the 'MIN" mark.

OIL AND FILTER CHANGE

▶ See Figures 41, 42, 43, 44 and 45

➡The factory maintenance intervals (every 7,500 miles) specify changing the oil filter at every second oil change after the initial service. We recommend replacing the oil filter with every oil change.

1. Run the engine until it reaches normal operating temperature.
2. Remove the oil filler cap.
3. Raise and safely support the front of the vehicle.
4. Remove the oil drain plug and allow the engine oil to drain.

➡The EPA warns that prolonged contact with used engine oil may cause a number of skin disorders, including cancer! You should make every effort to minimize your contact to used engine oil. Protective gloves should be worn when changing engine oil. Wash hands and any other exposed skin as soon as possible after contact.

5. Install the drain plug into the oil pan with new gasket installed. Tighten the drain plug to 26-32 ft. lbs. (35-45 Nm).

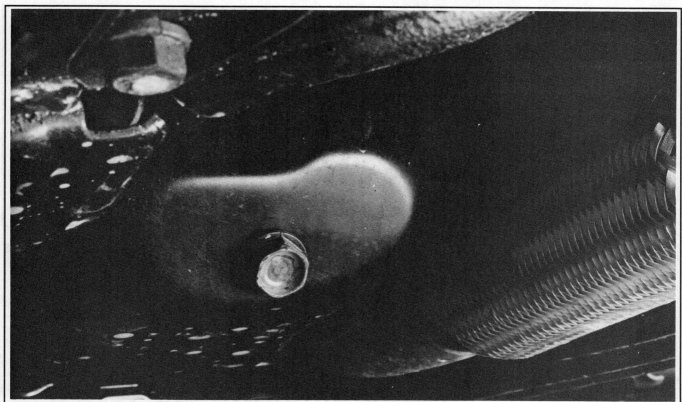

Fig. 42 Clean the area of the oil pan drain plug prior to removing from the oil pan. This will help to prevent the entrance of foreign material into the crankcase during oil plug installation.

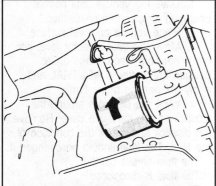

Fig. 43 Removing the oil filter from the engine

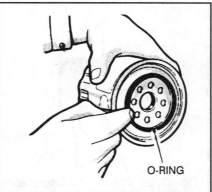

O-RING

Fig. 44 Lubrication of the oil filter O-ring prior to installation

6. Remove the engine oil filter using the appropriate sized oil filter wrench. Clean the oil filter mounting bracket.

➡**Make sure the O-ring from the old oil filter is on the old oil filter. If a new oil filter is installed with the old O-ring stuck on the engine filter head, a severe oil leak will result. This, if not noticed, could result in engine damage. The quality of replacement filters vary widely. Only high quality filters should be used to assure most**

efficient service. Install only oil filters capable of withstanding a pressure of 256 psi.

7. Coat the O-ring on the top of the new oil filter lightly with clean engine oil. Thread the filter onto the oil filter head by hand. Tighten the oil filter to 8-10 ft. lbs. (11-13 Nm).

8. Install new engine oil through the oil filler, filling to the appropriate level.

9. Start the engine and allow to run for a few minutes. Check for oil leaks.

10: Shut the engine OFF and check the oil level. Add as required to fill to the correct level.

Manual Transaxle

FLUID RECOMMENDATION

For manual transaxles, there are a variety of fluids available (depending upon the outside temperature); be sure to use Hypoid gear oil SAE 75W-85W conforming to API specifications GL-4 or higher.

FLUID LEVEL CHECK

▶ **See Figures 46 and 47**

Inspect each component for leaking. Check the oil level by removing the filler plug. If the oil is contaminated, it is necessary to replace it with new oil.

1. Park the vehicle on level surface.
2. Remove the filler plug and make sure the oil level is at the lower portion of the filler hose.

Fig. 45 Removing the oil filler cap from the engine

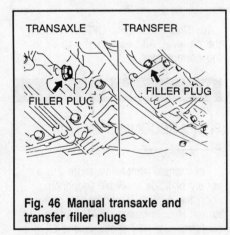

Fig. 46 Manual transaxle and transfer filler plugs

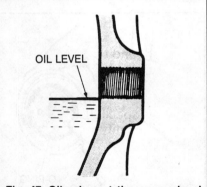

Fig. 47 Oil, when at the proper level, will register at the bottom of the oil drain hole opening

3. Check to be sure that the transaxle oil is not noticeably dirty and that it has a suitable viscosity.

DRAIN AND REFILL FLUID

1. Position the vehicle on a flat surface.
2. Remove the filler plug and the drain plug and allow the oil to drain.
3. Refill the transaxle to the proper level with Hypoid gear oil SAE 75W-85W

conforming to API specifications GL-4 or higher. The oil level should be at the bottom of the oil filler hole.

    F5M22 transaxle — 1.9 qts. (1.8 liters)
    F5M33 transaxle — 2.3 qts. (2.3 liters)
    Transfer case — 0.63 qts. (0.6 liters)

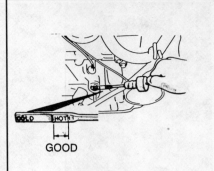

Fig. 48 Inspection of automatic transaxle fluid level using dipstick

Automatic Transaxle

FLUID RECOMMENDATION

Recommended automatic transaxle fluid is Mitsubishi Plus ATF or equivalent.

FLUID LEVEL CHECK

▶ **See Figure 48**

1. Drive the vehicle until normal operating temperature is reached.
2. Place vehicle on level surface.
3. Move the gear selector level into every position. Once this is done, position the shifter in NEUTRAL and apply the parking brake firmly.
4. Wipe the dirt from around the dipstick on the transaxle case. Remove the dipstick and check the condition of the fluid. The fluid should be changed if:
    The fluid smells burnt
    The fluid is discolored
    There is noticeable amounts of metal particles in the fluid.
5. Wipe the fluid from the end of the dipstick and reinsert it into the transaxle assembly. Pull out the dipstick and inspect the fluid level on the stick.
6. The fluid level should be in the HOT range on the stick. If not, add the appropriate fluid to fill to specifications.
7. Inspect the transaxle and related components for fluid leaks. Repair leaks as required.

➡**When adding fluid to the transaxle, do not overfill the transaxle. Too much fluid could result in fluid aeration and cause slipping and eventual damage.**

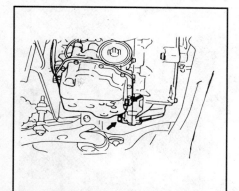

Fig. 49 Automatic transaxle drain plug locations

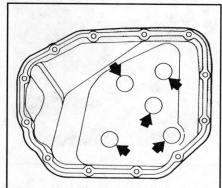

Fig. 50 Positioning of magnets inside of the automatic transaxle oil pan

FLUID DRAINING AND REFILLING

▶ **See Figures 49 and 50**

1. Position vehicle on level ground.
2. Remove the drain plug and allow the fluid to drain.
3. Remove the retainers and the transaxle oil pan.
4. Check the oil filter for damage or restrictions and replace as required.
5. Clean the inside of the pan and the magnets. Make sure the magnets are positioned to the concave part of the oil pan.
6. Clean all gasket mating surfaces.
To install:
7. Install the oil pan with new gasket in place. Install and tighten the retaining bolts to 8 ft. lbs. (12 Nm).
8. Install the oil drain plug with new gasket in place. Tighten the plug to 25 ft. lbs. (35 Nm).

9. Fill the transaxle assembly through the dipstick hole, adding 8.5 pints of specified ATF.
10. Start the engine and allow to idle for 2 minutes. Cycle the gear selector lever through all gears and then position in NEUTRAL. Add sufficient ATF to the transaxle to fill to the lower mark.
11. Allow the engine to reach normal operating temperature. Recheck the fluid level and add until fluid registers at the HOT mark on the dipstick.
12. Inspect for ATF leaks.

PAN AND FILTER SERVICE

1. Remove the transaxle oil pan.
2. Remove the oil filter retainers and the filter noting exact positioning of both.
To install:
3. Install the oil filter in the exact positioning as removed and secure with the retainers.
4. Install the transaxle oil pan.
5. Refill the transaxle assembly with the correct amount of ATF.
6. Inspect for fluid leaks.

Transfer Case

FLUID RECOMMENDATIONS

Recommended fluid for the transfer case is Hypoid gear oil SAE 75W-85W conforming to API specifications GL-4 or higher.

FLUID LEVEL CHECK

Inspect each component for leaking. Check the oil level by removing the filler plug. If the oil is contaminated, it is necessary to replace it with new oil.
1. Park the vehicle on level surface.
2. Remove the filler plug and make sure the oil level is at the lower portion of the filler hole.
3. Check to be sure that the oil is not noticeably dirty and that it has the proper viscosity.

DRAIN AND REFILL FLUID

1. Position the vehicle on a flat surface.

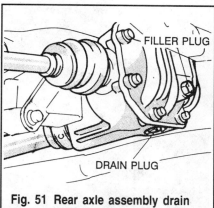

Fig. 51 Rear axle assembly drain plug — AWD models

2. Remove the filler and the drain plug and allow the oil to drain.
3. Refill the transfer to the proper level with Hypoid gear oil SAE 75W-85W conforming to API specifications GL-4 or higher. The oil level should be at the bottom of the oil filler hole.
Transfer case — 0.63 qts. (0.6 liters)

Rear Axle

FLUID RECOMMENDATIONS

AWD Vehicles

Recommended fluid for the rear axle is Hypoid gear oil SAE 80W-90 conforming to API specifications GL-5 or higher. Fluid viscosity range may vary depending on specific temperature range of operation.

FLUID LEVEL CHECK

1. Park the vehicle on level ground.
2. Remove the oil fill plug and check the oil level.
3. The oil level is sufficient if it reaches the lower portion of the filler plug hole. Add fluid as required.

DRAIN AND REFILL FLUID

▶ **See Figure 51**

1. Position the vehicle on a flat surface.
2. Remove the filler and the drain plug and allow the oil to drain.
3. Refill the transfer with 0.74 qt. Hypoid gear oil SAE 80W-90 conforming to API specifications GL-5 or higher. The

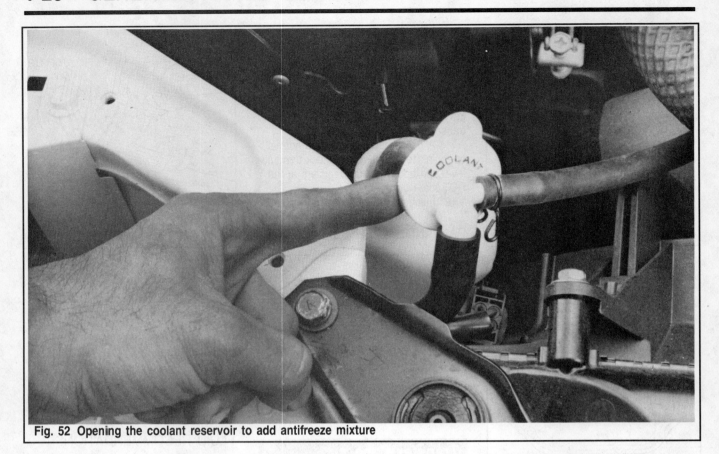

Fig. 52 Opening the coolant reservoir to add antifreeze mixture

oil level should be at the bottom of the oil filler hole.

Cooling System

FLUID RECOMMENDATIONS

The cooling fluid or antifreeze normally should be changed every 30,000 miles or 24 months. When replacing the fluid, use a mixture of 50% water and 50% ethylene glycol antifreeze.

LEVEL CHECK

▶ See Figure 52

Check the coolant level every 3,000 miles or once a month. In hot weather operation, it may be a good idea to check the level once a week. Check for loose connections and signs of deterioration of the coolant hoses. Check the coolant level in the reservoir when the engine is cold, making sure the level is at the **MAX** mark. If the bottle is empty, check the level in the radiator and refill as necessary. Then fill the bottle up to the MAX level.

❋❋CAUTION

Never remove the radiator cap when the vehicle is hot or overheated. Wait until it has cooled. Place a thick cloth over the radiator cap to shield yourself from the heat and turn the radiator cap, SLIGHTLY, until the sound of escaping pressure can be heard. DO NOT turn any more; allow the pressure to release gradually. When no more pressure can be heard escaping, remove the cap with the heavy cloth, CAUTIOUSLY.

➡Never add cold water to an overheated engine

After filling the radiator, run the engine until it reaches normal operating temperature, to make sure that the thermostat has opened and all the air is bled from the system.

DRAIN AND REFILL FLUID

Allow the engine to cool completely before draining the cooling system.

1. Remove the radiator cap. Place a drain plug under the draincock tube.
2. Loosen the draincock and allow the coolant mixture to drain from the system.
3. Drain the coolant from the reserve tank, as required.
4. Once all the coolant has drained from the system, close the draincock.
5. Pour the coolant/water mixture into the radiator until the fluid level is at the filler neck.
6. Pour the coolant/water mixture into the radiator reservoir bottle until fluid is at the proper level.
7. Start the engine and allow to idle until the thermostat opens. Carefully open the radiator cap and inspect the fluid level. Add as required to bring the fluid to the filler neck. Install the radiator cap securely.
8. Fill the reserve tank up to the FULL line. Inspect for coolant leaks.

Fig. 53 The correct fluid level in the clutch master cylinder reservoir (smaller) and the brake master cylinder reservoir (larger) is between the MAX and the MIN marks.

FLUSHING AND CLEANING THE SYSTEM

1. Drain the cooling system. Close the petcock slightly.
2. Remove the thermostat from the engine. Disconnect the upper radiator hose at the radiator neck.
3. Install a high pressure hose into the thermostat housing and allow the water pressure to back-flush the system.
4. Continue this procedure until the water coming from the hose is clean.
5. Reverse the removal procedure and refill the cooling system with fresh coolant.

Brake Master Cylinder

FLUID RECOMMENDATIONS

When adding or changing the fluid in the systems, use a quality brake fluid of the DOT 3 or DOT 4 specifications. Never reuse old brake fluid.

✻✻CAUTION

Be careful to avoid spilling any brake fluid on painted surfaces, because the paint coat will become discolored or damaged.

LEVEL CHECK

▶ **See Figure 53**

Check the levels of brake fluid in the brake and clutch master cylinder reservoirs once a month or every 3,000 miles. The fluid level should be maintained to a level between the Maximum (Max.) and the minimum (Min.) lines on the master cylinder reservoir. Never add a mixture of fluid. Any sudden decrease in the level in the reservoir indicates a leak in the brake system and should be checked out immediately. A gradual decreases in the fluid level in the master cylinder is normal due to brake pad wear.

Clutch Master Cylinder

FLUID RECOMMENDATIONS

When adding or changing the fluid in the systems, use a quality brake fluid conforming to DOT 3 specifications. Never reuse old brake fluid.

FLUID LEVEL CHECK

▶ **See Figure 53**

1. Remove the air cleaner assembly to gain access to the clutch master cylinder cap, if required.
2. Wipe the cap and the surrounding area clean with a shop towel.
3. Inspect the fluid in the reservoir making sure fluid is between the Max and the MIN marks
4. Remove the clutch master cylinder cover and add fresh fluid to fill to the top full mark on the reservoir, if required.

✻✻CAUTION

Be careful to avoid spilling any brake fluid on painted surfaces, because the paint coat will become discolored or damaged.

5. Reinstall the cover onto the clutch master cylinder.
6. Install the air cleaner assembly, if removed.

Power Steering Pump

FLUID RECOMMENDATIONS

When adding or changing the power steering fluid, use Dexron®II ATF (Automatic Transmission Fluid); the system uses approximately 0.95 qts. of fluid.

LEVEL CHECK

▶ **See Figure 54**

Like all other general maintenance items, check every 3,000 miles or once a month. Inspect the oil level in the reservoir by checking the position of the fluid against the mark on the dipstick.

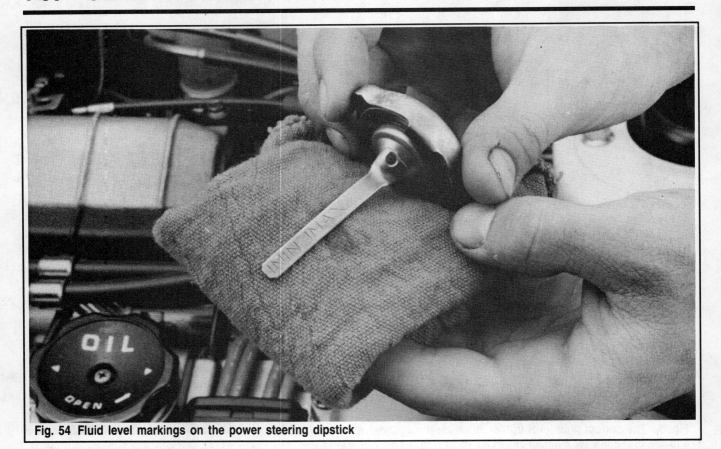

Fig. 54 Fluid level markings on the power steering dipstick

Add fluid to the reservoir if the fluid does not reach the appropriate full line.

Chassis Greasing

On most models, the manufacturer doesn't install lubrication fittings on lube points on the steering linkage or suspension. However, if the lubrication point does have a grease fitting, lubricate with multipurpose NLGI No. 2 (Lithium base) grease.

Body Lubrication and Maintenance

Body hinges and latches should be lubricated as needed to maintain smooth operation of doors, hoods and latches. When required, lubricate the hood lock latch, door lock strikers, seat adjusters, liftgate lock and the parking brake control cable mechanism with multipurpose NLGI No. 2 (Lithium base) grease. Wipe area to be lubricated clean with a shop rag and inspect for damage or misalignment. Adjust the component, if required, prior to lubrication.

Keeping your vehicle clean extends the beauty and the life of your vehicle.

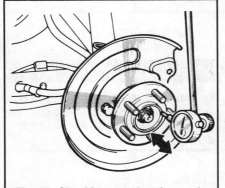

Fig. 55 Checking rear bearing end-play while moving the hub in and out

When washing your car with high pressure car washing equipment or steam car washing equipment, be sure to stay back away from the vehicle. Be sure to maintain the spray nozzle at a distance of at least 12 in. (300mm) from any plastic parts and all opening parts such as doors, luggage compartment, hood, etc. Do not clean aluminum wheels with pressure or steam cleaners or damage to the finish will occur.

Rear Wheel Bearings

▶ See Figure 55

REMOVAL & INSTALLATION

➡This section pertains to Front Wheel Drive vehicles only. For All Wheel Drive vehicles, please refer to the Drive Axle section.

1. Raise the vehicle and support safely.
2. Remove the tire and wheel assembly.
3. Remove the bolt(s) holding the speed sensor bracket to the knuckle and remove the assembly from the vehicle.

➡The speed sensor has a pole piece projecting from it. This exposed tip must be protected from impact or scratches. Do not allow the pole piece to contact the toothed wheel during removal or installation.

4. Remove the caliper from the brake disc and suspend with a wire.
5. Remove the brake disc.

➡ **The rear hub assembly can not be disassembled. If bearing replacement is required, replace the assembly as a unit.**

To install:

8. Install the hub and bearing assembly.

9. Install the tongued washer and a new self-locking nut.

10. Torque the nut to 144-188 ft. lbs. (200-260 Nm). Align with the indentation in the spindle and crimp nut down to lock in place.

11. Set up a dial indicator and measure the end-play while moving the hub in and out.

12. If the end-play exceeds 0.004 in. (0.01mm), retorque the nut and remeasure the end-play. If still beyond the limit, replace the hub and bearing unit.

13. Install the grease cap and brake parts.

14. Temporarily install the speed sensor to the knuckle; tighten the bolts only finger-tight.

15. Route the speed sensor cable correctly and loosely install the clips and retainers. All clips must be in their original position and the sensor cable must not be twisted. Improper installation may cause cable damage or system failure.

➡ **The wiring in the harness is easily damaged by twisting and flexing. Use the white stripe on the outer insulation to keep the sensor harness properly placed.**

16. Use a brass or other non-magnetic feeler gauge to check the air gap between the tip of the pole piece and the toothed wheel. Correct gap is 0.012-0.035 in. (0.3-0.9mm). Tighten the 2 sensor bracket bolts to 10 ft. lbs. (14 Nm) with the sensor located so the gap is the same at several points on the toothed wheel. If the gap is incorrect, it is likely that the toothed wheel is worn or improperly installed.

17. Install the tire and wheel assembly.

Trailer Towing

It is important to realize that towing places a substantial increase in load on the vehicle's transaxle, engine, braking system, cooling system, drive train and other systems. Your vehicle was primarily designed to carry passengers.

It is generally not considered to be a good choice in heavy applications.

PUSHING AND TOWING

▸ See Figures 56, 57 and 58

Front Wheel Drive Vehicles (FWD)

FRONT TOWING PICKUP

✳✳CAUTION

These vehicles can not be towed by a wrecker using sling-type equipment to prevent the bumper from deformation. If these vehicles require towing, use a wheel lift or flat bed equipment.

The vehicle may be towed on it's rear wheel, provided its parking brake is released. It is recommended that the vehicle be towed using the front pickup whenever possible.

REAR TOWING PICKUP

✳✳CAUTION

These vehicles can not be towed by a wrecker using sling-type equipment to prevent the bumper from deformation. If these vehicles require towing, use a wheel lift or flat bed equipment.

Manual transaxle vehicles may be towed on the front wheels provided that the transaxle is in NEUTRAL and the drive-line has not been damaged. The steering wheel must be clamped in the straight-ahead position with a steering wheel clamping device designed for towing service use.

✳✳CAUTION

Do not use the steering column lock to secure the front wheel position for towing.

Automatic transaxle vehicles may be towed on the front wheels at speeds not to exceed 30 mph (50 km/h) for a distance not to exceed 18 miles (30 km).

✳✳CAUTION

If these limits can not be met, then the front wheels must be placed on a tow dolly.

All Wheel Drive Models (AWD)

✳✳CAUTION

All Wheel Drive vehicles should only be towed with all 4 wheel lifted from the road surface. This means that the vehicle is to be towed with flat bed equipment or all 4 wheels off of the ground.

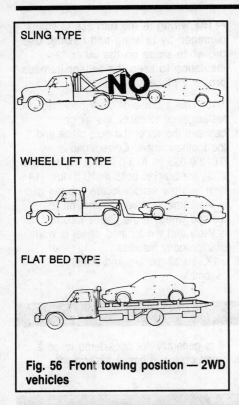

Fig. 56 Front towing position — 2WD vehicles

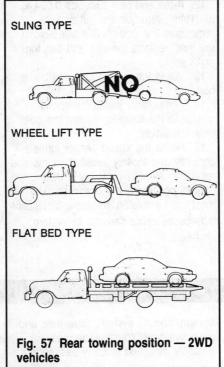

Fig. 57 Rear towing position — 2WD vehicles

On All Wheel Drive (AWD) models, the basic principal is that all 4 wheels are to be raised on tow dollies before towing. If equipped with manual transaxle, the shift lever is to be set in 1st Gear and the parking brake should be applied. If equipped with automatic transaxle, the selector lever should be set to P position and the parking brake should be applied.

The vehicle must not be towed by placing only its front wheels or its rear wheels on a rolling dolly because to do so will result in deterioration of the viscous coupling and result in the viscous coupling causing the vehicle to jump forward suddenly.

If only the front or the rear wheels are lifted for towing, the bumper will be damaged. In addition, lifting of the rear wheels causes the oil to flow forward, and may result in heat damage to the rear bushing of the transfer, and so should never be done.

JACKING THE VEHICLE

▶ **See Figures 59, 60, 61 and 62**

Jack receptacles are located at the body sills to accept the scissors jack supplied with the vehicle. Always block the opposite wheels and jack the vehicle on a level surface.

❋❋CAUTION

Climbing under a car supported by just the jack is extremely dangerous and should not be done.

The usual type of floor type jack is used in the following locations:
Front:
FWD — Under the mid point of the crossmember
AWD — Under the mid point of the crossmember
Rear:
FWD — Under the jack up bracket of the rear floor pan
AWD — Under the rear differential

Cautions

1. Never use a jack at the lateral rod or the rear suspension assembly on a FWD vehicle.

2. In order to prevent scaring of the centermember on FWD vehicle, or crossmember on AWD vehicle, place a piece of cloth on the jack's contact surface. This will prevent the formation of corrosion caused by damage to the protective coating.

3. A floor jack must never be used on any part of the underbody.

4. Do not attempt to raise one entire side of the vehicle by placing the jack midway between he front and the rear wheels. This practice may result in permanent damage to the body.

6. Remove the grease cap, self-locking nut and tongued washer.

7. Remove the rear hub and bearing assembly.

Towing methods	Remarks
If a tow truck is used Lifting method for 4 wheels — **Good** 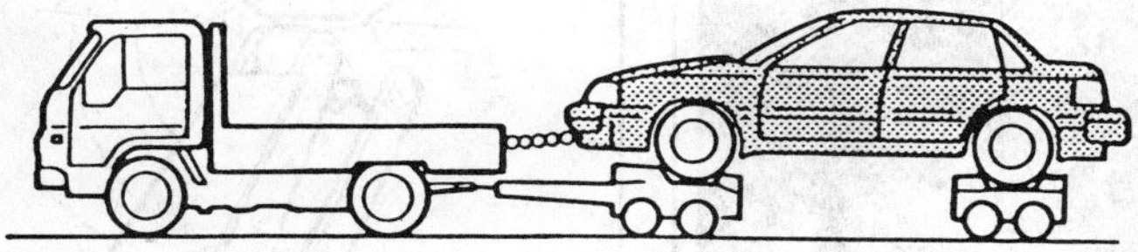	• For 4WD models, the basic principle is that all four wheels are to be raised before towing. • The shift lever should be set to 1st gear and the parking brake should be applied.
Front wheels lifted — **No good** 	• The vehicle must not be towed by placing only its front wheels or only the rear wheels on a rolling dolly, because to do so will result in deterioration of the viscous coupling and result in the viscous coupling causing the vehicle to jump forward suddenly.
Front wheels lifted — **No good** 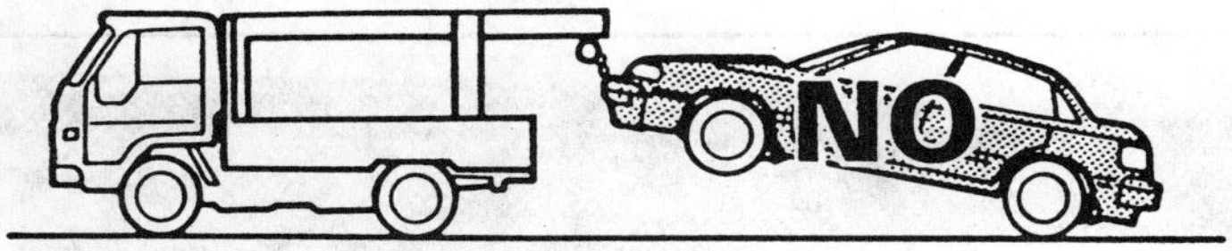	• If only the front wheels or only the rear wheels are lifted for towing, the bumper will be damaged. In addition, lifting of the rear wheels causes the oil to flow forward, and may result in heat damage to the rear bushing of the transfer, and so should never be done.
Rear wheels lifted — **No good** 	
Towing by rope or cable — **Good** 	• The front and rear wheels must rotate normally. • The various mechanisms must function normally. • The shift lever must be set to the neutral position and the ignition key must be set to "ACC".

Fig. 58 Towing instructions — 4WD models

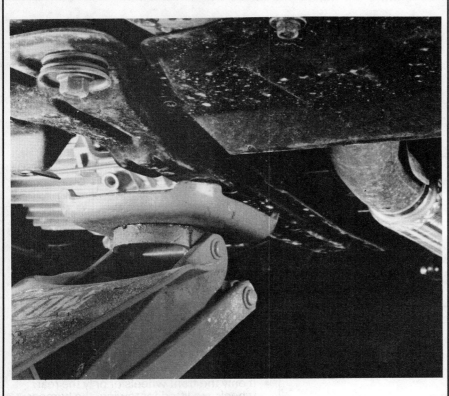

Fig. 60 Jack point on front of 2WD vehicle

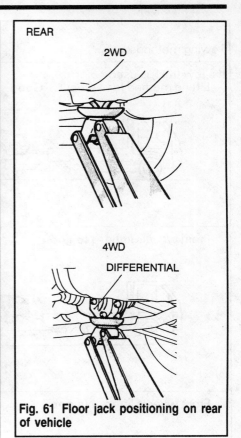

Fig. 61 Floor jack positioning on rear of vehicle

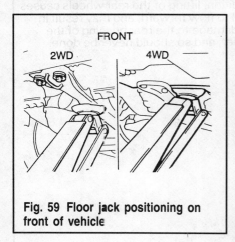

Fig. 59 Floor jack positioning on front of vehicle

Fig. 62 Jack point on rear of 2WD vehicle

GENERAL MAINTENANCE SERVICE SCHEDULE FOR NORMAL USAGE CONDITIONS

No.	General Maintenance		Service Intervals		Kilometers in Thousands	24	48	72	80	96
					Mileage in Thousands	15	30	45	50	60
5	Timing Belt (Including the Balancer Belt)		Replace	at						X
6	Drive Belt (for Water Pump and Alternator)		Inspect for tension	at			X			X
7	Engine Oil	Non-Turbo	Change Every Year	or	Every 12,000 km (7,500 miles)					
		Turbo	Change Every 6 Months		Every 8,000 km (5,000 miles)					
8	Engine Oil Filter	Non-Turbo	Change Every Year	or		X	X	X		X
		Turbo	Change Every Year		Every 16,000 km (10,000 miles)					
9	Manual Transaxle Oil		Inspect Oil Level	at			X			X
10	Automatic Transaxle Fluid		Inspect Fluid Level Every Year	or		X	X	X		X
			Change Fluid	at			X			X
11	Engine Coolant		Replace Every 2 Years	or			X			X
12	Disc Brake Pads		Inspect for Wear Every Year	or		X	X	X		X
13	Brake Hoses		Check for Deterioration or Leaks Every Year	or		X	X	X		X
14	Ball Joint and Steering Linkage Seals		Inspect for Grease Leaks and Damage Every 2 Years	or			X			X
15	Drive Shaft Boots		Inspect for Grease Leaks and Damage Every Year	or		X	X	X		X
16	Rear Axle <AWD>	With LSD	Change Oil				X			X
		Without LSD	Inspect Oil Level				X			X
17	Exhaust System (Connection Portion of Muffler, Pipings and Converter Heat Shields)		Check and Service as Required Every 2 Years	or			X			X

NOTE
LSD: Limited-slip differential

GENERAL MAINTENANCE SERVICE SCHEDULE FOR SEVERE USAGE CONDITIONS

Maintenance Item	Service to be Performed		Mileage Intervals Kilometers in Thousands (Miles in Thousands)									Severe Usage Conditions						
			12 (7.5)	24 (15)	36 (22.5)	48 (30)	60 (37.5)	72 (45)	80 (50)	84 (52.5)	96 (60)	A	B	C	D	E	F	G
Air Cleaner Element	Replace		More Frequently									X				X		
Spark Plugs	Replace			X		X		X			X		X		X			
Engine Oil	Change Every 3 Months	or	Every 4,800 km (3,000 miles)									X	X	X	X			X
Engine Oil Filter	Replace Every 6 Months	or	Every 9,600 km (6,000 miles)									X	X	X	X			X
Disc Brake Pads	Inspect for Wear		More Frequently									X					X	

Severe usage conditions
A—Driving in dusty conditions
B—Trailer towing or police, taxi, or commercial type operation
C—Extensive idling
D—Short trip operation at freezing temperatures (engine not thoroughly warmed up)
E—Driving in sandy areas
F—Driving in salty areas
G—More than 50% operation in heavy city trafic during hot weather above 32°C (90°F)

CAPACITIES

Year	Model	Engine ID/VIN	Engine Displacement Liters (cc)	Engine Crankcase with Filter	Transaxle (pts.) 4-Spd	Transaxle (pts.) 5-Spd	Transaxle (pts.) Auto.	Transfer Case (pts.)	Drive Axle Front (pts.)	Drive Axle Rear (pts.)	Fuel Tank (gal.)	Cooling System (qts.)
1990	Eclipse	T	1.8 (1755)	4.1	—	4.8	13.0	—	—	—	15.9	6.6
	Eclipse	R	2.0 (1997)	4.5	—	4.8	14.8	—	—	—	15.9	7.6
	Eclipse	U	2.0 (1997)	4.8	—	4.8	14.8	1.3	—	1.5	15.9	7.6
	Laser	T	1.8 (1755)	4.1	—	4.8	13.0	—	—	—	15.9	6.6
	Laser	R	2.0 (1997)	4.5	—	4.8	14.8	—	—	—	15.9	7.6
	Laser	U	2.0 (1997)	4.8	—	4.8	14.8	1.3	—	1.5	15.9	7.6
	Talon	R	2.0 (1997)	4.5	—	4.8	14.8	—	—	—	15.9	7.6
	Talon	U	2.0 (1997)	4.8	—	4.8	14.8	1.3	—	1.5	15.9	7.6
1991	Eclipse	T	1.8 (1755)	4.1	—	4.8	13.0	—	—	—	15.9	6.6
	Eclipse	R	2.0 (1997)	4.5	—	4.8	14.8	—	—	—	15.9	7.6
	Eclipse	U	2.0 (1997)	4.8	—	4.8	14.8	1.3	—	1.5	15.9	7.6
	Laser	T	1.8 (1755)	4.1	—	4.8	13.0	—	—	—	15.9	6.6
	Laser	R	2.0 (1997)	4.5	—	4.8	14.8	—	—	—	15.9	7.6
	Laser	U	2.0 (1997)	4.8	—	4.8	14.8	1.3	—	1.5	15.9	7.6
	Talon	R	2.0 (1997)	4.5	—	4.8	14.8	—	—	—	15.9	7.6
	Talon	U	2.0 (1997)	4.8	—	4.8	14.8	1.3	—	1.5	15.9	7.6
1992	Eclipse	T	1.8 (1755)	4.1	—	4.8	13.0	—	—	—	15.9	6.6
	Eclipse	R	2.0 (1997)	4.5	—	4.8	14.8	—	—	—	15.9	7.6
	Eclipse	U	2.0 (1997)	4.8	—	4.8	14.8	1.3	—	1.5	15.9	7.6
	Laser	T	1.8 (1755)	4.1	—	4.8	13.0	—	—	—	15.9	6.6
	Laser	R	2.0 (1997)	4.5	—	4.8	14.8	—	—	—	15.9	7.6
	Laser	U	2.0 (1997)	4.8	—	4.8	14.8	1.3	—	1.5	15.9	7.6
	Talon	R	2.0 (1997)	4.5	—	4.8	14.8	—	—	—	15.9	7.6
	Talon	U	2.0 (1997)	4.8	—	4.8	14.8	1.3	—	1.5	15.9	7.6
1993	Eclipse	B	1.8 (1755)	4.1	—	4.8	13.0	—	—	—	15.9	6.6
	Eclipse	E	2.0 (1997)	4.5	—	4.8	14.8	—	—	—	15.9	7.6
	Eclipse	F	2.0 (1997)	4.8	—	4.8	14.8	1.3	—	1.5	15.9	7.6
	Laser	B	1.8 (1755)	4.1	—	4.8	13.0	—	—	—	15.9	6.6
	Laser	E	2.0 (1997)	4.5	—	4.8	14.8	—	—	—	15.9	7.6
	Laser	F	2.0 (1997)	4.8	—	4.8	14.8	1.3	—	1.5	15.9	7.6
	Talon	B	1.8 (1755)	4.1	—	4.8	13.0	—	—	—	15.9	6.6
	Talon	E	2.0 (1997)	4.5	—	4.8	14.8	—	—	—	15.9	7.6
	Talon	F	2.0 (1997)	4.8	—	4.8	14.8	1.3	—	1.5	15.9	7.6

TORQUE SPECIFICATIONS

Component	U.S.	Metric
Fuel filter eye bolt	22 ft. lbs.	30 Nm
Fuel filter flare nut	25 ft. lbs.	35 Nm
Fuel filter mounting bolt	10 ft. lbs.	14 Nm
PCV valve	6 ft. lbs.	8 Nm
Battery cable terminals	10 ft. lbs.	15 Nm
Lug nuts	87-101 ft. lbs.	120-140 Nm
Oil drain plug	32 ft. lbs.	45 Nm
Oil filter	10 ft. lbs.	13 Nm
Automatic transaxle		
Oil pan retaining bolts	8 ft. lbs.	12 Nm
Oil drain plug	25 ft. lbs.	35 Nm
Rear wheel bearing nut	144-188 ft. lbs.	200-260 Nm
Speed sensor bracket bolt	10 ft. lbs.	14 Nm
Spark plug	18 ft. lbs.	24 Nm

Troubleshooting Basic Air Conditioning Problems

Problem	Cause	Solution
There's little or no air coming from the vents (and you're sure it's on)	• The A/C fuse is blown • Broken or loose wires or connections • The on/off switch is defective	• Check and/or replace fuse • Check and/or repair connections • Replace switch
The air coming from the vents is not cool enough	• Windows and air vent wings open • The compressor belt is slipping • Heater is on • Condenser is clogged with debris • Refrigerant has escaped through a leak in the system • Receiver/drier is plugged	• Close windows and vent wings • Tighten or replace compressor belt • Shut heater off • Clean the condenser • Check system • Service system
The air has an odor	• Vacuum system is disrupted • Odor producing substances on the evaporator case • Condensation has collected in the bottom of the evaporator housing	• Have the system checked/repaired • Clean the evaporator case • Clean the evaporator housing drains
System is noisy or vibrating	• Compressor belt or mountings loose • Air in the system	• Tighten or replace belt; tighten mounting bolts • Have the system serviced

Sight glass condition
  Constant bubbles, foam or oil streaks
  Clear sight glass, but no cold air
  Clear sight glass, but air is cold
  Clouded with milky fluid
	
	• Undercharged system
	
	• No refrigerant at all
	• System is OK
	• Receiver drier is leaking dessicant
		
		• Charge the system
		
		• Check and charge the system
		
		• Have system checked

Large difference in temperature of lines	• System undercharged	• Charge and leak test the system
Compressor noise	• Broken valves • Overcharged	• Replace the valve plate • Discharge, evacuate and install the correct charge

Troubleshooting Basic Air Conditioning Problems (cont.)

Problem	Cause	Solution
	• Piston slap	• Replace the compressor
	• Broken rings	• Replace the compressor
	• Drive belt pulley bolts are loose	• Tighten with the correct torque specification
Excessive vibration	• Incorrect belt tension	• Adjust the belt tension
	• Clutch loose	• Tighten the clutch
	• Overcharged	• Discharge, evacuate and install the correct charge
	• Pulley is misaligned	• Align the pulley
Condensation dripping in the passenger compartment	• Drain hose plugged or improperly positioned	• Clean the drain hose and check for proper installation
	• Insulation removed or improperly installed	• Replace the insulation on the expansion valve and hoses
Frozen evaporator coil	• Faulty thermostat	• Replace the thermostat
	• Thermostat capillary tube improperly installed	• Install the capillary tube correctly
	• Thermostat not adjusted properly	• Adjust the thermostat
Low side low—high side low	• System refrigerant is low	• Evacuate, leak test and charge the system
	• Expansion valve is restricted	• Replace the expansion valve
Low side high—high side low	• Internal leak in the compressor—worn	• Remove the compressor cylinder head and inspect the compressor. Replace the valve plate assembly if necessary. If the compressor pistons, rings or cylinders are excessively worn or scored replace the compressor
	• Cylinder head gasket is leaking	• Install a replacement cylinder head gasket
	• Expansion valve is defective	• Replace the expansion valve
	• Drive belt slipping	• Adjust the belt tension
Low side high—high side high	• Condenser fins obstructed	• Clean the condenser fins
	• Air in the system	• Evacuate, leak test and charge the system
	• Expansion valve is defective	• Replace the expansion valve
	• Loose or worn fan belts	• Adjust or replace the belts as necessary
Low side low—high side high	• Expansion valve is defective	• Replace the expansion valve
	• Restriction in the refrigerant hose	• Check the hose for kinks—replace if necessary
Low side low—high side high	• Restriction in the receiver/drier	• Replace the receiver/drier
	• Restriction in the condenser	• Replace the condenser
Low side and high normal (inadequate cooling)	• Air in the system	• Evacuate, leak test and charge the system
	• Moisture in the system	• Evacuate, leak test and charge the system

2

ENGINE
PERFORMANCE
AND
TUNE-UP

TUNE-UP PROCEDURES

GASOLINE ENGINE TUNE-UP SPECIFICATIONS

Year	Engine ID/VIN	Engine Displacement Liters (cc)	Spark Plugs Gap (in.)	Ignition Timing (deg.) MT	Ignition Timing (deg.) AT	Fuel Pump (psi)	Idle Speed (rpm) MT	Idle Speed (rpm) AT	Valve Clearance In.	Valve Clearance Ex.
1990	T	1.8 (1754)	0.039–0.043	5B	5B	38	750	750	Hyd.	Hyd.
	R	2.0 (2000)	0.039–0.043	5B	5B	38	750	750	Hyd.	Hyd.
	U	2.0 (2000)	0.028–0.031	5B	5B	27	750	750	Hyd.	Hyd.
1991	T	1.8 (1754)	0.039–0.043	5B	5B	38	750	750	Hyd.	Hyd.
	R	2.0 (2000)	0.039–0.043	5B	5B	38	750	750	Hyd.	Hyd.
	U	2.0 (2000)	0.028–0.031	5B	5B	27	750	750	Hyd.	Hyd.
1992	T	1.8 (1754)	0.039–0.043	5B	5B	38	750	750	Hyd.	Hyd.
	R	2.0 (2000)	0.039–0.043	5B	5B	38	750	750	Hyd.	Hyd.
	U	2.0 (2000)	0.028–0.031	5B	5B	27	750	750	Hyd.	Hyd.
1993	B	1.8 (1754)	0.039–0.043	5B	5B	38	750	750	Hyd.	Hyd.
	E	2.0 (2000)	0.039–0.043	5B	5B	38	750	750	Hyd.	Hyd.
	F	2.0 (2000)	0.028–0.031	5B	5B	27	750	750	Hyd.	Hyd.

NOTE: The lowest cylinder pressure should be within 75% of the highest cylinder pressure reading. For example, if the highest cylinder is 134 psi, the lowest should be 101. Engine should be at normal operating temperature with throttle valve in the wide open position.
The underhood specifications sticker often reflects tune-up specification changes in production. Sticker figures must be used if they disagree with those in this chart.
Hyd.—Hydraulic

Spark Plugs

▶ **See Figures 1, 2, 3, 4, 5 and 6**

Engine performance and fuel efficiency are of major concern to drivers today. One way to maximize your fuel economy and retain sound engine performance is to keep your car in good running order. One important component that should be periodically inspected is spark plugs.

A typical spark plug consists of a metal shell surrounding a ceramic insulator. A metal electrode extends downward through the center of the insulator and protrudes a small distance. Located at the end of the plug and attached to the side of the outer metal shell is the ground electrode. The ground electrode bends in at a 90° angle, so that its tip is even with and parallel to, the tip of the center electrode. The distance between these two electrodes (measured in thousandths of an inch) is called the spark plug gap.

Spark plugs, as many people know, ignite the air/fuel mixture that is inside each of the cylinders and thus causes an explosion. It is this explosion that forces the piston inside the cylinder downward. The downward motion of each piston results in a transfer of energy from inside each cylinder in the engine, through a number of components such as the crankshaft, transaxle and drive shafts (drivetrain) to the drive wheels of the vehicle. In order for this transfer of energy to take place efficiently, proper spark plug operation is essential. What many people do not know about spark plugs is that they come in many types. A spark plug with a long insulator nose retains heat longer. This type of plug can burn off oil and combustion deposits that accumulate on the electrode under light load conditions, resulting in less frequent fouling of the plug. Conversely, a short nose spark plug will dissipate heat rapidly and prevents pre-ignition and detonation (pinging) under heavy load conditions. Under normal driving conditions, the standard plug will be fine.

Spark plug life varies largely due to operating conditions. In order to maintain peak engine performance and fuel economy, the spark plugs should be removed and inspected at least once a year or every 12,000 miles. Faulty or excessively worn spark plugs should be replaced. It is helpful to check the plugs for various types of deposits as an indication of engine operating condition. Excessive or oily deposits can be an indication of engine component wear and should be investigated to find the source of the deposits.

Once the spark plug is removed from the engine, check that the electrodes are not burnt, insulators are not broken and the porcelain insulator is not burnt. Inspect the plug for the following:
Broken insulators
Wearing electrode
Deposited carbon
Damaged or broken gasket
Burnt condition of porcelain at spark plug gap Dark deposit of carbon on the spark plug indicates too rich a fuel mixture or extremely low air intake. Also, misfiring due to excessive spark gap is suspected. White burn indicates too lean fuel mixture or excessively advanced ignition timing. Also insufficient plug tightening can be suspected.

If the vehicle is driven under normal usage, the manufacturer recommends that new spark plugs be installed every 30,000 miles. If the vehicle is driven under severe usage, the manufacturer recommends that the spark plugs in the engine be changed every 15,000 miles.

It might be noted that when removing and inspecting the spark plugs, it is a good time to take a look around the engine compartment for problems in the

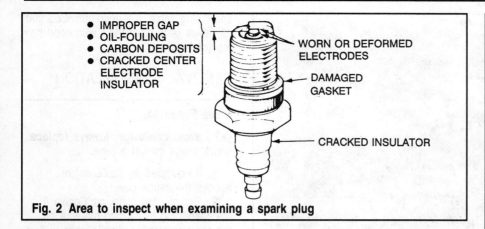

- IMPROPER GAP
- OIL-FOULING
- CARBON DEPOSITS
- CRACKED CENTER ELECTRODE INSULATOR

WORN OR DEFORMED ELECTRODES

DAMAGED GASKET

CRACKED INSULATOR

Fig. 2 Area to inspect when examining a spark plug

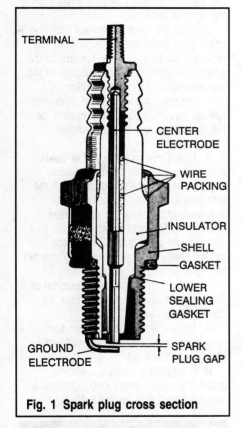

TERMINAL

CENTER ELECTRODE

WIRE PACKING

INSULATOR

SHELL

GASKET

LOWER SEALING GASKET

GROUND ELECTRODE

SPARK PLUG GAP

Fig. 1 Spark plug cross section

making, such as oil and fuel leaks, deteriorating radiator or heater hoses, loose and/or frayed belts etc.

REMOVAL & INSTALLATION

▶ **See Figures 7, 8, 9 and 10**

1. Label the location of the spark plug wires to assure correct installation.
2. Remove the center cover, if equipped with 2.0L engine.
3. Remove the spark plug cable from the spark plug by holding the wire on

Fig. 3 Clean between the outer shell and the insulator with a stiff wire

the cap, twist slightly and pull wire straight out.

4. Clean any dirt an foreign material from around the base of the spark plug. This will prevent dirt from entering the cylinder during spark plug removal.

5. Remove the spark plug from the cylinder head using spark plug socket. Make sure the rubber insert is inside of the spark plug socket or the plug will fall during removal.

To install:

6. Inspect the spark plug and replace if excess wear or damage is present.

7. Check and set the gap of the spark plug with a feeler gauge, as required. The spark plug gap is to be set as follows:

1.8L Engine — 0.039-0.043 in.
2.0L DOHC non-turbocharged engine — 0.039-0.043 in.
2.0L DOHC turbocharged engine — 0.028-0.031

8. Tighten the spark plug to 14-22 ft. lbs. (20-30 Nm).

➡**Do not overtighten the spark plug during installation.**

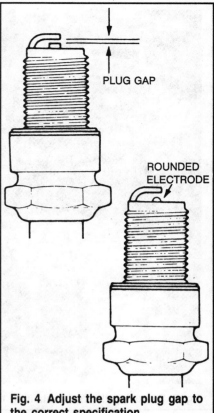

PLUG GAP

ROUNDED ELECTRODE

Fig. 4 Adjust the spark plug gap to the correct specification

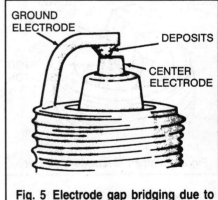

GROUND ELECTRODE

DEPOSITS

CENTER ELECTRODE

Fig. 5 Electrode gap bridging due to carbon deposit

9. Install the spark plug wire(s) onto the plug making sure the wire is securely seated on the plug.

10. If equipped with 2.0L engine, install the center cover and tighten the fasteners to 2.5 ft. lbs. (3.5 Nm).

Spark Plug Cables

Improper arrangement of the spark plug cables will induce voltage between the cables, causing miss firing and developing a surge at acceleration in

Fig. 7 Removing the bolts retaining the center cover

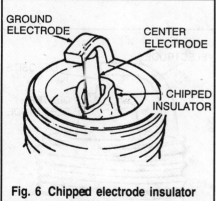

Fig. 6 Chipped electrode insulator

high-speed operation. Therefore, be careful to arrange the spark plug cables properly.

TESTING

▶ **See Figures 1, 12 and 13**

1. Remove the spark plug cable from the engine.

2. Check the cap and the coating of the cable for cracks. Replace the cable if cracks are present.

3. Using a volt ohmmeter, measure the resistance of the wire. Compare the measured resistance to following the desired values:

1.8L Engine
Spark plug cable No. 1 — 10.1K ohms
Spark plug cable No. 2 — 11.5K ohms
Spark plug cable No. 3 — 12.0K ohms
Spark plug cable No. 4 — 13.0K ohms
2.0L DOHC Engines
Spark plug cable No. 1 — 5.8 ohms
Spark plug cable No. 2 — 8.4 ohms
Spark plug cable No. 3 — 10.6 ohms
Spark plug cable No. 4 — 9.7 ohms

4. If the measures resistance differs from the desired values, replace the

spark plug cable. It is recommended that all cables be replaced if 1 wire need be replaced.

REMOVAL & INSTALLATION

▶ **See Figure 14**

➡**To avoid confusion, always replace spark plugs one at a time.**

1. If equipped with 2.0L engine, remove the center cover.
2. Remove the spark plug cable from the spark plug by holding the wire on the cap and twisting slightly while pulling the wire straight out.
3. With the cable removed from the spark plug, unfasten the wire from the retainers. Note the exact position of the wires to assure they are installed in the same position during installation.
4. Remove the cable end from the ignition coil on 2.0L DOHC engine, or the distributor cap on 1.8L engine.
   To install:
5. Install the cable over the spark plug and push until securely seated.
6. Install the cables securely in the retainers in the same position as removed. Make sure the cables are securely installed to avoid possible contact with metal parts. Install the cables neatly, ensuring that they are not too tight, loose, twisted or kinked.
7. Install cable end onto terminal of the ignition coil or distributor cap, as equipped.
8. Repeat this procedure on the remaining spark plug cables until all have been replaced.
9. If equipped with 2.0L engine, install the center cover and tighten the fasteners to 2.5 ft. lbs. (3.5 Nm).

Firing Orders

▶ **See Figures 15 and 16**

➡**To avoid confusion, remove and tag the wires one at a time, for replacement.**

DIAGNOSIS OF SPARK PLUGS

Problem	Possible Cause	Correction
Brown to grayish-tan deposits and slight electrode wear.	• Normal wear.	• Clean, regap, reinstall.
Dry, fluffy black carbon deposits.	• Poor ignition output.	• Check distributor to coil connections.
Wet, oily deposits with very little electrode wear.	• "Break-in" of new or recently overhauled engine. • Excessive valve stem guide clearances. • Worn intake valve seals.	• Degrease, clean and reinstall the plugs. • Refer to Section 3. • Replace the seals.
Red, brown, yellow and white colored coatings on the insulator. Engine misses intermittently under severe operating conditions.	• By-products of combustion.	• Clean, regap, and reinstall. If heavily coated, replace.
Colored coatings heavily deposited on the portion of the plug projecting into the chamber and on the side facing the intake valve.	• Leaking seals if condition is found in only one or two cylinders.	• Check the seals. Replace if necessary. Clean, regap, and reinstall the plugs.
Shiny yellow glaze coating on the insulator.	• Melted by-products of combustion.	• Avoid sudden acceleration with wide-open throttle after long periods of low speed driving. Replace the plugs.
Burned or blistered insulator tips and badly eroded electrodes.	• Overheating.	• Check the cooling system. • Check for sticking heat riser valves. Refer to Section 1. • Lean air-fuel mixture. • Check the heat range of the plugs. May be too hot. • Check ignition timing. May be over-advanced. • Check the torque value of the plugs to ensure good plug-engine seat contact.
Broken or cracked insulator tips.	• Heat shock from sudden rise in tip temperature under severe operating conditions. Improper gapping of plugs.	• Replace the plugs. Gap correctly.

ELECTRONIC IGNITION

▶ **See Figures 17 and 18**

Description & Operation

The ignition system on the 1.8L engine contains a contact pointless type distributor, whose advance mechanism is controlled by the Engine Control Module (ECM). The distributor houses a built in crankshaft position sensor, camshaft position sensor, ignition coil and ignition power transistor.

When the ignition switch is turned **ON**, battery voltage is applied to the ignition coil primary winding. As the shaft of the distributor rotates, signals are transmitted from the multi-port injection control unit to the ignition power transistor. These signals activate the power transistor to cause ignition coil primary winding current flow from the ignition coil negative terminal through the power transistor to ground repeatedly. This interruption induces high voltage in the ignition coil secondary windings, which is diverted through the distributor, spark plug cable and spark plug to ground, thus causing ignition in each cylinder.

The ignition system on the 2.0L DOHC engine is of distributorless type. The advance of this system, like the distributor type ignition system on the 1.8L engine, is controlled by the Engine Control Unit (ECU). The distributorless ignition system contains a crank angle sensor which detects the crank angle or position to each cylinder and convert this data into pulse signals. These signals are sent to the ECU, which will calculate the engine rpm and regulate the fuel injection and ignition timing accordingly. The system also contains a top dead center sensor which detects the top dead center position of each cylinder and converts this data into pulse signals. These signals are then sent to the ECU, which will calculate the sequence of fuel injection and engine rpm. Both sensors are located in a common housing on the cylinder head opposite the timing belt side. The power and ground for both sensors is supplied by the ECU.

When the ignition switch is turned **ON**, battery voltage is applied to the ignition coil primary winding. As the crank angle sensor shaft rotates, ignition signals are transmitted from the multi port injection control unit to the power transistor. These signals activate the power transistor to cause ignition coil primary winding current to flow from the ignition coil negative terminal through the power transistor to ground or be interrupted, repeatedly. This action induces high voltage in the secondary winding of the ignition coil. From the ignition coil, the secondary winding current produced flows through the spark plug to ground, thus causing ignition in each cylinder.

Diagnosis & Testing

SERVICE PRECAUTIONS

• Always turn the key **OFF** and isolate both ends of a circuit whenever testing for short or continuity.

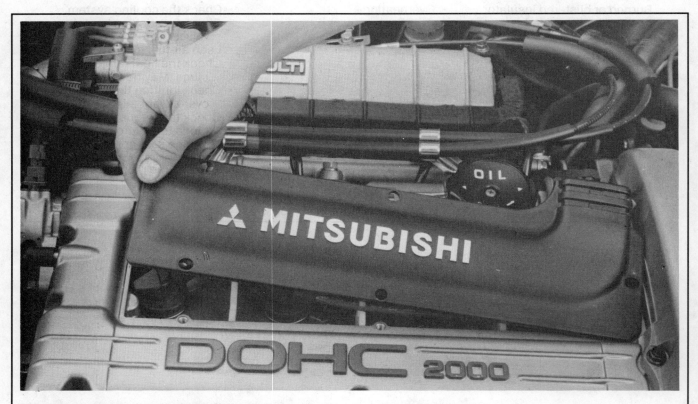

Fig. 8 Removing the center cover to access the spark plug wires and the spark plugs

Fig. 9 Disconnecting the spark plug wire. Always hold wire at the cap, do not disconnect by pulling on the wire.

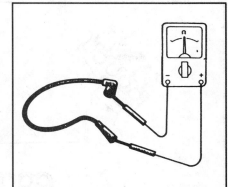

Fig. 11 Measuring the resistance of spark plug cable using a volt/ohmmeter

• Always disconnect solenoids and switches from the harness before measuring for continuity, resistance or energizing by way of a 12 volts source.

• When disconnecting connectors, inspect for damaged or pushed-out pins, corrosion, loose wires, etc. Service if required.

Fig. 10 Using the spark plug socket, remove the spark plug from the engine

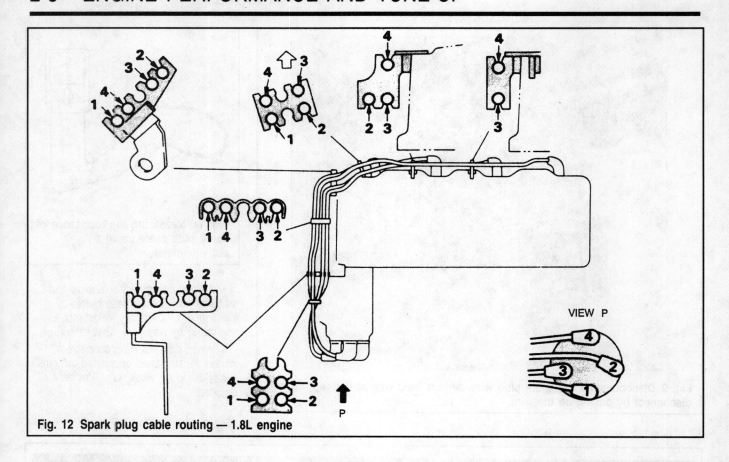

Fig. 12 Spark plug cable routing — 1.8L engine

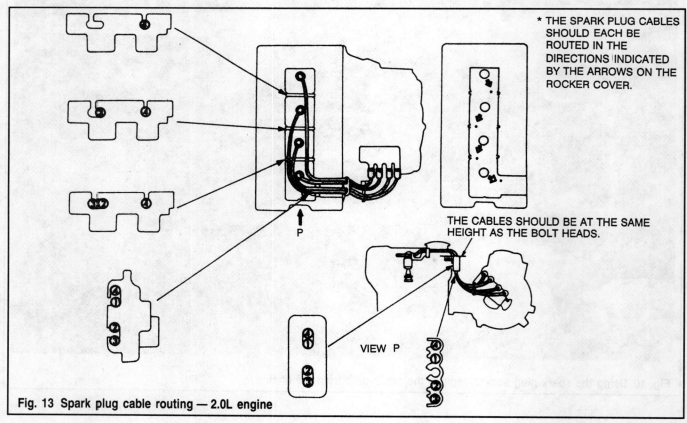

* THE SPARK PLUG CABLES SHOULD EACH BE ROUTED IN THE DIRECTIONS INDICATED BY THE ARROWS ON THE ROCKER COVER.

THE CABLES SHOULD BE AT THE SAME HEIGHT AS THE BOLT HEADS.

Fig. 13 Spark plug cable routing — 2.0L engine

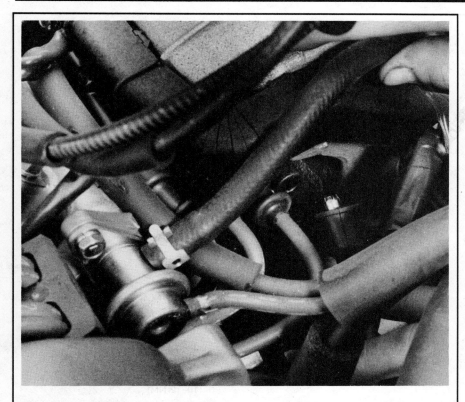

Fig. 14 Disconnecting the spark plug cables at the ignition coil — 2.0L engine

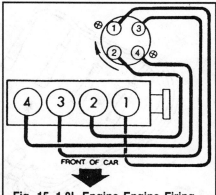

Fig. 15 1.8L Engine Engine Firing Order:1-3-4-2 Distributor Rotation: Clockwise

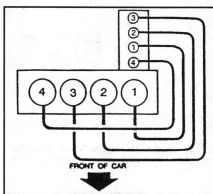

Fig. 16 2.0L DOHC Engine Engine Firing Order:1-3-4-2 Distributorless Ignition System

TROUBLESHOOTING HINTS

Testing

1. Engine cranks, but won't start:
   a. Spark is insufficient or does not occur at all, at spark plug:
   Check ignition coil.
   Check distributor (1.8L engine).
   Check crank angle sensor (2.0L engine).
   Check power transistor.
   Check spark plugs.
2. Spark is good:
   Check the ignition timing.
3. Engine Idles Roughly or Stalls:
   Check spark plugs.
   Check ignition timing.
   Check ignition coil.
   Check spark plug cables.
4. Poor Acceleration:
   Check ignition timing.
   Check ignition coil.
   Check spark plug cables.
5. Engine overheats or consumes excessive fuel:
   Check ignition timing.

DISTRIBUTOR CAP

1.8L Engine
▶ See Figure 19

1. Label and disconnect the ignition cables from the distributor cap.
2. Remove the distributor cap retainers and lift the cap from the distributor housing.
3. Inspect the cap thoroughly for cracks, damaged or worn electrodes.
4. Inspect the rotor for electrode wear.
5. Replace any component that shows excess wear or damage.
6. Prior to installing the cap onto the distributor housing, make sure the packing is in place.

IGNITION COIL

1.8L Engine
▶ See Figure 20

➡The ignition coil on the 1.8L engines is an integral part of the distributor.

1. Measure the resistance of the primary ignition coil as follows:
   a. Disconnect the electrical connector at the distributor. Using an ohmmeter, measure the resistance between terminal No. 1 and terminal No. 2. of the distributor connector.
   b. Compare the measured reading with the desired reading of 0.9-1.2 ohms.
   c. If the actual reading differs from the desired specification, replace the ignition coil.
   d. If the measured value is within standard allowance, there are no broken wires or short circuits.
2. Measure the resistance of the secondary ignition coil as follows:
   a. Insert 1 of the test leads into the secondary ignition coil terminal on top of the distributor cap.
   b. Touch the second test lead to terminal No. 1 or Terminal No. 2 of the distributor connector.

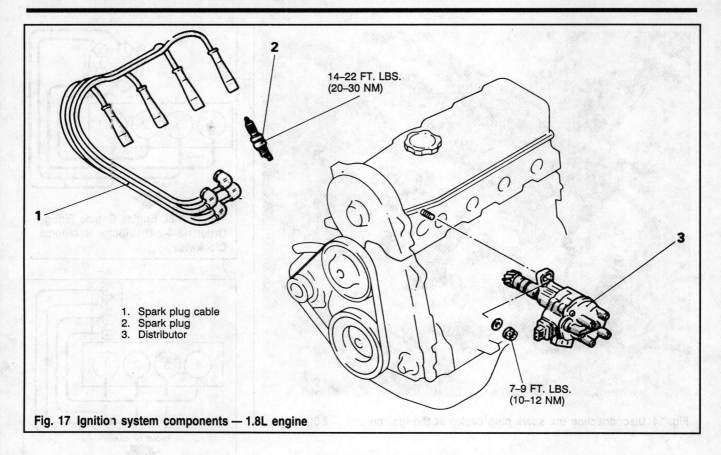

14–22 FT. LBS.
(20–30 NM)

7–9 FT. LBS.
(10–12 NM)

1. Spark plug cable
2. Spark plug
3. Distributor

Fig. 17 Ignition system components — 1.8L engine

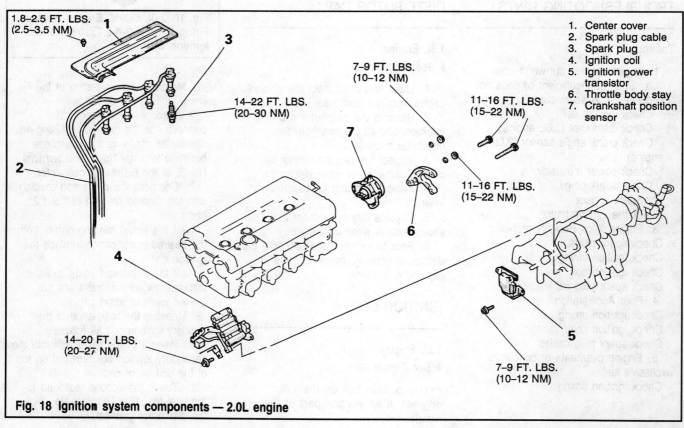

1.8–2.5 FT. LBS.
(2.5–3.5 NM)

14–22 FT. LBS.
(20–30 NM)

7–9 FT. LBS.
(10–12 NM)

11–16 FT. LBS.
(15–22 NM)

11–16 FT. LBS.
(15–22 NM)

14–20 FT. LBS.
(20–27 NM)

7–9 FT. LBS.
(10–12 NM)

1. Center cover
2. Spark plug cable
3. Spark plug
4. Ignition coil
5. Ignition power transistor
6. Throttle body stay
7. Crankshaft position sensor

Fig. 18 Ignition system components — 2.0L engine

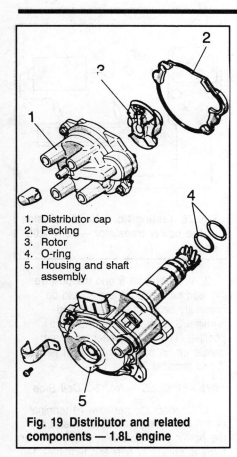

1. Distributor cap
2. Packing
3. Rotor
4. O-ring
5. Housing and shaft assembly

Fig. 19 Distributor and related components — 1.8L engine

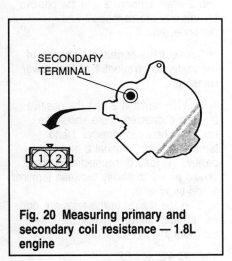

Fig. 20 Measuring primary and secondary coil resistance — 1.8L engine

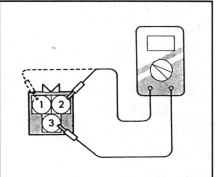

Fig. 21 Measuring primary coil resistance at the coil terminal connector — 1990 2.0L engine

c. Measure the resistance and compare to the desired specifications of 19-27 kohms.

d. If the measured value is within standard allowance, there are no broken wires or short circuits.

e. If the actual reading differs from the desired specification, replace the ignition coil.

2.0L Engine

1990 VEHICLES

▶ See Figure 21

1. Disconnect the negative battery cable and ignition coil harness connector.

2. Measure the primary coil resistance as follows:

a. Measure the resistance between terminals 4 and 2 (coils at the No. 1 and No. 4 cylinder sides) of the ignition coil, and between terminals 4 and 1 (coils at the No. 2 and No. 3 cylinder sides).

b. Compare reading to the desired primary coil resistance of 0.77-0.95 ohms.

3. Measure the coil secondary resistance as follows:

a. Disconnect the connector of the ignition coil.

b. Measure the resistance between the high-voltage terminals for the No. 1 and No. 4 cylinders, and between

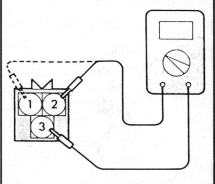

Fig. 22 Measuring primary coil resistance at the coil terminal connector — 1991-93 2.0L engine

the high-voltage terminals for the No. 2 and No. 3 cylinders.

c. Compare the measured resistance to the desired secondary coil resistance of 10.3-13.9 kilo-ohms.

4. If the readings are not within the specified value, replace the ignition coil.

EXCEPT 1990 VEHICLES

▶ See Figures 22 and 23

1. Disconnect the negative battery cable and ignition coil harness connector.

2. Measure the primary coil resistance as follows:

a. Measure the resistance between terminals 3 and 2 (coils at the No. 1 and No. 4 cylinder sides) of the ignition coil, and between terminals 3 and 1 (coils at the No. 2 and No. 3 cylinder sides).

b. Compare reading to the desired primary coil resistance of 0.70-0.86 ohms.

3. Measure the coil secondary resistance as follows:

a. Disconnect the connector of the ignition coil.

b. Measure the resistance between the high-voltage terminals for the No. 1 and No. 4 cylinders, and between the high-voltage terminals for the No. 2 and No. 3 cylinders.

c. The desired secondary coil resistance of 11.3-15.3 kilo-ohms.

4. If the readings are not within the specified value, replace the ignition coil.

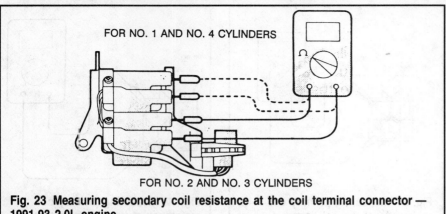

FOR NO. 1 AND NO. 4 CYLINDERS

FOR NO. 2 AND NO. 3 CYLINDERS

Fig. 23 Measuring secondary coil resistance at the coil terminal connector — 1991-93 2.0L engine

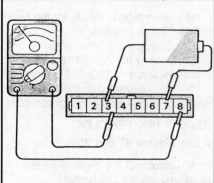

Fig. 24 Testing No. 1-No. 4 coil side of the power transistor — 1990 2.0L engine

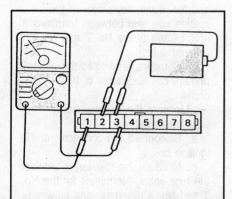

Fig. 25 Testing No. 2-No. 3 coil side of the power transistor — 1990 2.0L engine

POWER TRANSISTOR

2.0L Engine

➡When testing the power transistor(s), an analog-type circuit tester should be used.

1990 VEHICLES — No. 1-4 Coil Side
▶ See Figures 24 and 25

1. Select the ohm range on the analog tester and connect the tester between terminals 1 and 3.

2. Connect the positive (+) lead of a 1.5 volts dry cell between terminal 2 and the negative lead to terminal.

3. Continuity should be indicated when the dry cell is connected. No continuity should be indicated when the dry cell is disconnected.

1990 VEHICLES — No. 2-3 Coil Side

1. Select the ohm range on the analog tester and connect the tester between terminals 6 and 3.

2. Connect a 1.5 volts dry cell between terminals 3 and 5, with the positive (+) terminal and the negative (-) terminal connected to terminal 5 and terminal 3 respectively.

3. Continuity should be indicated when the dry cell is connected. No continuity should be indicated when the dry cell is disconnected.

4. If the results of the tests are not as indicated above, replace the power transistor(s).

1991 VEHICLES — No. 1-4 Coil Side
▶ See Figures 26 and 27

1. Connect the negative (-) terminal of a 1.5 volts dry cell to terminal 3 of the power transistor; then check whether there is continuity between terminals 7 and 3 when terminal 6 and the positive (+) terminal are connected and disconnected.

➡Connect the negative (-) probe of the tester to terminal 7 of the power transistor.

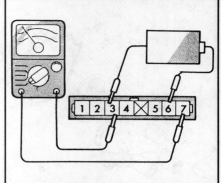

Fig. 26 Testing No. 1-No. 4 coil side of the power transistor — 1991 2.0L engine

2. With terminal 6 and the positive (+) lead connected, there should be continuity between terminal 7 and terminal 3. With terminal 6 and the positive (+) lead disconnected, there should be no continuity between terminal 7 and terminal 3.

1991 VEHICLES — No. 2-3 Coil Side

1. Connect the negative (-) terminal of a 1.5 volts dry cell to terminal 3 of the power transistor; then check whether there is continuity between terminals 1 and 3 when terminal 2 and the positive (+) terminal are connected and disconnected.

➡Connect the negative (-) probe of the tester to terminal 1 of the power transistor.

2. With terminal 2 and the positive (+) lead connected, there should be continuity between terminal 1 and terminal 3. With terminal 2 and the positive (+) lead disconnected, there should be no continuity between terminal 1 and terminal 3.

3. If the results of the tests are not as indicated above, replace the power transistor(s).

1992-93 VEHICLES

▶ See Figure 28
No. 1-4 Cylinder Transistor

1. Connect the negative (-) terminal of a 1.5 volts dry cell to terminal 3 of the power transistor; then check whether there is continuity between terminals 8 and 3 when terminal 7 and the positive

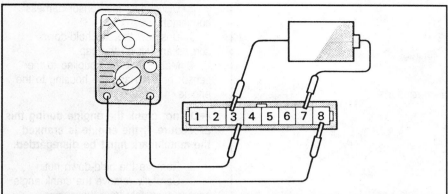

Fig. 28 Testing the power transistor for coil for No. 1 and No. 4 cylinders — 1992-93 2.0L engine

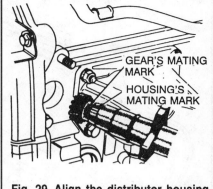

Fig. 29 Align the distributor housing and gear mating marks

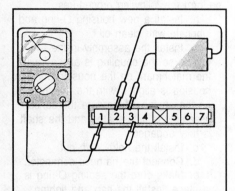

Fig. 27 Testing No. 2-No. 3 coil side of the power transistor — 1991 2.0L engine

Fig. 30 Alining the distributor housing mating mark with the center of the distributor installation stud

(+) terminal are connected and disconnected.

➡**Connect the negative (-) probe of the tester to terminal 8 of the power transistor.**

2. With terminal 7 and the positive (+) lead connected, there should be continuity between terminal 8 and terminal 3. With terminal 7 and the positive (+) lead disconnected, there should be no continuity between terminal 8 and terminal 3.

No. 2-3 Cylinder Transistor

3. Connect the negative (-) terminal of a 1.5 volts dry cell to terminal 3 of the power transistor; then check whether there is continuity between terminals 1 and 3 when terminal 2 and the positive (+) terminal are connected and disconnected.

➡**Connect the negative (-) probe of the tester to terminal 1 of the power transistor.**

4. With terminal 2 and the positive (+) lead connected, there should be continuity between terminal 1 and terminal 3. With terminal 2 and the positive (+) lead disconnected, there should be no continuity between terminal 1 and terminal 3.

5. If the results of the tests are not as indicated above, replace the power transistor(s).

Parts Replacement

DISTRIBUTOR CAP

1.8L Engine

1. Disconnect the negative battery cable. Remove the ignition wire cover, if equipped.

2. Unscrew the distributor cap hold-down screws or release the clips and lift off the distributor cap with all ignition cables still connected.

➡**To avoid confusion, remove and install the ignition cables from the distributor cap one at a time.**

3. Label and disconnect the ignition cables from the distributor cap.

To install:

4. Install the ignition cables into the new cap in the same orientation as removed.

5. Install the cap onto the housing and secure with the fasteners.

6. Install the ignition wire cover, if removed.

7. Reconnect the negative battery cable.

DISTRIBUTOR ASSEMBLY

1.8L Engine

▶ **See Figures 29 and 30**

1. Position the engine so No. 1 piston is at TDC on its compression stroke. Disconnect the negative battery cable.

2. Disconnect and tag the spark plug wires from the distributor cap.

3. Disconnect the distributor harness connector.

4. Loosen the distributor retaining bolt and remove the distributor from the engine.

To install:

5. If not already done, rotate the engine until No. 1 cylinder is at TDC on its compression stroke.

6. Align the distributor housing and gear mating marks, located at the base of the distributor shaft.

7. Install the distributor into the engine, while aligning the fine cut

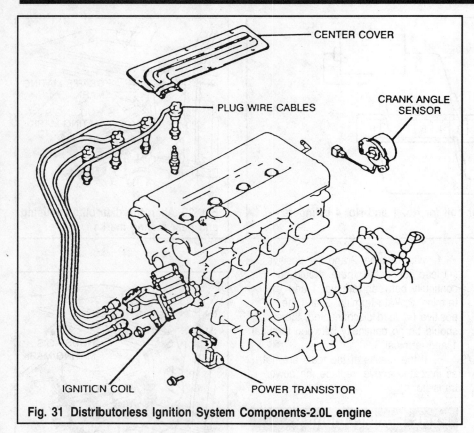

Fig. 31 Distributorless Ignition System Components-2.0L engine

(groove or projection) on the distributor flange with the center of the distributor mounting stud.

8. Install the distributor retaining bolt and tighten to 9 ft. lbs. (12 Nm).

9. Reconnect the distributor harness connector.

10. Install the distributor cap and spark plug wires.

11. Reconnect the negative battery cable.

POWER TRANSISTOR

1.8L Engine

1. Disconnect negative battery cable.
2. Disconnect the electrical connector from the transistor.
3. Remove the coil as required.
4. Remove the transistor mounting bolts and remove from the engine.
5. The installation is the reverse of the removal procedure.

2.0L Engine

1. Disconnect the negative battery cable.
2. Tag and disconnect the wires from the power transistor.

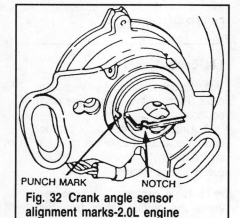

Fig. 32 Crank angle sensor alignment marks-2.0L engine

3. Remove the retaining screw and lift the power transistor from the engine.
4. Installation is the reverse of the removal procedure.

CRANK ANGLE SENSOR

2.0L Engine
▶ See Figures 31 and 32

1. Disconnect the negative battery cable.

2. Disconnect the sensor harness connector.

3. Unscrew the cap hold-down screws and lift off the cap.

4. Matchmark the coupling to the sensor housing and the housing to the engine.

➡**Do not crank the engine during this procedure. If the engine is cranked, the matchmark must be disregarded.**

5. Remove the hold-down nut.

6. Carefully remove the crank angle sensor assembly from the engine.

To install:

7. If the timing is not disturbed, perform the following procedures:

a. Install a new housing O-ring and lubricate with clean oil.

b. Install the assembly in the engine so the coupling is aligned with the matchmark on the housing and the housing is aligned with the matchmark on the engine. Make sure the sensor assembly is fully seated and the shaft is fully engaged.

c. Install the hold-down nut.

d. Connect the harness connector.

e. Make sure the sealing O-ring is in place, install the cap and tighten the screws.

f. Connect the negative battery cable.

g. Adjust the ignition timing, if applicable, and tighten the hold-down nut.

8. If the timing is disturbed, perform the following procedures:

a. Install a new housing O-ring and lubricate with clean oil.

b. Position the engine so the No. 1 piston is at TDC of its compression stroke and the mark on the vibration damper is aligned with 0 on the timing indicator.

c. Install the sensor in the engine so the factory matchmark on the coupling (notch) is aligned with the matchmark on the housing (punch mark) and the housing is aligned with the matchmark on the engine. Make sure the sensor assembly is fully seated and the shaft is fully engaged.

d. Install the hold-down nut.

e. Connect the harness connector.

f. Make sure the sealing O-ring is in place, install the cap and tighten the screws.

g. Connect the negative battery cable.

h. Adjust the ignition timing, if applicable, and tighten the hold-down nut.

IGNITION COIL

1. Disconnect the negative battery cable.

2. Tag and remove the spark plug wires from the ignition coil by gripping the boot and not the cable.

3. Remove the mounting screws and coil from engine.

4. Installation is the reverse of the removal procedure.

IGNITION TIMING

Timing

ADJUSTMENT

1.8L Engine
▶ **See Figures 33, 34, 35 and 36**

1. Apply the parking brake and block the wheels. Run the engine until the coolant reaches normal operating temperature.

2. Make certain all lights, cooling fan and accessories are **OFF**.

3. Position the steering wheel in straight ahead position and the gear selector lever in **P** or **N**.

4. Connect a timing light to the engine.

5. Insert a paper clip into the CRC filter connector (3-pole connector), located in the engine compartment of the vehicle.

6. Connect a tachometer to the inserted clip.

➡**During installation of the paper clip, do not separate the connector.**

7. Check the curb idle speed. It should be 600-800 rpm.

8. Turn the engine **OFF**. Connect a jumper wire to the terminal for ignition-timing adjustment (located in the engine compartment), and ground it.

9. Start and run the engine at curb idle speed.

10. Check the basic ignition timing and adjust, if necessary. Basic ignition timing should be 5 degrees BTDC.

11. If the timing is not within specifications, loosen the distributor hold-down bolt and turn the distributor to bring the timing within specifications. Turning the distributor to the right retards timing while to the left will advance timing.

12. Tighten the hold-down bolt after adjustment. Recheck the timing and adjust if necessary.

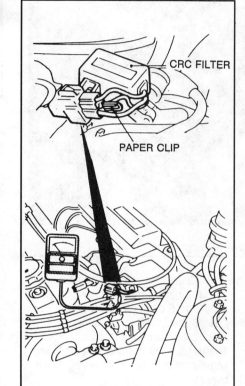

Fig. 33 Inserting a paper clip into the CRC filter connector — 1.8L engine

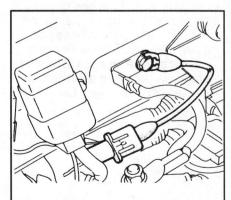

Fig. 34 Connecting a jumper wire to the terminal for ignition-timing adjustment — 1.8L engine

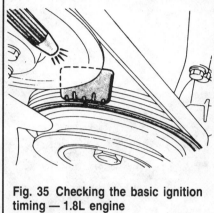

Fig. 35 Checking the basic ignition timing — 1.8L engine

13. Stop the engine and remove the ground for the ignition timing connector.

➡**Actual ignition timing may vary, depending on the control mode of the engine control unit. In such case, recheck the basic ignition timing. If there is no deviation, the ignition timing is functioning normally.**

14. Start the engine and run at curb idle. Check the actual ignition timing. Actual ignition timing should be 10 degrees BTDC.

➡**At altitudes, more than approximately 2,300 ft. (701m) above sea level, the actual ignition timing is further advanced to ensure good combustion.**

2.0L Engine
▶ **See Figures 37, 38, 39 and 40**

1. Apply the parking brake and block the wheels. Run the engine until it reaches normal operating temperature.

2. Make certain all lights, cooling fan and accessories are **OFF**.

3. Position the steering wheel in straight ahead position and the gear selector lever in **P** or **N**.

4. Connect a timing light to the engine.

5. Insert a paper clip into the engine revolution speed detection terminal (in

Fig. 36 Ignition timing scale — 2.0L engine

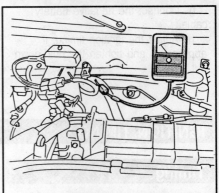

Fig. 37 Insert a paper clip into the engine revolution speed detection terminal and connect a tachometer to the inserted clip — 2.0L engine

engine compartment) and connect a tachometer to the inserted clip.

6. Check the curb idle speed. Should be 650-850 rpm.

7. Stop the engine and connect a jumper wire to the terminal for ignition-timing adjustment to ground.

8. Start and run the engine at curb idle speed.

9. Check the basic ignition timing and adjust, if necessary. Basic ignition timing should be 5 degrees BTDC.

10. If the timing is not within specification, loosen the crank angle sensor retaining nut and turn the crank angel sensor to bring the timing within specs.

11. Tighten the sensor retaining nut after adjustment. Recheck the timing and adjust if necessary.

12. Stop the engine and remove the ground for the ignition timing connector.

13. Start the engine and run at curb idle. Check the actual ignition timing. Actual ignition timing should be 8 degrees BTDC.

➡**Actual ignition timing may vary, depending on the control mode of the engine control unit. In such case, recheck the basic ignition timing. If there is no deviation, the ignition timing is functioning normally.   At altitudes, more than approximately 2,300 ft. (701m) above sea level, the actual ignition timing is further advanced to ensure good combustion.**

VALVE LASH

Adjustment

Both the 1.8L and the 2.0L DOHC engines are equipped with hydraulic valve adjusters. Because of this, there is no valve adjustment. Excess clearance between the tip of the valve and the rocker is taken up by engine oil pressure acting against a plunger in the adjuster.

1. If abnormal noise is heard from the lash adjusters check as follows:

a. Start the engine and allow to idle until normal operating temperature is reached. Shut the engine **OFF**.

b. While installed to the cylinder head, press the part of the rocker arm that contacts the lash adjuster. The part pressed will feel very hard if the lash adjuster condition is normal.

c. If, when pressed, it easily descends all the way downward, replace the lash adjuster.

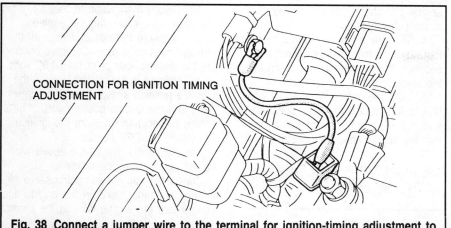

Fig. 38 Connect a jumper wire to the terminal for ignition-timing adjustment to ground — 2.0L engine

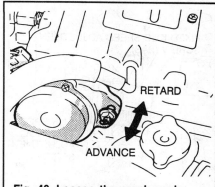

Fig. 40 Loosen the crank angle sensor retaining nut and turn the crank angel sensor to adjust the timing within specifications — 2.0L engine

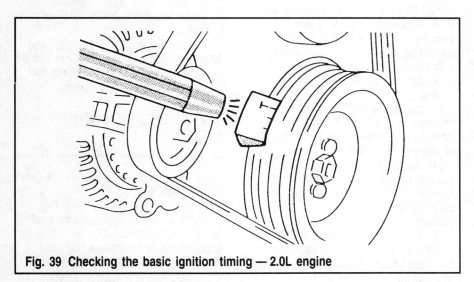

Fig. 39 Checking the basic ignition timing — 2.0L engine

d. If there is a spongy feeling when pressed, air is probably mixed in, and so the cause should be investigated. The cause is probably an insufficient amount of engine oil, or damage to the oil screen and/or gasket.

e. After finding the cause and taking the appropriate step, warm up the engine and drive at low speed for a short time. Then after stopping the engine and waiting a few minutes, drive again at low speed. Repeat this procedure a few times to bleed the air from the oil.

IDLE SPEED AND MIXTURE

Data from various sensors and switches are used by the ECU to determine the proper fuel/air mixture for optimal engine performance. The mixture setting is not adjustable.

Curb Idle Speed Adjustment

▶ See Figure 41

➡The idle speed is controlled electronically and adjustment is usually unnecessary.

1990 VEHICLES WITH 1.8L ENGINE

For this procedure, a multi-tester (scan tool) is required.

1. Warm the engine to operating temperature, leave lights, electric cooling fan and accessories **OFF**. The transaxle should be in **N** for manual transaxle or **P** for automatic transaxle. Place the steering wheel in a neutral (straight ahead) position for vehicles equipped with power steering.

2. Check the ignition timing and adjust, if necessary.

3. Connect a multi-use tester to the diagnostic connector, located beside the fuse block.

4. Set the ignition switch to **ON**, without starting the engine, and hold it in that position for 15 seconds or more. With the ignition in this position, the idle speed control motor will retract to the idle position. Turn the ignition switch **OFF**.

5. Uncouple the connector of the idle speed control servo to secure the idle speed servo at this position.

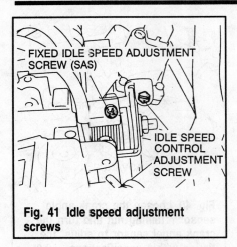

FIXED IDLE SPEED ADJUSTMENT SCREW (SAS)

IDLE SPEED CONTROL ADJUSTMENT SCREW

Fig. 41 Idle speed adjustment screws

6. In order to prevent the throttle valve from sticking, open it more than halfway 2 or 3 times and then release it to let it click shut. Loosen the fixed idle speed adjusting screw to allow for adjustment.

7. Start the engine and allow it to run at idle.

8. Check that the engine speed is at the desired reading of 650-750 rpm. The engine speed on a vehicle with 300 miles or less may be 20-100 rpm lower than specifications listed above, but adjustments may not be necessary.

9. If adjustment is required, turn the engine OFF and slacken the accelerator cable. Adjust the idle speed using the idle speed adjustment screw. When making the adjustment, use a hexagonal wrench in order to prevent play caused by backlash.

10. Once the engine rpm is set, screw in the fixed idle speed adjusting screw until the engine speed starts to rise. At this point return the fixed idle speed adjusting screw to find the point at which engine rpm does not change. Once at this point, turn the fixed idle speed adjusting screw in a half turn. Turn the engine OFF.

11. Switch the ignition to the ON position but do not start the engine.

12. Press code No. 14 on the scan tool and measure the output voltage of the throttle position sensor. Compare reading to the desired voltage of 0.48-0.52 volts. If the voltage is not correct, loosen the throttle position sensor mounting screws and turn the throttle position sensor to make the adjustment. Turn the ignition switch to the OFF position.

13. Adjust the play of the accelerator cable. Connect the idle speed control servo electrical connector. Start the

engine and check that the engine idles at the correct speed.

14. Turn the engine off and disconnect the battery terminals for longer than 10 seconds, then reconnect. By doing this, the memory data will be erased.

15. Start the engine once again and let idle for about 5 minutes. Check to be sure the idling condition is normal and that the engine speed is correct.

1991-92 LASER AND TALON WITH 1.8L ENGINE

1. Warm the engine to operating temperature, leave lights, electric cooling fan and accessories **OFF**. The transaxle should be in **N** for manual transaxle or **P** for automatic transaxle. Place the steering wheel in a neutral (straight ahead) position for vehicles equipped with power steering.

2. Connect a tachometer to the engine.

3. Connect a digital voltmeter between terminals 19 (throttle position sensor output voltage) and 24 (ground) of the engine control unit.

4. Set the ignition switch to the ON position but do not start the engine. Keep in this position for at least 15 seconds. Set the ignition to the OFF position.

5. Disconnect the idle speed control servo electrical connector. Back out the fixed idle speed adjusting screw enough to allow for adjustment.

6. Start the engine and run at idle.

7. Check that the engine speed is at the desired reading of 650-750 rpm. The engine speed on a vehicle with 300 miles or less may be 20-100 rpm lower than specifications listed above, but adjustments may not be necessary.

8. If adjustment is required, turn the engine OFF and slacken the accelerator cable. Adjust the idle speed using the idle speed adjustment screw. When making the adjustment, use a hexagonal wrench in order to prevent play caused by backlash.

9. Once the engine rpm is set, screw in the fixed idle speed adjusting screw until the engine speed starts to rise. At this point return the fixed idle speed adjusting screw to find the point at which engine rpm does not change. Once at this point, turn the fixed idle speed adjusting screw in a half turn. Turn the engine OFF.

10. Switch the ignition to the ON position but do not start the engine.

11. Measure the output voltage of the throttle position sensor. Compare reading to the desired voltage of 0.48-0.52 volts. If the voltage is not correct, loosen the throttle position sensor mounting screws and turn the throttle position sensor to make the adjustment. Turn the ignition switch to the OFF position.

12. Adjust the play of the accelerator cable and remove the voltmeter. Connect the idle speed control servo electrical connector. Start the engine and check that the engine idles at the correct speed.

13. Turn the engine OFF and disconnect the battery terminals for longer than 10 seconds, then reconnect. By doing this, the memory data will be erased.

14. Start the engine once again and let idle for about 5 minutes. Check to be sure the idling condition is normal and that the engine speed is correct.

1991-93 ECLIPSE WITH 1.8L ENGINE

For this procedure, a multi-tester (scan tool) is required.

1. Warm the engine to operating temperature, leave lights, electric cooling fan and accessories **OFF**. The transaxle should be in **N** for manual transaxle or **P** for automatic transaxle. Place the steering wheel in a neutral (straight ahead) position for vehicles equipped with power steering.

2. Slacken the accelerator cable to allow for adjustment. Connect the scan tool to the data link connector.

3. Switch the ignition to the ON position but do not start the engine. Leave in this position for 15 seconds or more. With the ignition in this position, the idle speed control motor will retract to the idle position. Turn the ignition switch OFF.

4. Uncouple the connector of the idle speed control servo to secure the idle speed servo at this position. In order to prevent the throttle valve from sticking, open it at least halfway 2 or more times and then release it so it will click shut.

5. Start the engine and let idle. Check the engine idle speed and compare to the desired specifications of 650-750 rpm.

6. If the idle speed is wrong, adjust with the idle speed control adjusting screw, using a hexagon wrench.

7. First loosen the fixed Speed Adjusting Screw (SAS). Then adjust the engine speed using the idle speed control adjusting screw until the desired engine speed is reached.

8. Once the engine rpm is set, screw in the fixed idle speed adjusting screw until the engine speed starts to rise. At this point return the fixed idle speed adjusting screw to find the point at which engine rpm does not change. Once at this point, turn the fixed idle speed adjusting screw in a half turn.

9. Turn the ignition switch OFF. Adjust the accelerator cable and the throttle position sensor.

10. Start the engine once again and let idle for about 5 minutes. Check to be sure the idling condition is normal and that the engine speed is correct.

1993 LASER AND TALON WITH 1.8L ENGINE

For this procedure, a multi-tester (scan tool) is required.

1. Warm the engine to operating temperature, leave lights, electric cooling fan and accessories **OFF**. The transaxle should be in **N** for manual transaxle or **P** for automatic transaxle. Place the steering wheel in a neutral (straight ahead) position for vehicles equipped with power steering.

2. Slacken the accelerator cable to allow for adjustment. Connect the scan tool to the data link connector.

3. Switch the ignition to the ON position but do not start the engine. Leave in this position for 15 seconds or more. With the ignition in this position, the idle speed control motor will retract to the idle position. Turn the ignition switch OFF.

4. Uncouple the connector of the idle speed control servo to secure the idle speed servo at this position. In order to prevent the throttle valve from sticking, open it at least halfway 2 or more times and then release it so it will click shut.

5. Start the engine and let idle. Check the engine idle speed and compare to the desired specifications of 650-750 rpm.

6. If the idle speed is wrong, adjust with the idle speed control adjusting screw, using a hexagon wrench.

7. First loosen the fixed Speed Adjusting Screw (SAS). Then adjust the engine speed using the idle speed control adjusting screw until the desired engine speed is reached.

8. Once the engine rpm is set, screw in the fixed idle speed adjusting screw until the engine speed starts to rise. At this point return the fixed idle speed adjusting screw to find the point at which engine rpm does not change. Once at this point, turn the fixed idle speed adjusting screw in a half turn.

9. Turn the ignition switch OFF. Adjust the accelerator cable and the throttle position sensor.

10. Start the engine once again and let idle for about 5 minutes. Check to be sure the idling condition is normal and that the engine speed is correct.

2.0L ENGINE

1. Warm the engine to operating temperature, leave lights, electric cooling fan and accessories **OFF**. The transaxle should be in **N**. Place the steering wheel in a neutral position for vehicles with power steering.

2. Check the ignition timing and adjust, if necessary.

3. Connect a tachometer to the 1 pin connector under the hood.

4. Locate the self-diagnosis terminal under the dashboard and connect terminal No. **10** to ground with a jumper wire.

5. Disconnect the waterproof female connector used for ignition timing adjustment. Connect this terminal to ground using a jumper wire.

6. Start the engine and allow to idle. Check that the basic idle speed is 700-800 rpm. Be aware that on some vehicles, the rpm reading may be half of the actual engine rpm. Adjust the engine rpm using the speed adjusting screw. If the idle speed still is difficult to adjust or deviates from the specification, note the following:

a. A new engine will idle more slowly. Break-in should take approximately 300 miles.

b. If the vehicle stalls or has a very low idle speed, suspect a deposit buildup on the throttle valve which must be cleaned.

c. If the idle speed is high even though the speed adjusting screw is fully closed, check that the idle position switch (fixed speed adjusting screw) position has changed. If so, adjust the idle position switch.

d. If after all these checks the idle is still out of specification, it may be that there is leakage resulting from deterioration of the Fast-Idle Air Valve (FIAV).

7. Turn the ignition switch **OFF**. Disconnect the jumper wire from the diagnosis connector, disconnect the jumper wire from the ignition timing connector and reconnect the waterproof connector. Disconnect the tachometer.

8. Restart the engine, allow to run for 5 minutes and check for good idle quality and correct idle speed.

TORQUE SPECIFICATIONS

Component	U.S.	Metric
Spark plug	22 ft. lbs.	30 Nm
Center cover mounting bolts	2.5 ft. lbs.	3.5 Nm
Distributor retaining bolt	9 ft. lbs.	12 Nm

3

ENGINE AND ENGINE OVERHAUL

ENGINE ELECTRICAL

➡ Disconnecting the negative battery cable on some vehicles may interfere with the functions of the on board computer systems and may require the computer to undergo a relearning process, once the negative battery cable is reconnected.

Ignition Coil

TESTING

▶ See Figures 1, 2 and 3

1.8L Engine

➡ The ignition coil on the 1.8L engine is an integral part of the distributor.

1. Measure the resistance of the primary ignition coil as follows:
   a. Disconnect the electrical connector at the distributor. Using an ohmmeter, measure the resistance between terminal No. 1 and terminal No. 2. of the distributor connector.
2. Compare the measured reading with the desired reading of 0.9-1.2 ohm.
3. If the actual reading differs from the desired specification, replace the ignition coil.
4. If the measured value is within standard allowance, there are no broken wires or short circuits.
5. Measure the resistance of the secondary ignition coil as follows:
   a. Insert 1 of the test leads into the secondary ignition coil terminal on top of the distributor cap.
   b. Touch the second test lead to terminal No. 1 or Terminal No. 2 of the distributor connector.
   c. Measure the resistance and compare to the desired specifications of 19-27 kohms.
   d. If the measured value is within standard allowance, there are no broken wires or short circuits.
   e. If the actual reading differs from the desired specification, replace the ignition coil.

2.0L Engine

1990 VEHICLES

1. Disconnect the negative battery cable and ignition coil harness connector.

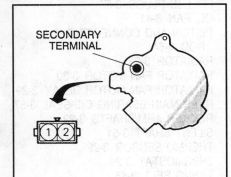

Fig. 1 Measuring resistance of primary and secondary ignition coils — 1.8L engine

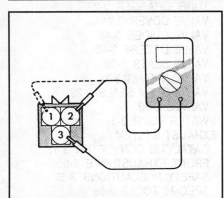

Fig. 2 Measuring primary ignition coil resistance — 2.0L engine

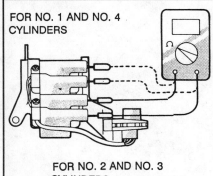

Fig. 3 measuring secondary ignition coil resistance — 2.0L engine

2. Measure the primary coil resistance as follows:
   a. Measure the resistance between terminals 4 and 2 (coils at the No. 1 and No. 4 cylinder sides) of the ignition coil, and between terminals 4 and 1 (coils at the No. 2 and No. 3 cylinder sides).

b. Compare reading to the desired primary coil resistance of 0.77-0.95 ohms.
3. Measure the coil secondary resistance as follows:
   a. Disconnect the connector of the ignition coil.
   b. Measure the resistance between the high-voltage terminals for the No. 1 and No. 4 cylinders, and between the high-voltage terminals for the No. 2 and No. 3 cylinders.
   c. Compare the measured resistance to the desired secondary coil resistance of 10.3-13.9 kilo-ohms.
4. If the readings are not within the specified value, replace the ignition coil.

EXCEPT 1990 VEHICLES

1. Disconnect the negative battery cable and ignition coil harness connector.
2. Measure the primary coil resistance as follows:
   a. Measure the resistance between terminals 3 and 2 (coils at the No. 1 and No. 4 cylinder sides) of the ignition coil, and between terminals 3 and 1 (coils at the No. 2 and No. 3 cylinder sides).
   b. Compare reading to the desired primary coil resistance of 0.70-0.86 ohms.
3. Measure the coil secondary resistance as follows:
   a. Disconnect the connector of the ignition coil.
   b. Measure the resistance between the high-voltage terminals for the No. 1 and No. 4 cylinders, and between the high-voltage terminals for the No. 2 and No. 3 cylinders.
   c. The desired secondary coil resistance of 11.3-15.3 kilo-ohms.
4. If the readings are not within the specified value, replace the ignition coil.

REMOVAL & INSTALLATION

1. Disconnect the negative battery cable.
2. Tag and remove the spark plug wires from the ignition coil by gripping the boot and not the cable.
3. Remove the mounting screws and coil from engine.
4. Installation is the reverse of the removal procedure.

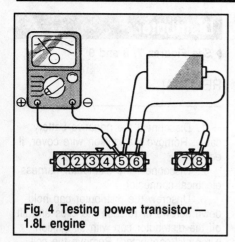

Fig. 4 Testing power transistor —
1.8L engine

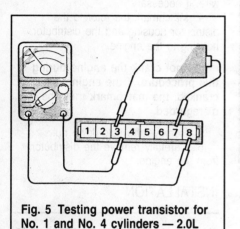

Fig. 5 Testing power transistor for
No. 1 and No. 4 cylinders — 2.0L
engine

Power Transistor

TESTING

◆ See Figures 4, 5 and 6

1.8L ENGINE

➡ When testing the power
transistor(s), an analog-type circuit
tester should be used.

1. Connect the negative terminal of a
1.5V power source to terminal No. 5 of
the power transistor connector.

2. Check whether there is continuity
between terminal No. 5 and terminal No.
8 when terminal 6 and the positive
terminal are connected and
disconnected.

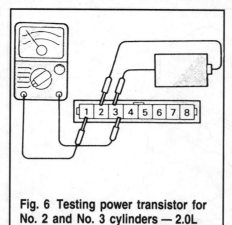

Fig. 6 Testing power transistor for
No. 2 and No. 3 cylinders — 2.0L
engine

➡ Connect the negative probe of the
tester to terminal No. 8 of the power
transistor connector.

3. With terminal No. 6 and the
positive terminal connected, there should
be continuity between terminal No. 5 and
terminal No. 8 of the connector.

4. With terminal No. 6 and the
positive terminal disconnected, there
should be no continuity between terminal
No. 5 and terminal No. 8 of the
connector.

5. If the results of the test are not as
specified above, replace the power
transistor.

2.0L Engine

➡ When testing the power
transistor(s), an analog-type circuit
tester should be used.

1990 VEHICLES
No. 1-4 Coil Side

1. Select the ohm range on the
analog tester and connect the tester
between terminals 1 and 3.

2. Connect the positive (+) lead of a
1.5 volts dry cell between terminal 2 and
the negative lead to terminal 3.

3. Continuity should be indicated
when the dry cell is connected. No
continuity should be indicated when the
dry cell is disconnected.

No. 2-3 Coil Side

4. Select the ohm range on the
analog tester and connect the tester
between terminals 6 and 3.

5. Connect a 1.5 volts dry cell
between terminals 3 and 5, with the
positive (+) terminal and the negative (-)

terminal connected to terminal 5 and
terminal 3 respectively.

6. Continuity should be indicated
when the dry cell is connected. No
continuity should be indicated when the
dry cell is disconnected.

7. If the results of the tests are not
as indicated above, replace the power
transistor(s).

1991 VEHICLES
No. 1-4 Coil Side

1. Connect the negative (-) terminal
of a 1.5 volts dry cell to terminal 3 of
the power transistor; then check whether
there is continuity between terminals 7
and 3 when terminal 6 and the positive
(+) terminal are connected and
disconnected.

➡ Connect the negative (-) probe of
the tester to terminal 7 of the power
transistor.

2. With terminal 6 and the positive
(+) lead connected, there should be
continuity between terminal 7 and
terminal 3. With terminal 6 and the
positive (+) lead disconnected, there
should be no continuity between terminal
7 and terminal 3.

No. 2-3 Coil Side

3. Connect the negative (-) terminal
of a 1.5 volts dry cell to terminal 3 of
the power transistor; then check whether
there is continuity between terminals 1
and 3 when terminal 2 and the positive
(+) terminal are connected and
disconnected.

➡ Connect the negative (-) probe of
the tester to terminal 1 of the power
transistor.

4. With terminal 2 and the positive
(+) lead connected, there should be
continuity between terminal 1 and
terminal 3. With terminal 2 and the
positive (+) lead disconnected, there
should be no continuity between terminal
1 and terminal 3.

5. If the results of the tests are not
as indicated above, replace the power
transistor(s).

1992-93 VEHICLES
No. 1-4 Cylinder Transistor

1. Connect the negative (-) terminal
of a 1.5 volts dry cell to terminal 3 of
the power transistor; then check whether
there is continuity between terminals 8
and 3 when terminal 7 and the positive

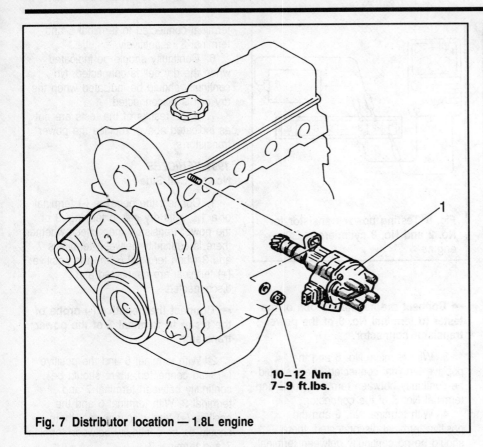

**10–12 Nm
7–9 ft.lbs.**

1

Fig. 7 Distributor location — 1.8L engine

(+) terminal are connected and disconnected.

➡ **Connect the negative (-) probe of the tester to terminal 8 of the power transistor.**

2. With terminal 7 and the positive (+) lead connected, there should be continuity between terminal 8 and terminal 3. With terminal 7 and the positive (+) lead disconnected, there should be no continuity between terminal 8 and terminal 3.

No. 2-3 Cylinder Transistor

3. Connect the negative (-) terminal of a 1.5 volts dry cell to terminal 3 of the power transistor; then check whether there is continuity between terminals 1 and 3 when terminal 2 and the positive (+) terminal are connected and disconnected.

➡ **Connect the negative (-) probe of the tester to terminal 1 of the power transistor.**

4. With terminal 2 and the positive (+) lead connected, there should be continuity between terminal 1 and terminal 3. With terminal 2 and the positive (+) lead disconnected, there should be no continuity between terminal 1 and terminal 3.

5. If the results of the tests are not as indicated above, replace the power transistor(s).

REMOVAL & INSTALLATION

1.8L Engine

1. Disconnect negative battery cable.
2. Disconnect the electrical connector from the transistor.
3. Remove the coil as required.
4. Remove the transistor mounting bolts and remove from the engine.
5. The installation is the reverse of the removal procedure.

2.0L Engine

1. Disconnect the negative battery cable.
2. Tag and disconnect the wires from the power transistor.
3. Remove the retaining screw and lift the power transistor from the engine.
4. Installation is the reverse of the removal procedure.

Distributor

➡ **See Figures 7, 8 and 9**

REMOVAL

1. Disconnect the negative battery cable. Remove the ignition wire cover, if equipped.
2. Disconnect the distributor harness electrical connector.
3. Unscrew the distributor cap hold-down screws or release the clips and lift off the distributor cap with all ignition wires still connected. Remove the coil wire, if necessary.
4. Matchmark the rotor to the distributor housing and the distributor housing to the engine.

➡ **Do not crank the engine during this procedure. If the engine is cranked, the matchmark must be disregarded.**

5. Remove the hold-down nut.
6. Carefully remove the distributor from the engine.

INSTALLATION

➡ **Some engines may be sensitive to the routing of the distributor sensor wires. If routed near the high-voltage coil wire or the spark plug wires, the electromagnetic field surrounding the high voltage wires could generate an occasional disruption of the ignition system operation.**

Timing Not Disturbed

1. Install a new distributor housing O-ring and lubricate with clean oil.
2. Install the distributor in the engine so the rotor is aligned with the matchmark on the housing and the housing is aligned with the matchmark on the engine. Make sure the distributor is fully seated and the distributor shaft is fully engaged.
3. Install the hold-down nut.
4. Connect the distributor harness connectors.
5. Make sure the sealing O-ring is in place, install the distributor cap and tighten the screws or secure the clips.
6. Connect the negative battery cable.
7. Adjust the ignition timing and tighten the hold-down nut.

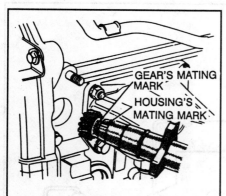

Fig. 8 Align the distributor housing and gear mating marks during installation — 1.8L engine

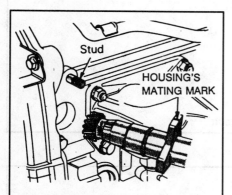

Fig. 9 Install the distributor in engine so the slot or groove of the distributor's installation flange aligns with the distributor installation stud in the engine block

Timing Disturbed

1. Install a new distributor housing O-ring and lubricate with clean oil.

2. Position the engine so the No. 1 piston is at TDC of its compression stroke and the mark on the vibration damper is aligned with **0** on the timing indicator.

3. Align the distributor housing and gear mating marks. Install the distributor in engine so the slot or groove of the distributor's installation flange aligns with the distributor installation stud in the engine block. Make sure the distributor is fully seated. Inspect alignment of the distributor rotor making sure the rotor is aligned with the position of the No. 1 ignition wire in the distributor cap.

➡ **Make sure the rotor is pointing to where the No. 1 runner originates inside the cap, if equipped, and not where the No. 1 ignition wire plugs into the cap.**

4. Install the hold-down nut.
5. Connect the distributor harness connectors.
6. Make sure the sealing O-ring is in place, install the distributor cap and tighten the screws or secure the clips.
7. Connect the negative battery cable.
8. Adjust the ignition timing and tighten the hold-down bolt.

Alternator

ALTERNATOR PRECAUTIONS

Several precautions must be observed with alternator-equipped vehicles to avoid damage to the unit.

• If the battery is removed for any reason, make sure it is reconnected with the correct polarity. Reversing the battery connections may result in damage to the 1-way rectifiers.

• When utilizing a booster battery as a starting aid, always connect the positive to positive terminals and the negative terminal from the booster battery to a good engine ground on the vehicle being started.

• Never use a fast charger as a booster to start vehicles.

• Disconnect the battery cables when charging the battery with a fast charger.

• Never attempt to polarize the alternator.

• Do not use test lamps of more than 12 volts when checking diode continuity.

• Do not short across or ground any of the alternator terminals.

• The polarity of the battery, alternator and regulator must be matched and considered before making any electrical connections within the system.

• Never separate the alternator on an open circuit. Make sure all connections within the circuit are clean and tight.

• Disconnect the battery ground terminal when performing any service on electrical components.

• Disconnect the battery if arc welding is to be done on the vehicle.

VOLTAGE DROP TESTING OF ALTERNATOR OUTPUT WIRE

◆ **See Figure 10**

This test will determine whether or not the wiring (including the fusible link) between the alternator **B** terminal and the battery positive terminal is sound by voltage drop method.

A clamp type ammeter that can measure current without disconnecting the harness is preferred for this test. The reason is that when checking a vehicle that has a low output current due to poor connection of the alternator **B** terminal, such poor connection is corrected as the **B** terminal is loosened and the test ammeter is connected in its place and as a result, causes for the trouble may not be determined.

Test Preparation

1. Turn the ignition **OFF** .
2. Disconnect the battery ground cable.
3. Disconnect the alternator output lead from the alternator **B** terminal.
4. Connect the positive lead of an ammeter to the **B** terminal and the negative lead to the disconnected output wire.
5. Connect a digital voltmeter between the alternator **B** terminal and the battery positive terminal. Connect the positive lead wire of the voltmeter to the **B** terminal and the negative lead wire to the battery positive terminal.
6. Connect the battery ground cable.
7. Leave the hood to the engine compartment open.

Test Procedure

1. Start the engine.
2. Turn the headlamps and small lamps **ON** and adjust the engine speed so that the ammeter reads 20A and read off the voltmeter indication under this condition.
3. It is OK if the voltmeter reads the standard value of 0.2V max.
4. If the voltmeter indicates a value larger than the standard value, poor wiring is suspected, in which case check the wiring from the alternator **B** terminal to the fuse link to battery positive terminal. Check for loose connection, color change due to overheating hardness, etc. and correct before testing again.

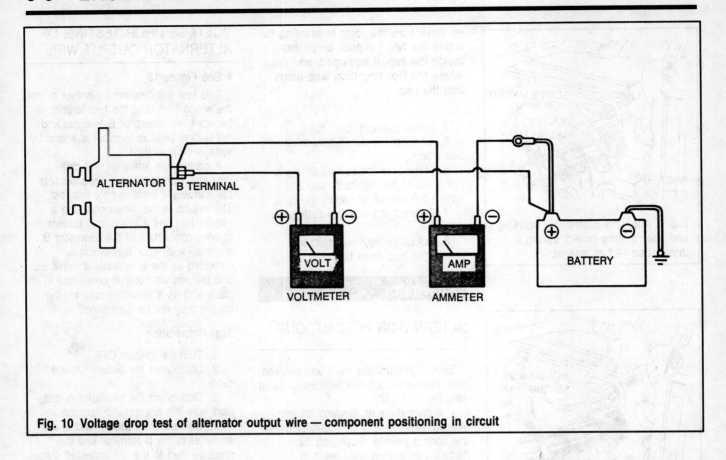

Fig. 10 Voltage drop test of alternator output wire — component positioning in circuit

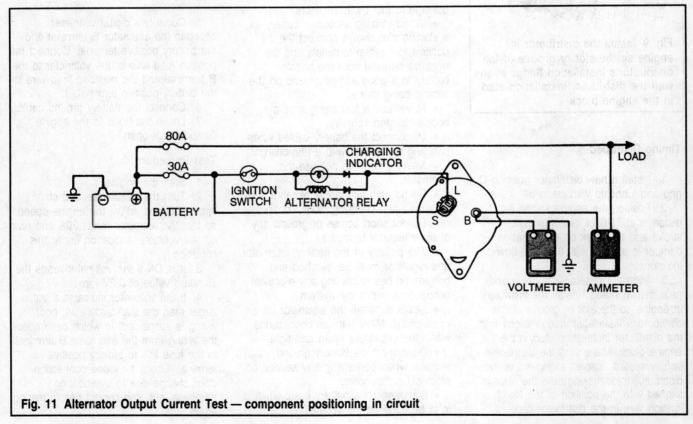

Fig. 11 Alternator Output Current Test — component positioning in circuit

5. Upon completion of the test, set the engine speed at idle.

6. Turn the lights and the ignition switch **OFF** .

7. Disconnect the battery ground cable.

8. Disconnect the ammeter and the voltmeter that have been connected for the purpose of the test.

9. Connect the alternator output wire to the alternator **B** terminal.

10. Connect the battery ground cable.

ALTERNATOR OUTPUT CURRENT TEST

▶ **See Figure 11**

This test is designed to judge whether or not the alternator gives an output current that is equivalent to the nominal output.

Test Preparation

1. Check the battery that installed in the vehicle, making sure it is in good sound state. Test the battery as required to assure this. The battery used to check output current should be one that has been rather discharged. With a fully discharged battery, the test may not be conducted correctly due to an insufficient load.

2. Check the alternator drive belt for proper tension as follows:

a. Place a straight-edge along the top edge of the belt and across 2 pulleys. Allow both ends of the straight-edge to rest on top of each pulley for support.

b. Measure the deflection of the belt from the straight-edge with a force of about 22 lbs. applied midway between the 2 pulleys. Deflection should measure as follows:

Alternator/Water Pump Drive Belt
1.8L engine — 0.315-0.433 in. (8.0-11.0mm)
2.0L engine — 0.354-0.453 in. (9.0-11.5mm)

3. Belt tension can also be checked with a tension gauge. Measure the belt tension between any 2 pulleys. The desired value should be 55-110 lbs. (250-500 N).

4. If the belt tension is not correct, adjust as follows:

a. Loosen the adjusting bolt or fixing bolt locknut on the alternator, alternator bracket or tension pulley.

b. Move the alternator or turn the adjusting bolt to adjust belt tension to the correct tension.

c. Once the desired value is reached, secure the bolt or locknut and recheck tension.

5. Make sure the ignition switch is turned **OFF** . Disconnect the battery ground cable.

6. Disconnect the alternator output lead from the alternator **B** terminal.

7. Connect the positive lead of an ammeter to the **B** terminal and the negative lead to the disconnected output wire.

➡ **Tighten each connection by bolt and nuts securely as a heavy current will flow through the wire. Do not relay on clips.**

8. Connect a voltmeter between the **B** terminal and ground. Connect the positive lead wire to the alternator **B** terminal and the negative lead wire to a sound ground.

9. Set the engine tachometer and connect the battery ground cable.

10. Leave the hood open.

Test Procedure

1. Check to see that the voltmeter reads the same value as the battery voltage. If the voltmeter reads 0 volts, an open circuit in the wire between the alternator **B** terminal and the battery negative terminal, a blown fuse link or poor grounding is suspected.

2. Turn ON the headlight switch and start the engine.

3. Set the headlights at high beam and the heater blower switch at high, quickly increase the engine speed to 2,500 rpm and read the maximum output current value indicated by the ammeter.

➡ **After engine start up, the charging current drops quickly, therefore, above operation must be done quickly to read the maximum current value correctly.**

4. The ammeter reading must be higher than the limit value. If it is lower than the limit value but the alternator output wire is normal, remove the alternator from the vehicle and check it further. The limit values are as follows:
65A alternator — 45.5A minimum
75A alternator — 52.5A minimum

➡ **The nominal output current value is shown on the nameplate affixed to the alternator body. The output**

current value changes with the electrical load and the temperature of the alternator itself. **Therefore, the nominal output current may not be obtained if the vehicle electrical load at the time of the test is small. In such case, keep the headlights on to cause discharge of the battery or use lights of another vehicle as a load to increase the electrical load. The nominal output current may not be obtained if the temperature of the alternator itself or ambient temperature is too high. In such case, reduce the temperature before testing again.**

5. Upon completion of the test, lower the engine speed to idle and turn the ignition switch OFF.

6. Disconnect the battery ground cable.

7. Disconnect and remove the engine tachometer, voltmeter and ammeter connected for the purpose of this test.

8. Connect the alternator output wire.

9. Connect the battery ground cable.

REGULATED VOLTAGE TEST

▶ **See Figure 12**

The purpose of this test is to determine that the electronic voltage regulator controls the voltage correctly.

Test Preparation

1. Check the battery that installed in the vehicle, making sure it is in good sound state. Test the battery as required to assure this. The battery must be fully charged.

2. Check the alternator drive belt for proper tension as follows:

a. Place a straight-edge along the top edge of the belt and across 2 pulleys. Allow both ends of the straight-edge to rest on top of each pulley for support.

b. Measure the deflection of the belt from the straight-edge with a force of about 22 lbs. applied midway between the 2 pulleys. Deflection should measure as follows:

Alternator/Water Pump Drive Belt
1.8L engine — 0.315-0.433 in. (8.0-11.0mm)
2.0L engine — 0.354-0.453 in. (9.0-11.5mm)

3. Belt tension can also be checked with a tension gauge. Measure the belt

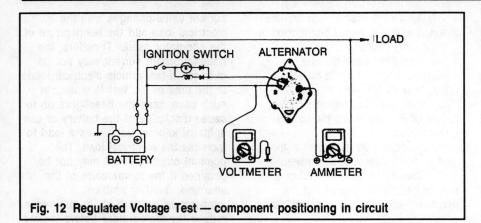

Fig. 12 Regulated Voltage Test — component positioning in circuit

tension between any 2 pulleys. The desired value should be 55-110 lbs. (250-500 N).

4. If the belt tension is not correct, adjust as follows:

a. Loosen the adjusting bolt or fixing bolt locknut on the alternator, alternator bracket or tension pulley.

b. Move the alternator or turn the adjusting bolt to adjust belt tension to the correct tension.

c. Once the desired value is reached, secure the bolt or locknut and recheck tension.

5. Make sure the ignition switch is turned **OFF** . Disconnect the battery ground cable.

6. Connect a digital voltmeter between the **S** terminal of the alternator and the alternator. inserting from the wire side of the 2-way connector. Connect the negative test lead to a sound ground or negative battery terminal.

7. Disconnect the alternator output lead from the alternator **B** terminal.

8. Connect the positive lead of an ammeter to the B terminal and the negative lead to the disconnected output wire.

9. Set the engine tachometer and connect the battery ground cable.

10. Leave the hood open.

Test Procedure

1. Turn the ignition switch **ON** . Check to see that the voltmeter reads the same value as the battery voltage. If the voltmeter reads 0 volts, an open circuit in the wire between the alternator **S** terminal and the battery positive terminal or the fuse link is blown.

2. Start the engine and keep all lights and accessories **OFF** .

3. Run the engine at 2,500 rpm and read the voltmeter when the alternator output current drops to 10A or less.

4. If the voltmeter readings agree with the desired readings listed below, the voltage regulator is functioning properly. If the measured reading disagrees with the desired value, the voltage regulator or the alternator is faulty. The desired voltages are as follows:

Voltage regulator ambient temperature at 68°F (20°C) — 13.9-14.9 volts

Voltage regulator ambient temperature at 140°F (68°C) — 13.4-14.6 volts.

Voltage regulator ambient temperature at 176°F (80°C) — 13.1-14.5 volts

5. Upon completion of test, set the engine speed at idle and turn the ignition switch **OFF** .

6. Disconnect the negative battery terminal.

7. Disconnect all test meters installed for this test procedure.

8. Connect the alternator output wire to the alternator **B** terminal.

9. Connect the negative battery terminal.

REMOVAL & INSTALLATION

1. Disconnect the negative battery cable. Remove the left side undercover from the vehicle.

2. If equipped with air conditioning, remove the condenser electric fan motor and shroud assembly.

3. Remove alternator, water pump and air conditioner compressor drive belts.

4. Remove both water pump pulleys and the alternator top brace.

5. Disconnect the alternator wiring and remove the alternator from the vehicle.

To install:

6. Install the alternator in position and connect the electrical harness.

7. Install the alternator top brace to the engine and tighten the mounting bolt to 20 ft. lbs. (27 Nm).

8. Install the alternator mounting bolt loosely.

9. Install the water pump pulleys and tighten retainer bolts to 7 ft. lbs. (10 Nm).

10. Install the drive belts as adjust until the proper tension is achieved. Secure the lower alternator through bolt nut to 18 ft. lbs. (25 Nm) and the upper alternator lock bolt to 11 ft. lbs. (15 Nm).

11. Install the condenser electric fan motor and shroud assembly.

12. Install the left side undercover from the vehicle, if removed.

13. Connect the negative battery cable. Start the engine and check the alternator for proper operation.

Regulator

All models use a regulator that is integral with the alternator and required no separate service.

Battery

REMOVAL & INSTALLATION

1. Make sure all accessories and the ignition key are in the **OFF** position. Raise the hood and support.

2. Remove the negative battery cable first and then the positive terminal.

3. Lubricate the hold-down with penetrating oil before loosening. Remove the battery clamp or hold-down.

4. Using an approved battery carrier, lift the battery out of the tray and place away from the vehicle.

5. Clean and inspect the battery tray.

6. Prepare a solution of baking soda and water, mixing to a thick paste. Paint the solution onto any corroded areas and allow to set for about one minute. Reapply the mixture a second time.

7. Rinse the area with warm water to remove baking soda mixture.

8. Apply a coat of rust converter to the affected areas and allow to dry. Apply a second application of rust converter once the first coat has dried.

9. Apply two coats of high quality metal primer to the affected areas and

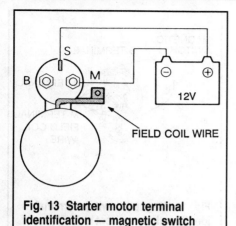

Fig. 13 Starter motor terminal identification — magnetic switch pull-in test

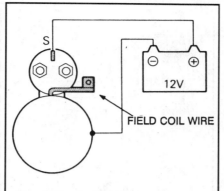

Fig. 14 Starter motor terminal identification — magnetic switch hold-in test

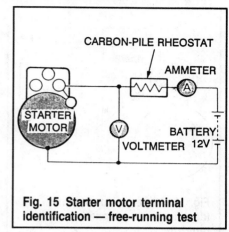

Fig. 15 Starter motor terminal identification — free-running test

allow to completely dry before going any further.

10. Apply a coat of undercoating to the battery tray area and allow to dry overnight, if possible.

11. Clean the battery with soap and water.

12. Install the battery into the vehicle and tighten the hold-down bolt or clamp.

13. Apply a light coat of petroleum jelly to the terminal posts.

14. Tighten the battery cable terminals to 10 ft. lbs. (15 Nm).

15. Clean the cable ends that connect to the body and starter motor, as required.

16. If a slow to no start problem still exists, it is possible that a battery cable or the starter motor may be defective.

Starter

TROUBLESHOOTING HINTS

1. If the starter motor does not operate at all, inspect the following:
Check the starter coil
Check for poor contact at the battery terminal

2. If the starter motor does not stop, inspect the following:
Check the starter magnetic switch

PULL-IN TEST OF MAGNETIC SWITCH

▶ See Figure 13

1. Disconnect the negative battery cable.

2. Remove the starter assembly from the vehicle.

3. Disconnect the field coil wire from the M-terminal of the magnetic switch.

4. Connect a 12V battery between S-terminal and M-terminal on the magnetic switch of the starter.

✳✳CAUTION

The test must be perform quickly, in less than 10 seconds, to prevent the coil from burning.

5. If the pinion moves out, the pull-in coil is functioning properly. If it doesn't, replace the magnetic switch.

HOLD-IN TEST OF MAGNETIC SWITCH

▶ See Figure 14

1. Disconnect the negative battery cable.

2. Remove the starter assembly from the vehicle.

3. Disconnect the field coil wire from the M-terminal of the magnetic switch.

4. Connect a 12V battery between S-terminal and body.

✳✳CAUTION

The test must be perform quickly, in less than 10 seconds, to prevent the coil from burning.

5. If the pinion remains out, everything is in order. If the pinion moves in, the hold-in circuit is open. Replace the magnetic switch.

FREE RUNNING TEST

▶ See Figure 15

1. Disconnect the negative battery cable.

2. Remove the starter assembly from the vehicle.

3. Place the starter assembly into a vise with soft jaws. Connect a fully charged 12 volt battery to the starter motor as follows:

a. Connect a test ammeter with 100 ampere scale and a carbon pile rheostat in series with battery positive post and starter motor terminal.

b. Connect a voltmeter across the starter motor.

c. Rotate the carbon pile to full-resistance position.

d. Connect the battery cable from the negative battery post to the starter motor body.

e. Adjust the rheostat until the battery voltage shown by the voltmeter is 11.5V for direct drive type starter, which are normally installed in 1.8L engines. On reduction-drive starters, which are normally installed in 2.0L engines, adjust the rheostat until the battery voltage shown by the voltmeter is 11.0 V.

f. Confirm that the maximum amperage is within the specifications listed below and the starter turns freely.

Direct drive starter — 60 Amps maximum

Reduction drive starter — 90 Amps maximum

4. If the starter is not within specifications, replace the unit.

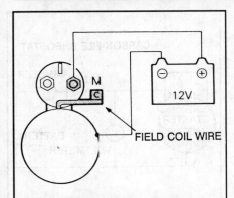

Fig. 16 Starter motor terminal identification — magnetic switch return test

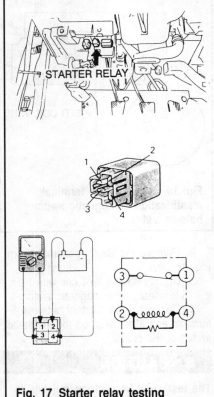

Fig. 17 Starter relay testing

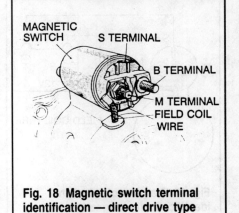

Fig. 18 Magnetic switch terminal identification — direct drive type starter motor

FREE RUNNING TEST

◆ **See Figure 16**

1. Disconnect the negative battery cable.

2. Remove the starter assembly from the vehicle.

3. Disconnect the field coil wire from the M-terminal of the magnetic switch.

4. Connect a 12 volt battery between the M-terminal of the starter and the starter body.

➡ **This test must be done quickly, in less than 10 seconds, to prevent coil from burning.**

5. Pull the pinion out and then release it. If the pinion quickly returns to its original position, everything is in order. If it doesn't, replace the magnetic switch.

Starter Motor

REMOVAL & INSTALLATION

1. Remove the battery and battery tray from the engine compartment.

2. Disconnect the speedometer cable connector at the transaxle end.

3. If equipped with 1.8L engine, remove the bracket on the lower side if the intake manifold.

4. Disconnect the starter motor electrical connections.

5. Remove the starter motor mounting bolts and remove the starter.

To install:

6. Clean both surfaces of starter motor flange and rear plate. Install the starter motor onto the engine and secure

with the retainer bolts. Tighten to 25 ft. lbs. (34 Nm).

7. Connect the electrical harness connector to the starter.

8. Install the intake manifold stay, if removed. Tighten the retainers to 18 ft. lbs. (25 Nm).

9. Install the speedometer cable at the transaxle. Install the battery tray and battery.

10. Operate the starter to assure proper operation.

Starter Relay

◆ **See Figure 17**

TESTING

1. Disarm the air bag as follows:

a. Position the front wheels in the straight-ahead position and place the key in the **LOCK** position. Remove the key from the ignition lock cylinder.

b. Disconnect the negative battery cable and insulate the cable end with high-quality electrical tape or similar non-conductive wrapping.

c. Wait at least 1 minute before working on the vehicle. The air bag system is designed to retain enough

voltage to deploy the air bag for a short period of time even after the battery has been disconnected.

d. If necessary, enter the vehicle from the passenger side and turn the key to unlock the steering column.

2. Remove the retainers and the knee protector from the vehicle.

3. Remove the starter relay from the underside of the relay box.

4. Connect battery voltage to terminal 2 and check continuity between the terminals with terminal 4 grounded.

With power supplied to terminal 2 and terminal 4 is grounded — there is no continuity between terminal 1 and 3.

When no power is supplied to terminal 2 — there is continuity between terminals 1 and 3 and also between terminals 2 and 4.

5. If the test results differ from the results listed above, replace the starter relay.

Magnetic Switch

REMOVAL & INSTALLATION

◆ **See Figures 18, 19, 20 and 21**

1. Disconnect the negative battery cable. Note routing and then disconnect all starter wiring.

2. Remove the starter from the engine.

3. Disconnect the field coil wire from the M-terminal of the magnetic switch.

4. Remove the 2 magnetic switch retainer bolts. Slide the solenoid to the rear slightly and grasp the shims that may be mounted between the solenoid and front housing.

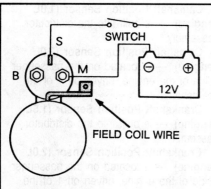

Fig. 19 Terminal connection for pinion gap adjustment

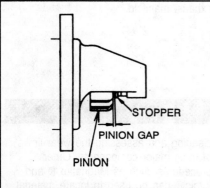

Fig. 20 Inspecting pinion to stopper clearance with a feeler gauge

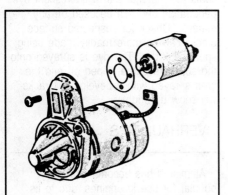

Fig. 21 Adjusting pinion gap by adding gaskets between magnetic switch and front bracket

5. Pull the solenoid assembly upward and to the rear so as to disengage the front end of the solenoid plunger from the shift lever. Once the plunger is free, remove the solenoid assembly.

To install:

6. Install the magnetic switch to the starter motor making sure the shims, if equipped, are in place and aligned. If necessary, turn the plunger so the slot will fit over the vertical top of the shift lever. Then, work the unit into position with the front end of the plunger near the top of the opening in the gear case

7. Position the opening in the front of the plunger over the shift lever and then lower the assembly so as to engage the plunger with the lever.

8. Install the magnetic switch retainer bolts. If necessary, turn the shims and solenoid assembly to line up the holes.

9. Check the pinion gap adjustment as follows:

   a. With the field coil wire disconnected from the M-terminal of the magnetic switch, connect a 12 volt battery supply between S-terminal and M-terminal. The pinion will move out.

✳✳CAUTION

This test must be done quickly, in less than 10 seconds, to prevent the coil from burning out.

   b. Check pinion to stopper clearance (pinion gap) with a feeler gauge and compare to the desired measurement of 0.020-0.079 in. (0.5-2.0mm).

   c. If the pinion gap is out of specifications, adjust by adding or removing gaskets between the magnetic switch and the front bracket.

10. Connect the field coil wire to the M-terminal of the magnetic switch.

11. Install the starter into position and connect the body electrical harness.

12. Install the starter motor retainer bolts and tighten to 25 ft. lbs. (24 Nm). Connect the negative battery cable.

13. Operate the starter assembly and check for proper operation.

Sending Units and Sensors

TESTING

Engine Coolant Temperature Gauge Unit

◗ See Figure 22

The coolant temperature sensor is used to operate the temperature gauge. Do not confuse this sensor with the other switches or sensors used to signal the engine control unit or air conditioning regarding temperature of the coolant. Usually, these other units are mounted near the coolant temperature sensor used for engine control.

1. Drain the engine coolant at least to a level below the intake manifold.

2. Disconnect the sensor wiring harness and remove the coolant temperature sensor.

3. Place the sensor tip in a pan of warm water. Use a thermometer to measure the water temperature.

4. Measure the resistance across the sensor terminals while the sensor is in the water.

5. Note the ohm reading and compare to the following specifications:

   Water temperature of 68°F (20°C) — 2.21-2.69 kohms resistance

   Water temperature of 158°F (70°C) — 90.5-117.5 ohms resistance

   Water temperature of 176°F (80°C) — 264-328 ohms resistance 6.

   If the resistance is not approximately accurate for the temperature, the sensor must be replaced.

Oil Pressure Switch

1. Remove the electrical harness connector from the switch and remove the switch from the oil filter head.

2. Connect an ohmmeter between the terminal and the sensor body cavity and check for conductivity. If there is no conductivity, replace the switch.

3. Next, insert a very thin wedge through the oil hole in the end of the sensor. Push the wedge in slightly and measure resistance. There should be no conductivity.

4. If there is conductivity, even when wedge is pushed, replace the switch.

5. If there is no conductivity when a 71 psi pressure is placed through the oil hole, the switch is operating properly.

6. Check to see that there is no air pressure leakage through the switch. If there is air pressure leakage, the diaphragm is broken and the switch will require replacement.

REMOVAL & INSTALLATION

Coolant Temperature Sensor

1. Drain the engine coolant at least to a level below the intake manifold.

2. Disconnect the sensor wiring harness and remove the coolant temperature sensor from the engine.

To install:

3. Coat the sensor threads with thread sealant (MOPAR 4318034 or equivalent). Install the engine coolant temperature gauge sensor into the bore in the engine and tighten to 9 ft. lbs. (12 Nm). If installing the larger engine coolant temperature sensor for engine control or the air conditioning engine coolant temperature sensor, tighten to 29 ft. lbs. (40 Nm).

4. Connect the electrical harness connector to the sensor and fill the cooling system to the proper level.

Oil Pressure Sensor

1. Disconnect the negative battery cable.

2. Raise and support the vehicle safely.

3. Disconnect the electrical harness connector from the switch and remove from the oil filter head.

To install:

4. Apply a thin bead of sealant to the threaded portion of the oil pressure sensor. Do not allow sealer to contact the end of the threaded portion of the sensor.

5. Install the sensor and tighten to 8 ft. lbs. (12 Nm). Do not overtighten the sensor.

6. Connect the electrical harness connector to the sensor.

There are many sensors located throughout these vehicles. Additional sensors and their locations are listed below. The service procedures are listed throughout this service manual, in the appropriate sections.

Air Flow Sensor — is located in the front of the air intake and has built into it, the air temperature and barometric pressure sensors.

Brake Fluid Level Sensor — is located in the brake master cylinder reservoir.

Camshaft Position Sensor (1.8L engine) — is built into the distributor assembly.

Camshaft Position Sensor (2.0L engine) — is located on the passenger side of the engine, driven off 1 of the camshafts.

Crankshaft Position Sensor (1.8L engine) — is built into the distributor assembly.

Crankshaft Position Sensor (2.0L engine) — is located on the passenger side of the engine, driven off 1 of the camshafts.

Knock Sensor (2.0L Turbocharged Only) — is located in the engine block near the starter/flywheel area.

Oxygen Sensor — is located in the exhaust manifold.

ENGINE MECHANICAL

Engine Overhaul Tips

Most engine overhaul procedures are fairly standard. In addition to specific parts replacement procedures and complete specifications for your individual engine, this chapter also is a guide to accepted rebuilding procedures. Examples of standard rebuilding practice are shown and should be used along with specific details concerning your particular engine.

Competent and accurate machine shop services will ensure maximum performance, reliability and engine life.

In most instances it is more profitable for the do-it-yourself mechanic to remove, clean and inspect the component, buy the necessary parts and deliver these to a shop for actual machine work.

On the other hand, much of the rebuilding work (crankshaft, block, bearings, piston rods, and other components) is well within the scope of the do-it-yourself mechanic.

TOOLS

The tools required for an engine overhaul or parts replacement will depend on the depth of your involvement. With a few exceptions, they will be the tools found in a mechanic's tool kit (see Chapter 1). More in-depth work will require any or all of the following:

a dial indicator (reading in thousandths) mounted on a universal base
micrometers and telescope gauges
jaw and screw-type pullers
scraper
valve spring compressor
ring groove cleaner
piston ring expander and compressor
ridge reamer
cylinder hone or glaze breaker
Plastigage®
engine stand

The use of most of these tools is illustrated in this chapter. Many can be rented for a one-time use from a local parts jobber or tool supply house specializing in automotive work.

Occasionally, the use of special tools is called for. See the information on Special Tools and Safety Notice in the front of this book before substituting another tool.

INSPECTION TECHNIQUES

Procedures and specifications are given in this chapter for inspecting, cleaning and assessing the wear limits of most major components. Other procedures such as Magnaflux® and Zyglo® can be used to locate material flaws and stress cracks. Magnaflux® is a magnetic process applicable only to ferrous materials. The Zyglo® process coats the material with a fluorescent dye penetrating and can be used on any material. Check for suspected surface cracks can be more readily made using spot check dye. The dye is sprayed onto the suspected area, wiped off and the area sprayed with a developer. Cracks will show up brightly.

OVERHAUL TIPS

Aluminum has become extremely popular for use in engines, due to its low weight. Observe the following precautions when handling aluminum parts:

• Never hot tank aluminum parts (the caustic hot tank solution will eat the aluminum.

• Remove all aluminum parts (identification tag, etc.) from engine parts prior to the tanking.

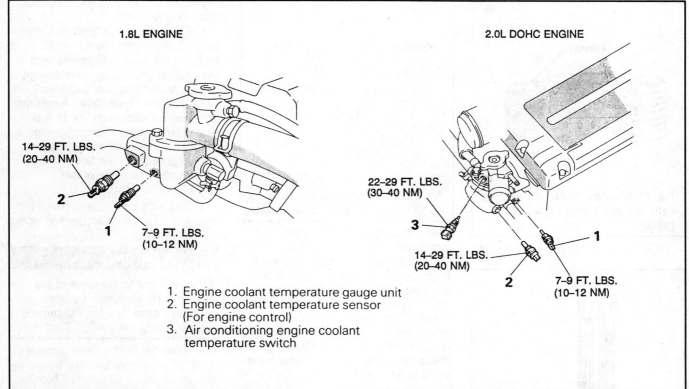

1.8L ENGINE

14–29 FT. LBS.
(20–40 NM)

2

1

7–9 FT. LBS.
(10–12 NM)

2.0L DOHC ENGINE

22–29 FT. LBS.
(30–40 NM)

3

1

14–29 FT. LBS.
(20–40 NM)

2

7–9 FT. LBS.
(10–12 NM)

1. Engine coolant temperature gauge unit
2. Engine coolant temperature sensor
   (For engine control)
3. Air conditioning engine coolant
   temperature switch

Fig. 22 Engine coolant temperature gauge unit, engine coolant temperature sensor and air conditioning engine coolant temperature switch

• Always coat threads lightly with engine oil or anti-seize compounds before installation, to prevent seizure.

• Never overtorque bolts or spark plugs especially in aluminum threads.

Stripped threads in any component can be repaired using any of several commercial repair kits (Heli-Coil®, Microdot®, Keenserts®, etc.).

When assembling the engine, any parts that will be frictional contact must be prelubed to provide lubrication at initial start-up. Any product specifically formulated for this purpose can be used, but engine oil is not recommended as a prelube.

When semi-permanent (locked, but removable) installation of bolts or nuts is desired, threads should be cleaned and coated with Loctite® or other similar, commercial non-hardening sealant.

REPAIRING DAMAGED THREADS

♦ **See Figures 23, 24, 25, 26 and 27**

Several methods of repairing damaged threads are available. Heli-Coil® (shown here), Keenserts® and Microdot® are

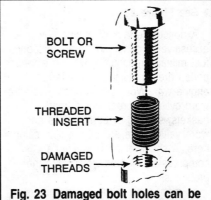

BOLT OR SCREW

THREADED INSERT

DAMAGED THREADS

Fig. 23 Damaged bolt holes can be repaired with thread repair inserts

among the most widely used. All involve basically the same principle — drilling out stripped threads, tapping the hole and installing a prewound insert — making welding, plugging and oversize fasteners unnecessary.

Two types of thread repair inserts are usually supplied: a standard type for most Inch Coarse, Inch Fine, Metric Course and Metric Fine thread sizes and a spark lug type to fit most spark plug port sizes. Consult the individual manufacturer's catalog to determine

Fig. 24 Drill out the damaged threads with specified drill. Drill completely through the hole or to the bottom of a blind hole.

exact applications. Typical thread repair kits will contain a selection of prewound threaded inserts, a tap (corresponding to the outside diameter threads of the insert) and an installation tool. Spark plug inserts usually differ because they require a tap equipped with pilot threads and a combined reamer/tap section. Most manufacturers also supply blister-packed thread repair inserts separately in addition to a master kit containing a

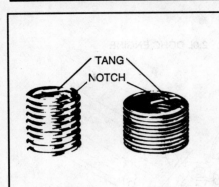

Fig. 25 Standard thread repair insert (left) and spark plug thread insert (right)

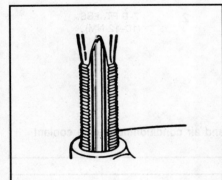

Fig. 26 With the tap supplied, tap the hole to receive the thread insert. Keep the tap well oiled and back it out frequently to avoid clogging the threads.

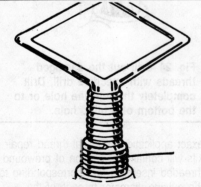

Fig. 27 Screw the threaded insert into the installation tool until the tang engages the slot. Screw the insert into the tapped hole until it is 1/4-1/2 turn below the top surface. After installation, break off the tang with a hammer.

Fig. 28 The screw-in type compression gauge is more accurate

variety of taps and inserts plus installation tools.

Before effecting a repair to a threaded hole, remove any snapped, broken or damaged bolts or studs. Penetrating oil can be used to free frozen threads. The offending item can be removed with locking pliers or with a screw or stud extractor. After the hole is clear, the thread can be repaired, as shown in the series of accompanying illustrations.

Checking Engine Compression

▶ See Figure 28

A noticeable lack of engine power, excessive oil consumption and/or poor fuel mileage measured over an extended period are all indicators of internal engine war. Worn piston rings, scored or worn cylinder bores, blown head gaskets, sticking or burnt valves and worn valve seats are all possible culprits here. A check of each cylinder's compression will help you locate the problems.

As mentioned in the Tools and Equipment part of Section 1, a screw-in type compression gauge is more accurate that the type you simply hold against the spark plug hole, although it takes slightly longer to use. It's worth it to obtain a more accurate reading. Follow the procedure listed below.

1. Before checking compression, make sure the engine oil, starter motor and the battery are all in good condition.
2. Start the engine and allow to idle until normal operating temperature of 185°F (85°C) is reached.
3. Stop the engine and remove all the spark plug cables.
4. Remove the spark plugs from the engine. Disconnect the high tension lead from the ignition coil.
5. Crank the engine to remove any foreign material from the cylinders.
➡ **Cover the spark plug holes with a shop rag to prevent expelled foreign objects from flying out, and keep away from the holes. When measuring compression with water, oil or fuel having entered the cylinder through a crack etc., these will come flying out of the spark plug hole hot and fast, so be sure to take the proper precautions**
6. Install the compression gauge into the spark plug hole to be checked, until the fitting is snug.

✳✳WARNING

Be careful not to crossthread the plug hole. On aluminum cylinder heads use extra care, as the threads in these heads are easily ruined.

7. Open the throttle valve completely either by operating the throttle linkage by hand or by having an assistant floor the accelerator pedal.
8. Ask the assistant to crank the engine using the ignition switch while you read the compression.
9. Compare the measured value to the standard values listed below:
1.8L Engine
Standard compression — 185 psi
Minimum compression — 130 psi
2.0L Non-Turbocharged Engine
Standard compression — 192 psi
Minimum compression — 145 psi
2.0L Turbocharged Engine
Standard compression — 164 psi
Minimum compression — 121 psi
10. Perform the test for all cylinders, ensuring that compression pressure differential for each of the cylinders is within the limit of 14 psi per cylinder maximum.
11. If the cylinder compression or the pressure differential exceeds the limit, add small amount of oil through the spark plug hole and repeat the compression test. If the addition of oil brings the compression value up, poor contact between the piston rings and the cylinder walls is a possible cause for low compression.
12. If the addition of oil into the cylinder doesn't bring the pressure up, valve seizure, poor valve seating or a compression leak from the gasket are all possible.

GENERAL ENGINE SPECIFICATIONS

Year	Engine ID/VIN	Engine Displacement Liters (cc)	Fuel System Type	Net Horsepower @ rpm	Net Torque @ rpm (ft. lbs.)	Bore × Stroke (in.)	Compression Ratio	Oil Pressure @ rpm
1990	T	1.8 (1754)	MPI	92 @ 5000	105 @ 3500	3.17 × 3.39	9.0:1	11.5 @ 750
	R	2.0 (2000)	MPI	135 @ 6000	125 @ 5000	3.35 × 3.47	9.0:1	11.5 @ 750
	U	2.0 (2000)	Turbo	190 @ 6000	203 @ 3000	3.35 × 3.47	7.8:1	11.5 @ 750
1991	T	1.8 (1754)	MPI	92 @ 5000	105 @ 3500	3.17 × 3.39	9.0:1	11.5 @ 750
	R	2.0 (2000)	MPI	135 @ 6000	125 @ 5000	3.35 × 3.47	9.0:1	11.5 @ 750
	U	2.0 (2000)	Turbo	190 @ 6000	203 @ 3000	3.35 × 3.47	7.8:1	11.5 @ 750
1992	T	1.8 (1754)	MPI	92 @ 5000	105 @ 3500	3.17 × 3.39	9.0:1	11.5 @ 750
	R	2.0 (2000)	MPI	135 @ 6000	125 @ 5000	3.35 × 3.47	9.0:1	11.5 @ 750
	U	2.0 (2000)	Turbo	190 @ 6000	203 @ 3000	3.35 × 3.47	7.8:1	11.5 @ 750
1993	B	1.8 (1754)	MPI	92 @ 5000	105 @ 3500	3.17 × 3.39	9.0:1	11.5 @ 750
	E	2.0 (2000)	MPI	135 @ 6000	125 @ 5000	3.35 × 3.47	9.0:1	11.5 @ 750
	F	2.0 (2000)	Turbo	195 @ 6000	203 @ 3000	3.35 × 3.47	7.8:1	11.5 @ 750

NOTE: Horsepower and torque are SAE net figures. They are measured at the rear of the transmission with all accessories installed and operating. Since the figures vary when a given engine is installed in different models, some are representative rather than exact.

MPI—Multi-Point Fuel Injection
Turbo—Turbocharged

CAMSHAFT SPECIFICATIONS

All measurements given in inches.

Year	Engine ID/VIN	Engine Displacement Liters (cc)	Journal Diameter 1	2	3	4	5	Elevation In.	Ex.	Bearing Clearance	Camshaft End Play
1990	T	1.8 (1754)	1.3360–1.3366	1.3360–1.3366	1.3360–1.3366	1.3360–1.3366	1.3360–1.3366	1.4138	1.4138	0.0020–0.0035	0.004–0.008
	R	2.0 (1997)	1.0217–1.0224	1.0217–1.0224	1.0217–1.0224	1.0217–1.0224	1.0217–1.0224	1.3974	1.3858	0.0020–0.0035	0.004–0.008
	U	2.0 (1997)	1.0217–1.0224	1.0217–1.0224	1.0217–1.0224	1.0217–1.0224	1.0217–1.0224	1.3974	1.3858	0.0020–0.0035	0.004–0.008
1991	T	1.8 (1754)	1.3360–1.3366	1.3360–1.3366	1.3360–1.3366	1.3360–1.3366	1.3360–1.3366	1.4138	1.4138	0.0020–0.0035	0.004–0.008
	R	2.0 (1997)	1.0217–1.0224	1.0217–1.0224	1.0217–1.0224	1.0217–1.0224	1.0217–1.0224	1.3974	1.3858	0.0020–0.0035	0.004–0.008
	U	2.0 (1997)	1.0217–1.0224	1.0217–1.0224	1.0217–1.0224	1.0217–1.0224	1.0217–1.0224	1.3974	1.3858	0.0020–0.0035	0.004–0.008
1992	T	1.8 (1754)	1.3360–1.3366	1.3360–1.3366	1.3360–1.3366	1.3360–1.3366	1.3360–1.3366	1.4138	1.4138	0.0020–0.0035	0.004–0.008
	R	2.0 (1997)	1.0217–1.0224	1.0217–1.0224	1.0217–1.0224	1.0217–1.0224	1.0217–1.0224	1.3974	1.3858	0.0020–0.0035	0.004–0.008
	U	2.0 (1997)	1.0217–1.0224	1.0217–1.0224	1.0217–1.0224	1.0217–1.0224	1.0217–1.0224	1.3974	1.3858	0.0020–0.0035	0.004–0.008
1993	B	1.8 (1754)	1.3360–1.3366	1.3360–1.3366	1.3360–1.3366	1.3360–1.3366	1.3360–1.3366	1.4138	1.4138	0.0020–0.0035	0.004–0.008
	E	2.0 (1997)	1.0217–1.0224	1.0217–1.0224	1.0217–1.0224	1.0217–1.0224	1.0217–1.0224	1.3974	1.3858	0.0020–0.0035	0.004–0.008
	F	2.0 (1997)	1.0217–1.0224	1.0217–1.0224	1.0217–1.0224	1.0217–1.0224	1.0217–1.0224	1.3974	1.3858	0.0020–0.0035	0.004–0.008

CRANKSHAFT AND CONNECTING ROD SPECIFICATIONS

All measurements are given in inches.

Year	Engine ID/VIN	Engine Displacement Liters (cc)	Crankshaft Main Brg. Journal Dia.	Crankshaft Main Brg. Oil Clearance	Crankshaft Shaft End-play	Crankshaft Thrust on No.	Connecting Rod Journal Diameter	Connecting Rod Oil Clearance	Connecting Rod Side Clearance
1990	T	1.8 (1754)	2.224	0.0008–0.0020	0.0020–0.0070	3	1.7700	0.0008–0.0020	0.0039–0.0098
	R	2.0 (1997)	2.243–2.244	0.0008–0.0020	0.0020–0.0070	3	1.7709–1.7715	0.0008–0.0020	0.0040–0.0098
	U	2.0 (1997)	2.243–2.244	0.0008–0.0020	0.0020–0.0070	3	1.7709–1.7715	0.0008–0.0020	0.0040–0.0098
1991	T	1.8 (1754)	2.224	0.0008–0.0020	0.0020–0.0070	3	1.7700	0.0008–0.0020	0.0039–0.0098
	R	2.0 (1997)	2.243–2.244	0.0008–0.0020	0.0020–0.0070	3	1.7709–1.7715	0.0008–0.0020	0.0040–0.0098
	U	2.0 (1997)	2.243–2.244	0.0008–0.0020	0.0020–0.0070	3	1.7709–1.7715	0.0008–0.0020	0.0040–0.0098
1992	T	1.8 (1754)	2.224	0.0008–0.0020	0.0020–0.0070	3	1.7700	0.0008–0.0020	0.0039–0.0098
	R	2.0 (1997)	2.243–2.244	0.0008–0.0020	0.0020–0.0070	3	1.7709–1.7715	0.0008–0.0020	0.0040–0.0098
	U	2.0 (1997)	2.243–2.244	0.0008–0.0020	0.0020–0.0070	3	1.7709–1.7715	0.0008–0.0020	0.0040–0.0098
1993	B	1.8 (1754)	2.224	0.0008–0.0020	0.0020–0.0070	3	1.7700	0.0008–0.0020	0.0039–0.0098
	E	2.0 (1997)	2.243–2.244	0.0008–0.0020	0.0020–0.0070	3	1.7709–1.7715	0.0008–0.0020	0.0040–0.0098
	F	2.0 (1997)	2.243–2.244	0.0008–0.0020	0.0020–0.0070	3	1.7709–1.7715	0.0008–0.0020	0.0040–0.0098

VALVE SPECIFICATIONS

Year	Engine ID/VIN	Engine Displacement Liters (cc)	Seat Angle (deg.)	Face Angle (deg.)	Spring Test Pressure (lbs. @ in.)	Spring Installed Height (in.)	Stem-to-Guide Clearance (in.)		Stem Diameter (in.)	
							Intake	Exhaust	Intake	Exhaust
1990	T	1.8 (1754)	44.0–44.5	45.0–45.5	68 @ ①	1.898–1.937	0.0012–0.0024	0.0020–0.0035	0.3100	0.3100
	R	2.0 (1997)	44.0–44.5	45.0–45.5	66 @ ①	1.469–1.516②	0.0008–0.0019	0.0020–0.0033	0.2585–0.2591	0.2671–0.2579
	U	2.0 (1997)	44.0–44.5	45.0–45.5	66 @ ①	1.469–1.516②	0.0008–0.0019	0.0020–0.0033	0.2585–0.2591	0.2671–0.2579
1991	T	1.8 (1754)	44.0–44.5	45.0–45.5	68 @ ①	1.898–1.937	0.0012–0.0024	0.0020–0.0035	0.3100	0.3100
	R	2.0 (1997)	44.0–44.5	45.0–45.5	66 @ ①	1.469–1.516②	0.0008–0.0019	0.0020–0.0033	0.2585–0.2591	0.2671–0.2579
	U	2.0 (1997)	44.0–44.5	45.0–45.5	66 @ ①	1.469–1.516②	0.0008–0.0019	0.0020–0.0033	0.2585–0.2591	0.2671–0.2579
1992	T	1.8 (1754)	44.0–44.5	45.0–45.5	68 @ ①	1.898–1.937	0.0012–0.0024	0.0020–0.0035	0.3100	0.3100
	R	2.0 (1997)	44.0–44.5	45.0–45.5	66 @ ①	1.469–1.516②	0.0008–0.0019	0.0020–0.0033	0.2585–0.2591	0.2671–0.2579
	U	2.0 (1997)	44.0–44.5	45.0–45.5	66 @ ①	1.469–1.516②	0.0008–0.0019	0.0020–0.0033	0.2585–0.2591	0.2671–0.2579
1993	B	1.8 (1754)	44.0–44.5	45.0–45.5	68 @ ①	1.898–1.937	0.0012–0.0024	0.0020–0.0035	0.3100	0.3100
	E	2.0 (1997)	44.0–44.5	45.0–45.5	66 @ ①	1.469–1.516②	0.0008–0.0019	0.0020–0.0033	0.2585–0.2591	0.2671–0.2579
	F	2.0 (1997)	44.0–44.5	45.0–45.5	66 @ ①	1.469–1.516②	0.0008–0.0019	0.0020–0.0033	0.2585–0.2591	0.2671–0.2579

① Installed height
② Free length

PISTON AND RING SPECIFICATIONS

All measurements are given in inches.

| Year | Engine ID/VIN | Engine Displacement Liters (cc) | Piston Clearance | Ring Gap | | | Ring Side Clearance | | |
				Top Compression	Bottom Compression	Oil Control	Top Compression	Bottom Compression	Oil Control
1990	T	1.8 (1754)	0.0004–0.0012	0.0018–0.0177	0.0079–0.0138	0.0080–0.0280	0.0018–0.0033	0.0008–0.0024	NA
	R	2.0 (1997)	0.0008–0.0016	0.0098–0.0177	0.0138–0.0197	0.0079–0.0276	0.0012–0.0028	0.0012–0.0028	NA
	U	2.0 (1997)	0.0012–0.0020	0.0118–0.0177	0.0138–0.0197	0.0079–0.0276	0.0012–0.0028	0.0012–0.0028	NA
1991	T	1.8 (1754)	0.0004–0.0012	0.0018–0.0177	0.0079–0.0138	0.0080–0.0280	0.0018–0.0033	0.0008–0.0024	NA
	R	2.0 (1997)	0.0008–0.0016	0.0098–0.0177	0.0138–0.0197	0.0079–0.0276	0.0012–0.0028	0.0012–0.0028	NA
	U	2.0 (1997)	0.0012–0.0020	0.0118–0.0177	0.0138–0.0197	0.0079–0.0276	0.0012–0.0028	0.0012–0.0028	NA
1992	T	1.8 (1754)	0.0004–0.0012	0.0018–0.0177	0.0079–0.0138	0.0080–0.0280	0.0018–0.0033	0.0008–0.0024	NA
	R	2.0 (1997)	0.0008–0.0016	0.0098–0.0177	0.0138–0.0197	0.0079–0.0276	0.0012–0.0028	0.0012–0.0028	NA
	U	2.0 (1997)	0.0012–0.0020	0.0118–0.0177	0.0138–0.0197	0.0079–0.0276	0.0012–0.0028	0.0012–0.0028	NA
1993	B	1.8 (1754)	0.0004–0.0012	0.0018–0.0177	0.0079–0.0138	0.0080–0.0280	0.0018–0.0033	0.0008–0.0024	NA
	E	2.0 (1997)	0.0008–0.0016	0.0098–0.0177	0.0138–0.0197	0.0079–0.0276	0.0012–0.0028	0.0012–0.0028	NA
	F	2.0 (1997)	0.0012–0.0020	0.0118–0.0177	0.0138–0.0197	0.0079–0.0276	0.0012–0.0028	0.0012–0.0028	NA

NA—Not available

TORQUE SPECIFICATIONS

All readings in ft. lbs.

Year	Engine ID/VIN	Engine Displacement Liters (cc)	Cylinder Head Bolts	Main Bearing Bolts	Rod Bearing Bolts	Crankshaft Damper Bolts	Flywheel Bolts	Manifold Intake	Manifold Exhaust	Spark Plugs	Lug Nut
1990	T	1.8 (1754)	51–54	37–39	24–25	80–94	94–101	13–18	13–18	18	87–101
	R	2.0 (1997)	65–72	47–51	36–38	80–94	94–101	18–22	18–22	18	87–101
	U	2.0 (1997)	65–72	47–51	36–38	80–94	94–101	18–22	18–22	18	87–101
1991	T	1.8 (1754)	51–54	37–39	24–25	80–94	94–101	13–18	13–18	18	87–101
	R	2.0 (1997)	65–72	47–51	36–38	80–94	94–101	18–22	18–22	18	87–101
	U	2.0 (1997)	65–72	47–51	36–38	80–94	94–101	18–22	18–22	18	87–101
1992	T	1.8 (1754)	51–54	37–39	24–25	80–94	94–101	13–18	13–18	18	87–101
	R	2.0 (1997)	65–72	47–51	36–38	80–94	94–101	18–22	18–22	18	87–101
	U	2.0 (1997)	65–72	47–51	36–38	80–94	94–101	18–22	18–22	18	87–101
1993	B	1.8 (1754)	51–54	37–39	24–25	80–94	94–101	13–18	13–18	18	87–101
	E	2.0 (1997)	①	③	②	80–94	94–101	18–22	18–22	18	87–101
	F	2.0 (1997)	①	③	②	80–94	94–101	18–22	18–22	18	87–101

① Step 1: 14.5 ft. lbs.
Step 2: Additional ¼ turn
Step 3: Additional ¼ turn
② Step 1: 14.5 ft. lbs.
Step 2: Additional ¼ turn
③ Step 1: 18 ft. lbs.
Step 2: Additional ¼ turn

Engine Assembly

REMOVAL & INSTALLATION

The following procedure can be used on all vehicles. Slight variations may occur due to extra connections, etc., but the basic procedure should cover all models.

1. Relieve fuel system pressure as follows:
a. Loosen the fuel filler cap to release fuel tank pressure.
b. Disconnect the fuel pump harness connector located at the rear of the fuel tank.
c. Start the vehicle and allow it to run until it stalls from lack of fuel. Turn the key to the OFF position.
d. Disconnect the negative battery cable, then reconnect the fuel pump connector and reinstall the fuel filler cap.
e. Wrap shop towels around the fitting that is being disconnected to absorb residual fuel in the lines.
2. Remove the engine under cover, if equipped.
3. Matchmark the hood and hinges and remove the hood assembly. Remove the air cleaner assembly and all adjoining air intake duct work.

4. Drain the engine coolant and remove the radiator assembly, coolant reservoir and intercooler.
5. Remove the transaxle and transfer case if equipped with AWD.
6. Disconnect and tag for assembly reference the connections for the accelerator cable, heater hoses, brake vacuum hose, connection for vacuum hoses, high pressure fuel line, fuel return line, oxygen sensor connection, coolant temperature gauge connection, coolant temperature sensor connector, connection for thermo switch sensor, if equipped with automatic transaxle, the connection for the idle speed control, the motor position sensor connector, the throttle position sensor connector, the EGR temperature sensor connection (California vehicles), the fuel injector connectors, the power transistor connector, the ignition coil connector, the condenser and noise filter connector, the distributor and control harness, the connections for the alternator and oil pressure switch wires.
7. Remove the air conditioner drive belt and the air conditioning compressor. Leave the hoses attached. Do not discharge the system. Wire the compressor aside.
8. Remove the power steering pump and wire aside.

9. Remove the exhaust manifold to head pipe nuts. Discard the gasket.
10. Attach a hoist to the engine and take up the engine weight. Remove the engine mount bracket. Remove any torque control brackets (roll stoppers). Note that some engine mount pieces have arrows on them for proper assembly. Double check that all cables, hoses, harness connectors, etc., are disconnected from the engine. Lift the engine slowly from the engine compartment.
To install:
11. Install the engine and secure in position. The front lower mount through bolt nut should not be tightened until the full weight of the engine is on the mount. Tightening the engine mount bolts as followings:
Upper mount to engine nuts and bolts — 36-47 ft. lbs. (50-65 Nm)
Upper mount through bolt nut — 43-58 ft. lbs. (60-80 Nm)
Lower mount through bolt nut — 33-47 ft. lbs. (45-65 Nm)
12. Install the exhaust pipe, power steering pump and air conditioning compressor.
13. Checking the tags installed during removal, reconnect all electrical and vacuum connections.
14. Install the transaxle to the vehicle and tighten the upper mounting bolts to

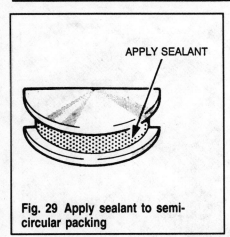

Fig. 29 Apply sealant to semi-circular packing

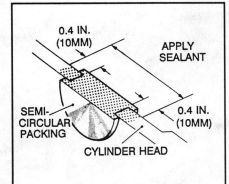

Fig. 30 Install semi-circular packing into cylinder head and apply sealant its top surface

65 ft. lbs (90 Nm). Install the starter assembly and tighten both mounting bolts to 54-65 ft. lbs. (75-90 Nm).

15. Install the radiator assembly and intercooler.

16. Install the air cleaner assembly. Install all control brackets, if not already done.

17. Fill the engine with the proper amount of engine oil. Connect the negative battery cable.

18. Refill the cooling system. Start the engine, allow it to reach normal operating temperature. Check for leaks.

19. Check the ignition timing and adjust, if necessary.

20. Install the hood making sure to align the matchmarks made during disassembly.

21. Road test the vehicle and check all functions for proper operation.

Valve Cover

REMOVAL & INSTALLATION

1.8L Engine
▶ **See Figures 29 and 30**

1. Relieve the fuel system pressure as follows:

a. Loosen the fuel filler cap to release fuel tank pressure.

b. Disconnect the fuel pump harness connector located at the rear of the fuel tank.

c. Start the vehicle and allow it to run until it stalls from lack of fuel. Turn the key to the **OFF** position.

d. Disconnect the negative battery cable, then reconnect the fuel pump connector and reinstall the fuel filler cap.

e. Wrap shop towels around the fitting that is being disconnected to absorb residual fuel in the lines.

2. Remove the air intake hose and the breather hose.

3. Remove the PCV hose from the valve.

4. Disconnect the accelerator cable. There will be 2 cables, if equipped with cruise-control.

5. Remove the spark plug cables as required. Label to aid in installation prior to removal.

6. Disconnect the vacuum line for the brake booster.

7. Remove the clamp that holds the power steering pressure hose to the engine mounting bracket, if required.

8. Remove the valve cover mounting bolt.

9. Remove the valve cover, gasket and half-round seal from the engine.

10. Clean all mating surfaces.

To install:

11. Apply sealer to the perimeter of the half-round seal and position in position in the cylinder head. Install a new valve cover gasket.

12. Install the valve cover to the engine and tighten to 5 ft. lbs. (7 Nm). Make sure the gasket remains in position during cover installation.

13. Connect or install all previously disconnected hoses, cables and electrical connections. Adjust the throttle cable(s).

14. Install the air intake hose. Connect the breather hose and PCV hose.

15. Connect the negative battery cable and let the engine idle. Inspect for leaks.

16. Check and adjust the idle speed as required.

2.0L Engine
▶ **See Figures 31 and 32**

1. Relieve the fuel system pressure as follows:

a. Loosen the fuel filler cap to release fuel tank pressure.

b. Disconnect the fuel pump harness connector located at the rear of the fuel tank.

c. Start the vehicle and allow it to run until it stalls from lack of fuel. Turn the key to the **OFF** position.

d. Disconnect the negative battery cable, then reconnect the fuel pump connector and reinstall the fuel filler cap.

e. Wrap shop towels around the fitting that is being disconnected to absorb residual fuel in the lines.

2. Disconnect the accelerator cable. There will be 2 cables if equipped with cruise-control.

3. Remove the air cleaner with the air intake hose.

4. Remove the spark plug cable center cover and remove the spark plug cable.

5. Reposition the electrical harness and vacuum hoses as not to interfere with valve cover removal. Remove the PCV hose, as required. Disconnect the brake booster vacuum hose.

6. Remove the valve cover retainer screw.

7. Remove the valve cover and the half-round seal from the engine.

8. Clean all gasket mating surfaces.

To install:

9. Apply sealer to the perimeter of the half-round seal and to the lower edges of the half-round portions of the belt-side of the new gasket. Install the valve cover.

10. Tighten the retainers evenly to 3 ft. lbs. (3.5 Nm).

11. Connect or install all previously disconnected hoses, cables and electrical connections. Adjust the throttle cable(s).

12. Install the spark plug cable center cover.

13. Install the air cleaner and intake hose. Connect the breather hose.

14. Connect the negative battery cable, run the vehicle until the thermostat opens and inspect for leaks.

15. Check and adjust the idle speed, as required.

Fig. 32 Removing the valve cover — 2.0L engine

Fig. 31 Removing the valve cover retainer screws — 2.0L engine

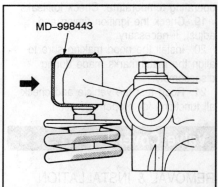

Fig. 33 Install lash adjuster retainer tools MD998443 to the rocker arms to hold lash adjusters in place

Rocker Arm/Shafts

REMOVAL & INSTALLATION

➡ The DOHC engines do not use rocker shafts. The valve are directly actuated by rocker arms. To remove the arms, the camshaft must first be removed. It is recommended that all rocker arms and lash adjusters are replaced together.

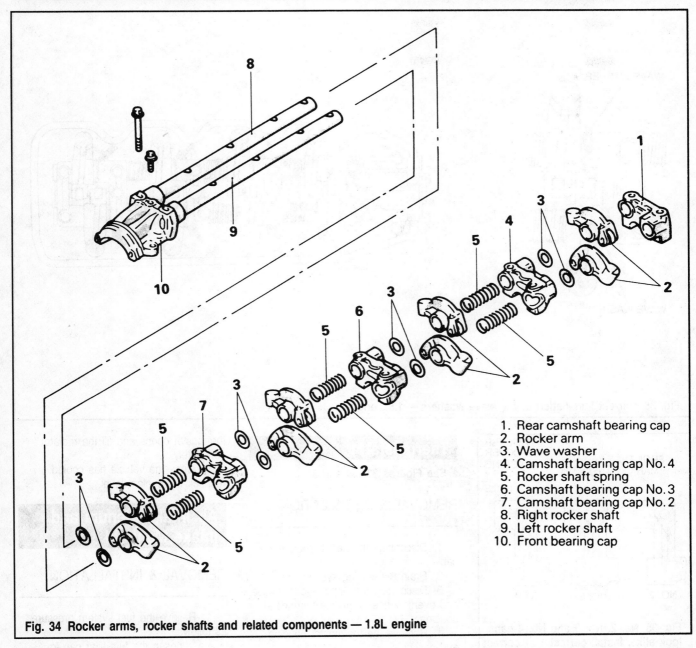

Fig. 34 Rocker arms, rocker shafts and related components — 1.8L engine

1. Rear camshaft bearing cap
2. Rocker arm
3. Wave washer
4. Camshaft bearing cap No. 4
5. Rocker shaft spring
6. Camshaft bearing cap No. 3
7. Camshaft bearing cap No. 2
8. Right rocker shaft
9. Left rocker shaft
10. Front bearing cap

1.8L Engine
▶ **See Figures 33, 34, 35 and 36**

1. Disconnect the negative battery cable.

2. Remove the valve cover. Install lash adjuster retainer tools MD998443 or equivalent, to the rocker arm.

3. Remove the distributor extension, if necessary.

4. Have a helper hold the rear of the camshaft down. If not, the timing belt will dislodge and valve timing will be lost.

5. Loosen the camshaft cap retaining bolts but don't remove them from the caps. Remove the rear bearing cap.

6. Loosen the remaining camshaft cap retaining bolts but don't remove them from the caps. Do not loosen the forward most camshaft bearing cap bolt.

7. At this point, the shafts can be removed as an assembly for service or service of individual components can be made by sliding component to be replaced off back end of shafts. If the later method of replacement is used, only disassembly as far as needed and keep all parts in order of removal. Installation of parts in the same location and orientation is necessary.

8. To remove the shafts as an assembly, remove bearing caps Nos. 2, 3 and 4, rocker arms, rocker shafts and

bolts. It is essential that all parts be kept in the same order and orientation for reinstallation. Remove the lash adjuster tools to replace the adjuster(s) as required. Inspect the roller surfaces of the rockers. Replace if there are any signs of damage or if the roller does not turn smoothly. Check the inside bore of the rockers and lifter for wear.

To install:

9. Apply a drop of sealant to the rear edges of the end caps.

10. Install the assembly into the front bearing cap making sure the notches in the rocker shafts are facing up. Insert the installation bolt but do not tighten at this point.

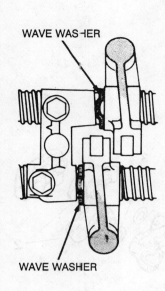

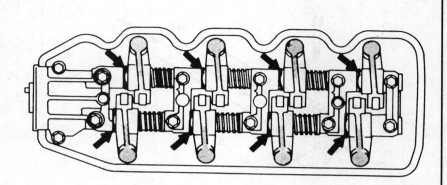

Fig. 36 Correct installation of the wave washers — 1.8L engine

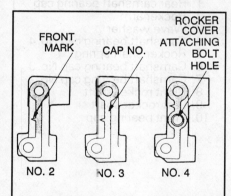

Fig. 35 No. 2, No. 3 and No. 4 caps look alike. Install correct 1 at correct position. They can be identified by the front mark, cap number and rocker cover attaching bolt hole — 1.8L engine

11. Install the remaining cap bolts. Tighten all bolts evenly and gradually to 15 ft. lbs. (20 Nm). Remove the lash adjuster retainers.

12. Install the distributor extension, if removed.

13. Install the valve cover with a new gasket in place.

14. Connect the negative battery cable.

Thermostat

▶ **See Figures 37, 38 and 39**

REMOVAL & INSTALLATION

1. Disconnect the negative battery cable.

2. Drain the cooling system.

3. Disconnect the upper radiator hose and overflow hose from the thermostat housing.

4. Remove the thermostat housing and gasket.

5. Remove the thermostat taking note of its original position in the housing.

To install:

6. Install the thermostat so its flange seats tightly in the machined recess in the thermostat housing. Refer to its location prior to removal.

7. Use a new gasket and reinstall the thermostat housing. Torque the housing mounting bolts to 12-14 ft. lbs. (17-20 Nm).

8. Connect the hoses and fill the system with coolant.

9. Connect the negative battery cable, run the vehicle until the thermostat opens and fill the radiator completely.

10. Once the vehicle has cooled, recheck the coolant level.

Air Intake Plenum and Intake Manifold

REMOVAL & INSTALLATION

1. Relieve the fuel system pressure as follows:

a. Loosen the fuel filler cap to release fuel tank pressure.

b. Disconnect the fuel pump harness connector located at the rear of the fuel tank.

c. Start the vehicle and allow it to run until it stalls from lack of fuel. Turn the key to the **OFF** position.

d. Disconnect the negative battery cable, then reconnect the fuel pump connector and reinstall the fuel filler cap.

e. Wrap shop towels around the fitting that is being disconnected to absorb residual fuel in the lines.

2. Drain the cooling system.

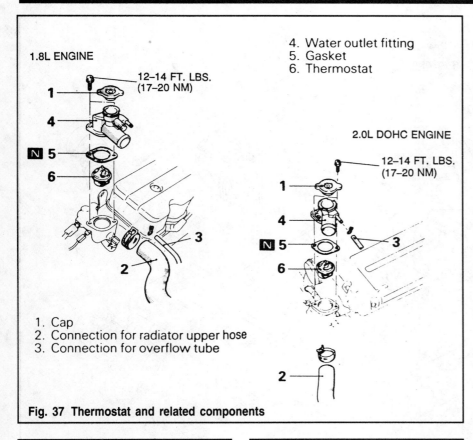

1.8L ENGINE

1 — 12–14 FT. LBS. (17–20 NM)

4

N 5

6

4. Water outlet fitting
5. Gasket
6. Thermostat

2.0L DOHC ENGINE

1 — 12–14 FT. LBS. (17–20 NM)

4

N 5

6

3

2

1. Cap
2. Connection for radiator upper hose
3. Connection for overflow tube

Fig. 37 Thermostat and related components

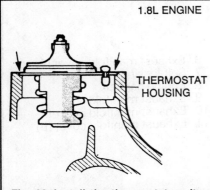

1.8L ENGINE

THERMOSTAT HOUSING

Fig. 38 Install the thermostat so its flange seats tightly in the machined recess in the thermostat housing — 1.8L engine

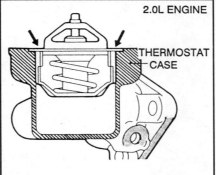

2.0L ENGINE

THERMOSTAT CASE

Fig. 39 Install the thermostat so its flange seats tightly in the machined recess in the thermostat housing — 2.0L engine

3. Disconnect the accelerator cable, breather hose and air intake hose.

4. Disconnect the upper radiator hose, heater hose and water bypass hose.

5. Remove all vacuum hoses and pipes as necessary, including the brake booster vacuum line.

6. Disconnect the high pressure fuel line, fuel return hose and remove throttle control cable brackets.

7. Tag and disconnect the electrical connectors from the oxygen sensor, coolant temperature sensor, thermo switch, idle speed control connection, EGR temperature sensor, spark plug wires, etc. that may interfere with the manifold removal procedure.

8. Remove the fuel rail, fuel injectors, pressure regulator and insulators from the engine.

9. If equipped with 1989-91 engines, remove the distributor from the engine if

it passes through the manifold. Matchmark the distributor shaft to the housing and the housing to the head or nearest accessory prior to removal.

10. If equipped with 2.0L engine, remove the ignition coil. Remove the intake manifold bracket.

11. Disconnect the water hose connections at the throttle body, water inlet, and heater assembly.

12. If the thermostat housing is preventing removal of the intake manifold, remove it.

13. Disconnect the vacuum connection at the power brake booster and the PCV valve if still connected.

14. Remove the intake manifold mounting bolts and remove the intake manifold assembly. Disassemble manifold from the intake plenum on a work bench as required.

To install:

15. Assemble the intake manifold assembly using all new gaskets. Torque air intake plenum bolts to 11-14 ft. lbs. (15-19 Nm).

16. Clean all gasket material from the cylinder head intake mounting surface and intake manifold assembly. Check both surfaces for cracks or other damage. Check the intake manifold water passages and jet air passages for clogging. Clean if necessary.

17. Install a new intake manifold gasket to the head and install the manifold. Torque the manifold in a crisscross pattern, starting from the inside and working outwards to 11-14 ft. lbs. (15-19 Nm).

18. Install the fuel delivery pipe, injectors and pressure regulator from the engine. Torque the retaining bolts to 7-9 ft. lbs. (10-13 Nm).

19. Install the thermostat housing, intake manifold brace bracket, distributor and throttle body stay bracket.

20. Connect or install all hoses, cables and electrical connectors that were removed or disconnected during the removal procedure.

21. Fill the system with coolant.

22. Connect the negative battery cable, run the vehicle until the thermostat opens, fill the radiator completely. Check for leaks.

23. Adjust the accelerator cable. Check and adjust the idle speed as required.

24. Once the vehicle has cooled, recheck the coolant level.

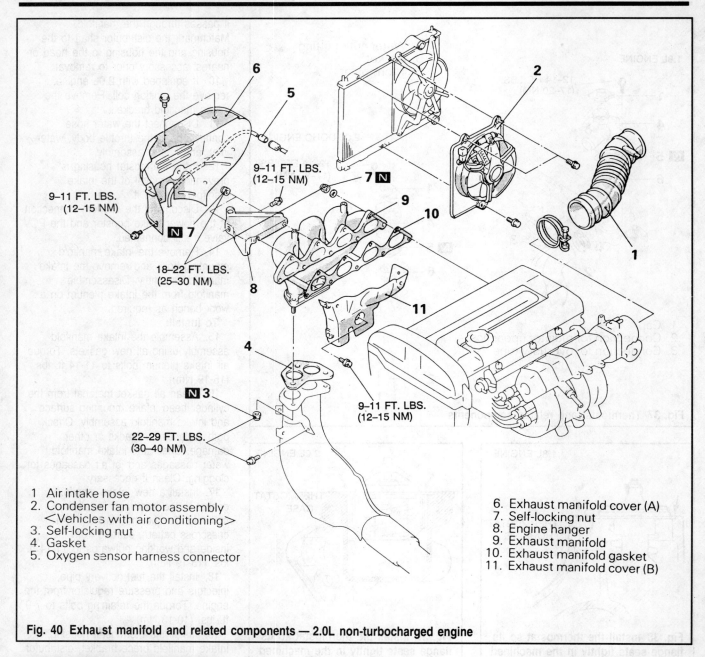

9–11 FT. LBS.
(12–15 NM)

9–11 FT. LBS.
(12–15 NM)

7 N

9–11 FT. LBS.
(12–15 NM)

18–22 FT. LBS.
(25–30 NM)

N 3

22–29 FT. LBS.
(30–40 NM)

9–11 FT. LBS.
(12–15 NM)

1 Air intake hose
2. Condenser fan motor assembly
   <Vehicles with air conditioning>
3. Self-locking nut
4. Gasket
5. Oxygen sensor harness connector

6. Exhaust manifold cover (A)
7. Self-locking nut
8. Engine hanger
9. Exhaust manifold
10. Exhaust manifold gasket
11. Exhaust manifold cover (B)

Fig. 40 Exhaust manifold and related components — 2.0L non-turbocharged engine

Exhaust Manifold

REMOVAL & INSTALLATION

Non-Turbocharged Engines

◆ See Figure 40

1. Disconnect battery negative cable.
2. Raise the vehicle and support safely.
3. Remove the exhaust pipe to exhaust manifold nuts and separate exhaust pipe. Discard gasket.
4. Lower vehicle.

5. Remove electric cooling fan assembly. Remove the oil dipstick tube.
6. Disconnect necessary EGR components. Disconnect and remove the oxygen sensor, as required.
7. Remove outer exhaust manifold heat shield and engine hanger.
8. Remove the exhaust manifold mounting bolts, the inner heat shield and the exhaust manifold from the engine.

To install:

9. Clean all gasket material from the mating surfaces and check the manifold for damage or cracking.

10. Install a new gasket and install the manifold. Tighten the nuts to in a crisscross pattern to:

1.8L engine — 11-14 ft. lbs. (15-20 Nm).

2.0L engine — 18-22 ft. lbs. (25-30 Nm).

11. Install the heat shields.
12. Connect EGR components.
13. Install the electric cooling fan assembly, dipstick tube and alternator, as required.
14. Install a new flange gasket and connect the exhaust pipe.
15. Connect the negative battery cable and check for exhaust leaks.

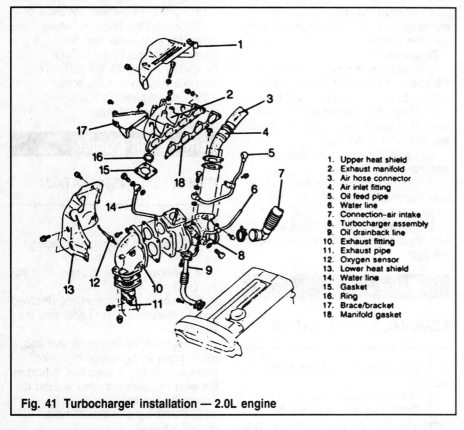

1. Upper heat shield
2. Exhaust manifold
3. Air hose connector
4. Air inlet fitting
5. Oil feed pipe
6. Water line
7. Connection-air intake
8. Turbocharger assembly
9. Oil drainback line
10. Exhaust fitting
11. Exhaust pipe
12. Oxygen sensor
13. Lower heat shield
14. Water line
15. Gasket
16. Ring
17. Brace/bracket
18. Manifold gasket

Fig. 41 Turbocharger installation — 2.0L engine

Turbocharger

REMOVAL & INSTALLATION

2.0L Engine
◆ See Figure 41

Many turbocharger failures are due to oil supply problems. Heat soak after hot shutdown can cause the engine oil in the turbocharger and oil lines to 'coke.' Often the oil feed lines will become partially or completely blocked with hardened particles of carbon, blocking oil flow. Check the oil feed pipe and oil return line for clogging. Clean these tubes well. Always use new gaskets above and below the oil feed eyebolt fitting. Do not allow particles of dirt or old gasket material to enter the oil passage hole and that no portion of the new gasket blocks the passage.

1. Disconnect the negative battery cable.
2. Drain the engine oil, cooling system and remove the radiator. On vehicles equipped with air conditioning, remove the condenser fan assembly with the radiator.
3. Disconnect the oxygen sensor connector and remove the sensor.
4. Remove the oil dipstick and tube.
5. Remove the air intake bellows hose, the wastegate vacuum hose, the connections for the air outlet hose, and the upper and lower heat shield.
6. Unbolt the power steering pump and bracket assembly and leaving the hoses connected, wire it aside.
7. Remove the self-locking exhaust manifold nuts, the triangular engine hanger bracket, the eyebolt and gaskets that connect the oil feed line to the turbo center section and the water cooling lines. The water line under the turbo has a threaded connection.
8. Remove the exhaust pipe nuts and gasket and lift off the exhaust manifold. Discard the gasket.
9. Remove the 2 through bolts and 2 nuts that hold the exhaust manifold to the turbocharger.
10. Remove the 2 capscrews from the oil return line (under the turbo). Discard the gasket. Separate the turbo from the exhaust manifold. The 2 water pipes and oil feed line can still be attached.
11. Visually check the turbine wheel (hot side) and compressor wheel (cold side) for cracking or other damage.

2.0L Turbocharged Engines

1. Disconnect the battery negative cable. Drain the cooling system.
2. If equipped with air conditioning, remove the condenser cooling fan from the vehicle. Remove the power steering pump and bracket and position aside.
3. Disconnect the oxygen sensor electrical harness.
4. Raise the vehicle and support safely.
5. Drain the oil from the crankcase and remove the oil level indicator and tube.
6. Remove the exhaust pipe to turbocharger nuts and separate the exhaust pipe. Discard the gasket.
7. Lower vehicle. Remove air intake and vacuum hose connections.
8. Remove the upper exhaust manifold and turbocharger heat shields. Remove the exhaust manifold to turbocharger attaching bolts and nut.
9. Remove the engine hanger, water and oil lines from the turbo.
10. Remove the exhaust manifold mounting nuts. Remove the exhaust manifold and gasket from the engine.

To install:

11. Clean all gasket material from the mating surfaces and check the manifold for damage.
12. Install new gaskets and install the manifold. Tighten the manifold to head nuts in a crisscross pattern to 18-22 ft. lbs. (25-30 Nm). Tighten the manifold to turbo nut and bolts to 40-47 ft. lbs. (55-65 Nm).
13. Install the engine hanger, water and oil lines to the turbocharger.
14. Install the heat shields.
15. Install the new gasket and connect the exhaust pipe.
16. Install the condenser cooling fan and power steering pump. Connect the oxygen sensor harness.
17. Install the oil level indicator and tube replacing O-ring as required.
18. Fill the crankcase with clean oil and refill the cooling system.
19. Connect the negative battery cable and check for exhaust leaks.

Check whether the turbine wheel and the compressor wheel can be easily turned by hand. Check for oil leakage. Check whether or not the wastegate valve remains open. If any problem is found, replace the part. Inspect oil passages for restriction or deposits and clean as required.

12. The wastegate can be checked with a pressure tester. Apply approximately 9 psi to the actuator and make sure the rod moves. Do not apply more than 10.3 psi or the diaphragm in the wastegate may be damaged. Vacuum applied to the wastegate actuator should be maintained, replace if leaks vacuum. Do not attempt to adjust the wastegate valve.

To install:

13. Prime the oil return line with clean engine oil. Replace all locking nuts. Before installing the threaded connection for the water inlet pipe, apply light oil to the inner surface of the pipe flange. Assemble the turbocharger and exhaust manifold.

14. Install the exhaust manifold using a new gasket.

15. Connect the water cooling lines, oil feed line and engine hanger.

16. If removed, install the power steering pump and bracket.

17. Install the heat shields, air outlet hose, wastegate hose and air intake bellows.

18. Install the oil dipstick tube and dipstick. Install the oxygen sensor.

19. Install the radiator assembly.

20. Fill the engine with oil, fill the cooling system and reconnect the negative battery cable.

Charge Air Cooler

REMOVAL & INSTALLATION

▶ See Figure 42

2.0L Turbocharged Engine

1. Disconnect the negative battery cable. Label and disconnect the vacuum hoses at the turbocharged by-pass valve.

2. Remove the air intake hoses and the air cleaner assembly from the engine compartment.

3. Remove the splash shield and extension from the inside of the right front wheel well. Disconnect and remove any remaining air plumbing.

4. Remove the charge air cooler mounting bolts and the charge air cooler from the vehicle.

To install:

5. Install the charge air cooler and the charge air cooler mounting bolts Tighten bolts to 11 ft. lbs. (15 Nm).

6. Install the lower air plumbing and the splash shield and extension inside of the right front wheel well.

7. Install the air cleaner assembly and the air intake hoses.

8. Connect the vacuum hoses at the turbocharged by-pass valve and the negative battery cable.

9. Confirm that all air hoses are securely installed and the connections are tight.

Radiator

REMOVAL & INSTALLATION

1. Disconnect the negative battery cable.

2. Drain the cooling system.

3. Disconnect the overflow tube. Some vehicles may also require removal of the overflow tank.

4. Disconnect upper and lower radiator hoses. Matchmark the upper radiator hose to assure installation in the orientation.

5. Disconnect electrical connectors for cooling fan and air conditioning condenser fan, if equipped. Remove the fan assembly from the engine compartment.

6. Disconnect thermo sensor wires.

7. Disconnect and plug automatic transaxle cooler lines, if equipped with automatic transaxle.

8. Remove the upper radiator mounts and lift out the radiator assembly.

9. Service the lower mounts, as required.

To install:

10. Install the radiator and fan assembly, if removed, as an assembly.

11. Connect the automatic transaxle cooler lines, if disconnected.

12. Connect the thermo wires.

13. Install the fan, if removed separately.

14. Install the radiator hoses. Make sure to position the upper radiator hose so the matchmarks made during disassembly are in proper alignment.

15. Install the overflow tube and reservoir, if removed.

16. Fill the system with coolant.

17. Connect the negative battery cable, run the vehicle until the thermostat opens, fill the radiator completely and check the automatic transaxle fluid level, if equipped.

18. Once the vehicle has cooled, recheck the coolant level.

Engine Oil Cooler

REMOVAL & INSTALLATION

▶ See Figure 43

1. Disconnect the negative battery cable.

2. Remove the front bumper from the vehicle.

3. Remove the left front splash shield and the left under cover panel from the vehicle.

4. Remove the engine oil feed and return tubes at the engine oil cooler. Make sure to use 2 wrenches, 1 holding the weld nut while the other is used to loosen the eye bolt.

5. Remove the engine oil tube mounting bracket.

6. Remove the engine oil cooler mounting bolts and the engine cooler from the vehicle.

7. Inspect the engine oil cooler line for bends, breaks or plugs. Check the engine oil cooler hoses for cracks, damage, restrictions or deterioration. Check the gaskets and the eye bolts for damage or deformation. Replace components as required.

To install:

8. Install the engine oil cooler onto the mounting bracket and install the mounting bolts. Tighten the mounting bolts to 9 ft. lbs. (12 Nm).

9. Install the oil feed and return lines to the engine oil cooler making sure to use new gaskets. Tighten fittings, using 2 wrenches, to 25 ft. lbs. (35 Nm). Do not put stress on the weld nut or leakage may occur.

10. Install the left front splash shield and the left under cover panel onto the vehicle.

11. Install the front bumper and connect the negative battery cable.

12. Inspect the oil level and add as required. Start the engine and check for leaks.

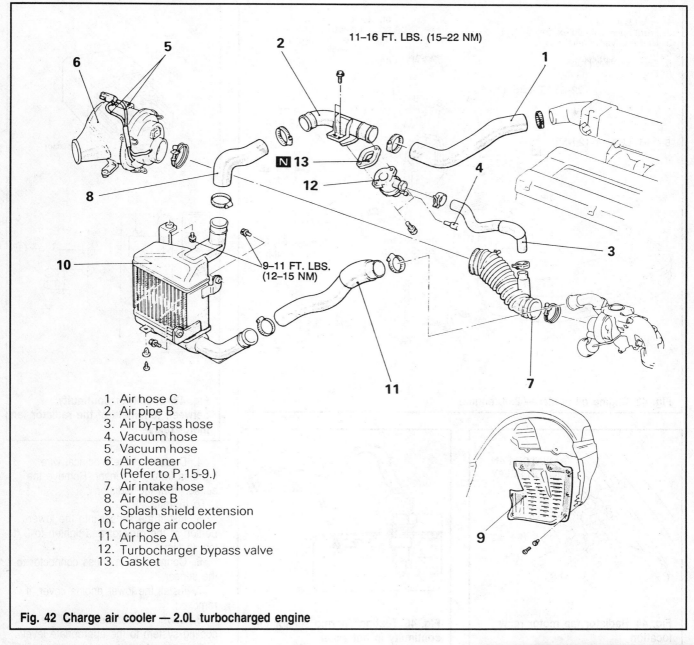

1. Air hose C
2. Air pipe B
3. Air by-pass hose
4. Vacuum hose
5. Vacuum hose
6. Air cleaner
   (Refer to P.15-9.)
7. Air intake hose
8. Air hose B
9. Splash shield extension
10. Charge air cooler
11. Air hose A
12. Turbocharger bypass valve
13. Gasket

11–16 FT. LBS. (15–22 NM)

9–11 FT. LBS.
(12–15 NM)

Fig. 42 Charge air cooler — 2.0L turbocharged engine

13. Turn the engine **OFF** and recheck the engine oil level. Add oil as required, to fill to the appropriate level.

Radiator Fan Motor Relay

TESTING

▶ See Figures 44 and 45

1. Disconnect the negative battery cable.

2. Remove the radiator fan motor relay from the relay box located at the right side in the engine compartment.

3. Check continuity between the terminals when the battery power supply is applied to terminal **2** , and terminal **4** is grounded. Inspect as follows:

a. When current flows between terminals **2** and **4** , continuity between terminals **1** and **3** is present.

b. When no current flows between terminals **2** and **4** , no continuity between terminals **1** and **3** is present.

4. If the above conditions are not present, replace the radiator motor relay.

Thermo Sensor

TESTING

▶ See Figure 46

1. Disconnect the negative battery cable. Drain the coolant from the radiator.

2. Disconnect the electrical connector from the thermo sensor, which is located at the base of the radiator. Removal of the fan or radiator hose may be required.

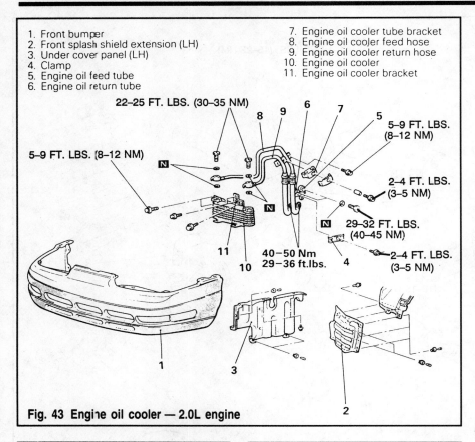

1. Front bumper
2. Front splash shield extension (LH)
3. Under cover panel (LH)
4. Clamp
5. Engine oil feed tube
6. Engine oil return tube
7. Engine oil cooler tube bracket
8. Engine oil cooler feed hose
9. Engine oil cooler return hose
10. Engine oil cooler
11. Engine oil cooler bracket

22–25 FT. LBS. (30–35 NM)

5–9 FT. LBS. (8–12 NM)

5–9 FT. LBS. (8–12 NM)

2–4 FT. LBS. (3–5 NM)

29–32 FT. LBS. (40–45 NM)

40–50 Nm
29–36 ft.lbs.

2–4 FT. LBS. (3–5 NM)

Fig. 43 Engine oil cooler — 2.0L engine

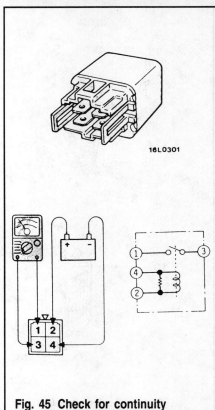

16L0301

Fig. 45 Check for continuity between terminals of the radiator fan motor relay

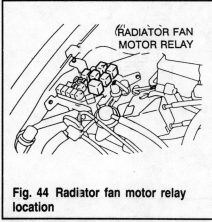

RADIATOR FAN MOTOR RELAY

Fig. 44 Radiator fan motor relay location

3. Remove the thermo sensor from the radiator.

➡ **For proper test results, make sure the water level is up to the sensor mounting threads.**

4. Place sensor is temperature controlled water and measure the resistance of the sensor. The desired resistance values are as follows:

a. 180-190°F (82-88°C) — continuity is present.

b. 172°F or less (78°C or less) — no continuity is present.

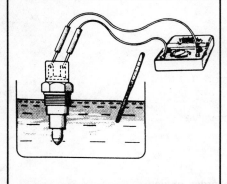

Fig. 46 Testing Thermo sensor for continuity in hot water

5. If the above conditions are not present, replace the thermo sensor.

REMOVAL & INSTALLATION

1. Disconnect the negative battery cable. Drain the radiator coolant.

2. Raise and safely support the vehicle.

3. Remove the engine under cover, if equipped.

4. Disconnect the electrical wire harness from the sensor. Remove the sensor from the radiator.

To install:

5. Install the sensor into the lower portion of the radiator and tighten to 3 ft. lbs. (4 Nm).

6. Connect the harness connector to the sensor.

7. Install the lower engine cover, if removed.

8. Lower the vehicle and fill the cooling system to the appropriate level.

9. Connect the negative battery cable and start the engine. Inspect for leaks.

10. Check for proper operation of the electric cooling fan.

Radiator Fan Motor

TESTING

▶ **See Figure 47**

1. Disconnect the negative battery cable.

2. Check to be sure the radiator fan rotates when spun by hand.

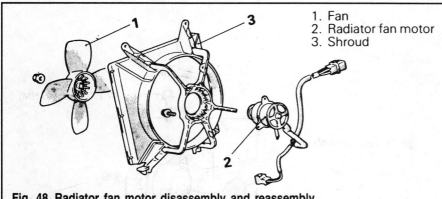

1. Fan
2. Radiator fan motor
3. Shroud

Fig. 48 Radiator fan motor disassembly and reassembly

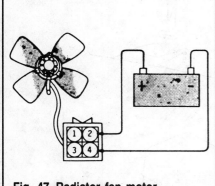

Fig. 47 Radiator fan motor inspection

3. Connect 1 end of a jumper wire to terminal **4** of the coolant fan motor connector. Connect the other end of the jumper wire to the negative battery terminal.

4. Connect 1 end of another jumper wire to terminal **2** of the coolant fan motor connector. Momentarily connect the other end of the jumper wire to the positive battery terminal.

5. The fan motor should run. Listen for abnormal noise while the motor is turning. If present, replace the coolant fan motor.

REMOVAL & INSTALLATION

▶ **See Figure 48**

1. Disconnect the negative battery cable.

2. Disconnect the electrical connector from the coolant fan motor.

3. Remove the mounting bolts, fan and shroud assembly from the vehicle.

4. Remove the fan blade retainer nut from the shaft on the fan motor and separate the fan from the motor.

5. Remove the motor to shroud attaching screws and the motor from the shroud.

To install:

6. Install the motor to the shroud and secure with the mounting bolts.

7. Install the fan to the motor shaft and secure with the retainer nut.

8. Install the fan and shroud assembly into the engine compartment and secure to the radiator. Reconnect the fan motor electrical connector.

9. Connect the negative battery cable and inspect the cooling fan for proper operation.

Condenser Fan Motor

REMOVAL & INSTALLATION

1. Disconnect the negative battery cable.

2. Disconnect the electrical connector from the fan.

3. Remove the fan shroud mounting screws and remove the fan motor and shroud as an assembly.

4. Installation is the reverse of the removal procedure. Tighten the motor and shroud assembly mounting bolts to 3 ft. lbs. (4 Nm).

Water Pump

REMOVAL & INSTALLATION

▶ **See Figures 49, 50, 51 and 52**

1. Disconnect the negative battery cable.

2. Drain the cooling system. Remove the accessory drive belt.

3. Remove the engine undercover.

4. Disconnect the clamp bolt from the power steering hose. Remove the tensioner pulley bracket.

5. Support the engine with the appropriate equipment and remove the engine mount bracket.

6. Remove both the outer and the inner timing belts from the front of the engine. If reusing old belt, mark the direction of rotation on the outer belt surface. This will assure belt rotation in the original direction and extend belt life.

7. Remove the alternator brace from the front of the water pump.

8. Remove the water pump and gasket from the engine.

9. Remove the O-ring where the water inlet pipe(s) joins the pump. Clean all mating surfaces and inspect for cracks or other damage. Replace components that are damaged or cracked.

To install:

10. Thoroughly clean and dry both gasket surfaces of the water pump and block.

11. Install a new O-ring into the groove on the front end of the water inlet pipe. Do not apply oils or grease to the O-ring. Wet with water only.

12. Install the gasket and pump assembly and tighten the bolts. Note the marks on the bolt heads. Those marked **4** should be torqued to 9-11 ft. lbs. (12-15 Nm). Those bolts marked **7** should be torqued from 15-19 ft. lbs. (20-27 Nm).

13. Connect the hoses to the pump.

14. Reinstall the timing belt and related parts.

15. Install the engine undercover.

16. Fill the system with coolant.

17. Connect the negative battery cable, run the vehicle until the thermostat opens and fill the radiator completely.

18. Once the vehicle has cooled, recheck the coolant level.

Cylinder Head

REMOVAL & INSTALLATION

1.8L Engine

▶ **See Figures 53, 54, 55 and 56**

1. Relieve the fuel system pressure as follows:

   a. Loosen the fuel filler cap to release fuel tank pressure.

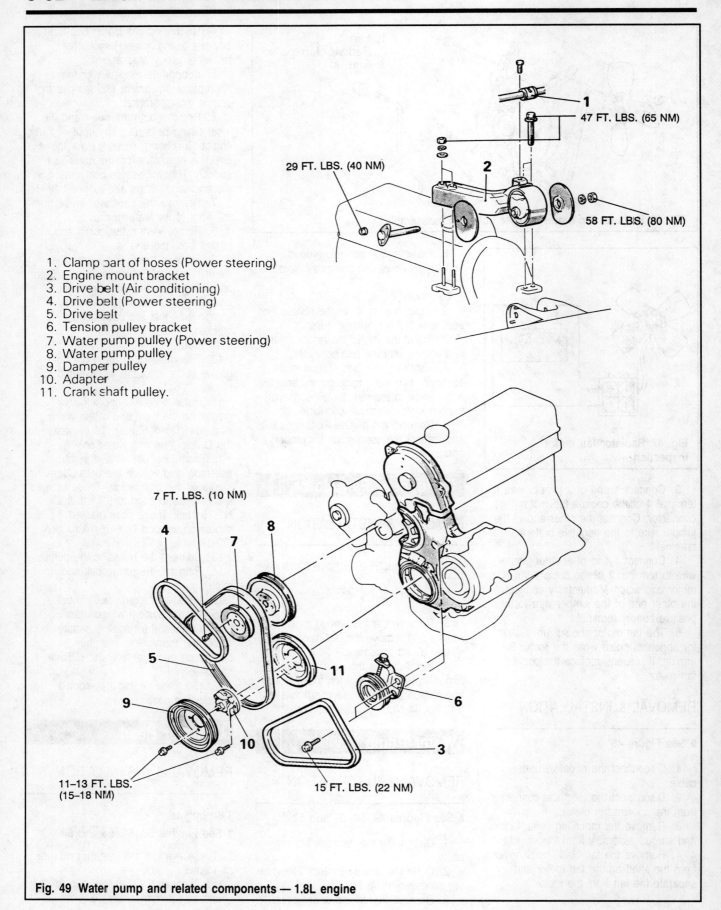

29 FT. LBS. (40 NM)

47 FT. LBS. (65 NM)

58 FT. LBS. (80 NM)

1. Clamp part of hoses (Power steering)
2. Engine mount bracket
3. Drive belt (Air conditioning)
4. Drive belt (Power steering)
5. Drive belt
6. Tension pulley bracket
7. Water pump pulley (Power steering)
8. Water pump pulley
9. Damper pulley
10. Adapter
11. Crank shaft pulley.

7 FT. LBS. (10 NM)

11–13 FT. LBS. (15–18 NM)

15 FT. LBS. (22 NM)

Fig. 49 Water pump and related components — 1.8L engine

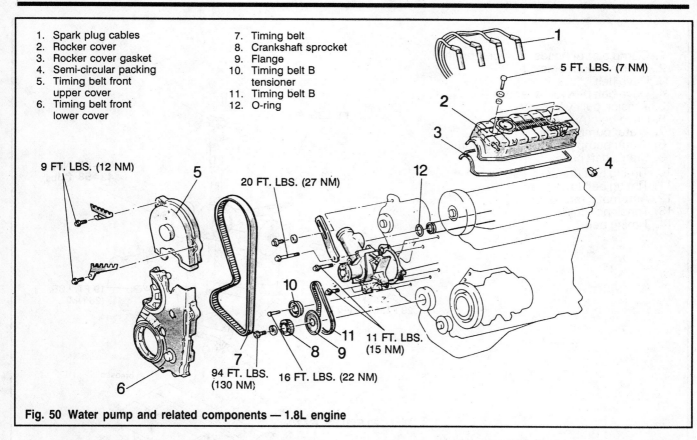

1. Spark plug cables
2. Rocker cover
3. Rocker cover gasket
4. Semi-circular packing
5. Timing belt front upper cover
6. Timing belt front lower cover
7. Timing belt
8. Crankshaft sprocket
9. Flange
10. Timing belt B tensioner
11. Timing belt B
12. O-ring

5 FT. LBS. (7 NM)

9 FT. LBS. (12 NM)

20 FT. LBS. (27 NM)

11 FT. LBS. (15 NM)

94 FT. LBS. (130 NM)

16 FT. LBS. (22 NM)

Fig. 50 Water pump and related components — 1.8L engine

b. Disconnect the fuel pump harness connector located at the rear of the fuel tank.

c. Start the vehicle and allow it to run until it stalls from lack of fuel. Turn the key to the **OFF** position.

d. Disconnect the negative battery cable, then reconnect the fuel pump connector and reinstall the fuel filler cap.

e. Wrap shop towels around the fitting that is being disconnected to absorb residual fuel in the lines.

2. Drain the cooling system.

3. Remove the air intake hose and the breather hose.

4. Disconnect the accelerator cable. There will be 2 cables, if equipped with cruise-control.

5. Place a shop towel around the high pressure fuel line to absorb any residual fuel remaining in the system. Disconnect the high pressure fuel line.

6. Remove the upper radiator hose, the water breather hose, the water bypass hose and the heater hose.

7. Disconnect the PCV hose.

8. Label and remove the spark plug cables. Make sure not to pull on the cable, when removing the spark plug wire.

9. Disconnect and plug the fuel return line.

10. Disconnect the vacuum line for the brake booster.

11. Disconnect the electrical connections for the oxygen sensor, engine coolant temperature gauge unit and the water temperature sensor.

12. Disconnect the electrical connections for the idle speed control motor, Throttle Position Sensor (TPS), distributor, motor position sensor connector, fuel injectors, EGR temperature sensor (California vehicles), power transistor, condenser and engine ground cable.

13. Disconnect the engine control wiring harness.

14. Remove the clamp that holds the power steering pressure hose to the engine mounting bracket.

15. Place a jack and wood block under the oil pan and carefully lift just enough to take the weight off the engine mounting bracket. Then remove the bracket.

16. Remove the valve cover, gasket and half-round seal. Remove the timing belt front upper cover.

17. If possible, rotate the crankshaft clockwise until the timing marks on the cam sprocket and belt align. This should

position the engine so No. 1 piston is at top dead center of it's compression stroke.

18. Remove the camshaft sprocket retainer bolt. Remove the camshaft sprocket with the timing belt attached, and allow to rest on the timing belt front lower cover. Do not allow the tension of the belt to slacken or engine timing may be lost.

19. Remove the timing belt rear upper cover. Remove the exhaust pipe self-locking nuts and separate the exhaust pipe from the exhaust manifold. Discard the gasket.

20. Loosen the cylinder head mounting bolts in 3 Steps, starting from the outside and working inward. Lift off the cylinder head assembly and remove the head gasket.

To install:

21. Thoroughly clean and dry the mating surfaces of the head and block. Check the cylinder head for cracks, damage or engine coolant leakage. Remove scale, sealing compound and carbon. Clean oil passages thoroughly. Check the head for flatness. End to end, the head should be within 0.002 in. normally with 0.008 in. the maximum allowed out of true. The total thickness

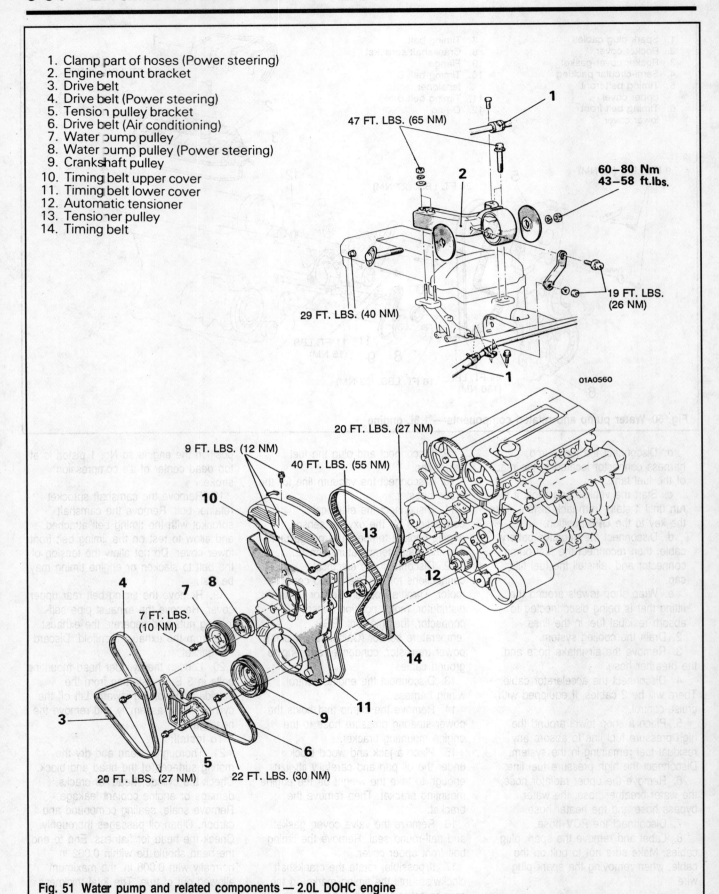

1. Clamp part of hoses (Power steering)
2. Engine mount bracket
3. Drive belt
4. Drive belt (Power steering)
5. Tension pulley bracket
6. Drive belt (Air conditioning)
7. Water pump pulley
8. Water pump pulley (Power steering)
9. Crankshaft pulley
10. Timing belt upper cover
11. Timing belt lower cover
12. Automatic tensioner
13. Tensioner pulley
14. Timing belt

47 FT. LBS. (65 NM)

60–80 Nm
43–58 ft.lbs.

19 FT. LBS. (26 NM)

29 FT. LBS. (40 NM)

01A0560

20 FT. LBS. (27 NM)

9 FT. LBS. (12 NM)

40 FT. LBS. (55 NM)

7 FT. LBS. (10 NM)

20 FT. LBS. (27 NM)

22 FT. LBS. (30 NM)

Fig. 51 Water pump and related components — 2.0L DOHC engine

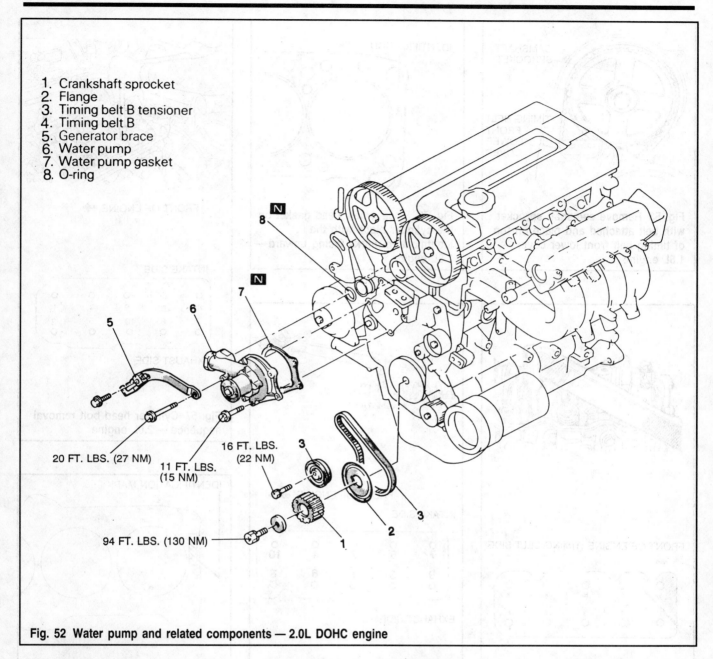

1. Crankshaft sprocket
2. Flange
3. Timing belt B tensioner
4. Timing belt B
5. Generator brace
6. Water pump
7. Water pump gasket
8. O-ring

20 FT. LBS. (27 NM)

11 FT. LBS. (15 NM)

16 FT. LBS. (22 NM)

94 FT. LBS. (130 NM)

Fig. 52 Water pump and related components — 2.0L DOHC engine

allowed to be removed from the head and block is 0.008 in. maximum.

22. Place a new head gasket on the cylinder block with the identification marks facing upward. Make sure the gasket has the proper identification mark for the engine. Do not use sealer on the gasket.

23. Carefully install the cylinder head on the block. Using 3 even Steps, torque the head bolts in sequence to 51-54 ft. lbs. (70-75 Nm).

24. Install a new exhaust pipe gasket and connect the exhaust pipe to the manifold. Install the upper rear timing cover.

25. Align the timing marks and install the cam sprocket. Torque the retaining bolt to 58-72 ft. lbs. (80-100 Nm). Check the belt tension and adjust, if necessary. Install the outer timing cover.

26. Apply sealer to the perimeter of the half-round seal. Install a new valve cover gasket. Install the valve cover.

27. Install the engine mount bracket. Once secure, remove the jack.

28. Install the clamp that holds the power steering pressure hose to the engine mounting bracket.

29. Connect or install all previously disconnected hoses, cables and electrical connections. Adjust the throttle cable(s).

30. Replace the O-rings and connect the fuel lines.

31. Install the air intake hose. Connect the breather hose.

32. Change the engine oil and oil filter.

33. Fill the system with coolant.

34. Connect the negative battery cable, run the vehicle until the thermostat opens, fill the radiator completely.

35. Check and adjust the idle speed and ignition timing.

36. Once the vehicle has cooled, recheck the coolant level.

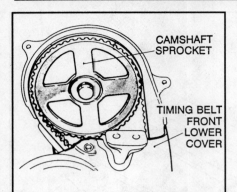

Fig. 53 Remove camshaft sprocket with belt attached and place on top of timing belt front lower cover — 1.8L engine

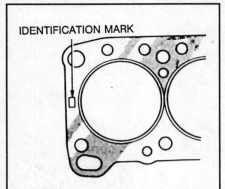

Fig. 55 Place a new head gasket on the cylinder block with the identification marks facing upward — 1.8L engine

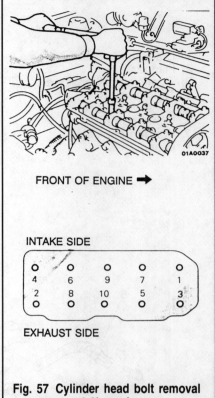

Fig. 57 Cylinder head bolt removal sequence — 2.0L engine

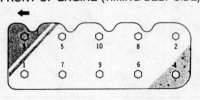

Fig. 54 Cylinder head bolt removal sequence — 1.8L engine

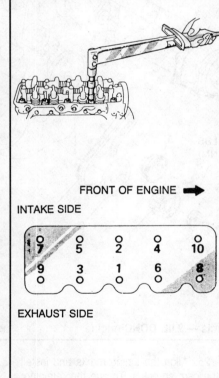

Fig. 56 Cylinder head bolt tightening sequence — 1.8L engine

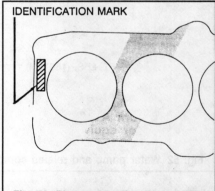

Fig. 58 Place a new head gasket on the cylinder block with the identification marks facing upward — 2.0L engine

2.0L Engine

▶ **See Figures 57, 58, 59 and 60**

1. Relieve the fuel system pressure as follows:

a. Loosen the fuel filler cap to release fuel tank pressure.

b. Disconnect the fuel pump harness connector located at the rear of the fuel tank.

c. Start the vehicle and allow it to run until it stalls from lack of fuel. Turn the key to the **OFF** position.

d. Disconnect the negative battery cable, then reconnect the fuel pump connector and reinstall the fuel filler cap.

e. Wrap shop towels around the fitting that is being disconnected to absorb residual fuel in the lines.

2. Drain the cooling system.

3. Disconnect the accelerator cable. There will be 2 cables if equipped with cruise-control.

4. Remove the air cleaner with the air intake hose.

5. Disconnect the oxygen sensor, engine coolant temperature sensor, the engine coolant temperature gauge unit and the engine coolant temperature switch on vehicles with air conditioning.

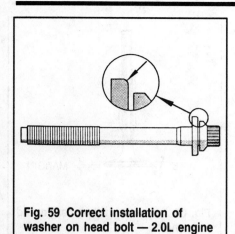

Fig. 59 Correct installation of washer on head bolt — 2.0L engine

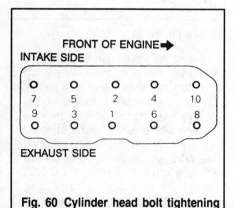

FRONT OF ENGINE →
INTAKE SIDE

| 7 | 5 | 2 | 4 | 10 |
| 9 | 3 | 1 | 6 | 8 |

EXHAUST SIDE

Fig. 60 Cylinder head bolt tightening sequence — — 2.0L engine

6. Disconnect the ISC motor, throttle position sensor, crankshaft angle sensor, fuel injectors, ignition coil, power transistor, noise filter, knock sensor on turbocharged engines, EGR temperature sensor (California vehicles), ground cable and engine control wiring harness.

7. Remove the upper radiator hose and the overflow tube.

8. Remove the spark plug cable center cover and remove the spark plug cable.

9. Disconnect and plug the high pressure fuel line.

10. Disconnect the small vacuum hoses.

11. Remove the heater hose and water bypass hose.

12. Remove the PCV hose.

13. If turbocharged, remove the vacuum hoses, water line and eyebolt connection for the oil line for the turbocharger.

14. Disconnect and plug the fuel return hose.

15. Disconnect the brake booster vacuum hose.

16. Remove the timing belt.

17. Remove the valve cover and the half-round seal.

18. On non-turbocharged engines, remove the exhaust pipe self-locking nuts and separate the exhaust pipe from the exhaust manifold. Discard the gasket.

19. On turbocharged engines, remove the sheet metal heat protector and remove the bolts that attach the turbocharger to the exhaust manifold.

20. Loosen the cylinder head mounting bolts in 3 steps, starting from the outside and working inward. Lift off the cylinder head assembly and remove the head gasket.

To install:

21. Thoroughly clean and dry the mating surfaces of the head and block. Check the cylinder head for cracks, damage or engine coolant leakage. Remove scale, sealing compound and carbon. Clean oil passages thoroughly. Check the head for flatness. End to end, the head should be within 0.002 in. normally with 0.008 in. the maximum allowed out of true. The total thickness allowed to be removed from the head and block is 0.008 in. maximum.

22. Place a new head gasket on the cylinder block with the identification marks at the front top (upward) position. Make sure the gasket has the proper identification mark for the engine. Do not use sealer on the gasket. Replace the turbo gasket and ring, if equipped.

23. If equipped with 1993 vehicle, inspect the cylinder head bolts shank length prior to installation. If the length exceeds 3.79 in. (96.4mm), the bolt must be replaced. Install the washer onto the bolt so the chamfer on the washer faces towards the head of the bolt.

24. Carefully install the cylinder head on the block.

25. On 1990-92 models, tighten the cylinder head bolts using 3 even steps, in sequence to 65-72 ft. lbs. (90-100 Nm). This torque applies to a cold engine. If checking cylinder head bolt torque on hot engine, the desired specification is 72-80 ft. lbs. (100-110 Nm).

26. On 1993 models, Tighten the cylinder head bolts as follows:

   a. Following the proper tightening sequence, tighten the cylinder head bolts to 54 ft. lbs. (75 Nm).

   b. Loosen all bolts completely.

   c. Torque bolts to 15 ft. lbs. (20 Nm).

   d. Tighten bolts an additional ¼ turn.

   e. Tighten bolts an additional ¼ turn.

27. On turbocharged engine, install the heat shield. On non-turbocharged engine, install a new exhaust pipe gasket and connect the exhaust pipe to the manifold.

28. Apply sealer to the perimeter of the half-round seal and to the lower edges of the half-round portions of the belt-side of the new gasket. Install the valve cover.

29. Install the timing belt and all related items.

30. Connect or install all previously disconnected hoses, cables and electrical connections. Adjust the throttle cable(s).

31. Install the spark plug cable center cover.

32. Replace the O-rings and connect the fuel lines.

33. Install the air cleaner and intake hose. Connect the breather hose.

34. Change the engine oil and oil filter.

35. Fill the system with coolant.

36. Connect the negative battery cable, run the vehicle until the thermostat opens, fill the radiator completely.

37. Check and adjust the idle speed and ignition timing.

38. Once the vehicle has cooled, recheck the coolant level.

CLEANING AND INSPECTION

▶ **See Figure 61**

Once the cylinder head is removed from the engine, thoroughly clean and dry the mating surfaces of the head and block. Check the cylinder head for cracks, damage or engine coolant leakage. Remove scale, sealing compound and carbon. Clean oil passages thoroughly. Check the head for flatness. End to end, the head should be within 0.002 in. normally with 0.008 in. the maximum allowed out of true. The total thickness allowed to be removed from the head and block is 0.008 in. maximum.

RESURFACING

Cylinder head resurfacing should be done only by a competent machine shop. If warpage exceeds the

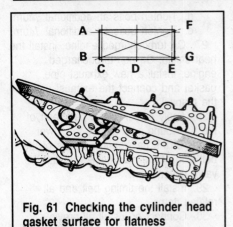

Fig. 61 Checking the cylinder head gasket surface for flatness

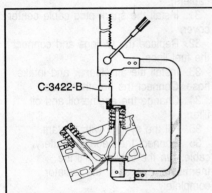

Fig. 62 Removing the valve retainer locks using spring compressor tool C-3422-B — 1.8L engine

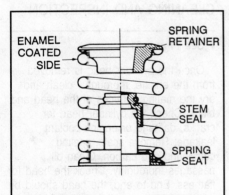

Fig. 63 Position valve spring with enamel coating toward valve spring retainers — 1.8L engine

manufacturer's tolerance, the cylinder head must be replaced.

Valves

REMOVAL & INSTALLATION

▶ **See Figures 62 and 63**

1. Remove the camshaft and the cylinder head from the engine.
2. Using spring compressor tool C-3422-B, compress the valve spring and remove the retainer locks from the grooves in the top of the valve.
3. Slowly release the tension on the tool and allow the spring to expand to it's relaxed state. Remove the spring compressor tool, valve spring, valve spring retainer and valve spring seat from the cylinder head.

➡ **Keep the removed retainers locks and valves together so they can be installed in their original locations.**

4. Remove the valve from the cylinder head.
5. Repeat this procedure on the remaining valves to be removed.
   To install:
6. Apply clean engine oil to each valve. Insert the valve into the valve guides, installing original valves in their original locations.

➡ **Avoid inserting the valve through the valve guide with force. Damage to the guide seal and possible oil consumption will result. After insertion, check to see that the valve moves smoothly in the guide and that there was no damage done to the valve seal.**

7. Install the valve guide spring with the enamel coated end facing upward toward the valve spring retainer.
8. Using spring compressor tool C-3422-B, compress the valve spring and install the retainer locks into the grooves in the top of the valve. Make certain that the retainer lock are positively installed.

➡ **When compressing the spring, check to see that the valve stem seal is not pressed to the bottom of the retainer.**

9. Repeat this procedure on the remaining valves to be installed.
10. Install the cylinder head and related components onto the engine.

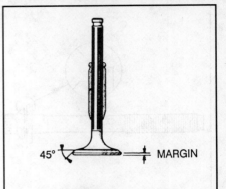

Fig. 64 Inspecting valve margin and face angle

INSPECTION

▶ **See Figure 64**

Before the valves can be properly inspected, the stem, the lower end of the stem, the entire valve face and head must be cleaned. An old valve works well for chipping carbon from the valve head, a wire brush, a gasket scraper or a putty knife can be used for cleaning the valve face and/or the area between the face and the lower stem. DO NOT scratch the valve face during cleaning. Clean the entire stem with a rag soaked in thinners to remove all of the varnish and gum.

Thorough inspection of the valves requires the use of a micrometer and a dial indicator. If these instruments are not available, the parts should be taken to a reputable machine shop for inspection.

1. Check the valve for wear, damage or deformation of the head and stem. Correct or replace excessively deformed, worn or damaged valves.
2. If the valve stem is pitted, correct with an oil stone or other similar means. The correction must be limited to a minimum. Also reface the valve.
3. Replace the valve if the margin has decreased to less than the limit. The margin limits are as follows:
   Intake valve — 0.028 in. (0.7mm)
   Exhaust valve — 0.039 in. (1.0mm)

REFACING

Valve refacing should only be handled by a reputable machine shop, as the experience and equipment needed to do the job are beyond that of the average owner/mechanic. During the course of a

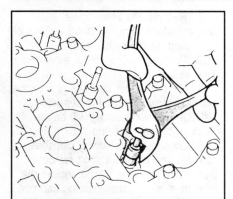

Fig. 65 Grasp the valve seal to be replaced firmly and pull from the valve guide

normal valve job, refacing is necessary when simply lapping the valves into their seats will not correct the seat and face wear. When the valves are reground (resurfaced), the valve seats must also be recut, again requiring special equipment and experience.

Valve Stem Seals

REPLACEMENT

▶ **See Figure 65**

With Cylinder Head Removed

1. Remove the valves and the valve springs from the cylinder head.
2. Using pliers, grasp the valve seal to be replaced firmly and pull from the valve guide. Be careful not to damage the cylinder head during seal removal. Do not reuse valve seals.

To install:

3. Install the valve spring seat, if removed.
4. Using valve seal installation tool MD998737 or exact equivalent, install a new valve stem seal to the valve guide. Tape tool to seat seal. Make sure the seal is seated against the valve spring seat.

➡ **Use tool MD998737 or exact equivalent to install valve stem seals. Improper installation can be a cause of leakage of oil down into the cylinder.**

5. Once the valve stem seal is installed and firmly seated, install the valve, valve spring and retainer. Compress the valve spring and install the retainer locks.

6. Repeat this procedure on the remaining valve seals to be replaced.
7. Install the cylinder head and related components onto the engine.

With Cylinder Head Installed

1. Remove the rocker arm (valve) cover as outlined in this chapter.
2. Remove the rocker arm assembly and the camshaft from the cylinder head.
3. Remove the spark plug of the affected cylinder.
4. Install a spark plug adapter and air chuck into the spark plug hole. Connect the air chuck to a compressed air source. This procedure fills the cylinder with compressed air so the valve does not fall into the cylinder when the valve spring retainer locks are removed.
5. Using a valve spring compressor, remove the valve spring retainer locks, spring retainer and valve spring as outlined in this chapter.
6. Using a pair of pliers, remove the valve seal from the cylinder head.

To install:

7. Lubricate the valve stem with engine oil and install the seal onto the valve guide, using tool MD998737 or exact equivalent. Make sure the seal is seated properly.
8. Install the valve spring, spring retainer and compress the spring using the spring compressor.
9. Install the spring retainer locks. Make sure they are properly seated before releasing the spring compressor.
10. Remove the air chuck and adapter.
11. Install the spark plug, camshaft and rocker arms.
12. Install the rocker arm shaft, as required.
13. Install the remaining components, start the engine and check for proper operation and leaks.

Valve Spring

REMOVAL & INSTALLATION

1. Using spring compressor tool C-3422-B, compress the valve spring and remove the retainer locks from the grooves in the top of the valve.
2. Slowly release the tension on the tool and allow the spring to expand to it's relaxed state. Remove the spring

compressor tool, valve spring, valve spring retainer and valve spring seat from the cylinder head. Keep the removed retainers locks and valve springs together so they can be installed in their original locations.

To install:

3. Install the valve guide spring with the enamel coated end facing upward toward the valve spring retainer.
4. Using spring compressor tool C-3422-B, compress the valve spring and install the retainer locks into the grooves in the top of the valve. Make certain that the retainer locks are positively installed.

➡ **When compressing the spring, check to see that the valve stem seal is not pressed to the bottom of the retainer.**

5. Repeat this procedure on the remaining valve springs to be installed.

INSPECTION

1. Check the valve springs for free-length and load. If they exceed the limits, replace the spring. The specifications are as follows:

1.8L Engine
Free length — 1.937 in. (49.2mm)
Load — 68 lbs. at installed height
Installed height — 1.469 in. (37.3mm)
2.0L Engine
Free length — 1.902 in. (48.3mm)

2. Using a straight edge, check spring for squareness and replace if the limit is exceeded. The limit for squareness is less than 1.5°, with a limit of 4°.

Valve Seats

CUTTING THE SEATS

➡ **To prevent damaging the other cylinder head components, completely disassemble the head.**

1. Before correcting the valve seat, check for clearance between the valve guide and the valve. If necessary, replace the valve guide.
2. Using the special tool or seat grinder, correct the valve seat to obtain a width of 0.035-0.051 in. (0.9-1.3mm) and the seat angle to 44 degrees.

3. After correction to the valve seat has been made, the valve and the valve seat should be lapped with a lapping compound.

REMOVAL & INSTALLATION

➡ **To prevent damaging the other cylinder head components, completely disassemble the head.**

1. Cut the valve seat to be replaced from the inside to thin the wall thickness.
2. Remove the seat from the head using punch of seat removal tool. Take extreme care not to mark or damage the cylinder head during removal.
3. Once the valve seat is removed, rebore the valve seat hole in the cylinder head to the appropriate oversized diameter.
4. Before fitting the valve seat, either heat the cylinder head up to approximately 482°F (250°C) or cool the valve seat in liquid nitrogen, to prevent the cylinder head bore from galling.
5. Using a valve seat cutter, correct the valve seat to a width of 0.035-0.051 in. (0.9-1.3mm) and the seat angle to 44 degrees.

Valve Guides

REMOVAL & INSTALLATION

➡ **To prevent damaging the other cylinder head components, completely disassemble the head.**

1. Using the special tool and press, remove the valve guide from the cylinder head, pressing toward the cylinder head gasket surface.
2. Rebore the valve guide hole to the new oversize valve guide outside diameter.

➡ **Do not try to install a replacement valve guide the same diameter as the original guide which is being replaced.**

3. Using the special tool, press fit the valve guide , working from the cylinder head top surface.
4. After installing the valve guides, insert new valves in them to check for a sliding of the guides.
5. When the valve guides have been replaced, check for valve contact and correct valve seats as necessary.

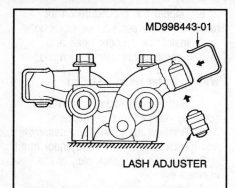

Fig. 66 Installation of lash adjuster retainer tool

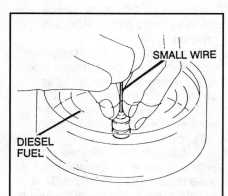

Fig. 67 Dip lash adjuster in clean diesel fuel and depress plunger to bleed

Valve Lifters

◆ **See Figures 66 and 67**

REMOVAL & INSTALLATION

1.8L Engine

1. Relieve the fuel system pressure as follows:
   a. Loosen the fuel filler cap to release fuel tank pressure.
   b. Disconnect the fuel pump harness connector located at the rear of the fuel tank.
   c. Start the vehicle and allow it to run until it stalls from lack of fuel. Turn the key to the **OFF** position.
   d. Disconnect the negative battery cable, then reconnect the fuel pump connector and reinstall the fuel filler cap.
   e. Wrap shop towels around the fitting that is being disconnected to absorb residual fuel in the lines.

2. Remove the valve cover. Install lash adjuster retainer tools MD998443 or equivalent, to the rocker arms.
3. Remove the distributor extension, if necessary.
4. Have a helper hold the rear of the camshaft down. If not, the belt will dislodge and valve timing will be lost. Remove the rear bearing cap.
5. Loosen the remaining camshaft cap retaining bolts but don't remove them from the caps. Do not loosen the forward most camshaft bearing cap bolt.
6. At this point, the shafts can be removed as an assembly for service or service of individual components can be made by sliding component to be replaced off back end of shafts. If the later method of replacement is used, keep all parts in order or removal and install parts in same location.
7. To remove the shafts as an assembly, remove bearing caps No. 2, 3 and 4, rocker arms, rocker shafts and bolts. It is essential that all parts be kept in the same order and orientation for reinstallation.
8. Remove the lash adjuster tools and the adjusters requiring replacement.
   To install:
9. If installing new lash adjuster, immerse the adjuster in clean diesel fuel. Using a small wire, move the plunger inside the lash adjuster up and down 4-5 times while pushing down lightly on the check ball in order to bleed air.
10. Install the bled lash adjusters in the rocker arms and install the adjuster retainer tools to hold them in place.
11. Apply a drop of sealant to the rear edges of the bearing end caps.
12. Install the rocker assembly into the front bearing cap making sure the notches in the rocker shafts are facing up. Insert the installation bolt but do not tighten at this point.
13. Install the remaining cap bolts and tighten evenly and gradually to 14 ft. lbs. (20 Nm). Remove the lash adjuster retainers.
14. Install the distributor extension, if removed.
15. Install the valve cover with a new gasket in place.
16. Connect the negative battery cable. Start the engine and allow to idle. Check for leaks and proper operation of the valve lifters.

2.0L Engine

1. Relieve the fuel system pressure as follows:

a. Loosen the fuel filler cap to release fuel tank pressure.

b. Disconnect the fuel pump harness connector located at the rear of the fuel tank.

c. Start the vehicle and allow it to run until it stalls from lack of fuel. Turn the key to the **OFF** position.

d. Disconnect the negative battery cable, then reconnect the fuel pump connector and reinstall the fuel filler cap.

e. Wrap shop towels around the fitting that is being disconnected to absorb residual fuel in the lines.

2. Disconnect the accelerator cable, PCV hoses, breather hoses, spark plug cables and the remove the valve cover.

3. Rotate the crankshaft clockwise and align the timing marks so No. 1 piston will be at TDC of the compression stroke. At this time the timing marks on the camshaft sprocket and the upper surface of the cylinder head should coincide, and the dowel pin of the camshaft sprocket should be at the upper side.

➡ **Always rotate the crankshaft in a clockwise direction. Make a mark on the back of the timing belt indicating the direction of rotation so it may be reassembled in the same direction if it is to be reused.**

4. Remove the timing belt upper and lower covers.

5. Remove the timing belt.

6. Remove the crank angle sensor.

7. Remove the camshafts, rocker arms and lash adjusters.

8. Visually inspect the rocker arm roller and replace if dent, damage or seizure is evident. Check the roller for smooth rotation. Replace if excess play or binding is present. Also, inspect valve contact surface for possible damage or seizure. It is recommended that all rocker arms and lash adjusters be replaced together.

To install:

9. Install the lash adjusters and rocker arms into the cylinder head. Lubricate lightly with clean oil prior to installation.

10. Apply engine oil to the lobes and journals of each camshaft. Install the camshafts into the cylinder head taking care not to confuse the intake and the exhaust camshaft; the intake camshaft has a slit on its rear end for driving the crank angle sensor. Align shafts so dowel pins on camshaft sprocket end are located on the top.

11. Install and tighten the camshaft bearing caps in the proper sequence torquing to specifications in 3 even progressions.

12. Replace the camshaft oil seals and install the sprockets.

13. Locate the dowel pin on the sprocket end of the intake camshaft at the top position, if not already done.

14. Align the punch mark on the crank angle sensor housing with the notch on the sensor plate. Install the crank angle sensor into the cylinder head.

15. Install the timing belt, covers and related components.

16. Install the valve cover using new gasket. Reconnect all related components.

17. Reconnect the negative battery cable.

OVERHAUL

Valve lifters usually exhibit external and well as internal wear. Because of this, overhaul of lifters is not recommended. It is more cost and time efficient to replace the worn or damaged valve lifters with new.

Oil Pan

REMOVAL & INSTALLATION

1. Disconnect the negative battery cable.

2. Raise the vehicle and support safely.

3. Remove the oil pan drain plug and drain the engine oil.

4. Disconnect and lower the exhaust pipe from the engine manifold.

5. If equipped with AWD, remove the transfer assembly and right drive shaft.

6. Using the appropriate equipment, support the weight of the engine. On AWD vehicles, remove the left member. On FWD vehicles, remove the retainer bolts and the center crossmember.

7. If equipped with turbocharger, disconnect the return pipe for the turbocharger from the side of the oil pan.

8. Remove the oil pan retainer bolts. Tap in thin prybar between the engine block and the oil pan.

➡ **Do not use a chisel, screwdriver or similar tool when removing the oil pan. Damage to engine components may occur.**

9. Inspect the oil pan for damage and cracks. Replace if faulty. While the pan is removed, inspect the oil screen for clogging, damage and cracks. Replace if faulty.

To install:

10. Using a wire brush or other tool, scrape clean all gasket surfaces of the cylinder block and the oil pan so that all loose material is removed. Clean sealing surfaces of all dirt and oil.

11. Apply sealant around the gasket surfaces of the oil pan in such a manner that all bolt holes are circled and there s a continuous bead of sealer around the entire perimeter of the oil pan.

➡ **The continuous bead of sealer should be applied in a bead approximately 0.16 in. (4mm) in diameter.**

12. Install the oil pan onto the cylinder block within 15 minutes after applying sealant. Install the fasteners and tighten to 4-6 ft. lbs. (6-8 Nm).

13. Install the oil return pipe using a new gasket, if removed. Tighten retainers to 5-7 ft. lbs. (7-10 Nm).

14. Install the crossmember. On FWD vehicles, tighten the crossmember mounting bolts to 72 ft. lbs. (100 Nm).

15. On AWD vehicles, install the left member and tighten the forward retainer bolts to 72 ft. lbs. (100 Nm). Tighten the rearward left member bolts to 58 ft. lbs. (80 Nm).

16. Install the transfer assembly and right drive shaft, if removed.

17. Connect the exhaust pipe from the engine manifold with new gasket in place. On turbocharged engine, tighten the exhaust pipe to manifold flange nuts to 43 ft. lbs. (60 Nm). On non-turbocharged engines, tighten the exhaust pipe to manifold flange nuts to 29 ft. lbs. (40 Nm). Install and tighten the support bolt to 29 ft. lbs. (40 Nm).

18. Install the oil drain plug and tighten to 33 ft. lbs., If not already done.

19. Lower the vehicle and fill the crankcase to the proper level with clean engine oil.

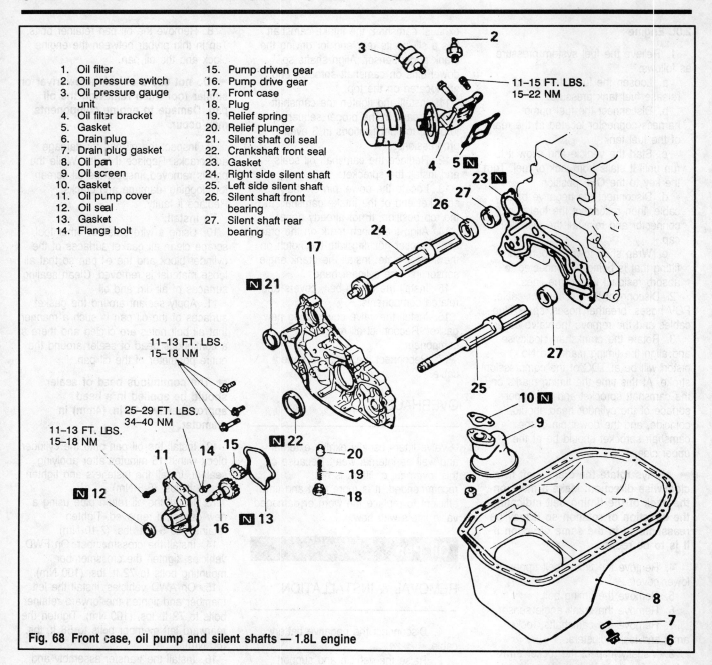

1. Oil filter
2. Oil pressure switch
3. Oil pressure gauge unit
4. Oil filter bracket
5. Gasket
6. Drain plug
7. Drain plug gasket
8. Oil pan
9. Oil screen
10. Gasket
11. Oil pump cover
12. Oil seal
13. Gasket
14. Flange bolt
15. Pump driven gear
16. Pump drive gear
17. Front case
18. Plug
19. Relief spring
20. Relief plunger
21. Silent shaft oil seal
22. Crankshaft front seal
23. Gasket
24. Right side silent shaft
25. Left side silent shaft
26. Silent shaft front bearing
27. Silent shaft rear bearing

11–15 FT. LBS.
15–22 NM

11–13 FT. LBS.
15–18 NM

25–29 FT. LBS.
34–40 NM

11–13 FT. LBS.
15–18 NM

Fig. 68 Front case, oil pump and silent shafts — 1.8L engine

20. Connect the negative battery cable. Start the engine and check for leaks.

Engine Oil Pump

▶ See Figures 68, 69, 70, 71 and 72

REMOVAL

➡ **Whenever the oil pump is disassembled or the cover removed, the gear cavity must be filled with petroleum jelly for priming purposes. Do not use grease.**

1.8L Engine

1. Disconnect the negative battery cable.
2. Remove the front engine mount bracket and accessory drive belts. Make sure to support the engine using the proper equipment prior to removing the front engine mount.
3. Remove timing belt upper and lower covers.
4. Remove the timing belt and crankshaft sprocket.
5. Drain the engine oil and remove the engine oil pan.
6. Remove the oil screen and gasket.

7. Remove the retainer bolts from the oil pump cover. Remove the oil pump cover from the front engine case cover.
8. Remove the flange bolt and the oil pump drive and oil pump driven gears.
9. Clean all mating surfaces of gasket material.

2.0L Engine

1. Disconnect the negative battery cable. Rotate the engine so No. 1 plug is on Top Dead Center (TDC) of it's compression stroke. The timing marks should be aligned at this point.
2. Raise and safely support the vehicle.

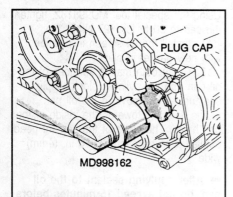

Fig. 69 Using special tool MD998162, remove the plug cap in the engine front cover — 2.0L engine

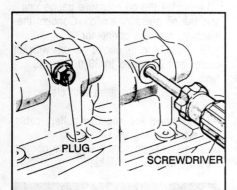

Fig. 70 Insert a Phillips screwdriver with a shank diameter of 0.32 in. (8mm) into the plug hole on the side of the engine block — 2.0L engine

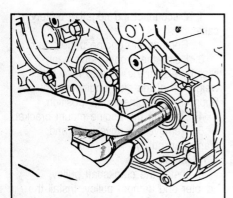

Fig. 71 Removing the driven gear bolt that secures the oil pump driven gear to the silent shaft — 2.0L engine

3. Drain the engine oil. Lower the vehicle.

4. Using the proper equipment, support the weight of the engine.

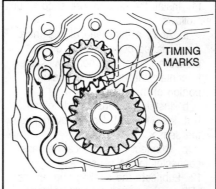

Fig. 72 Install the 2 oil pump gears into the case making sure to align the timing marks — 2.0L engine

Remove the front engine mount bracket and accessory drive belts.

5. Remove timing belt upper and lower covers.

6. Remove the timing belt and crankshaft sprocket.

7. Disconnect the electrical connector from the oil pressure sending unit. Remove the oil pressure sensor using special removal tool MD998054. Remove the oil filter and the oil filter bracket.

8. Remove the oil pan, oil screen and gasket.

9. Using special tool MD998162, remove the plug cap in the engine front cover. If the plug is too tight, hit the plug head with a hammer a few times and the plug should be easier to loosen.

10. Remove the plug on the side of the engine block. Insert a Phillips screwdriver with a shank diameter of 0.32 in. (8mm) into the plug hole. This will hold the silent shaft.

11. Remove the driven gear bolt that secures the oil pump driven gear to the silent shaft.

12. Remove and tag the front cover mounting bolts. Note the lengths of the mounting bolts as they are removed for proper installation.

13. Remove the front case cover and oil pump assembly. If necessary, the silent shaft can come out with the cover assembly.

14. Remove the oil pump cover, located on the back of the engine front cover. Remove the oil pump drive and driven gears.

INSPECTION

1.8L Engine

1. After disassembling the oil pump, clean all components.

2. Assemble the oil pump gears into the front case and rotate them to ensure smooth rotation and no looseness. Make sure there is no ridge wear on the contact surface between the front case and the gear surface of the oil pump front cover. Replace if any components are worn.

3. Measure the oil pump gears clearances as follows:

   a. With the drive and driven gears installed in the front case, measure the tip clearance of the gears. The distance between the tips of the drive gear's teeth and the case should be 0.0039-0.0079 in. with a limit of 0.071 in. The distance between the tips of the driven gear's teeth and the case should also be 0.0039-0.0079 in. with a limit of 0.071 in.

   b. Next measure the oil pump gear end-play. The end-play is checked by placing a straight-edge across the machined cover surface and measuring with a feeler gauge. The end-play for each gear should be 0.0024-0.0047 in. with a limit of 0.008 in.

4. If any measurement is beyond specification, replace the entire pump assembly.

2.0L DOHC Engine

1. After disassembling the oil pump, clean all components.

2. Assemble the oil pump gears into the front case and rotate it to ensure smooth rotation and no looseness. Make sure there is no ridge wear on the contact surface between the front case and the gear surface of the oil pump front cover.

3. The gear side clearance should be checked using the following procedure:

   a. With the drive and driven gears installed in the front case, measure the tip clearance of the gears. The distance between the tips of the drive gear's teeth and the case should be 0.0063-0.0083 in. with a limit of 0.0098 in. The distance between the tips of the driven gear's teeth and the case should be 0.0051-0.0071 in. with a limit of 0.0098 in.

b. Measure each oil pump gears end-play. The end-play is checked by placing a straight-edge across the machined cover surface and measuring with a feeler gauge. The end-play for the drive gear should be 0.0031-0.0055 in. with a limit of 0.0098 in. The end-play for the driven gear is 0.0024-0.0047 in. with a limit of 0.0098 in.

4. If any measurement is beyond specification, replace the entire pump assembly.

INSTALLATION

1.8L Engine

1. Clean all mating surfaces of gasket material.

2. Align the timing mark on the oil pump drive gear with that on the driven gear and install them into the engine front cover. Apply engine oil to the gears.

3. Insert a Phillips screwdriver with a shank diameter of 0.032 in. (8mm) into the plug hole on the left side of the cylinder to block the silent shaft. Tighten the flange bolt to 29 ft. lbs. (40 Nm).

4. Install a new oil pump gasket into the groove in the front case. When installing the gasket, face the round side to the oil pump cover.

5. Install a new oil seal into the oil pump cover, making sure the lip of the seal is in the correct direction.

6. Install the oil pump cover onto the engine front case and tighten the bolts to 13 ft. lbs. (18 Nm).

7. Install the oil screen in position with new gasket in place.

8. Clean both mating surfaces of the oil pan and the cylinder block. Apply sealant in the groove in the oil pan flange, keeping towards the inside of the bolt holes. The width of the sealant bead applied is to be about 0.016 in. (4mm) wide.

➡ **After applying sealant to the oil pan, do not exceed 15 minutes before installing the oil pan.**

9. Install the oil pan to the engine and secure with the retainers. Tighten bolts to 6 ft. lbs. (8 Nm).

10. Install the oil filter bracket to the engine with new gasket in place. Tighten the retainer bolts to 15 ft. lbs. (22 Nm).

11. Install the oil pressure sending unit as follows:

a. Apply a thin bead of sealant to the threaded portion of the oil pressure sensor. Do not allow sealer to contact the end of the threaded portion of the sensor.

b. Install the sensor and tighten to 8 ft. lbs. (12 Nm). Do not overtighten the sensor.

c. Connect the electrical harness connector to the sensor.

12. Lubricate the sealing ring on the oil filter with a small amount of clean engine oil. Install new oil filter, filled with clean oil, onto the filter bracket.

13. Fill the engine to the correct level with clean engine oil.

14. Connect the negative battery cable and start the engine. Check the oil pressure, making sure it is at the correct reading. Inspect for leaks.

2.0L DOHC Engine

1. Clean all mating surfaces of gasket material.

2. Align the timing mark on the oil pump drive gear with that on the driven gear and install them into the engine front case. Apply engine oil to the gears.

3. Install the oil pump cover and tighten the retainer bolts to 13 ft. lbs. (18 Nm).

4. Using the appropriate driver, install a new crankshaft seal into the front case.

5. Position new front case gasket in place. Set seal guide tool MD998285 on the front end of the crankshaft to protect the seal from damage. Apply a thin coat of oil to the outer circumference of the seal pilot tool.

6. Install the front case assembly through a new front case gasket and temporarily tighten the flange bolts.

7. Mount the oil filter on the bracket with new oil filter bracket gasket in place. Install the 4 bolts with washers and tighten to 25 ft. lbs. (34 Nm).

8. Insert a Phillips screwdriver into a hole in the left side of the engine block to lock the silent shaft in place.

9. Secure the oil pump drive gear onto the left silent shaft by installing and tightening the driven gear bolt to 29 ft. lbs. (40 Nm).

10. Install new O-ring to the groove in the front case and install the plug cap.

Using the special tool MD998162, tighten the cap to 20 ft. lbs. (27 Nm).

11. Install the oil screen in position with new gasket in place.

12. Clean both mating surfaces of the oil pan and the cylinder block. Apply sealant in the groove in the oil pan flange, keeping towards the inside of the bolt holes. The width of the sealant bead applied is to be about 0.016 in. (4mm) wide.

➡ **After applying sealant to the oil pan, do not exceed 15 minutes before installing the oil pan.**

13. Install the oil pan to the engine and secure with the retainers. Tighten bolts to 9 ft. lbs. (12 Nm).

14. Install the oil pressure gauge unit and the oil pressure switch. Connect the electrical harness connector.

15. Install the oil cooler secure with oil cooler bolt tightened to 33 ft. lbs. (45 Nm).

16. Install new oil filled with clean engine oil.

17. Connect the negative battery cable and start the engine. Check the oil pressure, making sure it is at the correct reading. Inspect for leaks.

Crankshaft and Damper Pulley

REMOVAL & INSTALLATION

1. Disconnect the negative battery cable.

2. Remove the engine undercover.

3. Using the proper equipment, slightly raise the engine to take the weight off of the side engine mount. Support the engine in this position.

4. Remove the engine mount bracket.

5. Remove the drive belts and crankshaft pulley.

To install:

6. Install the crankshaft pulley, adapter and damper pulley. Install the mounting bolts and tighten to to 13 ft. lbs. (18 Nm)on 1.8L engine or 22 ft. lbs. (30 Nm) on 2.0L DOHC engine.

7. Install the accessory drive belts and adjust to the correct tension.

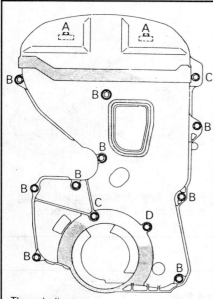

Thread diameter
× thread length
A: 6 × 16 (.24 × .63)
B: 6 × 18 (.24 × .70)
C: 6 × 25 (.24 × .98)
D: 6 × 28 (.24 × 1.10)   mm (in.)

Fig. 73 On 2.0L engine, be sure to install the bolts in their original locations. Their dimensions differ according to their locations.

8. Install the engine mount bracket. Tighten the bolts as follows:
Engine mount bracket through bolt — 58 ft. lbs. (80 Nm)
Upper mount bracket mounting nuts (2) — 47 ft. lbs. (65 Nm)
Upper mount bracket mounting bolt — 47 ft. lbs. (65 Nm)
9. Install the clamp for the power steering pressure hose.
10. Install the engine under cover.

Timing Belt Front Cover

REMOVAL & INSTALLATION

▶ See Figure 73

1. Disconnect the negative battery cable.
2. Remove the engine undercover.
3. Using the proper equipment, slightly raise the engine to take the weight off of the side engine mount. Support the engine in this position.
4. Remove the engine mount bracket.

5. Remove the drive belts, tension pulley brackets, water pump pulley and crankshaft pulley.
6. Remove all attaching screws and remove the upper and lower timing belt covers. If equipped with 2.0L engine, take note of the original locations of the cover fasteners prior to removal.

➡**On the 2.0L engine, take notice of the locations of each timing belt cover fastener during the removal procedure. Due to the difference in lengths, it is important that they are installed in their original locations. Refer to the illustration.**

7. The installation is the reverse of the removal procedure. Make sure all pieces of packing are positioned in the inner grooves of the covers when installing.

Timing Belt

It is recommended that the timing belt be replaced periodically to assure correct engine performance. Because of their composition, timing belts wear over a period of time and mileage. To avoid vehicle break down and possible engine damage, the manufacturer recommends timing belt replacement at 60,000 miles.

REMOVAL & INSTALLATION

1.8L Engine
▶ See Figures 74, 75 and 76

1. If possible, position the engine so the No. 1 piston is at TDC.
2. Disconnect the negative battery cable. On Summit Wagon with 2.4L engine, remove the coolant reservoir and the power steering and air conditioner hose clamp bolt.
3. Remove the timing belt covers.
4. Remove the timing (outer) belt tensioner and remove the outer timing belt.
5. Remove the outer crankshaft sprocket and flange.
6. Remove the silent shaft (inner) belt tensioner and remove the belt.
To install:
7. Align the timing marks of the silent shaft sprockets and the crankshaft sprocket with the timing marks on the front case. Wrap the timing belt around the sprockets so there is no slack in the

upper span of the belt and the timing marks are still aligned.
8. Install the tensioner pulley and move the pulley by hand so the long side of the belt deflects about ¼in.
9. Hold the pulley tightly so the pulley cannot rotate when the bolt is tightened. Tighten the bolt to 15 ft. lbs. (20 Nm) and recheck the deflection amount.
10. Install the timing belt tensioner fully toward the water pump and tighten the bolts. Place the upper end of the spring against the water pump body.
11. Align the timing marks of the camshaft, crankshaft and oil pump sprockets with their corresponding marks on the front case or rear cover.

➡ **There is a possibility to align all timing marks and have the oil pump sprocket and silent shaft out of time, causing an engine vibration during operation. If the following step is not followed exactly, there is a 50 percent chance that the silent shaft alignment will be 180 degrees off.**

12. Before installing the timing belt, ensure that the left side (rear) silent shaft (oil pump sprocket) is in the correct position as follows:
  a. Remove the plug from the rear side of the block and insert a tool with shaft diameter of 0.31 in. (8mm) into the hole.
  b. With the timing marks still aligned, the shaft of the tool must be able to go in at least 2 1/2 in. If the tool can only go in about 1 in., the shaft is not in the correct orientation and will cause a vibration during engine operation. Remove the tool from the hole and turn the oil pump sprocket 1 complete revolution. Realign the timing marks and insert the tool. The shaft of the tool must go in at least 2 1/3 in.
  c. Recheck and realign the timing mark.
  d. Leave the tool in place to hold the silent shaft while continuing.
13. Install the belt to the crankshaft sprocket, oil pump sprocket, then camshaft sprocket, in that order. While doing so, make sure there is no slack between the sprocket except where the tensioner is installed.
14. Recheck the timing marks' alignment. If all are aligned, loosen the tensioner mounting bolt and allow the tensioner to apply tension to the belt.

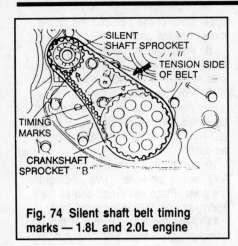

Fig. 74 Silent shaft belt timing marks — 1.8L and 2.0L engine

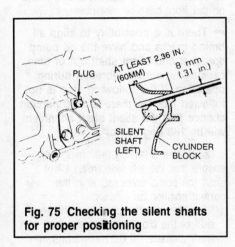

Fig. 75 Checking the silent shafts for proper positioning

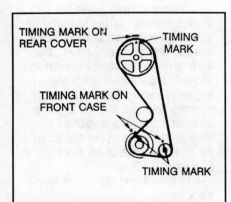

Fig. 76 Timing belt timing mark alignment — 1.8L engine

15. Remove the tool that is holding the silent shaft and rotate the crankshaft a distance equal to 2 teeth on the camshaft sprocket. This will allow the tensioner to automatically apply the proper tension on the belt. Do not manually overtighten the belt or it will howl.

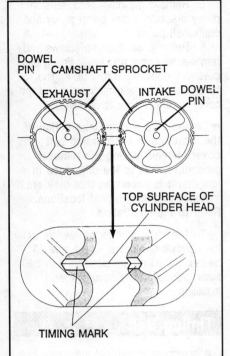

Fig. 77 Align the camshaft sprocket so marks face each other and are in alignment with the top surface of the cylinder head — 2.0L engine

16. Tighten the lower mounting bolt first, then the upper spacer bolt.

17. To verify correct belt tension, check that the deflection at the longest span of the belt is about ½in.

18. Install the timing belt covers and all related items.

19. Connect the negative battery cable.

2.0L DOHC Engine

▶ **See Figures 77, 78 and 79**

1. Disconnect the negative battery cable.

2. Remove the timing belt upper and lower covers.

3. Rotate the crankshaft clockwise and align the timing marks so No. 1 piston will be at TDC of the compression stroke. At this time the timing marks on the camshaft sprocket and the upper surface of the cylinder head should coincide, and the dowel pin of the camshaft sprocket should be at the upper side.

➡ **Always rotate the crankshaft in a clockwise direction. Make a mark on the back of the timing belt indicating**

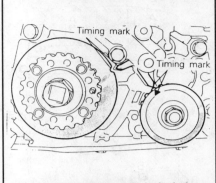

Fig. 78 Align the crankshaft timing mark and the oil pump sprocket timing mark — 2.0L engine

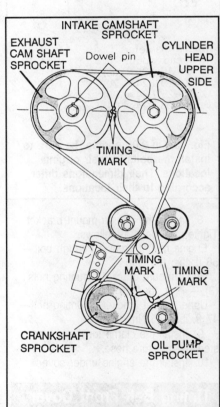

Fig. 79 Timing marks in alignment — 2.0L engine

the direction of rotation so it may be reassembled in the same direction if it is to be reused.

4. Remove the auto tensioner and remove the outermost timing belt.

5. Remove the timing belt tensioner pulley, tensioner arm, idler pulley, oil pump sprocket, special washer, flange and spacer.

6. Remove the silent shaft (inner) belt tensioner and remove the belt.

To install:

7. Align the timing marks on the crankshaft sprocket and the silent shaft sprocket. Fit the inner timing belt over the crankshaft and silent shaft sprocket. Ensure that there is no slack in the belt.

8. While holding the inner timing belt tensioner with your fingers, adjust the timing belt tension by applying a force towards the center of the belt, until the tension side of the belt is taut. Tighten the tensioner bolt.

➡ **When tightening the bolt of the tensioner, ensure that the tensioner pulley shaft does not rotate with the bolt. Allowing it to rotate with the bolt can cause excessive tension on the belt.**

9. Check belt for proper tension by depressing the belt on its' long side with your finger and noting the belt deflection. The desired reading is 0.20-0.28 in. (5-7mm). If tension is not correct, readjust and check belt deflection.

10. Install the flange, crankshaft and washer to the crankshaft. The flange on the crankshaft sprocket must be installed towards the inner timing belt sprocket. Tighten bolt to 80-94 ft. lbs. (110-130 Nm).

11. To install the oil pump sprocket, insert a Phillips screwdriver with a shaft 0.31 in. (8mm) in diameter into the plug hole in the left side of the cylinder block to hold the left silent shaft. Tighten the nut to 36-43 ft. lbs. (50-60 Nm).

12. Using a wrench, hold the camshaft at its' hexagon between journal No. 2 and 3 and tighten camshaft sprocket mounting bolt, if removed, to 58-72 ft. lbs. (80-100 Nm). If no hexagon is present between journal No. 2 and 3, hold the sprocket stationary with a spanner wrench while tightening the sprocket retainer bolt.

13. Carefully push the auto tensioner rod in until the set hole in the rod aligned up with the hole in the cylinder. Place a wire into the hole to retain the rod.

14. Install the tensioner pulley onto the tensioner arm. Locate the pinhole in the tensioner pulley shaft to the left of the center bolt. Then, tighten the center bolt finger-tight.

15. When installing the timing belt, turn the 2 camshaft sprockets so their dowel pins are located on top. Align the timing marks facing each other and with the top surface of the cylinder head. When you let go of the exhaust

camshaft sprocket, it will rotate 1 tooth in the counterclockwise direction. This should be taken into account when installing the timing belts on the sprocket.

➡ **Both camshaft sprockets are used for the intake and exhaust camshafts and are provided with 2 timing marks. When the sprocket is mounted on the exhaust camshaft, use the timing mark on the right with the dowel pin hole on top. For the intake camshaft sprocket, use the 1 on the left with the dowel pin hole on top.**

16. Align the crankshaft sprocket and oil pump sprocket timing marks.

17. After alignment of the oil pump sprocket timing marks, remove the plug on the cylinder block and insert a Phillips screwdriver with a shaft diameter of 0.31 in. (8mm) through the hole. If the shaft can be inserted 2.4 in. deep, the silent shaft is in the correct position. If the shaft of the tool can only be inserted 0.8-1.0 in. (20-25mm) deep, turn the oil pump sprocket 1 turn and realign the marks. Reinsert the tool making sure it is inserted 2.4 in. deep. Keep the tool inserted in hole for the remainder of this procedure.

➡ **The above step assures that the oil pump socket is in correct orientation to the silent shafts. This step must not be skipped or a vibration may develop during engine operation.**

18. Install the timing belt as follows:

a. Install the timing belt around the intake camshaft sprocket and retain it with 2 spring clips or binder clips.

b. Install the timing belt around the exhaust sprocket, aligning the timing marks with the cylinder head top surface using 2 wrenches. Retain the belt with 2 spring clips.

c. Install the timing belt around the idler pulley, oil pump sprocket, crankshaft sprocket and the tensioner pulley. Remove the 2 spring clips.

d. Lift upward on the tensioner pulley in a clockwise direction and tighten the center bolt. Make sure all timing marks are aligned.

e. Rotate the crankshaft ¼turn counterclockwise. Then, turn in clockwise until the timing marks are aligned again.

19. To adjust the timing (outer) belt, turn the crankshaft ¼turn

counterclockwise, then turn it clockwise to move No. 1 cylinder to TDC.

20. Loosen the center bolt. Using tool MD998738 or equivalent and a torque wrench, apply a torque of 1.88-2.03 ft. lbs. (2.6-2.8 Nm). Tighten the center bolt.

21. Screw the special tool into the engine left support bracket until its end makes contact with the tensioner arm. At this point, screw the special tool in some more and remove the set wire attached to the auto tensioner, if the wire was not previously removed. Then remove the special tool.

22. Rotate the crankshaft 2 complete turns clockwise and let it sit for approximately 15 minutes. Then, measure the auto tensioner protrusion (the distance between the tensioner arm and auto tensioner body) to ensure that it is within 0.15-0.18 in. (3.8-4.5mm). If out of specification, repeat Step 1-4 until the specified value is obtained.

23. If the timing belt tension adjustment is being performed with the engine mounted in the vehicle, and clearance between the tensioner arm and the auto tensioner body cannot be measured, the following alternative method can be used:

a. Screw in special tool MD998738 or equivalent, until its end makes contact with the tensioner arm.

b. After the special tool makes contact with the arm, screw it in some more to retract the auto tensioner pushrod while counting the number of turns the tool makes until the tensioner arm is brought into contact with the auto tensioner body. Make sure the number of turns the special tool makes conforms with the standard value of 2½-3 turns.

c. Install the rubber plug to the timing belt rear cover.

24. Install the timing belt covers and all related items.

25. Connect the negative battery cable.

Timing Belt and Tensioner

BELT TENSION

Adjustment

1.8L Engine

1. Disconnect the negative battery cable.

2. Remove the timing belt covers.

3. Adjust the silent shaft (inner) belt tension first. Loosen the idler pulley center bolt so the pulley can be moved.

4. Move the pulley by hand so the long side of the belt deflects about ¼in.

5. Hold the pulley tightly so the pulley cannot rotate when the bolt is tightened. Tighten the bolt to 15 ft. lbs. (20 Nm) and recheck the deflection amount.

6. To adjust the timing (outer) belt, first loosen the pivot side tensioner bolt and then the slot side bolt. Allow the spring to take up the slack.

7. Tighten the slot side tensioner bolt and then the pivot side bolt. If the pivot side bolt is tightened first, the tensioner could turn with bolt, causing over tension.

8. Turn the crankshaft clockwise. Loosen the pivot side tensioner bolt and then the slot side bolt. Tighten the slot bolt and then the pivot side bolt.

9. Check the belt tension by checking belt deflection. The deflection of the longest span of the belt should be about 0.40 in. Do not manually overtighten the belt or it will howl.

10. Install the timing belt covers and all related items.

11. Connect the negative battery cable.

2.0L DOHC Engine

1. Disconnect the negative battery cable.

2. Remove the timing belt covers.

3. Adjust the silent shaft (inner) belt tension first. Loosen the idler pulley center bolt so the pulley can be moved.

4. Move the pulley by hand so the long side of the belt deflects about ¼in.

5. Hold the pulley tightly so the pulley cannot rotate when the bolt is tightened. Tighten the bolt to 15 ft. lbs. (20 Nm) and recheck the deflection amount.

6. To adjust the timing (outer) belt, turn the crankshaft ¼turn counterclockwise, then turn it clockwise to move No. 1 cylinder to TDC.

7. Loosen the center bolt. Using tool MD998752 or equivalent and a torque wrench, apply a torque of 1.88-2.03 ft. lbs. (2.6-2.8 Nm) If the body of the vehicle interferes with the special tool and the torque wrench, use a jack and slightly raise the engine assembly. Holding the tensioner pulley, tighten the center bolt.

8. Screw special tool MD998738 or exact equivalent into the engine left support bracket until its end makes contact with the tensioner arm. At this point, screw the special tool in some more and remove the set wire attached to the auto tensioner, if wire was not previously removed. Then remove the special tool.

9. Rotate the crankshaft 2 complete turns clockwise and let it sit for approximately 15 minutes. Then, measure the auto tensioner protrusion (the distance between the tensioner arm and auto tensioner body) to ensure that it is within 0.15-0.18 in. (3.8-4.5mm). If out of specification, repeat Step 1-4 until the specified value is obtained.

10. If the timing belt tension adjustment is being performed with the engine mounted in the vehicle, and clearance between the tensioner arm and the auto tensioner body cannot be measured, the following alternative method can be used:

  a. Screw in special tool MD998738 or equivalent, until its end makes contact with the tensioner arm.

  b. After the special tool makes contact with the arm, screw it in some more to retract the auto tensioner pushrod while counting the number of turns the tool makes until the tensioner arm is brought into contact with the auto tensioner body. Make sure the number of turns the special tool makes conforms with the standard value of 2½-3 turns.

  c. Install the rubber plug to the timing belt rear cover.

11. Install the timing belt covers and all related items.

12. Connect the negative battery cable.

Timing Sprockets and Oil Seals

REMOVAL & INSTALLATION

1. Disconnect the negative battery cable.

2. Remove the valve cover(s) and timing belt(s).

3. Remove the crankshaft pulley retainer bolts and remove the pulley.

4. Remove the crankshaft sprocket retainer bolt and washer from the sprocket, if used, and remove sprocket.

If sprocket is difficult to remove, the appropriate puller may be used. If no bolts are used on the sprocket, use the appropriate puller to remove.

5. Hold the camshaft stationary using the hexagon cast between journals No. 2 and 3 and remove the retainer bolt. Remove the sprocket from the camshaft. If the camshaft does not have a hexagon cast between journals No. 2 and 3, use the appropriate spanner wrench to hold the shaft in position while removing the bolt.

6. Pry the seals from the bores and replace using the proper installation tools.

To install:

7. Install the sprockets to their appropriate shafts. Install the retainer bolts and torque the camshaft sprocket bolt to 58-72 ft. lbs. (80-100 Nm) on 1.8L and 2.0L engines.

8. Torque the crankshaft sprocket retaining bolt to 80-94 ft. lbs. (110-130 Nm).

9. Install the timing belt(s), timing covers, valve cover(s) and remaining components.

10. Install the engine under cover. Connect the negative battery cable, start the engine and check for leaks.

Camshaft and Bearings

REMOVAL & INSTALLATION

1.8L Engine
▶ See Figure 80

1. Disconnect the negative battery cable. Remove the air intake hose and the PCV hose.

2. Remove the valve covers and timing belt.

3. Install auto lash adjuster retainer tools MD998443 or equivalent, on the rocker arm.

4. Remove the camshaft bearing caps but do not remove the bolts from the carrier.

5. Remove the rocker arms, rocker shafts and bearing caps from the engine as an assembly.

6. Remove the camshaft from the cylinder head.

7. Inspect the bearing journals on the camshaft for excess wear or damage. Measure the cam lobe height and compare to the desired readings. Inspect the bearing surfaces in the cylinder

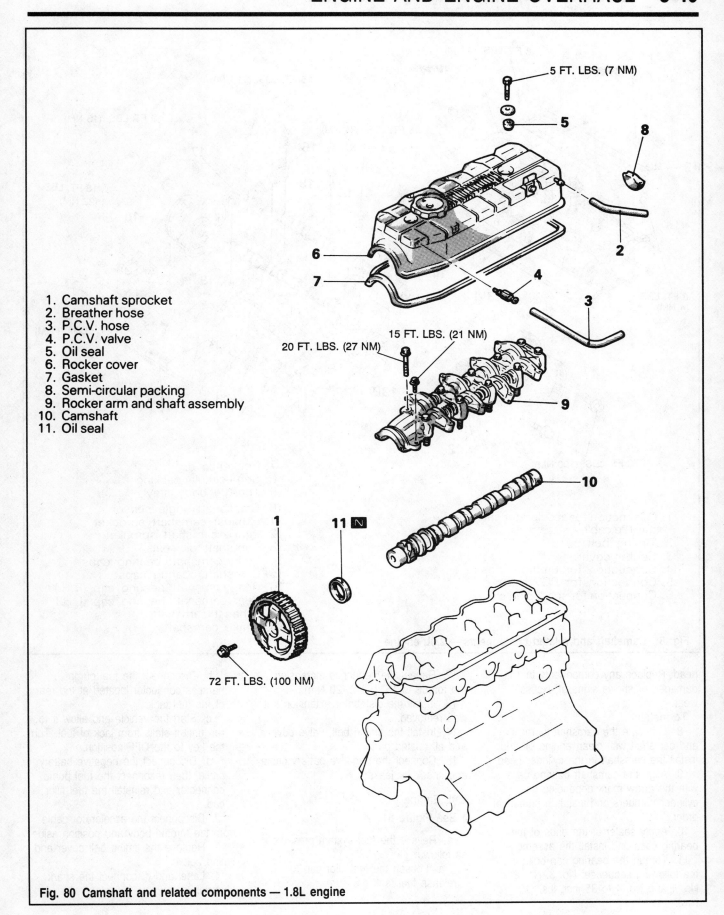

5 FT. LBS. (7 NM)

5

8

2

6

7

4

3

15 FT. LBS. (21 NM)

20 FT. LBS. (27 NM)

9

10

1. Camshaft sprocket
2. Breather hose
3. P.C.V. hose
4. P.C.V. valve
5. Oil seal
6. Rocker cover
7. Gasket
8. Semi-circular packing
9. Rocker arm and shaft assembly
10. Camshaft
11. Oil seal

1

11 N

72 FT. LBS. (100 NM)

Fig. 80 Camshaft and related components — 1.8L engine

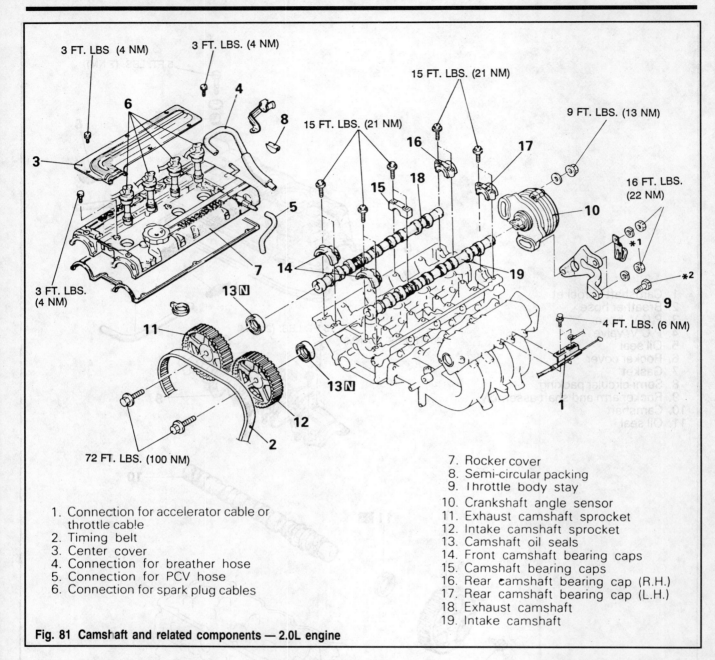

3 FT. LBS (4 NM)
3 FT. LBS. (4 NM)
15 FT. LBS. (21 NM)
9 FT. LBS. (13 NM)
15 FT. LBS. (21 NM)
16 FT. LBS. (22 NM)
3 FT. LBS. (4 NM)
4 FT. LBS. (6 NM)
72 FT. LBS. (100 NM)

1. Connection for accelerator cable or throttle cable
2. Timing belt
3. Center cover
4. Connection for breather hose
5. Connection for PCV hose
6. Connection for spark plug cables
7. Rocker cover
8. Semi-circular packing
9. Throttle body stay
10. Crankshaft angle sensor
11. Exhaust camshaft sprocket
12. Intake camshaft sprocket
13. Camshaft oil seals
14. Front camshaft bearing caps
15. Camshaft bearing caps
16. Rear camshaft bearing cap (R.H.)
17. Rear camshaft bearing cap (L.H.)
18. Exhaust camshaft
19. Intake camshaft

Fig. 81 Camshaft and related components — 2.0L engine

head. Replace any components that is damaged or shows signs of excess wear.

To install:

8. Lubricate the camshaft journals and camshaft with clean engine oil and install the camshaft in the cylinder head.

9. Align the camshaft bearing caps with the arrow mark depending on cylinder numbers and install in numerical order.

10. Apply sealer at the ends of the bearing caps and install the assembly.

11. Torque the bearing cap bolts in the following sequence: No. 3, No. 2, No. 1 and No. 4 to 85 inch lbs. (10 Nm).

12. Repeat the sequence increasing the torque to 15 ft. lbs. (20 Nm).

13. Install the distributor extension if it was removed.

14. Install the timing belt, valve cover and all related parts.

15. Connect the negative battery cable and check for leaks.

2.0L Engine

▶ **See Figure 81**

1. Relieve the fuel system pressure as follows:

a. Loosen the fuel filler cap to release fuel tank pressure.

b. Disconnect the fuel pump harness connector located at the rear of the fuel tank.

c. Start the vehicle and allow it to run until it stalls from lack of fuel. Turn the key to the **OFF** position.

d. Disconnect the negative battery cable, then reconnect the fuel pump connector and reinstall the fuel filler cap.

2. Disconnect the accelerator cable from the throttle body and position aside.

3. Remove the timing belt cover and timing belt.

4. Label and disconnect the spark plug cable.

5. Remove the center cover, breather and PCV hose.

6. Remove the rocker cover, semi-circular packing, throttle body stay, crankshaft angle sensor, both camshaft sprockets, and oil seal.

7. Loosen the bearing cap bolts in 2-3 steps. Label and remove all camshaft bearing caps.

➡ **If the bearing caps are difficult to remove, use a plastic hammer to gently tap the rear part of the camshaft.**

8. Remove the intake and exhaust camshafts.

9. Check the camshaft journals for wear or damage. Check the cam lobes for damage. Also, check the cylinder head oil holes for clogging.

To install:

10. Lubricate the camshafts with heavy engine oil and position the camshafts on the cylinder head.

➡ **Do not confuse the intake camshaft with the exhaust camshaft. The intake camshaft has a split on its rear end for driving the crank angle sensor.**

11. Make sure the dowel pin on both camshaft sprocket ends are located on the top.

12. Install the bearing caps. Tighten the caps in sequence and in 2 or 3 steps. No. 2 and 5 caps are of the same shape. Check the markings on the caps to identify the cap number and intake/exhaust symbol. Only **L** (intake) or **R** (exhaust) is stamped on No. 1 bearing cap. Also, make sure the rocker arm is correctly mounted on the lash adjuster and the valve stem end. Torque the retaining bolts to 15 ft. lbs. (20 Nm).

13. Apply a coating of engine oil to the oil seal. Using tool MD998307 or equivalent, press-fit the seal into the cylinder head.

14. Align the punch mark on the crank angle sensor housing with the notch in the plate. With the dowel pin on the sprocket side of the intake camshaft at top, install the crank angle sensor on the cylinder head.

➡ **Do not position the crank angle sensor with the punch mark positioned opposite the notch; this position will result in incorrect fuel injection and ignition timing.**

15. Install the timing belt, valve cover and all related parts.

16. Connect the negative battery cable and check for leaks.

Silent Shaft

REMOVAL & INSTALLATION

▸ **See Figure 82**

1.8L Engine

➡ **A special oil seal guide MD998285 or equivalent, is needed to complete this operation.**

1. Disconnect the negative battery cable.

2. Remove the oil filter, oil pressure switch, oil gauge sending unit, oil filter mounting bracket and gasket.

3. Raise and safely support the vehicle. Drain engine oil. Remove engine oil pan, oil screen and gasket.

4. Lower the vehicle. Remove the timing belts.

5. Remove the front engine cover. Different length bolts are used. Take note of their locations. If the cover sticks to the block, look for a special slot provided and pry with a flat bladed tool. Discard the shaft seal and gasket.

6. Remove the oil pump driven gear flange bolt. When loosening this bolt, first insert a tool approximately ⅜in. diameter into the plug hole on the left side of the cylinder block to hold the silent shaft. Remove the oil pump gears and remove the front case assembly. Remove the threaded plug, the oil pressure relief spring and plunger.

7. Remove the silent shaft oil seals, the crankshaft oil seal and front case gasket.

8. Remove the silent shafts and inspect as follows:

   a. Check the oil holes in the shaft for clogging.

   b. Check journals of the shaft for seizure, damage and contact with bearing. If there is anything wrong with the journal, replace the silent shaft bearing, silent shaft or front case.

   c. Check the silent shaft oil clearance. If the clearance is beyond the specifications, replace the silent shaft bearing, silent shaft or front

case. The specifications for oil clearances are as follows:

   Right shaft
   Front — 0.0008-0.0024 in.
   (0.02-0.06mm)
   Rear — 0.0020-0.0036 in.
   (0.05-0.09mm)
   Left shaft
   Front — 0.0008-0.0021 in.
   (0.02-0.05mm)
   Rear — 0.0020-0.0036 in.
   (0.05-0.09mm)

9. If bearing replacement is required, remove the front bearing from the engine block by using tool MD998282-01.

10. To remove the rear silent shaft bearing from the engine block, use tool MD998283-01.

To install:

11. Install the rear silent shaft bearing into the cylinder block as follows:

   a. Apply clean engine oil to the rear bearing outer circumference and bearing hole in the cylinder block.

   b. Using bearing installation tool MD998286-01 and a hammer, drive the rear bearing into the cylinder block.

➡ **Make sure the bearing oil holes align with the oil holes in the block, once the bearings are installed.**

12. Install the front silent shaft bearing into the cylinder block as follows:

   a. Install 2 guide pins, normally included in the special tool set with bearing installer MD998289-01, to the threaded holes in the cylinder block.

   b. Set the front baring on the installation tool so that the ratchet ball of the tool fits in the hole in the bearing.

   c. Apply engine oil to the bearing outer circumference and the bearing hole in the cylinder block.

   d. Set the installation tool on the guide pins and, using a hammer, drive the bearing into the cylinder block.

➡ **Make sure the bearing oil holes align with the oil holes in the block, once the bearings are installed.**

13. Lubricate the bearing surface of the shaft and the bearing journals with clean engine oil. Carefully install the silent shafts to the block.

14. Clean the gasket material from the mating surface of the cylinder block and the engine front cover. Install new gasket in place.

15. Using seal installation tool MD998304-01 or equivalent, install the crankshaft oil seal into the front engine cover.

16. Using the proper size socket wrench, press in the silent shaft oil seal into the front case.

17. Place pilot tool MD998285-01 or equivalent, onto the nose of the crankshaft. Apply clean engine oil to the outer circumference of the pilot tool.

18. Install the front case onto the engine block and install the retainer bolts in their original positions. Tighten retainers evenly to 12 ft. lbs. (17 Nm).

19. Install the oil pump relief plunger and spring into the bore in the front case and tighten to 33 ft. lbs. (45 Nm). Make sure a new gasket is in place.

20. Install the oil pump drive gear and driven gear to the front case, lining up the timing marks. Lubricate the gears with clean engine oil.

21. Inspect the orientation of the silent shaft as outlined in the timing belt section of this chapter. Insert the Phillips screwdriver into the hole on the side of the engine block. Install and tighten the flange bolt to 27 ft. lbs. (37 Nm).

22. Install a new oil pump cover gasket in the groove of the front case. When installing the gasket, make sure the round side of the gasket is towards the oil pump cover.

23. Install the oil pump seal into the oil pump cover, making sure the lip is facing the correct direction. The lip of the seal should be installed against the oil it is to stop.

➡ **The timing of the oil pump sprocket and connected silent shaft can be incorrect, even with the timing mark aligned. Incorrect orientation of the silent shaft will result in engine vibration during operation. Follow the alignment procedure in the timing belt section of this chapter.**

24. Install the timing belts and all related items. Make sure the timing and the orientation of the silent shafts is correct, using alignment tool in the hole in the left side of the engine block, as specified in the timing belt section of this chapter.

25. Install the oil pan, oil filter mounting bracket, oil switches and new oil filter to the engine. Fill the crankcase to the proper level with clean engine oil.

26. Connect the negative battery cable and start the engine. Check for proper timing and inspect for leaks.

2.0L Engine

➡ **A special oil seal guide MD998285 and a plug cap socket tool MD998162 or exact equivalents are needed to complete this operation.**

1. Disconnect the negative battery cable.

2. Remove the oil filter, oil pressure switch, oil gauge sending unit, oil filter mounting bracket and gasket. If equipped with turbocharged engine, remove the oil cooler bolt and oil cooler from the oil filter bracket.

3. Raise and safely support the vehicle. Drain engine oil. Remove engine oil pan, oil screen and gasket. Remove the relief plug, gasket, relief spring and relief plunger.

4. Lower the vehicle. Using the proper equipment, support the weight of the engine.

5. Remove the front engine mount bracket and accessory drive belt.

6. Disconnect the electrical connector from the oil pressure sending unit. Remove the oil pressure sensor using special removal tool MD998054. Remove the oil filter and the oil filter bracket.

7. Remove the oil pan, oil screen and gasket.

8. Using special tool MD998162, remove the plug cap in the engine front cover. If the plug is too tight, hit the plug head with a hammer a few times and the plug should be easier to loosen.

9. Remove the plug on the side of the engine block. Insert a Phillips screwdriver with a shank diameter of 0.32 in. (8mm) into the plug hole. This will hold the silent shaft.

10. Remove the driven gear bolt that secures the oil pump driven gear to the silent shaft.

11. Remove and tag the front cover mounting bolts. Note the lengths of the mounting bolts as they are removed for proper installation.

12. Remove the front case cover and oil pump assembly. If necessary, the silent shaft can come out with the cover assembly.

13. Remove the silent shaft oil seals, the crankshaft oil seal and front case gasket.

14. Remove the silent shafts and inspect as follows:
    a. Check the oil holes in the shaft for clogging.
    b. Check journals of the shaft for seizure, damage and contact with

bearing. If there is anything wrong with the journal, replace the silent shaft bearing, silent shaft or front case.

    c. Check the silent shaft oil clearance. If the clearance is beyond the specifications, replace the silent shaft bearing, silent shaft or front case. The specifications for oil clearances are as follows:
    Right shaft
    Front — 0.0012-0.0024 in. (0.03-0.06mm)
    Rear — 0.0008-0.0021 in. (0.02-0.05mm)
    Left shaft
    Front — 0.0020-0.0036 in. (0.05-0.09mm)
    Rear — 0.0017-0.0033 in. (0.04-0.08mm)

15. If bearing replacement is required, remove the front bearing from the engine block by using tool MD998371.

16. To remove the rear silent shaft bearing from the engine block, use tool MD998372 and guide plate MD998374.

To install:

17. Clean all mating surfaces of gasket material.

18. Install the rear silent shaft bearing into the cylinder block as follows:
    a. If installing the left rear bearing, install the guide plate MD9983774 to the cylinder block.
    b. Apply clean engine oil to the rear bearing outer circumference and bearing hole in the cylinder block.
    c. Using bearing installation tool MD998374, MD998373 and a hammer, drive the rear bearing into the cylinder block.

➡ **The left rear bearing has no oil holes. Make sure the oil hole in the right bearing align with the oil hole in the block, once the bearing is installed.**

19. Install the front silent shaft bearings into the cylinder block, using installation tool as follows:

20. Apply clean engine oil to the rear bearing outer circumference and bearing hole in the cylinder block.

21. Using bearing installation tool MD998373 and a hammer, drive the rear bearing into the cylinder block.

➡ **Make sure the bearing oil holes align with the oil holes in the block, once the bearings are installed. Also, make sure that the clinch of the bearing is in the upper most position.**

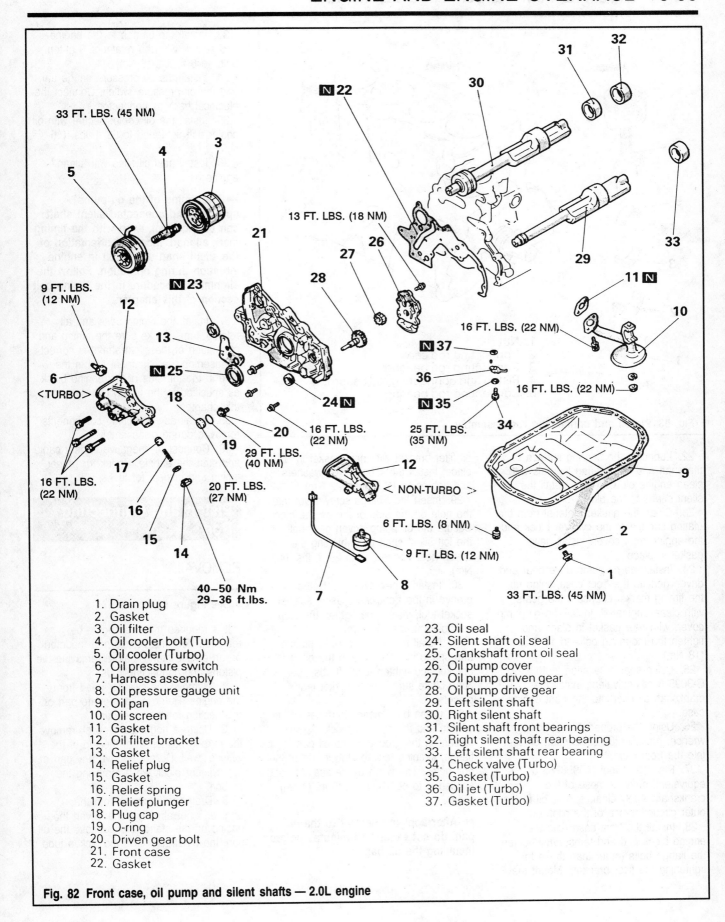

Fig. 82 Front case, oil pump and silent shafts — 2.0L engine

1. Drain plug
2. Gasket
3. Oil filter
4. Oil cooler bolt (Turbo)
5. Oil cooler (Turbo)
6. Oil pressure switch
7. Harness assembly
8. Oil pressure gauge unit
9. Oil pan
10. Oil screen
11. Gasket
12. Oil filter bracket
13. Gasket
14. Relief plug
15. Gasket
16. Relief spring
17. Relief plunger
18. Plug cap
19. O-ring
20. Driven gear bolt
21. Front case
22. Gasket
23. Oil seal
24. Silent shaft oil seal
25. Crankshaft front oil seal
26. Oil pump cover
27. Oil pump driven gear
28. Oil pump drive gear
29. Left silent shaft
30. Right silent shaft
31. Silent shaft front bearings
32. Right silent shaft rear bearing
33. Left silent shaft rear bearing
34. Check valve (Turbo)
35. Gasket (Turbo)
36. Oil jet (Turbo)
37. Gasket (Turbo)

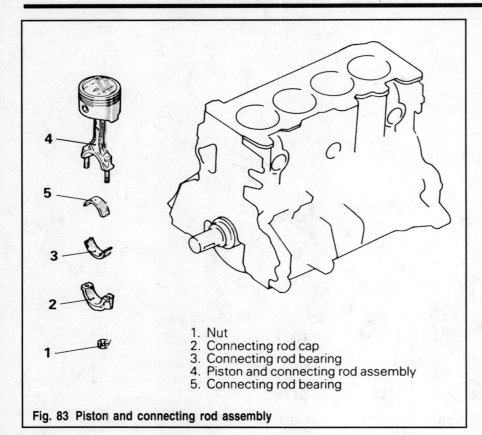

1. Nut
2. Connecting rod cap
3. Connecting rod bearing
4. Piston and connecting rod assembly
5. Connecting rod bearing

Fig. 83 Piston and connecting rod assembly

22. Lubricate the bearing surface of the shaft and the bearing journals with clean engine oil. Carefully install the silent shafts to the block.

23. Clean the gasket material from the mating surface of the cylinder block and the engine front cover. Install new gasket in place.

24. Install the oil pump drive gear and driven gear to the front case, lining up the timing marks. Lubricate the gears with clean engine oil. Install the oil pump cover, with new gasket in place and tighten the mounting bolts to 13 ft. lbs. (18 Nm).

25. Using seal installation tool C-3095-A or equivalent, install the crankshaft oil seal into the front engine case.

26. Using the proper size socket wrench, press in the silent shaft oil seal into the front case.

27. Place pilot tool MD998285 or equivalent, onto the nose of the crankshaft. Apply clean engine oil to the outer circumference of the pilot tool.

28. Install the front case onto the engine block and and temporarily tighten the flange bolts (other than those for tightening the filter bracket). Mount the oil filter bracket with new gasket in place. Install the 4 bolts with washers and tighten to 16 ft. lbs. (22 Nm).

29. Insert the Phillips screwdriver into the hole on the side of the engine block. Secure the oil pump driven gear onto the left silent shaft by tightening the driven gear flange bolt to 29 ft. lbs. (40 Nm).

30. Install a new O-ring onto the groove in the front case. Using special socket tool, install and tighten the plug cap to 20 ft. lbs. (27 Nm).

31. Install the oil pump relief plunger and spring into the bore in the oil filter bracket and tighten to 36 ft. lbs. (50 Nm). Make sure a new gasket is in place.

32. Clean both mating surfaces of the oil pan and the cylinder block. Apply sealant in the groove in the oil pan flange, keeping towards the inside of the bolt holes. The width of the sealant bead applied is to be about 0.016 in. (4mm) wide.

➡ **After applying sealant to the oil pan, do not exceed 15 minutes before installing the oil pan.**

33. Install the oil pan to the engine and secure with the retainers. Tighten bolts to 6 ft. lbs. (8 Nm).

34. Install the oil pressure gauge unit and the oil pressure switch. Connect the electrical harness connector.

35. Install the oil cooler secure with oil cooler bolt tightened to 33 ft. lbs. (45 Nm).

36. Install new oil filled with clean engine oil.

➡ **The timing of the oil pump sprocket and connected silent shaft can be incorrect, even with the timing mark aligned. Incorrect orientation of the silent shaft will result in engine vibration during operation. Follow the alignment procedure in the timing belt section of this chapter.**

37. Install the timing belts and all related items. Make sure the timing and orientation of the silent shafts is correct by inserting the alignment tool in the hole in the left side of the engine block, as specified in the timing belt section of this chapter.

38. Install any remaining components removed during disassembly.

39. Connect the negative battery cable and start the engine. Check for proper timing and inspect for leaks.

Piston and Connecting Rod

REMOVAL

▶ **See Figure 83**

It is required that the engine be removed from the vehicle and mounted on an engine stand, before removing the pistons and connecting rods.

1. Remove the cylinder head from the engine. Refer to appropriate part of this section for the procedure.

2. Using a ridge reamer tool, remove the carbon buildup from the top of the cylinder wall. Do not remove any part of the cylinder block while removing the carbon.

3. Drain the lubricant from the engine, if not already done. Invert the engine on the stand, then remove the oil pan, the oil strainer and the pickup tube.

4. Position the piston to be removed at the bottom of its stroke, so that the connecting rod bearing cap can be easily reached. Put the cylinder number on the side of the connecting rod big end to aid in identification during assembly.

5. Remove the connecting rod bearing cap nuts, cap and the lower half of the bearing. Cover the rod bolts with lengths of rubber tubing or hose to protect the crankshaft pin and the cylinder walls when the rod and piston assembly is removed.

6. Push the piston/connecting rod assembly out through the top of the cylinder block with a length of wood or a wooden hammer handle.

➡ When removing the piston/connecting rod assembly, be careful not to scratch the crank pin or the cylinder wall with the connecting rod.

IDENTIFICATION AND POSITIONING

Prior to removing the connecting rod cap, mark both the upper and the lower portion of the rod with the number cylinder it is being removed from. This will assure correct placement of the rod in its original cylinder and the correct cap on the rod.

The pistons are marked with an arrow stamped on the piston head. When installed in the engine, the arrow must be facing the front of the engine.

Connecting Rod, Piston and Piston Pin

CLEANING AND INSPECTION

1. Replace the piston pin if it has marks of streaks or seizure. Examine the outside thrust surface in particular.

2. Inspect for cracks and excess wear and replace if present.

3. If the piston pin can be inserted into the piston pin hole with snugly wit the thumb, it is reusable. If it can be inserted with no resistance or the is play, replace the piston and pin as a set.

4. Insert a new piston ring into the ring groove and check the side clearance. If the side clearance exceeds specifications, piston replacement is required.

5. Replace the connecting rod if it has damage on the thrust faces at either end. Also replace the connecting rod if it has ridge wear or severely rough surface on the inside diameter of the small end. When using new connecting rod, the cylinder number should be stamped on the cap and on the rod.

6. Using a connecting rod aligner, check the rod for bends and twist. Any connecting rod that has been bent or distorted should be replaced. Obviously, this should be checked by an experienced, reliable machine shop, who has the proper equipment.

PISTON RING REPLACEMENT

◆ See Figures 84, 85, 86, 87, 88, 89, 90, 91 and 92

Oil Control Ring

1. Fit the oil ring spacer into the bottom piston ring groove.

➡ The side rails may be installed in either direction.

2. Install the upper side rail as follows:

a. Fit 1 end of the rail into the piston groove.

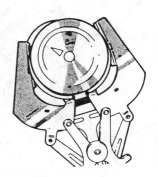

Fig. 84 Removing piston ring from piston using ring expander

b. Press the remaining portion of the rail into the groove using your finger.

➡ Using a piston ring expander to expand the side rail end gap can break the side rail, unlike on other rings. Do not use piston ring expander when installing side rail.

c. Install the lower side rail in the same manner.

d. Once both side rails are installed, make sure the side rails move smoothly in either direction.

3. Repeat this procedure for the installation of the reaming oil control rings.

Compression Rings

A piston ring expander is necessary for removing and installing the piston rings (to avoid damaging them). When the rings are removed, clean the ring grooves using an appropriate ring groove cleaning tool, using care not to cut too deeply. Use solvent to thoroughly remove all of the carbon and varnish deposits.

When installing the rings, make sure that the stamped mark on the ring is facing upwards. Install the bottom rings first, then the upper ones last. Be sure to use a ring expander, to keep from breaking the rings.

1. Remove the piston rings from the ring groove in each piston using a piston ring expander.

2. Clean the piston using the appropriate ring groove cleaner.

3. Insert a new piston ring into the ring groove and check the side clearance. If the side clearance exceeds specifications, piston replacement is required.

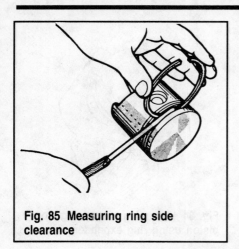

Fig. 85 Measuring ring side clearance

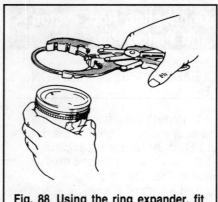

Fig. 88 Using the ring expander, fit No. 2, then No. 1 piston ring in position

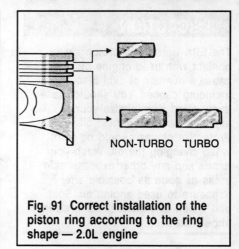

NON-TURBO   TURBO

Fig. 91 Correct installation of the piston ring according to the ring shape — 2.0L engine

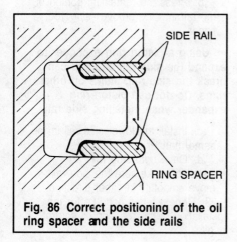

SIDE RAIL

RING SPACER

Fig. 86 Correct positioning of the oil ring spacer and the side rails

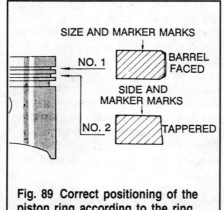

SIZE AND MARKER MARKS

NO. 1 — BARREL FACED

SIDE AND MARKER MARKS

NO. 2 — TAPPERED

Fig. 89 Correct positioning of the piston ring according to the ring markings — 1.8L engine

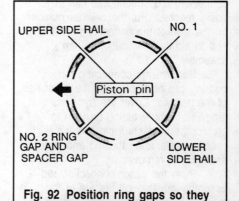

UPPER SIDE RAIL   NO. 1

Piston pin

NO. 2 RING GAP AND SPACER GAP   LOWER SIDE RAIL

Fig. 92 Position ring gaps so they are staggered as shown — 2.0L engine

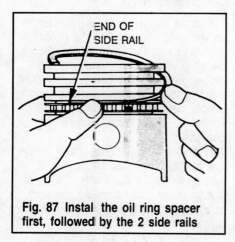

END OF SIDE RAIL

Fig. 87 Install the oil ring spacer first, followed by the 2 side rails

To install:

4. Using piston ring expander, fit the lower, then the upper piston ring into position. Make sure the piston rings are installed so the side of the ring with the marks on it is facing the top crown of the piston.

GAP OF LOWER SIDE RAIL   NO. 1 RING GAP

TIMING BELT SIDE   GAP UNDER UPPER SIDE RAIL

NO. 2 RING GAP AND SPACER GAP

Fig. 90 Position ring gaps so they are staggered as shown — 1.8L engine

ROD BEARING REPLACEMENT

1. Visually inspect the bearing surface for uneven contact, streaks, scratches and seizure. Replace if defects are evident. If streaks and seizure are excessive, check also the crankshaft. Replace or regrind the crankshaft as required.

2. Measure the connecting rod bearing inside diameter and the crankshaft pin outside diameter. If the oil clearance exceeds the limit, replace the bearing, and crankshaft if necessary. Or regrind the crankshaft to an undersize and replace bearing with an undersize one.

3. If the clearances are not within the tolerances, replaced the bearings.When installing new bearings, make sure that the tabs fit correctly into the notch of the bearing cap and rod. Lubricate the face of each insert before installing them onto the crankshaft.

INSTALLATION

1. Liberally coat engine oil on the circumference of the piston, piston ring and oil control ring.

2. Arrange the piston ring and oil ring gaps so they are staggered.

3. Rotate the crankshaft so the piston pin of the piston to the installed is on center of the cylinder bore.

4. Using a approved piston ring compressor, compress the piston rings. Install rubber hoses over the bolts on the connecting rod to protect the cylinder wall and the crankshaft pin. Care must be taken not to nick the crank pin.

5. Place the assembly into the cylinder with the alignment mark (arrow) on the top of the piston facing the front of the engine (timing belt side).

6. Using a wooden hammer handle, tap the piston assembly into the cylinder. Make sure the connecting rod engages the crankshaft.

7. Install the bearing cap making sure the direction of orientation is correct. If a new connecting rod is being installed, position the notches for locking the bearings on the same side.

8. Examine the connecting rod bolts and nuts prior to installation. Inspect the bolt threads for necking (stretching). Necking can be determined by running a nut with the fingers, down the full length of the bolt. If the nut does not turn smoothly, the bolt and nut should be replaced.

9. Install the connecting rod cap onto the big end of the connecting rod. Oil the threads using clean engine oil and install the retaining nuts.

10. Tighten both nuts finger-tight, then alternately torque each nut to specifications in 2 or 3 steps.

Freeze Plugs

REMOVAL & INSTALLATION

The freeze plugs are located on the sides of the engine block. The freeze plugs serve 2 functions. The first function is to fill the hole where the casting sand is removed during the manufacturing process. The second function is to release internal pressure if the engine coolant freezes. The plug is supposed to pop out when the coolant freezes, reducing the chance of engine damage.

1. Drain the engine coolant.

2. Remove all components that are in the way.

3. If accessible, drill a hole in the plug and remove it with a dent puller. If not accessible, drive a chisel into the plug and pry out of the block.

4. Clean the block sealing surface with sandpaper.

➡ **Expandable rubber plugs are made for tight places were the standard freeze plug can not be installed. Consult the local parts warehouse for the rubber plugs**

5. Use a ratchet socket that fits into the freeze plug to install. Coat the sealing surface with RTV sealer.

6. Install the plug far enough to make a good seal.

Rear Main Bearing Oil Seal

REMOVAL & INSTALLATION

1. Disconnect the negative battery cable.

2. Remove the transaxle and transfer, if equipped, from the vehicle.

3. If equipped with automatic transaxle, remove the flywheel/ring gear assembly from the crankshaft.

4. If equipped with manual transaxle, remove the rear engine plate and the bellhousing cover.

5. If the crankshaft rear oil seal case is leaking, remove it. Otherwise, just remove the oil seal. Some engines have a separator that should also be removed.

To install:

6. Lubricate the inner diameter of the new seal with clean engine oil.

7. Install the oil seal in the crankshaft rear oil seal case using tool MD998376 or equivalent. Press the seal all the way in without tilting it. Force the oil separator into the oil seal case so the oil hole in the separator is downward.

8. Install the seal case to the crankshaft with a new gasket, if removed.

9. Install the flywheel or drive plate, transfer and transaxle.

10. Connect the negative battery cable and check for leaks.

11. Fill the crankcase with clean oil to the proper level. Start the engine and check for leaks.

Crankshaft and Main Bearings

REMOVAL & INSTALLATION

1. Relieve the fuel system pressure as follows:

   a. Loosen the fuel filler cap to release fuel tank pressure.

   b. Disconnect the fuel pump harness connector located at the rear of the fuel tank.

   c. Start the vehicle and allow it to run until it stalls from lack of fuel. Turn the key to the **OFF** position.

   d. Disconnect the negative battery cable, then reconnect the fuel pump connector and reinstall the fuel filler cap.

   e. Wrap shop towels around the fitting that is being disconnected to absorb residual fuel in the lines.

2. Remove the transaxle from the vehicle.

3. Remove the engine from the vehicle. Place the engine in a suitable workstand.

4. If equipped with manual transaxle, remove the flywheel from the crankshaft. If equipped with automatic transaxle, remove the adapter, drive plate, adapter plate and the crankshaft bushing from the crankcase.

5. Remove the front timing cover and timing belts.

6. Remove the oil pan and oil pump screen. On 2.0L turbocharged engine, remove the oil jets from the engine.

7. Remove the rear seal carrier.

8. Remove the connecting rod bearing caps and place rubber hoses over the connecting rod bolts to protect the crankshaft journals.

9. Remove the crankshaft bearing caps.

10. Push the connecting rods into the cylinder slightly.

11. Carefully lift the crankshaft out of the engine, turning slightly to disengage the connecting rods.

To install:

12. Install oil pan bolts adjacent to the connecting rods. Place rubber bands around the connecting rod bolt to hold

the rod in place during crankshaft installation.

13. Lubricate all bearing surfaces with assembly lube. Make sure all bearings are in place before assembly.

14. Install the crankshaft with the help of an assistant.

15. Install the crankshaft main bearing caps as follows:

a. Install the bearing caps so the that their arrows are positioned facing the timing belt end of the block. All bearing caps must be positioned in numerical order.

➡ **When installing the bearing caps on 1993 2.0L engines, check that the shank length of each bolt meets the limit prior to installation. If the limit is exceeded, replace the bolt.**

b. If equipped with 1993 2.0L engine, inspect the shank length of the bolt and compare to the limit of 2.79 in. (71.1mm). If the limit is exceeded, replace the bolt.

c. Install the bolts and tighten to the proper torque value.

d. Once the bearing caps are installed and the bolts are torqued, check that the crankshaft can be freely turned. Inspect that the crankshaft end-play is within specifications, and if not, replace the bearings. The end-play limit is 0.0098 in. (0.25mm).

16. Install the connecting rod bearing caps and torque to specifications.

17. Install the rear seal carrier and new rear seal.

18. Install the oil jets, making sure the nozzles are pointing towards the piston.

19. Install the timing belts the covers.

20. Install the remaining components onto the engine.

21. Install the engine into the vehicle.

22. Confirm that all items are connected securely.

23. Start the engine and check for proper oil pressure and leaks.

CLEANING AND INSPECTION

The crankshaft inspection and servicing should be handled exclusively by a reputable machinist because all measurements require a dial indicator, fixing jigs and a large micrometer; also machine tools, such as: crankshaft grinder. The crankshaft should be thoroughly cleaned (especially the oil passages), Magnafluxed (to check for cracks) and the following checks made: Main journal diameter, crank pin

(connecting rod journal) diameter, taper, out-of-round and run-out. Wear, beyond the specification limits, in any of these areas means the crankshaft must be reground or replaced.

MAIN BEARING CLEARANCE INSPECTION

The crankshaft oil clearance can be measured by using a plastic gauge, as follows:

1. Remove the oil, grease and other foreign materials from he crankshaft journals and bearing inner surface.

2. Install the crankshaft.

3. Cut the plastic gauge to the same length as the width of the bearing and place it on the journal in parallel with the axis.

4. Gently place the crankshaft bearing cap over it and tighten the bolts to the specified torque.

5. Remove the bolts and carefully remove the bearing cap.

6. Measure the width of the smashed plastic gauge at its widest point by using the scale printed on the plastic gauge bag.

7. Repeat the procedure for the remaining bearings. If the bearing journal appears to be in good shape (with no unusual wear visible) and are within tolerances, no further main bearing service is required. If unusual wear is evident and/or the clearances are outside specifications, the bearings must be replaced and the cause of their wear determined.

BEARING REPLACEMENT

If the oil clearance is not within specifications and signs of wear are evident of the bearing, replacement of the bearings is required. When replacement is 1 set of bearings is required, it is recommended that all bearings be replaced. Remove the bearing from the engine block. Install the new bearings as follows:

1. Install the upper crankshaft main bearing into the cylinder block. The upper bearing will have an oil groove in the shell.

2. Install the crankshaft.

3. Install the upper bearing into each bearing cap. The upper bearing has no oil groove.

4. Once installed, lubricate the bearing surfaces with clean engine oil.

5. Install the bearing caps so the that their arrows are positioned facing the timing belt end of the block.

➡ **When installing the bearing caps on 1993 2.0L engines, check that the shank length of each bolt meets the limit prior to installation. If the limit is exceeded, replace the bolt.**

6. If equipped with 1993 2.0L engine, inspect the shank length of the bolt and compare to the limit of 2.79 in. (71.1mm). If the limit is exceeded, replace the bolt.

7. Install the bolts and tighten to the proper torque value.

8. Once the bearing caps are installed and the bolts are torqued, check that the crankshaft can be freely turned. Inspect that the crankshaft end-play is within specifications, and if not, replace the bearings. The end-play limit is 0.0098 in. (0.25mm).

Flywheel and Ring Gear

REMOVAL & INSTALLATION

1. If equipped with a manual transaxle, refer to the Clutch, Removal and Installation procedures in Chapter 7, then remove the transaxle and the clutch assembly. If equipped with an automatic transaxle, refer to the Automatic Transaxle, Removal and Installation procedures in Chapter 7, then remove the transaxle and the torque converter.

2. For manual transaxles, remove the flywheel-to-crankshaft bolts and the flywheel. For automatic transaxles, remove the drive plate-to-crankshaft bolts and the drive plate.

3. To install, reverse the removal procedures. Torque the flywheel-to-crankshaft bolts to specifications.

RING GEAR REPLACEMENT

Manual Transaxle

To remove the ring gear from the flywheel, strike the ring gear at several points on its outer circumference. The ring gear can not be removed if it is heated. To install the ring gear, heat the ring gear to 572°F (300°C) for a shrink fit and install.

EXHAUST SYSTEM

▶ **See Figures 93, 94 and 95**

Safety Precautions

For a number of reasons, exhaust system work can be dangerous. Always observe the following precautions:

1. Support the vehicle securely by using jackstands or equivalent under the frame of the vehicle.

2. Wear safety goggles to protect your eyes from metal chips that may become air born while working on the exhaust system.

3. If you are using a torch, be careful not to come close to any fuel lines or components that may burn.

4. Always use the proper tool for the job.

5. Once the exhaust system has been repaired, make sure all connections are tight and do not leak exhaust fumes. If entered into the passenger compartment of the vehicle, exhaust gases can cause serious personal injury or death.

Special Tools

A number of special exhaust tools can be rented or bought from a local auto parts store. It may also be quite helpful to use solvents designed to loosen rusted nuts or bolts. Remember that these products are often flammable, apply only to parts after they are cool.

Front Exhaust Pipe

REMOVAL & INSTALLATION

1. Disconnect the negative battery cable.

➡ **Allow the engine and exhaust system to cool completely prior to starting any repairs on the exhaust system.**

2. Raise and support the vehicle safely.

3. Apply penetrating solvent to the front exhaust pipe to catalytic converter flange bolts and the manifold to exhaust pipe nuts. This will aid in fastener removal.

4. Disconnect the mounts and hangers located on the front portion of the system only. This will allow for some movement of the exhaust system yet still provide support for the system.

5. Remove the bolts retaining the front exhaust pipe to the front catalytic converter flange. Once removed, move the rear section of the exhaust out of the way as best as possible.

6. Remove the front exhaust pipe to manifold retainers and remove the front pipe and gasket from the vehicle.

To install:

7. Hold the replacement pipe against the original and make sure the pipe is of the same shape. If the replacement pipe does not match the original pipe exactly,

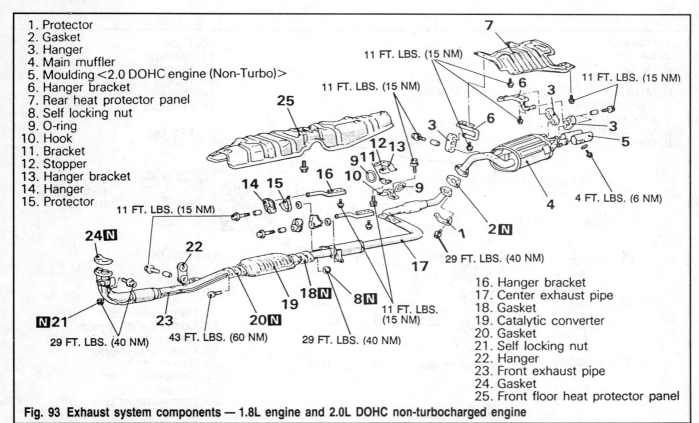

1. Protector
2. Gasket
3. Hanger
4. Main muffler
5. Moulding <2.0 DOHC engine (Non-Turbo)>
6. Hanger bracket
7. Rear heat protector panel
8. Self locking nut
9. O-ring
10. Hook
11. Bracket
12. Stopper
13. Hanger bracket
14. Hanger
15. Protector
16. Hanger bracket
17. Center exhaust pipe
18. Gasket
19. Catalytic converter
20. Gasket
21. Self locking nut
22. Hanger
23. Front exhaust pipe
24. Gasket
25. Front floor heat protector panel

Fig. 93 Exhaust system components — 1.8L engine and 2.0L DOHC non-turbocharged engine

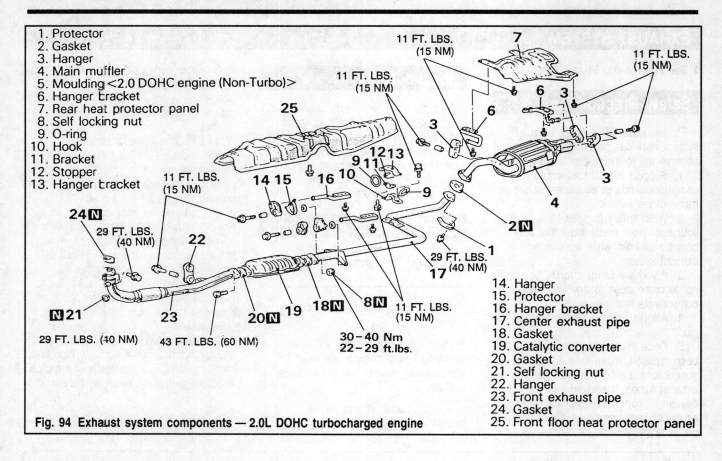

1. Protector
2. Gasket
3. Hanger
4. Main muffler
5. Moulding <2.0 DOHC engine (Non-Turbo)>
6. Hanger bracket
7. Rear heat protector panel
8. Self locking nut
9. O-ring
10. Hook
11. Bracket
12. Stopper
13. Hanger bracket

14. Hanger
15. Protector
16. Hanger bracket
17. Center exhaust pipe
18. Gasket
19. Catalytic converter
20. Gasket
21. Self locking nut
22. Hanger
23. Front exhaust pipe
24. Gasket
25. Front floor heat protector panel

Fig. 94 Exhaust system components — 2.0L DOHC turbocharged engine

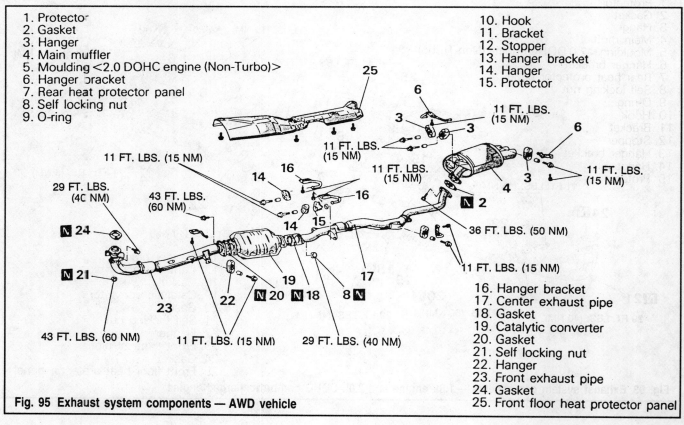

1. Protector
2. Gasket
3. Hanger
4. Main muffler
5. Moulding <2.0 DOHC engine (Non-Turbo)>
6. Hanger bracket
7. Rear heat protector panel
8. Self locking nut
9. O-ring

10. Hook
11. Bracket
12. Stopper
13. Hanger bracket
14. Hanger
15. Protector
16. Hanger bracket
17. Center exhaust pipe
18. Gasket
19. Catalytic converter
20. Gasket
21. Self locking nut
22. Hanger
23. Front exhaust pipe
24. Gasket
25. Front floor heat protector panel

Fig. 95 Exhaust system components — AWD vehicle

return the pipe to place of purchase for the correct replacement part.

➡ **Do not try to install an exhaust component not manufactured for your particular application. If the replacement part is not of the same shape, it is more than likely the wrong part. Do not try to install a component and then modify it to make it work. This practice usually does more harm than good and is not worth the time or the aggravation .**

8. Install new gasket on the front pipe to manifold flange and position the pipe. Secure with new retainers, started by hand. Do not fully tighten the fasteners at this point.

9. Install the hangers and mounts that were disconnected during disassembly. Inspect the alignment of the pipe, making sure there is clearance on all sides, the entire length of the pipe.

10. Tighten the new self-locking nuts on the manifold to pipe connection to 29 ft. lbs. (40 Nm) on FWD non-turbocharged engine or 43 ft. lbs. on turbocharged or AWD vehicles.

11. Install new fasteners at the converter to front pipe flange and tighten to 43 ft. lbs. (60 Nm).

12. Inspect the entire exhaust system making sure there is clearance between the exhaust system and all components of the vehicle.

13. Start the engine and make sure there is no leakage from the exhaust system.

14. If a leak of exhaust fumes is found, shut the engine **OFF** and allow the system to cool completely. Then repair the leak as required.

Catalytic Converter

REMOVAL & INSTALLATION

1. Disconnect the negative battery cable.

➡ **Allow the engine and exhaust system to cool completely prior to starting any repairs on the exhaust system.**

2. Raise and support the vehicle safely.

3. Apply penetrating solvent to the front exhaust pipe to catalytic converter flange bolts. This will aid in fastener removal.

4. Disconnect the mounts and hangers located on the rear portion of the system only. This will allow for some movement of the exhaust system yet still provide support for the system.

5. Remove the bolts retaining the front exhaust pipe to the front catalytic converter flange. Once removed, move the rear section of the exhaust out of the way as best as possible.

6. Remove the bolts at the rear converter flange and the converter from the vehicle. Remove the old gasket.

To install:

7. Inspect the heat dam material mounted to the underside of the passenger compartment. Make sure it is fully intact and securely fastened. Clean all mating surfaces to assure leakproof connection.

8. Install the catalytic converter to the front exhaust pipe and secure using new retainers and gasket, if used. Tighten the retainers to 29-43 ft. lbs. (40-60 Nm).

9. Connect the rear part of the exhaust system to the catalytic converter with new gasket and fasteners in place. Tighten the nuts to 29 ft. lbs. (40 Nm).

10. Reconnect all mounts and hangers disconnected during this procedure.

11. Inspect the entire exhaust system making sure there is clearance between the exhaust system and all components of the vehicle.

12. Start the engine and make sure there is no leakage from the exhaust system.

13. If an exhaust leak is found, shut the engine **OFF** and allow the system to cool completely. Then repair the leak as required.

Tailpipe And Muffler

REMOVAL & INSTALLATION

1. Remove tailpipe connection at catalytic converter.

2. Remove all brackets and exhaust clamps.

3. Remove tailpipe from muffler. On some models the tailpipe and muffler are one piece.

4. To install reverse the removal procedures. Always use new clamps and exhaust seals, start engine and check for leaks.

ENGINE MECHANICAL SPECIFICATIONS

Component	US	Metric
Cylinder head		
Surface warpage		
1.8L engine	0.002-0.008 in.	0.05-0.20mm
2.0L engine	0.002-0.008 in.	0.05-0.20mm
Camshaft		
Cam height		
1.8L engine	1.3941-1.4138 in.	35.41-35.91mm
2.0L engine		
intake (ID mark A, D):	1.3972-1.3776 in.	35.49-34.99mm
intake (ID mark B, C, E, F):	1.3858-1.3661 in.	35.20-34.70mm
exhaust (ID mark A):	1.3858-1.3661 in.	35.20-34.70mm
exhaust (ID mark C):	1.3972-1.3776 in.	35.49-34.99mm
exhaust (ID mark E, F):	1.3744-1.3547 in.	35.91-34.41mm
Fuel pump driving cam diameter		
1.8L engine	1.55-1.57 in.	39.5-40.0mm
Journal diameter		
1.8L engine	1.336-1.337 in.	33.94-33.95mm
2.0L engine	1.0217-1.0224 in.	25.95-25.97mm
Oil clearance		
1.8L engine	0.0020-0.0035 in.	0.05-0.09mm
2.0L engine	0.0020-0.0035 in.	0.05-0.09mm
Valve		
Stem diameter		
1.8L engine		
intake:	0.313-0.314 in.	7.960-7.980mm
exhaust:	0.312-0.313 in.	7.930-7.950mm
2.0L engine		
intake:	0.2587-0.2591 in.	6.47-6.58
exhaust:	0.2571-0.2579 in.	6.53-6.66mm
Face angle		
1.8L engine	45 degrees-45 degrees 30 min.	45 degrees-45 degrees 30 min.
2.0L engine	45 degrees-45 degrees 30 min.	45 degrees-45 degrees 30 min.
Valve margin		
1.8L engine		
intake:	0.047-0.028 in.	1.2-0.7mm
exhaust:	0.059-0.039 in.	1.5-1.0mm
2.0L engine		
intake:	0.039-0.028 in.	0.039-0.028mm
exhaust:	0.059-0.039 in.	1.5-1.0mm
Valve stem to guide clearance		
1.8L engine		
intake:	0.0012-0.0039 in.	0.03-0.10mm
exhaust:	0.0020-0.0059 in.	0.05-0.15mm
2.0L engine		
intake:	0.0008-0.0040 in.	0.02-0.004mm
exhaust:	0.002-0.006 in.	0.05-0.15mm
Valve Spring		
Free height		
1.8L engine	1.898-1.937 in.	48.2-49.2mm
2.0L engine	1.866-1.902 in.	47.4-48.3
Installed height		
1.8L engine	1.469 in.	37.3mm
2.0L engine	1.57 in.	40mm
Out of squareness		
1.8L engine	2-4 degrees	2-4 degrees
2.0L engine	1.5-4 degrees	1.5-4 degrees

ENGINE MECHANICAL SPECIFICATIONS

Component	US	Metric
Valve guide		
Overall length		
1.8L engine		
intake:	1.730 in.	44.0mm
exhaust:	1.890 in.	48.0mm
2.0L engine		
intake:	1.791 in.	45.5mm
exhaust:	1.988 in.	50.5mm
Inside diameter		
1.8L engine	0.2835-0.2854 in.	7.20-7.25mm
2.0L engine	0.2598-0.2606 in.	6.60-6.62mm
Outside diameter		
1.8L engine	0.5142-0.5146 in.	13.06-13.07mm
2.0L engine	0.4748-0.4752 in.	12.06-12.07mm
Valve seat		
Seat angle		
1.8L engine	43 degrees 30 min.-44 degrees	43 degrees 30 min.-44 degrees
2.0L engine	43 degrees 30 min.-44 degrees	43 degrees 30 min.-44 degrees
Valve contact width		
1.8L engine	0.035-0.051 in.	0.90-1.30mm
2.0L engine	0.035-0.051 in.	0.90-1.30mm
Valve seat sinkage (max.)		
1.8L engine	0.008 in.	0.20mm
2.0L engine	0.008 in.	0.20mm
Rocker arm		
Inside diameter		
1.8L engine	0.7444-0.7453 in.	18.91-18.93mm
Rocker arm to shaft clearance		
1.8L engine	0.0004-0.0040 in.	0.01-0.10mm
Rocker shaft		
Outside diameter		
1.8L engine	0.7437-0.7440 in.	18.89-18.90mm
Overall length		
1.8L engine		
intake:	14.035 in.	365.5mm
exhaust:	13.780 in.	350.0mm
Silent shaft		
Journal diameter		
1.8L engine		
Right front	1.5339-1.5346 in.	38.96-38.98mm
Right rear	1.4154-1.4161 in.	35.95-35.97mm
Left front	0.7272-0.7276 in.	18.47-18.48mm
Left rear	1.4154-1.4161 in.	35.95-35.97mm
2.0L engine		
Right front	1.6520-1.6528 in.	41.96-41.98
Right rear	1.6122-1.6130 in.	40.95-40.97mm
Left front	0.7272-0.7276 in.	18.47-18.48mm
Left rear	1.6122-1.6130 in.	40.95-40.97mm
Oil clearance		
1.8L engine		
Right front	0.0008-0.0024 in.	0.02-0.06mm
Right rear	0.0020-0.0035 in.	0.05-0.09mm
Left front	0.0008-0.0020 in.	0.02-0.05mm
Left rear	0.0020-0.0035 in.	0.05-0.09mm
2.0L engine		
Right front	0.0012-0.0024 in.	0.03-0.06mm
Right rear	0.0020-0.0036 in.	0.05-0.09mm
Left front	0.0008-0.0020 in.	0.02-0.05mm
Left rear	0.0020-0.0036 in.	0.05-0.09mm

ENGINE MECHANICAL SPECIFICATIONS

Component	US	Metric
Piston		
Outside diameter		
1.8L engine	3.1720-3.1732 in.	80.57-80.60mm
2.0L engine		
Non-turbocharged:	3.3453-3.3465 in.	84.07-85.00mm
Turbocharged:	3.3449-3.3461 in.	84.96-84.99mm
Piston to cylinder clearance		
1.8L engine	0.0008-0.0016 in.	0.02-0.04mm
2.0L engine		
Non-turbocharged:	0.0008-0.0016 in.	0.02-0.04mm
Turbocharged:	0.0012-0.0020 in.	0.03-0.05mm
Piston ring		
End gap		
1.8L engine		
No. 1 ring	0.0118-0.0310 in.	0.30-0.80mm
No. 2 ring	0.0079-0.0310 in.	0.20-0.80mm
Oil ring	0.0079-0.0390 in.	0.02-1.0mm
2.0L engine		
No. 1 ring	0.0098-0.0310 in.	0.25-0.80mm
No. 2 ring	0.0177-0.0310 in.	0.45-0.80mm
Oil ring	0.0079-0.0390 in.	0.20-1.00mm
Ring to groove clearance		
1.8L engine		
No. 1 ring	0.0020-0.0035 in.	0.05-0.09mm
No. 2 ring	0.0008-0024 in.	0.02-0.06mm
2.0L engine		
No. 1 ring	0.0012-0.0040 in.	0.03-0.10mm
No. 2 ring	0.0012-0.0040 in.	0.03-0.10mm
Piston pin		
Outside diameter		
1.8L engine	0.7480-0.7484 in.	19.00-19.01mm
2.0L engine	0.8268-0.8272 in.	21.00-21.01mm
Press in load		
1.8L engine	1102-3307 lbs.	5000-15,000 N
2.0L engine	1653-3858 lbs.	7500-17,500 N
Press in temperature		
1.8L engine	Room tempreature	Room tempreature
2.0L engine	Room tempreature	Room tempreature
Connecting rod		
Big end center to small end center length		
1.8L engine	6.047-6.051 in.	153.6-153.7mm
2.0L engine	5.902-5.906 in.	149.9-150.0mm
Bend		
1.8L engine	0.0020 in.	0.05mm
2.0L engine	0.0020 in.	0.05mm
Twist		
1.8L engine	0.004 in.	0.10mm
2.0L engine	0.004 in.	0.10mm
Big end side clearance		
1.8L engine	0.0039-0.0160 in.	0.10-0.40mm
2.0L engine	0.0039-0.0160 in.	0.10-0.40mm

ENGINE MECHANICAL SPECIFICATIONS

Component	US	Metric
Crankshaft		
End play		
1.8L engine	0.0020-0.120 in.	0.05-0.30mm
2.0L engine	0.0020-0.0098 in.	0.05-0.25mm
Journal outside diameter		
1.8L engine	2.24 in.	57.0mm
2.0L engine	2.2433-2.2441 in.	56.98-57.00mm
Pin outside diameter		
1.8L engine	1.77 in.	45.0mm
2.0L engine	1.7709-1.7717 in.	44.98-45.00mm
Out of roundness and taper of journal and pin		
1.8L engine	0.0004 in.	0.01mm
2.0L engine	0.0004 in.	0.01mm
Oil clearance of journal		
1.8L engine	0.0008-0.0040 in.	0.02-0.10mm
2.0L engine	0.0008-0.0040 in.	0.02-0.10mm
Oil clearance of pin		
1.8L engine	0.0008-0.0040 in.	0.01-0.10mm
2.0L engine	0.0008-0.0040 in.	0.01-0.10mm
Cylinder block		
Cylinder inside diameter		
1.8L engine	3.1732-3.1744 in.	80.60-80.63mm
2.0L engine	3.3465-3.3476 in.	85.00-85.03mm
Flatness of gasket surface		
1.8L engine	0.02-0.0040 in.	0.05-0.10mm
2.0L engine	0.0020-0.0040 in.	0.05-0.10mm
Oil pump		
Side cleaance		
1.8L engine		
Drive gear	0.0031-0.0055 in.	0.08-0.14mm
Driven gear	0.0024-0.0047 in.	0.06-0.12mm
2.0L engine		
Drive gear	0.0031-0.0055 in.	0.08-0.140mm
Driven gear	0.0024-0.0047 in.	0.06-0.120mm

TORQUE SPECIFICATIONS

Component	US	Metric
Accelerator cable to intake manifold plenum		
1.8L Engine	4 ft. lbs.	6 Nm
2.0L Engine	4 ft. lbs.	6 Nm
Accelerator cable adjusting bolts		
1.8L Engine	4 ft. lbs.	6 Nm
2.0L Engine	4 ft. lbs.	6 Nm
Camshaft bearing cap bolt		
1.8L Engine		
— —6 x 20:	4 ft. lbs.	6 Nm
— —8 x 65:	15 ft. lbs.	21 Nm
2.0L Engine	15 ft. lbs.	21 Nm
Camshaft sprocket bolt		
1.8L Engine	58–72 ft. lbs.	80–100 Nm
2.0L Engine	58–72 ft. lbs.	80–100 Nm
Center cover mounting bolts		
2.0L Engine	3 ft. lbs.	4 Nm
Connecting rod nuts		
1.8L Engine	25 ft. lbs.	35 Nm
2.0L Engine	38 ft. lbs.	53 Nm
Crankshaft damper pulley retaining bolt(s)		
1.8L Engine	13 ft. lbs.	18 Nm
2.0L Engine	22 ft. lbs.	30 Nm
Crankshaft pulley bolt		
1.8L Engine	13 ft. lbs.	18 Nm
2.0L Engine	22 ft. lbs.	30 Nm
Crankshaft sprocket bolt		
1.8L Engine	80–94 ft. lbs.	110–130 Nm
2.0L Engine	80–94 ft. lbs.	110–130 Nm
Cylinder head bolts		
1.8L Engine		
Step 1:	32 ft. lbs.	42 Nm
Step 2:	54 ft. lbs.	75 Nm
2.0L Engine		
1990-92		
Step 1:	33–41 ft. lbs.	45–55 Nm
Step 2:	65–72 ft. lbs.	90–100 Nm
1993		
Step 1:	14.5 ft. lbs.	20 Nm
Step 2:	additional ¼ turn	additional ¼ turn
Step 3:	additional ¼ turn	additional ¼ turn
EGR valve mounting bolts		
1.8L Engine	11 ft. lbs.	15 Nm
2.0L Engine	16 ft. lbs.	22 Nm
Engine mount insulator nut (large)		
1.8L Engine	43–58 ft. lbs.	60–80 Nm
2.0L Engine	43–58 ft. lbs.	60–80 Nm
Engine mount insulator nut (small)		
1.8L Engine	22–29 ft. lbs.	30–40 Nm
2.0L Engine	22–29 ft. lbs.	30–40 Nm
Exhaust manifold bolts		
1.8L Engine	11–14 ft. lbs.	15–20 Nm
2.0L Engine	18 – 22 ft. lbs.	25–30 Nm
Exhaust pipe clamp bolts		
1.8L Engine	22–29 ft. lbs.	30–40 Nm
2.0L Engine	22–29 ft. lbs.	30–40 Nm
Exhaust pipe-to-manifold nuts		
1.8L Engine with FWD	22–29 ft. lbs.	30–40 Nm
2.0L Engine with FWD	22–29 ft. lbs.	30–40 Nm
2.0L Engine with AWD	29–43 ft. lbs.	40–60 Nm

TORQUE SPECIFICATIONS

Component	US	Metric
Flywheel to crankshaft mounting bolts		
1.8L Engine	94–101 ft. lbs.	130–140 Nm
2.0L Engine	94–101 ft. lbs.	130–140 Nm
Fuel rail mounting bolts		
1.8L Engine	9 ft. lbs.	13 Nm
2.0L Engine	9 ft. lbs.	13 Nm
Front case		
2.0L Engine	16 ft. lbs.	22 Nm
Front roll stopper to insulator nut		
1.8L Engine	47 ft. lbs.	65 Nm
2.0L Engine	36–47 ft lbs.	50–65 Nm
Front roll stopper bracket to centermember		
1.8L Engine	36 ft. lbs.	50 Nm
2.0L Engine	29–36 ft lbs.	40–50 Nm
Rear roll stopper to insulator nut		
1.8L Engine	47 ft. lbs.	65 Nm
2.0L Engine	36 ft. lbs.	50 Nm
Ground cable to intake manifold plenum bolt		
1.8L Engine	9 ft. lbs.	12 Nm
2.0L Engine	4 ft. lbs.	6 Nm
Intake manifold to engine		
1.8L Engine	11–14 ft. lbs.	15–20 Nm
2.0L Engine		
M8 bolt:	14 ft. lbs.	20 Nm
M10 nut and bolt:	30 ft. lbs.	42 Nm
Intake manifold plenum-to-intake manifold bolts		
1.8L Engine	11–14 ft. lbs.	15–20 Nm
Main bearing cap bolts		
1.8L Engine	39–55 ft. lbs.	82–88 Nm
2.0L Engine	51 ft. lbs.	70 Nm

TORQUE SPECIFICATIONS

Component	US	Metric
Oil pan drain plug		
1.8L Engine	33 ft. lbs.	45 Nm
2.0L Engine	33 ft. lbs.	45 Nm
Oil pan bolts		
1.8L Engine	6 ft. lbs.	8 Nm
2.0L Engine	6 ft. lbs.	8 Nm
Oil pan nuts		
1.8L Engine	5 ft. lbs.	7 Nm
2.0L Engine	5 ft. lbs.	7 Nm
Oil pump cover bolts		
2.0L Engine	11 ft. lbs.	15 Nm
Oil pump sprocket bolt		
1.8L Engine	29 ft. lbs.	40 Nm
2.0L Engine	43 ft. lbs.	60 Nm
Heated oxygen sensor		
1.8L Engine	36 ft. lbs.	50 Nm
2.0L Engine	36 ft. lbs.	50 Nm
Rear roll stopper insulator nut	39 ft. lbs.	50 Nm
Rear roll stopper bracket to centermember	36 ft. lbs.	50 Nm
Rocker arm cover bolts		
1.8L Engine	5 ft. lbs.	7 Nm
2.0L Engine	3 ft. lbs.	4 Nm
Silent shaft sprocket bolt		
1.8L Engine	29 ft. lbs.	40 Nm
2.0L Engine	31–35 ft. lbs.	43–49 Nm
Thermostat housing (water outlet fitting) bolts		
1.8L Engine	12–14 ft. lbs.	17–20 Nm
2.0L Engine	12–14 ft. lbs.	17–20 Nm
Throttle body bolts		
1.8L Engine	16 ft. lbs.	22 Nm
2.0L Engine	16 ft. lbs.	22 Nm
Timing belt front cover bolts	9 ft. lbs.	12 Nm
Timing belt rear cover bolts		
1.8L Engine	9 ft. lbs.	12 Nm
2.0L Engine	9 ft. lbs.	12 Nm
Inner timing belt (B) tensioner bolt	16 ft. lbs.	22 Nm
Outer timing belt tensioner bolt	22 ft. lbs.	30 Nm
Timing belt tensioner pulley		
2.0L Engine	31–40 ft. lbs.	43–55 Nm
Water pump installation bolts		
Bolt head mark 4T	9–11 ft. lbs.	12–15 Nm
Bolt head mark 7T	14–20 ft. lbs.	20–27 Nm
Water pump pulley bolts	7 ft. lbs.	10 Nm

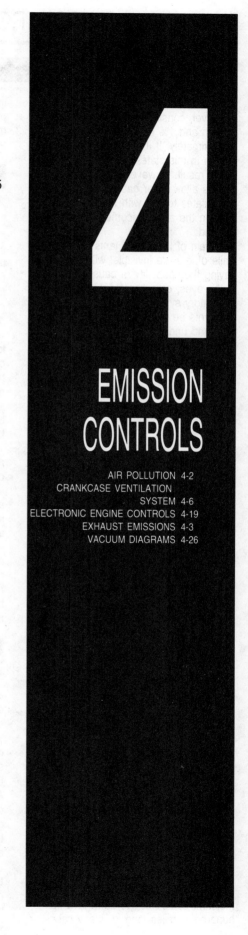

4
EMISSION CONTROLS

AIR POLLUTION

The earth's atmosphere, at or near sea level, consists of 78% nitrogen, 21% oxygen and 1% other gases, approximately. If it were possible to remain in this state, 100% clean air would result. However, many varied causes allow other gases and particulates to mix with the clean air, causing the air to become unclean or polluted.

Certain of these pollutants are visible while others are invisible, with each having the capability of causing distress to the eyes, ears, throat, skin and respiratory system. Should these pollutants be concentrated in a specific area and under the right conditions, death could result due to the displacement or chemical change of the oxygen content in the air. These pollutants can cause much damage to the environment and to the many man made objects that are exposed to the elements.

To better understand the causes of air pollution, the pollutants can be categorized into 3 separate types, natural, industrial and automotive.

Natural Pollutants

Natural pollution has been present on earth before man appeared and is still a factor to be considered when discussing air pollution, although it causes only a small percentage of the present overall pollution problem existing in our country. It is the direct result of decaying organic matter, wind born smoke and particulates from such natural events as plains and forest fires (ignited by heat or lightning), volcanic ash, sand and dust which can spread over a large area of the countryside.

Such a phenomenon of natural pollution has been recent volcanic eruptions, with the resulting plume of smoke, steam and volcanic ash blotting out the sun's rays as it spreads and rises higher into the atmosphere, where the upper air currents catch and carry the smoke and ash, while condensing the steam back into water vapor. As the water vapor, smoke and ash traveled on their journey, the smoke dissipates into the atmosphere while the ash and moisture settle back to earth in a trail hundred of miles long. In many cases,

lives are lost and millions of dollars of property damage result, and ironically, man can only stand by and watch it happen.

Industrial Pollution

Industrial pollution is caused primarily by industrial processes, the burning of coal, oil and natural gas, which in turn produces smoke and fumes. Because the burning fuels contain much sulfur, the principal ingredients of smoke and fumes are sulfur dioxide and particulate matter. This type of pollutant occurs most severely during still, damp and cool weather, such as at night. Even in its less severe form, this pollutant is not confined to just cities. Because of air movements, the pollutants move for miles over the surrounding countryside, leaving in its path a barren and unhealthy environment for all living things.

Working with Federal, State and Local mandated rules, regulations and by carefully monitoring the emissions, industries have greatly reduced the amount of pollutant emitted from their industrial sources, striving to obtain an acceptable level. Because of the mandated industrial emission clean up, many land areas and streams in and around the cities that were formerly barren of vegetation and life, have now begun to move back in the direction of nature's intended balance.

Automotive Pollutants

The third major source of air pollution is the automotive emissions. The emissions from the internal combustion engine were not an appreciable problem years ago because of the small number of registered vehicles and the nation's small highway system. However, during the early 1950's, the trend of the American people was to move from the cities to the surrounding suburbs. This caused an immediate problem in the transportation areas because the majority of the suburbs were not afforded mass transit conveniences. This lack of transportation created an attractive market for the automobile manufacturers, which resulted in a dramatic increase in the number of vehicles produced and

sold, along with a marked increase in highway construction between cities and the suburbs. Multi-vehicle families emerged with much emphasis placed on the individual vehicle per family member. As the increase in vehicle ownership and usage occurred, so did the pollutant levels in and around the cities, as the suburbanites drove daily to their businesses and employment in the city and its fringe area, returning at the end of the day to their homes in the suburbs.

It was noted that a fog and smoke type haze was being formed and at times, remained in suspension over the cities and did not quickly dissipate. At first this "smog", derived from the words "smoke" and "fog", was thought to result from industrial pollution but it was determined that the automobile emissions were largely to blame. It was discovered that as normal automobile emissions were exposed to sunlight for a period of time, complex chemical reactions would take place.

It was found the smog was a photo chemical layer and was developed when certain oxides of nitrogen (NOx) and unburned hydrocarbons (HC) from the automobile emissions were exposed to sunlight and was more severe when the smog would remain stagnant over an area in which a warm layer of air would settle over the top of a cooler air mass at ground level, trapping and holding the automobile emissions, instead of the emissions being dispersed and diluted through normal air flows. This type of air stagnation was given the name "Temperature Inversion".

Temperature Inversion

In normal weather situations, the surface air is warmed by the heat radiating from the earth's surface and the sun's rays and will rise upward, into the atmosphere, to be cooled through a convection type heat expands with the cooler upper air. As the warm air rises, the surface pollutants are carried upward and dissipated into the atmosphere.

When a temperature inversion occurs, we find the higher air is no longer cooler but warmer than the surface air, causing the cooler surface air to become trapped and unable to move. This warm air blanket can extend from above ground level to a few hundred or even a few thousand feet into the air. As the surface air is trapped, so are the pollutants, causing a severe smog condition. Should this stagnant air mass extend to a few thousand feet high, enough air movement with the inversion takes place to allow the smog layer to rise above ground level but the pollutants still cannot dissipate. This inversion can remain for days over an area, with only the smog level rising or lowering from ground level to a few hundred feet high. Meanwhile, the pollutant levels increases, causing eye irritation, respiratory problems, reduced visibility, plant damage and in some cases, cancer type diseases.

This inversion phenomenon was first noted in the Los Angeles, California area. The city lies in a basin type of terrain and during certain weather conditions, a cold air mass is held in the basin while a warmer air mass covers it like a lid.

Because this type of condition was first documented as prevalent in the Los Angeles area, this type of smog was named Los Angeles Smog, although it occurs in other areas where a large concentration of automobiles are used and the air remains stagnant for any length of time.

Internal Combustion Engine Pollutants

Consider the internal combustion engine as a machine in which raw materials must be placed so a finished product comes out. As in any machine operation, a certain amount of wasted material is formed. When we relate this to the internal combustion engine, we find that by putting in air and fuel, we obtain power from this mixture during the combustion process to drive the vehicle. The by-product or waste of this power is, in part, heat and exhaust gases with which we must concern ourselves.

HEAT TRANSFER

The heat from the combustion process can rise to over 4000°F (2204°C). The dissipation of this heat is controlled by a ram air effect, the use of cooling fans to cause air flow and having a liquid coolant solution surrounding the combustion area and transferring the heat of combustion through the cylinder walls and into the coolant. The coolant is then directed to a thin-finned, multi-tubed radiator, from which the excess heat is transferred to the outside air by

1 or all of the 3 heat transfer methods, conduction, convection or radiation.

The cooling of the combustion area is an important part in the control of exhaust emissions. To understand the behavior of the combustion and transfer of its heat, consider the air/fuel charge. It is ignited and the flame front burns progressively across the combustion chamber until the burning charge reaches the cylinder walls. Some of the fuel in contact with the walls is not hot enough to burn, thereby snuffing out or Quenching the combustion process. This leaves unburned fuel in the combustion chamber. This unburned fuel is then forced out of the cylinder along with the exhaust gases and into the exhaust system.

Many attempts have been made to minimize the amount of unburned fuel in the combustion chambers due to the snuffing out or "Quenching", by increasing the coolant temperature and lessening the contact area of the coolant around the combustion area. Design limitations within the combustion chambers prevent the complete burning of the air/fuel charge, so a certain amount of the unburned fuel is still expelled into the exhaust system, regardless of modifications to the engine.

EXHAUST EMISSIONS

Composition Of The Exhaust Gases

The exhaust gases emitted into the atmosphere are a combination of burned and unburned fuel. To understand the exhaust emission and its composition review some basic chemistry.

When the air/fuel mixture is introduced into the engine, we are mixing air, composed of nitrogen (78%), oxygen (21%) and other gases (1%) with the fuel, which is 100% hydrocarbons, in a semi-controlled ratio. As the combustion process is accomplished, power is produced to move the vehicle while the heat of combustion is transferred to the cooling system. The exhaust gases are then composed of nitrogen, the same as was introduced in the engine, carbon dioxide, the same gas that is used in beverage carbonation and water vapor. The nitrogen, for the most part passes

through the engine unchanged, while the oxygen reacts (burns) with the hydrocarbons and produces the carbon dioxide and the water vapors. If this chemical process would be the only process to take place, the exhaust emissions would be harmless. However, during the combustion process, other pollutants are formed and are considered dangerous. These pollutants are carbon monoxide, hydrocarbons, oxides of nitrogen oxides of sulfur and engine particulates.

Lead is considered 1 of the particulates and is present in the exhaust gases whenever leaded fuels are used. Lead does not dissipate easily. Levels can be high along roadways when it is emitted from vehicles and can pose a health threat. Since the increased usage of unleaded gasoline and the phasing out of leaded gasoline for fuel, this pollutant is gradually diminishing. While not considered a

major threat lead is still considered a dangerous pollutant.

HYDROCARBONS

Hydrocarbons are essentially unburned fuel that have not been successfully burned during the combustion process or have escaped into the atmosphere through fuel evaporation. The main sources of incomplete combustion are rich air/fuel mixtures, low engine temperatures and improper spark timing. The main sources of hydrocarbon emission through fuel evaporation come from the vehicle's fuel tank.

To reduce combustion hydrocarbon emission, engine modifications were made to minimize dead space and surface area in the combustion chamber. In addition the air/fuel mixture was made more lean through improved fuel injection and by the addition of external controls to aid in further combustion of the hydrocarbons outside the engine. One such method was the installation of a catalytic converter, a unit that is able to burn traces of hydrocarbons without affecting the internal combustion process or fuel economy.

To control hydrocarbon emissions through fuel evaporation, modifications were made to the fuel tank to allow storage of the fuel vapors during periods of engine shut-down, and at specific times during engine operation, to purge and burn these same vapors by blending them with the air/fuel mixture.

CARBON MONOXIDE

Carbon monoxide is formed when not enough oxygen is present during the combustion process to convert carbon to carbon dioxide. An increase in the carbon monoxide emission is normally accompanied by an increase in the hydrocarbon emission because of the lack of oxygen to completely burn all of the fuel mixture.

Carbon monoxide also increases the rate at which the photo chemical smog is formed by speeding up the conversion of nitric oxide to nitrogen dioxide. To accomplish this, carbon monoxide combines with oxygen and nitrogen dioxide to produce carbon dioxide and nitrogen dioxide.

The dangers of carbon monoxide, which is an odorless, colorless toxic gas are many. When carbon monoxide is inhaled into the lungs and passed into the blood stream, oxygen is replaced by the carbon monoxide in the red blood cells, causing a reduction in the amount of oxygen being supplied to the many parts of the body. This lack of oxygen causes headaches, lack of coordination, reduced mental alertness and should the carbon monoxide concentration be high enough, death could result.

NITROGEN

Normally, nitrogen is an inert gas. When heated to approximately 2500°F (1371°C) through the combustion process, this gas becomes active and causes an increase in the nitric oxide (NOx) emission.

Oxides of nitrogen are composed of approximately 97-98% nitric oxide. Nitric

oxide is a colorless gas but when it is passed into the atmosphere, it combines with oxygen and forms nitrogen dioxide. The nitrogen dioxide then combines with chemically active hydrocarbons and when in the presence of sunlight, causes the formation of photo chemical smog.

OZONE

To further complicate matters, some of the nitrogen dioxide is broken apart by the sunlight to form nitric oxide and oxygen. This single atom of oxygen then combines with diatomic (meaning 2 atoms) oxygen to form ozone. Ozone is 1 of the smells associated with smog. It has a pungent and offensive odor, irritates the eyes and lung tissues, affects the growth of plant life and causes rapid deterioration of rubber products. Ozone can be formed by sunlight as well as electrical discharge into the air.

The most common discharge area on the automobile engine is the secondary ignition electrical system, especially when inferior quality spark plug cables are used. As the surge of high voltage is routed through the secondary cable, the circuit builds up an electrical field around the wire, acting upon the oxygen in the surrounding air to form the ozone. The faint glow along the cable with the engine running that may be visible on a dark night, is called the "corona discharge". It is the result of the electrical field passing from a high along the cable, to a low in the surrounding air, which forms the ozone gas. The combination of corona and ozone has been a major cause of cable deterioration. Recently, different types and better quality insulating materials have lengthened the life of the electrical cables.

Although ozone at ground level can be harmful, ozone is beneficial to the earth's inhabitants. By having a concentrated ozone layer called the 'ozonosphere", between 10 and 20 miles (16-32km) up in the atmosphere much of the ultra violet radiation from the sun's rays are absorbed and screened. If this ozone layer were not present, much of the earth's surface would be burned, dried and unfit for human life.

There is much discussion concerning the ozone layer and its density. A feeling exists that this protective layer of ozone

is slowly diminishing and corrective action must be directed to this problem. Much experimenting is presently being conducted to determine if a problem exists and if so, the short and long term effects of the problem and how it can be remedied.

OXIDES OF SULFUR

Oxides of sulfur were initially ignored in the exhaust system emissions, since the sulfur content of gasoline as a fuel is less than $\frac{1}{2}$ of 1%. Because of this small amount, it was felt that it contributed very little to the overall pollution problem. However, because of the difficulty in solving the sulfur emissions in industrial pollutions and the introduction of catalytic converter to the automobile exhaust systems, a change was mandated. The automobile exhaust system, when equipped with a catalytic converter, changes the sulfur dioxide into the sulfur trioxide.

When this combines with water vapors, a sulfuric acid mist is formed and is a very difficult pollutant to handle and is extremely corrosive. This sulfuric acid mist that is formed, is the same mist that rises from the vents of an automobile storage battery when an active chemical reaction takes place within the battery cells.

When a large concentration of vehicles equipped with catalytic converters are operating in an area, this acid mist will rise and be distributed over a large ground area causing land, plant, crop, paints and building damage.

PARTICULATE MATTER

A certain amount of particulate matter is present in the burning of any fuel, with carbon constituting the largest percentage of the particulates. In gasoline, the remaining percentage of particulates is the burned remains of the various other compounds used in its manufacture. When a gasoline engine is in good internal condition, the particulate emissions are low but as the engine wears internally, the particulate emissions increase. By visually inspecting the tail pipe emissions, a determination can be made as to where an engine defect may exist. An engine with light gray smoke emitting from the tail pipe normally indicates an increase in the oil consumption through burning due to internal engine wear. Black smoke would indicate a defective fuel delivery system, causing the engine to operate in a rich mode. Regardless of the color of the smoke, the internal part of the engine or the fuel delivery system should be repaired to a "like new" condition to prevent excess particulate emissions.

Diesel and turbine engines emit a darkened plume of smoke from the exhaust system because of the type of fuel used. Emission control regulations are mandated for this type of emission and more stringent measures are being used to prevent excess emission of the particulate matter. Electronic components are being introduced to control the injection of the fuel at precisely the proper time of piston travel, to achieve the optimum in fuel ignition and fuel usage. Other particulate after-burning components are being tested to achieve a cleaner particular emission.

Good grades of engine lubricating oils should be used, meeting the manufacturers specification. "Cut-rate" oils can contribute to the particulate emission problem because of their low "flash" or ignition temperature point. Such oils burn prematurely during the combustion process causing emissions of particulate matter.

The cooling system is an important factor in the reduction of particulate matter. With the cooling system operating at a temperature specified by the manufacturer, the optimum of combustion will occur. The cooling system must be maintained in the same manner as the engine oiling system, as each system is required to perform properly in order for the engine to operate efficiently for a long time.

Other Automobile Emission Sources

Before emission controls were mandated on the internal combustion engines, other sources of engine pollutants were discovered, along with the exhaust emission. It was determined the engine combustion exhaust produced 60% of the total emission pollutants, fuel evaporation from the fuel tank and carburetor vents produced 20%, with the another 20% being produced through the

crankcase as a by-product of the combustion process.

CRANKCASE EMISSIONS

Crankcase emissions are made up of water, acids, unburned fuel, oil fumes and particulates. The emissions are classified as hydrocarbons and are formed by the small amount of unburned, compressed air/fuel mixture entering the crankcase from the combustion area during the compression and power strokes, between the cylinder walls and piston rings. The head of the compression and combustion help to form the remaining crankcase emissions.

Since the first engines, crankcase emissions were allowed to go into the air through a road draft tube, mounted on the lower side of the engine block. Fresh air came in through an open oil filler cap or breather. The air passed through the crankcase mixing with blow-by gases. The motion of the vehicle and the air blowing past the open end of the road draft tube caused a low pressure area at the end of the tube. Crankcase emissions were simply drawn out of the road draft tube into the air.

To control the crankcase emission, the road draft tube was deleted. A hose and/or tubing was routed from the crankcase to the intake manifold so the blow-by emission could be burned with the air/fuel mixture. However, it was found that intake manifold vacuum, used to draw the crankcase emissions into the manifold, would vary in strength at the wrong time and not allow the proper emission flow. A regulating type valve was needed to control the flow of air through the crankcase.

Testing, showed the removal of the blow-by gases from the crankcase as quickly as possible, was most important to the longevity of the engine. Should large accumulations of blow-by gases remain and condense, dilution of the engine oil would occur to form water, soots, resins, acids and lead salts, resulting in the formation of sludge and varnishes. This condensation of the blow-by gases occur more frequently on vehicles used in numerous starting and stopping conditions, excessive idling and when the engine is not allowed to attain normal operating temperature through short runs. The crankcase purge control

or PCV system will be described in detail later in this section.

FUEL EVAPORATIVE EMISSIONS

Gasoline fuel is a major source of pollution, before and after it is burned in the automobile engine. From the time the fuel is refined, stored, pumped and transported, again stored until it is pumped into the fuel tank of the vehicle, the gasoline gives off unburned hydrocarbons into the atmosphere. Through redesigning of the storage areas and venting systems, the pollution factor has been diminished but not eliminated, from the refinery standpoint. However, the automobile still remained the primary source of vaporized, unburned hydrocarbon emissions.

Fuel pumped form an underground storage tank is cool but when exposed to a Warner ambient temperature, will expand. Before controls were mandated, an owner would fill the fuel tank with fuel from an underground storage tank and park the vehicle for some time in warm area, such as a parking lot. As the fuel would warm, it would expand and should no provisions or area be provided for the expansion, the fuel would spill out the filler neck and onto the ground, causing hydrocarbon pollution and creating a severe fire hazard. To correct this condition, the vehicle manufacturers added overflow plumbing and/or gasoline tanks with built in expansion areas or domes.

However, this did not control the fuel vapor emission from the fuel tank and the carburetor bowl. It was determined that most of the fuel evaporation occurred when the vehicle was stationary and the engine not operating. Most vehicles carry 5-25 gallons (19-95 liters) of gasoline. Should a large concentration of vehicles be parked in one area, such as a large parking lot, excessive fuel vapor emissions would take place, increasing as the temperature increases.

To prevent the vapor emission from escaping into the atmosphere, the fuel system is designed to trap the fuel vapors while the vehicle is stationary, by sealing the fuel system from the atmosphere. A storage system is used to collect and hold the fuel vapors from the carburetor and the fuel tank when the

EMISSION CONTROL SYSTEM—TROUBLESHOOTING GUIDE

Symptom	Probable cause	Remedy
Engine will not start or hard to start	Vacuum hose disconnected or damaged The EGR valve is not closed Malfunction of the evaporative emission purge solenoid	Repair or replace
Rough idle or engine stalls	The EGR valve is not closed. Vacuum hose disconnected or damaged	Repair or replace
	Malfunction of the positive crankcase ventilation valve	Replace
	Malfunction of the purge control system	Check the system; if there is a problem, check its component parts
Engine hesitates or poor acceleration	Malfunction of the exhaust gas recirculation system	Check the system; if there is a problem, check its component parts
Excessive oil consumption	Positive crankcase ventilation line clogged	Check positive crankcase ventilation system
Poor fuel mileage	Malfunction of the exhaust gas recirculation system	Check the system; if there is a problem, check its component parts

engine is not operating. When the engine is started, the storage system is then purged of the fuel vapors, which are drawn into the engine and burned with the air/fuel mixture.

The components of the fuel evaporative system will be described in detail later in this section.

The majority of emission and fuel related items are warranted by the manufacturer for an extended mileage and time. Before replacing any fuel or emission related components, check with the manufacturer for component replacement under warranty.

CRANKCASE VENTILATION SYSTEM

Positive Crankcase Ventilation (PCV) System

OPERATION

All engines are equipped with the Positive Crankcase Ventilation (PCV) system. The PCV system vents crankcase gases into the engine air intake where they are burned with the fuel and air mixture. The PCV system keeps pollutants from being released into the atmosphere, and also helps to keep the engine oil clean, by ridding the crankcase of moisture and corrosive fumes. The PCV system consists of the rocker arm cover mounted PCV valve, the nipple in the air intake and the connecting hoses.

The PCV valve regulates the amount of ventilating air and blow-by gas to the intake manifold. It also prevents backfire from traveling into the crankcase, avoiding the explosion of crankcase gases.

PCV Valve

TESTING

◗ See Figures 1, 2, 3 and 4

1. Disconnect the ventilation hose from the PCV valve. Remove the PCV valve from the engine. Once removed, reconnect the ventilation hose to the valve.
2. Start the engine and allow to idle. Place a finger over open end of the PCV valve. Make sure intake manifold vacuum is felt on finger.
3. If vacuum is not felt, the PCV valve may be restricted.
4. Turn the engine **OFF** and remove the PCV valve from the hose.

5. Insert a thin stick into the threaded end of the PCV valve. Push on the inner plunger and inspect for movement.
6. If plunger inside the PCV valve is not free to move back and forth, the valve is clogged and will require replacement.

➡**It is possible to clean the valve using the appropriate solvent, but replacement is recommended.**

REMOVAL & INSTALLATION

1. Disconnect ventilation hose from the PCV valve. Unthread the PCV valve from the cylinder head cover.
2. Inspect hose for restriction, cracks or blockage and replace as required.
   To install:
3. Install new valve into the opening in the valve cover. Tighten the PCV valve to 7 ft. lbs. (10 Nm).

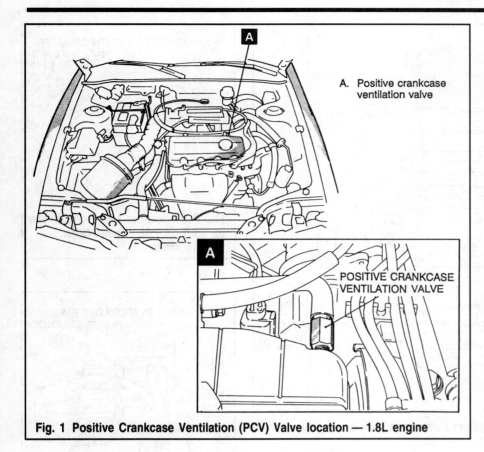

A. Positive crankcase ventilation valve

POSITIVE CRANKCASE VENTILATION VALVE

Fig. 1 Positive Crankcase Ventilation (PCV) Valve location — 1.8L engine

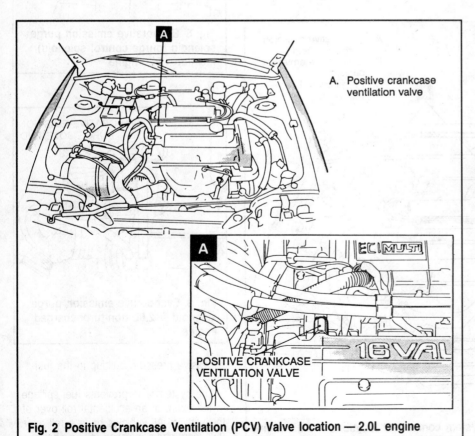

A. Positive crankcase ventilation valve

POSITIVE CRANKCASE VENTILATION VALVE

Fig. 2 Positive Crankcase Ventilation (PCV) Valve location — 2.0L engine

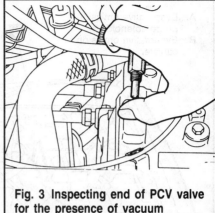

Fig. 3 Inspecting end of PCV valve for the presence of vacuum

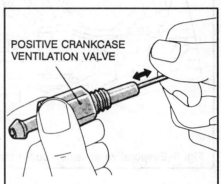

POSITIVE CRANKCASE VENTILATION VALVE

Fig. 4 Inspecting the PCV valve for inner plunger movement. If bound or sticking, replace the PCV valve.

4. Install ventilation hose to the valve. Inspect the fit of the ventilation hose over the PCV valve. If the hose fits loosely on valve stem, replace the ventilation hose.

Fuel Evaporative Emission Control System

OPERATION

▶ **See Figures 5 and 6**

The function of this control system is to prevent the emissions of gasoline vapors from the fuel tank into the atmosphere. When fuel evaporates in the fuel tank, the vapors pass through vent hoses or tubes to a charcoal canister. There they are temporarily held until they can be drawn into the intake manifold when the engine is running and burned the combustion process during engine operation. This action prevents

A. Evaporative emission purge solenoid
B. Evaporative emission canister

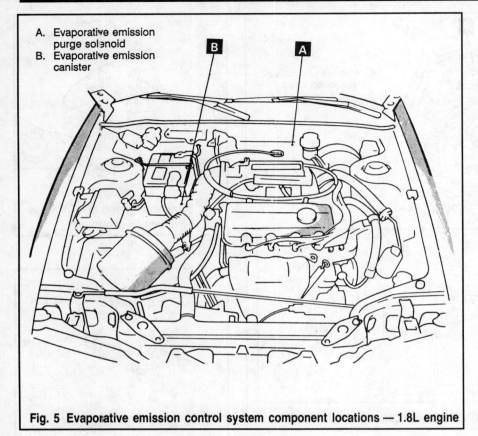

Fig. 5 Evaporative emission control system component locations — 1.8L engine

A. Evaporative emission purge solenoid
B. Evaporative emission canister
C. Purge control valve (turbo)

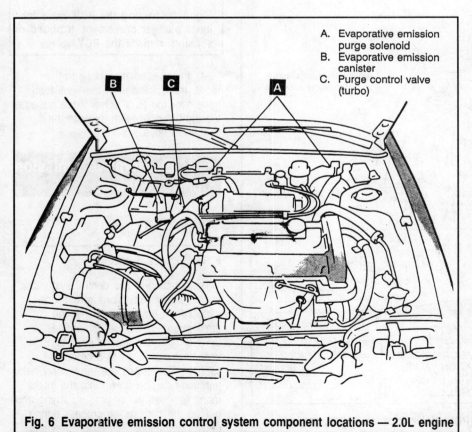

Fig. 6 Evaporative emission control system component locations — 2.0L engine

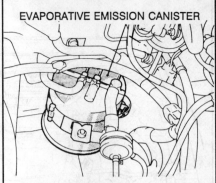

Fig. 7 Evaporative emission canister as seen on 1.8L engine. Other models similar.

Fig. 8 Evaporative emission purge solenoid (purge control solenoid) location-1.8L engine

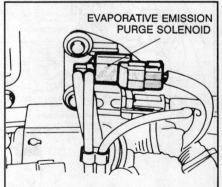

Fig. 9 Evaporative emission purge solenoid — 2.0L non-turbocharged engine

excessive pressure buildup in the fuel tank.

The system also prevents fuel spillage in the event of an accidental roll over of the vehicle. All vehicles have a roll over (two-way) valve installed in-line above

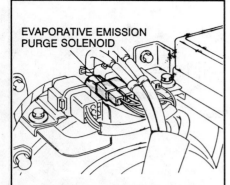

Fig. 10 Evaporative emission purge solenoid — 2.0L turbocharged engine

the tank to release fuel tank pressure to the canister and to prevent fuel from leaking in the event of an accidental vehicle roll over.

Charcoal Canister
▶ See Figure 7

A sealed, maintenance free charcoal canister is used on all vehicles. The fuel tank's vents lead to the canister. Fuel vapors are temporarily held in the canister's activated charcoal until they can be drawn into the intake manifold and burned in the combustion chamber. There is no scheduled maintenance interval on the charcoal canister.

Purge Control System
▶ See Figures 8, 9, 10, 11 and 12

The canister is connected to the engine via a purge control solenoid; a purge control valve is added to the system on turbocharged engines. The purge control solenoid is located as follows:

- Eclipse with 1.8L engine — on the firewall just to the right of the brake fluid reservoir.
- Eclipse with 2.0L non-turbocharged engine — on the firewall, slightly right of center.
- Eclipse with 2.0L turbocharged engine — the innermost solenoid in the solenoid cluster at the left rear corner of the engine compartment.
- Laser and Talon with 1.8L engine — on the firewall just to the right of the brake fluid reservoir.
- Laser and Talon with 2.0L non-turbocharged engine — on the firewall, slightly right of center.

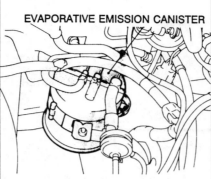

Fig. 11 Evaporative emission canister — 2.0L non-turbocharged engine

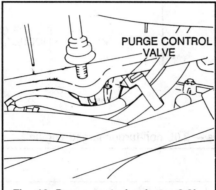

Fig. 12 Purge control valve — 2.0L turbocharged engine

- Laser and Talon with 2.0L turbocharged engine — the innermost solenoid in the solenoid cluster at the left rear corner of the engine compartment.

FUEL EVAPORATIVE EMISSION CONTROL SYSTEM TESTING

Non-Turbocharged Engines
▶ See Figure 13

1. Disconnect red striped vacuum hose from the throttle body and connect it to a hand held vacuum pump.
2. Plug the open nipple on the throttle body.
3. Using the hand pump, apply vacuum while the engine is idling. Check that vacuum is maintained or released as outlined below:
   a. With engine coolant temperature of 140°F (60°C) or less — 14.8 in. Hg of vacuum is maintained.

b. With engine coolant temperature of 158°F (70°C) or higher — 14.8 in. Hg of vacuum is maintained.
4. With the engine coolant temperature of 158°F (70°C) or higher, run the engine at 3000 rpm within 3 minutes of starting vehicle. Try to apply vacuum using the hand held pump. Vacuum should leak.
5. With the engine coolant temperature of 158°F (70°C) or higher, run the engine at 3000 rpm after 3 minutes have elapsed after starting vehicle. Apply 14.8 in. Hg of vacuum. The vacuum should be maintained momentarily, after which it should leak.

➡ **The vacuum will leak continuously if the altitude is 7,200 ft. or higher, or the intake air temperature is 122°F (50°C) of higher.**

6. If the test results differ from the desired results, the purge control system is not operating properly.

1990-92 Turbocharged Engine
▶ See Figure 14

1. Disconnect the purge air hose from the intake hose and plug the air intake hose.
2. Connect a hand vacuum pump to the purge air hose.
3. Under various engine conditions, inspect the system operation:
   a. Allow the engine to cool to a temperature of 140°F (60°C) or below.
   b. Start the engine and run at idle.
   c. Using the hand pump, apply 14.8 in. Hg of vacuum. In this condition, the vacuum should be maintained.
   d. Raise the engine speed to 3000 rpm.
   e. Using the hand pump, apply 14.8 in. Hg of vacuum. In this condition, the vacuum should be maintained.
4. Run the engine until the coolant temperature reaches 158°F (70°C). Inspect system operations as follows:
   a. Using the hand pump, apply 14.8 in. Hg of vacuum with the engine at idle. In this condition, vacuum should be maintained.
   b. Increase the engine speed to 3000 rpm within 3 minutes of starting the engine. Try applying vacuum. The vacuum should leak.
   c. After 3 minutes have elapsed after starting engine, raise the engine speed to 3000 rpm. Apply 14.8 in. Hg of vacuum. Vacuum should be

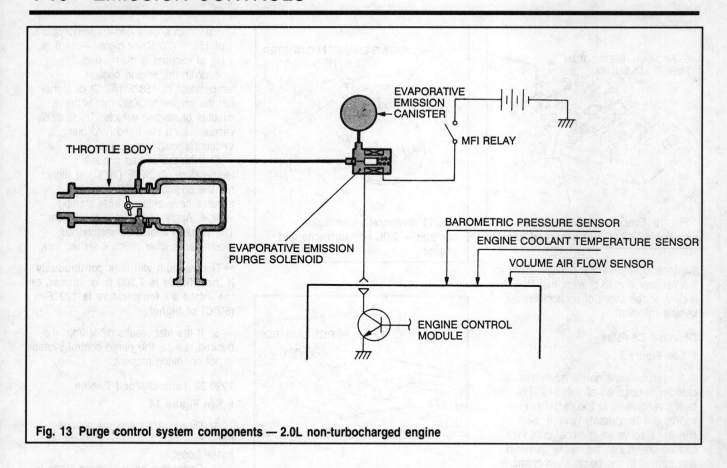

Fig. 13 Purge control system components — 2.0L non-turbocharged engine

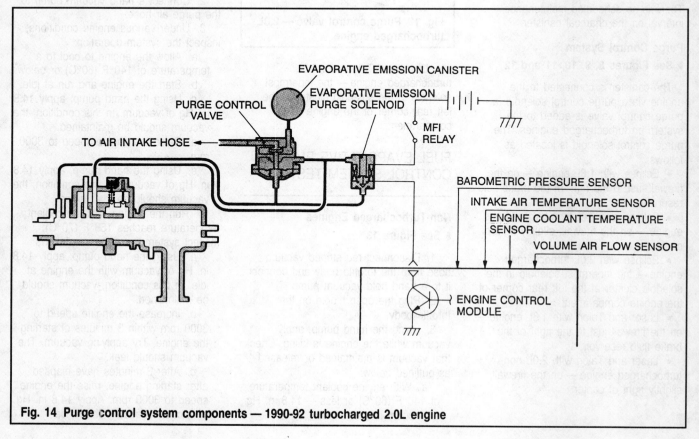

Fig. 14 Purge control system components — 1990-92 turbocharged 2.0L engine

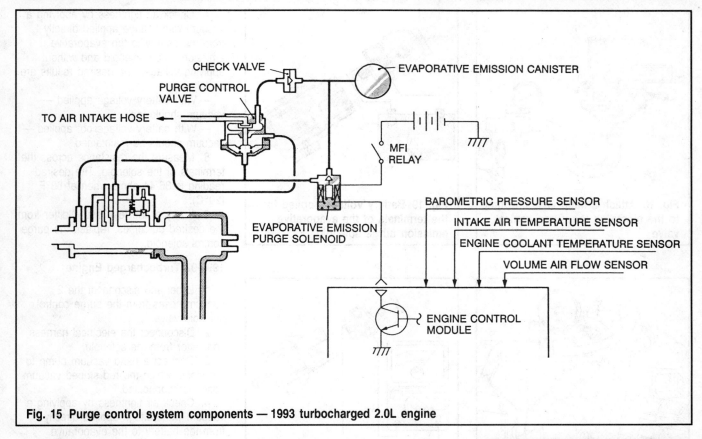

Fig. 15 Purge control system components — 1993 turbocharged 2.0L engine

maintained momentarily, after which it will leak.

➡ **The vacuum will leak continuously if the altitude is 7200 ft. or higher or the air temperature is 122°F (50°C) or higher.**

5. If the results of either test differs from specifications, the system is not functioning properly and will require further diagnosis.

1993 Turbocharged Engine
▶ **See Figure 15**

1. Disconnect red striped vacuum hose from the throttle body and connect it to a hand held vacuum pump.
2. Plug nipple from which the vacuum hose was disconnected.
3. Allow the engine coolant to cool below 104°F (60°C) or below. Check system operation as follows:
   a. Start the engine and run at idle speed. Apply 14.8 in. Hg of vacuum. The vacuum should be maintained.
   b. Run the engine at 3000 rpm. Apply 14.8 in. Hg of vacuum. The vacuum should be maintained.
4. Run the engine until the coolant temperature reaches 158°F (70°C) of

above. Inspect system operation as follows:
   a. With the engine at idle, apply 14.8 in. Hg of vacuum using the hand pump. The vacuum should be maintained.
   b. Run the engine at 3000 rpm within 3 minutes after starting the engine, and try applying vacuum. The vacuum should leak.
   c. Run the engine at 3000 rpm after 3 minutes have elapsed after starting the vehicle and apply 14.8 in Hg of vacuum. Vacuum will be maintained momentarily, then it will leak.

➡ **The vacuum will leak continuously if the altitude is 7200 ft. or higher or the air temperature is 122°F (50°C) or higher.**

5. If any of the test results differ from the specifications, there is a fault in the operation of the system and further diagnosis is required.

Purge Control Valve

TESTING

Turbocharged 2.0L Engine
▶ **See Figure 16**

1. The purge control valve is located to the right side of the battery. Remove the purge control valve from the engine compartment.
2. Connect a hand vacuum pump to the vacuum nipple of the purge control valve.
3. Apply 15.7 in. Hg of vacuum and check air tightness. Blow in air lightly from the evaporative emission canister side nipple and check conditions as follows:
   — No vacuum applied to the valve — air will not pass.
   — 8.0 in. Hg of vacuum applied to the valve — air will pass through.
4. Connect a hand vacuum pump to the positive pressure nipple of the purge control valve.
5. Apply a vacuum of 15.7 in. Hg and check for air tightness. The valve should be air tight.

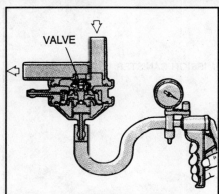

Fig. 16 Attach a hand vacuum pump to the nipple on the purge control valve

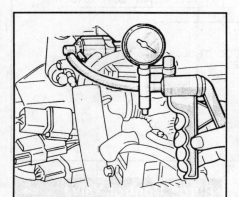

Fig. 17 Apply a vacuum to the evaporative emission purge solenoid and check for air-tightness with and with out voltage — 1.8L engine

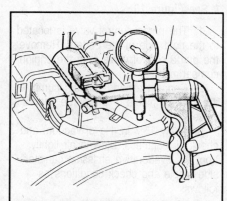

Fig. 18 Apply a vacuum to the evaporative emission purge solenoid and check for air-tightness — 2.0L engine

6. If the results differ from the desired outcomes, replace the purge control valve.

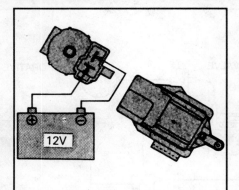

Fig. 19 Battery voltage applied to the terminals of the evaporative emission purge solenoid

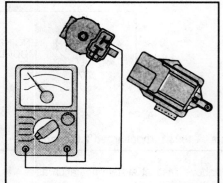

Fig. 20 Measuring the resistance between the terminals of the evaporative emission purge solenoid

Evaporative Emission Purge Solenoid

The evaporative emission purge solenoid is an ON/OFF type switch which controls the induction of purge air from the canister into the intake manifold plenum.

TESTING

1990-93 Non-Turbocharged and 1993 Turbocharged Engines

▶ **See Figures 17, 18, 19 and 20**

1. Label and disconnect the 2 vacuum hoses from the purge control solenoid valve.
2. Disconnect the electrical harness connector from the solenoid.
3. Connect a hand vacuum pump to the nipple which the red striped vacuum hose was connected.

4. Check air tightness by applying a vacuum with voltage applied directly from the battery to the evaporative emission purge solenoid and without applying voltage. The desired results are as follows:
   — With battery voltage applied — vacuum should leak
   — With battery voltage not applied — vacuum should be maintained
5. Measure the resistance across the terminals of the solenoid. The desired reading is 36-44 ohms when at 68°F (20°C).
6. If any of the test results differ from the desired outcomes, replace the purge control solenoid.

1990-92 Turbocharged Engine

1. Label and disconnect the 2 vacuum hoses from the purge control solenoid valve.
2. Disconnect the electrical harness connector from the solenoid.
3. Connect a hand vacuum pump to the nipple which the red striped vacuum hose was connected.
4. Check air tightness by applying a vacuum with voltage applied directly from the battery to the evaporative emission purge solenoid and without applying voltage. The desired results are as follows:
   — With battery voltage applied — vacuum should be maintained
   — With battery voltage not applied — vacuum should leak
5. Measure the resistance across the terminals of the solenoid. The desired reading is 36-44 ohms when at 68°F (20°C).
6. If any of the test results differ from the specifications, replace the emission purge control solenoid.

REMOVAL & INSTALLATION

1. Disconnect the negative battery cable.
2. Label and remove the vacuum and electrical harness connections from the purge control solenoid.
3. Remove the solenoid and mounting bracket from the engine compartment.
4. Installation is the reverse of the removal procedure.

A. EGR valve
B. EGR temperature sensor
C. EGR solenoid

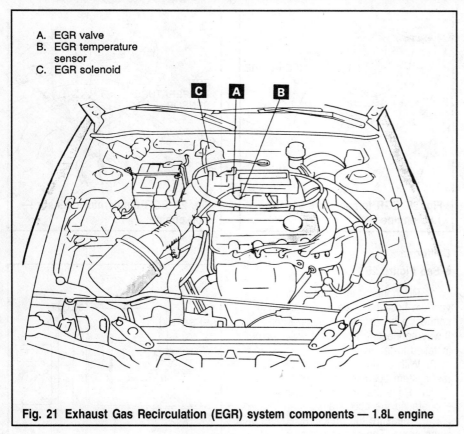

Fig. 21 Exhaust Gas Recirculation (EGR) system components — 1.8L engine

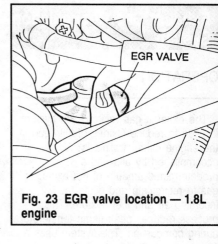

Fig. 23 EGR valve location — 1.8L engine

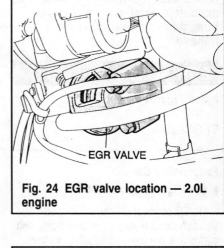

Fig. 24 EGR valve location — 2.0L engine

A. Thermal vacuum valve (Federal)
B. EGR valve
C. EGR temperature sensor (California)
D. EGR solenoid (California)

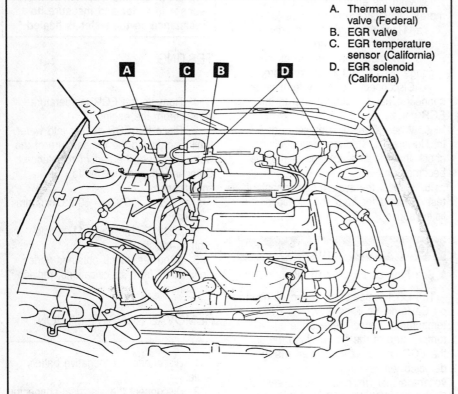

Fig. 22 Exhaust Gas Recirculation (EGR) system components — 2.0L engine

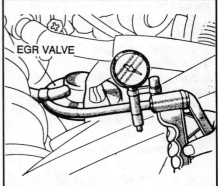

Fig. 25 Applying vacuum to the EGR valve to inspect system operation-California vehicle

Exhaust Gas Recirculation System

OPERATION

The exhaust gas recirculation system is used to reduce Oxides of Nitrogen in the engine exhaust. This is accomplished by allowing a predetermined amount of hot exhaust gas to recirculate and dilute the incoming air and fuel mixture. This process reduces peak flame temperature during combustion. The system uses a vacuum-controlled Exhaust Gas Recirculation (EGR) valve, in order to modulate exhaust gas flow from the exhaust manifold into the intake manifold.

EGR System

▶ **See Figures 21 and 22**

TESTING

Federal and Canada
▶ **See Figures 23 and 24**

1. Disconnect the green striped vacuum hose from the throttle body and connect a hand vacuum pump to the vacuum hose.
2. Plug the nipple from which the vacuum hose was disconnected.
3. Under the engine conditions listed below, inspect the system operation by applying vacuum from a hand held vacuum pump.
4. With the engine temperature cold, 104°F (40°C) or below, the response should be as follows:
   a. Engine at idle — vacuum should leak
5. With the engine at temperature of 176°F (80°C) or higher, the response should be as follows:
   a. With 1.8 in. Hg vacuum applied — the engine idles and vacuum is maintained.
   b. With 8.5 in. Hg vacuum applied — the engine changes from idling to slightly unstable and vacuum is maintained.
6. Inspect the system components if test results differ from above.

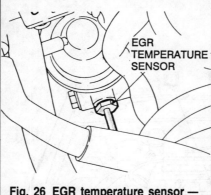

Fig. 26 EGR temperature sensor — 1.8L engine

California
▶ **See Figure 25**

1. Disconnect the green striped vacuum hose from the EGR valve and connect a hand vacuum pump through a 3-way connector. The pump will now be installed in the line.
2. With engine cold (below 68°F), test system operation as follows:
   a. Race the engine by rapidly operating the accelerator.
   b. Measure the pressure reading on the pump. The negative pressure at the valve should not change.
3. With the engine warm (68°F or more), test system operation as follows:
   a. Race the engine by rapidly operating the accelerator.
   b. The negative pressure at the gauge rises to 3.9 in. Hg or more.
4. Disconnect the 3-way terminal and connect a hand vacuum pump to the EGR valve.
5. When a negative pressure of 8.5 in. Hg is applied during engine idling, check that the engine stops or the idle becomes unstable.
6. Inspect the system components if test results differ from specifications listed above.

EGR Temperature Sensor

▶ **See Figures 26, 27 and 28**

The EGR temperature sensor is used on California vehicles only. The EGR temperature sensor detects the temperature of the gas passing through the EGR control valve. It converts the detected temperature into an electrical voltage signal which is sent the vehicles Engine Control Unit (ECU). If the circuit of the EGR temperature sensor is broken, the warning light will come on.

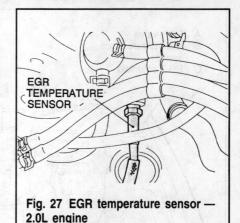

Fig. 27 EGR temperature sensor — 2.0L engine

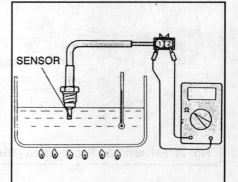

Fig. 28 Place the EGR temperature sensor in water and measure its resistance as the water is heated

TESTING

1. Remove the EGR temperature sensor from the engine.
2. Place the EGR sensor into water. While increasing the temperature of the water, measure the sensor resistance. Compare obtained values to specifications:
   a. 122°F (50°C) — 60-83 kilo-ohms resistance
   b. 212°F (100°C) — 11-14 kilo-ohms resistance
3. If the resistance obtained varies significantly from specifications, replace the sensor.

REMOVAL & INSTALLATION

1. Disconnect the negative battery cable.
2. Disconnect the electrical connector from the sensor.

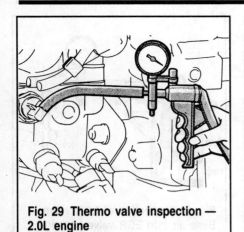

**Fig. 29 Thermo valve inspection —
2.0L engine**

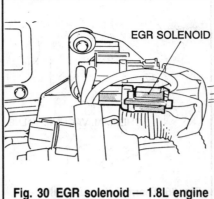

Fig. 30 EGR solenoid — 1.8L engine

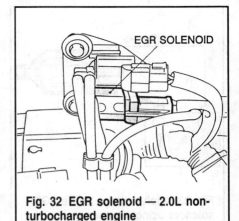

**Fig. 32 EGR solenoid — 2.0L non-
turbocharged engine**

3. Remove the sensor from the
engine.
To install:
4. Install the sensor to the engine
am\and tighten to 8 ft. lbs. (12 Nm).
5. Install the electrical connector to
the sensor.
6. Connect the negative battery
cable.

Thermo Valve

▶ **See Figure 29**

TESTING

Federal and Canadian Vehicles.

1. Label and disconnect the vacuum
hose at the thermo valve.
2. Connect a hand held vacuum
pump to the vacuum hose on the thermo
valve.
3. Apply a vacuum and check the air
passage through the thermo valve.
Compare results to specifications:
   a. Engine coolant temperature of
122°F (50°C) or less — vacuum leaks
   b. Engine coolant temperature of
176°F (80°C) or more — vacuum is
maintained
4. If the results differ from the
desired specifications, replace the valve.

REMOVAL & INSTALLATION

1. Disconnect the negative battery
cable.
2. Disconnect the vacuum line from
the thermo valve.

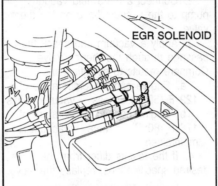

**Fig. 31 EGR solenoid — 2.0L
turbocharged engine**

3. Using a wrench, remove the valve
from the engine.

➡**When removing or installing the
valve, do not allow wrenches or other
tool to contact the resin part of the
valve. Damage to the valve may
occur.**

4. Inspect the vacuum hose for
cracks and replace as required.
To install:
5. Apply sealant to the threads of the
thermo valve and install into the engine.
6. Tighten the valve to 15-30 ft. lbs.
(20-40 Nm). When installing the valve,
do not allow the wrench to come in
contact with the resin part of the valve.
7. Reconnect the vacuum hose to the
valve.
8. Reconnect the negative battery
cable.

EGR Control Solenoid

TESTING

▶ **See Figures 30, 31, 32, 33 and 34**

On California vehicles, the EGR
control solenoid is located:
• Vehicles with non-turbocharged
engine — on the firewall, slightly right of
center.
• Vehicles with turbocharged
engine — the outermost solenoid in the
cluster at the left rear corner of the
engine compartment.
1. Label and disconnect the yellow
and green striped vacuum hose from the
EGR solenoid.
2. Disconnect the electrical harness
connector.
3. Connect a hand vacuum pump to
the nipple to which the green-striped
vacuum hose was connected.
4. Apply a vacuum and check for air-
tightness when voltage is applied and
discontinued. When voltage is applied,
the vacuum should be maintained. When
voltage is discontinued, vacuum should
leak.
5. Measure the resistance between
the terminals of the solenoid valve. The
resistance should be 36-44 ohms at
68°F (20°C).
6. If the test results differ from the
specifications, replace the EGR solenoid.

REMOVAL & INSTALLATION

1. Disconnect the negative battery
cable.
2. Label and disconnect the vacuum
connections at the EGR solenoid.

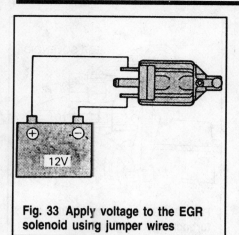

Fig. 33 Apply voltage to the EGR solenoid using jumper wires

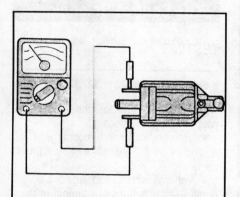

Fig. 34 Measure the resistance between the terminals of the EGR solenoid

3. Disconnect the electrical harness from the solenoid.

4. Remove the solenoid from the mounting bracket and replace as required.

To install:

5. Install the solenoid to the mounting bracket and secure in position.

6. Install the electrical connector.

7. Connect the vacuum hoses to the solenoid making sure they are installed in their original location.

8. Connect the negative battery cable.

Thermal Vacuum Valve

▶ See Figure 35

TESTING

1. Label and disconnect the vacuum hose at the thermo valve.

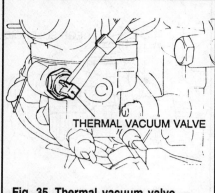

Fig. 35 Thermal vacuum valve — 2.0L engine (Federal)

2. Connect a hand held vacuum pump to the vacuum hose on the thermo valve.

3. Apply a vacuum and check the air passage through the thermo valve. Compare results to specifications:

　a. Engine coolant temperature of 122°F (50°C) or less — vacuum leaks

　b. Engine coolant temperature of 176°F (80°C) or more — vacuum is maintained

4. If the results differ from the desired specifications, replace the valve.

REMOVAL & INSTALLATION

1. Disconnect the negative battery cable.

2. Disconnect the vacuum line from the thermo valve.

3. Using a wrench, remove the valve from the engine.

➡When removing or installing the valve, do not allow wrenches or other tool to contact the resin part of the valve. Damage to the valve may occur.

4. Inspect the vacuum hose for cracks and replace as required.

To install:

5. Apply sealant to the threads of the thermo valve and install into the engine.

6. Tighten the valve to 15-30 ft. lbs. (20-40 Nm). When installing the valve, do not allow the wrench to come in contact with the resin part of the valve.

7. Reconnect the vacuum hose to the valve.

8. Reconnect the negative battery cable.

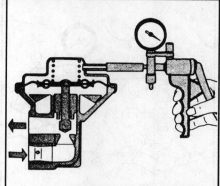

Fig. 36 Inspection of EGR valve. Blow air into EGR valve passage with vacuum applied and compare results to specifications.

Exhaust Gas Recirculation (EGR) Valve

▶ See Figures 21, 22, 23, 24 and 36

TESTING

1. Remove the EGR valve and check for sticking of plunger caused by excess carbon deposits. If such a condition exists, clean with appropriate solvent so valve seats correctly.

2. Connect a vacuum pump to the valve and apply 19.8 in. Hg of vacuum.

3. Check for air tightness. If the valve has 2 vacuum ports; pick one and plug the other. The vacuum must be retained.

4. Blow air from 1 passage of the EGR to check condition as follows:

　a. With 1.8 in. Hg of vacuum or less applied to the valve, air should not pass through the valve.

　b. With 8.5 in. Hg of vacuum or more applied to the valve, air should pass through the valve.

5. If the results are not as described, replace the EGR valve.

REMOVAL & INSTALLATION

1. Disconnect the negative battery cable.

2. Remove the air cleaner and intake hoses as required.

3. Disconnect the vacuum hose at the EGR valve.

4. Remove the mounting bolts and the EGR valve from the engine.

5. Clean the mating surfaces on the valve and the engine. Make sure to remove all gasket material.

6. Inspect the valve for a sticking plunger, caused by excess carbon deposits. If such a condition exists, clean with appropriate solvent so valve seats correctly.

To install:

7. Install EGR valve onto engine with new gasket in place.

8. Install the mounting bolts and tighten as follows:
- Vehicles with 1.8L engine — 7-10 ft. lbs. (10 — 15 Nm)
- Vehicles with 2.0L DOHC engine — 10-15 ft. lbs. (15 — 22 Nm)

9. Connect the vacuum hose to the EGR valve.

10. Install the air cleaner and air intake hoses as required.

11. Reconnect the negative battery cable.

Oxygen Sensor

The oxygen sensor is usually located on the gathering area of the exhaust manifolds. All exhaust gas leaving the engine flows past the oxygen sensor. The oxygen sensor produces an electrical voltage when exposed to oxygen present in the exhaust gases. Where there is a large amount of oxygen present (lean mixture), the sensor produces a low voltage. When there is a lesser amount of oxygen present (rich mixture), the sensor produces a higher voltage. By monitoring the oxygen content and converting it to electrical voltage, the sensor acts as a rich/lean switch. The voltage from the sensor is transmitted to the Engine Control Unit (ECU), which changes the fuel injection ratio accordingly. On later models, the oxygen sensor may be electrically heated internally. This allows for faster switching during cold engine operation.

TESTING

➡If the oxygen sensor has failed, the driveability of the vehicle may not be influenced. Since the air fuel mixture ratio shifts toward the rich side, the driveability of the vehicle may become better. However, levels of hazardous components such as HC, CO, and NOx emitted out the tailpipe

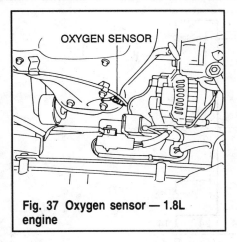

Fig. 37 Oxygen sensor — 1.8L engine

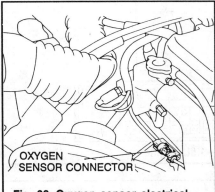

Fig. 38 Oxygen sensor electrical connector — 1.8L engine

will be elevated. This is because the ECU is not controlling the air fuel mixture.

1.8L Engine
▶ See Figures 37, 38 and 39

1. Before testing, make certain the engine is fully warm. Coolant temperature must be 185-205°F.

2. Disconnect the oxygen sensor connector.

3. Connect the positive probe of a digital ohmmeter to terminal 1 of the sensor connector. Connect the other meter probe to chassis ground.

4. Repeatedly race the engine; measure the voltage output of the sensor. As the mixture becomes richer from repeated racing of the engine, the sensor output voltage should register 0.6-1.0 volts.

5. If the voltage output is incorrect, the sensor will require replacement.

6. Shut the ignition OFF, disconnect the test equipment and reconnect the sensor to the wiring harness.

2.0L Engine
▶ See Figures 40, 41 and 42

1. Disconnect the oxygen sensor connector.

2. Measure the resistance between terminal Nos. 3 and 4. Correct resistance is approximately 12 ohms at 68°F.

3. If there is no continuity or if the resistance is not approximately correct, the sensor must be replaced.

4. Operate the engine until fully warmed up. Coolant temperature must be at least 176°F.

5. Using jumper wires, carefully connect terminal 3 to battery positive voltage and connect terminal No. 4 to a ground (-).

➡Use extreme care when connecting the jumpers. Incorrect circuiting will destroy the sensor.

6. Connect the probes of a digital voltmeter across to terminal No. 1 and terminal No. 2.

7. Repeatedly race the engine; measure the voltage output of the sensor. As the mixture becomes richer from repeated racing of the engine, the sensor output voltage should become 0.6-1.0 volt.

8. If the voltage output is incorrect, the sensor must be replaced.

9. Shut the ignition OFF, disconnect the test equipment and reconnect the sensor to the wiring harness.

REMOVAL & INSTALLATION

❊❊CAUTION

The temperature of the exhaust system is extremely high after the engine has been run. To prevent personal injury, allow the exhaust system to cool completely before removing sensor from the exhaust system.

1. Disconnect negative battery cable.

2. Raise and safely support the vehicle.

3. Disconnect the electrical connector at the oxygen sensor.

4. Using socket MD998770, or equivalent, remove the oxygen sensor.

To install:

5. If installing old oxygen sensor, coat the threads with anti-seize

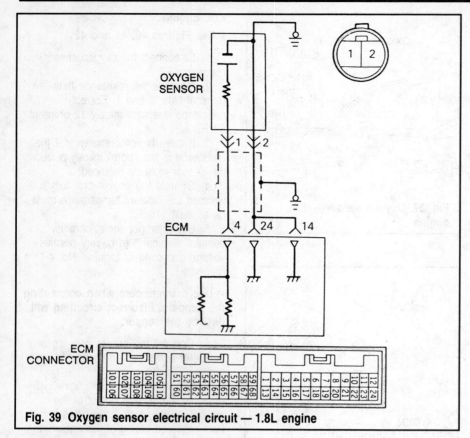

Fig. 39 Oxygen sensor electrical circuit — 1.8L engine

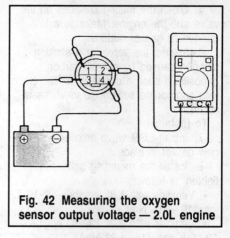

Fig. 42 Measuring the oxygen sensor output voltage — 2.0L engine

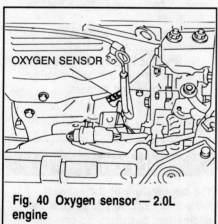

Fig. 40 Oxygen sensor — 2.0L engine

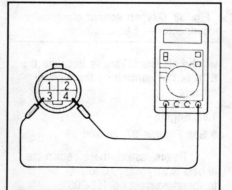

Fig. 41 Testing continuity of oxygen sensor — 2.0L engine

compound. New sensors are already coated. Take care not to contaminate the oxygen sensor probe with the anti-seize compound.

6. Install the oxygen sensor into the exhaust manifold. Tighten the sensor, using the correct tool, to 33 ft. lbs. (45 Nm)

7. Connect the wiring to the sensor.

8. Connect the negative battery cable.

Catalytic Converter

Engine exhaust consists mainly of nitrogen, however, it also contains carbon monoxide, carbon dioxide, water vapor, oxygen, nitrogen oxides and hydrogen, as well as various, unburned hydrocarbons. Three of these exhaust components, carbon monoxide, oxides of nitrogen and hydrocarbons, are major air pollutants, so their emission to the atmosphere must be controlled.

The catalytic converter, mounted in the engine exhaust stream, plays a major role in the emission control system. The converter works as a gas reactor and it's catalytic function is to speed up the heat producing chemical reaction between the exhaust gas components in order to reduce the air pollutants in the engine exhaust.

Heat shields are used to protect both the vehicle and environment from the high temperature developed in the vicinity of the catalytic converter during engine operation. There should be no trace of undercoating or rust proofing materials on the shield; these substances greatly reduce the heat dissipating efficiency of the shield.

An elevation in exhaust temperature can occur when the engine misfires or otherwise does not operate at peak efficiency. Spark plug wires should not be disconnected from the spark plugs while the engine is running, if the exhaust system is equipped with a catalytic converter. Failure of the converter can occur due to temperature increases caused by unburned fuel passing through the converter.

There is no regularly scheduled maintenance on any catalytic converters. Inspect catalytic converters for damage, cracking or deterioration. If damaged in any way, the converter should be replaced.

Catalytic Converter

REMOVAL & INSTALLATION

➡**Allow the exhaust system on the vehicle to cool completely before starting repairs on the vehicle. When working on exhaust system components, always wear eye protection. Personal injury from flying debris or hot exhaust components could occur.**

1. Disconnect negative battery cable.
2. Remove the catalytic converter flange bolts or clamps as required.
3. Separate and remove the converter from the exhaust system.

To install:

4. Install catalytic converter and remaining exhaust components onto the vehicle. Install new clamps and retainers and tighten just enough to hold exhaust components in position.
5. Working from the front of the vehicle and moving towards the rear, align each component and then tighten the appropriate retainer. Tighten the converter flange mounting bolts to 36-43 ft. lbs. (50-60 Nm).
6. Once all components are installed, inspect the system. Make sure the exhaust system is free of all other components.
7. Lower the vehicle and start the engine. Inspect the exhaust system for leaks.

ELECTRONIC ENGINE CONTROLS

Fuel Injection System

The majority of fuel injection and emission related items are warranted by the manufacturer for an extended mileage and time. Before replacing any fuel or emission related components, check with the manufacturer for possible component replacement under warranty.

SERVICE PRECAUTIONS

Safety is an important factor when servicing the fuel system. Failure to conduct maintenance and repairs in a safe manner may result in serious personal injury. Maintenance and testing of the vehicle's fuel system components can be accomplished safely and effectively by adhering to the following rules and guidelines.

• To avoid the possibility of fire and personal injury, always disconnect the negative battery cable unless the repair or test procedure requires that battery voltage be applied.

• Always relieve the fuel system pressure prior to disconnecting any fuel system component (injector, fuel rail, pressure regulator, etc.), fitting or fuel line connection. Exercise extreme caution whenever relieving fuel system pressure to avoid exposing skin, face and eyes to fuel spray. Be advised that fuel under pressure may penetrate the skin or any part of the body that it contacts.

• Always place a shop towel or cloth around the fitting or connection prior to loosening to absorb any excess fuel due to spillage. Ensure that all fuel spillage is quickly removed from engine surfaces.

Ensure that all fuel soaked cloths or towels are deposited into a suitable waste container.

• Always keep a dry chemical (Class B) fire extinguisher near the work area.

• Do not allow fuel spray or fuel vapors to come into contact with a spark or open flame.

• Always use a backup wrench when loosening and tightening fuel line connection fittings. This will prevent unnecessary stress and torsion to fuel line piping. Always follow the proper torque specifications.

• Always replace worn fuel fitting O-rings. Do not substitute fuel hose where fuel pipe is installed.

Fuel Pressure Relief Procedure

Because of the high pressure present in the multi-point fuel system of these engines, prior to opening the fuel system, the pressure in the system must be released. Failure to release the pressure prior to opening the system may result in personal and/or property damage caused by fuel spray under pressure.

1. Loosen the fuel filler cap to release the fuel tank pressure.
2. Disconnect the fuel pump harness connector, located at the rear of the fuel tank.
3. Start the vehicle and allow it to run until it stalls from lack of fuel. Turn the key to the **OFF** position.
4. Disconnect the negative battery cable, then reconnect the fuel pump connector and reinstall the fuel filler cap.

Diagnosis and Testing

PRELIMINARY CHECKS

1. Visually inspect the engine compartment to ensure that all vacuum lines and wires are properly routed and securely connected.
2. Examine all wiring harnesses and connectors for insulation damage, burned, overheated, loose or broken conditions.
3. Be certain that the battery is fully charged and that all accessories are **OFF** during diagnosis.
4. Check the mounting bolts holding the throttle body to the manifold.
5. In the case of a vehicle which will not start, check the most obvious items first. Is there fuel in the tank? Are the battery cables clean and tight at the battery and starter solenoid? Are components such as the fuel pump and/or relays heard to operated when the ignition switch is turned **ON**?
6. Pay particular attention to the condition and position of vacuum line(s). Check the ends and undersides of each vacuum line for splitting or melting.

It is recommended that all system and/or component testing be performed with the DRB-II or equivalent scan tool. The number of tests and functions performed by the tool can eliminate extended diagnostic time. Use of the scan tool requires use of the MMC Adapter, an interface device allowing the DRB-II to communicate with the on-board ECU.

In the event that a scan tool is not available, many components of the fuel system may be tested using a

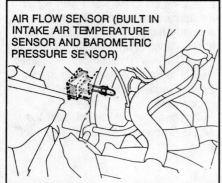

Fig. 43 Air flow sensor, intake air temperature sensor and barometric pressure sensor — 2.0L engine. Air flow sensor, intake air temperature sensor and barometric pressure sensor on 1.8L engine is similar.

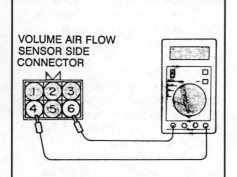

Fig. 44 Measuring the intake air temperature sensor resistance at the volume air flow sensor harness connector-Non-turbocharged engine

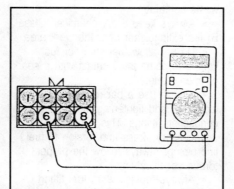

Fig. 45 Measuring the intake air temperature sensor resistance at the volume air flow sensor harness connector-Turbocharged engine

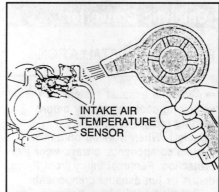

Fig. 46 Measuring the intake air temperature sensor resistance while heating with hair drier

volt/ohmmeter. Additionally, stored fault codes may be retrieved with an ohmmeter but this is not the preferred method. The test procedures outlines in this section use volt/ohmmeter testing.

For testing and code retrieval, both a digital and an analog meter will be required. The digital meter must be high-impedance; use of other meters may yield faulty readings and/or cause damage to the ECU.

Air Flow Sensor

♦ See Figure 43

The air flow sensor, including the air temperature sensor and barometric pressure sensor, is located inside the air filter housing. Care must be used when removing the air filter housing cover to avoid damaging the sensor assembly.

REMOVAL & INSTALLATION

Replacing the sensor requires disconnecting the electrical connector, then carefully removing the lid of the air filter housing. On Stealth, remove the bolts holding the sensor to the top of the air cleaner cover and disconnect the rubber boot leading to the intake tube.

Handle the sensor assembly carefully, protecting it from impact, extremes of temperature and/or exposure to shop chemicals.

Intake Air Temperature Sensor

The intake air temperature sensor, provided on the air flow sensor, is a resistor which measures the intake air temperature. The ECU will determine the intake air temperature according to the output voltage from the sensor, and compensates the fuel injection amount according to the intake air temperature.

When the intake air temperature sensor is faulty, it controls the fuel injection amount based on an intake air temperature default of 77°F (25°C). The driveability of the vehicle may become poor during cold temperature operation. The trouble may not be noticeable when the ambient temperature is around 77°F (25°C).

TESTING

♦ See Figures 44, 45 and 46

1. Disconnect the air flow sensor electrical connector.
2. If equipped with non-turbocharged engines, measure the resistance between terminals No. 4 and No. 6 of the electrical connector.
3. If equipped with turbocharged engine, measure the resistance between terminals No. 6 and No. 8 of the sensor electric connector.
4. Compare test readings to specifications:
   a. Sensor temperature of 32°F (0°C) — 6.0 kilo-ohms
   b. Sensor temperature of 68°F (20°C) — 2.7 kilo-ohms
   c. Sensor temperature of 176°F (80°C) — 0.4 kilo-ohms
5. Measure the sensor resistance while heating the sensor area with a hair dryer. As the temperature of the sensor increases, sensor resistance should become smaller.
6. If the measured resistance deviates from the standard value or the resistance remains unchanged, replace the air flow sensor assembly.

Barometric Pressure Sensor

The barometric pressure sensor is provided on the air flow sensor. It detects the barometric pressure of the surrounding environment and converts it to a signal. This signal is sent to the Engine Control Unit (ECU), which judges the present altitude of the vehicle. According to the altitude, the ECU

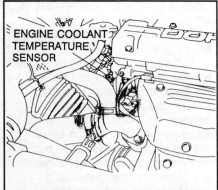

Fig. 47 Engine coolant temperature sensor as seen on 2.0L engine

compensates the fuel injection amount to gain a suitable air fuel mixture, thus improving the driveability of the vehicle at higher altitudes. If a defect in the barometric pressure sensor is found, replacement of the air flow sensor is required.

Coolant Temperature Sensor

▶ **See Figure 47**

The engine coolant temperature sensor is located on the thermostat case. It a resistor type sensor which detect the engine coolant temperature. The warm-up state of the engine is judged from the sensor output voltage. When a cold signal is sent to the ECU, the fuel injection amount, idle revolution speed and the injection timing are suitably controlled.

When the coolant temperature sensor is faulty, the engine coolant temperature is regarded as being 176°F (80°C). Because of this, engine starting may be difficult, driveability may become poor and the idle quality may be poor during cold operating conditions. The trouble may not be noticed when the engine is at normal operating temperature.

TESTING

1. Drain the engine coolant to a level below the intake manifold.
2. Disconnect the sensor wiring harness and remove the coolant temperature sensor from the engine.
3. Place the temperature sensing portion of the sensor into a pan of hot water. Use a thermometer to monitor the water temperature.

4. Measure the resistance across the sensor terminals while the sensor is in the water. Compare obtained reading to specifications:
   a. Water temperature of 32°F (0°C) — 5.9 kilo-ohms present
   b. Water temperature of 68°F (20°C) — 2.5 kilo-ohms present
   c. Water temperature of 104°F (40°C) — 2.7 kilo-ohms present
   d. Water temperature of 176°F (80°C) — 0.3 kilo-ohms present
5. If the resistance differs greatly from standard value, replace the sensor.

REMOVAL & INSTALLATION

1. Disconnect the negative battery cable.
2. Drain the engine coolant to a level below the intake manifold.
3. Disconnect the sensor wiring harness.
4. Unthread and remove the sensor from the engine.
To install:
5. Coat the threads of the sensor with sealant (MOPAR 4318034 or equivalent) and install onto engine.
6. Tighten the sensor to 14-29 ft. lbs. (20-40 Nm) torque.
7. Refill the cooling system to the proper level.
8. Reconnect the electrical connector to the sensor securely.

Idle Position Switch

The idle position switch senses accelerator operation. The switch is located at the tip of the Idle Speed Control (ISC) servo.

When a short circuit or wire breakage occurs in the idle position switch circuit, the dashpot control is not executed. Therefore, the revolution speed of the engine rapidly drops according to the position of the accelerator during deceleration.

TESTING

1990 Vehicle with 1.8L Engine

1. With the ignition **OFF**, disconnect the idle position switch connector.
2. With the accelerator released and fully closed, measure resistance between

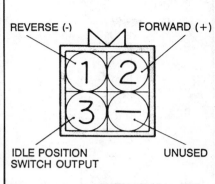

Fig. 48 Idle position switch connector terminal identification — 1990 vehicle

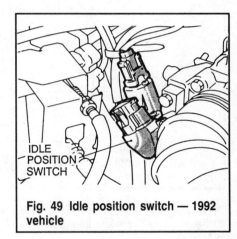

Fig. 49 Idle position switch — 1992 vehicle

terminal No. 3 and a good body ground. There should be 0 ohms resistance.
3. Depress the accelerator to the fully open position and measure the resistance. In this position, there should be no conductivity.
4. If the test results differ from specifications, the idle speed control servo assembly must be replaced. Refer to Section 5 of this manual for service procedure.

1991-92 Vehicles with 1.8L Engines
▶ **See Figures 48, 49 and 50**

1. With the ignition **OFF**, disconnect the idle position switch connector.
2. With the accelerator released and fully closed, measure resistance between terminal No. 4 and a good body ground. There should be 0 ohms resistance.
3. Depress the accelerator to the fully open position and measure the resistance. In this position, there should be no conductivity.
4. If the test results differ from specifications, the idle speed control

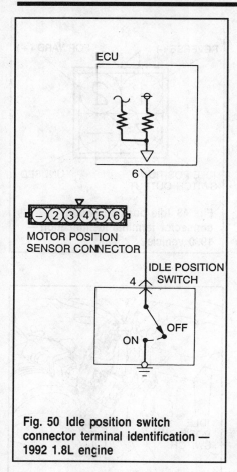

Fig. 50 Idle position switch connector terminal identification — 1992 1.8L engine

servo assembly must be replaced. Refer to Section 5 of this manual for service procedure.

1993 Vehicle With 1.8L Engine

1. With the ignition **OFF**, disconnect the idle position switch connector.
2. Measure the resistance between terminal No. 2 and No. 3 of the motor position sensor connector. Compare the obtained value to the specifications of 4-6 kilo-ohms.
3. Disconnect the idle speed control motor connector.
4. Connect a 6 volt DC power source between terminal No. 1 and No. 2 of the idle speed control motor connector. Measure the resistance between terminals No. 3 and No. 5 of the idle speed control motor position sensor connector when the idle speed control motor is activated. The desired reaction is a smooth increase/decrease in accordance with expansion and

contraction of the idle speed control motor plunger.

➡**Only apply a 6 volt DC or lower to the idle speed control motor. Application of higher voltage to the motor could cause locking of the servo gears.**

5. If there is a deviation from the standard value or if the change is not smooth, replace the idle speed control motor assembly.

1990-92 Vehicles With 2.0L Engine

1. Disconnect the electrical connector from the idle speed control servo.
2. Check the continuity between terminal No. 1 and body ground with the accelerator depressed and released. Compare the reading to specifications:
   a. Throttle depressed — Non-conductive
   b. Throttle released — Conductive
3. If results differ from the specifications, replace the idle speed control servo assembly. Refer to Section 5 for removal procedure.

Motor Position Sensor (MPS)

The motor position sensor is a variable resistor, integrated in the Idle Speed Control (ISC) servo system. The sliding pin of the MPS is in contact with the end of the idle speed control plunger. As the ISC plunger moves, the internal resistance of the MPS varies. This will cause a variation in the output voltage of the MPS. The MPS detects the plunger position of the ISC system, and sends the signal to the ECU. The ECU processes this signal, along with the signals from the coolant temperature sensor, idle signal, load signal and vehicle speed signal, to control the opening angle of the throttle valve and revolution speed during engine idling.

TESTING

1990 Vehicles With 1.8L Engine

1. With the ignition **OFF**, disconnect the 4 terminal motor position sensor connector.
2. Measure the resistance between terminal No. 1 and No. 3 of the motor position sensor connector. Compare the

obtained value to the specifications of 4-6 kilo-ohms.
3. Disconnect the idle speed control servo connector.
4. Connect a DC 6 volt power source between terminal No. 1 and No. 2 of the idle speed control motor connector. Measure the resistance between terminals No. 1 and No. 2 of the idle speed control motor position sensor connector when the idle speed control motor is activated. The desired reaction is a smooth increase/decrease in accordance with expansion and contraction of the idle speed control motor plunger.

➡**Only apply a 6 volt DC or lower to the idle speed control motor. Application of higher voltage to the motor could cause locking of the servo gears.**

5. If there is a deviation from the standard value or if the change is not smooth, replace the idle speed control motor assembly.

1991-92 Vehicles With 1.8L Engine
▶ **See Figure 50**

1. With the ignition **OFF**, disconnect the idle position switch connector.
2. Measure the resistance between terminal No. 2 and No. 3 of the motor position sensor connector. Compare the obtained value to the specifications of 4-6 kilo-ohms.
3. Disconnect the idle speed control motor connector.
4. Connect a 6 volt DC power source between terminal No. 1 and No. 2 of the idle speed control motor connector. Measure the resistance between terminals No. 3 and No. 5 of the idle speed control motor position sensor connector when the idle speed control motor is activated. The desired reaction is a smooth increase/decrease in accordance with expansion and contraction of the idle speed control motor plunger.

➡**Only apply a 6 volt DC or lower current to the idle speed control motor. Application of higher voltage to the motor could cause locking of the servo gears.**

5. If there is a deviation from the standard value or if the change is not smooth, replace the idle speed control motor assembly.

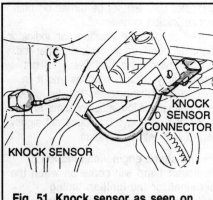

Fig. 51 Knock sensor as seen on 1993 Laser/Talon equipped with 2.0L turbocharged engine

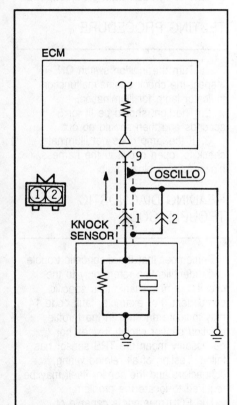

Fig. 52 Knock sensor circuit. Note positioning of oscilloscope during testing.

Knock Sensor

▶ **See Figures 51 and 52**

The knock sensor is used on turbocharged engines only. The sensor is normally mounted to the rear side of the engine block. It reacts to the ping or knock caused during detonation by sending a signal to the ECU. The ECU

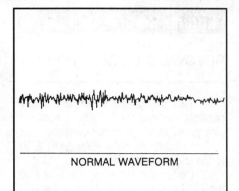

Fig. 53 Ideal test patter of knock sensor as seen on oscilloscope

retards the timing to eliminate the detonation.

If the knock sensor is experiencing functional problems, a slight reduction in engine power may be felt. This is because the ignition timing will be forcibly retarded.

TESTING

▶ **See Figure 53**

1. Position test probe of oscilloscope lead to the oscilloscope pick-up point as shown in the illustration.
2. Start the engine and allow to idle. Accelerate the engine to a maximum speed of 5,000 rpm and check the waveform on the scope.
3. If the test pattern obtained differentiates from the desired pattern shown in the illustration, replace the knock sensor.

REMOVAL & INSTALLATION

1. Disconnect the negative battery cable.
2. Disconnect the electrical connector at the sensor.
3. Remove the sensor from the engine block.
   To install:
4. Install the knock sensor in the opening in the engine block and tighten the sensor retainer.
5. Reconnect electrical connector to the sensor.
6. Connect the negative battery cable.

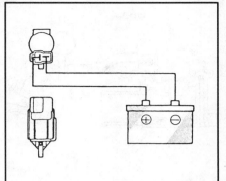

Fig. 54 Inspecting turbocharger waste gate solenoid for air-tightness with voltage applied and disconnected from the solenoid valve terminals

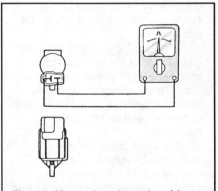

Fig. 55 Measuring the solenoid valve terminal resistance

Turbocharger Wastegate Solenoid

The turbocharger wastegate control solenoid (supercharging pressure relief solenoid valve) is an ON/OFF type solenoid valve, controlling the leakage rate of the supercharging pressure to the turbocharger waste gate actuator.

TESTING

▶ **See Figures 54 and 55**

1. Label and disconnect the vacuum hoses from the valve. Disconnect the electrical connector from the solenoid terminals.
2. Using a hand vacuum pump, apply a negative pressure to the solenoid valve nipple to which the white vacuum hose was connected, and check air-tightness when voltage is applied and

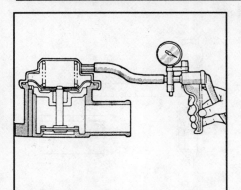

Fig. 56 Apply a negative pressure to the turbocharger by-pass valve and check for air tightness

disconnected from the solenoid terminals.

a. Apply battery voltage while covering the other nipple of the solenoid valve — vacuum should be maintained.

b. Apply battery voltage while the other nipple of the solenoid valve is open — vacuum should leak.

c. Battery voltage disconnected while the nipple of the solenoid valve is open — vacuum should be maintained.

3. Check the continuity of the solenoid coil by measuring the resistance across the 2 solenoid terminals. Compare the measured value to the desired reading of 36-44 ohms at 68°F (20°C).

4. If any test result differs from specifications, replace the turbocharger waste gate solenoid.

Turbocharger Bypass Valve

TESTING

▸ **See Figure 56**

1. Remove the turbocharger bypass valve from the engine.

2. Connect a vacuum pump to the nipple on the turbocharger bypass valve.

3. Apply a negative pressure and check the operation and air-tightness of the valve. The valve should start to open at 7.7 psi of vacuum and should maintain vacuum.

4. If results differ from specifications, replace the turbocharger bypass valve.

Engine Control Unit (ECU)

REMOVAL & INSTALLATION

1. Turn the ignition switch to the **OFF** position. Remove the left and/or right side panel from the center console.

2. Remove the bolts holding the ECU to the mounting bracket.

3. Disconnect the wiring harness from the ECU and remove ECU from the vehicle.

To install:

4. Connect the electrical harness to the ECU. Make certain the multi-pin connector is firmly and squarely seated to the ECU.

5. Install the ECU in the mounting bracket and secure in position.

6. Install the side panels to the center console.

7. Connect the negative battery cable.

Self-Diagnostics

The Engine Control Unit (ECU) monitors the signals of input and output sensors, some all the time and others at certain times and processes each signal. When the ECU has noticed an irregularity has continued for a specified time or longer from when the irregular signal was initially monitored, the ECU judges that a malfunction has occurred and will memorize the malfunction code. The code is then stored in the memory of the ECU and is accessible through the data link (diagnostic connector) with the use of an electronic scan tool or an voltmeter.

Check Engine/Malfunction Indicator Light

Among the on-board diagnostic items, a check engine/malfunction indicator light comes on to notify the driver of a emission control component irregularity. If the irregularity detected returns to normal or the engine control module judges that the component has returned to normal, the check check engine/malfunction indicator light will be turned off. Moreover, if the ignition is turned OFF and then the engine is restarted, the check engine/malfunction

indicator light will not be turned on until a malfunction is detected.

The check engine/malfunction indicator light will come on immediately after the ignition switch is turned ON. The light should stay lit for 5 seconds and then will go off. This indicates that the check engine/malfunction indicator lamp is operating normally. This does not signify a problem with the system.

➡**The check engine/malfunction indicator lamp will come on when the terminal for the ignition timing adjustment is shorted to ground. Therefore, it is not abnormal that the light comes on even when the terminal for ignition timing is shorted at time of ignition timing adjustment.**

TESTING PROCEDURE

1. Turn the ignition switch **ON**. Inspect the check engine/malfunction indicator lamp for illumination.

2. The light should be lit for 5 seconds and then should go out.

3. If the lamp does not illuminate, check for open circuit in the harness, blown fuse or blown bulb.

READING DIAGNOSTIC TROUBLE CODE

Remember that the diagnostic trouble code identification refers only to the circuit, not necessarily to a specific component. For example, fault code 14 may indicate an error in the throttle position sensor circuit; it does not necessarily mean the TPS sensor has failed. Testing of all related wiring, connectors and the sensor itself may be required to locate the problem.

The ECU memory is capable of storing multiple codes. During diagnosis the codes will be transmitted in numerical order from lowest to highest, regardless of the order of occurrence. If multiple codes are stored, always begin diagnostic work with the lowest numbered code.

Precautions for Operation

1. When battery voltage is low, no detection of failure is possible. Be sure to check the battery voltage and other conditions before starting the test.

2. Diagnostic items are erased if the battery or the engine controller connection is disconnected. Do not disconnect either of these components until the diagnostic material present in the engine control module has been read completely.

3. Be sure to connect and disconnect the scan tool to the data link connector with the ignition key **OFF**. If the scan tool in connected or disconnected with the ignition key **ON**, ABS diagnostic trouble codes may be falsely stored and the ABS warning light may be illuminated.

Reading Using Scan Tool
▶ **See Figure 57**

The procedure listed below is to be used only as a guide, when using Chrysler's DRB II scan tool. For specific operating instructions, follow the directions supplied with the particular scan tool being used.

1. Remove the under dash cover, if equipped. Connect the scan tool to the data link connector, located on the left underside of the instrument panel.

2. Using the scan tool, read and record the on-board diagnostic output.

3. Diagnose and repair the faulty components as required.

4. Turn the ignition switch **OFF** and then turn it **ON**.

5. Erase the diagnostic trouble code.

6. Recheck the diagnostic trouble code and make sure that the normal code is output.

Reading Using Voltmeter
▶ **See Figure 58**

1. Remove the under dash cover, if equipped.

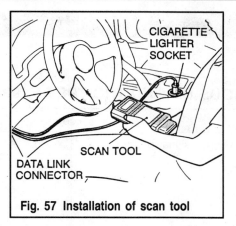

Fig. 57 Installation of scan tool

2. Connect an analog voltmeter between the on-board diagnostic output terminal of the data link connector and the ground terminal.

3. Turn the ignition switch **ON**.

4. Read the on-board diagnostic output pattern from the voltmeter and record.

5. Diagnose and repair the faulty components as required.

6. Erase the trouble code.

7. Turn **ON** the ignition switch and read the diagnostic trouble codes checking that a normal code is output.

TROUBLE CODE ERASURE

Scan Tool Not Available

1. Turn the ignition switch **OFF**.

2. Disconnect the negative battery cable from the battery for 10 seconds or more and then reconnect it.

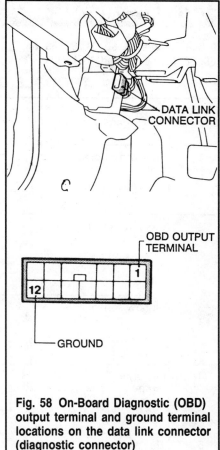

Fig. 58 On-Board Diagnostic (OBD) output terminal and ground terminal locations on the data link connector (diagnostic connector)

3. Turn **ON** the ignition switch and read the diagnostic trouble codes checking that a normal code is output.

Diagnostic Trouble Codes

VACUUM DIAGRAMS

▶ **See Figures 59, 60, 61, 62, 63, 64, 65, 66, 67, 68, 69 and 70**

TROUBLE CODE DIAGNOSTIC TREE—1990–92 1.8L ENGINE

Output preference order	Diagnosis item	Diagnosis code			Check item (Remedy)
		Output signal pattern	No.	Memory	
1	Engine control unit	H ⎍ L	—	—	(Replace engine control unit)
2	Oxygen sensor	H ⎍ L	11	Retained	• Harness and connector • Fuel pressure • Injectors (Replace if defective) • Intake air leaks • Oxygen sensor
3	Air flow sensor	H ⎍ L	12	Retained	• Harness and connector (If harness and connector are normal, replace air flow sensor assembly.)
4	Intake air temperature sensor	H ⎍ L	13	Retained	• Harness and connector • Intake air temperature sensor
5	Throttle position sensor	H ⎍ L	14	Retained	• Harness and connector • Throttle position sensor • Idle position switch
6	Motor position sensor	H ⎍ L	15	Retained	• Harness and connector • Motor position sensor • Throttle position sensor

TROUBLE CODE DIAGNOSTIC TREE — 1990–92 1.8L ENGINE, CONT.

Output preference order	Diagnosis item	Diagnosis code			Check item (Remedy)
		Output signal pattern	No.	Memory	
7	Engine coolant temperature sensor		21	Retained	• Harness and connector • Engine coolant temperature sensor
8	Crank angle sensor		22	Retained	• Harness and connector (If harness and connector are normal, replace distributor assembly.)
9	No. 1 cylinder top dead center sensor		23	Retained	• Harness and connector (If harness and connector are normal, replace distributor assembly.)
10	Vehicle speed sensor (reed switch)		24	Retained	• Harness and connector • Vehicle speed sensor (reed switch)
11	Barometric pressure sensor		25	Retained	• Harness and connector (If harness and connector are normal, replace barometric pressure sensor assembly.)
12	Ignition timing adjustment signal		36	–	• Harness and connector
13	Injector		41	Retained	• Harness and connector • Injector coil resistance
14	Fuel pump		42	Retained	• Harness and connector • Control relay
15	EGR <California>		43	Retained	• Harness and connector • EGR temperature sensor • EGR valve • EGR valve control solenoid valve • EGR valve control vacuum
16	Normal state		–	–	–

NOTE
Replace the engine control unit if a malfunction code is output although the inspection reveals that there is no problem with the check items.

TROUBLE CODE DIAGNOSTIC TREE — 1993 1.8L ENGINE

Diagnostic trouble code		Diagnostic item	Check item (Remedy)	Memory
No.	Output signal pattern			
—	H ⎍ L	Engine control module	(Replace engine control module)	—
11	H ⎍ L	Oxygen sensor	• Harness and connector • Oxygen sensor • Fuel pressure • Injectors (Replace if defective.) • Intake air leaks	Retained
12	H ⎍ L	Volume air flow sensor	• Harness and connector (If harness and connector are normal, replace volume air flow sensor assembly.)	Retained
13	H ⎍ L	Intake air temperature sensor	• Harness and connector • Intake air temperature sensor	Retained
14	H ⎍ L	Throttle position sensor	• Harness and connector • Throttle position sensor • Closed throttle position switch	Retained
15	H ⎍ L	Idle speed control motor position sensor	• Harness and connector • Idle speed control motor position sensor • Throttle position sensor	Retained
21	H ⎍ L	Engine coolant temperature sensor	• Harness and connector • Engine coolant temperature sensor	Retained
22	H ⎍ L	Crankshaft position sensor	• Harness and connector (If harness and connector are normal, replace distributor assembly.)	Retained

TROUBLE CODE DIAGNOSTIC TREE — 1993 1.8L ENGINE, CONT.

No.	Output signal pattern	Diagnostic item	Check item (Remedy)	Memory
Diagnostic trouble code				
23	H L	Camshaft position sensor	• Harness and connector (If harness and connector are normal, replace distributor assembly.)	Retained
24	H L	Vehicle speed sensor (reed switch)	• Harness and connector • Vehicle speed sensor (reed switch)	Retained
25	H L	Barometric pressure sensor	• Harness and connector (If harness and connector are normal, replace barometric pressure sensor assembly.)	Retained
36	H L	Ignition timing adjustment signal	• Harness and connector	–
41	H L	Injector	• Harness and connector • Injector coil resistance	Retained
42	H L	Fuel pump	• Harness and connector • MFI relay	Retained
43	H L	EGR <California>	• Harness and connector • EGR temperature sensor • EGR valve • EGR solenoid • EGR valve control vacuum	Retained
–	H L	Normal state	–	–

NOTE
1. Replace the engine control module if a diagnostic trouble code is output although the inspection reveals that there is no problem with the check items.

TROUBLE CODE DIAGNOSTIC TREE—1990–92 2.0L ENGINE

Output preference order	Diagnosis item	Diagnosis code			Check item (Remedy)
		Output signal pattern	No.	Memory	
1	Engine control unit	H ⎡‾‾‾‾‾ L ⎦	–	–	(Replace engine control unit)
2	Oxygen sensor	H ⎡ ⎤ L ⎦	11	Retained	• Harness and connector • Oxygen sensor • Fuel pressure • Injectors (Replace if defective) • Intake air leaks
3	Air flow sensor	H ⎡ ⎤ L ⎦	12	Retained	• Harness and connector (If harness and connector are normal, replace air flow sensor assembly.)
4	Intake air temperature sensor	H ⎡ ⎤ L ⎦	13	Retained	• Harness and connector • Intake air temperature sensor
5	Throttle position sensor	H ⎡ ⎤ L ⎦	14	Retained	• Harness and connector • Throttle position sensor • Idle position switch
6	Engine coolant temperature sensor	H ⎡ ⎤ L ⎦	21	Retained	• Harness and connector • Engine coolant temperature sensor

TROUBLE CODE DIAGNOSTIC TREE — 1990–92 2.0L ENGINE, CONT.

Output preference order	Diagnosis item	Diagnosis code			Check item (Remedy)
		Output signal pattern	No.	Memory	
7	Crank angle sensor	H ⎍ (waveform) L	22	Retained	• Harness and connector (If harness and connector are normal, replace crank angle sensor assembly.)
8	Top dead center sensor (No.1 and No.4 cylinder)	H ⎍ (waveform) L	23	Retained	• Harness and connector (If harness and connector are normal, replace crank angle sensor assembly.)
9	Vehicle speed sensor (reed switch)	H ⎍ (waveform) L	24	Retained	• Harness and connector • Vehicle speed sensor (reed switch)
10	Barometric pressure sensor	H ⎍ (waveform) L	25	Retained	• Harness and connector (If harness and connector are normal, replace barometric pressure sensor assembly.)
11	Detonation sensor <Turbo>	H ⎍ (waveform) L	31	Retained	• Harness and connector • Detonation sensor
12	Injector	H ⎍ (waveform) L	41	Retained	• Harness and connector • Injector coil resistance
13	Fuel pump	H ⎍ (waveform) L	42	Retained	• Harness and connector • Control relay
14	EGR <California>	H ⎍ (waveform) L	43	Retained	• Harness and connector • EGR temperature sensor • EGR valve • EGR valve control solenoid valve • EGR valve control vacuum
15	Ignition coil	H ⎍ (waveform) L	44	Retained	• Harness and connector • Ignition coil • Power transistor
16	Normal state	H ⎍ (waveform) L	–	–	–

NOTE
Replace the engine control unit if a malfunction code is output although the inspection reveals that there is no problem with the check items.

TROUBLE CODE DIAGNOSTIC TREE—1993 2.0L ENGINE

Diagnostic trouble code		Diagnostic item	Check item (Remedy)	Memory
No.	Output signal pattern			
—	H ⎍ L	Engine control module	(Replace engine control module)	—
11	H ⎍ L	Heated oxygen sensor	• Harness and connector • Heated oxygen sensor • Fuel pressure • Injectors (Replace if defective.) • Intake air leaks	Retained
12	H ⎍ L	Volume air flow sensor	• Harness and connector (If harness and connector are normal, replace volume air flow sensor assembly.)	Retained
13	H ⎍ L	Intake air temperature sensor	• Harness and connector • Intake air temperature sensor	Retained
14	H ⎍ L	Throttle position sensor	• Harness and connector • Throttle position sensor • Closed throttle position switch	Retained
21	H ⎍ L	Engine coolant temperature sensor	• Harness and connector • Engine coolant temperature sensor	Retained
22	H ⎍ L	Crankshaft position sensor	• Harness and connector (If harness and connector are normal, replace crankshaft position assembly.)	Retained
23	H ⎍ L	Camshaft position sensor	• Harness and connector (If harness and connector are normal, replace crankshaft position assembly.)	Retained

TROUBLE CODE DIAGNOSTIC TREE – 1993 2.0L ENGINE, CONT.

No.	Diagnostic trouble code — Output signal pattern	Diagnostic item	Check item (Remedy)	Memory
24		Vehicle speed sensor (reed switch)	• Harness and connector • Vehicle speed sensor (reed switch)	Retained
25		Barometric pressure sensor	• Harness and connector (If harness and connector are normal, replace barometric pressure sensor assembly.)	Retained
31		Knock sensor <Turbo>	• Harness and connector (If harness and connector are normal, replace knock sensor.)	Retained
41		Injector	• Harness and connector • Injector coil resistance	Retained
42		Fuel pump	• Harness and connector • MFI relay	Retained
43		EGR <California>	• Harness and connector • EGR temperature sensor • EGR valve • EGR solenoid • EGR valve control vacuum	Retained
44		Ignition coil, Ignition power transistor unit	• Harness and connector • Ignition coil • Ignition power transistor	Retained
–		Normal state	–	–

NOTE
1. Replace the engine control module if a diagnostic trouble code is output although the inspection reveals that there is no problem with the check items.

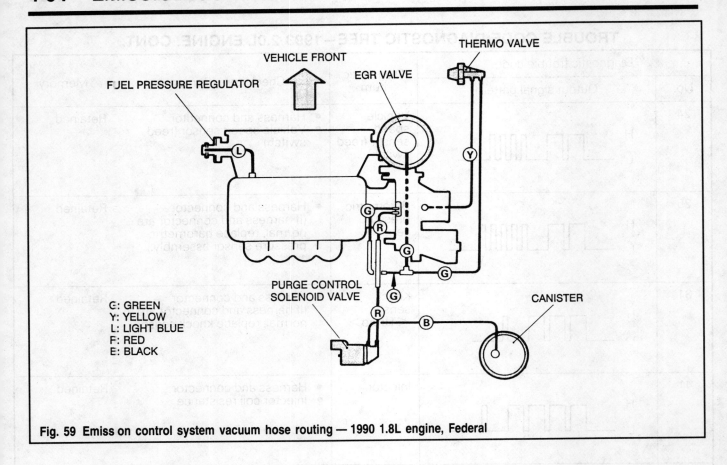

Fig. 59 Emission control system vacuum hose routing — 1990 1.8L engine, Federal

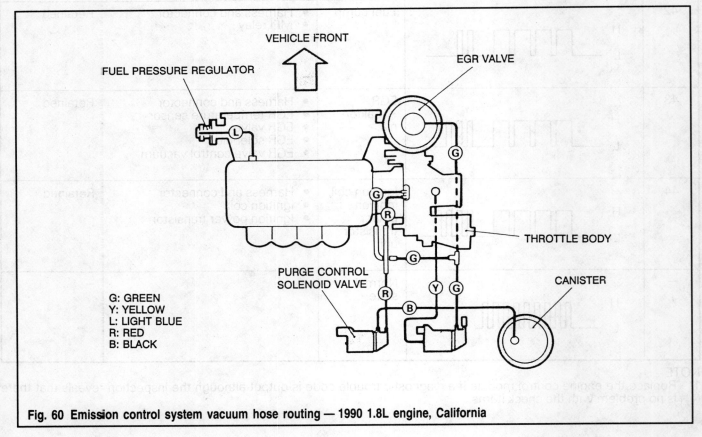

Fig. 60 Emission control system vacuum hose routing — 1990 1.8L engine, California

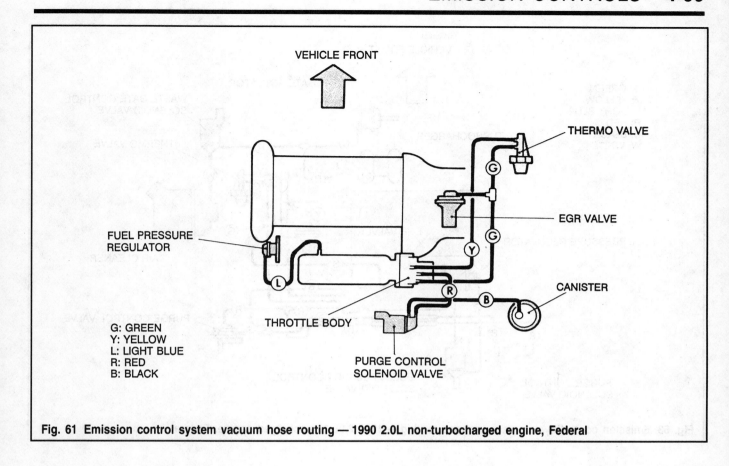

VEHICLE FRONT

THERMO VALVE

EGR VALVE

CANISTER

FUEL PRESSURE
REGULATOR

THROTTLE BODY

PURGE CONTROL
SOLENOID VALVE

G: GREEN
Y: YELLOW
L: LIGHT BLUE
R: RED
B: BLACK

Fig. 61 Emission control system vacuum hose routing — 1990 2.0L non-turbocharged engine, Federal

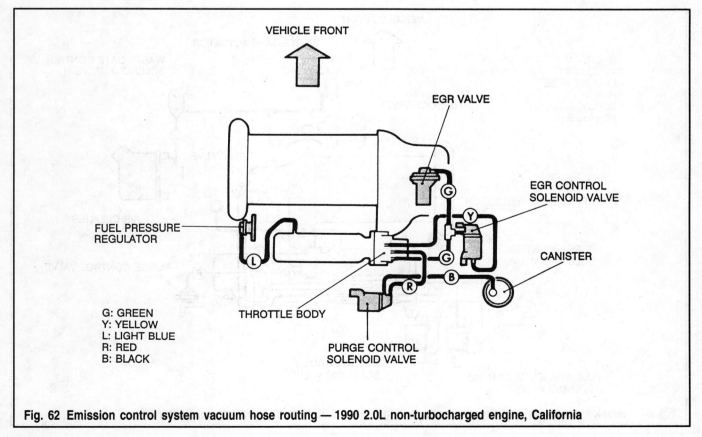

VEHICLE FRONT

EGR VALVE

EGR CONTROL
SOLENOID VALVE

CANISTER

FUEL PRESSURE
REGULATOR

THROTTLE BODY

PURGE CONTROL
SOLENOID VALVE

G: GREEN
Y: YELLOW
L: LIGHT BLUE
R: RED
B: BLACK

Fig. 62 Emission control system vacuum hose routing — 1990 2.0L non-turbocharged engine, California

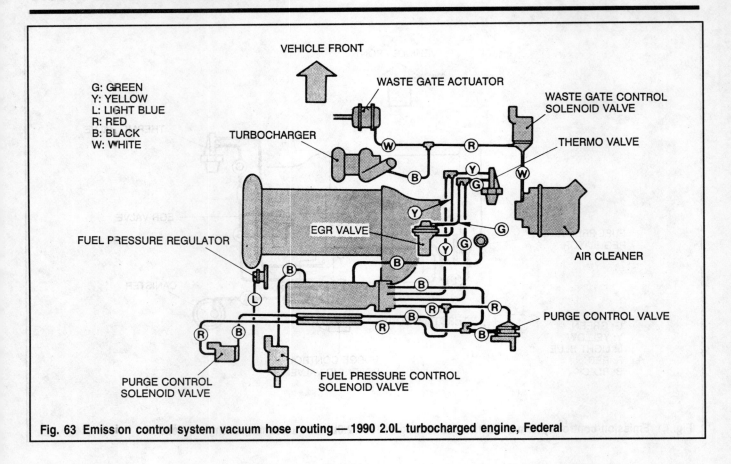

Fig. 63 Emission control system vacuum hose routing — 1990 2.0L turbocharged engine, Federal

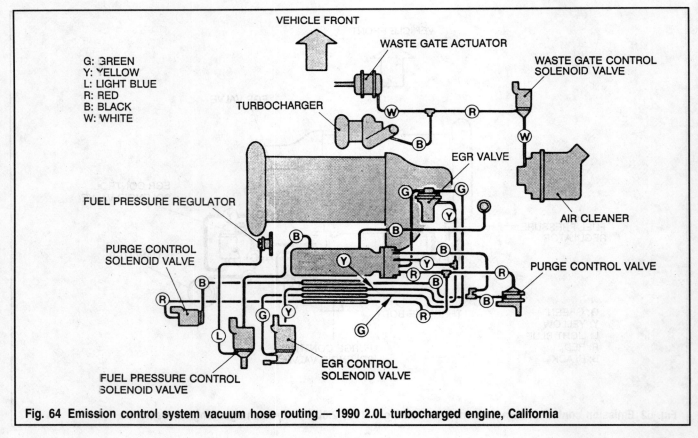

Fig. 64 Emission control system vacuum hose routing — 1990 2.0L turbocharged engine, California

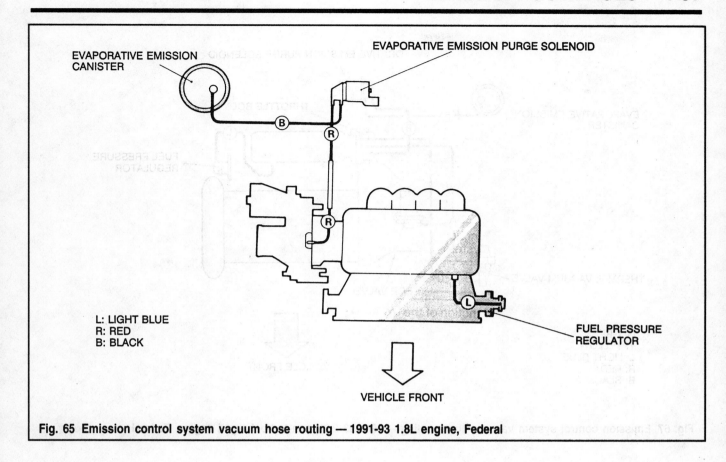

Fig. 65 Emission control system vacuum hose routing — 1991-93 1.8L engine, Federal

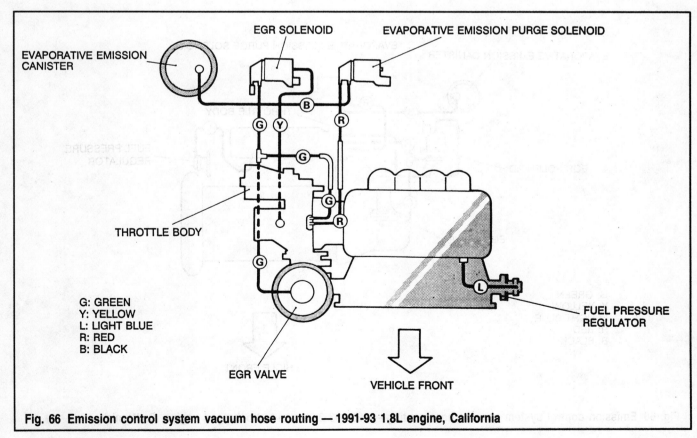

Fig. 66 Emission control system vacuum hose routing — 1991-93 1.8L engine, California

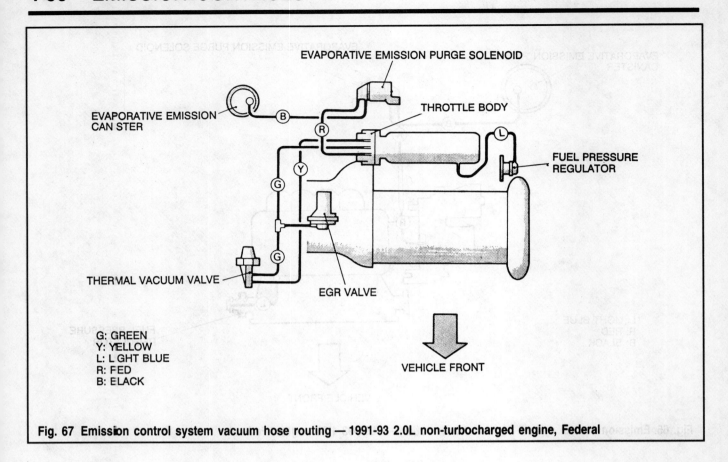

G: GREEN
Y: YELLOW
L: LIGHT BLUE
R: RED
B: BLACK

Fig. 67 Emission control system vacuum hose routing — 1991-93 2.0L non-turbocharged engine, Federal

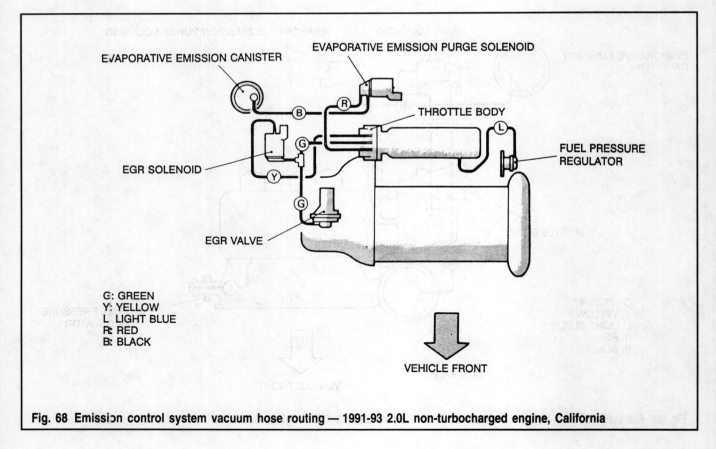

G: GREEN
Y: YELLOW
L: LIGHT BLUE
R: RED
B: BLACK

Fig. 68 Emission control system vacuum hose routing — 1991-93 2.0L non-turbocharged engine, California

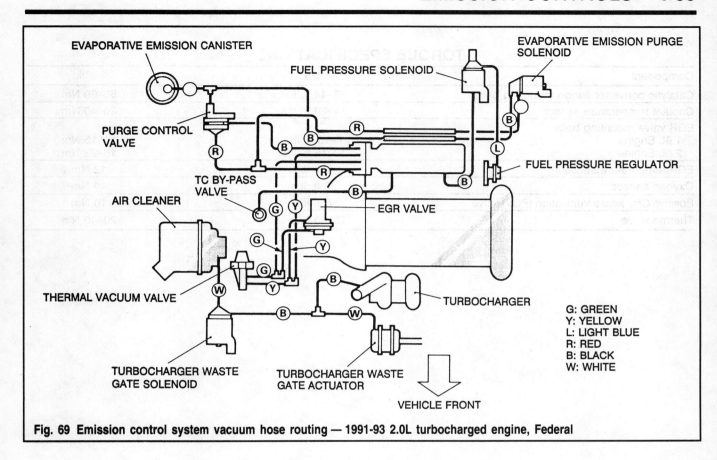

Fig. 69 Emission control system vacuum hose routing — 1991-93 2.0L turbocharged engine, Federal

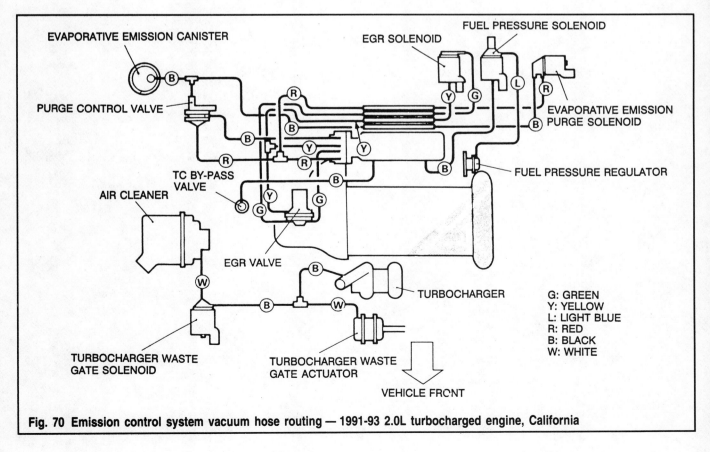

Fig. 70 Emission control system vacuum hose routing — 1991-93 2.0L turbocharged engine, California

TORQUE SPECIFICATIONS

Component	US	Metric
Catalytic converter flange mounting bolts	36–43 ft. lbs.	50–60 Nm
Coolant temperature sensor	14–29 ft. lbs.	20–40 Nm
EGR valve mounting bolts		
1.8L Engine	7–10 ft. lbs.	10–15 Nm
2.0L Engine	10–15 ft. lbs.	15–22 Nm
EGR temperature sensor	8 ft. lbs.	12 Nm
Oxygen sensor	33 ft. lbs.	45 Nm
Positive Crankcase Ventilation (PCV) valve	7 ft. lbs.	10 Nm
Thermo valve	15–30 ft. lbs.	20–40 Nm

5

FUEL SYSTEM

GASOLINE FUEL INJECTION SYSTEM

Description of System

The Multi-Point Injection (MPI) system is electronically controlled by the Engine Control Module (ECM), based on data from various sensors. The ECM controls the fuel flow, idle speed and ignition timing.

Fuel is supplied to the injectors by an electric in-tank fuel pump and is distributed to the respective injectors via the main fuel pipe. The fuel pressure applied to the injector is constant and higher than the pressure in the intake manifold. The pressure is controlled by the fuel pressure regulator. The excess fuel is returned to the fuel tank through the fuel return pipe.

When an electric current flows in the injector, the injector valve is fully opened to supply fuel. Since the fuel pressure is constant, the amount of the fuel injected from the injector into the manifold is increased or decreased in proportion to the time the electric current flows. Based on ECU signals, the injectors inject fuel to the cylinder manifold ports in firing order.

The flow rate of the air drawn through the air cleaner is measured by the air flow sensor. The air enters the air intake plenum or manifold through the throttle body.

In the intake manifold, the air is mixed with the fuel from the injectors and is drawn into the cylinder. The air flow rate is controlled according to the degree of the throttle valve and the servo motor openings.

The system is monitored through a number of sensors which feed information on engine conditions and requirements to the ECM. The ECM calculates the injection time and rate according to the signals from the sensors.

Fuel System Service Precaution

Safety is an important factor when servicing the fuel system. Failure to conduct maintenance and repairs in a safe manner may result in serious personal injury. Maintenance and testing of the vehicle's fuel system components

Fig. 1 Fuel pump harness connector is located at the rear of the fuel tank

can be accomplished safely and effectively by adhering to the following rules and guidelines.

• To avoid the possibility of fire and personal injury, always disconnect the negative battery cable unless the repair or test procedure requires that battery voltage be applied.

• Always relieve the fuel system pressure prior to disconnecting any fuel system component (injector, fuel rail, pressure regulator, etc.), fitting or fuel line connection. Exercise extreme caution whenever relieving fuel system pressure to avoid exposing skin, face and eyes to fuel spray. Please be advised that fuel under pressure may penetrate the skin or any part of the body that it contacts.

• Always place a shop towel or cloth around the fitting or connection prior to loosening to absorb any excess fuel due to spillage. Ensure that all fuel spillage is quickly removed from engine surfaces. Ensure that all fuel soaked cloths or towels are deposited into a suitable waste container.

• Always keep a dry chemical (Class B) fire extinguisher near the work area.

• Do not allow fuel spray or fuel vapors to come into contact with a spark or open flame.

• Always use a backup wrench when loosening and tightening fuel line connection fittings. This will prevent unnecessary stress and torsion to fuel line piping. Always follow the proper torque specifications.

• Always replace worn fuel fitting O-rings. Do not substitute fuel hose where fuel pipe is installed.

Relieving Fuel System Pressure

▶ See Figure 1

As mentioned above, in MPI fuel systems, fuel under high pressure is supplied to the fuel rail and injectors. Because of the high pressure present, it is essential that the pressure be released prior to loosening any connections where fuel flows. If the pressure is not released and a line or fitting is loosened, fuel will leak out the line under pressure, possibly causing serious personal as well as property damage.

1. Loosen the fuel filler cap to release fuel tank pressure.

2. Disconnect the fuel pump harness connector located at the rear of the fuel tank.

3. Start the vehicle and allow it to run until it stalls from lack of fuel. Turn the key to the **OFF** position.

4. Disconnect the negative battery cable, then reconnect the fuel pump connector and reinstall the fuel filler cap.

✳✳WARNING

Always wrap shop towels around a fitting that is being disconnected to absorb residual fuel in the lines.

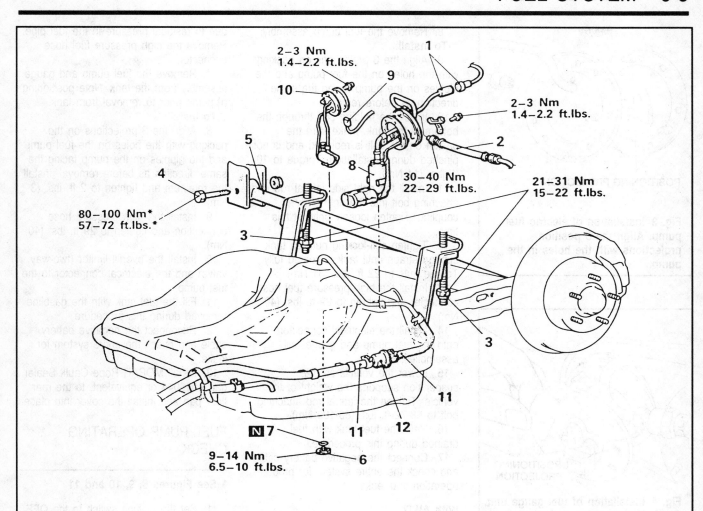

2–3 Nm
1.4–2.2 ft.lbs.

2–3 Nm
1.4–2.2 ft.lbs.

30–40 Nm
22–29 ft.lbs.

21–31 Nm
15–22 ft.lbs.

80–100 Nm*
57–72 ft.lbs.*

9–14 Nm
6.5–10 ft.lbs.

1. Connection for fuel pump connector
2. High pressure fuel hose
3. Self locking nut
4. Lateral rod attaching bolt
5. Lateral rod and body connection
6. Bolt
7. O-ring
8. Electric fuel pump
9. Connection for fuel gauge unit connector
10. Fuel gauge unit
11. Connection for vapor hose
12. Fuel tank pressure control valve

Fig. 2 Fuel pump, fuel gauge and related components — Front Wheel Drive (FWD) vehicle

Electric Fuel Pump

REMOVAL & INSTALLATION

With FWD

▶ **See Figures 2, 3, 4 and 5**

1. Relieve the fuel system pressure as follows:

a. Loosen the fuel filler cap to release fuel tank pressure.

b. Disconnect the fuel pump harness connector located at the rear of the fuel tank.

c. Start the vehicle and allow it to run until it stalls from lack of fuel. Turn the key to the **OFF** position.

d. Disconnect the negative battery cable, then reconnect the fuel pump connector and reinstall the fuel filler cap.

2. Raise and safely support the vehicle.

3. Drain the fuel from the fuel tank into an approved gasoline container.

Remove the electrical connectors at the fuel pump. Make sure there is enough slack in the electrical harness of the fuel gauge unit to allow for the fuel tank to be lowered slightly. If not, label

and disconnect the electrical harness at the fuel gauge unit.

❋❋CAUTION

Cover the high pressure fuel hose with rags to prevent splash of fuel caused by residual pressure in the fuel pipe line.

4. Disconnect the high pressure fuel line connector at the pump.

5. Loosen self-locking nuts on tank support straps to the end of the stud bolts.

6. Remove the right side lateral rod attaching bolt and disconnect the arm

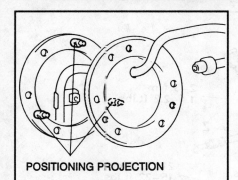

POSITIONING PROJECTION

Fig. 3 Installation of electric fuel pump. Align the 3 position projections with the holes in the pump.

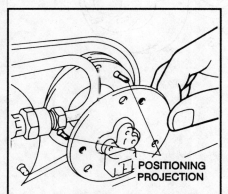

POSITIONING PROJECTION

Fig. 4 Installation of fuel gauge unit. Align the 2 position projections with the holes in the fuel gauge unit.

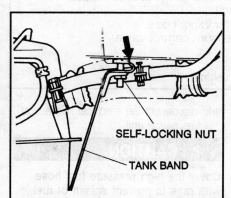

SELF-LOCKING NUT

TANK BAND

Fig. 5 Tighten the self-locking nuts until the rear end of the tank band contacts the body of the vehicle

from the right body coupling. Lower the lateral rod and suspend from the axle beam using wire.

7. Remove the holding bolt and gasket from the base of the tank.

8. Remove the fuel pump assembly.

To install:

9. Align the 3 projections on packing with the holes on the fuel pump and the nipples on the pump facing the same direction as before removal.

10. Install the holding bolt through the bottom of the tank. Make sure the gasket on the bolt is replaced and is not pinched during installation. Torque to 10 ft. lbs. (14 Nm).

11. Install the right side lateral rod and attaching bolt into the right body coupling. Tighten loosely only, at this time.

12. Tighten self-locking nuts on tank support straps until tank is seated fully. Torque nuts to 22 ft. lbs. (31 Nm).

13. Install the high pressure fuel hose connector and tighten to 29 ft. lbs. (40 Nm).

14. Install the electrical connectors onto the fuel pump and gauge unit assemblies.

15. Lower the vehicle so the suspension supports the weight of the vehicle. Tighten the lateral rod attaching bolt to 58-72 ft. lbs. (80-100 Nm).

16. Refill the fuel tank with fuel drained during this procedure.

17. Connect the negative battery cable and check the entire system for proper operation and leaks.

With AWD

▶ **See Figures 6 and 7**

1. Relieve the fuel system pressure as follows:

   a. Loosen the fuel filler cap to release fuel tank pressure.

   b. Disconnect the fuel pump harness connector located at the rear of the fuel tank.

   c. Start the vehicle and allow it to run until it stalls from lack of fuel. Turn the key to the **OFF** position.

   d. Disconnect the negative battery cable, then reconnect the fuel pump connector and reinstall the fuel filler cap.

2. The fuel pump is located in the fuel tank. Remove the hole cover located in the rear floor pan.

3. Partially drain the fuel tank into an approved gasoline container.

4. Remove the electrical connector from the fuel pump.

5. Remove the overfill limiter (two-way valve), as required.

6. Cover the hose connection with a shop towel to prevent any splash of fuel

due to residual pressure in the fuel pipe. Remove the high pressure fuel hose connector.

7. Remove the fuel pump and gauge assembly from the tank. Note positioning of pump prior to removal from tank.

To install:

8. Align the 3 projections on the packing with the holes on the fuel pump and the nipples on the pump facing the same direction as before removal. Install the retainers and tighten to 2 ft. lbs. (3 Nm).

9. Install the high pressure hose connection and tighten to 29 ft. lbs. (40 Nm).

10. Install the overfill limiter (two-way valve) and the electrical connector to the fuel pump.

11. Fill the fuel tank with the gasoline removed during this procedure.

12. Reconnect the negative battery cable and check the entire system for leaks.

13. Install MOPAR Rope Caulk Sealer part 4026044 or equivalent, to the rear floor pan and install the cover into place.

FUEL PUMP OPERATING CHECK

▶ **See Figures 8, 9, 10 and 11**

1. Set the ignition switch to the **OFF** position.

2. Check that when battery voltage is directly applied to the fuel pump check terminal located in the engine compartment, the operating sound of the fuel pump can be heard.

➡ **Since the fuel pump is located in the fuel tank, its operating sound cannot be readily heard. Remove the fuel tank cap and listen to the operating sound through the filler port.**

3. Hold the high pressure fuel hose between your fingers and check that the fuel pressure in the lines can be felt.

PRESSURE TESTING

▶ **See Figures 12, 13, 14 and 15**

1. Relieve the fuel system pressure as follows:

   a. Loosen the fuel filler cap to release fuel tank pressure.

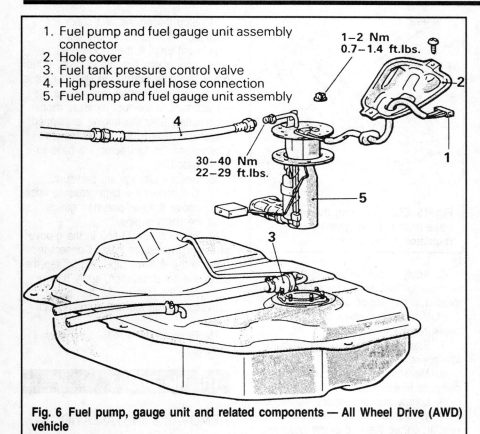

1. Fuel pump and fuel gauge unit assembly connector
2. Hole cover
3. Fuel tank pressure control valve
4. High pressure fuel hose connection
5. Fuel pump and fuel gauge unit assembly

1–2 Nm
0.7–1.4 ft.lbs.

30–40 Nm
22–29 ft.lbs.

Fig. 6 Fuel pump, gauge unit and related components — All Wheel Drive (AWD) vehicle

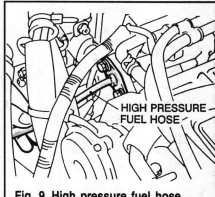

HIGH PRESSURE FUEL HOSE

Fig. 9 High pressure fuel hose location — 2.0L engine

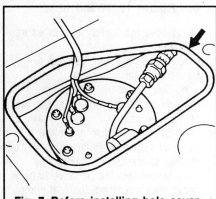

Fig. 7 Before installing hole cover, apply sealant to the rear floor pan

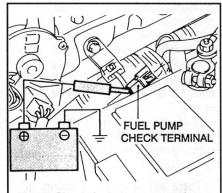

FUEL PUMP CHECK TERMINAL

Fig. 8 Check fuel pump operation with battery voltage applied to the fuel pump check terminal — 2.0L engine

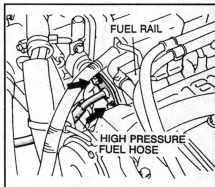

FUEL RAIL

HIGH PRESSURE FUEL HOSE

Fig. 10 High pressure fuel hose to fuel rail mounting bolts — 2.0L engine

b. Disconnect the fuel pump harness connector located at the rear of the fuel tank.

c. Start the vehicle and allow it to run until it stalls from lack of fuel. Turn the key to the **OFF** position.

d. Disconnect the negative battery cable, then reconnect the fuel pump connector and reinstall the fuel filler cap.

2. Disconnect the high pressure fuel line at the fuel rail.

❊❊CAUTION

Cover the hose connection with a shop towel to prevent the splash of fuel that can be caused by residual pressure in the fuel pipe line.

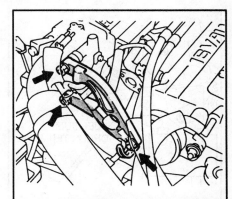

Fig. 11 Removing the throttle body stay — 2.0L engine

3. On 2.0L engine, remove the throttle body stay.

4. Connect a fuel pressure gauge to tools MD998709 and MD998742 or exact equivalent, with appropriate adaptors, seals and/or gaskets to prevent leaks during the test. Install the gauge and

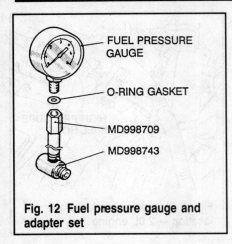

Fig. 12 Fuel pressure gauge and adapter set

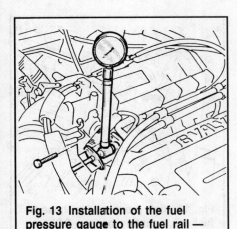

Fig. 13 Installation of the fuel pressure gauge to the fuel rail — 2.0L engine

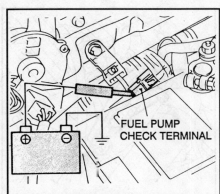

Fig. 14 Applying battery voltage to the fuel pump check terminal to activate the fuel pump. Inspect the gauge and hose connections for fuel leakage.

adapter between the delivery pipe and high pressure hose.

5. Connect the negative battery cable.

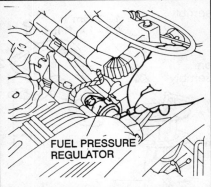

Fig. 15 Disconnecting the vacuum hose from the fuel pressure regulator — 2.0L engine

6. Apply battery voltage to the terminal for fuel pump check terminal located in the engine compartment. Run the fuel pump and check for leaks at the pressure gauge connection.

7. Start the engine and run at curb idle speed.

8. Measure the fuel pressure and compare to specification.

9. Locate and disconnect the vacuum hose running to the fuel pressure regulator. Plug the end of the hose and record the fuel pressure again. The fuel pressure should have increased approximately 10 psi.

10. Reconnect the vacuum hose the fuel pressure regulator. After the fuel pressure stabilizes, race the engine 2-3 times and check that the fuel pressure does not fall when the engine is running at idle.

11. Check to be sure there is fuel pressure in the return hose by gently pressing the fuel return hose with fingers while racing the engine.

➡ **There will be no fuel pressure in the return hose when the volume of fuel flow is low.**

12. If fuel pressure is too low, check for a clogged fuel filter, a defective fuel pressure regulator or a defective fuel pump, any of which will require replacement.

13. If fuel pressure is too high, the fuel pressure regulator is defective and will have to be replaced or the fuel return is bent or clogged. If the fuel pressure reading does not change when the vacuum hose is disconnected, the hose is clogged or the valve is stuck in the fuel pressure regulator and it will have to be replaced.

14. Stop the engine and check for changes in the fuel pressure gauge. It should not drop. If the gauge reading does drop, watch the rate of drop. If fuel pressure drops slowly, the likely cause is a leaking injector which will require replacement. If the fuel pressure drops immediately after the engine is stopped, the check valve in the fuel pump isn't closing and the fuel pump will have to be replaced.

15. Relieve fuel system pressure.

16. Disconnect the high pressure hose and remove the fuel pressure gauge from the delivery pipe.

17. Install a new O-ring in the groove of the high pressure hose. Connect the hose to the delivery pipe and tighten the screws. After installation, apply battery voltage to the terminal for fuel pump activation to run the fuel pump. Check for leaks.

18. Reinstall the throttle body stay, if removed.

Throttle Body

REMOVAL & INSTALLATION

▶ **See Figures 16, 17 and 18**

1. Relieve the fuel system pressure as follows:

a. Loosen the fuel filler cap to release fuel tank pressure.

b. Disconnect the fuel pump harness connector located at the rear of the fuel tank.

c. Start the vehicle and allow it to run until it stalls from lack of fuel. Turn the key to the **OFF** position.

d. Disconnect the negative battery cable, then reconnect the fuel pump connector and reinstall the fuel filler cap.

2. Matchmark the location of the adjuster bolt on the cable mounting flange. This will assure that the cable is installed in its original location. Remove the throttle cable adjusting bolt and disconnect the cable from the lever on the throttle body. Position cable aside.

3. Remove the connection for the breather hose and the air intake hose from the throttle body and position aside.

4. Disconnect the electrical connectors at the throttle body. Label and disconnect the electrical harness connectors from the Idle Air Control (IAC) motor connection, ISC motor

1. Connection for accelerator cable (Refer to P.14FE-17.)
2. Connection for breather hose
3. Connection for air intake hose
4. Connection for vacuum hose
5. Connection for ISC motor connector
6. Connection for ISC motor position sensor connector
7. Connection for TPS connector
8. Connection for water hose
9. Connection for water by-pass hose
10. Throttle body
11. Gasket

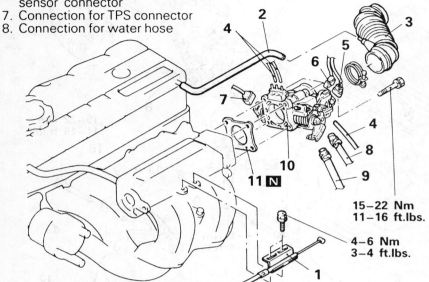

15–22 Nm
11–16 ft.lbs.

4–6 Nm
3–4 ft.lbs.

Fig. 16 Throttle body and related components — 1.8L engine

connection, ISC motor position sensor, Throttle Position Sensor (TPS), closed throttle position switch, etc., as equipped.

5. Drain the engine cooling system.

6. Disconnect the water and water by-pass hoses at the base of the throttle body.

7. Remove the throttle body stay and ground plate from the engine.

8. Remove the air fitting and gasket.

9. Remove the throttle body mounting bolts and throttle body from the engine. Remove old gasket and discard.

To install:

10. Clean all old gasket material from the both throttle body mounting surfaces. Install new gasket onto the intake manifold plenum mounting surface so the projection on the gasket is as illustrated.

➡**Poor idling quality and poor performance may be experienced if the gasket is installed incorrectly.**

11. Install the throttle body to the intake manifold plenum and tighten the mounting bolts and nuts to 16 ft. lbs. (22 Nm).

12. Install the air fitting, if equipped, making sure new gasket is in place.

13. Install the throttle body stay and ground plate. Secure with retainers tightened to 11-16 ft. lbs. (15-22 Nm). Install the ground plate mounting screw.

14. Install the water hoses to the throttle body. Install new hose clamps if required.

15. Connect the electrical and vacuum connectors at the throttle body.

16. Install the accelerator cable to the throttle body and install the adjusting bolt in original position. Check adjustment of cable.

17. Install the air intake and breather hoses. Connect the negative battery cable. Refill the cooling system.

IDLE SPEED CONTROL AND THROTTLE POSITION SENSOR ADJUSTMENTS

1.8L Engine

1990-91 VEHICLES AND 1992 LASER AND TALON

▶ **See Figures 19, 20 and 21**

1. Warm the engine to operating temperature, leave lights, electric cooling fan and accessories **OFF** . The transaxle should be in **N** or **P** for automatic

transaxle. The steering wheel should be in a neutral (straight ahead) position for vehicles with power steering.

2. Check the ignition timing and adjust, if necessary.

3. Connect a tachometer to the CRC filter connector. Use a paper clip for a tach adapter.

4. Run the engine for more than 10 seconds at 2000-3000 rpm. Allow the engine to idle for 2 minutes. Check the idle rpm. Curb idle should be 750 ± 100 rpm.

5. If adjustment is required, turn the engine **OFF** . Loosen the accelerator cable adjusting bolts and slacken the accelerator cable.

6. Connect a digital voltmeter between terminal **19** throttle position sensor output voltage) of the engine control unit and terminal **24** (ground).

7. Set the ignition switch to **ON** , without starting the engine, and hold it in that position for 15 seconds or more. Turn the ignition switch **OFF** .

8. Disconnect the connectors of the idle speed control servo and lock the idle speed control plunger at the initial position. Back out the fixed Speed Adjusting Screw (SAS).

9. Start the engine and allow to idle. Basic idle speed should be at specification. A new engine may idle a little lower. If the vehicle stalls or has a very low idle speed, suspect a deposit buildup on the throttle valve which must be cleaned.

10. If the idle speed is wrong, adjust with the idle speed control adjusting screw. Use a hexagon wrench if possible. Turn in the fixed SAS until the engine speed rises. Then back out the fixed SAS until the Touch Point where the engine speed does not fall any longer, is found. Back out the fixed SAS an additional ½turn from the touch point.

11. Stop the engine. Turn the ignition switch to **ON** but do not start engine. Check that the output voltage from the throttle position sensor is 0.48-0.52 volts. If it is out of specification, adjust by loosening the throttle position sensor mounting screws and rotating the throttle position sensor. Turning the throttle position sensor clockwise increases the output voltage. After adjustment, tighten screws firmly.

12. Turn the ignition switch **OFF** .

13. Adjust the free-play of the accelerator cable, reconnect the connectors of the idle speed control servo and remove the voltmeter.

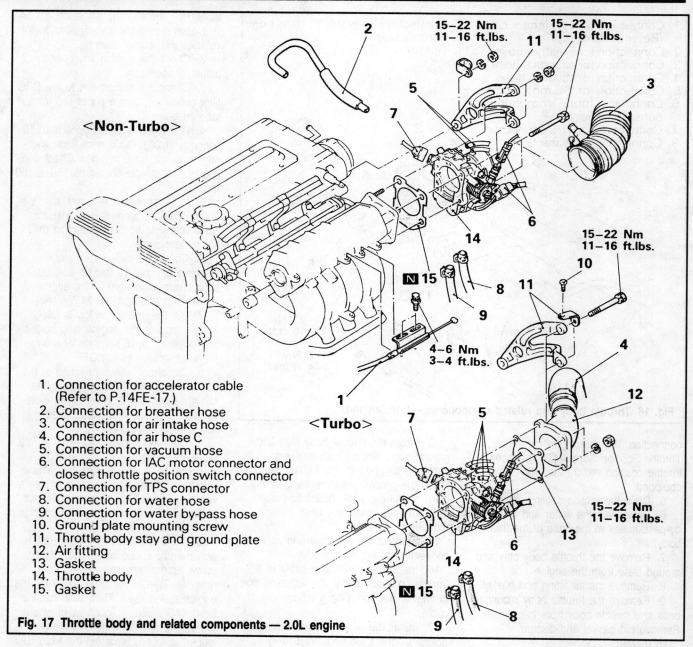

<Non-Turbo>

15–22 Nm
11–16 ft.lbs.

15–22 Nm
11–16 ft.lbs.

15–22 Nm
11–16 ft.lbs.

4–6 Nm
3–4 ft.lbs.

<Turbo>

15–22 Nm
11–16 ft.lbs.

1. Connection for accelerator cable
   (Refer to P.14FE-17.)
2. Connection for breather hose
3. Connection for air intake hose
4. Connection for air hose C
5. Connection for vacuum hose
6. Connection for IAC motor connector and
   closed throttle position switch connector
7. Connection for TPS connector
8. Connection for water hose
9. Connection for water by-pass hose
10. Ground plate mounting screw
11. Throttle body stay and ground plate
12. Air fitting
13. Gasket
14. Throttle body
15. Gasket

Fig. 17 Throttle body and related components — 2.0L engine

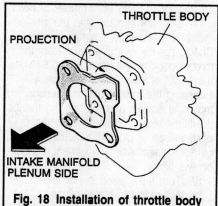

THROTTLE BODY

PROJECTION

INTAKE MANIFOLD
PLENUM SIDE

**Fig. 18 Installation of throttle body
base gasket**

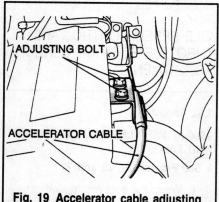

ADJUSTING BOLT

ACCELERATOR CABLE

**Fig. 19 Accelerator cable adjusting
bolts — 1.8L engine**

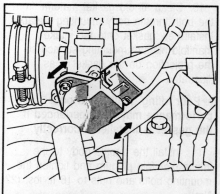

**Fig. 20 Adjusting Throttle Position
Sensor (TPS) — 1.8L engine**

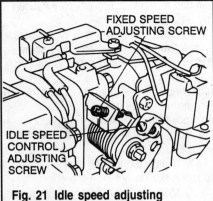

Fig. 21 Idle speed adjusting screws — 1.8L engine

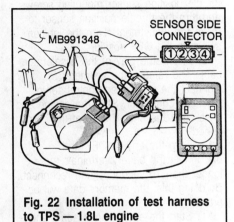

Fig. 22 Installation of test harness to TPS — 1.8L engine

14. Start the engine and check the curb idle. It should be 700 ± 100 rpm.

15. Turn the ignition switch to **OFF**, disconnect the negative battery cable for more than 10 seconds and reconnect. This clears any trouble codes introduced during testing.

16. Restart the engine, allow to run for 5 minutes and check for good idle quality.

Throttle Position Sensor

ADJUSTMENT

1.8L Engine

1992-93 VEHICLES EXCEPT 1992 LASER AND TALON

‣ See Figure 22

➡Complete the basic idle speed adjustment prior to adjusting the throttle position sensor output voltage.

1. Disconnect the throttle position sensor connectors.

2. Install Test Harness Set tool MB991348 or equivalent, between the disconnected connectors of the throttle position sensor.

3. Connect a digital voltmeter between the throttle position sensor terminal **4** and terminal **2**.

4. Turn the ignition key **ON**, but do not start the engine. Hold in this position for 15 seconds or more.

➡Turning the ignition switch ON will extend the idle speed control plunger to the fast idle position. In 15 seconds, the plunger will contract to stop at the idle position. In this position, the idle speed control motor position sensor output voltage of 0.9 volts is present.

5. Check the throttle position sensor output voltage. The throttle position sensor output voltage should read between 0.48-0.52 volts. If measured reading deviates from the desired specifications, adjustment of the throttle position sensor is required.

6. Adjust the throttle position sensor as follows:

  a. Loosen the throttle position sensor mounting screws.

  b. Turn the throttle position sensor body to make the necessary adjustment of the output voltage.

➡Turning the sensor clockwise will increase the output voltage.

  c. Once the adjustment has been made, tighten the throttle position sensor mounting bolts.

➡During adjustment of the throttle position sensor output voltage, the diagnostic trouble code for the idle speed control motor may have been set. This is not idle speed control motor trouble, but a result of deviation of throttle position sensor output voltage from the normal, during the sensor body rotation. Erase the fault code if set.

2.0L Engine

1990-91 VEHICLES AND 1992 LASER AND TALON

1. Loosen the accelerator cable adjusting bolts and slacken the accelerator cable.

2. Connect a digital voltmeter between terminal **19** (throttle position sensor output voltage) of the engine control unit and terminal **24** (ground).

3. Set the ignition switch to **ON**, without starting the engine, and Turn the ignition switch to **ON** but do not start engine. Check that the output voltage from the throttle position sensor is 0.48-0.52 volts.

4. If it is out of specification, adjust the output voltage by loosening the throttle position sensor mounting screws and rotating the throttle position sensor.

➡Turning the throttle position sensor clockwise increases the output voltage.

5. Once the desired adjustment has been made, tighten the sensor mounting screws firmly.

6. Turn the ignition switch **OFF**.

7. Adjust the free-play of the accelerator cable and remove the voltmeter.

8. Start the engine and check the curb idle. It should be 700 ± 100 rpm.

9. Turn the ignition switch to **OFF**, disconnect the negative battery cable for more than 10 seconds and reconnect. This clears any trouble codes introduced during testing.

10. Restart the engine, allow to run for 5 minutes and check for good idle quality.

1992-93 VEHICLES EXCEPT 1992 LASER AND TALON

‣ See Figure 22

➡Complete the basic idle speed adjustment prior to adjusting the throttle position sensor output voltage.

1. Disconnect the throttle position sensor connectors.

2. Install Test Harness Set tool MB991348 or equivalent, between the disconnected connectors of the throttle position sensor.

3. Connect a digital voltmeter between the throttle position sensor terminal **4** and terminal **2**.

4. Turn the ignition key **ON**, but do not start the engine. Hold in this position for 15 seconds or more.

➡Turning the ignition switch ON will extend the idle speed control plunger to the fast idle position. In 15 seconds, the plunger will contract to stop at the idle position. In this position, the idle speed control motor position sensor output voltage of 0.9 volts is present.

5. Check the throttle position sensor output voltage. The throttle position sensor output voltage should read between 0.48-0.52 volts. If measured reading deviates from the desired specifications, adjustment of the throttle position sensor is required.

6. Adjust the throttle position sensor as follows:

   a. Loosen the throttle position sensor mounting screws.

   b. Turn the throttle position sensor body to make the necessary adjustment of the output voltage.

➡Turning the sensor clockwise will increase the output voltage.

   c. Once the adjustment has been made, tighten the throttle position sensor mounting bolts.

➡During adjustment of the throttle position sensor output voltage, the diagnostic trouble code for the idle speed control motor may have been set. This is not idle speed control motor trouble, but a result of deviation of throttle position sensor output voltage from the normal, during the sensor body rotation. Erase the fault code if set.

Idle Position Switch

ADJUSTMENT

The idle position switch has been adjusted by the manufacturer. Do not, therefore, disturb the setting of the idle position switch.

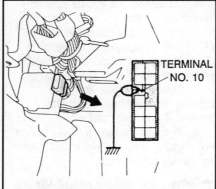

Fig. 23 Using jumper wire to ground the diagnostic test mode control terminal — 2.0L engine

Basic Idle Speed

ADJUSTMENT

1.8L Engine

1991-92 LASER AND TALON

1. Warm the engine to operating temperature, leave lights, electric cooling fan and accessories **OFF**. The transaxle should be in **N** for manual transaxle or **P** for automatic transaxle. Place the steering wheel in a neutral (straight ahead) position for vehicles equipped with power steering.

2. Connect a tachometer to the engine.

3. Connect a digital voltmeter between terminals 19 (throttle position sensor output voltage) and 24 (ground) of the engine control unit.

4. Set the ignition switch to the ON position but do not start the engine. Keep in this position for at least 15 seconds. Set the ignition to the OFF position.

5. Disconnect the idle speed control servo electrical connector. Back out the fixed idle speed adjusting screw enough to allow for adjustment.

6. Start the engine and run at idle.

7. Check that the engine speed is at the desired reading of 650-750 rpm. The engine speed on a vehicle with 300 miles or less may be 20-100 rpm lower than specifications listed above, but adjustments may not be necessary.

8. If adjustment is required, turn the engine OFF and slacken the accelerator cable. Adjust the idle speed using the idle speed adjustment screw. When making the adjustment, use a hexagonal

wrench in order to prevent play caused by backlash.

9. Once the engine rpm is set, screw in the fixed idle speed adjusting screw until the engine speed starts to rise. At this point return the fixed idle speed adjusting screw to find the point at which engine rpm does not change. Once at this point, turn the fixed idle speed adjusting screw in a half turn. Turn the engine OFF.

10. Switch the ignition to the ON position but do not start the engine.

11. Measure the output voltage of the throttle position sensor. Compare reading to the desired voltage of 0.48-0.52 volts. If the voltage is not correct, loosen the throttle position sensor mounting screws and turn the throttle position sensor to make the adjustment. Turn the ignition switch to the OFF position.

12. Adjust the play of the accelerator cable and remove the voltmeter. Connect the idle speed control servo electrical connector. Start the engine and check that the engine idles at the correct speed.

13. Turn the engine OFF and disconnect the battery terminals for longer than 10 seconds, then reconnect. By doing this, the memory data will be erased.

14. Start the engine once again and let idle for about 5 minutes. Check to be sure the idling condition is normal and that the engine speed is correct.

2.0L Engine

▶ **See Figures 23, 24, 25, 26 and 27**

1. Warm the engine to operating temperature, leave lights, electric cooling fan and accessories **OFF**. The transaxle should be in **N** for manual transaxles or **P**, if equipped with automatic transaxle.

2. If equipped with power steering, position the steering wheel in a neutral position (straight ahead).

3. Insert a paper clip into the harness side of the 1 pin connector. Connect a tachometer to the paper clip and read the engine rpm.

➡Half (1/2) of the actual engine rpm is indicated on the tachometer. In order to measure the actual engine rpm, multiply the reading on the tachometer by 2.

4. Use a jumper wire to ground the diagnostic test mode control terminal (terminal 10), of the data link connector.

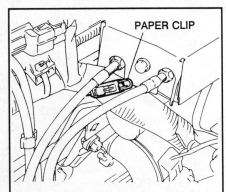

Fig. 24 Insertion of paper clip into the single pin connector — 2.0L engine

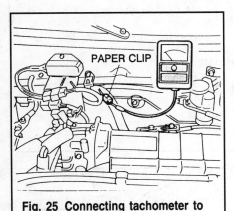

Fig. 25 Connecting tachometer to paper clip — 2.0L engine

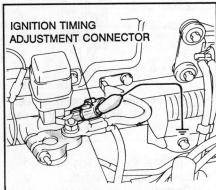

Fig. 26 Using jumper wire to ground the terminal for adjustment of ignition timing — 2.0L engine

5. Use a jumper wire to ground the terminal for adjustment of the ignition timing.

6. Start the engine and allow to idle. Check the standard idle speed and

Fig. 27 Adjustment of the engine speed adjustment screw — 2.0L engine

compare to the desired reading of 750 ± 50 rpm.

a. A new engine will idle more slowly. The engine speed may be 20-100 rpm low for new vehicles (300 miles or less). If this is the case, adjustment is not necessary.

b. If the vehicle stalls or has a very low idle speed, suspect a deposit buildup on the throttle valve which must be cleaned.

7. If not within specifications, turn the engine speed adjusting screw to make the necessary adjustment.

8. Turn the ignition switch to the **OFF** position. Disconnect the jumper wire from the diagnostic test mode control terminal and the terminal for adjustment of ignition timing. Return the connectors to their original condition.

9. Start the engine and allow to idle for about 10 minutes. Check to be sure that the idle condition is normal.

THROTTLE BODY CLEANING

1. Warm the engine to operating temperature and then turn the ignition **OFF** .

2. Remove the air intake hose from the throttle body.

3. Plug the by-pass inlet in the throttle body.

4. Spray cleaning solvent into the valve through the throttle body intake port and let is stand for approximately 5 minutes.

5. Start the engine and race it several times. Allow the engine to idle for about 1 minute. If the idling speed becomes unstable or if the engine stalls, slightly open the throttle valve to keep the engine running.

6. If the deposits on the throttle valve are not removed, repeat Steps 4 and 5.

7. Unplug the bypass passage inlet. Attach the air intake hose.

8. Disconnect the negative battery terminal for 10 seconds or more and then reconnect it.

9. Adjust the basic idle speed, if required.

➡**If the engine hunts (surges) at idle after the basic idle speed has been adjusted, disconnect the negative battery cable from the battery for at least 10 seconds and then, restart the engine and allow to idle.**

Fuel Filter

A replaceable fuel filter is located in the fuel line in the engine compartment.

REMOVAL & INSTALLATION

✳✳CAUTION

Do not use conventional fuel filters, hoses or clamps when servicing fuel injection systems. They are not compatible with the injection system and could fail, causing personal injury or damage to the vehicle. Use only hoses and clamps specifically designed for fuel injection.

1. Relieve the fuel system pressure as follows:

a. Loosen the fuel filler cap to release fuel tank pressure.

b. Disconnect the fuel pump harness connector located at the rear of the fuel tank.

c. Start the vehicle and allow it to run until it stalls from lack of fuel. Turn the key to the **OFF** position.

d. Disconnect the negative battery cable, then reconnect the fuel pump connector and reinstall the fuel filler cap.

➡**Wrap shop towels around the fitting that is being disconnected to absorb residual fuel in the lines.**

2. Hold the fuel filter nut securely with a backup or spanner wrench. Cover the hoses with shop towels and remove the eye bolt. Discard the gaskets.

3. Separate the flare nut connection at the filter. Discard the gaskets.

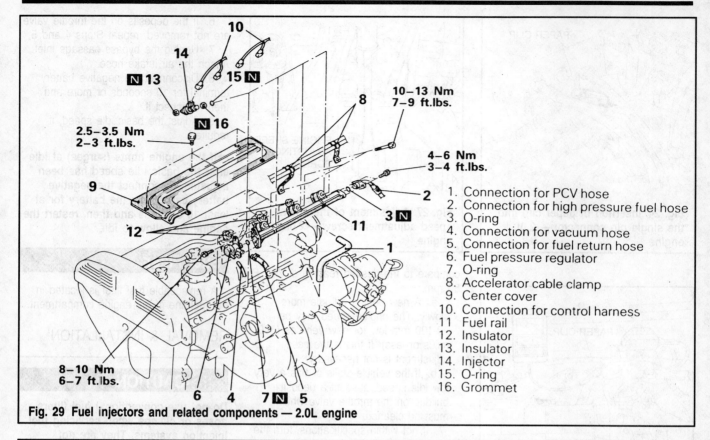

2.5–3.5 Nm
2–3 ft.lbs.

10–13 Nm
7–9 ft.lbs.

4–6 Nm
3–4 ft.lbs.

8–10 Nm
6–7 ft.lbs.

1. Connection for PCV hose
2. Connection for high pressure fuel hose
3. O-ring
4. Connection for vacuum hose
5. Connection for fuel return hose
6. Fuel pressure regulator
7. O-ring
8. Accelerator cable clamp
9. Center cover
10. Connection for control harness
11. Fuel rail
12. Insulator
13. Insulator
14. Injector
15. O-ring
16. Grommet

Fig. 29 Fuel injectors and related components — 2.0L engine

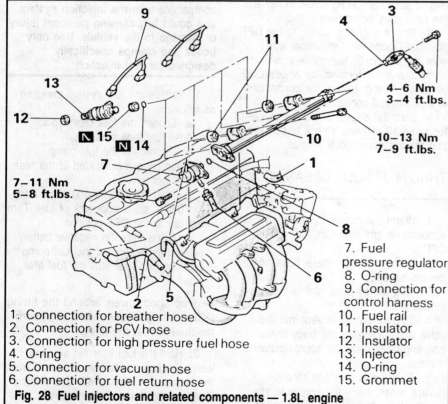

7–11 Nm
5–8 ft.lbs.

4–6 Nm
3–4 ft.lbs.

10–13 Nm
7–9 ft.lbs.

1. Connection for breather hose
2. Connection for PCV hose
3. Connection for high pressure fuel hose
4. O-ring
5. Connection for vacuum hose
6. Connection for fuel return hose
7. Fuel pressure regulator
8. O-ring
9. Connection for control harness
10. Fuel rail
11. Insulator
12. Insulator
13. Injector
14. O-ring
15. Grommet

Fig. 28 Fuel injectors and related components — 1.8L engine

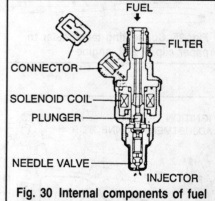

Fig. 30 Internal components of fuel injector — 1.8L engine. Other models similar.

4. Remove the mounting bolts and the fuel filter from the vehicle.

To install:

5. If equipped with flare fitting, install a new O-ring and tighten the fitting by hand before installing the filter to the vehicle.

6. Install the filter to its bracket only finger-tight. Movement of the filter will ease attachment of the fuel lines.

7. Install new O-rings and connect the high pressure hose and eye bolt, then the main pipe and eye bolt. While

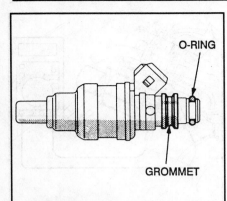

Fig. 31 Positioning of O-ring and grommet on fuel injector

holding the fuel filter nut, tighten the eye bolts to 22 ft. lbs. (30 Nm). Tighten the flare nut to 25 ft. lbs. (35 Nm).

8. Tighten the filter mounting bolts fully.

9. Install the air cleaner assembly, if removed.

10. Connect the negative battery cable, install the fuel filler cap, turn the key to the **ON** position to pressurize the fuel system and check for leaks.

11. Release the fuel pressure and repair leaks as required.

Fuel Injectors

REMOVAL & INSTALLATION

▶ **See Figures 28, 29, 30 and 31**

1. Relieve the fuel system pressure as follows:

a. Loosen the fuel filler cap to release fuel tank pressure.

b. Disconnect the fuel pump harness connector located at the rear of the fuel tank.

c. Start the vehicle and allow it to run until it stalls from lack of fuel. Turn the key to the **OFF** position.

d. Disconnect the negative battery cable, then reconnect the fuel pump connector and reinstall the fuel filler cap.

2. Disconnect the PCV hose from the valve cover. On 1.8L engines, also disconnect the breather hose at the opposite end of the valve cover.

3. Remove the bolts holding the high pressure fuel line to the fuel rail and disconnect the line. Be prepared to contain fuel spillage; plug the line to keep out dirt and debris.

4. Remove the vacuum hose from the fuel pressure regulator.

5. Disconnect the fuel return hose from the pressure regulator. Remove the fuel pressure regulator mounting bolts and remove from the fuel rail.

6. On 2.0L engines, remove the clamps holding the accelerator cable.

7. On 2.0L engines, remove the center engine cover from between the cam cover.

8. Label and disconnect the electrical connector from each injector.

9. Remove the bolt(s) holding the fuel rail to the manifold. Carefully lift the rail up and remove it with the injectors attached. Take great care not to drop an injector. Place the rail and injectors in a safe location on the workbench; protect the tips of the injectors from dirt and/or impact.

10. Remove and discard the injector insulators from the intake manifold. The insulators are not reusable.

11. Remove the injectors from the fuel rail by pulling gently in a straight outward motion. Make certain the grommet and O-ring come off with the injector.

To install:

12. Install a new insulator in each injector port in the manifold.

13. Remove the old grommet and O-ring from each injector. Install a new grommet and O-ring; coat the O-ring lightly with clean, thin oil.

14. If the fuel pressure regulator was removed, replace the O-ring with a new one and coat it lightly with clean, thin oil. Insert the regulator straight into the rail, then check that it can be rotated freely. If it does not rotate smoothly, remove it and inspect the O-ring for deformation or jamming. When properly installed, align the mounting holes and tighten the retaining bolts to 8 ft. lbs. (11 Nm). This procedure must be followed even if the fuel rail was not removed.

15. Install the injector into the fuel rail, constantly turning the injector left and right during installation. When fully installed, the injector should still turn freely in the rail. If it does not, remove the injector and inspect the O-ring for deformation or damage.

16. Install the delivery pipe and injectors to the engine. Make certain that each injector fits correctly into its port and that the rubber insulators for the fuel rail mounts are in position.

17. Install the fuel rail retaining bolts and tighten them to 8 ft. lbs. (11 Nm).

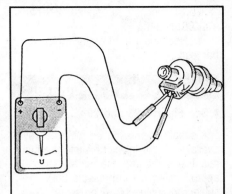

Fig. 32 Measuring resistance of fuel injector

On 2.0L engines, install the accelerator cable retaining clips before the rail retaining bolts.

18. Connect the wiring harnesses to the appropriate injector.

19. On 2.0L engines, reinstall the center cover. Tighten the retaining bolts only to 3 ft. lbs. (4 Nm).

20. Connect the fuel return hose to the pressure regulator, then connect the vacuum hose.

21. Replace the O-ring on the high pressure fuel line, coat the O-ring lightly with clean, thin oil and install the line to the fuel rail. Tighten the mounting bolts to 4 ft. lbs. (6 Nm).

22. Connect the PCV hose and the breather hose if they were disconnected.

23. Connect the negative battery cable. Pressurize the fuel system and inspect all connections for leaks.

TESTING

▶ **See Figure 32**

With the engine running at idle, use a stethoscope or similar tool to listen to the individual injectors. With the stethoscope, each injector should exhibit a distinct clicking as it functions. The speed of the clicking should increase with engine speed. Note that other injectors may be heard through a non-functioning injector.

To check the resistance of each injector:

1. Turn the ignition **OFF** .

2. Disconnect the injector harness from the injector to be tested.

3. Measure the resistance across the injector terminals. Reference resistances are:

a. Non-turbocharged engines: 13-16 ohms at 68 degrees F.

b. Turbocharged engines: 2-3 ohms at 68 degrees F.

4. Reconnect the injector wiring harness.

Fuel Pressure Regulator

▶ See Figures 28 and 29

A supply of pressurized fuel is always available to the injectors within the fuel rail. The fuel enters the rail at approximately pump pressure but is subject to fluctuation. A pressure regulator is used to provide accurate, constant pressure. The regulator contains a spring valve and diaphragm; vacuum is supplied to the chamber behind the diaphragm. If the fuel pressure rises above the compression point of the spring, fuel is allowed to bleed off into the fuel return line to the tank. Changes in vacuum, determined by engine operation, repositions the spring within the regulator, slightly altering the amount of pressure necessary to open the passage.

REMOVAL & INSTALLATION

1. Relieve the fuel system pressure as follows:

   a. Loosen the fuel filler cap to release fuel tank pressure.

   b. Disconnect the fuel pump harness connector located at the rear of the fuel tank.

   c. Start the vehicle and allow it to run until it stalls from lack of fuel. Turn the key to the **OFF** position.

   d. Disconnect the negative battery cable, then reconnect the fuel pump connector and reinstall the fuel filler cap.

2. Disconnect the PCV hose from the valve cover. On 1.8L engines, also disconnect the breather hose at the opposite end of the valve cover.

3. Remove the vacuum hose from the fuel pressure regulator.

4. Disconnect the fuel return hose from the pressure regulator.

5. Remove the fuel regulator retainer bolts and the fuel regulator from the fuel rail.

To install:

6. Replace the O-ring on fuel pressure regulator with a new one and coat it lightly with clean, thin oil.

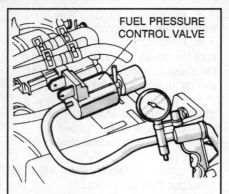

Fig. 33 Testing fuel pressure control valve with vacuum and no voltage applied — 1991 2.0L turbocharged engine

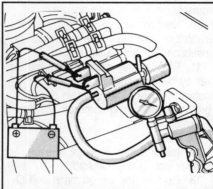

Fig. 34 Testing fuel pressure control valve with vacuum and voltage applied — 1991 2.0L turbocharged engine

7. Insert the regulator straight into the rail, then check that it can be rotated freely.

➡**If it does not rotate smoothly, remove it and inspect the O-ring for deformation or damage.**

8. When properly installed, align the mounting holes. Install and tighten the retaining bolts to 8 ft. lbs. (11 Nm).

9. Connect the fuel return hose to the pressure regulator.

10. Install the vacuum hose to the fuel pressure regulator.

11. Connect the PCV hose to the valve cover. On 1.8L engines, also connect the breather hose at the opposite end of the valve cover.

12. Connect the negative battery cable and pressurize the fuel system. Inspect for leaks.

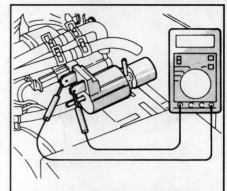

Fig. 35 Measuring the resistance of the fuel pressure control valve — 1991 2.0L turbocharged engine

Fuel Pressure Control Valve

▶ See Figures 33, 34 and 35

OPERATIONAL CHECK

1990-92 2.0L Turbocharged Engine

1. Label and disconnect the vacuum hoses from the solenoid valve.

2. Disconnect the electrical harness connector from the valve.

3. Apply vacuum with a hand pump to the nipple to which the black vacuum hose was connected. Check for air tightness both when a voltage is applied to the solenoid valve terminal and when the voltage is removed. Test and compare the results to specifications as follows:

   a. With battery voltage applied to the solenoid and the other nipple on the solenoid valve open — vacuum should be retained.

   b. When battery voltage is removed and the other nipple is blocked off with finger — Vacuum should be retained.

   c. When battery voltage is removed and the other nipple is not blocked off with finger — Vacuum should not be retained.

4. If the test results differ from specifications, replace the fuel pressure control valve.

5. Measure the resistance across the 2 terminals of the valve. The desired value is 36-46 ohms at 68°F (20°C).

6. If the measured resistance differs from specifications, replace the fuel pressure control valve.

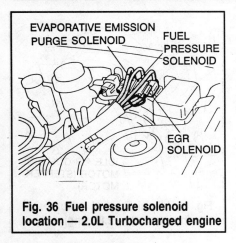

Fig. 36 Fuel pressure solenoid location — 2.0L Turbocharged engine

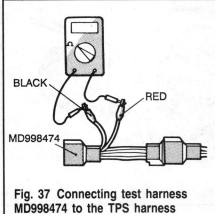

Fig. 37 Connecting test harness MD998474 to the TPS harness connector

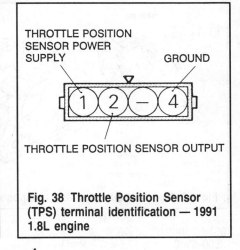

Fig. 38 Throttle Position Sensor (TPS) terminal identification — 1991 1.8L engine

REMOVAL & INSTALLATION

1. Disconnect the negative battery cable.
2. Label and disconnect the vacuum hose connections from the valve.
3. Disconnect the electrical connections from the valve.
4. Remove the retainer and the valve from the vehicle.

To install:

5. Install the valve to the vehicle and secure.
6. Install the electrical and vacuum connections to the valve in their original locations.
7. Connect the negative battery cable.

Fuel Pressure Solenoid

▶ **See Figure 36**

1993 2.0L turbocharged engines incorporate an electric fuel pressure solenoid to provide more precise fuel pressure control. The solenoid is located in the engine compartment, to the left of the brake reservoir at the firewall.

The negative pressure in the intake manifold is usually applied to the fuel pressure regulator. This was to assure the fuel injection amount to be proportional to the injector drive time by maintaining the fuel pressure to be constant to the pressure in the intake manifold. When the engine is started with high engine coolant temperature or the high intake air temperature is high, the ECM flows electricity through the fuel pressure solenoid to apply atmospheric pressure to the fuel pressure regulator. This increases the fuel pressure to prevent vaporization of fuel at the higher

temperatures, thus maintaining the stability of idling soon after restarting the engine.

If the fuel pressure solenoid is not turned on because of wire breakage in the circuit, restartability after hot shutdown may be poor.

REMOVAL & INSTALLATION

1993 2.0L Turbocharged Engine

1. Disconnect the negative battery cable.
2. Label and disconnect the vacuum connectors at the group of solenoids.
3. Remove the fuel pressure solenoid from the mounting bracket and replace as required.
4. Installation is the reverse of the removal procedure.

Throttle Position Sensor

The Throttle Position Sensor (TPS) is an electrical resistor which is activated by the movement of the throttle shaft. It is mounted on the throttle body and senses the angle of the throttle blade opening. The voltage that the sensor produces increases or decreases according to the throttle blade opening. This voltage is transmitted to the ECU where it is used along with data from other sensors to adjust the air/fuel ratio to varying conditions and during acceleration, deceleration, idle, with wide open throttle operations.

TESTING

▶ **See Figures 37 and 38**

1990 Vehicles

1. With the ignition **OFF** , disconnect the sensor connector from the harness.
2. If equipped with 1.8L engine, connect test harness MD998474 or equivalent, to the TPS harness. If equipped with 2.0L engine, connect test harness MD998464 or equivalent, to the TPS harness. The test harness will prevent damage to the female connector. If the test is done without the test harness, do not insert the test probe into the female connector of the sensor.
3. On 1.8L engine, using an analog (needle type) ohmmeter, measure the resistance across the sensor power supply terminal (black clip) and the sensor ground terminal (red clip). Normal resistance is 3500-6500 ohms (3.5-6.5 Kohm;).
4. If equipped with 2.0L engine, using an analog (needle type) ohmmeter, measure the resistance across the sensor power supply terminal (white clip) and the sensor ground terminal (red clip). Normal resistance is 3500-6500 ohms (3.5-6.5 Kohm;).
5. If the measured value differs from the desired readings, the sensor must be replaced.

Except 1990 Vehicles

1. With the ignition **OFF** , disconnect the sensor connector from the harness.
2. Using an analog (needle type) ohmmeter, measure the resistance across terminal No. **1** (sensor power supply) and terminal No. **4** (sensor

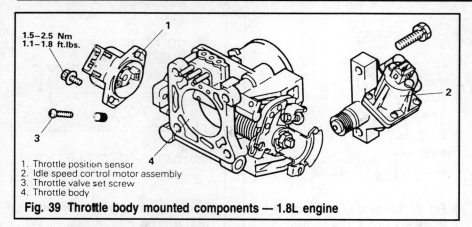

1.5–2.5 Nm
1.1–1.8 ft.lbs.

1. Throttle position sensor
2. Idle speed control motor assembly
3. Throttle valve set screw
4. Throttle body

Fig. 39 Throttle body mounted components — 1.8L engine

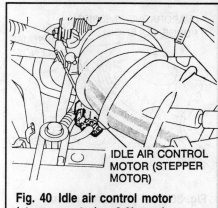

IDLE AIR CONTROL MOTOR (STEPPER MOTOR)

Fig. 40 Idle air control motor (stepper motor) — 2.0L engine

ground terminal. Normal resistance is 3500-6500 ohms (3.5-6.5 Kohm;).

3. Move the ohmmeter probes to test across terminal **2** and terminal **4** . Slowly operate the throttle from idle to wide open; the resistance shown on the meter must change evenly and in proportion to throttle movement.

4. If resistance is out of specification or fails to change smoothly, the sensor must be replaced.

REMOVAL & INSTALLATION

▶ **See Figure 39**

1. Disconnect negative battery cable.
2. Disconnect the electrical connector from the throttle position sensor.
3. Remove the mounting screws from the sensor taking care not to round the Phillips screw head.
4. Remove the sensor from the throttle body.
   To install:
5. Install the throttle position sensor onto the throttle body and rotate the sensor counterclockwise on the throttle shaft and temporarily tighten the screws.
6. Connect the electrical harness to the sensor.
7. Adjust the throttle position sensor. Torque the retainer screws to 1.8 ft. lbs. (2.5 Nm).

Idle Speed Control Servo

▶ **See Figures 39, 40, 41 and 42**

The idle speed control servo is mounted on the throttle body. Data from various sensors and switches are used by the ECU to tell the servo to adjust engine idle correctly. The servo adjusts the air portion of the air/fuel mixture through the throttle body.

TESTING

1.8L Engine

1. With the ignition **OFF** , disconnect the idle speed servo connector.
2. Use an ohmmeter to check continuity and resistance between the terminals. Reference resistance is 5-35 ohms at 68 degrees F.
3. Apply +6 volts DC to terminal 1; connect terminal 2 to the 6 volt ground (-).

➡ **Use only 6 volts DC or lower voltage for this test. Higher voltage may cause the servo gears to lock.**

4. With voltage applied the servo should operate.
5. If either the resistance test or the operation test do not give proper results, the servo must be replaced as an assembly.

2.0L Engine

▶ **See Figures 41 and 42**

1. Turn the ignition **ON** but do not start the engine. Listen for the sound of the stepper motor. If no sound is heard, inspect the motor drive circuit wiring.
2. Turn the ignition switch **OFF** . Disconnect the idle speed control servo connector.
3. Measure the resistance between terminal 2 and either terminal 1 or 3. Measure the resistance between terminal 5 and either terminal 4 or 6.
4. Reference resistance for each circuit is 28-33 ohms at 68°F. All circuits must pass the resistance test; if any resistance is significantly out of specification the servo must be replaced as an assembly.
5. To perform the operational test, remove the throttle body from the

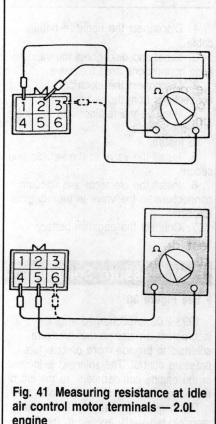

Fig. 41 Measuring resistance at idle air control motor terminals — 2.0L engine

vehicle. Remove the stepper motor from the throttle body.

➡ **Use only 6 volts DC or lower voltage for this test. Higher voltage may cause the servo gears to lock.**

6. Connect a 6 volt DC power supply so that the positive terminal is connected to terminals 2 and 5 of the servo connector. The test is performed by connecting the negative side of the power supply to pairs of connector

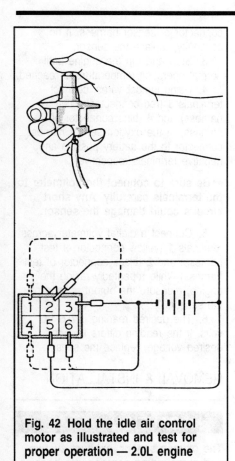

Fig. 42 Hold the idle air control motor as illustrated and test for proper operation — 2.0L engine

terminals. When connection is made, the stepper motor should be felt to vibrate slightly when held in the hand. Connect the negative side of the power supply to the following connector terminal pairs:

    a. Terminals 3 and 6
    b. Terminals 1 and 6
    c. Terminals 1 and 4
    d. Terminals 3 and 4
    e. Terminals 3 and 6
    f. Repeat the sequence in reverse order: 3-6, 3-4, 1-4, etc.

    7. The motor must respond correctly to all tests. If any test is failed, the stepper motor must be replaced.

REMOVAL & INSTALLATION

    1. Disconnect the negative battery cable.
    2. Drain the engine coolant.
    3. Label and disconnect the electrical connectors at the throttle body.
    4. Remove the mounting bolts and the throttle body from the engine.

➡**The Phillips screws are tightly installed; use a screwdriver of the correct size to avoid damage to the**

heads. **Additionally, some screws are coated with adhesive or thread-locking compound.**

    5. Unbolt the idle speed control servo and remove from the throttle body.
    To install:
    6. Clean the throttle body and removed servo, but the idle speed control servo must not be sprayed or immersed in cleaner. The insulation of these units will be damaged and they will not function properly; clean them only with a soft cloth. Vacuum ports should be inspected and cleaned with compressed air.
    7. Install the idle speed control servo to the throttle body. Install the retaining bolts and tighten to 1.5 ft. lbs (2 Nm).
    8. Correctly position a new gasket between the throttle body and the air plenum. Install the throttle body. On turbocharged engines, remember to install the air fitting and gasket on the intake side before the retainers.
    9. Reconnect the electrical connectors in their original location. Install the accelerator cable and remaining components disconnected during removal of the throttle body.
    10. Reconnect the negative battery cable and fill the cooling system.

Engine Control Unit (ECU)

REMOVAL & INSTALLATION

    1. Remove the left and/or right side panel from the center console.
    2. Remove the bolts holding the ECU to the mounting bracket.
    3. Disconnect the wiring harness from the ECU.
    4. Reinstall in reverse order. Make certain the multi-pin connector is firmly and squarely seated to the ECU. Secure any retaining clips or locks holding the connector.
    5. Reinstall the console side panels.

Heated Oxygen Sensor

TESTING

    The oxygen sensor, mounted in the exhaust gas flow, is a device which produces an electrical voltage when exposed to oxygen present in the engine

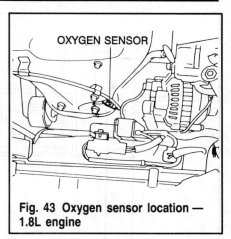

Fig. 43 Oxygen sensor location — 1.8L engine

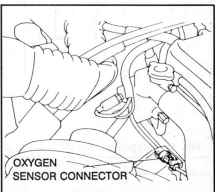

Fig. 44 Oxygen sensor connector location — 1.8L engine

exhaust gases. The electrical voltage produced by the oxygen sensor is proportional to the amount of oxygen present in the exhaust and therefore, representative of the air/fuel ratio. The voltage produced by the sensor is sent to the ECM. The ECM will check whether the actual air/fuel mixture ratio is richer or leaner than the optimal (theoretical) ratio, and adjust accordingly. The oxygen sensor is electrically heated internally for faster switching during cold engine operation.

    If the oxygen sensor has malfunctioned, the driveability of the vehicle may not be influenced. However, hazardous components (HC, CO, NOx) in the exhaust gas will increase.

1.8L Engine
▶ **See Figures 43, 44 and 45**

    1. Warm the engine to operating temperature.
    2. Use an accurate digital voltmeter for testing. Disconnect the oxygen sensor connector and connect the voltmeter.

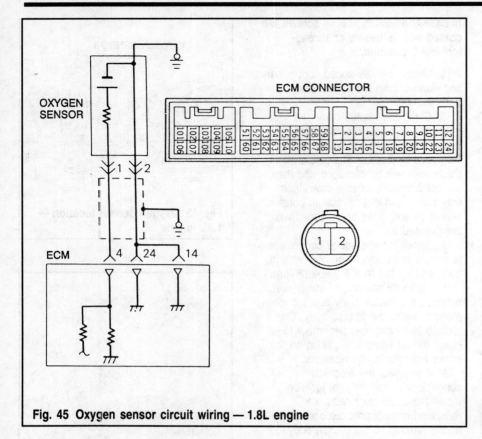

Fig. 45 Oxygen sensor circuit wiring — 1.8L engine

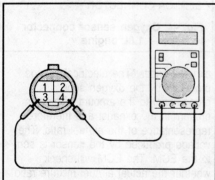

Fig. 46 Testing oxygen sensor for continuity at connector — 2.0L engine

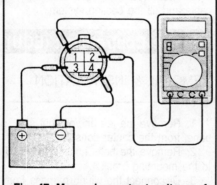

Fig. 47 Measuring output voltage of oxygen sensor — 2.0L engine

3. While repeatedly racing the the engine to richen the air-fuel mixture, measure the oxygen sensor output voltage. The voltage should be 0.6-1.0 volts.

4. If the voltage reading is out of specification, replace the oxygen sensor.

2.0L Engine
▶ **See Figures 46 and 47**

1. Disconnect the oxygen sensor connector. Connect test harness

MD998460 or equivalent, to the sensor harness. If the test harness is not available, perform the test procedure using the sensor harness terminals listed below. The color codes may not be the same on the sensor harness as they are for the test harness.

2. Measure the resistance across terminals **4** (blue connector of test harness) and **3** (red connector of test harness) of the oxygen sensor test

connector or sensor harness. If no continuity, replace the sensor.

3. Start and run the engine until normal operating temperature is reached.

4. Using jumper wires, connect terminals **3** (red connector of test harness) and **4** (blue connector of test harness) of the oxygen sensor harness connector to the battery positive and negative terminals respectively.

➡**Be sure to connect the voltmeter to the terminals carefully. Any short circuits could damage the sensor.**

5. Connect a digital voltmeter across terminals **1** (yellow connector of test harness) and **2** (black connector of test harness).While repeatedly racing the engine, measure the output voltage of the sensor.

6. The desired reading is 0.6-1.0 volts. If the reading differs from the desired voltage, replace the sensor.

REMOVAL & INSTALLATION

✷✷CAUTION

The oxygen sensor, exhaust system and surrounding components become very hot during engine operation. Avoid personal injury by waiting for a cooled motor or wearing protective clothing and gloves.

1. Disconnect negative battery cable.
2. Disconnect the electrical connector from the sensor.
3. Raise and safely support the vehicle.
4. Remove the oxygen sensor from the exhaust manifold.
   To install:
5. If not already done, coat the threads of the replacement sensor with anti-seize compound. New sensors are already coated with the substance. Take great care not to contaminate the oxygen sensor probe with the anti-seize compound.
6. Install the oxygen sensor into the exhaust manifold. Tighten the sensor to 33 ft. lbs. (45 Nm).
7. Reconnect the wiring harness to the sensor. Connect the negative battery cable.

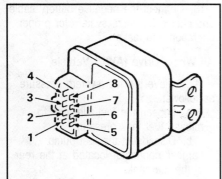

Fig. 48 Control relay terminal identification — 1991 1.8L engine

Control Relay

TESTING

1.8L Engine

▶ See Figure 48

1. With the ignition switch **OFF** , disconnect and remove the control relay.

2. Use an ohmmeter to check continuity and resistance between terminals as follows:

  a. Terminals 3 and 5: Approximately 95 ohms

  b. Terminals 2 and 5: Approximately 95 ohms

  c. Terminals 6 and 7: Approximately 35 ohms

  d. Terminals 6 and 8: Continuity in one direction only

➡The following steps require the application of 12 volts and ground to relay pins. Take great care to connect the jumpers correctly; the relay will be damaged if a mistake is made.

3. Connect relay terminal 7 to +12 volt power supply. Connect relay terminal 6 to ground (-). Check continuity between relay terminals 1 and 4; continuity should exist when the 12 volt circuit is connected. There should be no continuity when the 12 volt circuit is open or disconnected.

4. Omit this Step on 1990 vehicles. Disconnect the 12 volt circuit. Connect relay terminal 2 to +12 volt and connect relay terminal 5 to ground (-). Check continuity between relay terminals 1 and 4; continuity should exist when the 12 volt circuit is connected. There should be no continuity when the 12 volt circuit is open or disconnected.

5. Disconnect the 12 volt circuit. Connect relay terminal 8 to +12 volt and connect relay terminal 6 to ground (-). Check continuity between relay terminals 2 and 4; continuity should exist when the 12 volt circuit is connected. There should be no continuity when the 12 volt circuit is open or disconnected.

6. The relay must pass all tests; if any circuit is found faulty, the relay must be replaced.

2.0L Engine

▶ See Figure 49

1. With the ignition switch **OFF** , disconnect and remove the control relay.

➡The following steps require the application of 12 volts and ground to relay pins. Take great care to connect the jumpers correctly; the relay will be damaged if a mistake is made.

2. Using jumper wire, connect control relay terminal No. 10 to the positive battery terminal (+12 volts) and connect relay terminal No. 8 to the battery negative terminal (-). Check voltage between relay terminals 4 and 5; voltage

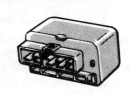

Fig. 49 Control Relay terminal identification — 1991 2.0L engine

should exist when the 12 volt circuit is connected. There should be no voltage present when the 12 volt circuit is open or disconnected.

3. Disconnect the 12 volt circuit. Connect relay terminal No. 9 to battery positive terminal (+12 volt) and connect relay terminal No. 6 to the battery negative terminal (-). Check continuity between relay terminals No. 2 and No. 3; continuity should exist when the 12 volt circuit is connected. There should be no continuity when the 12 volt circuit is open or disconnected.

4. Disconnect the 12 volt power supply from the circuit. Connect relay terminal No. 3 to +12 volt power supply and connect relay terminal No. 7 to battery negative terminal (-). Check for voltage at relay terminal No. 2; battery voltage should be present when the 12 volt circuit is connected. There should be no voltage present when the 12 volt circuit is open or disconnected.

5. The relay must pass all tests; if any circuit is found faulty, the relay must be replaced.

FUEL TANK

Tank Assembly

REMOVAL & INSTALLATION

▶ See Figures 50 and 51

1. Relieve the fuel system pressure as follows:

  a. Loosen the fuel filler cap to release fuel tank pressure.

  b. Disconnect the fuel pump

harness connector located at the rear of the fuel tank.

  c. Start the vehicle and allow it to run until it stalls from lack of fuel. Turn the key to the **OFF** position.

  d. Disconnect the negative battery cable, then reconnect the fuel pump connector and reinstall the fuel filler cap.

➡Wrap shop towels around the fitting that is being disconnected to absorb residual fuel in the lines.

2. Raise the vehicle and support safely.

3. Drain the fuel from the fuel tank into an approved container.

4. Disconnect the return hose, high pressure hose and vapor hoses from the fuel pump.

5. Disconnect the electrical connectors at the pump/sending unit.

❊❊CAUTION

Cover all fuel hose connections with a shop towel, prior to disconnecting, to prevent splash of fuel that could be caused by residual pressure remaining in the fuel line.

6. Disconnect the filler and vent hoses.

7. Place a transmission jack under the center of the fuel tank and apply a slight upward pressure. Remove the fuel tank strap retaining nut.

8. Lower the tank slightly and disconnect any remaining electrical or hose connectors at the fuel tank.

9. Remove the fuel tank from the vehicle.

To install:

10. Install the fuel tank onto the transmission jack. Raise the tank in position under the vehicle. Leave enough clearance to attach the electrical and hose connections to the top of the fuel pump.

11. Attach all connections to the top of the tank.

12. Raise the tank completely and position the retainer straps around the fuel tank. Install new fuel tank self-locking nuts and tighten to 22 ft. lbs. (31 Nm).

13. Connect the return hose and high pressure hoses.

14. Install the vapor hose and the filler hose. Install the filler hose retainer screws to the fender, if removed.

15. Lower the vehicle and pour the drained fuel into the gas tank.

16. Connect the negative battery cable. Check the fuel pump for proper pressure and inspect the entire system for leaks.

FUEL SENDING UNIT REPLACEMENT

Front Wheel Drive Vehicle

1. Relieve the fuel system pressure as follows:

a. Loosen the fuel filler cap to release fuel tank pressure.

b. Disconnect the fuel pump harness connector located at the rear of the fuel tank.

c. Start the vehicle and allow it to run until it stalls from lack of fuel. Turn the key to the **OFF** position.

d. Disconnect the negative battery cable, then reconnect the fuel pump connector and reinstall the fuel filler cap.

➡**Wrap shop towels around the fitting that is being disconnected to absorb residual fuel in the lines.**

2. Raise the vehicle and support safely.

3. Drain the fuel from the fuel tank into an approved container.

4. Loosen the self locking nuts until they are at the bottom of the stud bolts.

5. Remove the right side lateral rod attaching bolt and disconnect the arm from the right body coupling. Lower the lateral rod and suspend from the axle beam using wire.

6. Disconnect the electrical connector from the fuel gauge sending unit.

7. Remove the sending unit retainer nuts and sending unit from the fuel tank.

To install:

8. Install the fuel gauge sending unit into the fuel tank while aligning the 2 positioning projections on the packing with the holes in the fuel gauge unit.

❊❊CAUTION

When mounting the fuel gauge unit, incline the end of the float towards the left and install it into the tank. Since the reservoir cap is provided in the fuel tank, the fuel gauge unit will contact the reservoir cap if the fuel gauge unit is inclined rightward during insertion.

9. Connect the electrical harness to the fuel gauge unit.

10. Install the right side lateral rod and attaching bolt into the right body coupling. Tighten loosely only, at this time.

11. Tighten self-locking nuts on tank support straps until tank is seated fully. Torque nuts to 22 ft. lbs. (31 Nm).

12. Install the high pressure fuel hose connector and tighten to 29 ft. lbs. (40 Nm), if disconnected.

13. Install the electrical connectors onto the fuel pump and gauge unit assemblies.

14. Lower the vehicle so the suspension supports the weight of the vehicle. Tighten the lateral rod attaching bolt to 58-72 ft. lbs. (80-100 Nm).

15. Refill the fuel tank with fuel drained during this procedure.

16. Connect the negative battery cable and check the entire system for proper operation and leaks.

All Wheel Drive (AWD) Vehicle

1. Relieve the fuel system pressure as follows:

a. Loosen the fuel filler cap to release fuel tank pressure.

b. Disconnect the fuel pump harness connector located at the rear of the fuel tank.

c. Start the vehicle and allow it to run until it stalls from lack of fuel. Turn the key to the **OFF** position.

d. Disconnect the negative battery cable, then reconnect the fuel pump connector and reinstall the fuel filler cap.

2. The fuel pump/gauge unit is located in the fuel tank. Remove the hole cover located in the rear floor pan.

3. Partially drain the fuel tank into an approved gasoline container.

4. Remove the electrical connector from the fuel pump.

5. Remove the overfill limiter (two-way valve), as required.

6. Cover the hose connection with a shop towel to prevent any splash of fuel due to residual pressure in the fuel pipe. Remove the high pressure fuel hose connector.

7. Remove the fuel pump and gauge assembly from the tank. Note positioning of pump prior to removal from tank.

To install:

8. Align the 3 projections on the packing with the holes on the fuel pump and the nipples on the pump facing the same direction as before removal. Install the retainers and tighten to 2 ft. lbs. (3 Nm).

9. Install the high pressure hose connection and tighten to 29 ft. lbs. (40 Nm).

10. Install the overfill limiter (two-way valve) and the electrical connector to the fuel pump assembly, if removed.

11. Fill the fuel tank with the gasoline removed during this procedure.

12. Reconnect the negative battery cable and check the entire system for leaks and proper operation.

13. Install MOPAR Rope Caulk Sealer part 4026044 or equivalent, to the rear floor pan and install the cover into place.

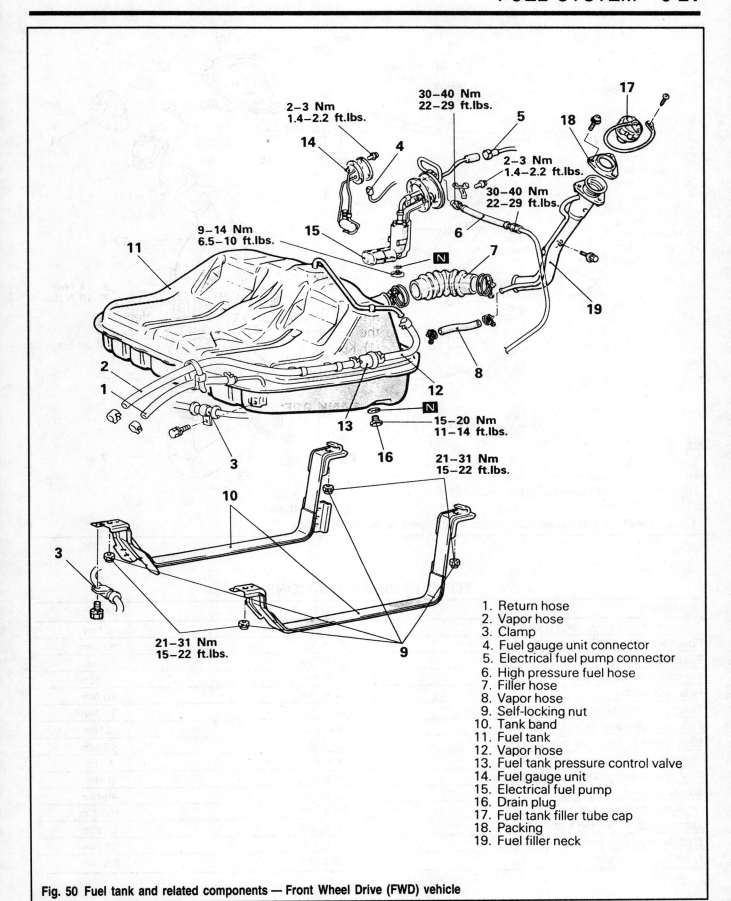

2–3 Nm
1.4–2.2 ft.lbs.

30–40 Nm
22–29 ft.lbs.

2–3 Nm
1.4–2.2 ft.lbs.

30–40 Nm
22–29 ft.lbs.

9–14 Nm
6.5–10 ft.lbs.

15–20 Nm
11–14 ft.lbs.

21–31 Nm
15–22 ft.lbs.

21–31 Nm
15–22 ft.lbs.

1. Return hose
2. Vapor hose
3. Clamp
4. Fuel gauge unit connector
5. Electrical fuel pump connector
6. High pressure fuel hose
7. Filler hose
8. Vapor hose
9. Self-locking nut
10. Tank band
11. Fuel tank
12. Vapor hose
13. Fuel tank pressure control valve
14. Fuel gauge unit
15. Electrical fuel pump
16. Drain plug
17. Fuel tank filler tube cap
18. Packing
19. Fuel filler neck

Fig. 50 Fuel tank and related components — Front Wheel Drive (FWD) vehicle

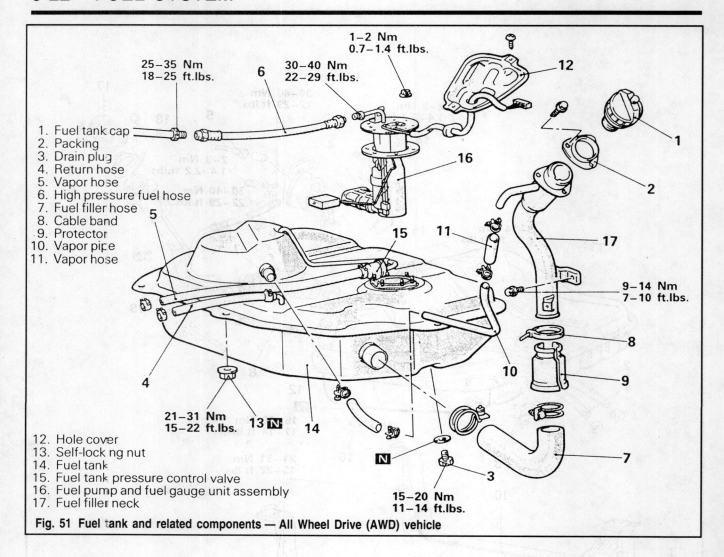

25–35 Nm
18–25 ft.lbs.

1–2 Nm
0.7–1.4 ft.lbs.

30–40 Nm
22–29 ft.lbs.

1. Fuel tank cap
2. Packing
3. Drain plug
4. Return hose
5. Vapor hose
6. High pressure fuel hose
7. Fuel filler hose
8. Cable band
9. Protector
10. Vapor pipe
11. Vapor hose

12. Hole cover
13. Self-locking nut
14. Fuel tank
15. Fuel tank pressure control valve
16. Fuel pump and fuel gauge unit assembly
17. Fuel filler neck

9–14 Nm
7–10 ft.lbs.

21–31 Nm
15–22 ft.lbs.

15–20 Nm
11–14 ft.lbs.

Fig. 51 Fuel tank and related components — All Wheel Drive (AWD) vehicle

TORQUE SPECIFICATIONS

Component	US	Metric
Accelerator cable adjusting bolts	4 ft. lbs.	6 Nm
Fuel filter eye bolts	22 ft. lbs.	30 Nm
Fuel filter flare nut	25 ft. lbs.	35 Nm
Fuel rail mounting bolts	8 ft. lbs.	11 Nm
Fuel tank drain plug	14 ft. lbs.	20 Nm
Fuel tank self-locking nuts	22 ft. lbs.	31 Nm
Fuel pump/module to tank retainer nuts	2 ft. lbs.	3 Nm
High pressure fuel hose to fuel pump	29 ft. lbs.	40 Nm
High pressure fuel hose to fuel rail	4 ft. lbs.	6 Nm
Fuel pressure regulator mounting bolts	6 ft. lbs.	8 Nm
Oxygen sensor	33 ft. lbs.	45 Nm
Right rear lateral rod and attaching bolt	72 ft. lbs.	100 Nm
Throttle body mounting bolts/nuts	16 ft. lbs.	22 Nm
Throttle body stay bolts	16 ft. lbs.	22 Nm
Throttle position sensor	1.8 ft. lbs.	2.5 Nm

6

CHASSIS ELECTRICAL

UNDERSTANDING AND TROUBLESHOOTING ELECTRICAL SYSTEMS

With the rate at which both import and domestic manufacturers are incorporating electronic control systems into their production lines, it won't be long before every new vehicle is equipped with one or more on-board computer, like the unit installed on your car. These electronic components (with no moving parts) should theoretically last the life of the vehicle, provided nothing external happens to damage the circuits or memory chips.

While it is true that electronic components should never wear out, in the real world malfunctions do occur. It is also true that any computer-based system is extremely sensitive to electrical voltages and cannot tolerate careless or haphazard testing or service procedures. An inexperienced individual can literally do major damage looking for a minor problem by using the wrong kind of test equipment or connecting test leads or connectors with the ignition switch ON. When selecting test equipment, make sure the manufacturers instructions state that the tester is compatible with whatever type of electronic control system is being serviced. Read all instructions carefully and double check all test points before installing probes or making any test connections.

The following section outlines basic diagnosis techniques for dealing with computerized automotive control systems. Along with a general explanation of the various types of test equipment available to aid in servicing modern electronic automotive systems, basic repair techniques for wiring harnesses and connectors is given. Read the basic information before attempting any repairs or testing on any computerized system, to provide the background of information necessary to avoid the most common and obvious mistakes that can cost both time and money. Although the replacement and testing procedures are simple in themselves, the systems are not, and unless one has a thorough understanding of all components and their function within a particular computerized control system, the logical test sequence these systems demand cannot be followed. Minor malfunctions can make a big difference, so it is important to know how each component

affects the operation of the overall electronic system to find the ultimate e cause of a problem without replacing good components unnecessarily. It is not enough to use the correct test equipment; the test equipment must be used correctly.

Safety Precautions

✳✳CAUTION

Whenever working on or around any computer based microprocessor control system, always observe these general precautions to prevent the possibility of personal injury or damage to electronic components.

- Never install or remove battery cables with the key ON or the engine running. Jumper cables should be connected with the key OFF to avoid power surges that can damage electronic control units. Engines equipped with computer controlled systems should avoid both giving and getting jump starts due to the possibility of serious damage to components from arcing in the engine compartment when connections are made with the ignition ON.
- Always remove the battery cables before charging the battery. Never use a high output charger on an installed battery or attempt to use any type of 'hot shot' (24 volt) starting aid.
- Exercise care when inserting test probes into connectors to insure good connections without damaging the connector or spreading the pins. Always probe connectors from the rear (wire) side, NOT the pin side, to avoid accidental shorting of terminals during test procedures.
- Never remove or attach wiring harness connectors with the ignition switch ON, especially to an electronic control unit.
- Do not drop any components during service procedures and never apply 12 volts directly to any component (like a solenoid or relay) unless instructed specifically to do so. Some component electrical windings are designed to safely handle only 4 or 5 volts and can be destroyed in seconds if 12 volts are applied directly to the connector.

- Remove the electronic control unit if the vehicle is to be placed in an environment where temperatures exceed approximately 176°F (80°C), such as a paint spray booth or when arc or gas welding near the control unit location in the car.

ORGANIZED TROUBLESHOOTING

When diagnosing a specific problem, organized troubleshooting is a must. The complexity of a modern automobile demands that you approach any problem in a logical, organized manner. There are certain troubleshooting techniques that are standard:

1. Establish when the problem occurs. Does the problem appear only under certain conditions? Were there any noises, odors, or other unusual symptoms?

2. Isolate the problem area. To do this, make some simple tests and observations; then eliminate the systems that are working properly. Check for obvious problems such as broken wires, dirty connections or split or disconnected vacuum hoses. Always check the obvious before assuming something complicated is the cause.

3. Test for problems systematically to determine the cause once the problem area is isolated. Are all the components functioning properly? Is there power going to electrical switches and motors? Is there vacuum at vacuum switches and/or actuators? Is there a mechanical problem such as bent linkage or loose mounting screws? Doing careful, systematic checks will often turn up most causes on the first inspection without wasting time checking components that have little or no relationship to the problem.

4. Test all repairs after the work is done to make sure that the problem is fixed. Some causes can be traced to more than one component, so a careful verification of repair work is important to pick up additional malfunctions that may cause a problem to reappear or a different problem to arise. A blown fuse, for example, is a simple problem that may require more than another fuse to repair. If you don't look for a problem

that caused a fuse to blow, for example, a shorted wire may go undetected.

Experience has shown that most problems tend to be the result of a fairly simple and obvious cause, such as loose or corroded connectors or air leaks in the intake system; making careful inspection of components during testing essential to quick and accurate troubleshooting. Special, hand held computerized testers designed specifically for diagnosing the system are available from a variety of aftermarket sources, as well as from the vehicle manufacturer, but care should be taken that any test equipment being used is designed to diagnose that particular computer controlled system accurately without damaging the control unit (ECU) or components being tested.

➡**Pinpointing the exact cause of trouble in an electrical system can sometimes only be accomplished by the use of special test equipment. The following describes commonly used test equipment and explains how to put it to best use in diagnosis. In addition to the information covered below, the manufacturer's instructions booklet provided with the tester should be read and clearly understood before attempting any test procedures.**

TEST EQUIPMENT

Jumper Wires

Jumper wires are simple, yet extremely valuable, pieces of test equipment. Jumper wires are merely wires that are used to bypass sections of a circuit. The simplest type of jumper wire is merely a length of multistrand wire with an alligator clip at each end. Jumper wires are usually fabricated from lengths of standard automotive wire and whatever type of connector (alligator clip, spade connector or pin connector) that is required for the particular vehicle being tested. The well equipped tool box will have several different styles of jumper wires in several different lengths. Some jumper wires are made with three or more terminals coming from a common splice for special purpose testing. In cramped, hard-to-reach areas it is advisable to have insulated boots over the jumper wire terminals in order to prevent accidental grounding, sparks,

and possible fire, especially when testing fuel system components.

Jumper wires are used primarily to locate open electrical circuits, on either the ground (-) side of the circuit or on the hot (+) side. If an electrical component fails to operate, connect the jumper wire between the component and a good ground. If the component operates only with the jumper installed, the ground circuit is open. If the ground circuit is good, but the component does not operate, the circuit between the power feed and component is open. You can sometimes connect the jumper wire directly from the battery to the hot terminal of the component, but first make sure the component uses 12 volts in operation. Some electrical components, such as fuel injectors, are designed to operate on about 4 volts and running 12 volts directly to the injector terminals can burn out the wiring. By inserting an in-line fuseholder between a set of test leads, a fused jumper wire can be used for bypassing open circuits. Use a 5 amp fuse to provide protection against voltage spikes. When in doubt, use a voltmeter to check the voltage input to the component and measure how much voltage is being applied normally. By moving the jumper wire successively back from the lamp toward the power source, you can isolate the area of the circuit where the open is located. When the component stops functioning, or the power is cut off, the open is in the segment of wire between the jumper and the point previously tested.

✳✳CAUTION

Never use jumpers made from wire that is of lighter gauge than used in the circuit under test. If the jumper wire is of too small gauge, it may overheat and possibly melt. Never use jumpers to bypass high resistance loads (such as motors) in a circuit. Bypassing resistances, in effect, creates a short circuit which may, in turn, cause damage and fire. Never use a jumper for anything other than temporary bypassing of components in a circuit.

12 Volt Test Light

The 12 volt test light is used to check circuits and components while electrical current is flowing through them. It is used for voltage and ground tests.

Twelve volt test lights come in different styles but all have three main parts; a ground clip, a probe, and a light. The most commonly used 12 volt test lights have pick-type probes. To use a 12 volt test light, connect the ground clip to a good ground and probe wherever necessary with the pick. The pick should be sharp so that it can penetrate wire insulation to make contact with the wire, without making a large hole in the insulation. The wrap-around light is handy in hard to reach areas or where it is difficult to support a wire to push a probe pick into it. To use the wrap around light, hook the wire to probed with the hook and pull the trigger. A small pick will be forced through the wire insulation into the wire core.

✳✳CAUTION

Do not use a test light to probe electronic ignition spark plug or coil wires. Never use a pick-type test light to probe wiring on computer controlled systems unless specifically instructed to do so. Any wire insulation that is pierced by the test light probe should be taped and sealed with silicone after testing.

Like the jumper wire, the 12 volt test light is used to isolate opens in circuits. But, whereas the jumper wire is used to bypass the open to operate the load, the 12 volt test light is used to locate the presence of voltage in a circuit. If the test light glows, you know that there is power up to that point; if the 12 volt test light does not glow when its probe is inserted into the wire or connector, you know that there is an open circuit (no power). Move the test light in successive steps back toward the power source until the light in the handle does glow. When it does glow, the open is between the probe and point previously probed.

➡**The test light does not detect that 12 volts (or any particular amount of voltage) is present; it only detects that some voltage is present. It is advisable before using the test light to touch its terminals across the battery posts to make sure the light is operating properly.**

Self-Powered Test Light

The self-powered test light usually contains a 1.5 volt penlight battery. One type of self-powered test light is similar in design to the 12 volt test light. This type has both the battery and the light in the handle and pick-type probe tip. The second type has the light toward the open tip, so that the light illuminates the contact point. The self-powered test light is dual purpose piece of test equipment. It can be used to test for either open or short circuits when power is isolated from the circuit (continuity test). A powered test light should not be used on any computer controlled system or component unless specifically instructed to do so. Many engine sensors can be destroyed by even this small amount of voltage applied directly to the terminals.

Open Circuit Testing

To use the self-powered test light to check for open circuits, first isolate the circuit from the vehicle's 12 volt power source by disconnecting the battery or wiring harness connector. Connect the test light ground clip to a good ground and probe sections of the circuit sequentially with the test light. (start from either end of the circuit). If the light is out, the open is between the probe and the circuit ground. If the light is on, the open is between the probe and end of the circuit toward the power source.

Short Circuit Testing

By isolating the circuit both from power and from ground, and using a self-powered test light, you can check for shorts to ground in the circuit. Isolate the circuit from power and ground. Connect the test light ground clip to a good ground and probe any easy-to-reach test point in the circuit. If the light comes on, there is a short somewhere in the circuit. To isolate the short, probe a test point at either end of the isolated circuit (the light should be on). Leave the test light probe connected and open connectors, switches, remove parts, etc., sequentially, until the light goes out. When the light goes out, the short is between the last circuit component opened and the previous circuit opened.

➡The 1.5 volt battery in the test light does not provide much current. A weak battery may not provide enough power to illuminate the test light even when a complete circuit is made (especially if there are high resistances in the circuit). Always make sure that the test battery is strong. To check the battery, briefly touch the ground clip to the probe; if the light glows brightly the battery is strong enough for testing. Never use a self-powered test light to perform checks for opens or shorts when power is applied to the electrical system under test. The 12 volt vehicle power will quickly burn out the 1.5 volt light bulb in the test light.

Voltmeter

A voltmeter is used to measure voltage at any point in a circuit, or to measure the voltage drop across any part of a circuit. It can also be used to check continuity in a wire or circuit by indicating current flow from one end to the other. Voltmeters usually have various scales on the meter dial and a selector switch to allow the selection of different voltages. The voltmeter has a positive and a negative lead. To avoid damage to the meter, always connect the negative lead to the negative (-) side of circuit (to ground or nearest the ground side of the circuit) and connect the positive lead to the positive (+) side of the circuit (to the power source or the nearest power source). Note that the negative voltmeter lead will always be black and that the positive voltmeter will always be some color other than black (usually red). Depending on how the voltmeter is connected into the circuit, it has several uses.

A voltmeter can be connected either in parallel or in series with a circuit and it has a very high resistance to current flow. When connected in parallel, only a small amount of current will flow through the voltmeter current path; the rest will flow through the normal circuit current path and the circuit will work normally. When the voltmeter is connected in series with a circuit, only a small amount of current can flow through the circuit. The circuit will not work properly, but the voltmeter reading will show if the circuit is complete or not.

Available Voltage Measurement

Set the voltmeter selector switch to the 20V position and connect the meter negative lead to the negative post of the battery. Connect the positive meter lead to the positive post of the battery and turn the ignition switch ON to provide a load. Read the voltage on the meter or digital display. A well charged battery should register over 12 volts. If the meter reads below 11.5 volts, the battery power may be insufficient to operate the electrical system properly. This test determines voltage available from the battery and should be the first step in any electrical trouble diagnosis procedure. Many electrical problems, especially on computer controlled systems, can be caused by a low state of charge in the battery. Excessive corrosion at the battery cable terminals can cause a poor contact that will prevent proper charging and full battery current flow.

Normal battery voltage is 12 volts when fully charged. When the battery is supplying current to one or more circuits it is said to be 'under load". When everything is off the electrical system is under a 'no-load" condition. A fully charged battery may show about 12.5 volts at no load; will drop to 12 volts under medium load; and will drop even lower under heavy load. If the battery is partially discharged the voltage decrease under heavy load may be excessive, even though the battery shows 12 volts or more at no load. When allowed to discharge further, the battery's available voltage under load will decrease more severely. For this reason, it is important that the battery be fully charged during all testing procedures to avoid errors in diagnosis and incorrect test results.

Voltage Drop

When current flows through a resistance, the voltage beyond the resistance is reduced (the larger the current, the greater the reduction in voltage). When no current is flowing, there is no voltage drop because there is no current flow. All points in the circuit which are connected to the power source are at the same voltage as the power source. The total voltage drop always equals the total source voltage. In a long circuit with many connectors, a series of small, unwanted voltage drops due to corrosion at the connectors can add up to a total loss of voltage which impairs the operation of the normal loads in the circuit.

INDIRECT COMPUTATION OF VOLTAGE DROPS

1. Set the voltmeter selector switch to the 20 volt position.
2. Connect the meter negative lead to a good ground.
3. Probe all resistances in the circuit with the positive meter lead.

4. Operate the circuit in all modes and observe the voltage readings.

DIRECT MEASUREMENT OF VOLTAGE DROPS

1. Set the voltmeter switch to the 20 volt position.
2. Connect the voltmeter negative lead to the ground side of the resistance load to be measured.
3. Connect the positive lead to the positive side of the resistance or load to be measured.
4. Read the voltage drop directly on the 20 volt scale.

Too high a voltage indicates too high a resistance. If, for example, a blower motor runs too slowly, you can determine if there is too high a resistance in the resistor pack. By taking voltage drop readings in all parts of the circuit, you can isolate the problem. Too low a voltage drop indicates too low a resistance. If, for example, a blower motor runs too fast in the MED and/or LOW position, the problem can be isolated in the resistor pack by taking voltage drop readings in all parts of the circuit to locate a possibly shorted resistor. The maximum allowable voltage drop under load is critical, especially if there is more than one high resistance problem in a circuit because all voltage drops are cumulative. A small drop is normal due to the resistance of the conductors.

HIGH RESISTANCE TESTING

1. Set the voltmeter selector switch to the 4 volt position.
2. Connect the voltmeter positive lead to the positive post of the battery.
3. Turn on the headlights and heater blower to provide a load.
4. Probe various points in the circuit with the negative voltmeter lead.
5. Read the voltage drop on the 4 volt scale. Some average maximum allowable voltage drops are:
FUSE PANEL — 7 volts
IGNITION SWITCH — 5 volts
HEADLIGHT SWITCH — 7 volts
IGNITION COIL (+) — 5 volts
ANY OTHER LOAD — 1.3 volts

➡**Voltage drops are all measured while a load is operating; without current flow, there will be no voltage drop.**

Ohmmeter

The ohmmeter is designed to read resistance (ohms) in a circuit or component. Although there are several different styles of ohmmeters, all will usually have a selector switch which permits the measurement of different ranges of resistance (usually the selector switch allows the multiplication of the meter reading by 10, 100, 1000, and 10,000). A calibration knob allows the meter to be set at zero for accurate measurement. Since all ohmmeters are powered by an internal battery (usually 9 volts), the ohmmeter can be used as a self-powered test light. When the ohmmeter is connected, current from the ohmmeter flows through the circuit or component being tested. Since the ohmmeter's internal resistance and voltage are known values, the amount of current flow through the meter depends on the resistance of the circuit or component being tested.

The ohmmeter can be used to perform continuity test for opens or shorts (either by observation of the meter needle or as a self-powered test light), and to read actual resistance in a circuit. It should be noted that the ohmmeter is used to check the resistance of a component or wire while there is no voltage applied to the circuit. Current flow from an outside voltage source (such as the vehicle battery) can damage the ohmmeter, so the circuit or component should be isolated from the vehicle electrical system before any testing is done. Since the ohmmeter uses its own voltage source, either lead can be connected to any test point.

➡**When checking diodes or other solid state components, the ohmmeter leads can only be connected one way in order to measure current flow in a single direction. Make sure the positive (+) and negative (-) terminal connections are as described in the test procedures to verify the one-way diode operation.**

In using the meter for making continuity checks, do not be concerned with the actual resistance readings. Zero resistance, or any resistance readings, indicate continuity in the circuit. Infinite resistance indicates an open in the circuit. A high resistance reading where there should be none indicates a problem in the circuit. Checks for short circuits are made in the same manner as checks for open circuits except that the circuit must be isolated from both power and normal ground. Infinite resistance indicates no continuity to ground, while zero resistance indicates a dead short to ground.

RESISTANCE MEASUREMENT

The batteries in an ohmmeter will weaken with age and temperature, so the ohmmeter must be calibrated or 'zeroed" before taking measurements. To zero the meter, place the selector switch in its lowest range and touch the two ohmmeter leads together. Turn the calibration knob until the meter needle is exactly on zero.

➡**All analog (needle) type ohmmeters must be zeroed before use, but some digital ohmmeter models are automatically calibrated when the switch is turned on. Self-calibrating digital ohmmeters do not have an adjusting knob, but its a good idea to check for a zero readout before use by touching the leads together. All computer controlled systems require the use of a digital ohmmeter with at least 10 megohms impedance for testing. Before any test procedures are attempted, make sure the ohmmeter used is compatible with the electrical system or damage to the on-board computer could result.**

To measure resistance, first isolate the circuit from the vehicle power source by disconnecting the battery cables or the harness connector. Make sure the key is OFF when disconnecting any components or the battery. Where necessary, also isolate at least one side of the circuit to be checked to avoid reading parallel resistances. Parallel circuit resistances will always give a lower reading than the actual resistance of either of the branches. When measuring the resistance of parallel circuits, the total resistance will always be lower than the smallest resistance in the circuit. Connect the meter leads to both sides of the circuit (wire or component) and read the actual measured ohms on the meter scale. Make sure the selector switch is set to the proper ohm scale for the circuit being tested to avoid misreading the ohmmeter test value.

Ammeters

An ammeter measures the amount of current flowing through a circuit in units called amperes or amps. Amperes are units of electron flow which indicate how fast the electrons are flowing through the circuit. Since Ohms Law dictates that current flow in a circuit is equal to the circuit voltage divided by the total circuit resistance, increasing voltage also increases the current level (amps). Likewise, any decrease in resistance will increase the amount of amps in a circuit. At normal operating voltage, most circuits have a characteristic amount of amperes, called "current draw" which can be measured using an ammeter. By referring to a specified current draw rating, measuring the amperes, and comparing the two values, one can determine what is happening within the circuit to aid in diagnosis. An open circuit, for example, will not allow any current to flow so the ammeter reading will be zero. More current flows through a heavily loaded circuit or when the charging system is operating.

An ammeter is always connected in series with the circuit being tested. All of the current that normally flows through the circuit must also flow through the ammeter; if there is any other path for the current to follow, the ammeter reading will not be accurate. The ammeter itself has very little resistance to current flow and therefore will not affect the circuit, but it will measure current draw only when the circuit is closed and electricity is flowing. Excessive current draw can blow fuses and drain the battery, while a reduced current draw can cause motors to run slowly, lights to dim and other components to not operate properly. The ammeter can help diagnose these conditions by locating the cause of the high or low reading.

Multimeters

Different combinations of test meters can be built into a single unit designed for specific tests. Some of the more common combination test devices are known as Volt/Amp testers, Tach/Dwell meters, or Digital Multimeters. The Volt/Amp tester is used for charging system, starting system or battery tests and consists of a voltmeter, an ammeter and a variable resistance carbon pile. The voltmeter will usually have at least two ranges for use with 6, 12 and 24 volt systems. The ammeter also has more than one range for testing various levels of battery loads and starter current draw and the carbon pile can be adjusted to offer different amounts of resistance. The Volt/Amp tester has heavy leads to carry large amounts of current and many later models have an inductive ammeter pickup that clamps around the wire to simplify test connections. On some models, the ammeter also has a zero-center scale to allow testing of charging and starting systems without switching leads or polarity. A digital multimeter is a voltmeter, ammeter and ohmmeter combined in an instrument which gives a digital readout. These are often used when testing solid state circuits because of their high input impedance (usually 10 megohms or more).

The tach/dwell meter combines a tachometer and a dwell (cam angle) meter and is a specialized kind of voltmeter. The tachometer scale is marked to show engine speed in rpm and the dwell scale is marked to show degrees of distributor shaft rotation. In most electronic ignition systems, dwell is determined by the control unit, but the dwell meter can also be used to check the duty cycle (operation) of some electronic engine control systems. Some tach/dwell meters are powered by an internal battery, while others take their power from the car battery in use. The battery powered testers usually require calibration much like an ohmmeter before testing.

Special Test Equipment

A variety of diagnostic tools are available to help troubleshoot and repair computerized engine control systems. The most sophisticated of these devices are the console type engine analyzers that usually occupy a garage service bay, but there are several types of

aftermarket electronic testers available that will allow quick circuit tests of the engine control system by plugging directly into a special connector located in the engine compartment or under the dashboard. Several tool and equipment manufacturers offer simple, hand held testers that measure various circuit voltage levels on command to check all system components for proper operation. Although these testers usually cost about $300-$500, consider that the average computer control unit (or ECM) can cost just as much and the money saved by not replacing perfectly good sensors or components in an attempt to correct a problem could justify the purchase price of a special diagnostic tester the first time it's used.

These computerized testers can allow quick and easy test measurements while the engine is operating or while the car is being driven. In addition, the on-board computer memory can be read to access any stored trouble codes; in effect allowing the computer to tell you where it hurts and aid trouble diagnosis by pinpointing exactly which circuit or component is malfunctioning. In the same manner, repairs can be tested to make sure the problem has been corrected. The biggest advantage these special testers have is their relatively easy hookups that minimize or eliminate the chances of making the wrong connections and getting false voltage readings or damaging the computer accidentally.

➡It should be remembered that these testers check voltage levels in circuits; they don't detect mechanical problems or failed components if the circuit voltage falls within the preprogrammed limits stored in the tester PROM unit. Also, most of the hand held testers are designed to work only on one or two systems made by a specific manufacturer.

A variety of aftermarket testers are available to help diagnose different computerized control systems. Owatonna Tool Company (OTC), for example, markets a device called the OTC Monitor which plugs directly into the assembly line diagnostic link (ALDL). The OTC tester makes diagnosis a simple matter of pressing the correct buttons and, by changing the internal PROM or inserting a different diagnosis cartridge, it will work on any model from full size to subcompact, over a wide range of years.

An adapter is supplied with the tester to allow connection to all types of ALDL links, regardless of the number of pin terminals used. By inserting an updated PROM into the OTC tester, it can be easily updated to diagnose any new modifications of computerized control systems.

Wiring Harnesses

INFORMATION

The average automobile contains about ½ mile of wiring, with hundreds of individual connections. To protect the many wires from damage and to keep them from becoming a confusing tangle, they are organized into bundles, enclosed in plastic or taped together and called wire harnesses. Different wiring harnesses serve different parts of the vehicle. Individual wires are color coded to help trace them through a harness where sections are hidden from view.

A loose or corroded connection or a replacement wire that is too small for the circuit will add extra resistance and an additional voltage drop to the circuit. A ten percent voltage drop can result in slow or erratic motor operation, for example, even though the circuit is complete. Automotive wiring or circuit conductors can be in any one of three forms:

1. Single strand wire
2. Multistrand wire
3. Printed circuitry

Single strand wire has a solid metal core and is usually used inside such components as alternators, motors, relays and other devices. Multistrand wire has a core made of many small strands of wire twisted together into a single conductor. Most of the wiring in an automotive electrical system is made up of multistrand wire, either as a single conductor or grouped together in a harness. All wiring is color coded on the insulator, either as a solid color or as a colored wire with an identification stripe. A printed circuit is a thin film of copper or other conductor that is printed on an insulator backing. Occasionally, a printed circuit is sandwiched between two sheets of plastic for more protection and flexibility. A complete printed circuit, consisting of conductors, insulating material and connectors for lamps or other components is called a printed circuit board. Printed circuitry is used in place of individual wires or harnesses in places where space is limited, such as behind instrument panels.

Wire Gauge

Since computer controlled automotive electrical systems are very sensitive to changes in resistance, the selection of properly sized wires is critical when systems are repaired. The wire gauge number is an expression of the cross section area of the conductor. The most common system for expressing wire size is the American Wire Gauge (AWG) system.

Wire cross section area is measured in circular mils. A mil is 0.001 inch; a circular mil is the area of a circle one mil in diameter. For example, a conductor ¼ inch in diameter is 0.250 in. or 250 mils. The circular mil cross section area of the wire is 250 squared or 62,500 circular mils. Imported car models usually use metric wire gauge designations, which is simply the cross section area of the conductor in square millimeters.

Gauge numbers are assigned to conductors of various cross section areas. As gauge number increases, area decreases and the conductor becomes smaller. A 5 gauge conductor is smaller than a 1 gauge conductor and a 10 gauge is smaller than a 5 gauge. As the cross section area of a conductor decreases, resistance increases and so does the gauge number. A conductor with a higher gauge number will carry less current than a conductor with a lower gauge number.

➡Gauge wire size refers to the size of the conductor, not the size of the complete wire. It is possible to have two wires of the same gauge with different diameters because one may have thicker insulation than the other.

12 volt automotive electrical systems generally use 10, 12, 14, 16 and 18 gauge wire. Main power distribution circuits and larger accessories usually use 10 and 12 gauge wire. Battery cables are usually 4 or 6 gauge, although 1 and 2 gauge wires are occasionally used. Wire length must also be considered when making repairs to a circuit. As conductor length increases, so does resistance. An 18 gauge wire, for example, can carry a 10 amp load for 10 feet without excessive voltage drop; however if a 15 foot wire is required for the same 10 amp load, it must be a 16 gauge wire.

An electrical schematic shows the electrical current paths when a circuit is operating properly. It is essential to understand how a circuit works before trying to figure out why it doesn't. Schematics break the entire electrical system down into individual circuits and show only one particular circuit. In a schematic, no attempt is made to represent wiring and components as they physically appear on the vehicle; switches and other components are shown as simply as possible. Face views of harness connectors show the cavity or terminal locations in all multi-pin connectors to help locate test points.

If you need to backprobe a connector while it is on the component, the order of the terminals must be mentally reversed. The wire color code can help in this situation, as well as a keyway, lock tab or other reference mark.

WIRING REPAIR

Soldering is a quick and efficient method of joining metals permanently. Everyone who has the occasion to make wiring repairs should know how to solder. Electrical connections that are soldered are far less likely to come apart and will conduct electricity much better than connections that are only 'pig-tailed" together. The most popular (and preferred) method of soldering is with an electrical soldering gun. Soldering irons are available in many sizes and wattage ratings. Irons with higher wattage ratings deliver higher temperatures and recover lost heat faster. A small soldering iron rated for no more than 50 watts is recommended, especially on electrical systems where excess heat can damage the components being soldered.

There are three ingredients necessary for successful soldering; proper flux, good solder and sufficient heat. A soldering flux is necessary to clean the metal of tarnish, prepare it for soldering and to enable the solder to spread into tiny crevices. When soldering, always use a resin flux or resin core solder which is non-corrosive and will not attract moisture once the job is finished. Other types of flux (acid core) will leave a residue that will attract moisture and cause the wires to corrode. Tin is a unique metal with a low melting point. In a molten state, it dissolves and alloys easily with many metals. Solder is made by mixing tin with lead. The most common proportions are 40/60, 50/50 and 60/40, with the percentage of tin listed first. Low priced solders usually contain less tin, making them very difficult for a beginner to use because more heat is required to melt the solder. A common solder is 40/60 which is well suited for all-around general use, but 60/40 melts easier, has more tin f or a better joint and is preferred for electrical work.

Soldering Techniques

Successful soldering requires that the metals to be joined be heated to a temperature that will melt the solder — usually 360-460°F (182-238°C). Contrary to popular belief, the purpose of the soldering iron is not to melt the solder itself, but to heat the parts being soldered to a temperature high enough to melt the solder when it is touched to the work. Melting flux-cored solder on the soldering iron will usually destroy the effectiveness of the flux.

➡**Soldering tips are made of copper for good heat conductivity, but must be ''tinned'' regularly for quick transference of heat to the project and to prevent the solder from sticking to the iron. To 'tin' the iron, simply heat it and touch the flux-cored solder to the tip; the solder will flow over the hot tip. Wipe the excess off with a clean rag, but be careful as the iron will be hot.**

After some use, the tip may become pitted. If so, simply dress the tip smooth with a smooth file and 'tin' the tip again. An old saying holds that 'metals well cleaned are half soldered.'' Flux-cored solder will remove oxides but rust, bits of insulation and oil or grease must be removed with a wire brush or emery cloth. For maximum strength in soldered parts, the joint must start off clean and tight. Weak joints will result in gaps too wide for the solder to bridge.

If a separate soldering flux is used, it should be brushed or swabbed on only those areas that are to be soldered. Most solders contain a core of flux and separate fluxing is unnecessary. Hold the work to be soldered firmly. It is best to solder on a wooden board, because a metal vise will only rob the piece to be

soldered of heat and make it difficult to melt the solder. Hold the soldering tip with the broadest face against the work to be soldered. Apply solder under the tip close to the work, using enough solder to give a heavy film between the iron and the piece being soldered, while moving slowly and making sure the solder melts properly. Keep the work level or the solder will run to the lowest part and favor the thicker parts, because these require more heat to melt the solder. If the soldering tip overheats (the solder coating on the face of the tip burns up), it should be retinned. Once the soldering is completed, let the soldered joint stand until cool. Tape and seal all soldered wire splices after the repair has cooled.

Wire Harness and Connectors

The on-board computer (ECM) wire harness electrically connects the control unit to the various solenoids, switches and sensors used by the control system. Most connectors in the engine compartment or otherwise exposed to the elements are protected against moisture and dirt which could create oxidation and deposits on the terminals. This protection is important because of the very low voltage and current levels used by the computer and sensors. All connectors have a lock which secures the male and female terminals together, with a secondary lock holding the seal and terminal into the connector. Both terminal locks must be released when disconnecting ECM connectors.

These special connectors are weatherproof and all repairs require the use of a special terminal and the tool required to service it. This tool is used to remove the pin and sleeve terminals. If removal is attempted with an ordinary pick, there is a good chance that the terminal will be bent or deformed. Unlike standard blade type terminals, these terminals cannot be straightened once they are bent. Make certain that the connectors are properly seated and all of the sealing rings in place when connecting leads. On some models, a hinge-type flap provides a backup or secondary locking feature for the terminals. Most secondary locks are used to improve the connector reliability by retaining the terminals if the small terminal lock tangs are not positioned properly.

Molded-on connectors require complete replacement of the connection.

This means splicing a new connector assembly into the harness. All splices in on-board computer systems should be soldered to insure proper contact. Use care when probing the connections or replacing terminals in them as it is possible to short between opposite terminals. If this happens to the wrong terminal pair, it is possible to damage certain components. Always use jumper wires between connectors for circuit checking and never probe through weatherproof seals.

Open circuits are often difficult to locate by sight because corrosion or terminal misalignment are hidden by the connectors. Merely wiggling a connector on a sensor or in the wiring harness may correct the open circuit condition. This should always be considered when an open circuit or a failed sensor is indicated. Intermittent problems may also be caused by oxidized or loose connections. When using a circuit tester for diagnosis, always probe connections from the wire side. Be careful not to damage sealed connectors with test probes.

All wiring harnesses should be replaced with identical parts, using the same gauge wire and connectors. When signal wires are spliced into a harness, use wire with high temperature insulation only. With the low voltage and current levels found in the system, it is important that the best possible connection at all wire splices be made by soldering the splices together. It is seldom necessary to replace a complete harness. If replacement is necessary, pay close attention to insure proper harness routing. Secure the harness with suitable plastic wire clamps to prevent vibrations from causing the harness to wear in spots or contact any hot components.

➡**Weatherproof connectors cannot be replaced with standard connectors. Instructions are provided with replacement connector and terminal packages. Some wire harnesses have mounting indicators (usually pieces of colored tape) to mark where the harness is to be secured.**

In making wiring repairs, it's important that you always replace damaged wires with wires that are the same gauge as the wire being replaced. The heavier the wire, the smaller the gauge number. Wires are color-coded to aid in identification and whenever possible the same color coded wire should be used for replacement. A wire stripping and crimping tool is necessary to install solderless terminal connectors. Test all crimps by pulling on the wires; it should not be possible to pull the wires out of a good crimp.

Wires which are open, exposed or otherwise damaged are repaired by simple splicing. Where possible, if the wiring harness is accessible and the damaged place in the wire can be located, it is best to open the harness and check for all possible damage. In an inaccessible harness, the wire must be bypassed with a new insert, usually taped to the outside of the old harness.

When replacing fusible links, be sure to use fusible link wire, NOT ordinary automotive wire. Make sure the fusible segment is of the same gauge and construction as the one being replaced and double the stripped end when crimping the terminal connector for a good contact. The melted (open) fusible link segment of the wiring harness should be cut off as close to the harness as possible, then a new segment spliced in as described. In the case of a damaged fusible link that feeds two harness wires, the harness connections should be replaced with two fusible link wires so that each circuit will have its own separate protection.

➡**Most of the problems caused in the wiring harness are due to bad ground connections. Always check all vehicle ground connections for corrosion or looseness before performing any power feed checks to eliminate the chance of a bad ground affecting the circuit.**

Repairing Hard Shell Connectors

Unlike molded connectors, the terminal contacts in hard shell connectors can be replaced. Weatherproof hard-shell connectors with the leads molded into the shell have non-replaceable terminal ends. Replacement usually involves the use of a special terminal removal tool that depress the locking tangs (barbs) on the connector terminal and allow the connector to be removed from the rear of the shell. The connector shell should be replaced if it shows any evidence of burning, melting, cracks, or breaks. Replace individual terminals that are burnt, corroded, distorted or loose.

➡**The insulation crimp must be tight to prevent the insulation from sliding back on the wire when the wire is pulled. The insulation must be visibly compressed under the crimp tabs, and the ends of the crimp should be turned in for a firm grip on the insulation.**

The wire crimp must be made with all wire strands inside the crimp. The terminal must be fully compressed on the wire strands with the ends of the crimp tabs turned in to make a firm grip on the wire. Check all connections with an ohmmeter to insure a good contact. There should be no measurable resistance between the wire and the terminal when connected.

Mechanical Test Equipment

INFORMATION

Vacuum Gauge

Most gauges are graduated in inches of mercury (in.Hg), although a device called a manometer reads vacuum in inches of water (in. H_2O). The normal vacuum reading usually varies between 18 and 22 in.Hg at sea level. To test engine vacuum, the vacuum gauge must be connected to a source of manifold vacuum. Many engines have a plug in the intake manifold which can be removed and replaced with an adapter fitting. Connect the vacuum gauge to the fitting with a suitable rubber hose or, if no manifold plug is available, connect the vacuum gauge to any device using manifold vacuum, such as EGR valves, etc. The vacuum gauge can be used to determine if enough vacuum is reaching a component to allow its actuation.

Hand Vacuum Pump

Small, hand-held vacuum pumps come in a variety of designs. Most have a built-in vacuum gauge and allow the component to be tested without removing it from the vehicle. Operate the pump lever or plunger to apply the correct amount of vacuum required for the test specified in the diagnosis routines. The level of vacuum in inches of Mercury (in.Hg) is indicated on the pump gauge. For some testing, an additional vacuum gauge may be necessary.

Intake manifold vacuum is used to operate various systems and devices on late model vehicles. To correctly diagnose and solve problems in vacuum control systems, a vacuum source is necessary for testing. In some cases, vacuum can be taken from the intake manifold when the engine is running, but vacuum is normally provided by a hand vacuum pump. These hand vacuum pumps have a built-in vacuum gauge that allow testing while the device is still attached to the component. For some tests, an additional vacuum gauge may be necessary.

HEATER

Blower Motor

REMOVAL & INSTALLATION

▶ **See Figure 1**

1. Disconnect battery negative cable.
2. Remove the right side duct, if equipped.

3. Remove the molded hose from the blower assembly.
4. Remove the blower motor assembly.
5. Remove the packing seal.
6. Remove the fan retaining nut and fan in order to renew the motor.
To install:
7. Check that the blower motor shaft is not bent and that the packing is in good condition. Clean all parts of dust, etc.

8. Assemble the motor and fan. Install the blower motor then connect the motor terminals to battery voltage. Check that the blower motor operates smoothly. Then, reverse the polarity and check that the blower motor operates smoothly in the reverse direction.
9. Install the molded hose and duct, if removed.
10. Connect the negative battery cable and check the entire climate control system for proper operation.

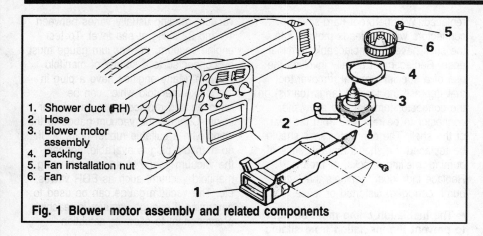

1. Shower duct (RH)
2. Hose
3. Blower motor assembly
4. Packing
5. Fan installation nut
6. Fan

Fig. 1 Blower motor assembly and related components

Fig. 3 Set the air selector damper to the position that permits outside air introduction

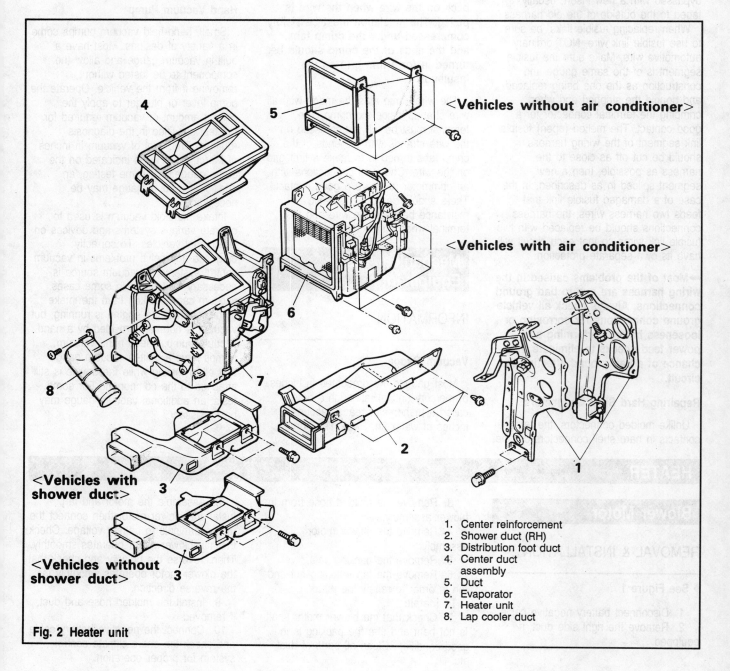

<Vehicles without air conditioner>

<Vehicles with air conditioner>

<Vehicles with shower duct>

<Vehicles without shower duct>

1. Center reinforcement
2. Shower duct (RH)
3. Distribution foot duct
4. Center duct assembly
5. Duct
6. Evaporator
7. Heater unit
8. Lap cooler duct

Fig. 2 Heater unit

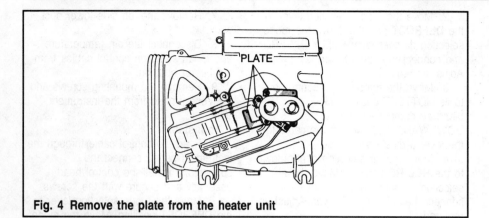

Fig. 4 Remove the plate from the heater unit

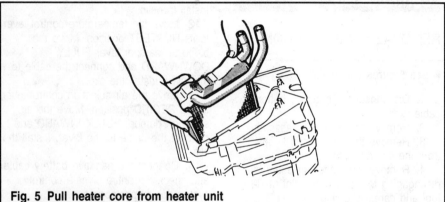

Fig. 5 Pull heater core from heater unit

Heater Core

REMOVAL & INSTALLATION

◆ **See Figures 2, 3, 4 and 5**

➡**The evaporator housing can be removed by itself, without removing the console, instrument panel or heater core. The heater core, though, cannot be removed without removing the evaporator.**

1. Disconnect the negative battery cable.

2. Drain the cooling system and properly discharge the air conditioning system and disconnect the refrigerant lines from the evaporator, if equipped. Cover the exposed ends of the lines to minimize contamination.

3. Remove the floor console by first removing the plugs, then the screws retaining the side covers and the small cover piece in front of the shifter. Remove the shifter knob, manual transmission, and the cup holder. Remove both small pieces of upholstery to gain access to retainer screws. Disconnect both electrical connectors from the front of the console. Remove the shoulder harness guide plates and the console assembly.

4. Remove the instrument panel assembly by performing the following:

a. Locate the rectangular plugs in the knee protector on either side of the steering column. Pry these plugs out, remove the screws. Remove the screws from the hood lock release lever and the knee protector.

b. Remove the upper and lower column covers.

c. Remove the narrow panel covering the instrument cluster cover screws and the cover.

d. Remove the radio panel and the radio.

e. Remove the center air outlet assembly by reaching through the grille and pushing the side clips out with a small flat-tipped tool while carefully prying the outlet free.

f. Pull the heater control knobs off and remove the heater control panel assembly.

g. Open the glove box, remove the plugs from the sides and the glove box assembly.

h. Remove the instrument gauge cluster and the speedometer adapter by disconnecting the speedometer cable from the transaxle, pulling the cable sightly towards the vehicle interior, then giving a slight twist on the adapter to release it.

i. Remove the left and right speaker covers from the top of the instrument panel.

j. Remove the center plate below the heater controls.

k. Remove the heater control assembly installation screws.

l. Remove the lower air ducts.

m. Drop the steering column by removing the bolts.

n. Remove the instrument panel mounting screws, bolts and the instrument panel assembly.

5. Remove both stamped steel reinforcement pieces.

6. Remove the lower ductwork from the heater box.

7. Remove the upper center duct.

8. Vehicles without air conditioning will have a square duct in place of the evaporator. Remove this duct if present. If the vehicle is equipped with air conditioning, remove the evaporator assembly:

a. Remove the wiring harness connectors and the electronic control unit.

b. Remove the drain hose and lift out the evaporator unit.

c. If servicing the assembly, disassemble the housing and remove the expansion valve and evaporator.

9. With the evaporator removed, remove the heater unit. To prevent bolts from falling inside the blower assembly,

set the inside/outside air-selection damper to the position that permits outside air introduction.

10. Remove the cover plate around the heater tubes and remove the core fastener clips. Pull the heater core from the heater box, being careful not to damage the fins or tank ends.

To install:

11. Install the heater core to the heater box. Install the clips and cover.

12. Install the heater box and connect the duct work.

13. Assemble the housing, evaporator and expansion valve, making sure the gaskets are in good condition. Install the evaporator housing.

14. Using new lubricated O-rings, connect the refrigerant lines to the evaporator.

15. Install the electronic transmission ELC box. Connect all wires and control cables.

16. Install the instrument panel assembly and the console by reversing their removal procedures.

17. Evacuate and recharge the air conditioning system. If the evaporator was replaced, add 2 oz. of refrigerant oil during the recharge.

18. Connect the negative battery cable and check the entire climate control system for proper operation. Check the system for leaks.

Control Cables

ADJUSTMENT

1. Disconnect the negative battery cable. Remove the glove box, if necessary.

2. Move the mode selection lever to the **DEFROST** position. Move the mode selection damper lever FULLY INWARD and connect the cable to the lever. Adjust as required.

3. Move the temperature control lever to its HOTTEST position. Move the blend air damper lever FULLY DOWNWARD and connect the cable to the lever. Adjust as required.

4. Move the air selection control lever to the **RECIRC** position. Move the air selection damper FULLY INWARD and connect the cable to the lever. Adjust as required.

Control Head

REMOVAL & INSTALLATION

▶ **See Figures 6 and 7**

1. Disconnect the negative battery cable.

2. Remove the glove box assembly.

3. Remove the dial control knobs from the control head.

4. Remove the center air outlet by disengaging the tabs with a flat blade tool and carefully prying out.

5. Remove the instrument cluster bezel and radio bezel.

6. Remove the knee protector and lower the hood lock release handle.

7. Remove the left side lower duct work.

8. Disconnect the air, temperature and mode selection control cables from the heater housing.

9. Remove the mounting screws and the control head from the instrument panel.

To install:

10. Feed the control cable through the instrument panel, connect the connectors, install the control head assembly and secure with the screws.

11. Move the mode selection lever to the **DEFROST** position. Move the mode selection damper lever FULLY INWARD and connect the cable to the lever. Install the clip.

12. Move the temperature control lever to its HOTTEST position. Move the blend air damper lever FULLY DOWNWARD and connect the cable to the lever. Install the clip.

13. Move the air selection control lever to the **RECIRC** position. Move the air selection damper FULLY INWARD and connect the cable to the lever. Install the clip.

14. Connect the negative battery cable and check the entire climate control system for proper operation.

15. If everything is satisfactory, install the remaining interior pieces.

AIR CONDITIONER

Discharging the air conditioning system is required before the system is opened to the atmosphere during a service procedure. When discharging the system, an approved R-12 Recovery/Recycling machine that meets SAE standard J1991 should be employed. If such a recovery/recycling station is not used and the R-12 is allowed to vent into the atmosphere, depletion of the ozone layer in the upper atmosphere will be increased.

Service Valve Location

The suction (low pressure) port is normally located on the compressor itself. The discharge (high pressure) port is located on the discharge line at the left front corner of the engine compartment.

System Discharging

The R-12 refrigerant is a chlorofluorocarbon which, when mishandled, can contribute to the depletion of the ozone layer in the upper atmosphere. Ozone filters out harmful radiation from the sun. In order to protect the ozone layer, an approved R-12 Recovery/Recycling machine that meets SAE standard J1991 should be employed when discharging the system. Follow the operating instructions provided with the approved equipment exactly to properly discharge the system. Refer to Section 1 in this manual for more information.

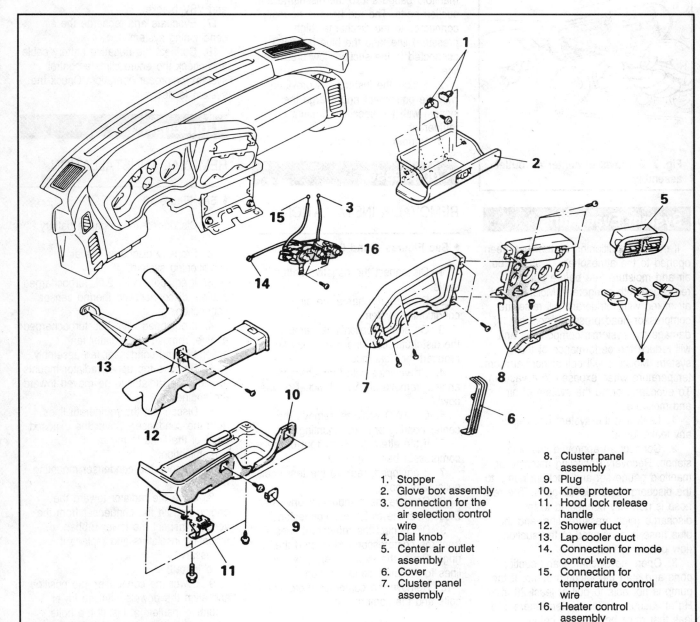

1. Stopper
2. Glove box assembly
3. Connection for the air selection control wire
4. Dial knob
5. Center air outlet assembly
6. Cover
7. Cluster panel assembly
8. Cluster panel assembly
9. Plug
10. Knee protector
11. Hood lock release handle
12. Shower duct
13. Lap cooler duct
14. Connection for mode control wire
15. Connection for temperature control wire
16. Heater control assembly

Fig. 6 Heater control assembly

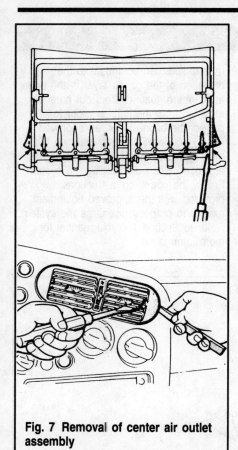

Fig. 7 Removal of center air outlet assembly

System Evacuating

If the air conditioning system has been opened to the atmosphere, it should be air and moisture free before being recharged with refrigerant. Moisture and air mixed with refrigerant will raise the compressor head pressure, possibly damage the system's components and will reduce the performance of the system. Moisture will boil at normal room temperature when exposed to a vacuum. To evacuate, or rid the system of air and moisture:

1. Leak test the system and repair any leaks found.
2. Connect an approved charging station, Recovery/Recycling machine or manifold gauge set and vacuum pump to the discharge and suction ports. The red hose is normally connected to the discharge (high pressure) line, and the blue hose is connected to the suction (low pressure) line
3. Open the discharge and suction ports and start the vacuum pump. If the pump is not able to pull at least 26 in. Hg of vacuum on the system, there is a leak that must be repaired before evacuation can occur.

4. Once the system has reached at least 26-28 in. Hg of vacuum, allow the system to evacuate for at least 15 minutes. The longer the system is evacuated, the more contaminants will be removed.
5. Close all valves and turn the pump OFF. If the system loses more than 2 in. Hg of vacuum after 10 minutes, there is a leak that should be repaired.

System Charging

1. Connect an approved charging station, Recovery/Recycling machine or manifold gauge set to the discharge and suction ports. The red hose is normally connected to the discharge (high pressure) line, and the blue hose is connected to the suction (low pressure) line.
2. Follow the instructions provided with the equipment and charge the system with the specified amount of refrigerant.
3. Perform a leak test.

Compressor

REMOVAL & INSTALLATION

▶ **See Figures 8 and 9**

1. Disconnect the negative battery cable.
2. Properly discharge the air conditioning system.
3. On some applications, removal of the distributor cap and wires may be required for clearance.
4. On vehicles with turbocharged engine, remove the VSV bracket on the cowl.
5. On AWD vehicles, remove the center bearing bracket mounting bolt.
6. If the alternator is in front of the compressor belt, remove it.
7. If equipped, remove the tensioner pulley assembly.
8. Remove the compressor drive belt. Disconnect the clutch coil connector.
9. Disconnect the refrigerant lines from the compressor and discard the O-rings. Cover the exposed ends of the lines to minimize contamination.
10. Remove the compressor mounting bolts and the compressor.

To install:
11. Install the compressor and torque the mounting bolts. Connect the clutch coil connector.
12. Using new lubricated O-rings, connect the refrigerant lines to the compressor.
13. Install the belt and tensioner pulley, if removed. Adjust the belt to specifications.
14. Install the alternator belt, if removed.
15. Install the center bearing bracket mounting bolts and VSV bracket, if removed. Torque the center bearing bracket mounting bolts.
16. Install the distributor cap, wires and VSV bracket.
17. Evacuate and recharge the air conditioning system.
18. Connect the negative battery cable and check the entire climate control system for proper operation. Check the system for leaks.

Condenser

REMOVAL & INSTALLATION

▶ **See Figure 10**

1. Disconnect the negative battery cable.
2. Properly discharge the air conditioning system.
3. If equipped with 2.0L turbocharged engine, disconnect the thermo sensor connection.
4. If equipped with 2.0L turbocharged engine, remove the radiator fan assembly and condenser fan assembly.
5. Remove the upper radiator mounts to allow the radiator to be moved toward the engine.
6. Disconnect the refrigerant lines from the condenser. Cover the exposed ends of the lines to minimize contamination.
7. Remove the condenser mounting bolt.
8. Move the radiator toward the engine and lift the condenser from the vehicle. Inspect the lower rubber mounting insulators and replace if necessary.

To install:
9. Lower the condenser into position and align the dowels with the lower mounting insulators. Install the bolts.

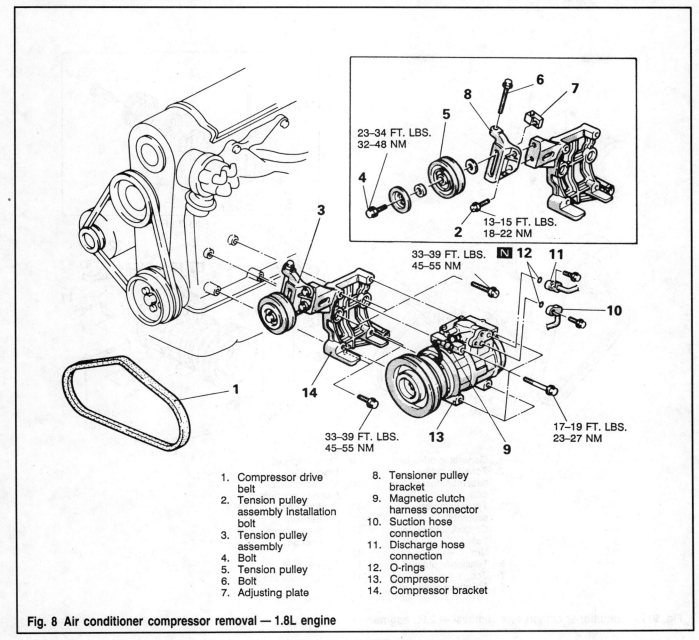

Fig. 8 Air conditioner compressor removal — 1.8L engine

23–34 FT. LBS.
32–48 NM

13–15 FT. LBS.
18–22 NM

33–39 FT. LBS.
45–55 NM

17–19 FT. LBS.
23–27 NM

33–39 FT. LBS.
45–55 NM

1. Compressor drive belt
2. Tension pulley assembly installation bolt
3. Tension pulley assembly
4. Bolt
5. Tension pulley
6. Bolt
7. Adjusting plate
8. Tensioner pulley bracket
9. Magnetic clutch harness connector
10. Suction hose connection
11. Discharge hose connection
12. O-rings
13. Compressor
14. Compressor bracket

10. Using new O-rings, connect the refrigerant lines.

11. Install the radiator mounts and cooling fans.

12. Install the grille, alternator, windshield washer reservoir, battery tray and battery, if removed.

13. Evacuate and recharge the air conditioning system. Add 2 oz. of refrigerant oil during the recharge.

14. Connect the negative battery cable and check the entire climate control system for proper operation. Check the system for leaks.

Evaporator Core

REMOVAL & INSTALLATION

▶ See Figures 11 and 12

1. Disconnect the negative battery cable.

2. Drain the cooling system and properly discharge the air conditioning system and disconnect the refrigerant lines from the evaporator. Cover the exposed ends of the lines to minimize contamination.

3. Pull the heater control knobs off and remove the control panel assembly.

4. Open the glove box, remove the plugs from the sides and the glove box assembly.

5. Remove the lower air ducts and the lower frame.

6. Remove the wiring harness connectors and the electronic air conditioning control unit.

7. Remove the drain hose and lift out the evaporator unit.

8. If servicing the assembly, disassemble the housing and remove the expansion valve and evaporator.

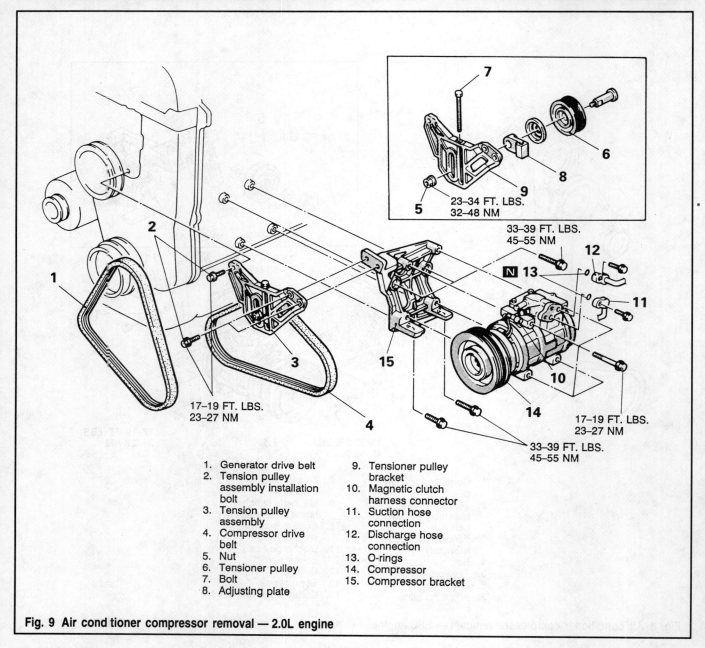

23–34 FT. LBS.
32–48 NM

33–39 FT. LBS.
45–55 NM

17–19 FT. LBS.
23–27 NM

17–19 FT. LBS.
23–27 NM

33–39 FT. LBS.
45–55 NM

1. Generator drive belt
2. Tension pulley assembly installation bolt
3. Tension pulley assembly
4. Compressor drive belt
5. Nut
6. Tensioner pulley
7. Bolt
8. Adjusting plate
9. Tensioner pulley bracket
10. Magnetic clutch harness connector
11. Suction hose connection
12. Discharge hose connection
13. O-rings
14. Compressor
15. Compressor bracket

Fig. 9 Air cond tioner compressor removal — 2.0L engine

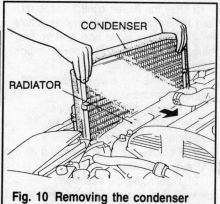

CONDENSER

RADIATOR

Fig. 10 Removing the condenser

To install:

9. Assemble the housing, evaporator and expansion valve, making sure the gaskets are in good condition. Install the evaporator housing.

10. Install the electronic air conditioning control unit. Connect all wires and control cables.

11. Install the lower air ducts, lower frame and the drain hose.

12. Install the glove box.

13. Using new lubricated O-rings, connect the refrigerant lines to the evaporator.

14. Evacuate and recharge the air conditioning system, adding 2 oz. of refrigerant oil during the recharge.

15. Connect the negative battery cable and check the entire climate control system for proper operation. Check the system for leaks.

Air Conditioning Switch

REMOVAL & INSTALLATION

▶ See Figures 13 and 14

1. Disconnect the negative battery cable.

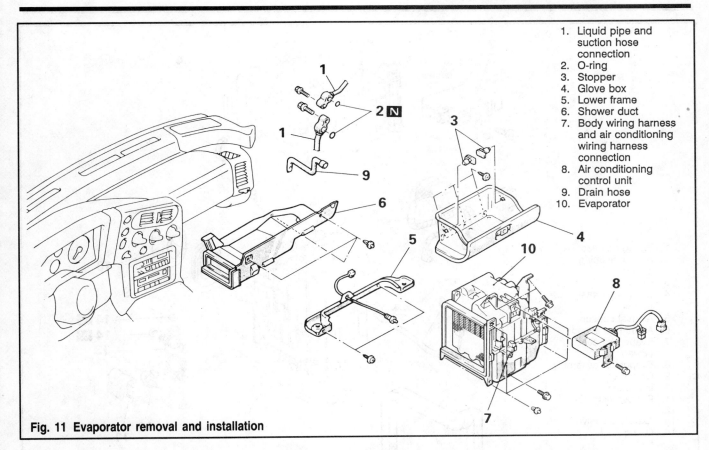

1. Liquid pipe and suction hose connection
2. O-ring
3. Stopper
4. Glove box
5. Lower frame
6. Shower duct
7. Body wiring harness and air conditioning wiring harness connection
8. Air conditioning control unit
9. Drain hose
10. Evaporator

Fig. 11 Evaporator removal and installation

2. Remove the instrument cluster bezel and radio bezel. Remove the radio and tape player, as equipped.

3. Insert your hand through the produced opening to the back of the cluster panel. Catching the air conditioning switch on the right and the left sides, push it towards you for removal.

4. Installation is the reverse of the removal procedure.

Expansion Valve

REMOVAL & INSTALLATION

▶ **See Figures 15, 16 and 17**

1. Disconnect the negative battery cable.

2. Properly discharge the air conditioning system.

3. Remove the evaporator housing and separate the upper and lower case.

4. Remove the expansion valve from the evaporator line.

5. The installation is the reverse of the removal procedures. Use new lubricated O-rings when assemblies.

6. Evacuate and recharge the air conditioning system.

7. Connect the negative battery cable and check the entire climate control system for proper operation. Check the system for leaks.

Receiver/Drier

The receiver/drier is located forward of the condenser.

REMOVAL & INSTALLATION

1. Disconnect the negative battery cable. Properly discharge the air conditioning system.

2. Disconnect the electrical connector from the switch on the receiver/drier, if equipped.

3. Disconnect the refrigerant lines from the receiver/drier assembly. Discard the O-rings. Cover the exposed ends of the lines to minimize contamination.

4. Open or remove the mounting strap and the receiver/drier from its bracket.

5. The installation is the reverse of the removal procedures. Use new lubricated O-rings when assembling.

6. Evacuate and recharge the air conditioning system. Add 1 oz. of refrigerant oil during the recharge.

7. Connect the negative battery cable and check the entire climate control system for proper operation. Check the system for leaks.

Refrigerant Lines

REMOVAL & INSTALLATION

▶ **See Figure 18**

1. Disconnect the negative battery cable. Remove the reserve tank and the battery as required.

2. Properly discharge the air conditioning system.

3. Remove the nuts or bolts that attach the refrigerant lines sealing plates to the adjoining components. If the line is not equipped with a sealing plate, separate the flare connection.

4. Remove the line and discard the O-rings.

To install:

5. Coat the new O-rings with refrigerant oil and install. Connect the refrigerant lines to the adjoining

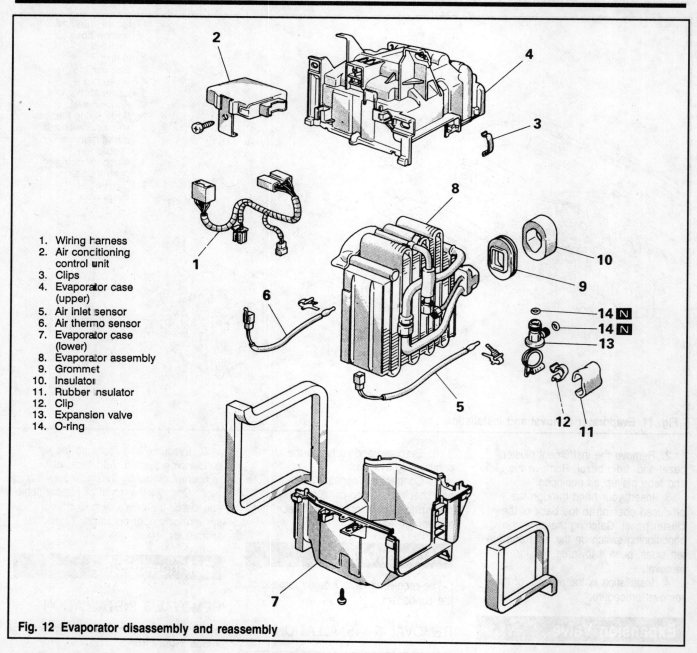

1. Wiring harness
2. Air conditioning control unit
3. Clips
4. Evaporator case (upper)
5. Air inlet sensor
6. Air thermo sensor
7. Evaporator case (lower)
8. Evaporator assembly
9. Grommet
10. Insulator
11. Rubber insulator
12. Clip
13. Expansion valve
14. O-ring

Fig. 12 Evaporator disassembly and reassembly

components and tighten the nuts, bolts or flare connections.

6. Evacuate and recharge the air conditioning system.

7. Connect the negative battery cable and check the entire climate control system for proper operation. Check the system for leaks.

Power Relays

OPERATION

These vehicles use relays to control the compressor clutch coil, heater,

condenser fan and speed of the condenser fan. The compressor coil, condenser fan and condenser fan speed relays are in a small relay block located in the left rear corner of the engine compartment. The heater relay is located on top of the interior junction block under the left side of the instrument panel.

TESTING

◆ See Figure 19

Compressor Coil, Heater and Condenser Fan Relays

1. Remove the relay.

2. Use jumper wires to connect the positive battery terminal to terminal **2** of the relay and the negative terminal to terminal **4**.

3. With 12 volts applied, there should be continuity across terminals **1** and **3**.

4. When the voltage is disconnected:

   a. Terminals **1** and **3** should be open.

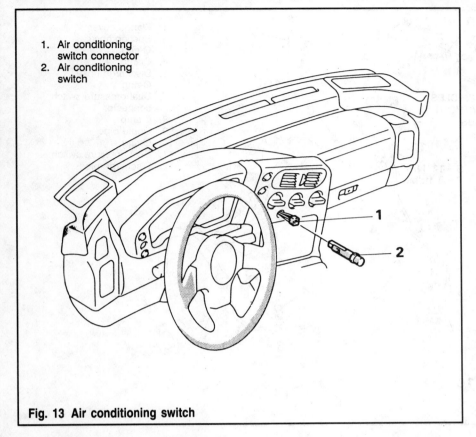

1. Air conditioning switch connector
2. Air conditioning switch

Fig. 13 Air conditioning switch

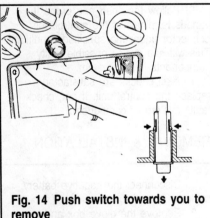

Fig. 14 Push switch towards you to remove

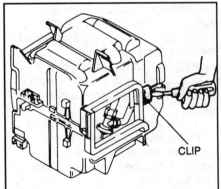

Fig. 15 Remove the clips from the heater case

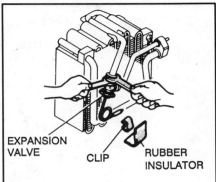

EXPANSION VALVE   CLIP   RUBBER INSULATOR

Fig. 16 Loose the flare nut on the expansion valve using 2 wrenches

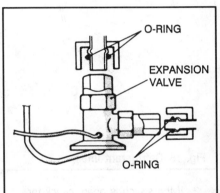

O-RING

EXPANSION VALVE

O-RING

Fig. 17 Apply compressor oil to the expansion valve and O-rings

b. Terminals **2** and **4** should have continuity.

5. Replace the relay if faulty.

Condenser Fan Control Relay

▶ **See Figure 20**

1. Remove the relay.

2. Use jumper wires to connect the positive battery terminal to terminal **3** of the relay and the negative terminal to terminal **5**.

3. With 12 volts applied, there should be continuity across terminals **1** and **2**.

4. When the voltage is disconnected:

a. Terminals **1** and **4** should have continuity.

b. Terminals **3** and **5** should have continuity.

c. Terminals **1** and **2** should be open.

5. Replace the relay if faulty.

Dual Pressure Switch

OPERATION

The dual pressure switch is a combination of a low pressure cut off switch and high pressure cut off switch. The function of this switch is to prevent the operation of the compressor in the event of either high of low refrigerant charge, preventing damage to the system. The switch is located near the sight glass on the refrigerant line.

The dual pressure switch is designed to cut off voltage to the compressor coil when the pressure either drops below 30 psi or rises above 384 psi.

TESTING

1. Check for continuity through the switch, using a jumper wire at the switch harness and an ohmmeter on the switch terminals.

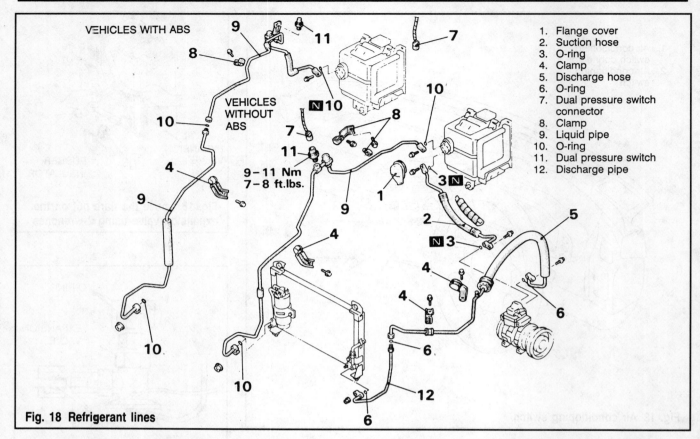

VEHICLES WITH ABS

VEHICLES WITHOUT ABS

9 – 11 Nm
7 – 8 ft.lbs.

1. Flange cover
2. Suction hose
3. O-ring
4. Clamp
5. Discharge hose
6. O-ring
7. Dual pressure switch connector
8. Clamp
9. Liquid pipe
10. O-ring
11. Dual pressure switch
12. Discharge pipe

Fig. 18 Refrigerant lines

2. If the switch is open, check for insufficient refrigerant charge or excessive pressures.

3. If neither of the above conditions exist and the switch is open, replace the switch.

REMOVAL & INSTALLATION

1. Disconnect the negative battery cable.

2. Properly discharge the air conditioning system.

3. Remove the switch from the refrigerant line.

4. The installation is the reverse of the removal installation.

5. Evacuate and recharge the air conditioning system.

6. Connect the negative battery cable and check the entire climate control system for proper operation. Check the system for leaks.

A/C Control Unit

The A/C control unit is an electronic control unit used to process information received from various sensors and switches to control the air conditioning compressor. The unit is located behind the glove box on top of the evaporator housing. The function of the control unit is to send current to the dual pressure switch when the following conditions are met:

1. The air conditioning switch is in either the **ECONO** or **A/C** mode.

2. The refrigerant temperature sensor, if equipped, is reading 347°F (175°C) or less.

3. The air thermo and air inlet sensors are both reading at least 39°F (4°C).

TESTING

1. Disconnect the control unit connector.

2. Turn the ignition switch **ON**.

3. Turn the air conditioning switch **ON**.

4. Turn the temperature control lever too its COOLEST position.

5. Turn the blower switch to its HIGHEST position.

6. Follow the chart and probe the various terminals of the control unit connector under the specified conditions. This will rule out all possible faulty components in the system.

7. If all checks are satisfactory, replace the control unit. If not, check the faulty system or component.

REMOVAL & INSTALLATION

1. Disconnect the negative battery cable.

2. Remove the glove box and locate the control module.

3. Disconnect the connector to the module and remove the mounting screws.

4. Remove the module from the evaporator housing.

5. The installation is the reverse of the removal installation.

6. Connect the negative battery cable and check the entire climate control system for proper operation.

CRUISE CONTROL

Control Switch

The cruise control switch is part of the column switch.

REMOVAL & INSTALLATION

▶ See Figure 21

1. Remove the horn pad screw at the back of the steering wheel. Push up on the pad to remove it.
2. Remove the nut and use a puller to remove the steering wheel.
3. Remove the upper and lower steering column covers. On some models, the lower knee pad and air conditioning duct must also be removed.
4. Disconnect the wiring and remove the screws to remove the column switches as an assembly.
5. Installation is the reverse of removal. Torque the steering wheel nut to 30 ft. lbs. (40 Nm).

Speed Sensor

REMOVAL & INSTALLATION

1. The instrument panel and speedometer must be removed to replace the speed sensor. Make sure all fault codes have been read from control unit memory and disconnect the negative battery cable.
2. Remove the bezel screws and the instrument panel bezel.
3. Remove the screws and move the top of the instrument panel down to

CONDENSER FAN MOTOR RELAY, MAGNET CLUTCH RELAY, BLOWER MOTOR HIGH RELAY

HEATER RELAY

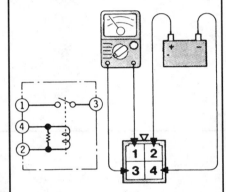

Fig. 19 Testing condenser fan motor relay, magnetic clutch relay, blower motor high relay and heater relay

disconnect the speedometer cable and wiring. Remove the instrument panel.
4. Installation is the reverse of removal.

Actuator Assembly

REMOVAL & INSTALLATION

1. Disconnect the negative battery cable.
2. Remove the retainer clip and fastener holding the cruise control cable to the cruise control actuator.
3. Remove the vacuum hose from the rear side of the cruise control actuator.
4. Remove the mounting bolts and the cruise control actuator.
5. Installation is the reverse of the removal procedure.
6. Connect the negative battery cable and confirm proper operation of the cruise control system.

Control Unit

REMOVAL & INSTALLATION

1. Disconnect the negative battery cable.
2. Remove the lower interior trim from the left side of the instrument panel.
3. Remove the junction block to access the cruise control unit.
4. Remove the mounting screws and the control unit from the vehicle.
5. Installation is the reverse of the removal procedure.

ENTERTAINMENT SYSTEMS

Radio/Tape Player/CD Player

REMOVAL & INSTALLATION

▶ See Figures 22 and 23

1. Disconnect the negative battery cable.
2. Using a plastic trim tool, remove the radio panel. Pry the lower part of the radio panel away first, then separate and remove the pan.
3. Remove the mounting bolts and pull the radio receiver, tape player or CD player out slightly. Disconnect all harness connectors and remove unit from the vehicle.
4. Installation is the reverse of the removal procedure.

Speakers

REMOVAL & INSTALLATION

▶ See Figures 24, 25 and 26

Front Speaker

1. Disconnect the negative battery cable.
2. Remove the front speaker garnish.

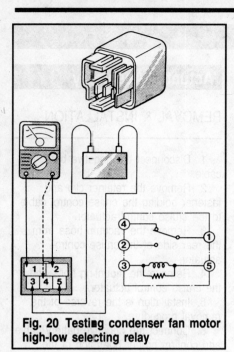

Fig. 20 Testing condenser fan motor high-low selecting relay

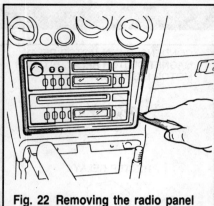

Fig. 22 Removing the radio panel trim plate

3. Remove the retainers, disconnect the harness connector and remove the front speaker.

✵✵WARNING

Handle the speaker carefully to avoid damaging the cone during removal and installation.

4. Installation is the reverse of the removal procedure

Door Speaker

1. Disconnect the negative battery cable.

2. Remove the door trim. Refer to the procedure on Section 10.

3. Remove the mounting screws, disconnect the harness connector and remove the front speaker.

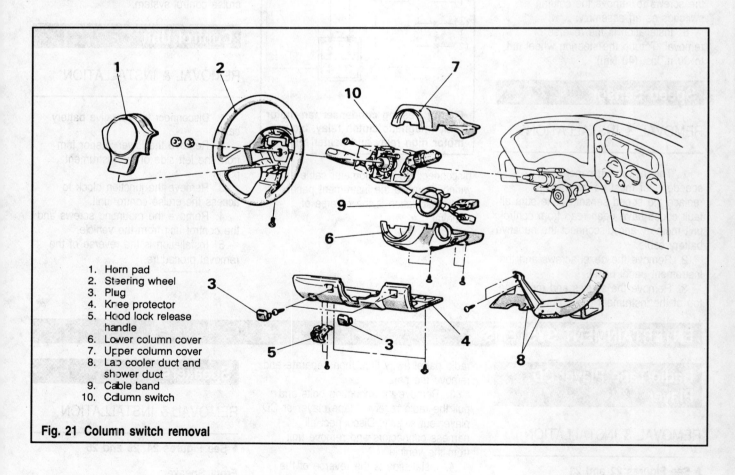

1. Horn pad
2. Steering wheel
3. Plug
4. Knee protector
5. Hood lock release handle
6. Lower column cover
7. Upper column cover
8. Lap cooler duct and shower duct
9. Cable band
10. Column switch

Fig. 21 Column switch removal

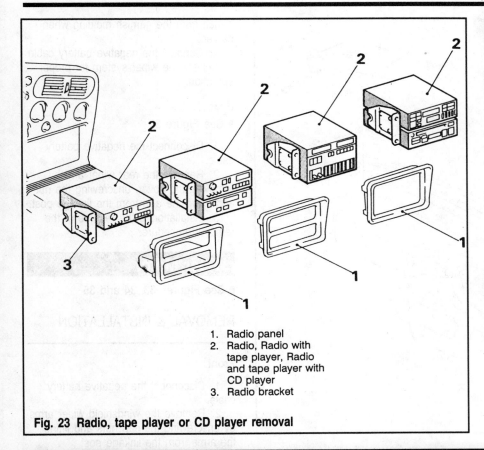

1. Radio panel
2. Radio, Radio with tape player, Radio and tape player with CD player
3. Radio bracket

Fig. 23 Radio, tape player or CD player removal

Handle the speaker carefully to avoid damaging the cone during removal and installation.

4. Installation is the reverse of the removal procedure

Rear Speaker

1. Disconnect the negative battery cable.
2. Remove the luggage compartment side tray.
3. Remove the speaker retainers and lift the speaker from the speaker cover. Disconnect the harness connector.

⁕⁕WARNING

Handle the speaker carefully to avoid damaging the cone during removal and installation.

4. Installation is the reverse of the removal procedure.
5. Connect the negative battery cable.

WINDSHIELD WIPER AND WASHERS

The windshield wipers and washers can be operated with the wiper switch only when the ignition switch is in the **ACC** or **ON** positions. A circuit breaker, integral with the wiper switch or fuse in the fuse box, protects the circuitry of the wiper system and the vehicle.

Wiper blades, exposed to the weather for a long period of time, tend to lose their wiping effectiveness. Periodic cleaning of the the wiper blade element is suggested. The wiper blade element, arms and windshield should be cleaned with a sponge or cloth and a mild detergent or non-abrasive cleaner. If the the wiper element continues to streak or smear, replace the wiper blade element on or arm.

Windshield Wiper Blade

REMOVAL & INSTALLATION

♦ **See Figures 27, 28 and 29**

1. Lift the wiper arm away from the glass.
2. Depress the release lever on the bridge and remove the blade assembly from the arm. On later model vehicles, remove the blade assembly from the arm by inserting a small tool into the release slot of the wiper blade and push downward, or by pushing the release button. Refer to the necessary illustration.
3. To remove the wiping element from the blade assembly: Lift the tab and pinch the end bridge to release it from the center bridge.
4. Slide the end bridge from the blade element and the element from the opposite end bridge.
5. Assembly is the reverse of removal. Make sure that the element

locking tabs are securely locked in position.

Windshield Wiper Blade and Arm

REMOVAL & INSTALLATION

♦ **See Figures 30, 31 and 32**

Front

1. Disconnect the negative battery cable.
2. Remove the windshield wiper arms by unscrewing the cap nuts and lifting the arms from the linkage posts.
   To install:
3. Install the wiper blades. Note that the driver's side wiper arm should be marked **D** or **Dr** and the passenger's side wiper arm should be marked **A** or **As**. The identification marks should be located at the base of the arm, near the pivot. Install the arms so the blades are

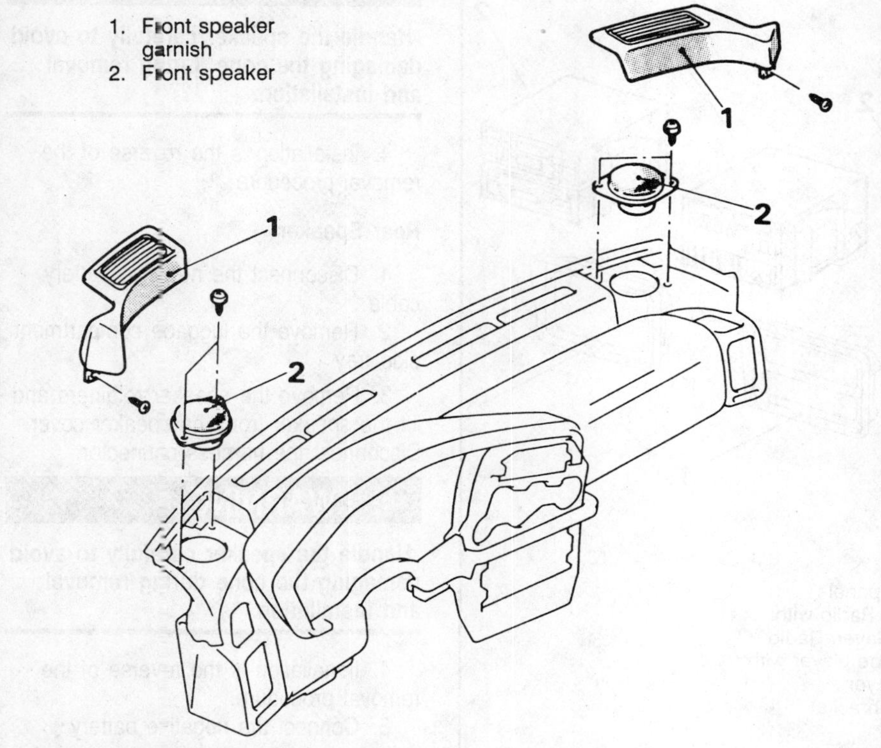

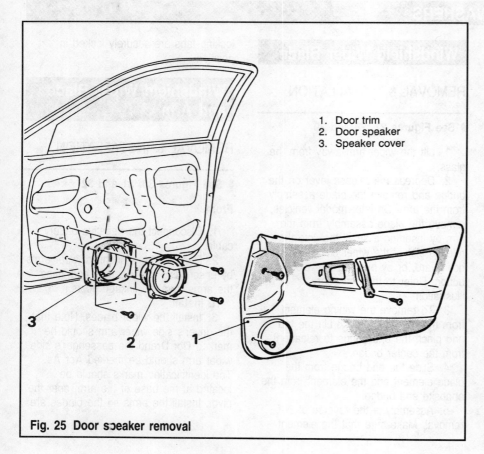

1. Front speaker garnish
2. Front speaker

Fig. 24 Front speaker removal

1. Door trim
2. Door speaker
3. Speaker cover

Fig. 25 Door speaker removal

1 inch from the garnish molding when parked.

4. Connect the negative battery cable and check the wiper system for proper operation.

Rear

▶ **See Figure 33**

1. Disconnect the negative battery cable.

2. Remove the rear wiper arm by removing the cover, unscrewing the nut and lifting the arm from the linkage post.

3. Installation is the reverse of the removal procedure.

Wiper Motor

▶ **See Figures 33, 34 and 35**

REMOVAL & INSTALLATION

Front

1. Disconnect the negative battery cable.

2. Remove the windshield wiper arms by unscrewing the cap nuts and lifting the arms from the linkage post.

3. Remove the front deck garnish panel.

4. Remove the air inlet trim pieces.

5. Remove the hole cover.

6. Remove the wiper motor by loosening the mounting bolts, removing the motor assembly, then disconnecting the linkage.

➡**The installation angle of the crank arm and motor has been factory set; do not remove them unless it is necessary to do so. If they must be removed, remove them only after marking their mounting positions.**

To install:

7. Install the windshield wiper motor and connect the linkage.

8. Reinstall all trim pieces.

9. Reinstall the wiper blades. Note that the driver's side wiper arm should be marked **D** or **Dr** and the passenger's side wiper arm should be marked **A** or **As**. The identification marks should be located at the base of the arm, near the pivot. Install the arms so the blades are 1 inch from the garnish molding when parked.

10. Connect the negative battery cable and check the wiper system for proper operation.

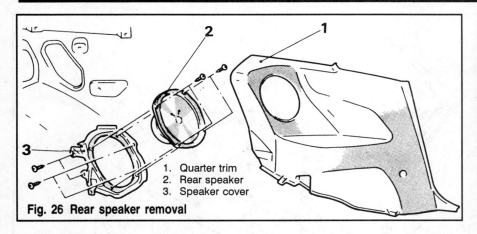

1. Quarter trim
2. Rear speaker
3. Speaker cover

Fig. 26 Rear speaker removal

Fig. 27 Removing the blade assembly from the wiper arm

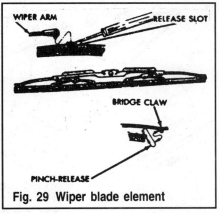

Fig. 29 Wiper blade element

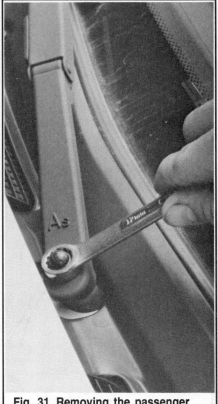

Fig. 31 Removing the passenger side wiper arm

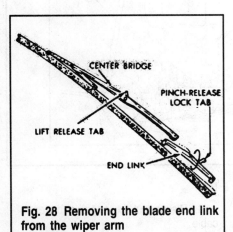

Fig. 28 Removing the blade end link from the wiper arm

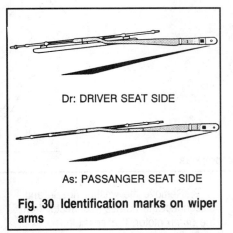

Dr: DRIVER SEAT SIDE

As: PASSANGER SEAT SIDE

Fig. 30 Identification marks on wiper arms

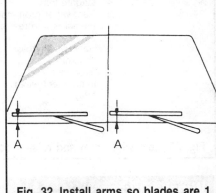

Fig. 32 Install arms so blades are 1 in. from the garnish molding when parked

Rear

1. Disconnect the negative battery cable.

2. Remove the rear wiper arm by removing the cover, unscrewing the nut and lifting the arm from the linkage post.

3. Remove the large interior trim panel. Use a plastic trim stick to unhook the trim clips of the liftgate trim.

4. If equipped with rear air spoiler, remove the wiper grommet.

5. Remove the rear wiper assembly. Do not loosen the grommet for the wiper post.

To install:

6. Install the motor and grommet. Mount the grommet so the arrow on the grommet is pointing upward.

7. Install the wiper arm.

8. Connect the negative battery cable and check the rear wiper for proper operation.

9. If operation is satisfactory, fit the tabs on the upper part of the liftgate trim into the liftgate clips and secure the liftgate trim.

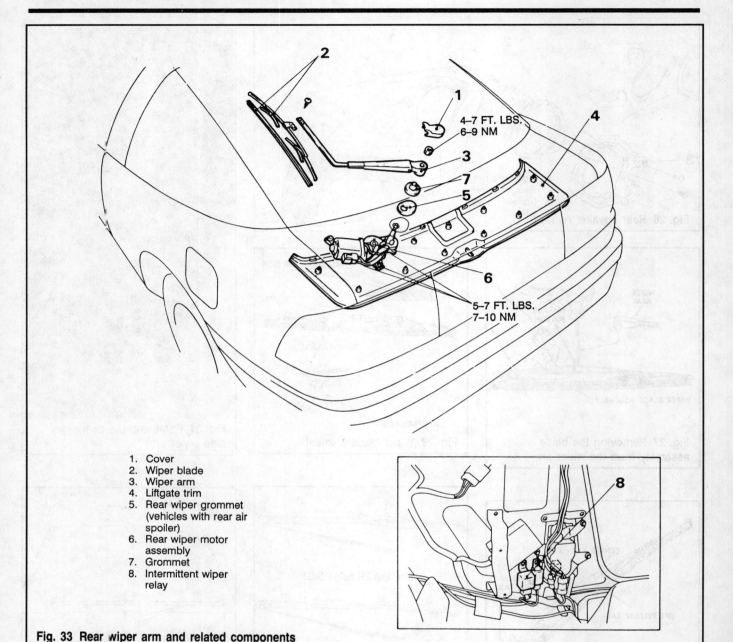

1. Cover
2. Wiper blade
3. Wiper arm
4. Liftgate trim
5. Rear wiper grommet (vehicles with rear air spoiler)
6. Rear wiper motor assembly
7. Grommet
8. Intermittent wiper relay

Fig. 33 Rear wiper arm and related components

Wiper Linkage

On these vehicles, the front wiper linkage is removed along with the front windshield wiper motor. Refer to the appropriate procedure above for instructions.

Windshield Washer Motor and Fluid Reservoir

REMOVAL & INSTALLATION

▶ **See Figure 36**

Front

1. Disconnect the negative battery cable.

2. Remove the washer nozzle and disconnect the harness connector from the pump motor, if accessible.

3. Disconnect the washer tube.

4. Remove the front splash shield extension from the left front wheel well area. Disconnect the pump motor harness, if still connected.

5. Remove the tank retainers and the tank from the vehicle. Remove the washer motor assembly from the tank and replace as required.

6. Installation is the reverse of the removal procedure.

Rear

1. Disconnect the negative battery cable.

2. Remove the rear side trim. Refer to Section 10 for detained instructions.

3. Disconnect the electrical harness at the reservoir.

4. Remove the reservoir retainers. Disconnect the fluid feed tubes at the tank and remove.

5. Installation is the reverse of the removal procedure.

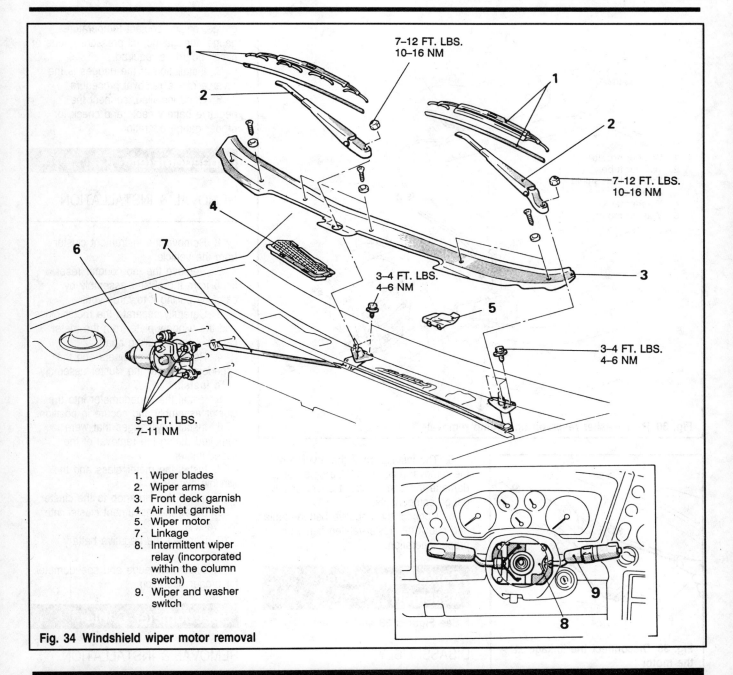

1. Wiper blades
2. Wiper arms
3. Front deck garnish
4. Air inlet garnish
5. Wiper motor
7. Linkage
8. Intermittent wiper relay (incorporated within the column switch)
9. Wiper and washer switch

Fig. 34 Windshield wiper motor removal

INSTRUMENTS AND SWITCHES

The removal and installation procedures in this section covers gauges in the dash assembly. For sending unit service procedures, see Section 3 or Section 5 in this manual.

Instrument Cluster

REMOVAL & INSTALLATION

♦ See Figure 37

1. Disconnect the negative battery cable.

2. Remove the screw cover on the side of the cluster panel assembly.

3. Remove the front instrument cluster bezel.

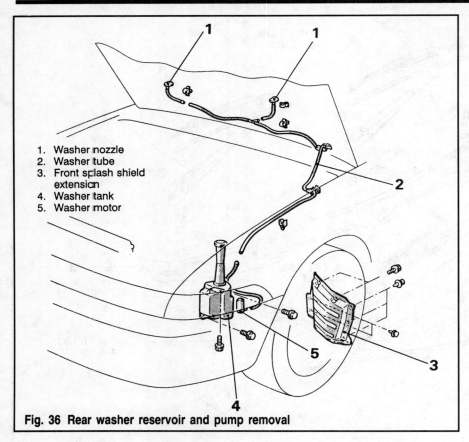

1. Washer nozzle
2. Washer tube
3. Front splash shield extension
4. Washer tank
5. Washer motor

Fig. 36 Rear washer reservoir and pump removal

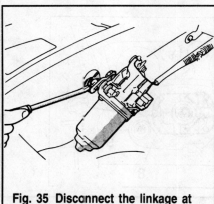

Fig. 35 Disconnect the linkage at the motor

4. Remove the instrument cluster. Disassemble and remove gauges or the speedometer as required.

➡️If the speedometer cable adapter requires service, disconnect the cable at the transaxle end. Pull the cable slightly toward the vehicle interior, release the lock by turning the adapter to the right or left and remove the adapter.

5. The installation is the reverse of the removal procedure. Use care not to damage the printed circuit board or any gauge components.

6. Connect the negative battery cable and check all cluster-related items for proper operation.

Combination Meter Assembly

♦ See Figures 38 and 39

DISASSEMBLY

1. Remove the instrument cluster from the vehicle.

2. Remove the trip counter reset knob from the cluster assembly by carefully pulling it towards you.

3. Carefully separate the meter glass and the window plate from the meter case by gently pulling apart.

4. Remove the retainers and the tachometer or tachometer and pressure

gauge, engine coolant temperature gauge, fuel gauge, oil pressure gauge or circuit board as required.

5. Installation of the gauges is the reverse of the removal procedure.

6. Once installed, connect the negative battery cable and check for proper gauge operation.

Speedometer

REMOVAL & INSTALLATION

1. Remove the instrument cluster from the vehicle.

2. Remove the trip counter reset knob from the cluster assembly by carefully pulling it towards you.

3. Carefully separate the meter glass and the window plate from the meter case by gently pulling apart.

4. Remove the retainers and the speedometer from the cluster assembly.

To install:

5. Install the speedometer into the cluster assembly and secure in position.

6. Install any gauges that were removed during the removal of the speedometer.

7. Install the meter glass and the window plate.

8. Install the trip knob to the cluster.

9. Install the instrument cluster into the vehicle.

10. Connect the negative battery cable.

11. Check all gauge and speedometer for proper operation.

Speedometer Cable

REMOVAL & INSTALLATION

1. Disconnect the negative battery cable.

2. Remove the instrument cluster from the vehicle.

3. Raise and support the vehicle safely.

4. Loosen the mounting nut and extract the speedometer cable from the transaxle assembly and adapter. Remove the cable from the vehicle.

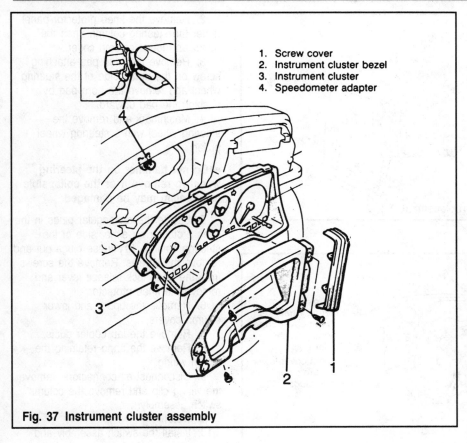

1. Screw cover
2. Instrument cluster bezel
3. Instrument cluster
4. Speedometer adapter

Fig. 37 Instrument cluster assembly

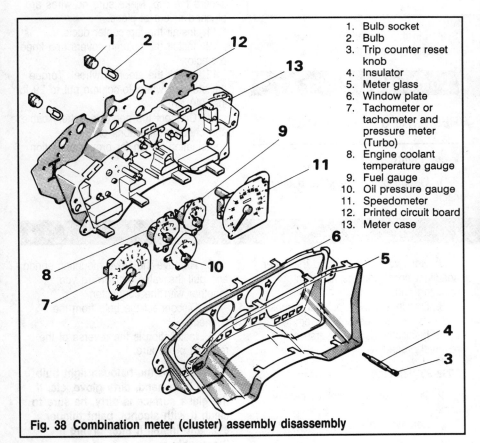

1. Bulb socket
2. Bulb
3. Trip counter reset knob
4. Insulator
5. Meter glass
6. Window plate
7. Tachometer or tachometer and pressure meter (Turbo)
8. Engine coolant temperature gauge
9. Fuel gauge
10. Oil pressure gauge
11. Speedometer
12. Printed circuit board
13. Meter case

Fig. 38 Combination meter (cluster) assembly disassembly

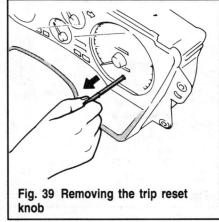

Fig. 39 Removing the trip reset knob

To install:

5. Install the cable on the vehicle, routing in the same location as the original.

➡The cable arrangement should be made so that the radius of cable bends is 5.9 in. (150mm) or more. The arrangement of the speedometer cable should be such that it does not interfere with other parts.

6. Connect the cable to the adapter, then install the adapter to the instrument cluster.

7. At the transaxle end of the cable, the key joint is to be inserted in the transaxle and the nut should be securely tightened.

Windshield Wiper Switch

The windshield wiper switch and the intermittent wiper relay is built into 1 multi-function combination switch that is mounted on the steering column. See Combination Switch in this section for removal and installation procedure.

Rear Window Wiper Switch

REMOVAL & INSTALLATION

▶ See Figure 40

1. Disconnect the negative battery cable.

2. Remove the cluster panel containing the hazard switch, rear window defogger switch and rear wiper switch. See Section 10 for procedure.

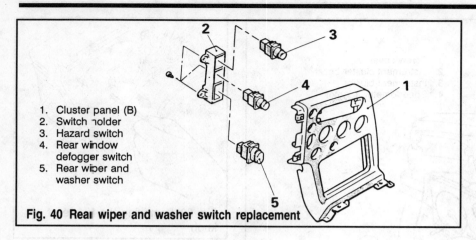

1. Cluster panel (B)
2. Switch holder
3. Hazard switch
4. Rear window defogger switch
5. Rear wiper and washer switch

Fig. 40 Rear wiper and washer switch replacement

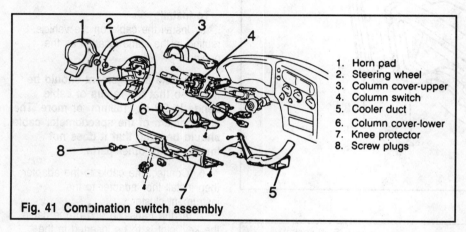

1. Horn pad
2. Steering wheel
3. Column cover-upper
4. Column switch
5. Cooler duct
6. Column cover-lower
7. Knee protector
8. Screw plugs

Fig. 41 Combination switch assembly

3. Remove the switch holder from the back of the panel. Remove the rear wipe switch and replace as required.

4. Installation is the reverse of the removal procedure.

Headlight Switch

The headlight, turn signal and dimmer switches are built into 1 multi-function combination switch that is mounted on the steering column. See Combination Switch in this section for removal and installation procedure.

Combination Switch

REMOVAL & INSTALLATION

▶ See Figure 41

1. Disconnect the negative battery cable.

2. Remove the knee protector panel under the steering column, then the upper and lower column cover.

3. Remove the horn pad attaching screw on the under side of the steering wheel and remove the horn pad by pushing the pad upward.

4. Matchmark and remove the steering wheel with a steering wheel puller.

➡**Do not hammer on the steering wheel to remove it or the collapsible mechanism may be damaged.**

5. Locate the rectangular plugs in the knee protector on either side of the steering column. Pry these plugs out and remove the screws. Remove the screws from the hood lock release lever and remove the knee protector.

6. Remove the upper and lower column covers.

7. Remove the lap cooler ducts.

8. Remove the band retaining the switch wiring.

9. Disconnect all connectors, remove the wiring clip and remove the column switch assembly.

To install:

10. Install the switch assembly and secure the clip. Make sure no wires are pinched or out of place.

11. Install the lap cooler ducts.

12. Install the column covers and knee protector.

13. Install the steering wheel. Torque the steering wheel-to-column nut to 29 ft. lbs. (40 Nm).

14. Connect the negative battery cable and check all functions of the combination switch for proper operation.

LIGHTING

Headlight Bulb (Sealed Beam)

▶ See Figures 42, 43, 44, 45 and 46

REMOVAL & INSTALLATION

1990-91 Vehicles

1. Raise the headlights using the pop-up switch. Disconnect the negative battery cable.

2. Remove the upper and the lower headlight bezel and the headlight retaining ring.

3. Disconnect the electrical harness and remove the headlight assembly.

4. Installation is the reverse of the removal procedure.

1992-93 Vehicles

1. Disconnect the negative battery cable.

2. Remove the socket cover.

3. Remove the valve mounting spring and pull the valve out toward you, together with the connector.

4. Disconnect the bulb from the connector.

5. Installation is the reverse of the removal procedure.

➡**Never hold the halogen light bulb with a bare hand, dirty glove, etc. If the glass surface is dirty, be sure to clean it with alcohol, paint thinner, etc., and install it after drying if thoroughly.**

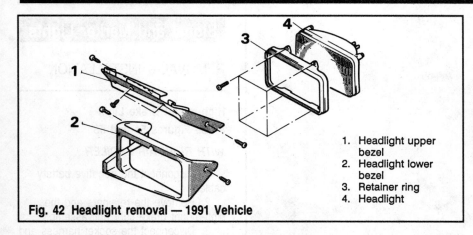

1. Headlight upper bezel
2. Headlight lower bezel
3. Retainer ring
4. Headlight

Fig. 42 Headlight removal — 1991 Vehicle

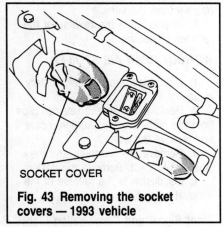

Fig. 43 Removing the socket covers — 1993 vehicle

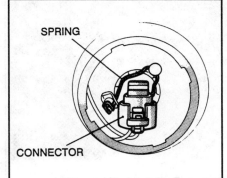

Fig. 44 Remove the valve mounting spring and pull the valve out toward you with the connector

Headlight (Composite)

REMOVAL & INSTALLATION

1992-93 Vehicle
▶ See Figures 47, 48, 49 and 50

1. Disconnect the negative battery cable.
2. Remove the mounting screws and the front side marker lamp.
3. Remove the mounting screws and the headlamp.
   To install:
4. Install the headlamp into the front of the vehicle and secure in place.
5. Insert the boss of the front side marker light into the clip areas of the front fender. Insert the ribs of the front side marker light into the mounting holes on the headlight side. Then mount the front side marker light with the mounting screws.
6. Connect the negative battery cable and check lamp operation.

Concealed Headlights

MANUAL OPERATION

If the headlight covers will not raise electrically, remove the fusible link from the relay box, then remove the boot on the rear area of the pop-up motor and turn the manual knob clockwise until the cover is open. Perform this procedure on both the left and right sides.

HEADLIGHT ALIGNMENT PREPARATION

HEADLAMPS SHOULD BE ADJUSTED WITH A SPECIAL ALIGNMENT TOOL, STATE REGULATIONS MAY VARY THIS PROCEDURE, USE THIS PROCEDURE BELOW FOR TEMPORARY ADJUSTMENTS.

1. Verify the headlamp dimmer switch operation.
2. Test operation of high beam indicator, mounted on instrument panel.
3. Inspect fro badly rusted or faulty headlight assemblies. These conditions must be corrected before the headlights can be properly adjusted.
4. Place vehicle on a level surface.
5. Bounce the front suspension thoroughly.
6. Inspect tire inflation and adjust as required.
7. Rock the vehicle sideways to allow the vehicle to assume tits proper resting position.
8. If the fuel tank is not full, place a weight in the trunk of the vehicle to simulate the weight of a full gas tank. (approx. 6.5 lbs. Per gallon of missing fuel).
9. There should be not other weight in the vehicle aside from the driver, or substitute, weighing approximately 150 lbs.
10. Adjust the headlights as required.

HEADLAMP ALIGNMENT PROCEDURE1990-91 Vehicles

A rough adjustment can be made while shining the headlights on a wall or on the rear of another vehicle, but headlight adjustment should really be made using proper headlight aiming equipment. To adjust headlamps, adjust the alignment screws to achieve the proper beam positioning.

1991-93 Vehicle
▶ See Figures 51 and 52

VERTICAL ADJUSTING

Insert the screwdriver into the vertical adjusting hole and turn the screw clockwise or counterclockwise to bring the bubble of the vertical angle gauge to the center.

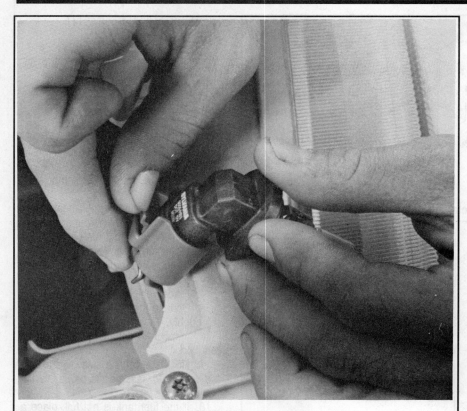

Fig. 45 Disconnecting the halogen headlight bulb from the connector.

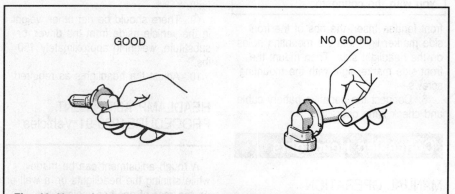

GOOD   NO GOOD

Fig. 46 Never hold the halogen light bulb with a bare hand

HORIZONTAL ADJUSTING

1. After pulling out the stopper upward, press the larger gear forward to disengage it from the smaller gear.
2. Insert the screwdriver into the horizontal adjusting hole and adjust with the headlight test.
3. After adjustment, align the 0 mark of the gear to the centerline and press in the stopper for locking.

Fog Lights

AIMING

▶ See Figure 53

1. Place the vehicle on a level surface.
2. Use the adjusting screws, located on the upper right and middle left of the mounting rim of the light, to adjust the beam of light to the desired area.

Signal and Marker Lights

REMOVAL & INSTALLATION

High-mount Brake Light
▶ See Figures 54 and 55

WITH REAR AIR SPOILER

1. Disconnect the negative battery cable.
2. Remove the retainers and the high-mount brake light unit assembly.
3. Disconnect the socket harness and remove from the vehicle.
4. Installation is the reverse of the removal procedure.

WITHOUT REAR AIR SPOILER

1. Disconnect the negative battery cable.
2. Remove the square retainer clips from the high-mount brake light cover and remove the cover.
3. Remove the fasteners and the lens and bracket. Remove the gasket.
4. Installation is the reverse of the removal procedure.

Front Signal and Combination Lights
▶ See Figure 56

1. Remove the front garnish.
2. Raise the headlights using the pop-up switch. Disconnect the negative battery cable.
3. Disconnect the front turn signal light. Remove the retainers and the turn signal light.
4. Remove the retainers and the front combination light.
5. Installation is the reverse of the removal procedure.

Rear Turn Signal, Brake and Parking Light
▶ See Figures 57, 58, 59 and 60

1. Disconnect the negative battery cable.
2. Remove the back-up lights and license plate light from the rear panel garnish, if equipped.
3. Remove the rear deck garnish. Refer to Section 10.
4. Remove the mounting must from inside the rear compartment and remove the combination light. Turn the socket to remove from the light.
5. Installation is the reverse of the removal procedure. During installation,

tighten the mounting nuts in the sequence illustrated.

Front Side Marker Light

1. Disconnect the negative battery cable.
2. Remove the mounting screws and the front side marker lamp.

To install:

3. Insert the boss of the front side marker light into the clip areas of the front fender. Insert the ribs of the front side marker light into the mounting holes on the headlight side. Then mount the front side marker light with the mounting screws.

4. Connect the negative battery cable and check lamp operation.

License Plate Light

The license plate lights are located in the rear bumper fascia. For service procedures, refer to Section 10.

TRAILER WIRING

➡ **The vehicles covered in this manual were not designed with trailer towing in mind. Proper consideration should be given concerning the above normal load that will be placed on the your vehicle during trailer towing. These vehicles may not be the best choice for such conditions.**

Wiring the car for towing is fairly easy. There are a number of good wiring kits available and these should be used, rather than trying to design your own. All trailers will need brake lights and turn signals as well as tail lights and side marker lights. Most states require extra marker lights for overly wide trailers. Also, most states have recently required back-up lights for trailers, and most trailer manufacturers have been building trailers with back-up lights for several years.

Additionally, some Class I, most Class II and just about all Class III trailers will have electric brakes.

Add to this number an accessories wire, to operate trailer internal equipment or to charge the trailer's battery, and you can have as many as seven wires in the harness.

Determine the equipment on your trailer and buy the wiring kit necessary. The kit will contain all the wires needed, plus a plug adapter set which included the female plug, mounted on the bumper or hitch, and the male plug, wired into, or plugged into the trailer harness.

When installing the kit, follow the manufacturer's instructions. The color coding of the wires is standard throughout the industry.

One point to note, some domestic vehicles, and most imported vehicles, have separate turn signals. On most domestic vehicles, the brake lights and rear turn signals operate with the same bulb. For those vehicles with separate

turn signals, you can purchase an isolation unit so that the brake lights won't blink whenever the turn signals are operated, or, you can go to your local electronics supply house and buy four diodes to wire in series with the brake and turn signal bulbs. Diodes will isolate the brake and turn signals. The choice is yours. The isolation units are simple and quick to install, but far more expensive than the diodes. The diodes, however, require more work to install properly, since they require the cutting of each bulb's wire and soldering in place of the diode.

One final point, the best kits are those with a spring loaded cover on the vehicle mounted socket. This cover prevents dirt and moisture from corroding the terminals. Never let the vehicle socket hang loosely. Always mount it securely to the bumper or hitch.

CIRCUIT PROTECTION

Fuses

▶ **See Figures 61, 62, 63 and 64**

There are a number of fuse blocks on these vehicles. The multi-purpose fuse block is located inside the passenger compartment of the vehicle, under the instrument panel, towards the left of the steering column. The dedicated fuses are located in the engine compartment at the left rear corner (if equipped with A/C) and also in front of the right strut tower. Blade type fuses are used.

A blade type fuse has test taps provided to allow checking of the fuse itself, without removing it from the fuse block. A test light can be used. The fuse is okay if the test light comes ON when its one test lead is connected to the test

taps (one at a time) and the other lead is grounded. Change the position of the ignition switch so the fuse circuit being tested has voltage applied to it.

When a fuse is blown, there are two probable causes as follows: One is that the fuse was blown due to a flow of current exceeding its rating. The other is that it is blown due to a repeated ON/OFF current flowing through it. Which of the two causes is responsible can be easily determined by a visual check as described below:

1. Fuse blown due to current exceeding rating:

As seen in the illustration, a fuse that is blown due to an exceeding current has a section missing. If the blown fuse looks like this, do not replace the fuse with a new one hastily since a current heavy enough to blow the fuse had

flowed through it. First, check the circuit for shorting and check for abnormal electrical parts. Only after the correction of such a shorting condition, the fuse is to be replaced with one of the same capacity. Never use a fuse of a larger capacity than the one blown. If such a fuse is used, electrical components or wiring can be damaged before the fuse blows in the event that an overcurrent occurs again.

2. Fuse blown due to repeated current ON/OFF flow:

The figure shows a fuse blown due to repeated current ON/OFF. Normally, this type of problem occurs after fairly long period of use and hence is less frequent than the above type. In this case, you may simply replace the fuse with a new fuse of the same capacity.

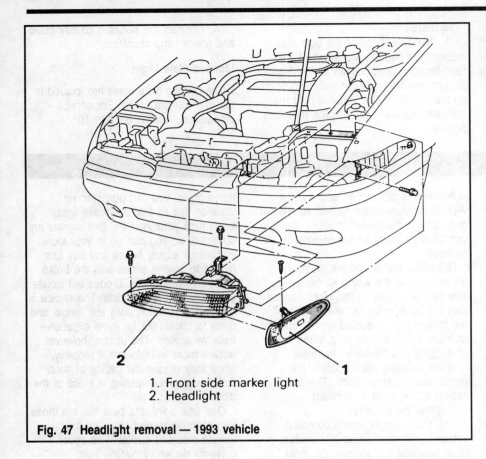

1. Front side marker light
2. Headlight

Fig. 47 Headlight removal — 1993 vehicle

Fusible Links

✴✴CAUTION

Do not replace blown fusible links with standard wire. Only fusible type wire with hypalon insulation can be used, or damage to the electrical system will occur!

A number of fuse links are used on these vehicles to protect wiring and electrical components. There is a collection of fuse links located near the battery. These are referred to as the main fuse links. A second group of links are located in the box with the dedicated fuses. If replacement of a fuse link is required, use the exact same link as removed.

When a fusible link blows it is very important to find out why. They are placed in the electrical system for protection against dead shorts to ground, which can be caused by electrical

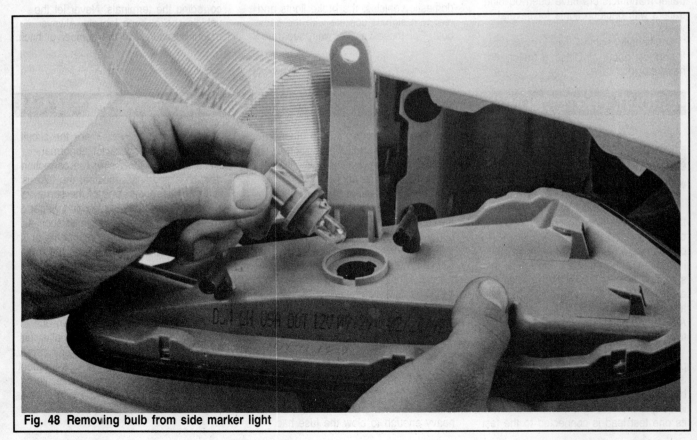

Fig. 48 Removing bulb from side marker light

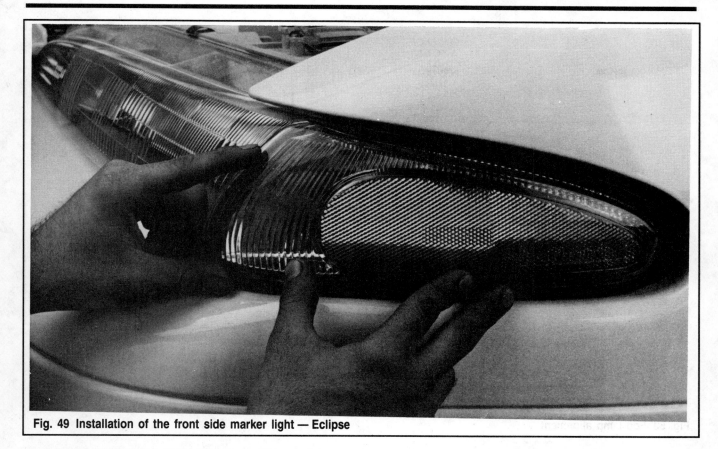

Fig. 49 Installation of the front side marker light — Eclipse

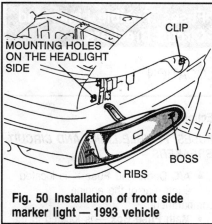

Fig. 50 Installation of front side marker light — 1993 vehicle

component failure or various wiring failures.

✳✳CAUTION

Do not just replace the fusible link to correct a problem!

When replacing all fusible links, they are to be replaced with the same type of prefabricated link available from your vehicle manufacturer.

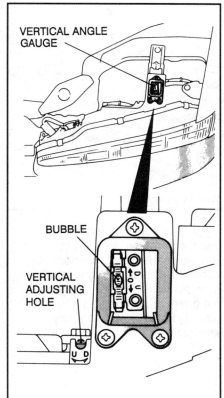

Fig. 51 Vertical adjustment of the headlights — 1992 vehicle

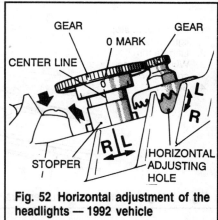

Fig. 52 Horizontal adjustment of the headlights — 1992 vehicle

REPLACEMENT

The following procedure applies to main fuse links that are installed in the vehicle wiring harness.

1. Cut the fusible link including the connection insulator from the main harness wire.

2. Remove 1 inch (25mm) of insulation from both new fusible link, and the main harness, and wrap together.

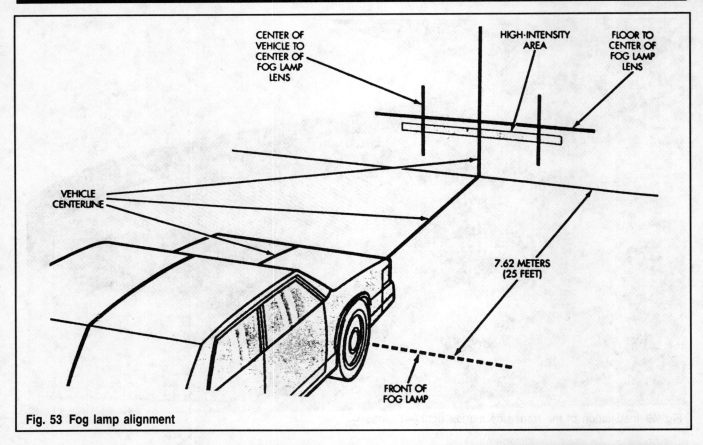

Fig. 53 Fog lamp alignment

1. High mount stop light unit
2. Liftgate trim
3. Wiring harness connector
4. Socket

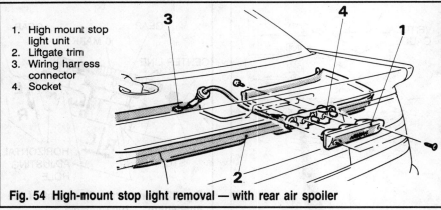

Fig. 54 High-mount stop light removal — with rear air spoiler

3. Heat the splice with a soldering gun, and apply rosin type solder.

➡**Do not use acid core solder.**

4. Allow the connection to cool, and wrap the new splice with at least 3 layers of electrical tape.

Flashers

The turn signal and hazard flasher unit is located in the multi-purpose fuse panel located under the driver's left side knee protector. They are replaced by simply pulling them straight out. Note that the prongs are arranged in such a way that the flasher must be properly oriented before attempting to install it. Turn the flasher until the orientation of the prongs is correct and simply push it firmly in until the prongs are fully engaged.

Fuses, Fusible Links and Circuit Breakers

LOCATION

Eclipse

FUSES, FUSIBLE LINKS AND CIRCUIT BREAKERS

• **A/C Dedicated Fuse** — is located at the left rear of the engine compartment.

• **Main Fuse Box** — is located in the engine compartment, at the left fender well.

• **Multi-purpose Fuse Box** — is located under the left side dash.

• **Main Fusible Links** — are located in the engine compartment next to the battery.

• **ABS Circuit Diode** — is located in the trunk, in front of the right side wheel well.

• **Door Ajar Warning Circuit Diode** — is located behind the passenger side rear quarter interior panel.

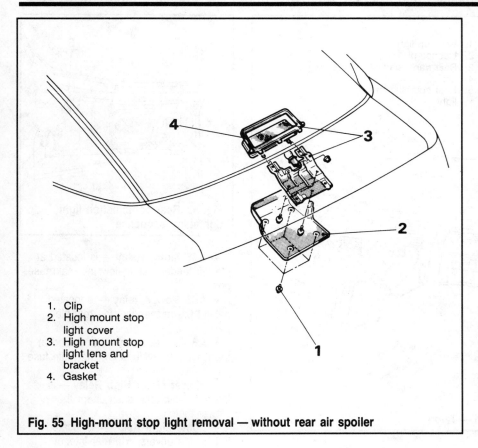

1. Clip
2. High mount stop light cover
3. High mount stop light lens and bracket
4. Gasket

Fig. 55 High-mount stop light removal — without rear air spoiler

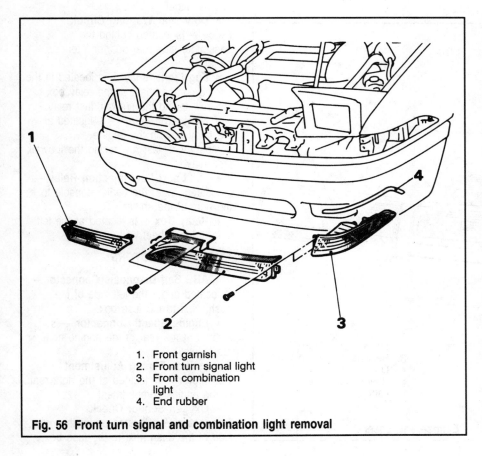

1. Front garnish
2. Front turn signal light
3. Front combination light
4. End rubber

Fig. 56 Front turn signal and combination light removal

RELAYS

- **Automatic Seat Belt Motor Relay** — is located at the left quarter panel behind the left seat.
- **Condenser Fan Motor Changeover Relay** — is located at the relay box on the right side of the engine compartment, largest square relay in that row.
- **Condenser Fan Motor Relay** — is located at the relay box on the right side of the engine compartment, middle relay.
- **Defogger Timer** — is located at the relay box under the left side of the dash, third relay in.
- **Door Lock Relay** — is located at the relay box under the left side of the dash, first relay.
- **Headlight Relay** — is located at the relay box on the left side of the engine compartment, first row, second relay in.
- **Heater Relay** — is located under the left side of the dash above the junction block.
- **Intermittent Front Wiper Relay** — is located inside the steering column switch.
- **Intermittent Rear Wiper Relay** — is located at the left quarter panel behind the left seat.
- **Magnet Clutch Relay** — is located at the right rear of the engine compartment, against the fire wall, last relay at the fire wall.
- **Power Window Relay** — is located at the relay box on the left side of the engine compartment, last relay last row.
- **Radiator Fan Motor Relay** — is located at the relay box on the left side of the engine compartment, middle relay last row.
- **Starter Relay** — is located at the relay box under the left side of the dash, last relay towards the console.
- **Taillight Relay** — is located at the relay box on the left side of the engine compartment, first row of relays, last relay in that row.
- **Theft Alarm Relay** — is located at the relay box under the left side of the dash, first relay.
- **Theft Alarm Starter Relay (automatic transmission)** — is located under the left side of the dash above the fuse box.
- **Theft Alarm Starter Relay (manual transmission)** — is located under the front of the console.
- **Warning Buzzer** — is located under the dash at the steering column.

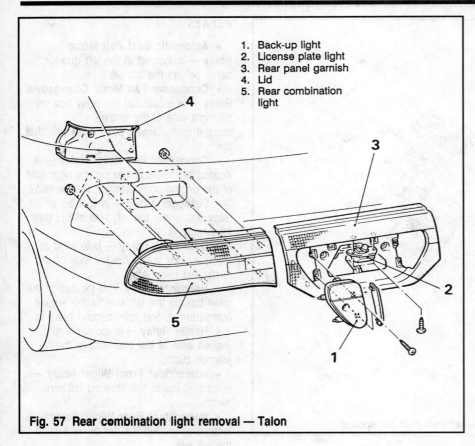

1. Back-up light
2. License plate light
3. Rear panel garnish
4. Lid
5. Rear combination light

Fig. 57 Rear combination light removal — Talon

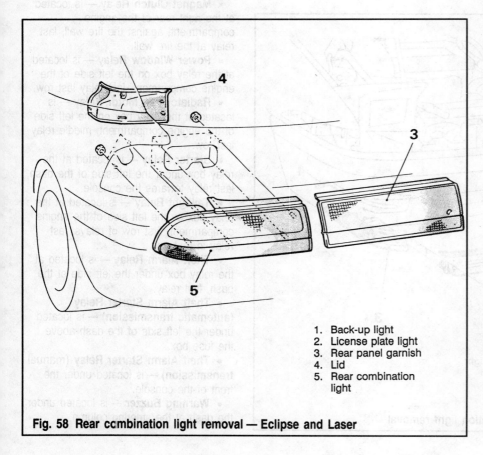

1. Back-up light
2. License plate light
3. Rear panel garnish
4. Lid
5. Rear combination light

Fig. 58 Rear combination light removal — Eclipse and Laser

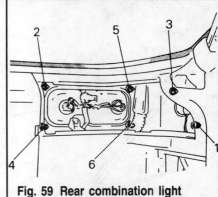

Fig. 59 Rear combination light tightening sequence

- **ABS Motor Relay** — is located at the left fender well, below the main fuse box.
- **ABS Power Relay** — is located behind the passenger side rear side panel.
- **ABS Valve Relay** — is located at the left fender well, below the main fuse box.
- **Blower Motor High Relay** — is located behind the dash under the blower motor.
- **Dome Light Relay** — is located behind the drivers side rear interior quarter panel.
- **Door Ajar Warning Circuit Diode** — is located behind the passenger side rear quarter interior panel.
- **Fog Light Relay** — is located in the engine compartment at the relay box, last row facing the firewall, last relay.
- **Generator Relay** — is located at the relay box, in the engine compartment, last row facing the firewall, first relay.
- **Multi-port Fuel Injection Relay** — is located under the center console to the right of the shifter.
- **Relay Box** — is located at the left side fender wheel well.

CHECK CONNECTORS

- **ABS Self Diagnosis Connector** — is located under the left side of the dash, next to the fuse box.
- **Engine Speed Connector** — is located at the rear of the engine near the firewall.
- **Ignition Timing Adjustment Connector** — is located at the right rear of the engine compartment.
- **Oxygen Sensor Check Connector** — is located under the right side of the dash next to the fuse box.

Fig. 60 Removing the socket and bulb from the rear combination light

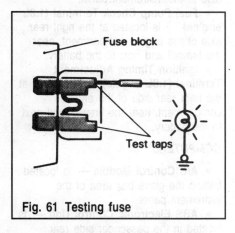

Fig. 61 Testing fuse

- **Self-Diagnosis Connector** — is located under the left side of the dash next to the fuse box.
- **Data Link Connector** — is located under the left side dash next to the fuse box.
- **Fuel Pump Check Connector** — is located in the engine compartment, in front of the wiper transmission.

COMPUTERS

- **ABS Control Unit** — is located behind the rear seat.
- **Automatic Seat Belt Control Unit** — is located at the left quarter panel behind the left seat.
- **Automatic Transaxle Control Unit** — is located at the front of the console under the dash.
- **Cruise Control Actuator** — is located at the right shock tower in the engine compartment.
- **Door Lock Control Unit** — is located at the right kick panel.
- **Engine Control Unit** — is located at the front of the console under the dash.
- **Theft Alarm Control Unit** — is located at the right kick panel.
- **4-Speed Automatic Transaxle Control Module** — is located behind the center console towards the firewall.
- **A/C Control Unit** — is located under the right side of the dash near the A/C unit.

Laser and Talon

FUSES, FUSIBLE LINKS AND CIRCUIT BREAKERS

Junction Block, Electrical System — is located under the instrument panel on the left side.

Fuses, Multi-Purpose Fuse Block — is located under the instrument panel on the left side.

Fuses, Underhood, Dedicated — are located at the left rear corner of the engine compartment and also near the right strut tower.

Fusible Link, Subsystem — are located underhood next to the passenger side strut tower.

Fusible Links, Main — are located next to the battery.

RELAYS

- **A/C Clutch Relay** — is located in the underhood relay panel next to the left side strut tower.
- **A/C Compressor Clutch Relay** — is located underhood on the relay bracket next to the strut tower.
- **A/C Condenser Fan Motor High-Low Selecting Relay** — is located in the underhood relay panel next to the left side strut tower.
- **A/C Condenser Fan Motor Relay** — is located in the underhood relay panel next to the left side strut tower.
- **ABS Power Relay** — is located on the ABS control unit bracket.
- **Blower Motor High Speed Relay** — is located under the passenger side instrument panel on the heater case.
- **Central Door Locking System Relay** — is located under the driver's side instrument panel above the pedal pivots.
- **Cooling Fan Relay** — is located in the underhood relay bank near the right strut tower.
- **Door Locking System Relay** — is located under the driver's side instrument panel above the pedal pivots.
- **Fuel Injection Relay** — is located under the center front of the console.
- **Head Light Relay** — is located in the underhood fuse/relay box next to the right side strut tower.
- **Heater Relay** — is located on the top of the junction block that is under the driver's side instrument panel.
- **Heater Resistor** — is located in the heater ducting near the motor on the

Fig. 62 Fuse block and cover

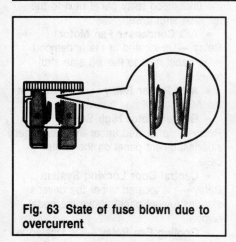

Fig. 63 State of fuse blown due to overcurrent

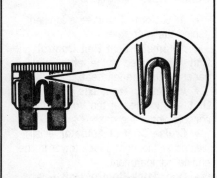

Fig. 64 State of fuse blown due to thermal fatigue

passenger side, behind the glove box assembly.
• **Radiator Fan Motor Relay** — is located in the underhood relay bank near the right strut tower.
• **Relay, A/C Clutch** — is located in the underhood relay panel next to the left side strut tower.
• **Relay, A/C Compressor Clutch** — is located underhood on the relay bracket next to the strut tower.
• **Relay, A/C Condenser Fan Motor High-Low Selecting** — is located in the

underhood relay panel next to the left side strut tower.
• **Relay, A/C Condenser Fan Motor** — is located in the underhood relay panel next to the left side strut tower.
• **Relay, ABS Power** — is located on the ABS control unit bracket.
• **Relay, Blower Motor High Speed** — is located under the passenger side instrument panel on the heater case.

• **Relay, Head Light** — is located in the underhood fuse/relay box next to the right side strut tower.
• **Relay, Heater** — is located on the top of the junction block that is under the driver's side instrument panel.
• **Relay, Multi-Point Fuel Injection** — is located inside the center console toward the front, under the lower instrument panel.
• **Relay, Multi-port Fuel Injection** — is located under the center front of the console.
• **Relay, Radiator Fan Motor** — is located in the underhood relay bank near the right strut tower.
• **Relay, Tail Light** — is located in the underhood fuse/relay box next to the right side strut tower.

CHECK CONNECTORS

• **Connector, Data Link** — is located under the instrument panel to the left of the pedal assemblies.
• **Diagnostic Connector** — is located under the instrument panel to the left of the pedal assemblies.
• **On-Board Diagnostic Connector** — is located under the left side of the instrument panel.
• **Fuel Pump Check Terminal (1.8L engine)** — is located at the right rear side of the engine compartment, near the firewall and next to the battery.
• **Ignition Timing Adjustment Terminal (1.8L engine)** — is located at the right rear side of the engine compartment, near the firewall and next to the battery.

COMPUTERS

• **A/C Control Module** — is located behind the glove box area of the instrument panel.
• **ABS Electronic Control Unit** — is located in the passenger side rear quarter panel under the quarter trim.
• **ABS System G-Sensor** — is located under the rear seat.
• **Automatic Seat Belt Control Unit** — is located in the quarter panel at the base of the door pillar, behind the trim panel.
• **Central Door Locking System Control Unit** — is located inside the passenger side cowl behind the trim panel.
• **Control Unit, Automatic Seat Belt** — is located in the quarter panel at the base of the door pillar, behind the trim panel.

- **ECU (Electronic Control Unit), Multi-Point Fuel Injection** — is located inside the center console toward the front, under the lower instrument panel.
- **Electronic Control Module, 4-Speed Automatic Transaxle** — is located under the instrument panel forward of the console and forward of the engine control module.

- **Electronic Control Unit, ABS** — is located in the passenger side rear quarter panel under the quarter trim.
- **Engine Control Module** — is located under the center front of the console.

- **Multi-Point Fuel Injection Electronic Control Unit (ECU)** — is located inside the center console toward the front, under the lower instrument panel.

WIRING DIAGRAMS

TORQUE SPECIFICATIONS

Component	US	Metric
Air conditioning compressor mounting bolt	33 ft. lbs.	45 Nm
Air conditioning dual pressure switch	8 ft. lbs.	11 Nm
Steering wheel nut	30 ft. lbs.	40 Nm
Steering wheel-to-column nut to	30 ft. lbs.	40 Nm
Tensioner pulley center nut (bolt)	24 ft. lbs.	32 Nm
Wheel lug nut	87–101 ft. lbs.	120–140 Nm

HOW TO READ CIRCUIT DIAGRAMS

The circuit of each system from the fuse (or fusible link) to ground is shown. The power supply is shown at the top and the ground at the bottom to facilitate understanding of how the current flows.

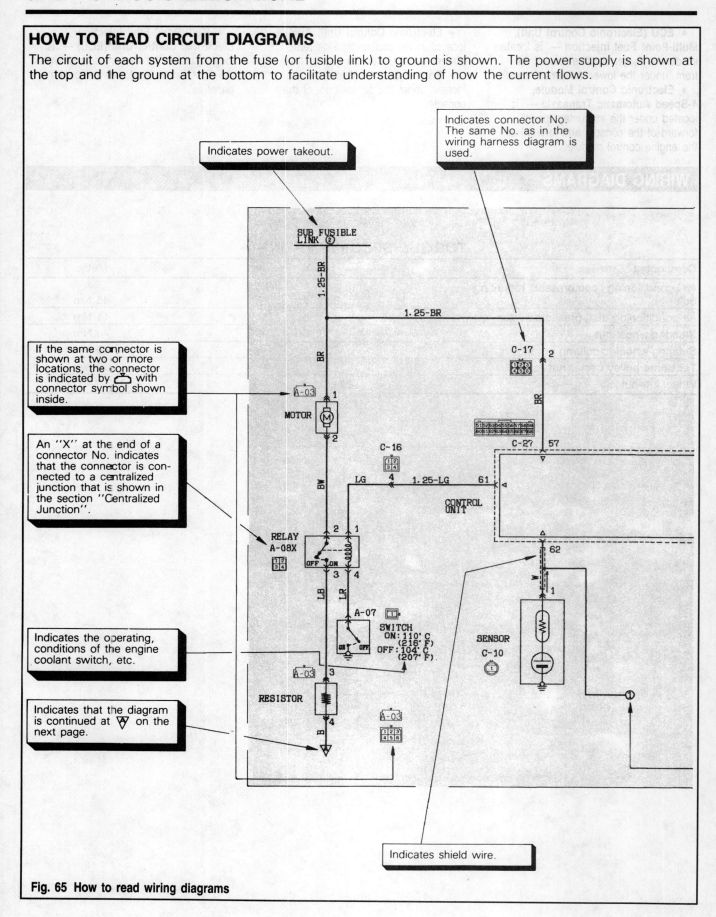

Fig. 65 How to read wiring diagrams

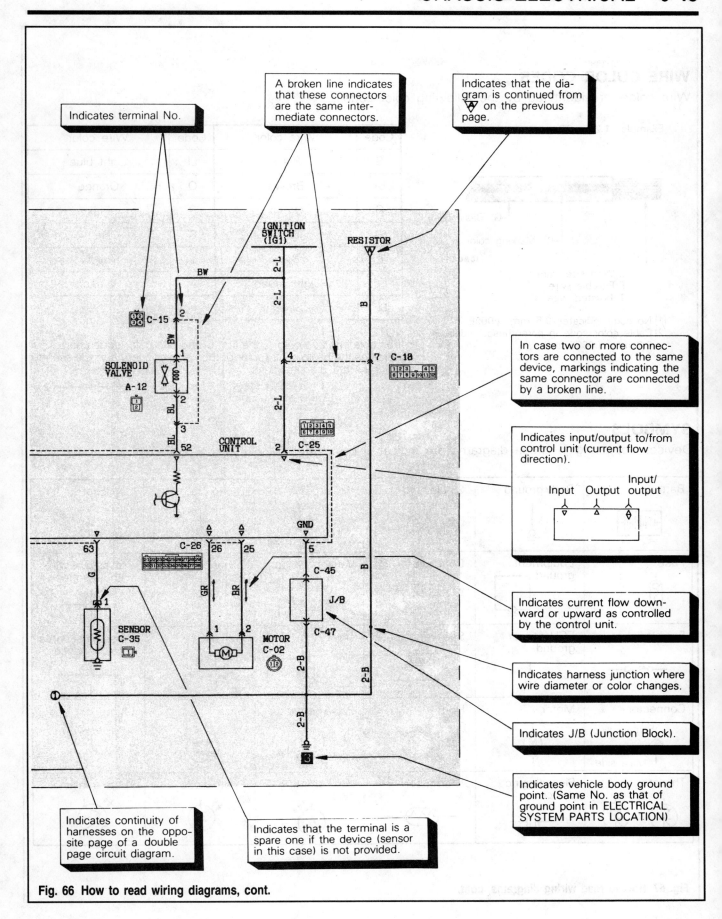

Indicates terminal No.

A broken line indicates that these connectors are the same intermediate connectors.

Indicates that the diagram is continued from ▽ on the previous page.

IGNITION SWITCH (IG1)

RESISTOR

BW

2-L

2-L

C-15 2

BW

SOLENOID VALVE
A-12

4

7 C-18

C-25

CONTROL UNIT

GND

C-26

63

SENSOR
C-35

GR BR

MOTOR
C-02

J/B

C-45

C-47

In case two or more connectors are connected to the same device, markings indicating the same connector are connected by a broken line.

Indicates input/output to/from control unit (current flow direction).

Input Output Input/output

Indicates current flow downward or upward as controlled by the control unit.

Indicates harness junction where wire diameter or color changes.

Indicates J/B (Junction Block).

Indicates vehicle body ground point. (Same No. as that of ground point in ELECTRICAL SYSTEM PARTS LOCATION)

Indicates continuity of harnesses on the opposite page of a double page circuit diagram.

Indicates that the terminal is a spare one if the device (sensor in this case) is not provided.

Fig. 66 How to read wiring diagrams, cont.

WIRE COLOR CODES

Wire colors are identified by the following color codes.

Example: 1.25F—GB

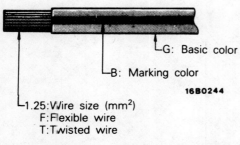

- G: Basic color
- B: Marking color

16B0244

- 1.25: Wire size (mm²)
  F: Flexible wire
  T: Twisted wire

(1) No code indicates 0.5 mm² (.0008 in.²).
(2) Cable color code in parentheses indicates 0.3 mm² (.0005 in.²).

Code	Wire color	Code	Wire color
B	Black	Ll	Light blue
Br	Brown	O	Orange
G	Green	P	Pink
Gr	Gray	R	Red
L	Blue	Y	Yellow
Lg	Light green	W	White
Sb	Silver		

NOTE
If a cable has two colors, the first of the two color code characters indicates the basic color (color of the cable coating) and the second indicates the marking color.

SYMBOLS

Devices appearing in circuit diagrams are indicated by the following symbols.

Battery	Body ground	Single bulb	Resistor	Diode	Capacitor
Fuse	Equipment ground	Dual bulb	Variable resistor	Zener diode	Crossing of wires without connection
Fusible link	ECU interior ground	Speaker	Coil	Transistor	Crossing of wires with connection
Connector Female side Male side	Motor	Horn	Pulse generator	Buzzer	Chime
Thyristor	Piezoelectric device	Thermistor	Light emitting diode	Photo diode	Photo transistor

Fig. 67 How to read wiring diagrams, cont.

	No.	Layout indications	Symbol	Description
Connector indications	①	Male 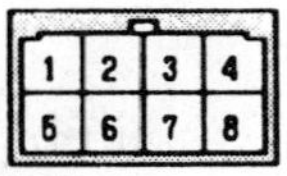		Male and female terminals are distinguished one from the other as shown in the illustration: connectors framed by a double line are male terminals, and those framed by a single line are female terminals.
	–	Female 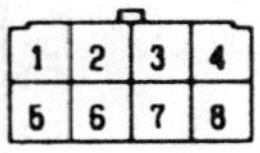		
Connector symbol indications	②	Equipment  / Intermediate connector 16A0333	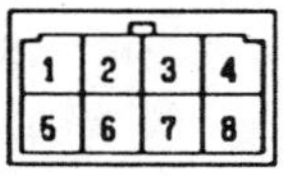	Symbols are shown as facing in the direction indicated in the illustration. For connections to the equipment is shown; for intermediate connectors, the symbol for the connector at the male side is shown.
Connector connection indications	③	Direct-connect type 	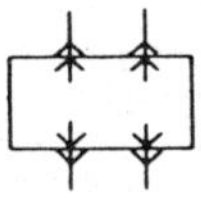	There are two types of connection between the equipment and the connector at the harness: the type by which there is direct plug-in to the equipment (the direct-connect type), and the type by which connection is with the harness connector at the equipment (the type with harness); these are individually identified as shown in the illustration.
	④	Type with harness  16A0334	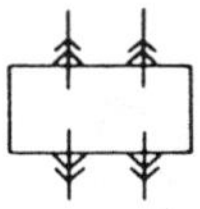	
	⑤	Intermediate connector  16A0339		
Ground indications	⑥	Chassis ground 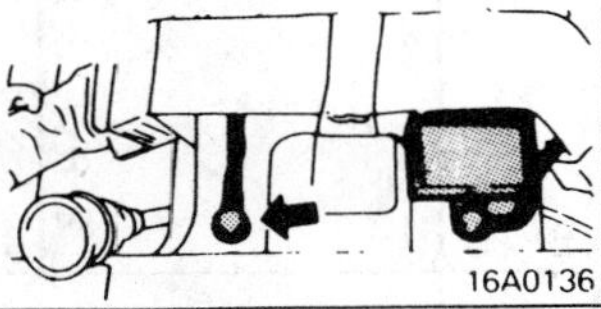 16A0136		There are three types of grounds: the chassis ground, the equipment ground, and the ground within the control unit; these are individually identified as shown in the illustration.
	⑦	Equipment ground 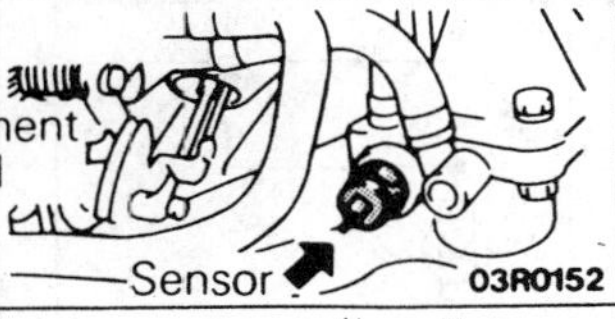 — Sensor 03R0152		
	⑧	Ground within control unit 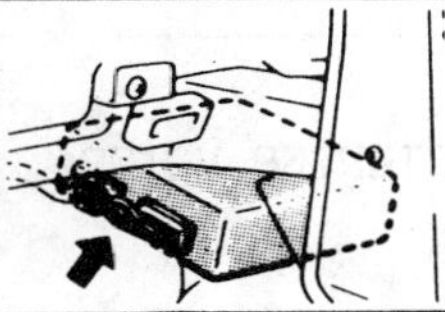 16A0109		

Fig. 68 How to read wiring diagrams, cont.

2. STARTING CIRCUIT

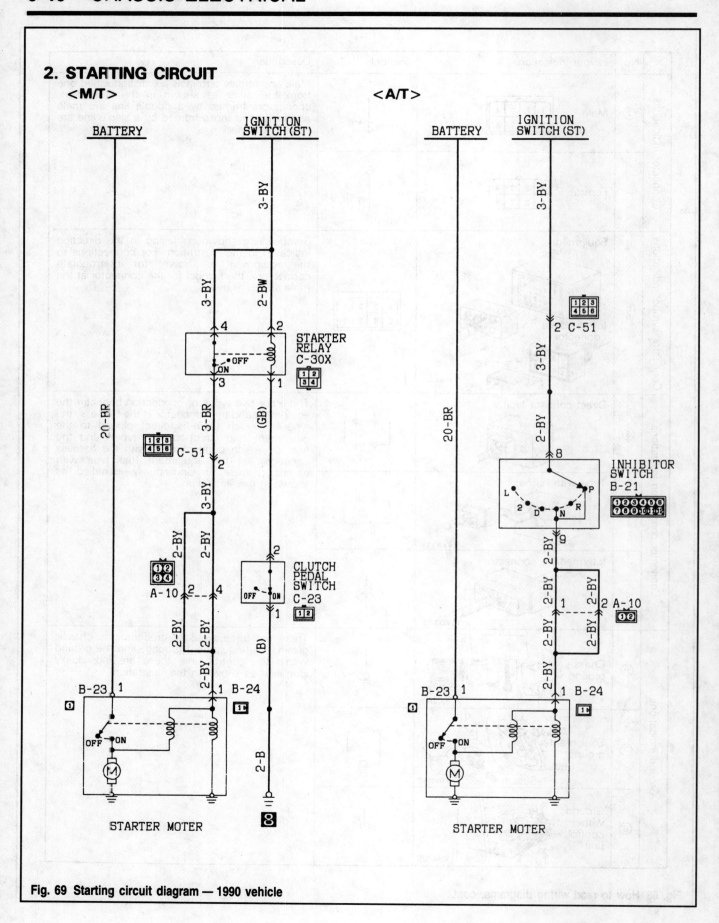

Fig. 69 Starting circuit diagram — 1990 vehicle

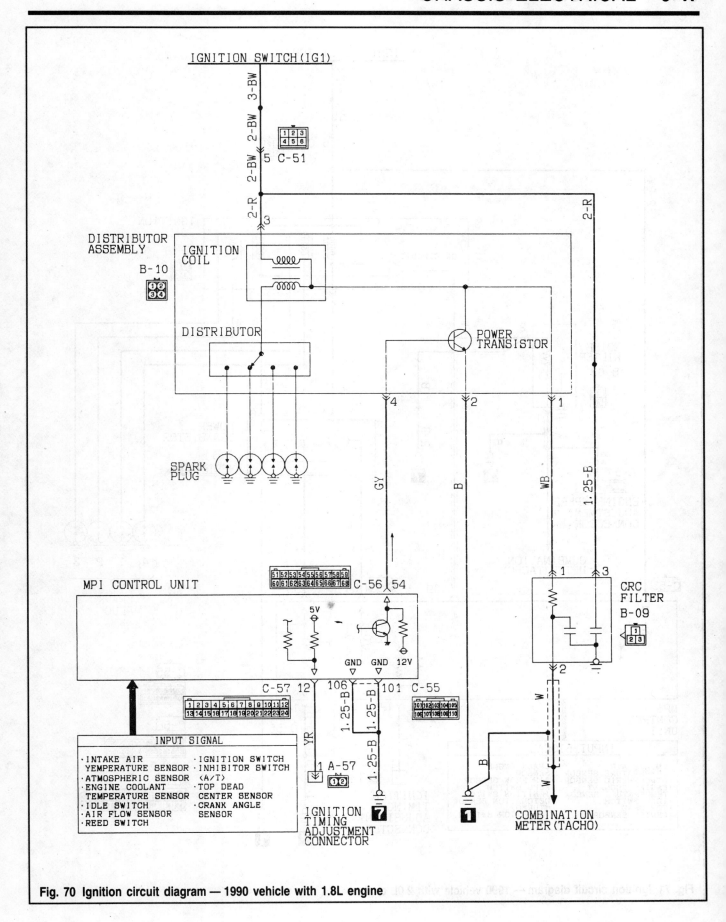

Fig. 70 Ignition circuit diagram — 1990 vehicle with 1.8L engine

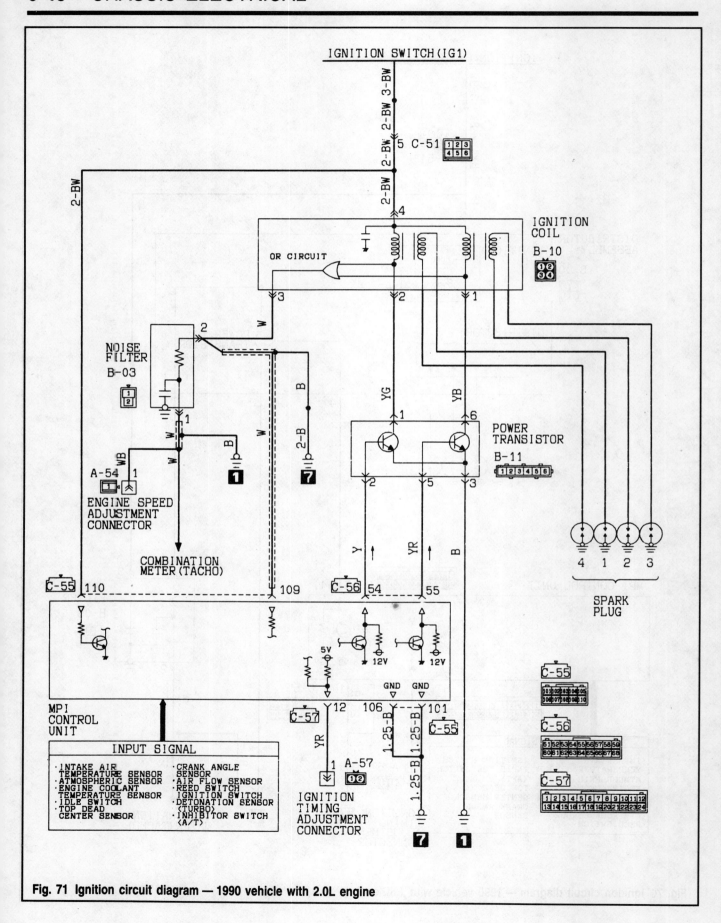

Fig. 71 Ignition circuit diagram — 1990 vehicle with 2.0L engine

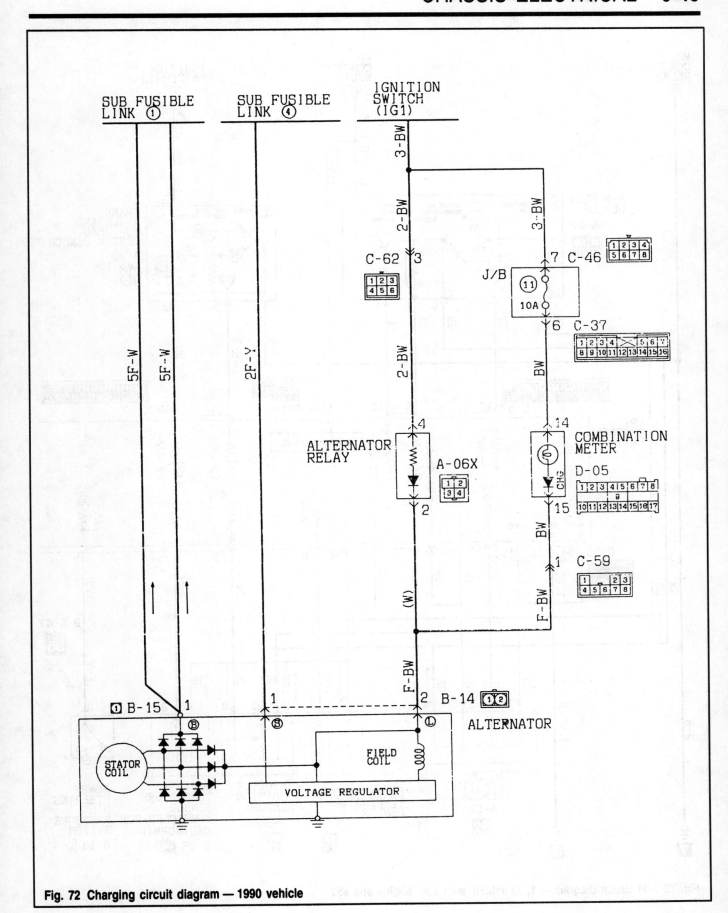

Fig. 72 Charging circuit diagram — 1990 vehicle

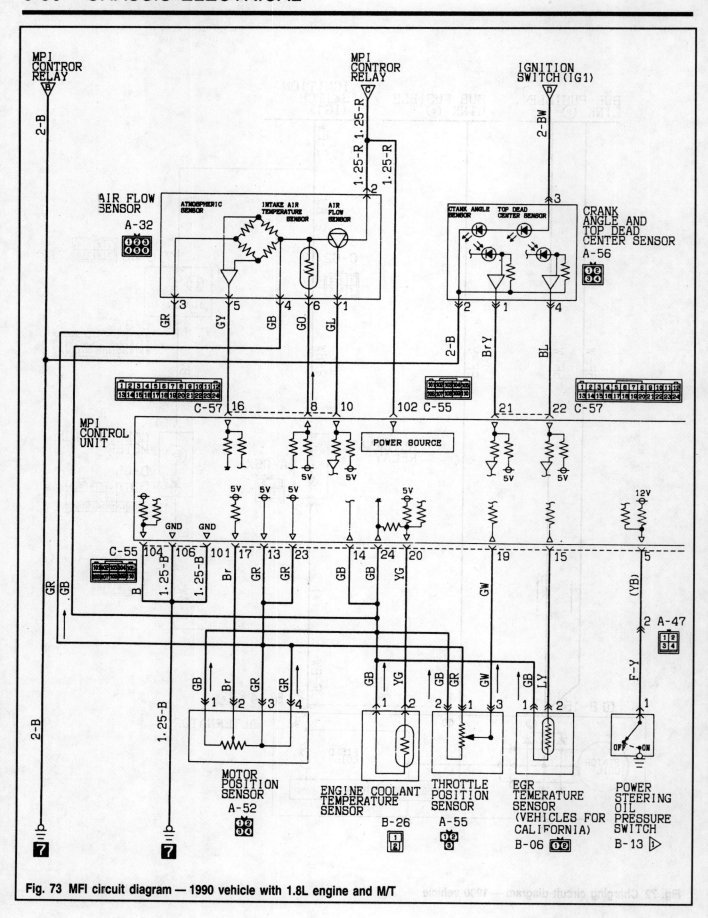

Fig. 73 MFI circuit diagram — 1990 vehicle with 1.8L engine and M/T

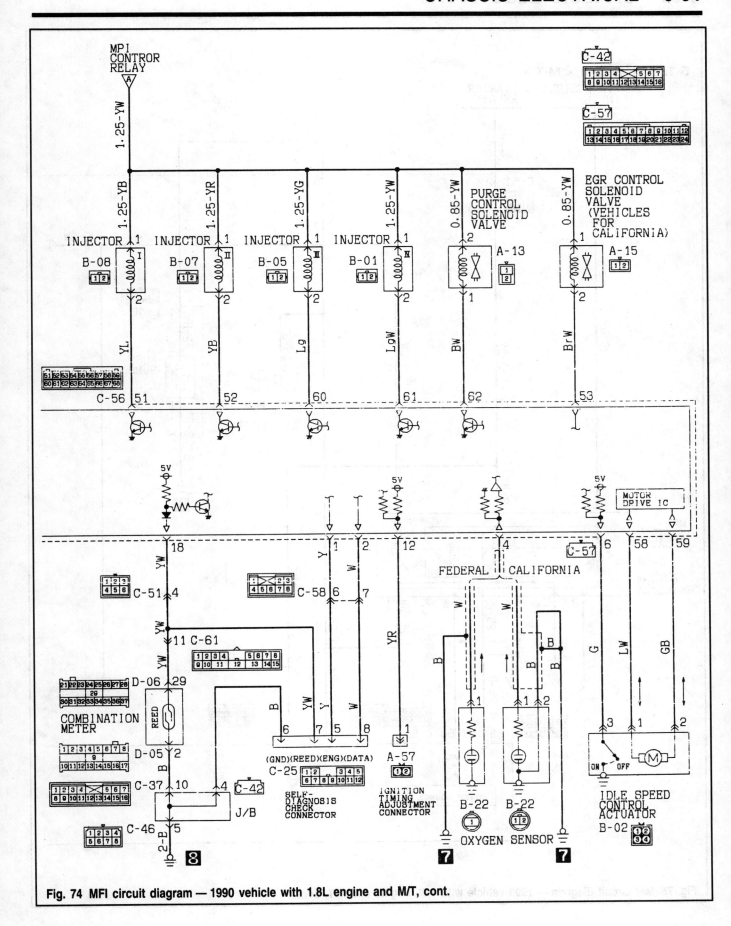

Fig. 74 MFI circuit diagram — 1990 vehicle with 1.8L engine and M/T, cont.

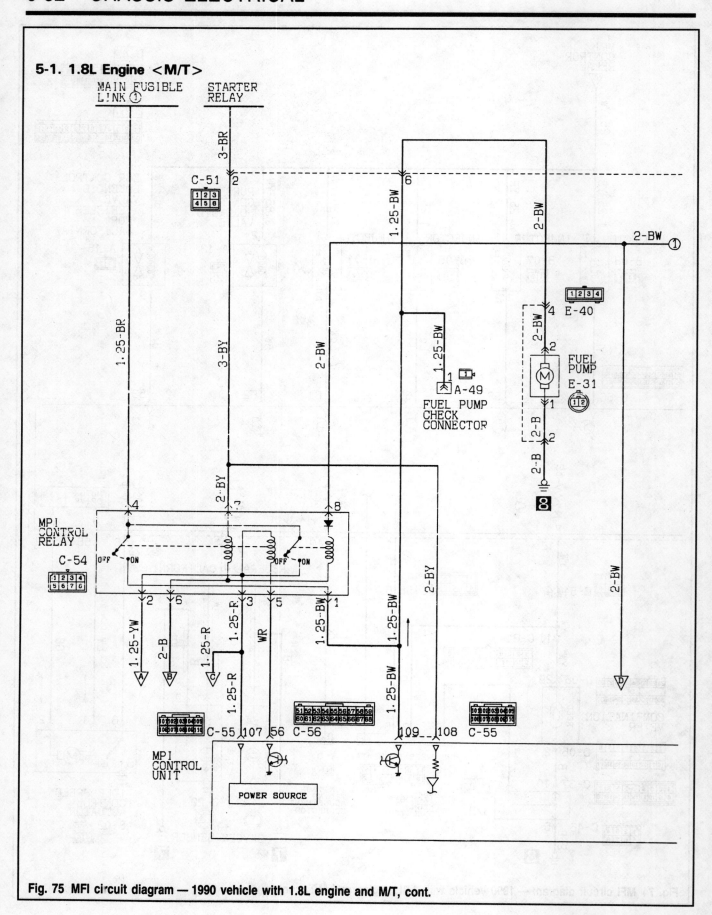

Fig. 75 MFI circuit diagram — 1990 vehicle with 1.8L engine and M/T, cont.

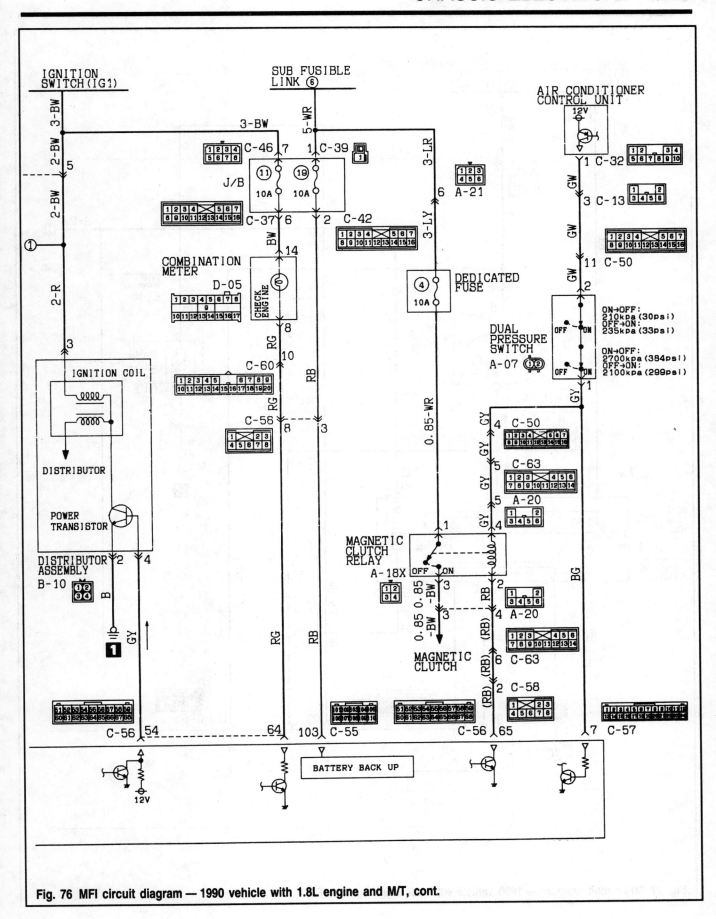

Fig. 76 MFI circuit diagram — 1990 vehicle with 1.8L engine and M/T, cont.

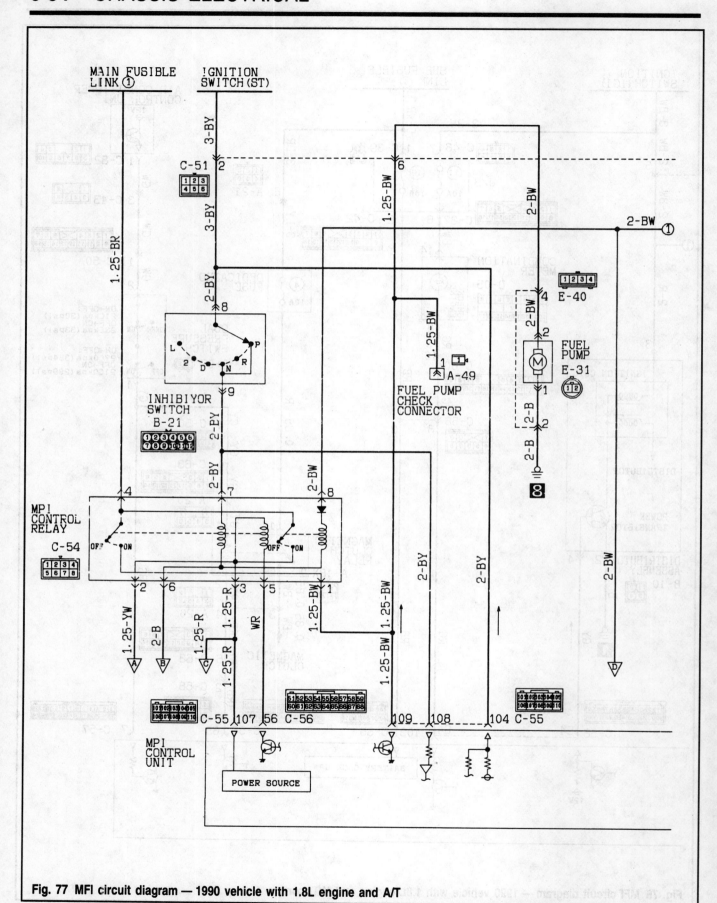

Fig. 77 MFI circuit diagram — 1990 vehicle with 1.8L engine and A/T

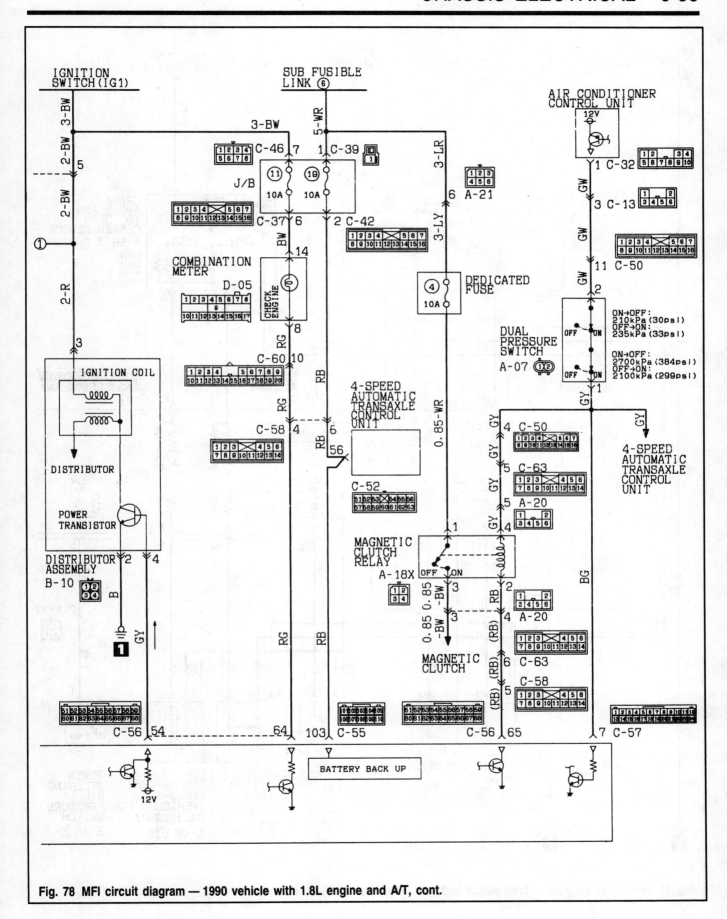

Fig. 78 MFI circuit diagram — 1990 vehicle with 1.8L engine and A/T, cont.

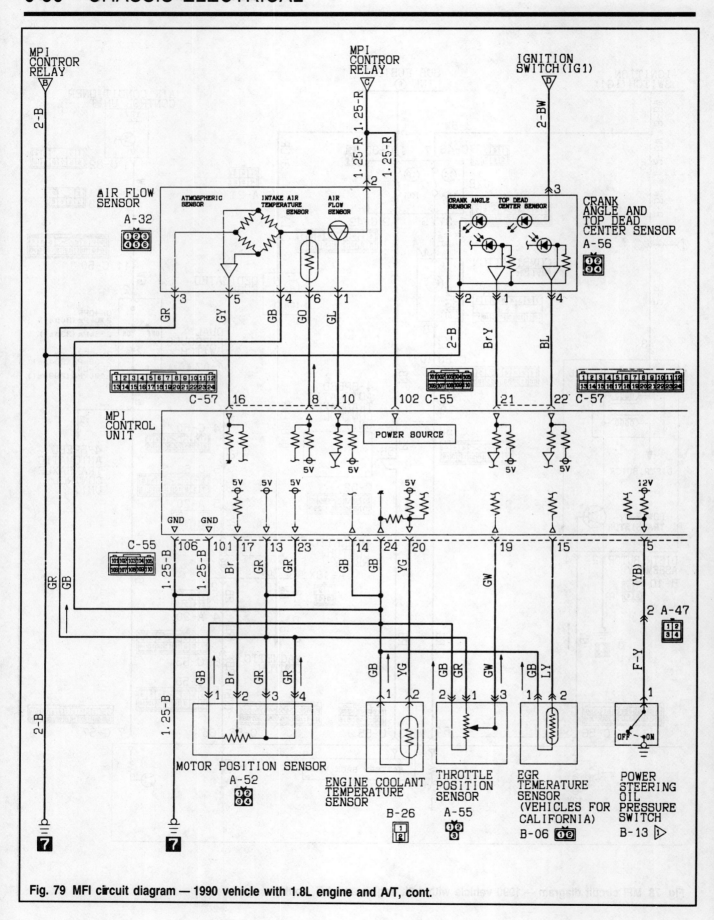

Fig. 79 MFI circuit diagram — 1990 vehicle with 1.8L engine and A/T, cont.

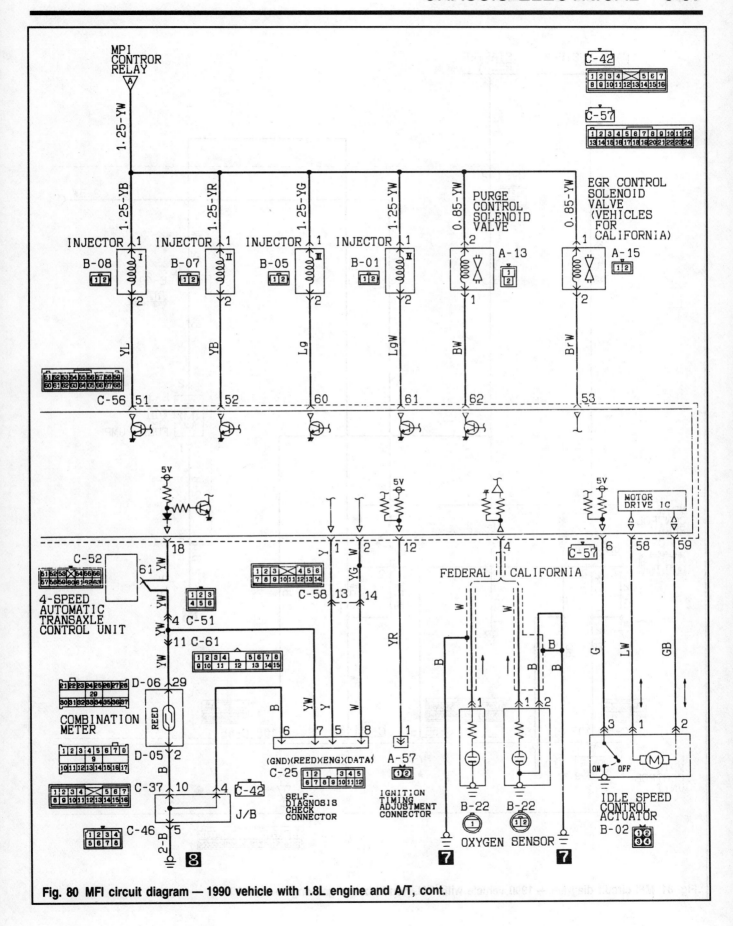

Fig. 80 MFI circuit diagram — 1990 vehicle with 1.8L engine and A/T, cont.

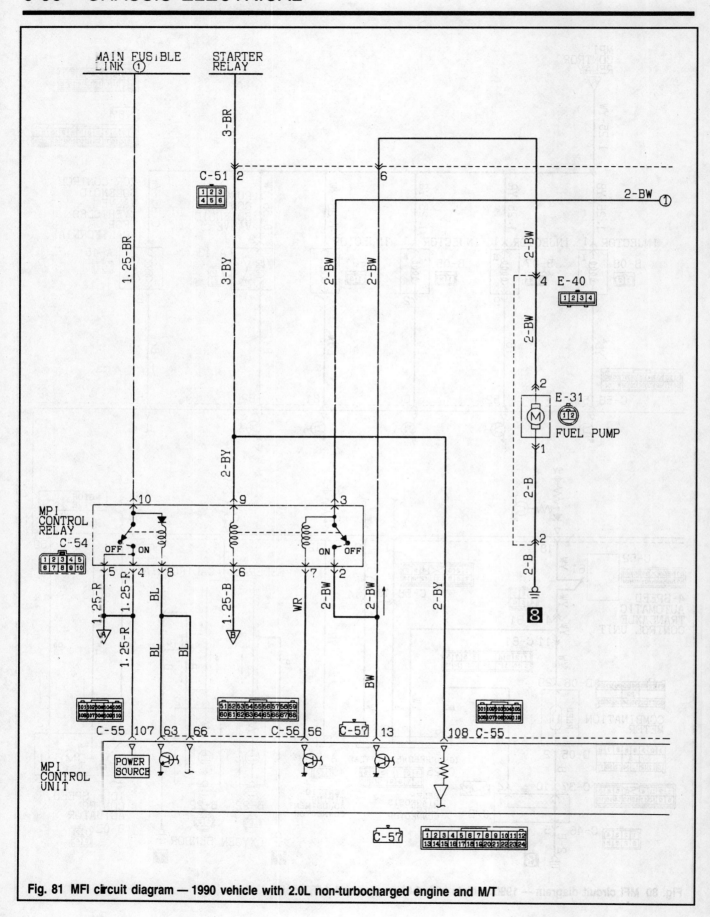

Fig. 81 MFI circuit diagram — 1990 vehicle with 2.0L non-turbocharged engine and M/T

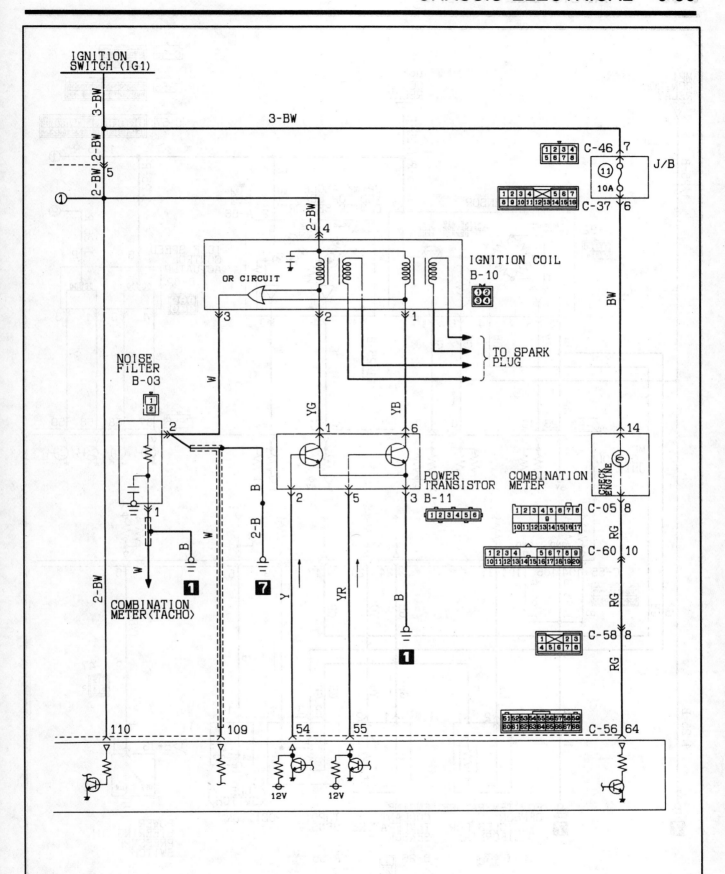

Fig. 82 MFI circuit diagram — 1990 vehicle with 2.0L non-turbocharged engine and M/T, cont.

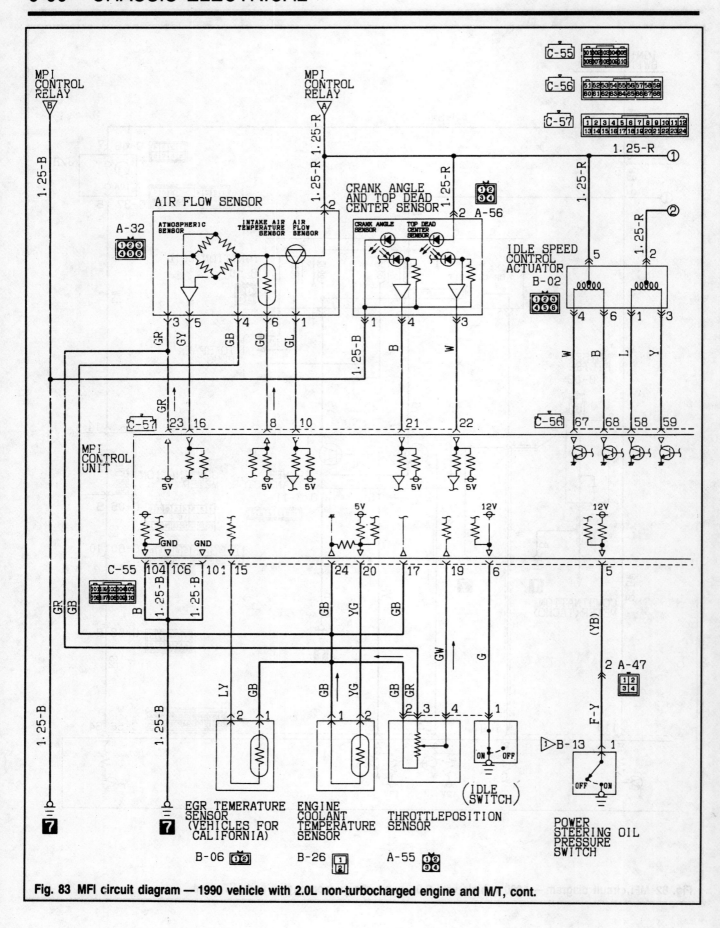

Fig. 83 MFI circuit diagram — 1990 vehicle with 2.0L non-turbocharged engine and M/T, cont.

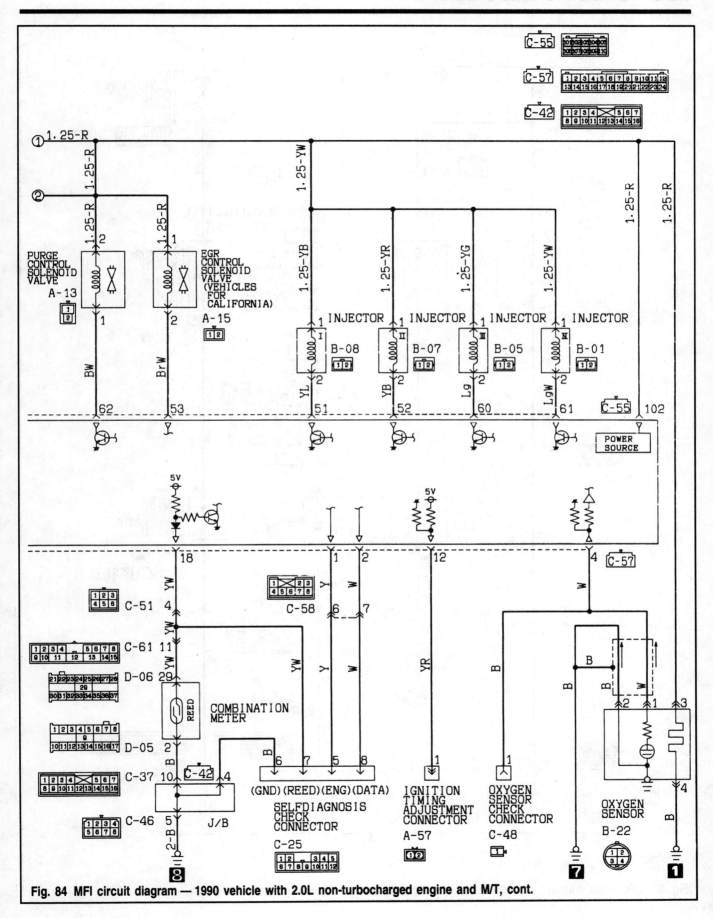

Fig. 84 MFI circuit diagram — 1990 vehicle with 2.0L non-turbocharged engine and M/T, cont.

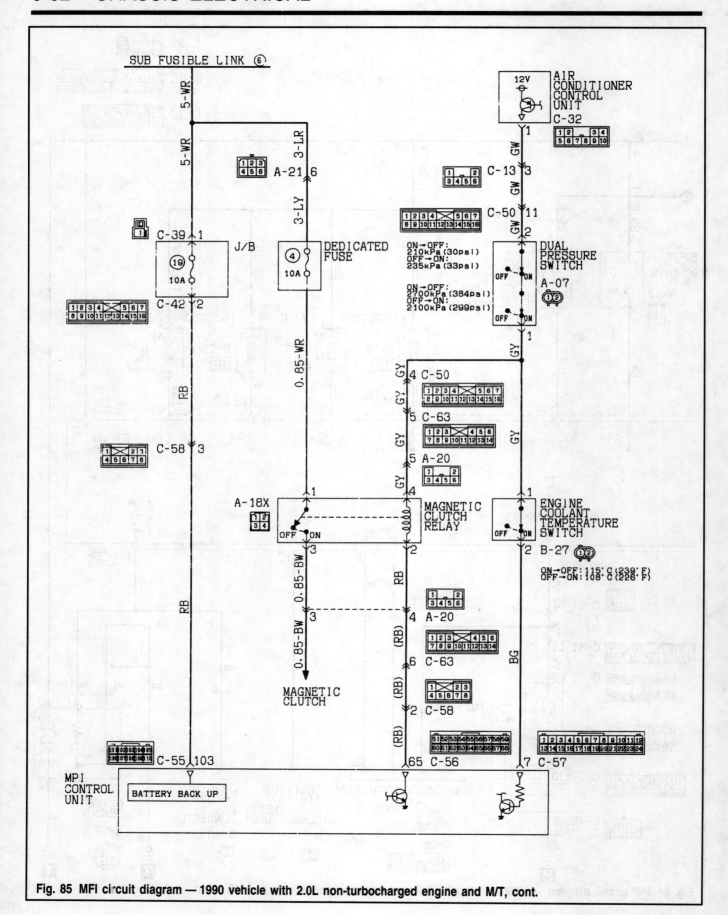

Fig. 85 MFI circuit diagram — 1990 vehicle with 2.0L non-turbocharged engine and M/T, cont.

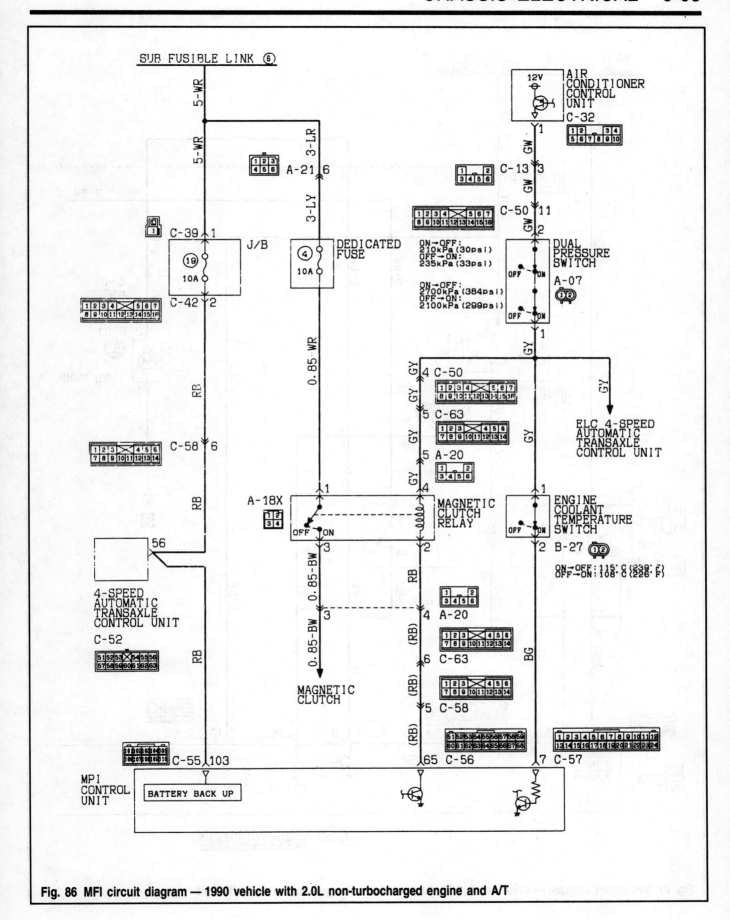

Fig. 86 MFI circuit diagram — 1990 vehicle with 2.0L non-turbocharged engine and A/T

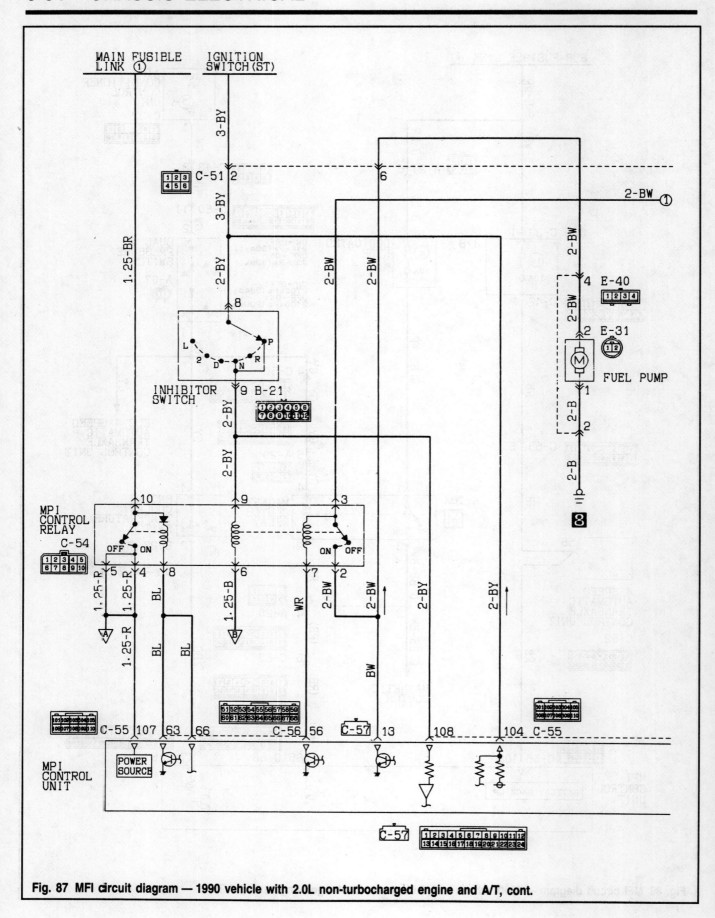

Fig. 87 MFI circuit diagram — 1990 vehicle with 2.0L non-turbocharged engine and A/T, cont.

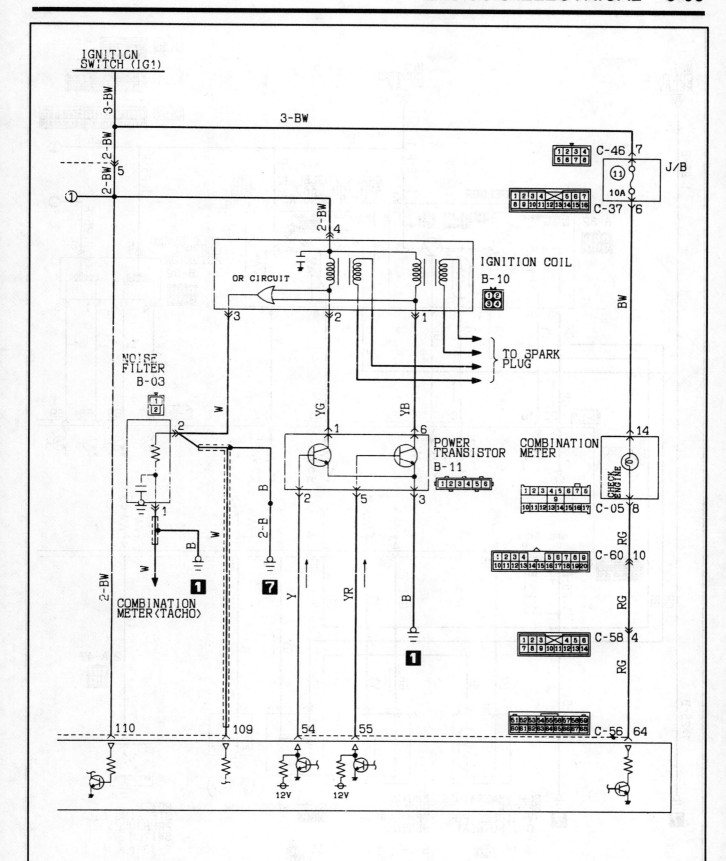

Fig. 88 MFI circuit diagram — 1990 vehicle with 2.0L non-turbocharged engine and A/T, cont.

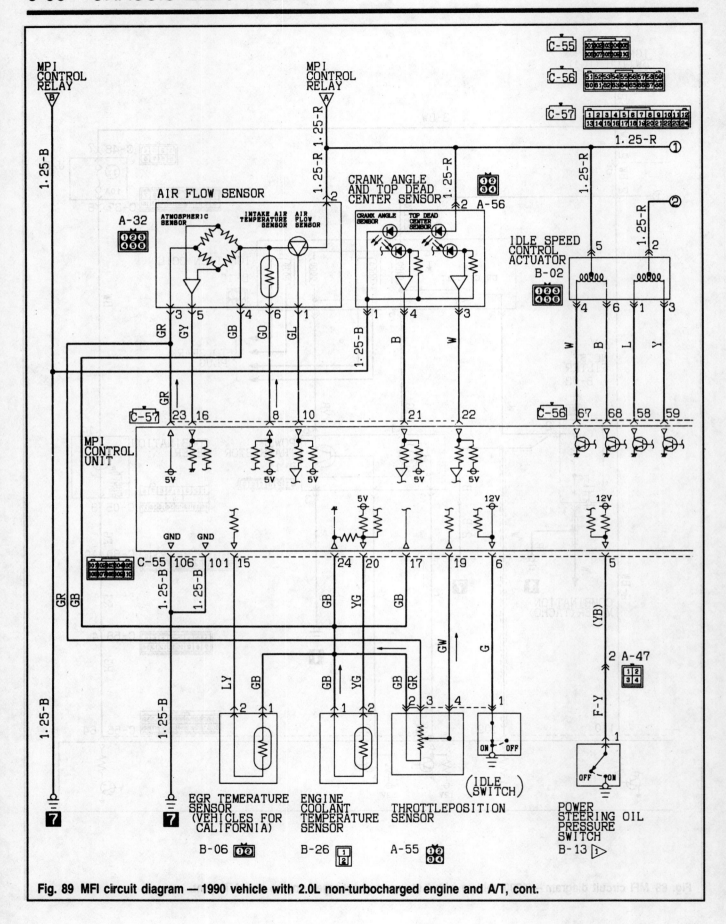

Fig. 89 MFI circuit diagram — 1990 vehicle with 2.0L non-turbocharged engine and A/T, cont.

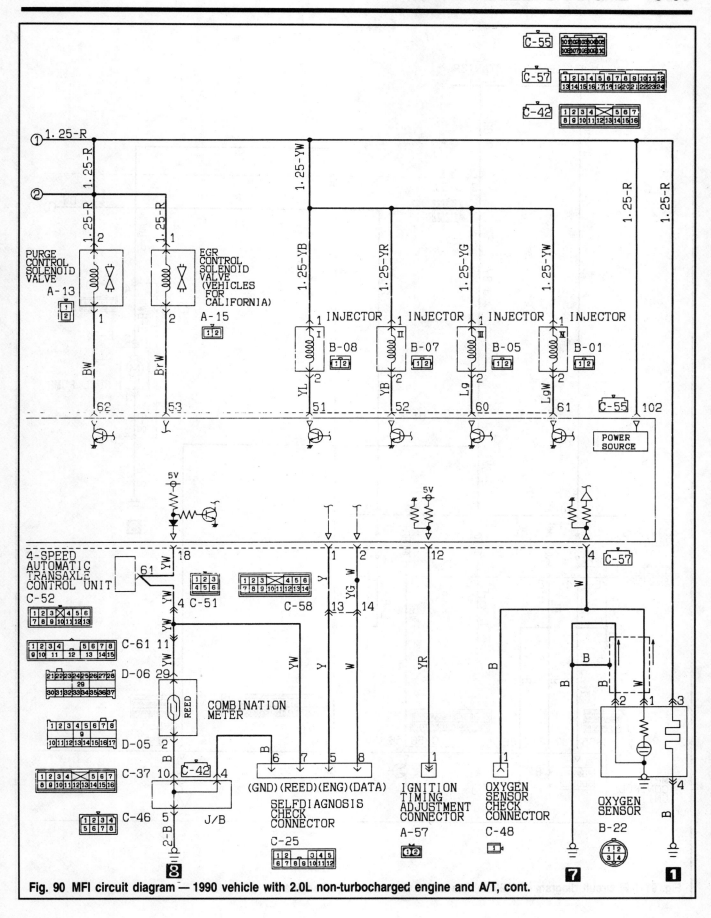

Fig. 90 MFI circuit diagram — 1990 vehicle with 2.0L non-turbocharged engine and A/T, cont.

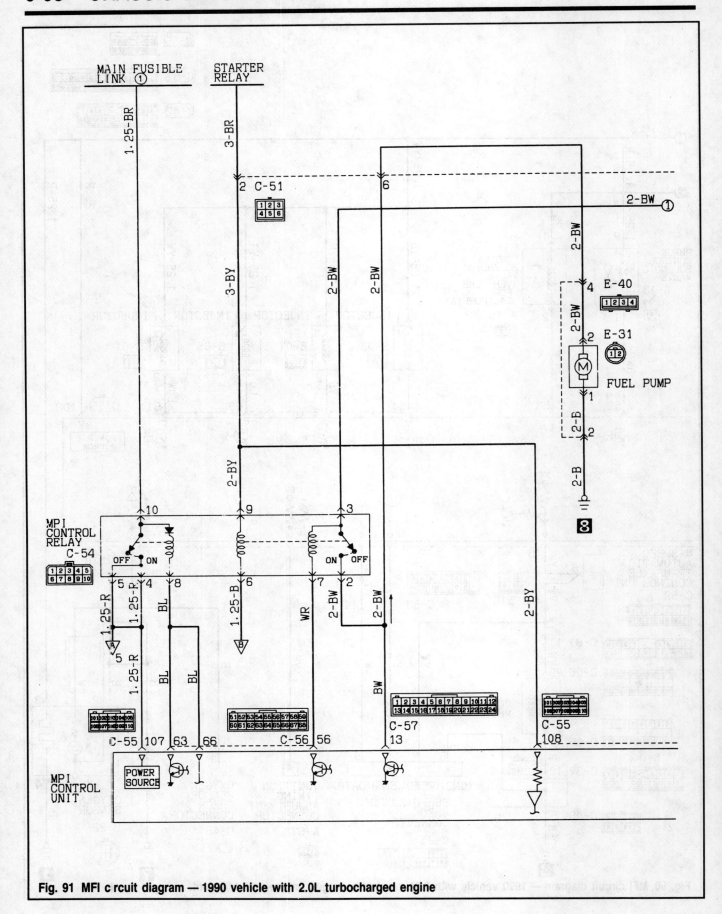

Fig. 91 MFI circuit diagram — 1990 vehicle with 2.0L turbocharged engine

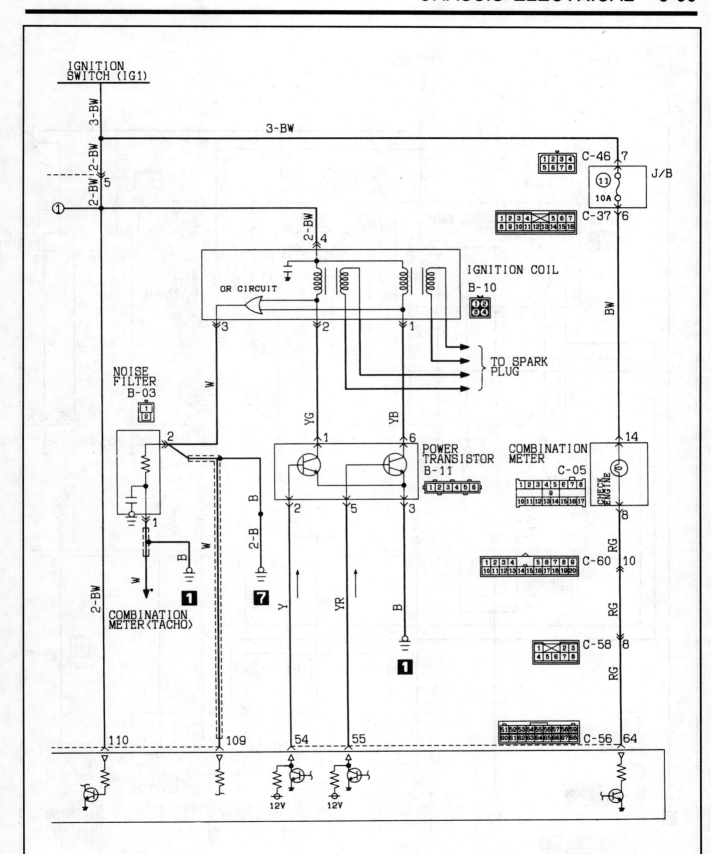

Fig. 92 MFI circuit diagram — 1990 vehicle with 2.0L turbocharged engine, cont.

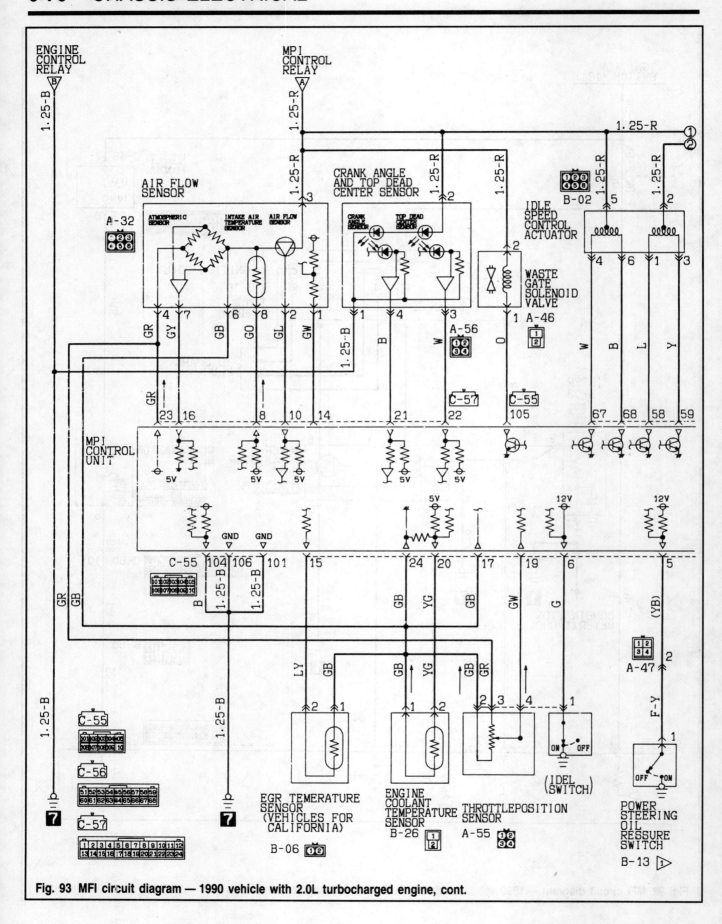

Fig. 93 MFI circuit diagram — 1990 vehicle with 2.0L turbocharged engine, cont.

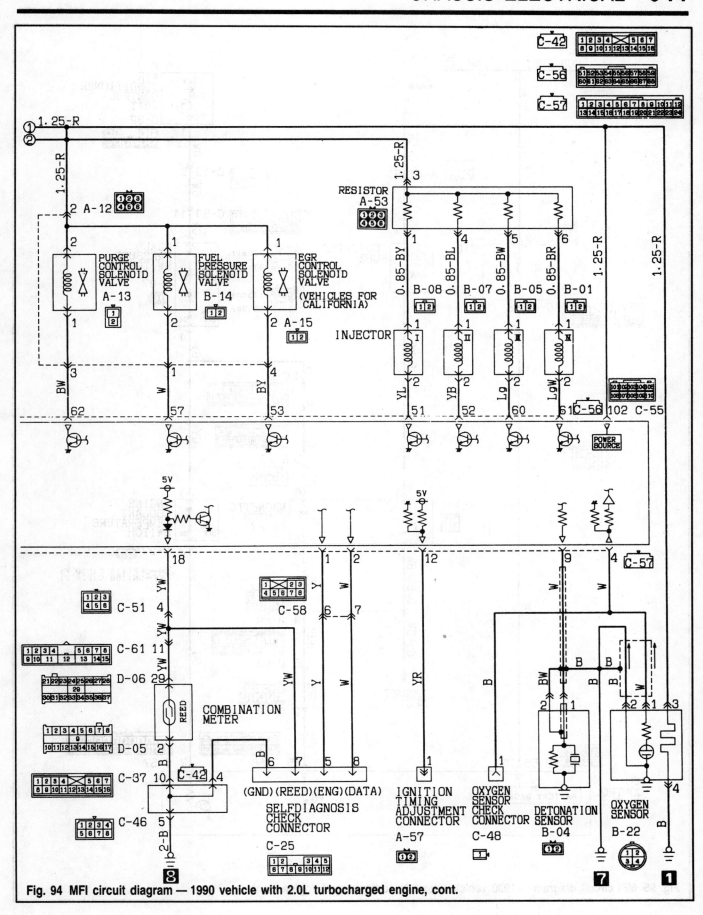

Fig. 94 MFI circuit diagram — 1990 vehicle with 2.0L turbocharged engine, cont.

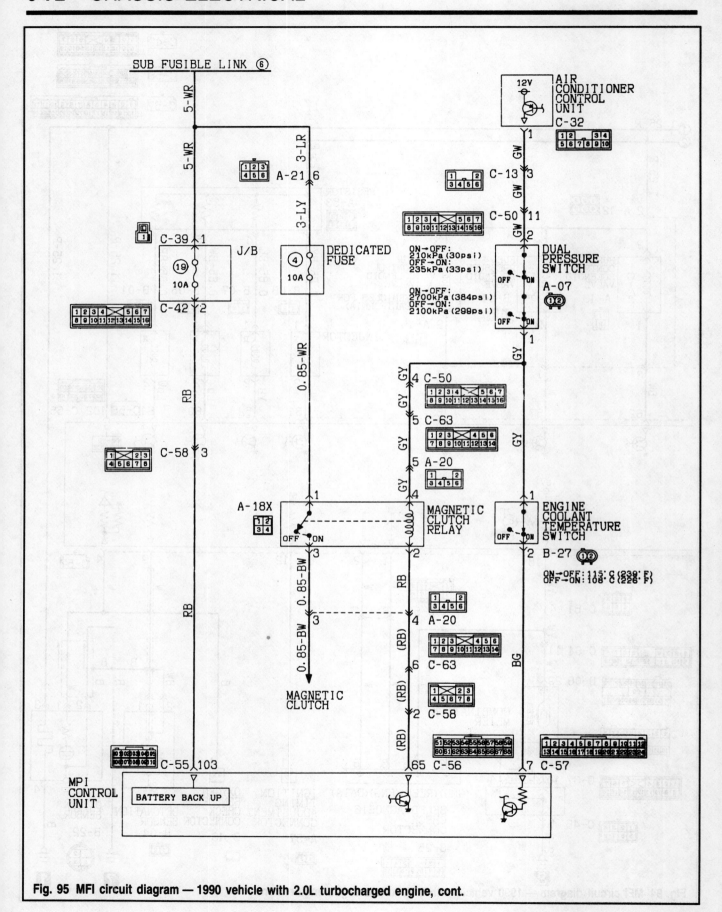

Fig. 95 MFI circuit diagram — 1990 vehicle with 2.0L turbocharged engine, cont.

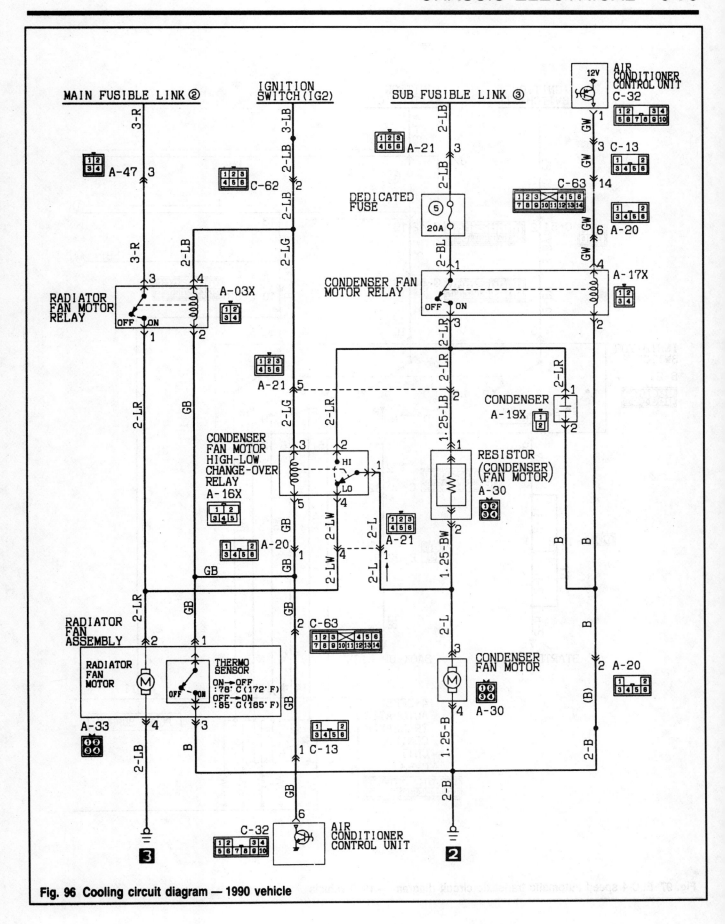

Fig. 96 Cooling circuit diagram — 1990 vehicle

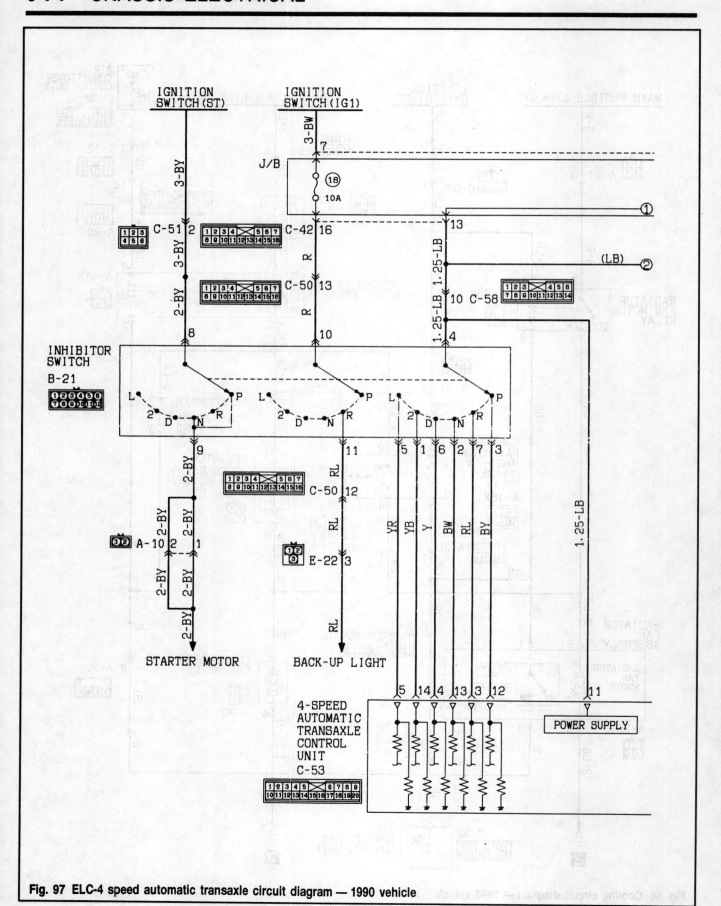

Fig. 97 ELC-4 speed automatic transaxle circuit diagram — 1990 vehicle

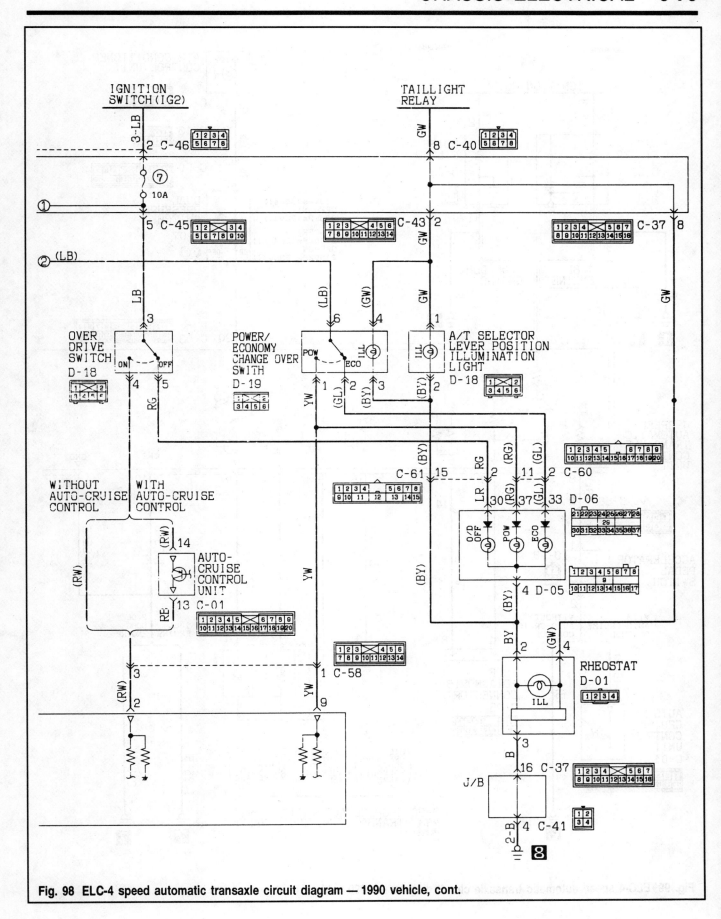

Fig. 98 ELC-4 speed automatic transaxle circuit diagram — 1990 vehicle, cont.

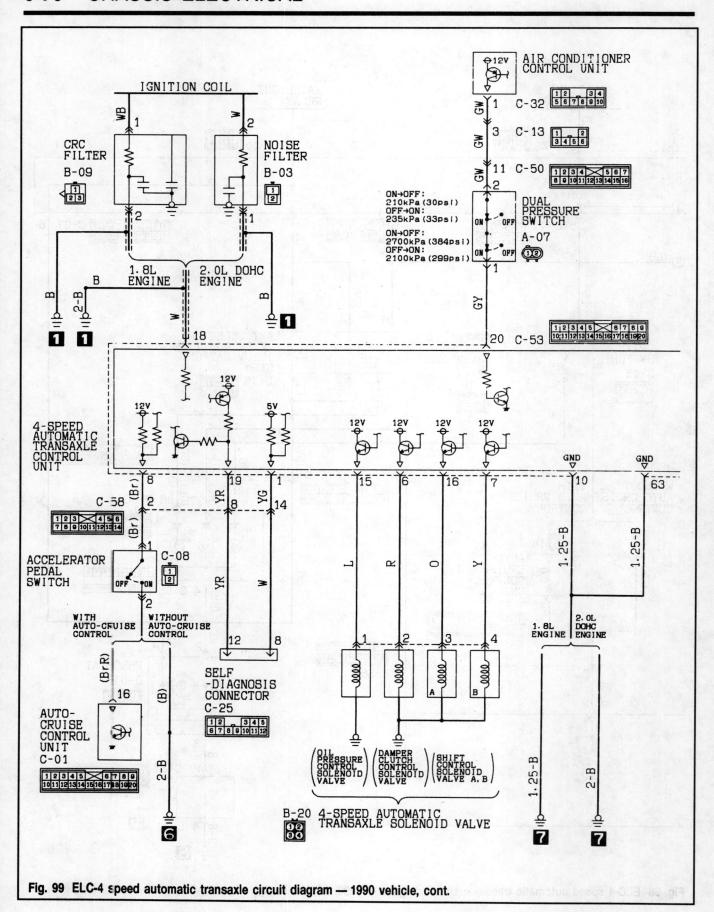

Fig. 99 ELC-4 speed automatic transaxle circuit diagram — 1990 vehicle, cont.

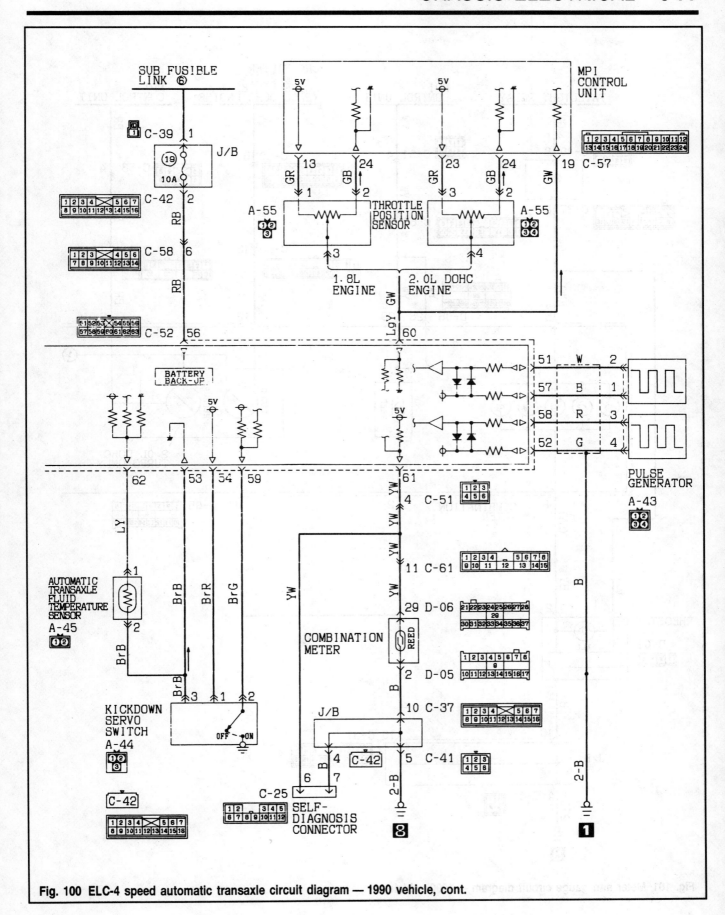

Fig. 100 ELC-4 speed automatic transaxle circuit diagram — 1990 vehicle, cont.

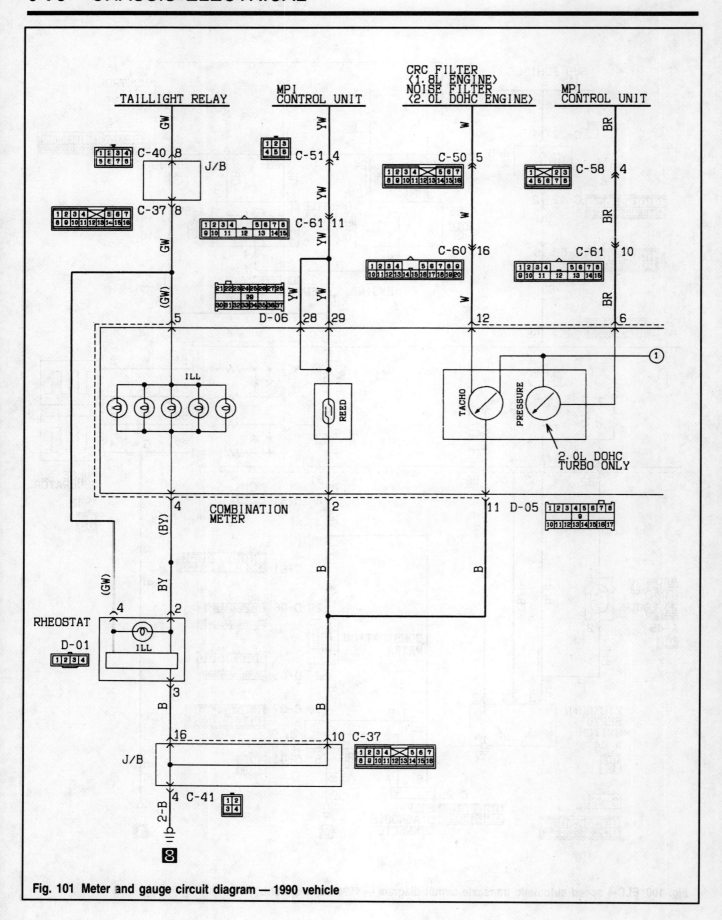

Fig. 101 Meter and gauge circuit diagram — 1990 vehicle

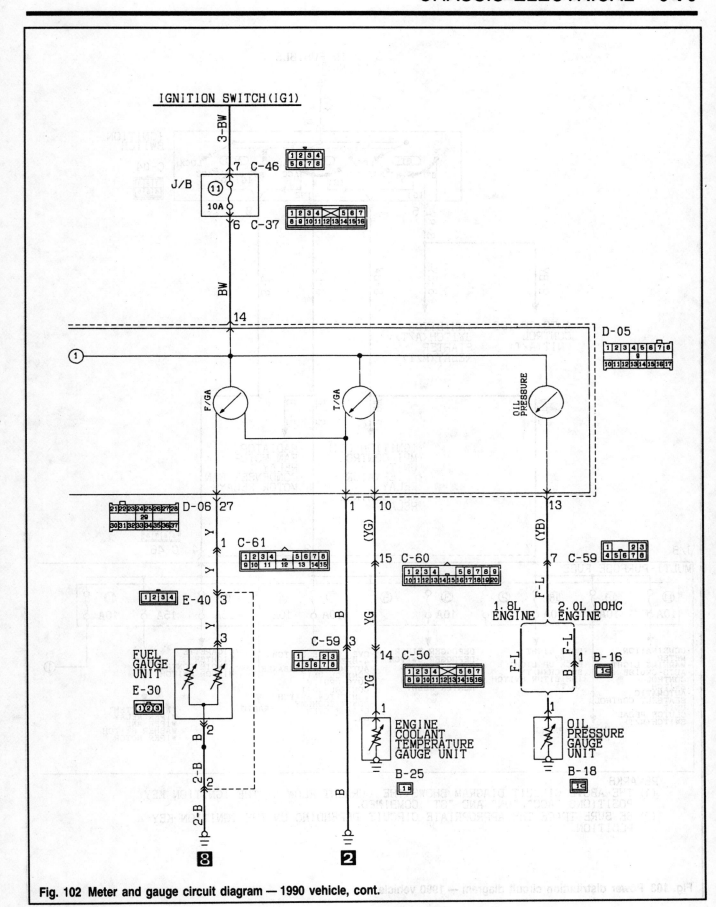

Fig. 102 Meter and gauge circuit diagram — 1990 vehicle, cont.

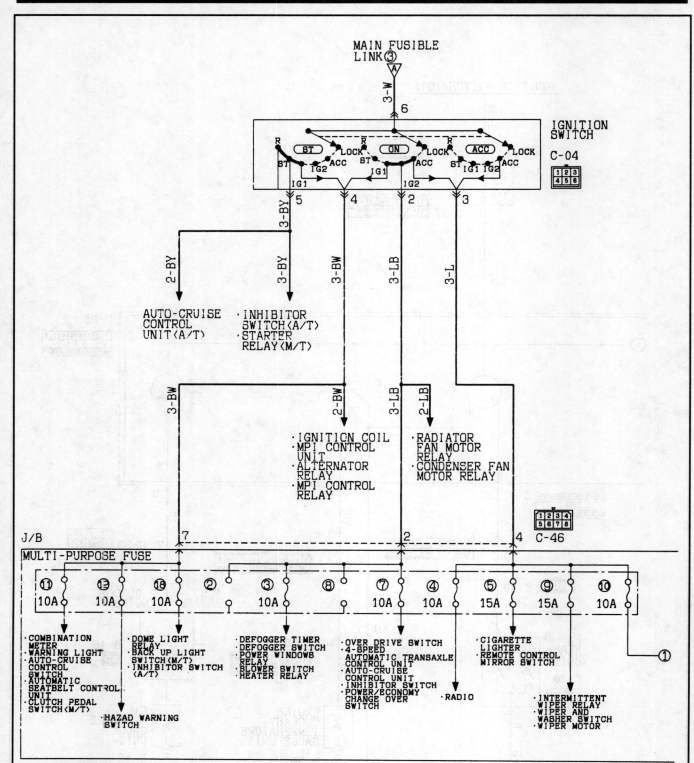

Fig. 103 Power distribution circuit diagram — 1990 vehicle

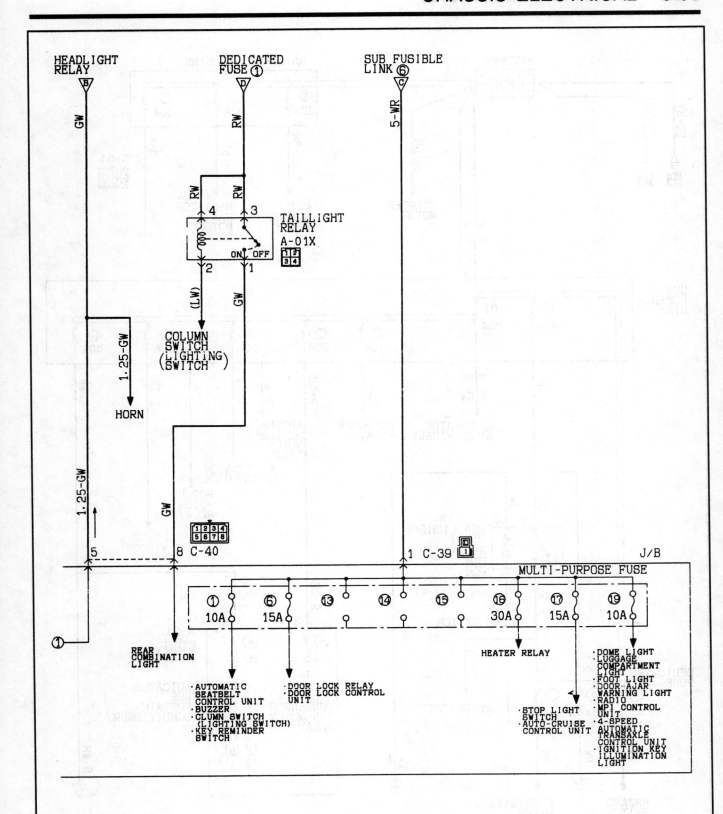

Fig. 104 Power distribution circuit diagram — 1990 vehicle, cont.

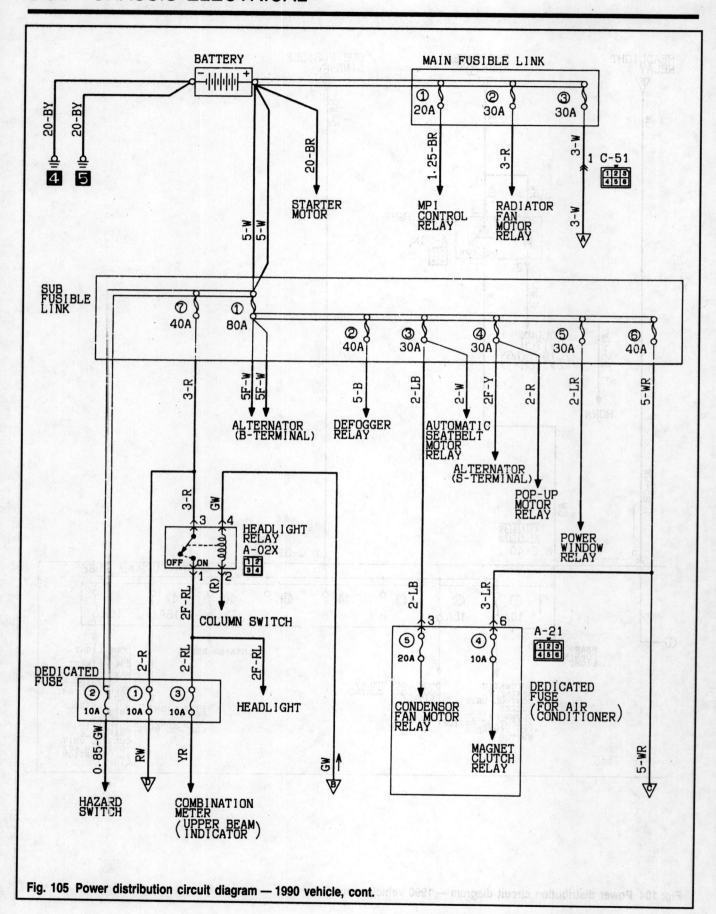

Fig. 105 Power distribution circuit diagram — 1990 vehicle, cont.

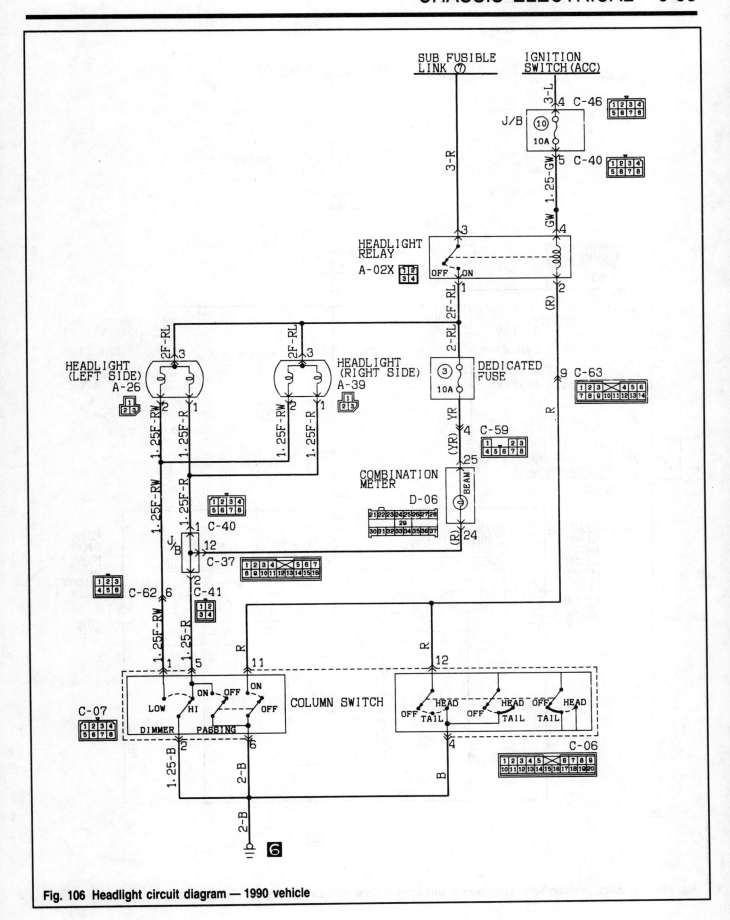

Fig. 106 Headlight circuit diagram — 1990 vehicle

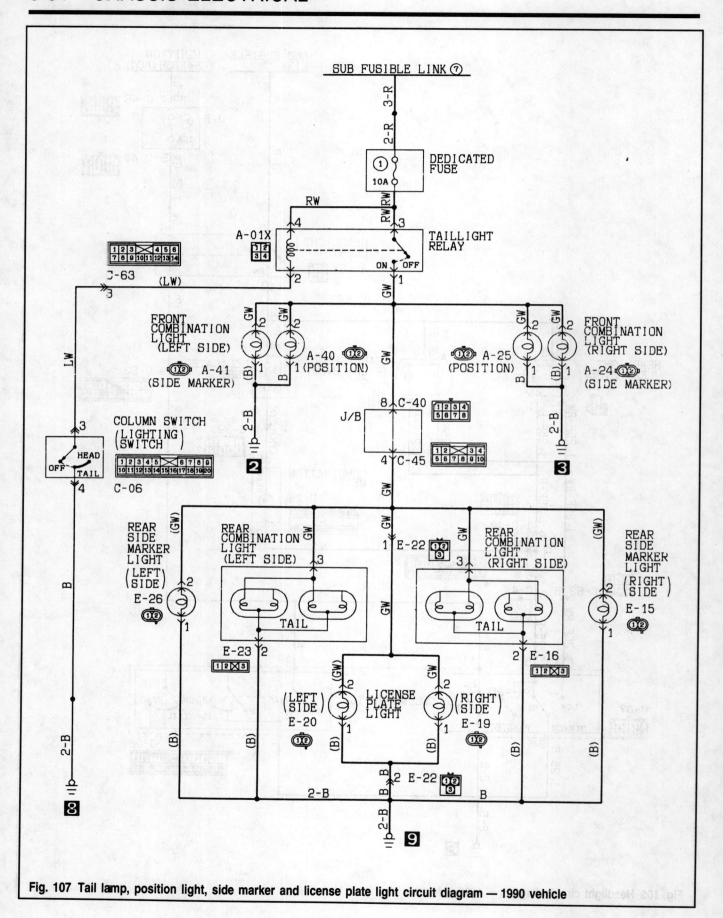

Fig. 107 Tail lamp, position light, side marker and license plate light circuit diagram — 1990 vehicle

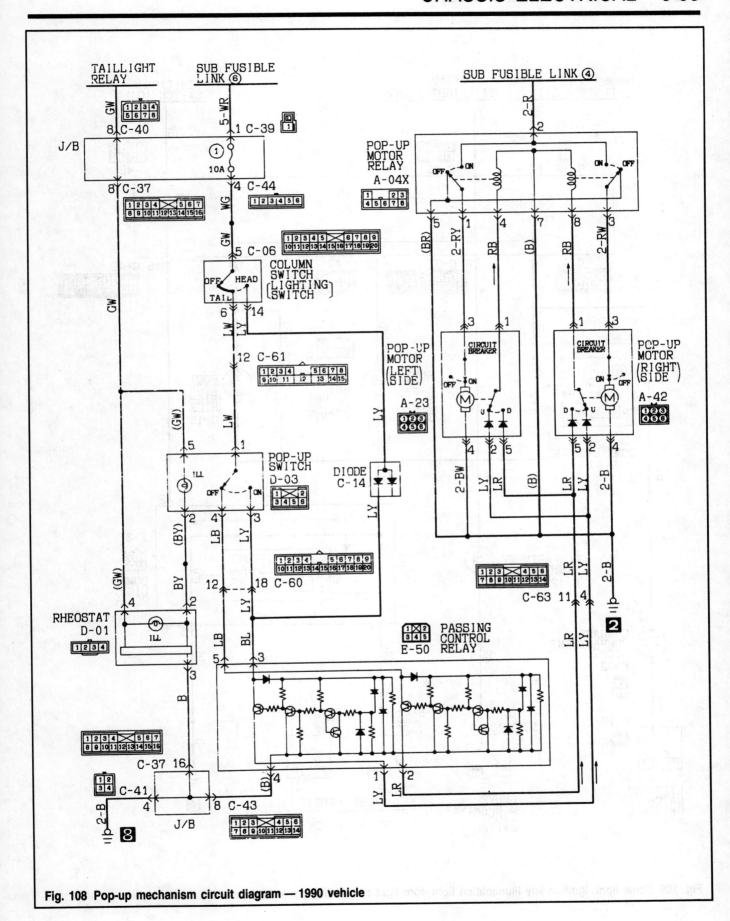

Fig. 108 Pop-up mechanism circuit diagram — 1990 vehicle

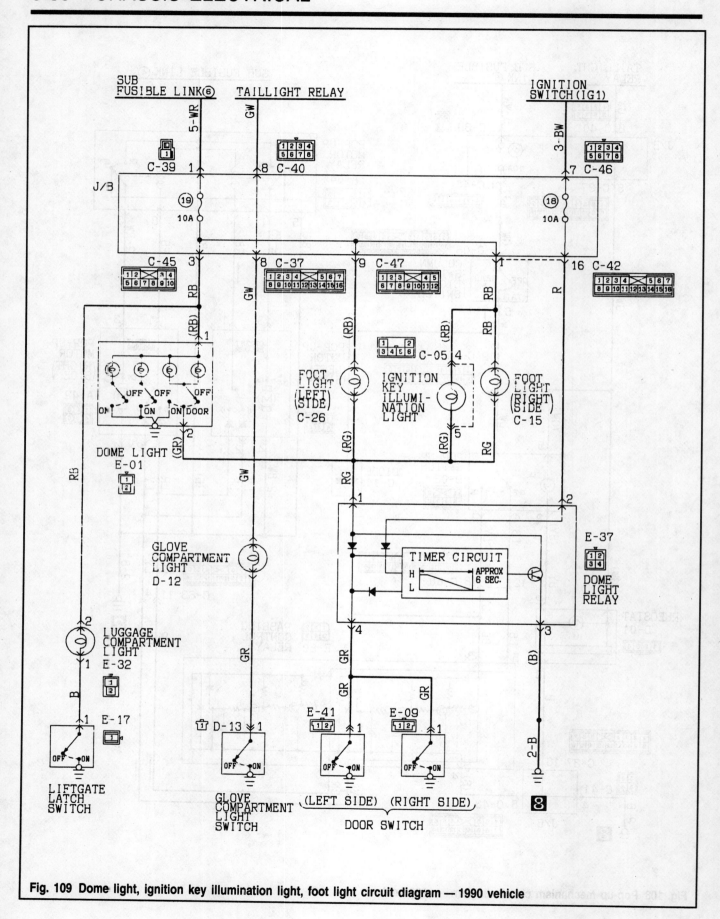

Fig. 109 Dome light, ignition key illumination light, foot light circuit diagram — 1990 vehicle

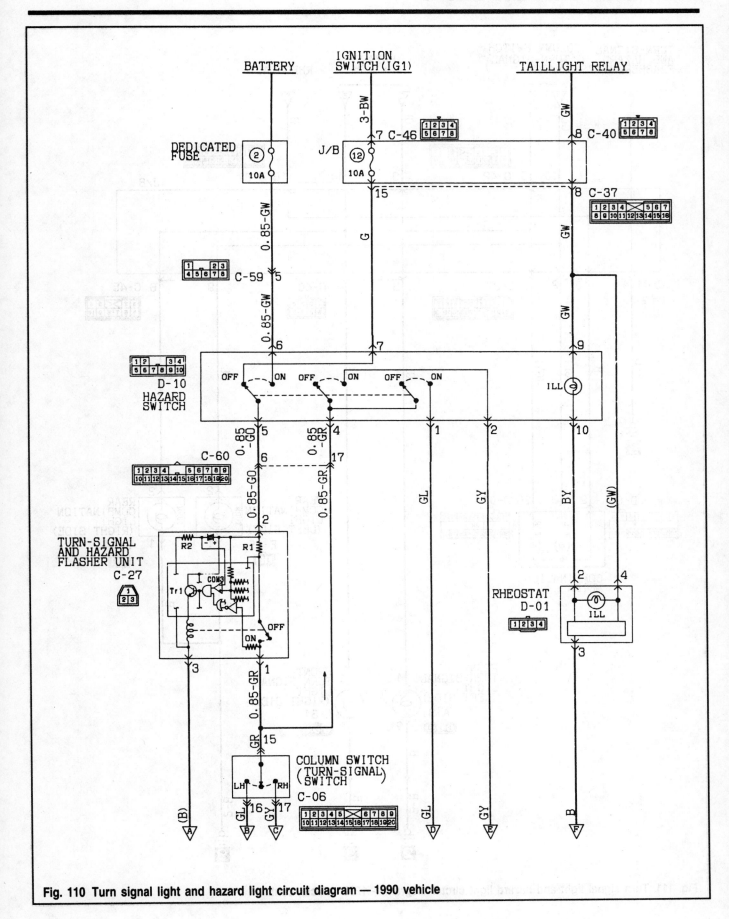

Fig. 110 Turn signal light and hazard light circuit diagram — 1990 vehicle

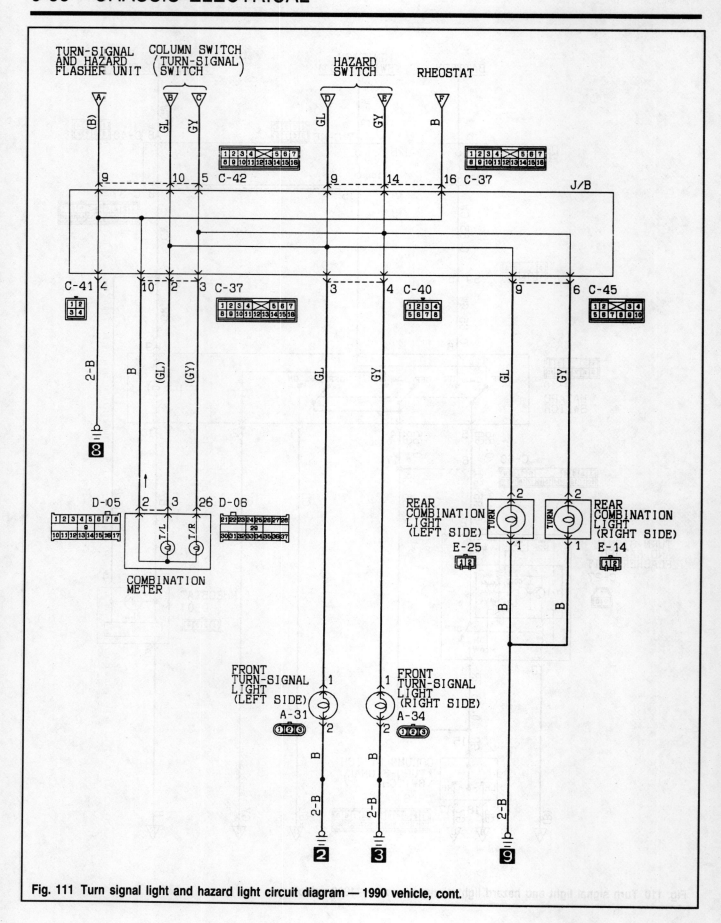

Fig. 111 Turn signal light and hazard light circuit diagram — 1990 vehicle, cont.

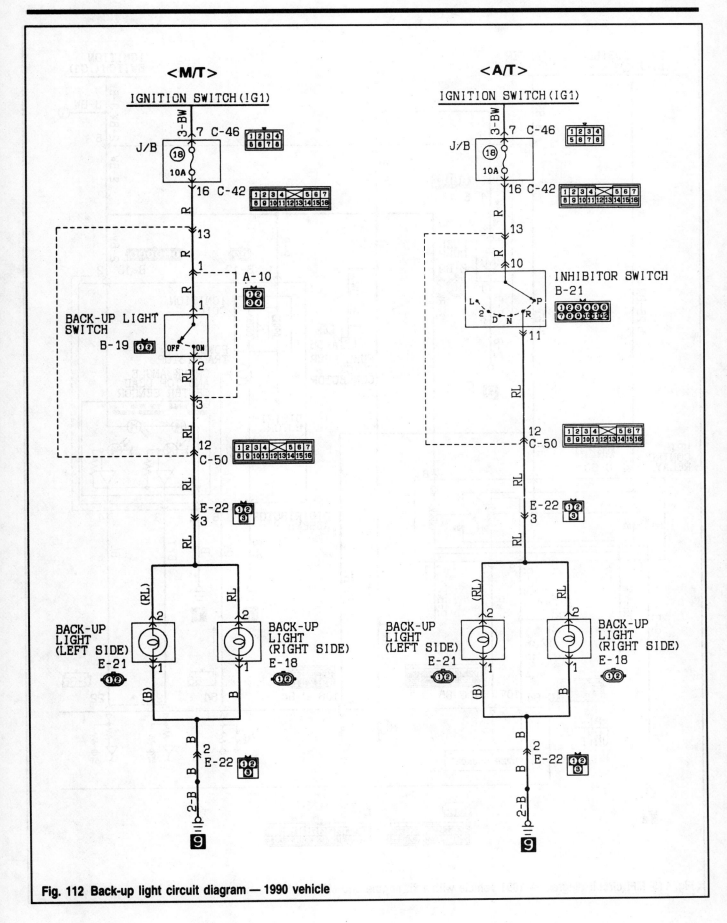

Fig. 112 Back-up light circuit diagram — 1990 vehicle

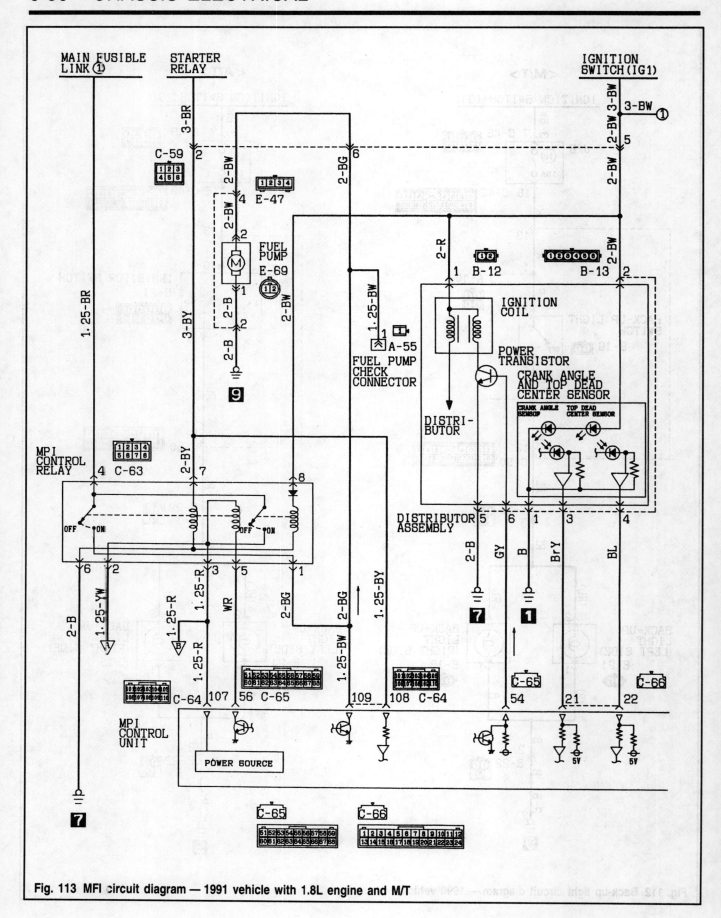

Fig. 113 MFI circuit diagram — 1991 vehicle with 1.8L engine and M/T

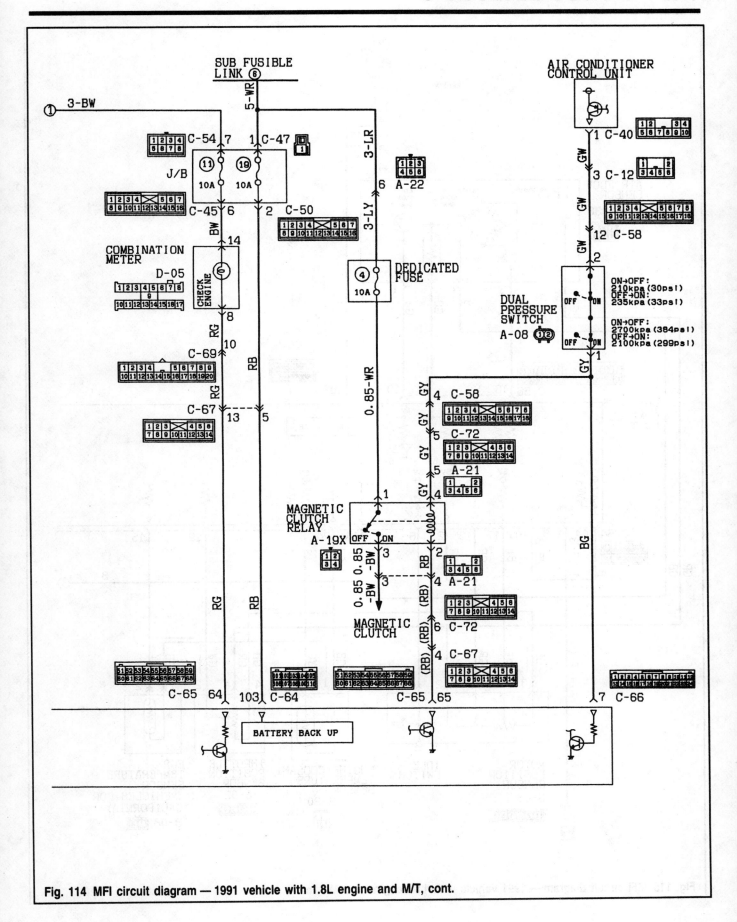

Fig. 114 MFI circuit diagram — 1991 vehicle with 1.8L engine and M/T, cont.

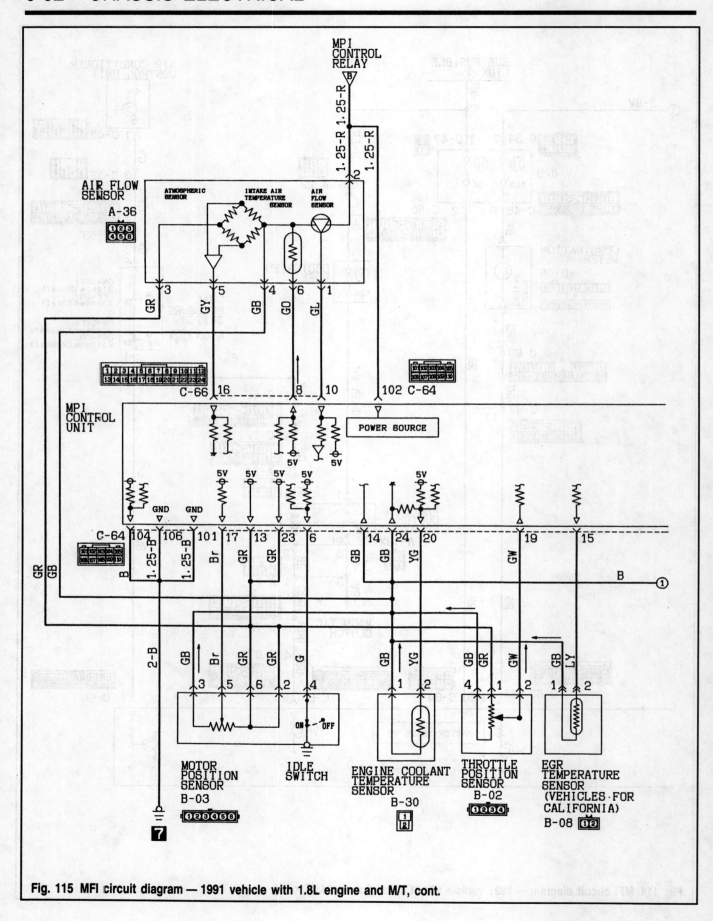

Fig. 115 MFI circuit diagram — 1991 vehicle with 1.8L engine and M/T, cont.

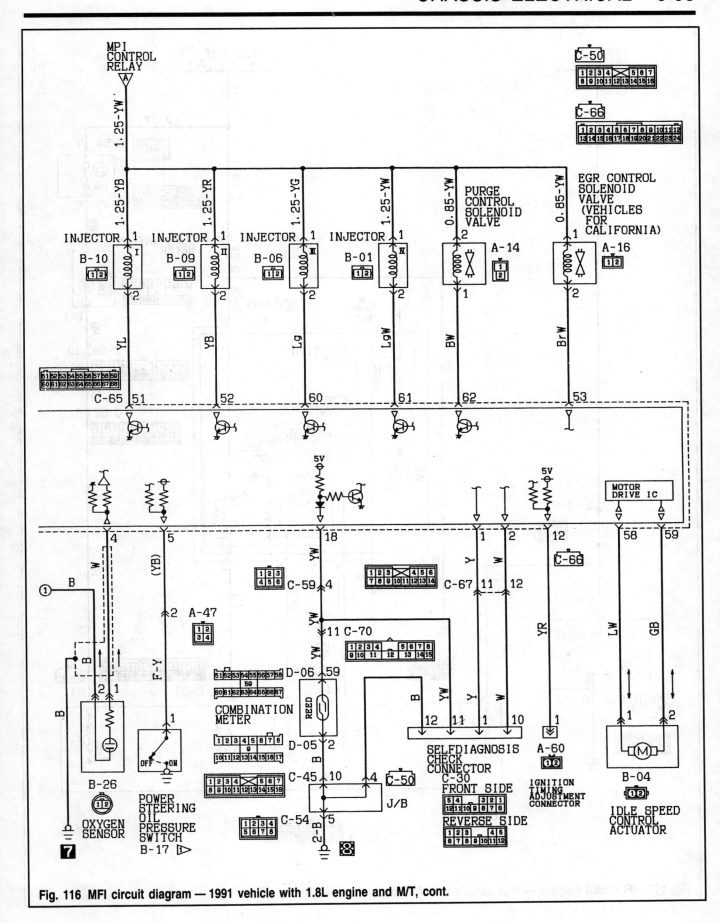

Fig. 116 MFI circuit diagram — 1991 vehicle with 1.8L engine and M/T, cont.

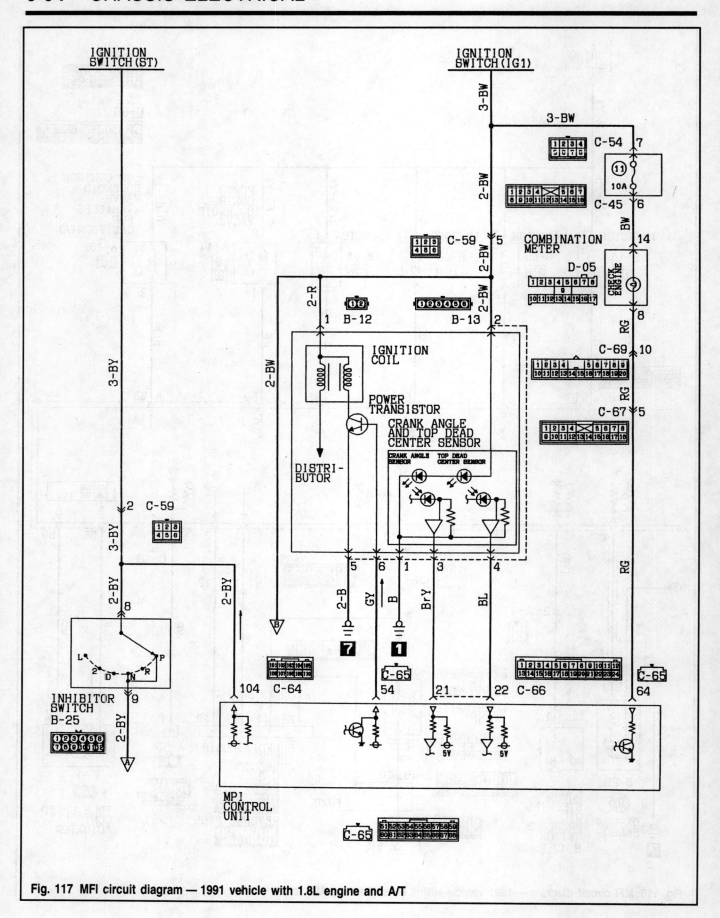

Fig. 117 MFI circuit diagram — 1991 vehicle with 1.8L engine and A/T

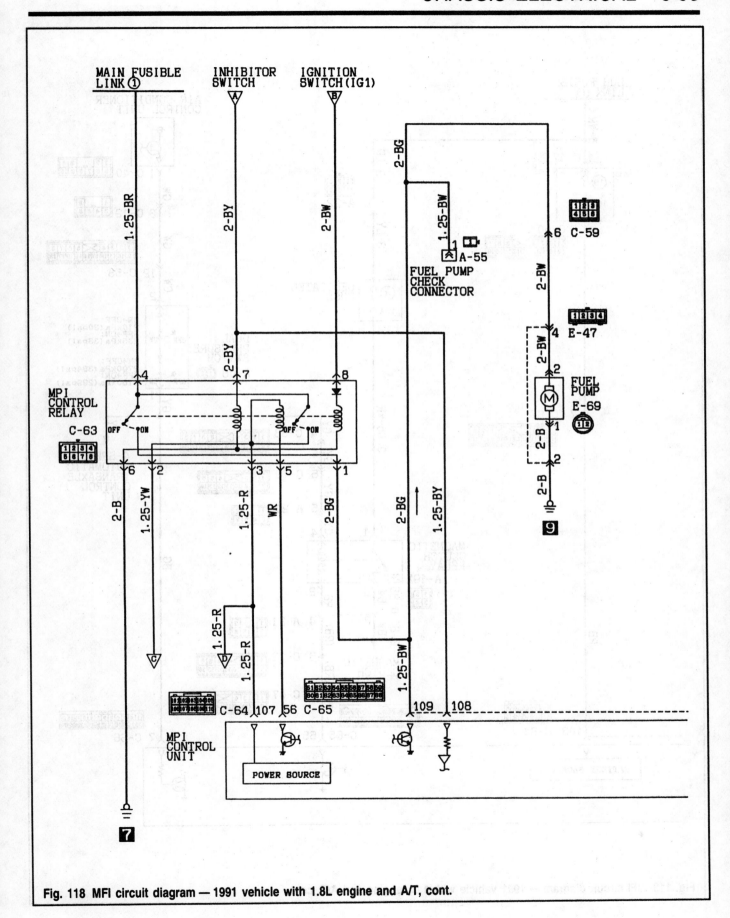

Fig. 118 MFI circuit diagram — 1991 vehicle with 1.8L engine and A/T, cont.

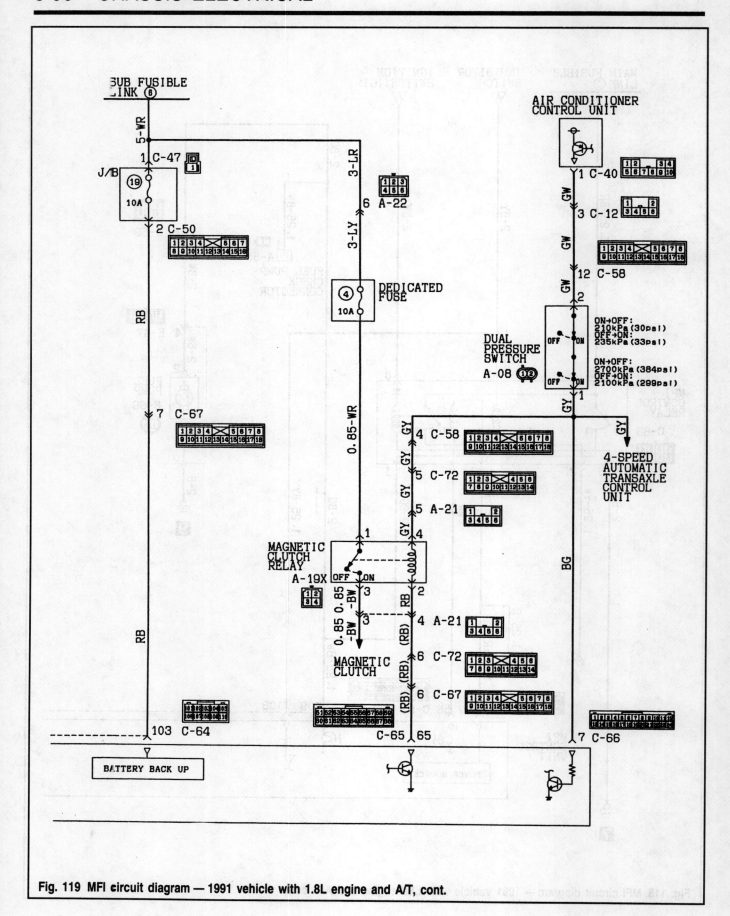

Fig. 119 MFI circuit diagram — 1991 vehicle with 1.8L engine and A/T, cont.

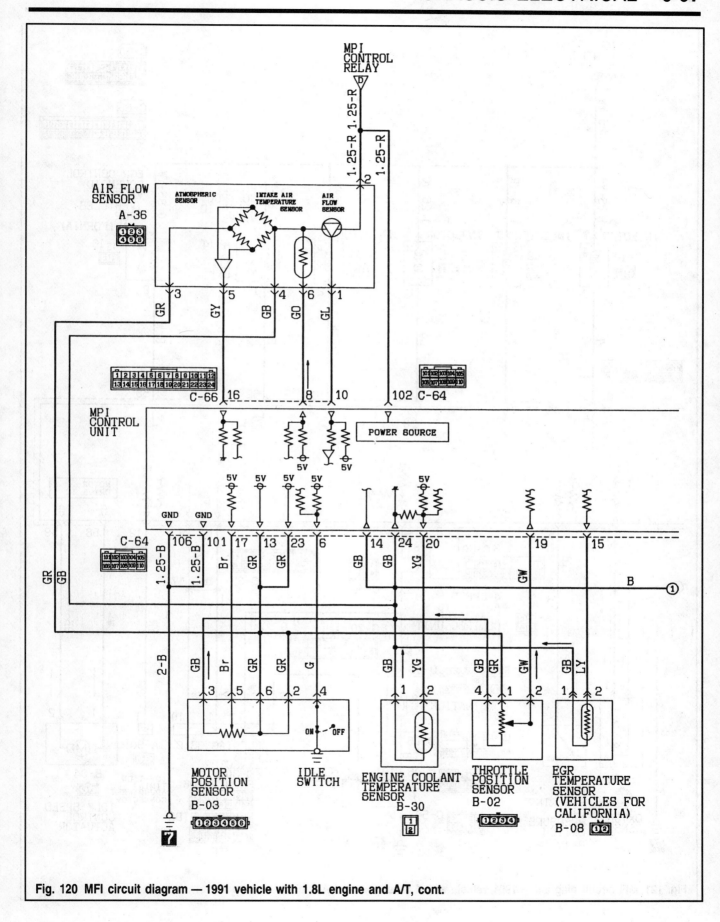

Fig. 120 MFI circuit diagram — 1991 vehicle with 1.8L engine and A/T, cont.

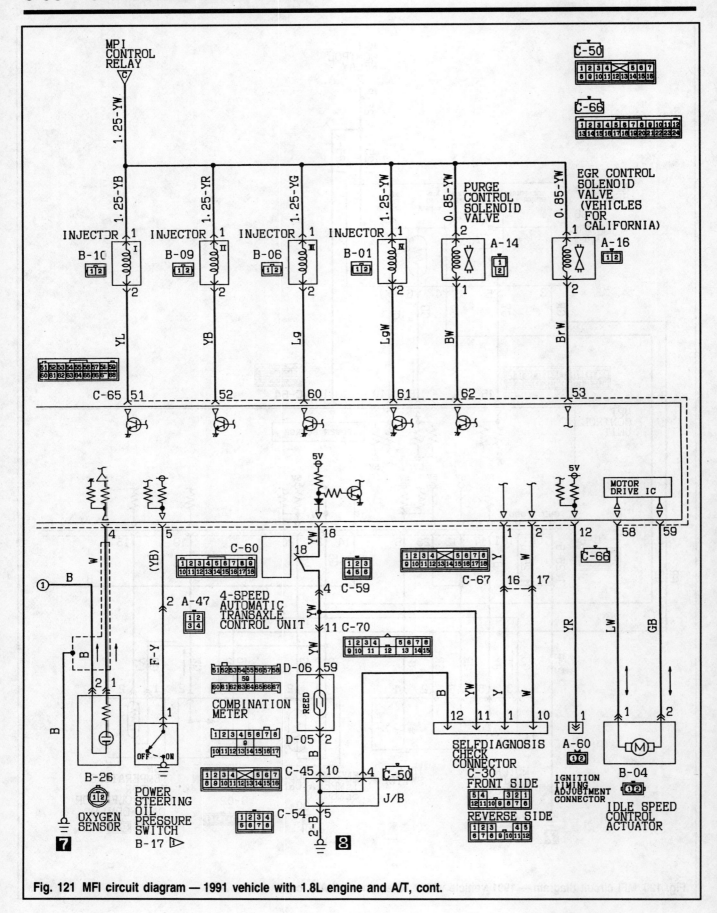

Fig. 121 MFI circuit diagram — 1991 vehicle with 1.8L engine and A/T, cont.

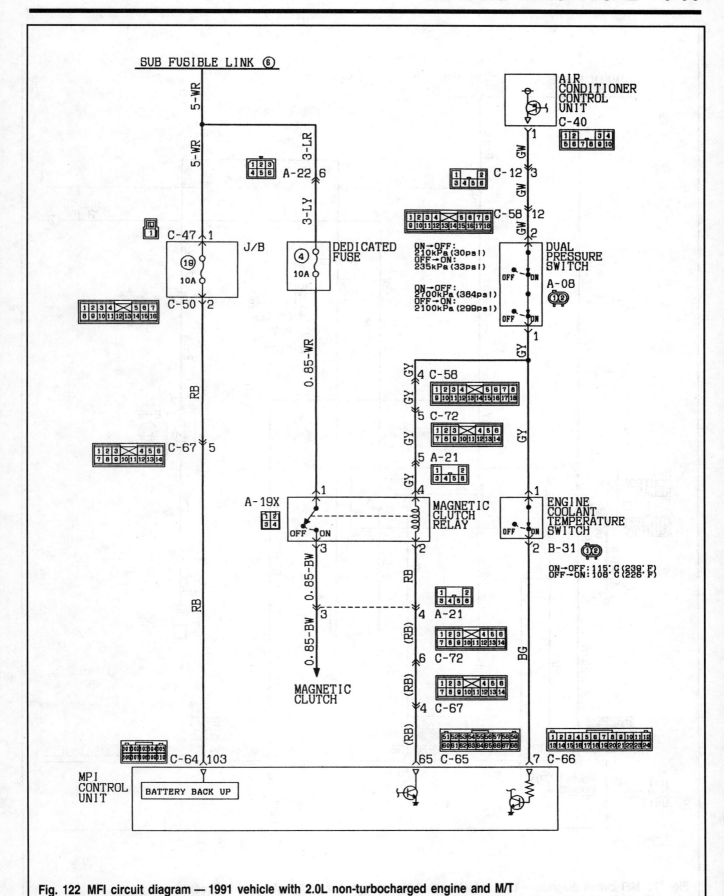

Fig. 122 MFI circuit diagram — 1991 vehicle with 2.0L non-turbocharged engine and M/T

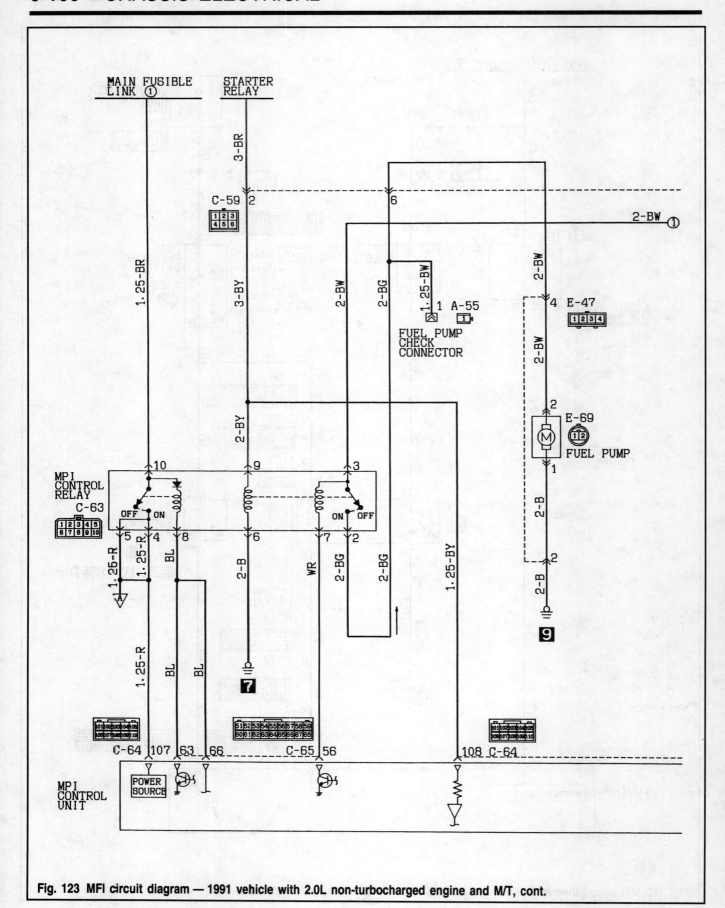

Fig. 123 MFI circuit diagram — 1991 vehicle with 2.0L non-turbocharged engine and M/T, cont.

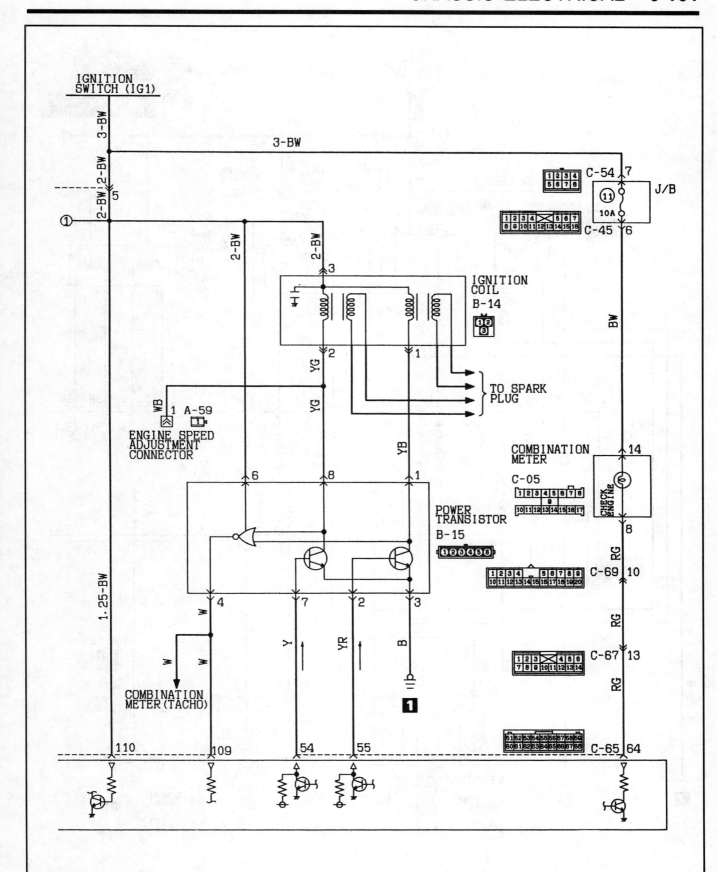

Fig. 124 MFI circuit diagram — 1991 vehicle with 2.0L non-turbocharged engine and M/T, cont.

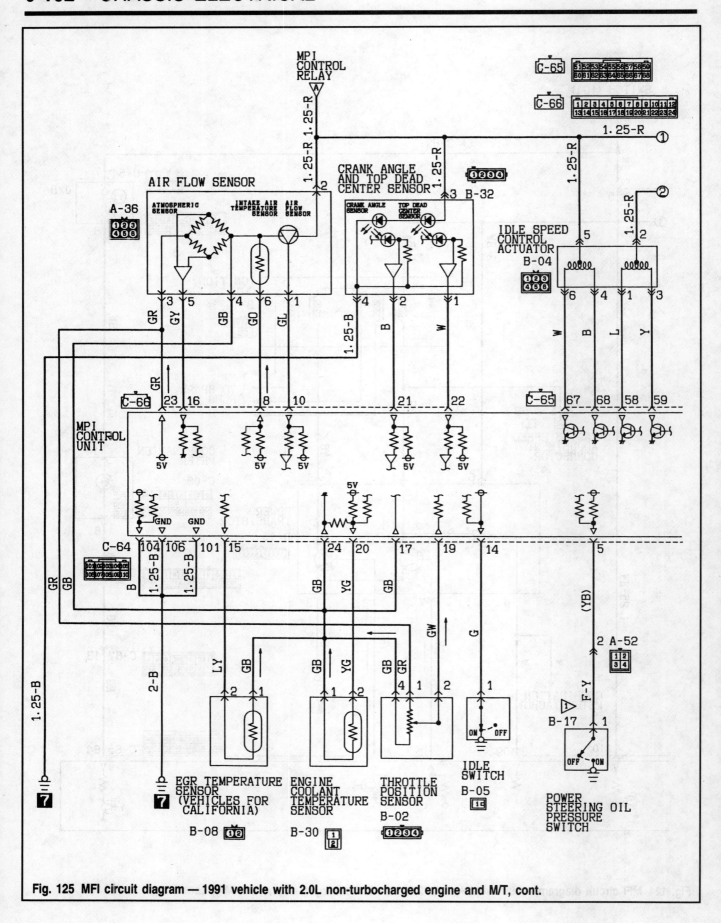

Fig. 125 MFI circuit diagram — 1991 vehicle with 2.0L non-turbocharged engine and M/T, cont.

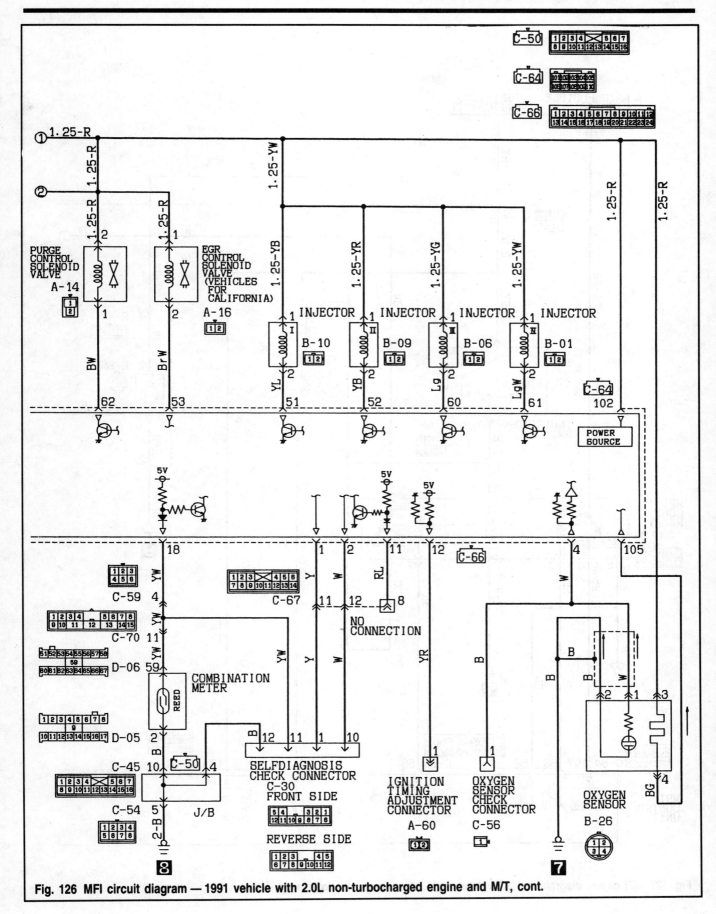

Fig. 126 MFI circuit diagram — 1991 vehicle with 2.0L non-turbocharged engine and M/T, cont.

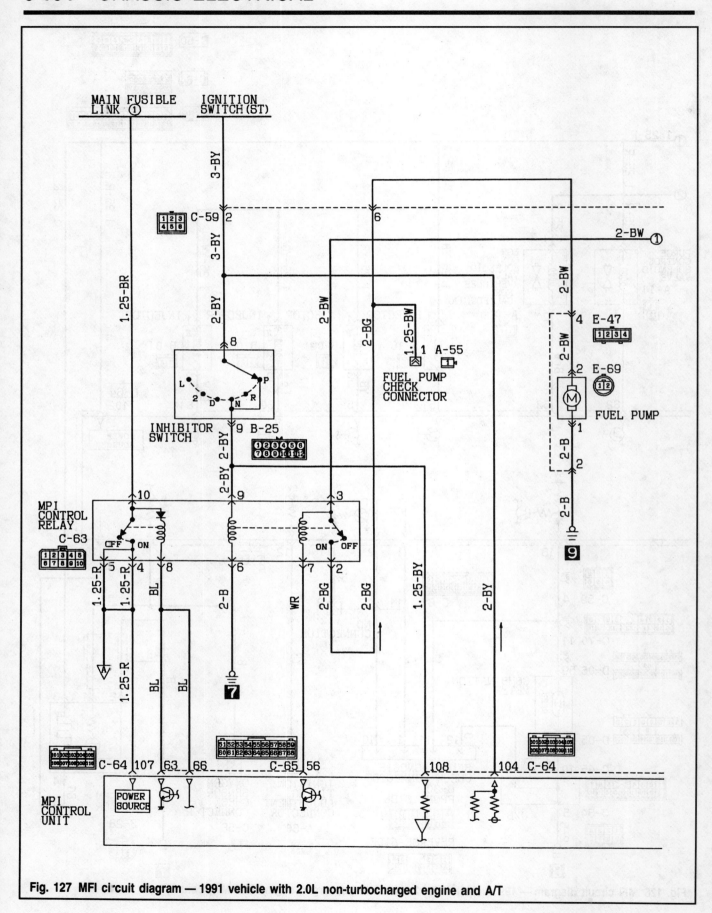

Fig. 127 MFI circuit diagram — 1991 vehicle with 2.0L non-turbocharged engine and A/T

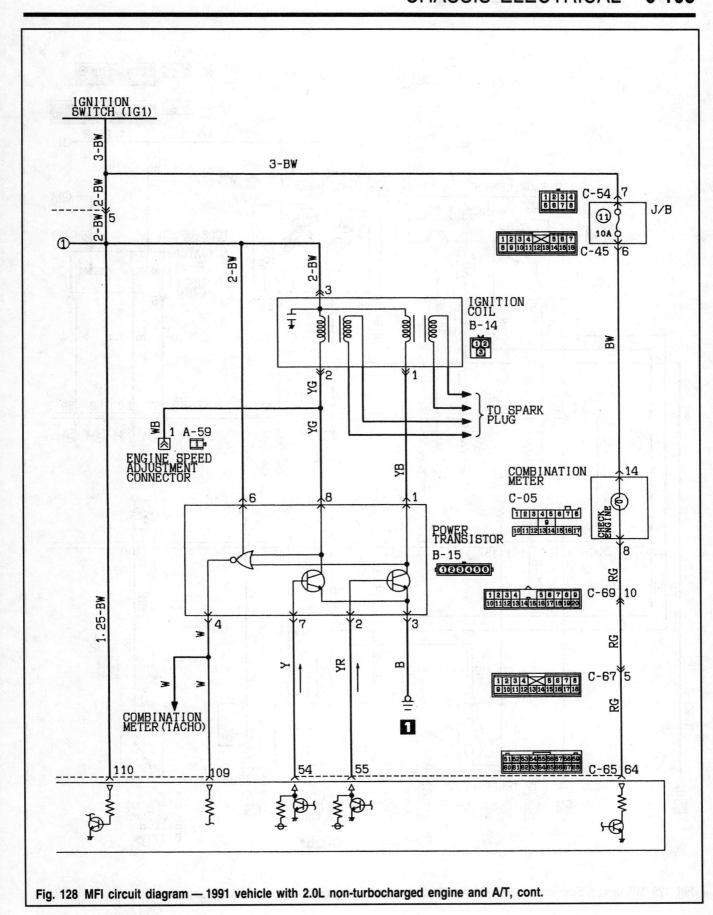

Fig. 128 MFI circuit diagram — 1991 vehicle with 2.0L non-turbocharged engine and A/T, cont.

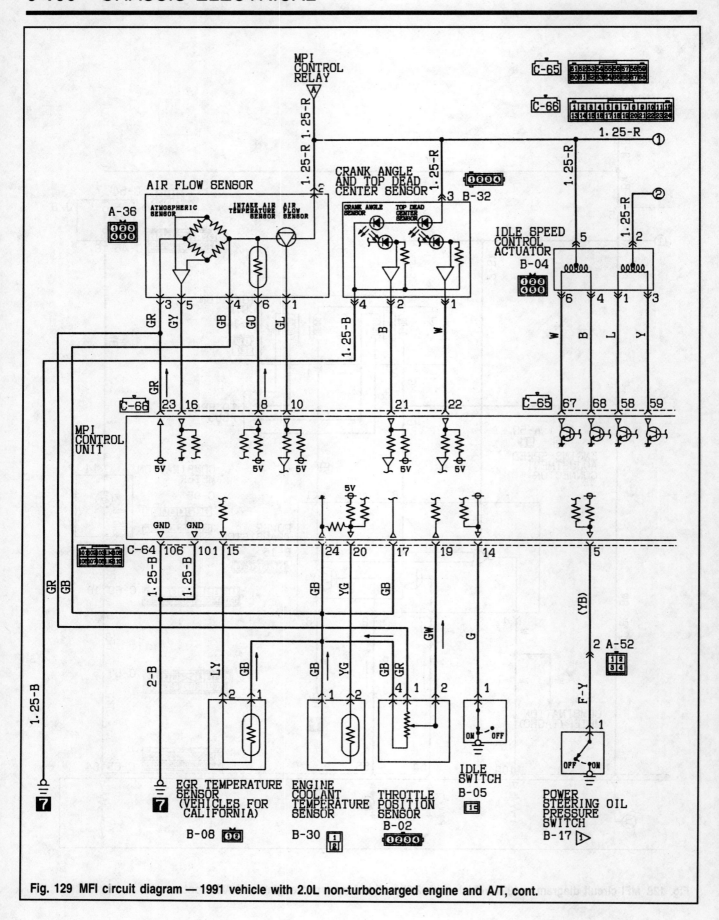

Fig. 129 MFI circuit diagram — 1991 vehicle with 2.0L non-turbocharged engine and A/T, cont.

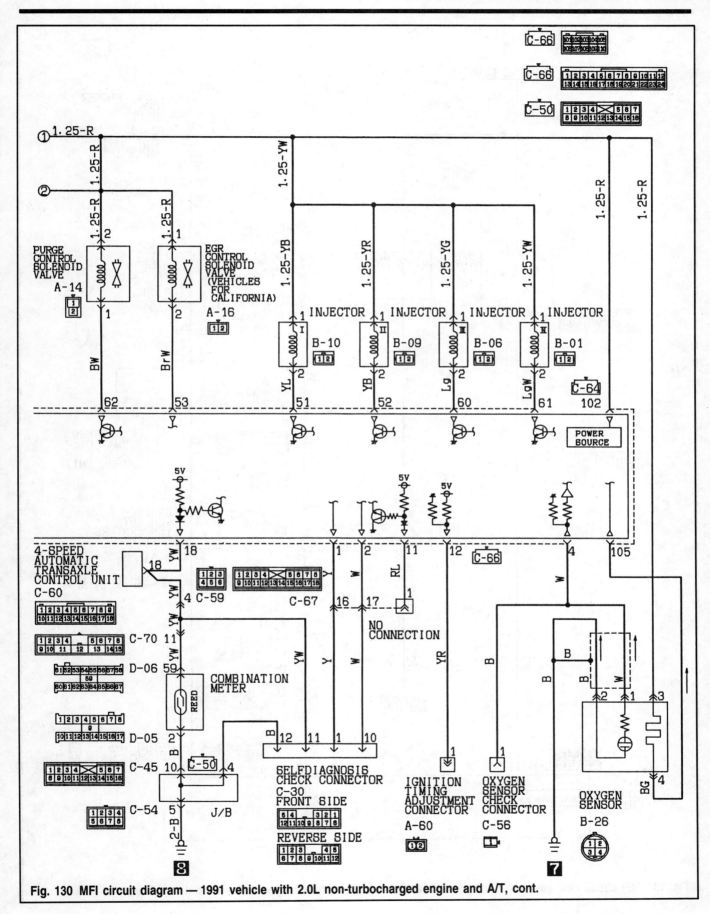

Fig. 130 MFI circuit diagram — 1991 vehicle with 2.0L non-turbocharged engine and A/T, cont.

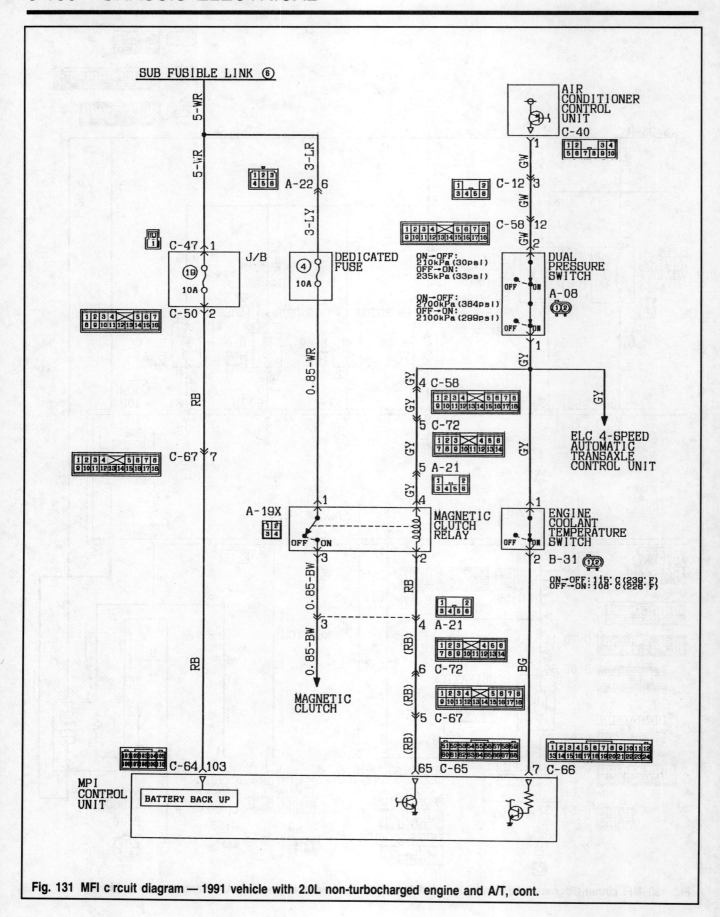

Fig. 131 MFI circuit diagram — 1991 vehicle with 2.0L non-turbocharged engine and A/T, cont.

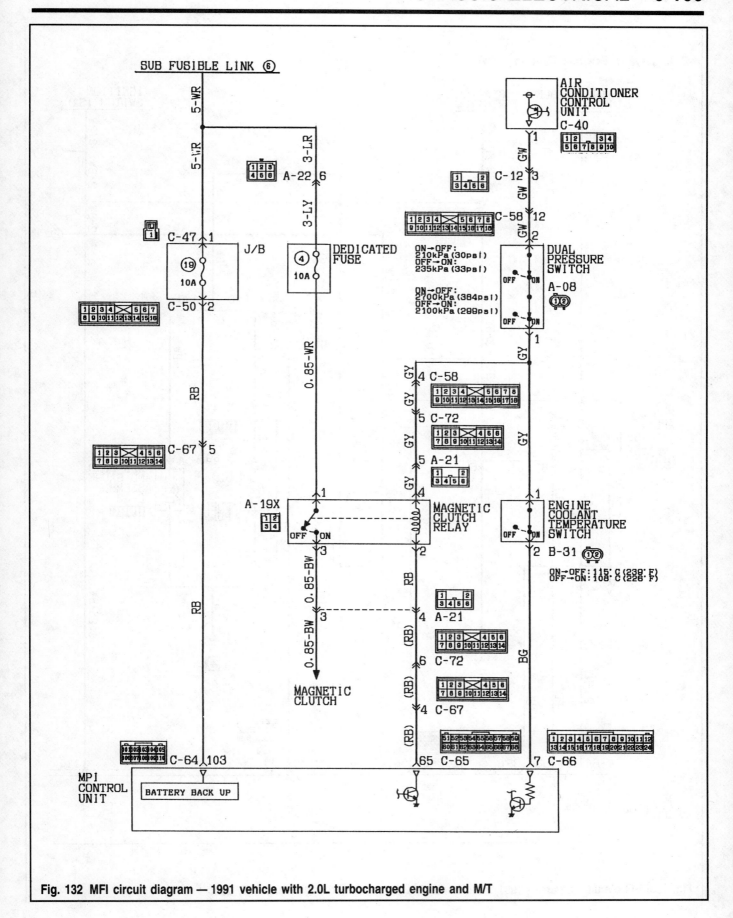

Fig. 132 MFI circuit diagram — 1991 vehicle with 2.0L turbocharged engine and M/T

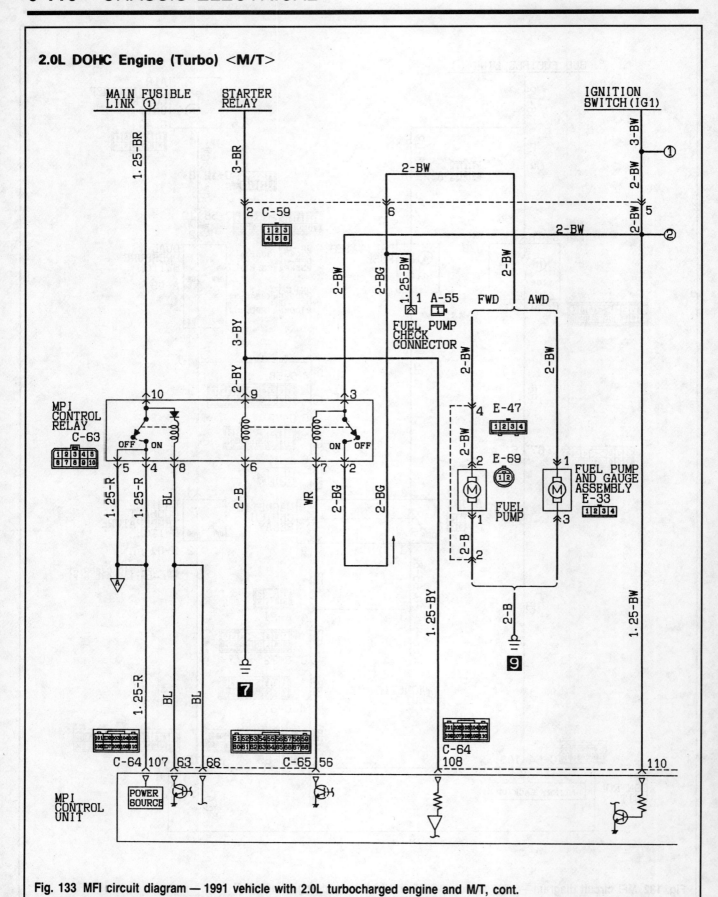

Fig. 133 MFI circuit diagram — 1991 vehicle with 2.0L turbocharged engine and M/T, cont.

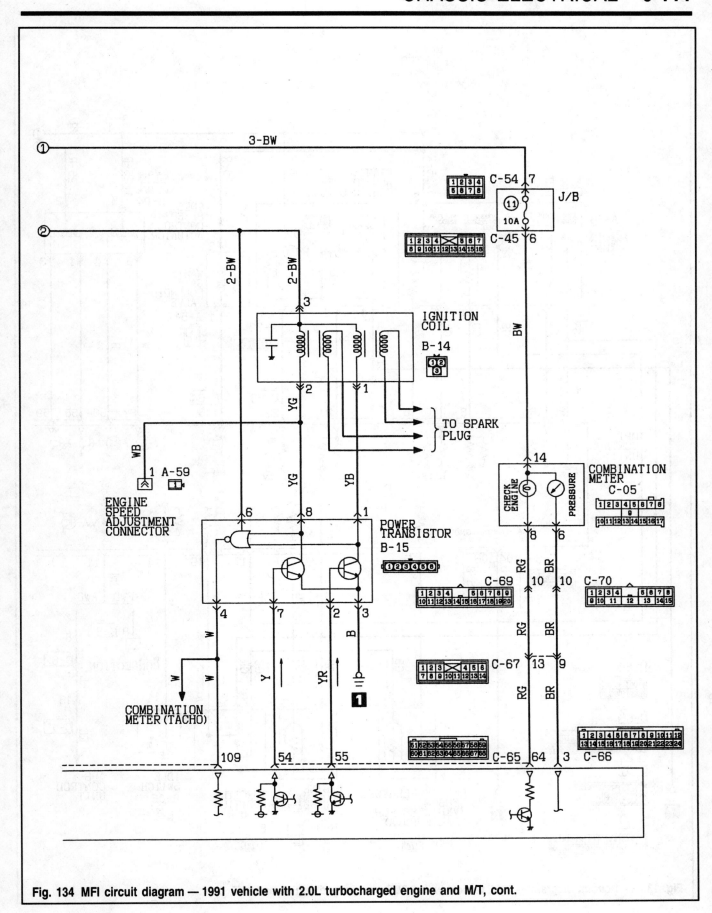

Fig. 134 MFI circuit diagram — 1991 vehicle with 2.0L turbocharged engine and M/T, cont.

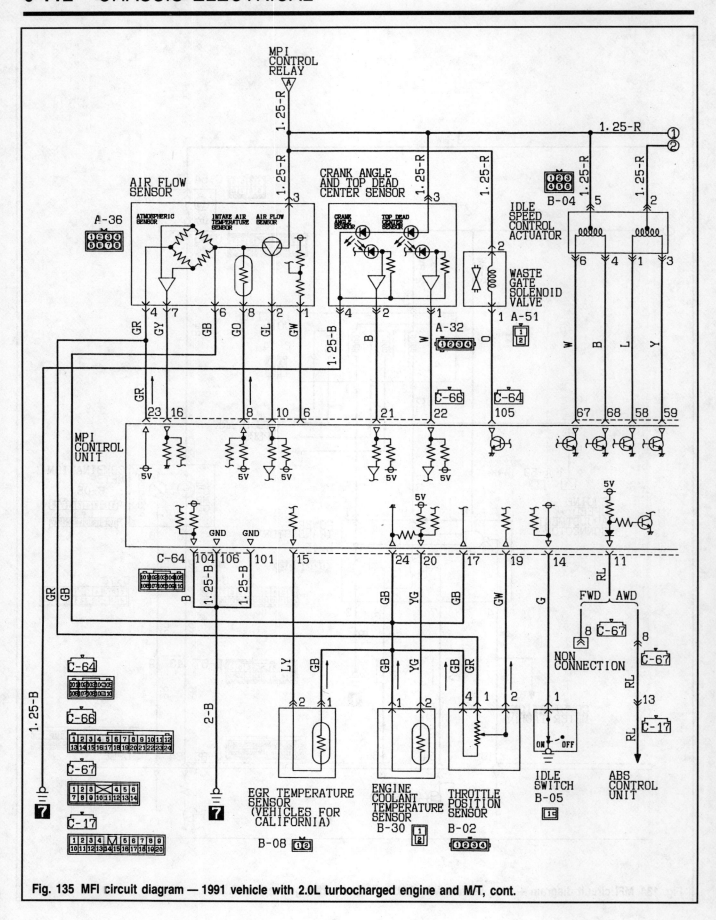

Fig. 135 MFI circuit diagram — 1991 vehicle with 2.0L turbocharged engine and M/T, cont.

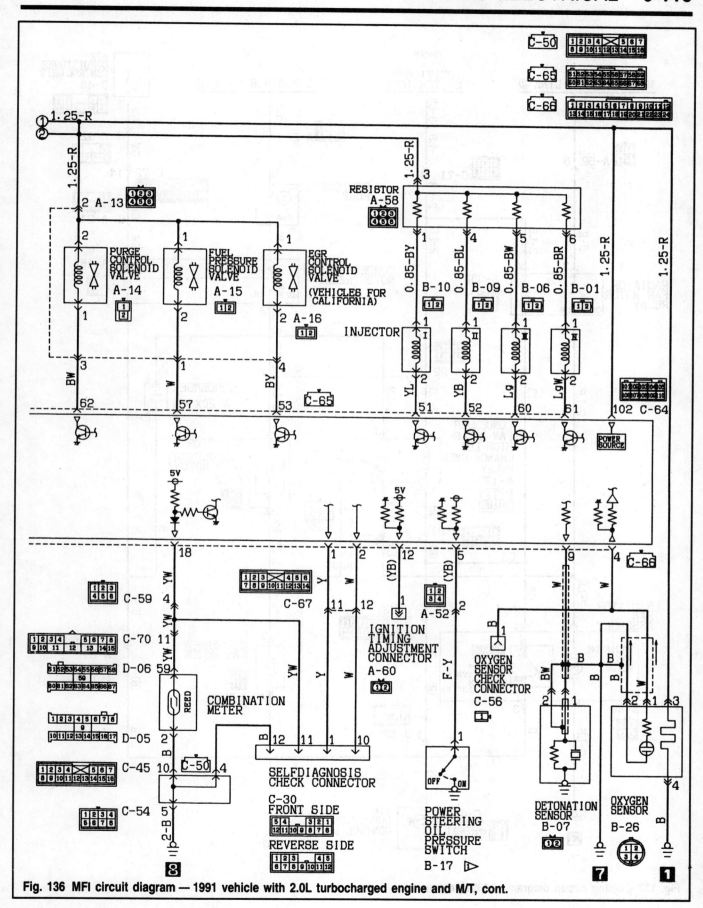

Fig. 136 MFI circuit diagram — 1991 vehicle with 2.0L turbocharged engine and M/T, cont.

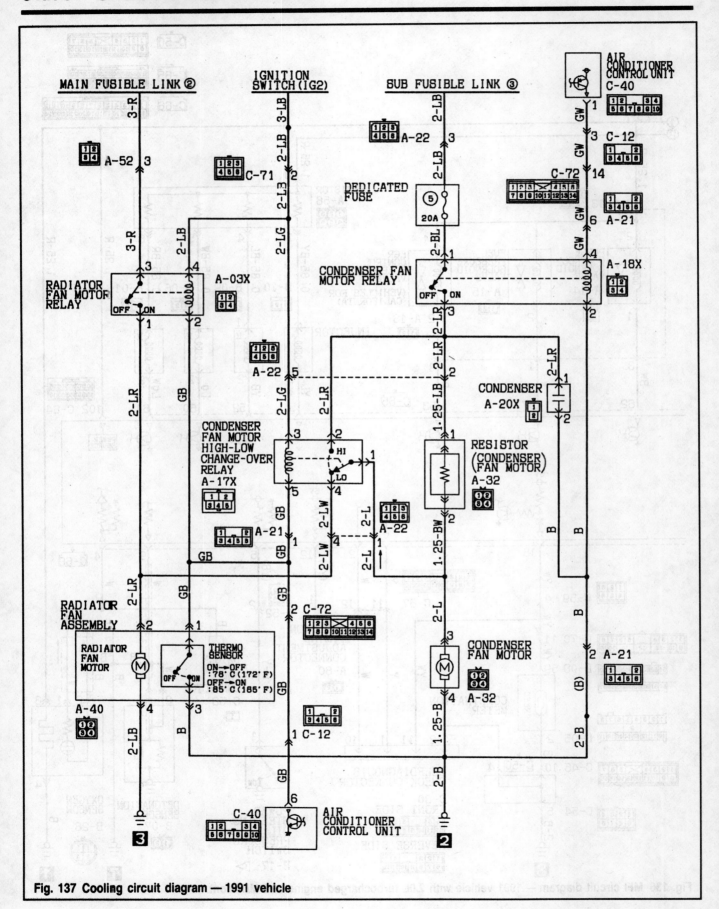

Fig. 137 Cooling circuit diagram — 1991 vehicle

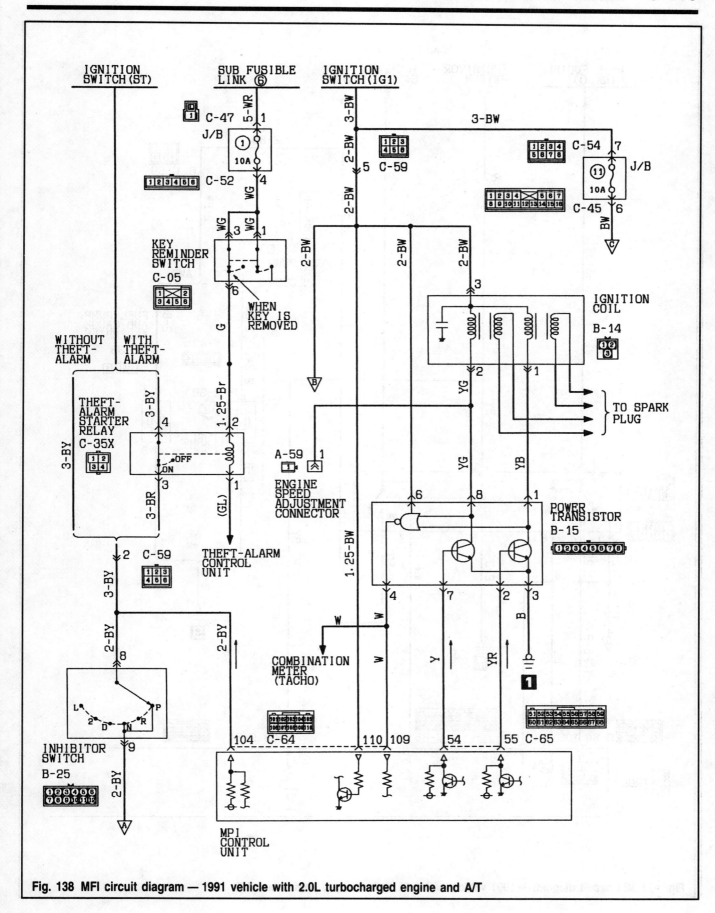

Fig. 138 MFI circuit diagram — 1991 vehicle with 2.0L turbocharged engine and A/T

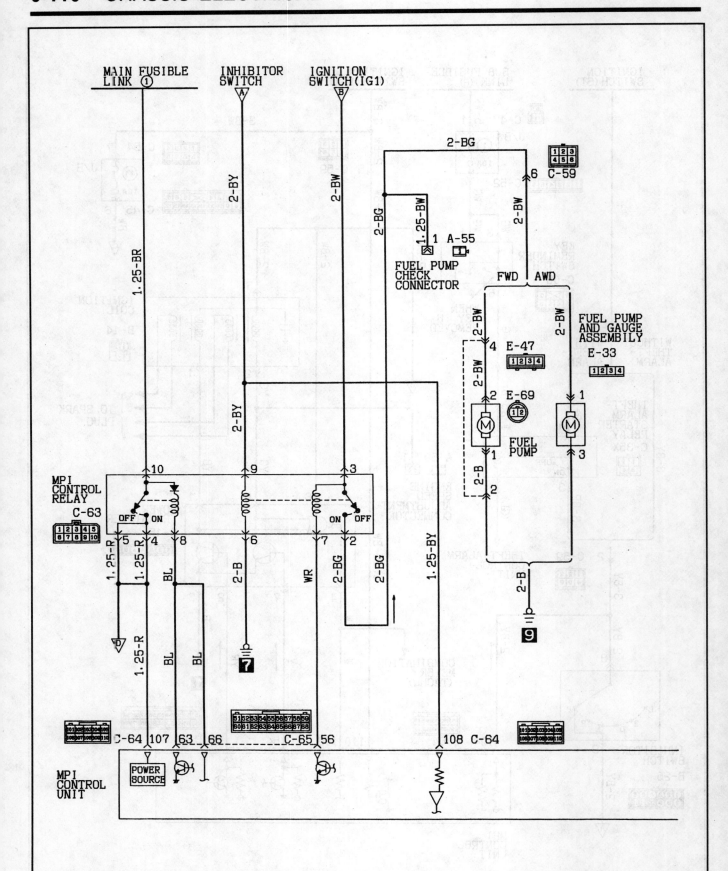

Fig. 139 MFI circuit diagram — 1991 vehicle with 2.0L turbocharged engine and A/T

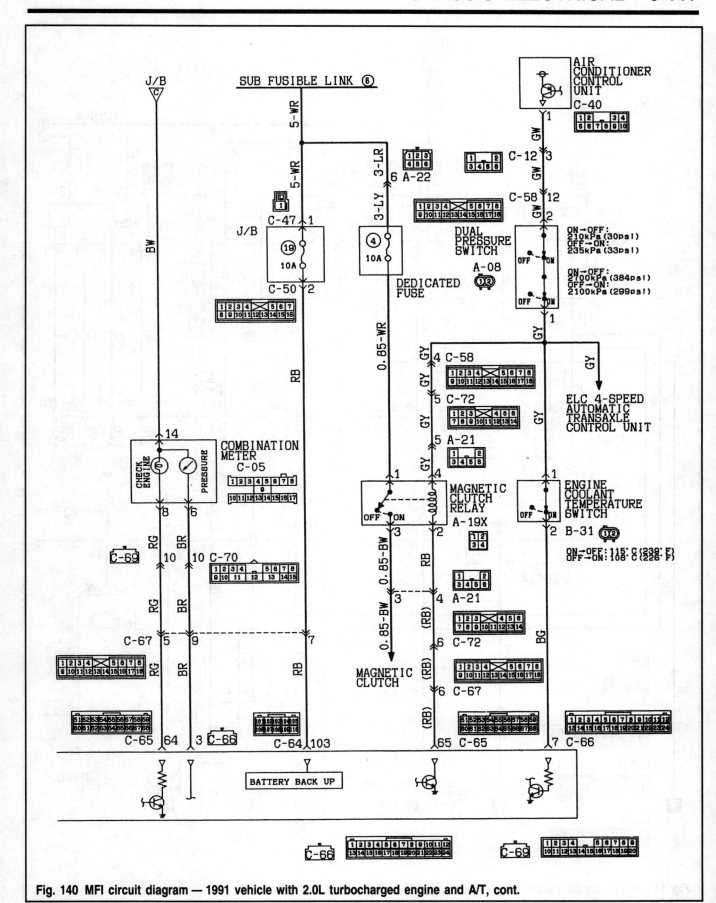

Fig. 140 MFI circuit diagram — 1991 vehicle with 2.0L turbocharged engine and A/T, cont.

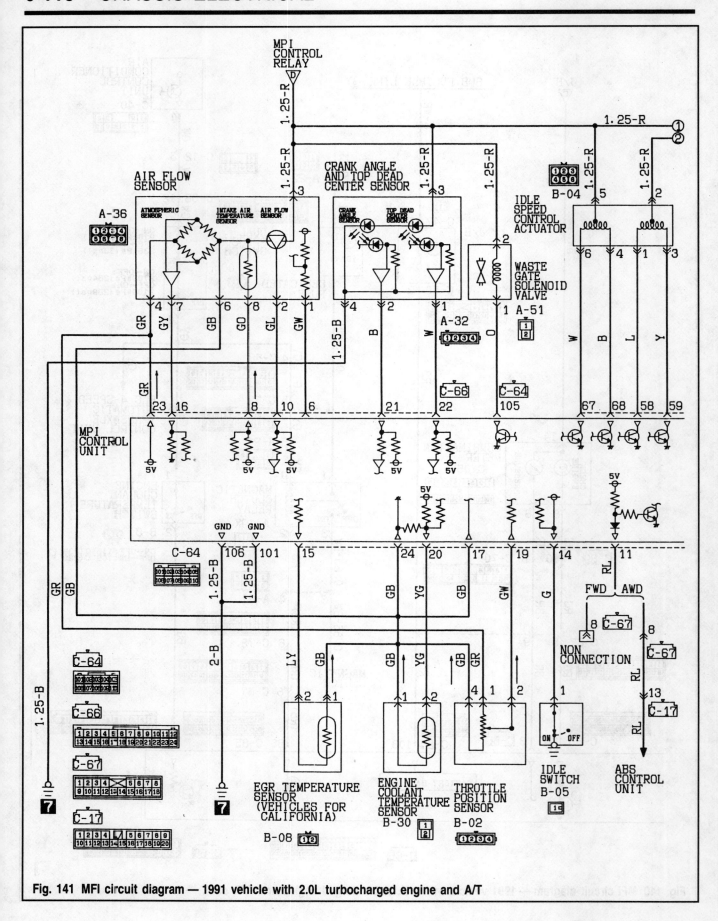

Fig. 141 MFI circuit diagram — 1991 vehicle with 2.0L turbocharged engine and A/T

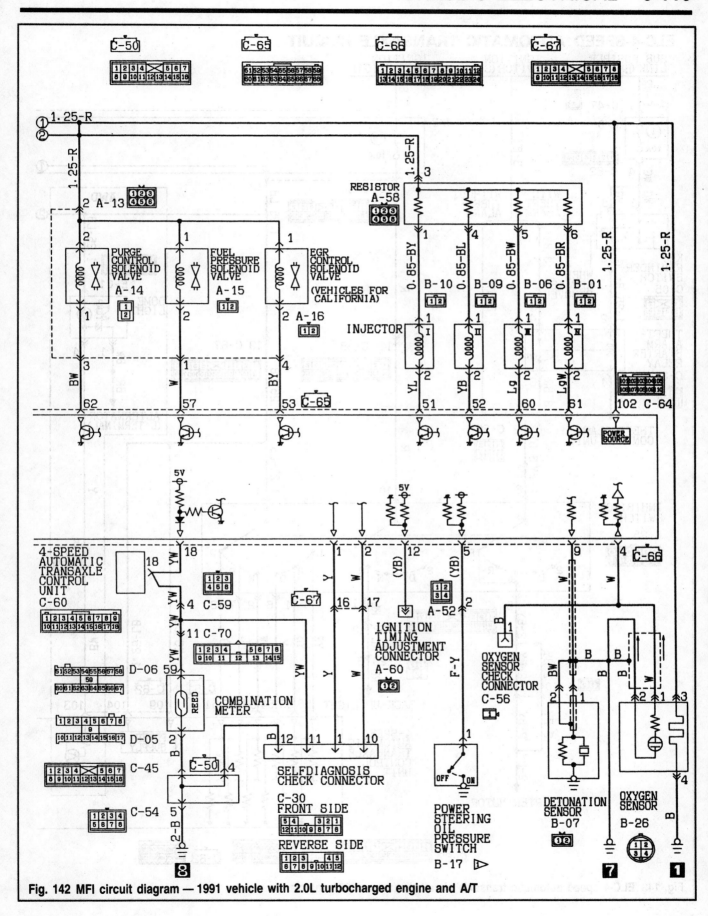

Fig. 142 MFI circuit diagram — 1991 vehicle with 2.0L turbocharged engine and A/T

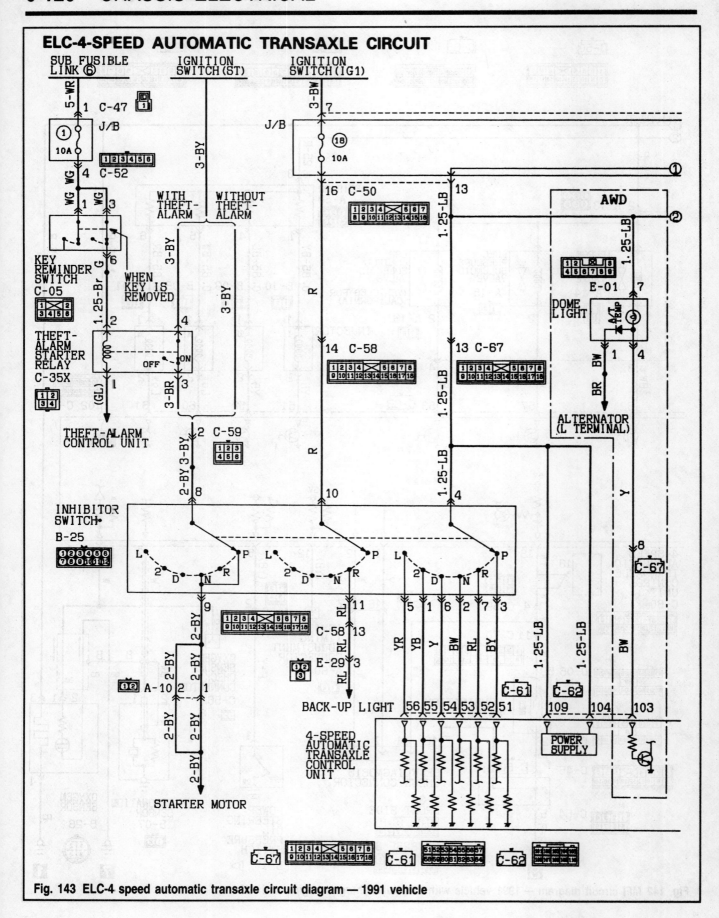

Fig. 143 ELC-4 speed automatic transaxle circuit diagram — 1991 vehicle

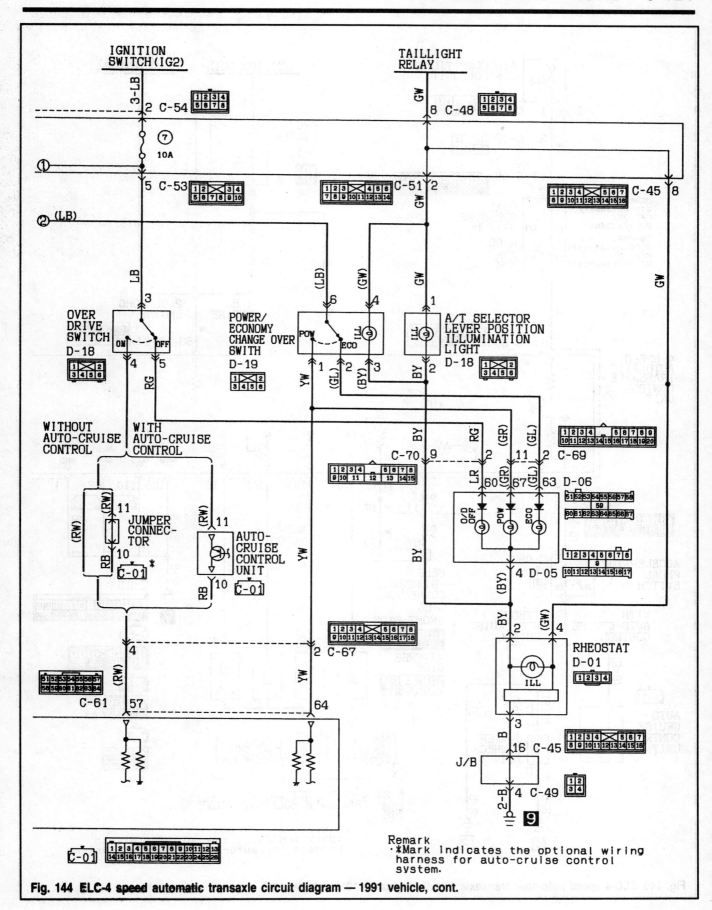

Fig. 144 ELC-4 speed automatic transaxle circuit diagram — 1991 vehicle, cont.

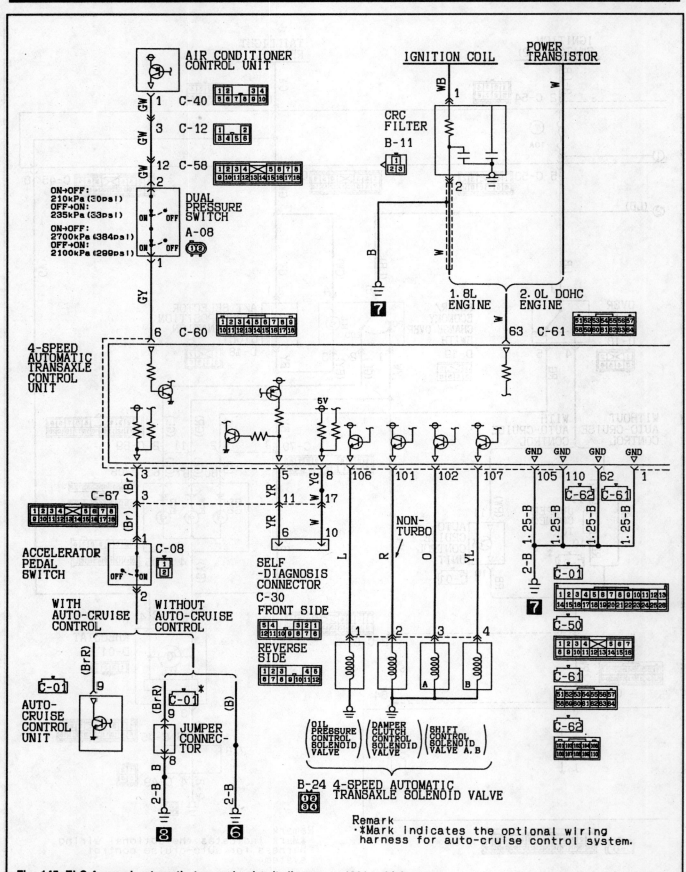

Fig. 145 ELC-4 speed automatic transaxle circuit diagram — 1991 vehicle, cont.

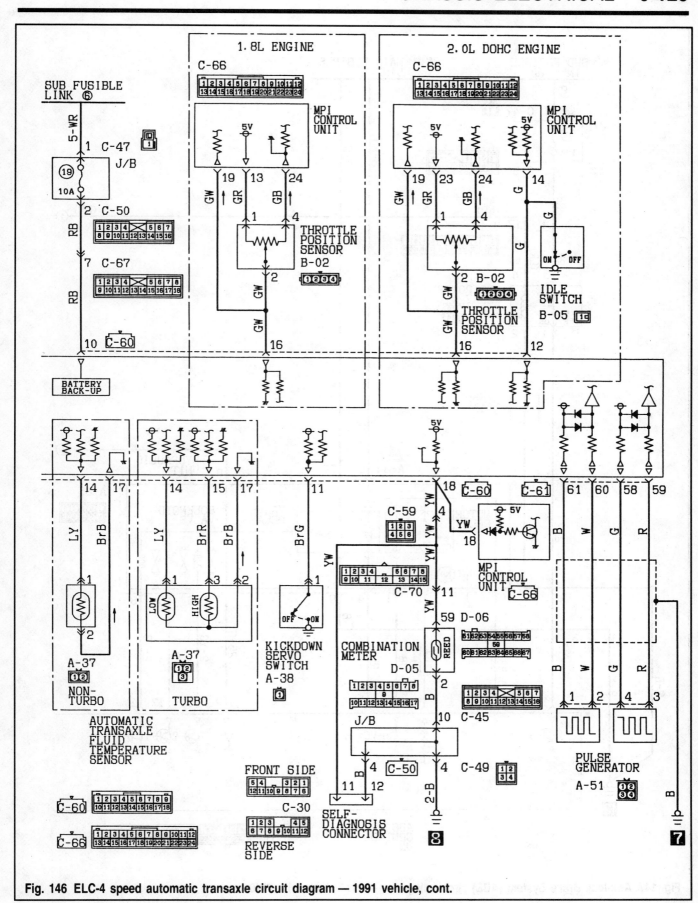

Fig. 146 ELC-4 speed automatic transaxle circuit diagram — 1991 vehicle, cont.

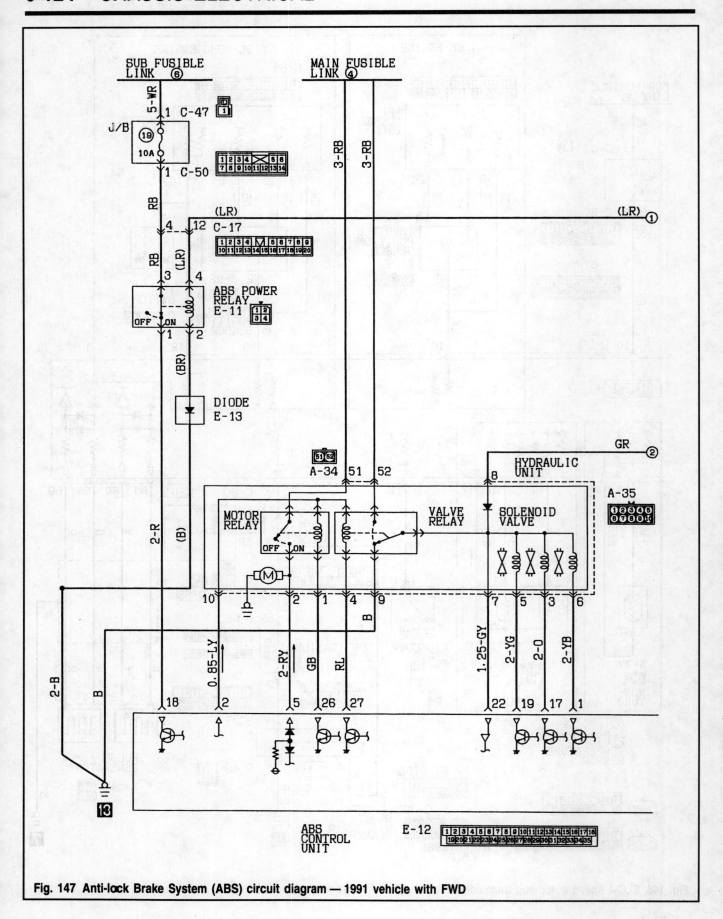

Fig. 147 Anti-lock Brake System (ABS) circuit diagram — 1991 vehicle with FWD

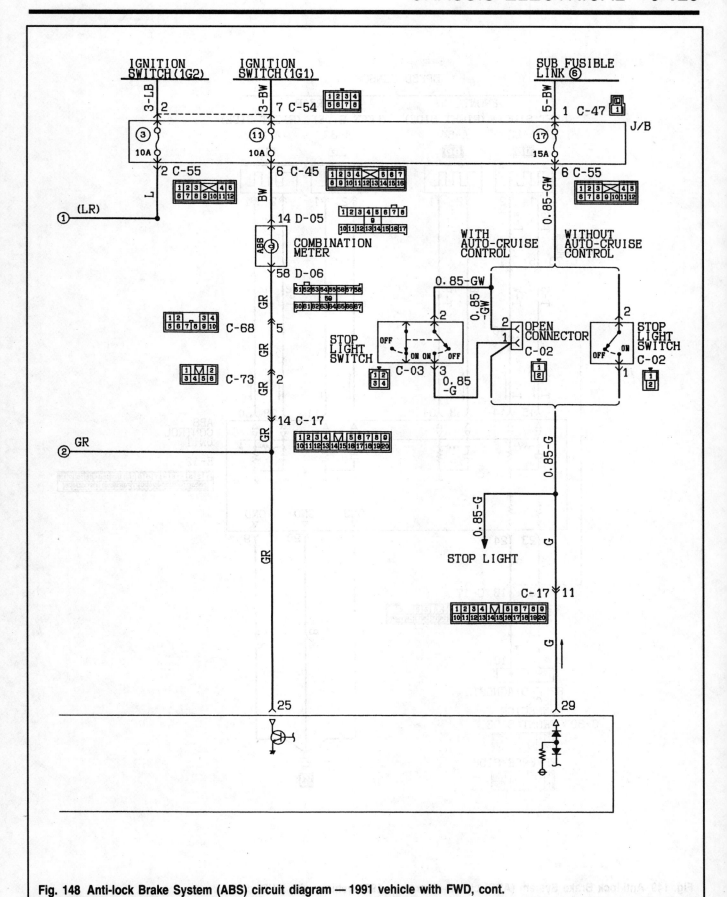

Fig. 148 Anti-lock Brake System (ABS) circuit diagram — 1991 vehicle with FWD, cont.

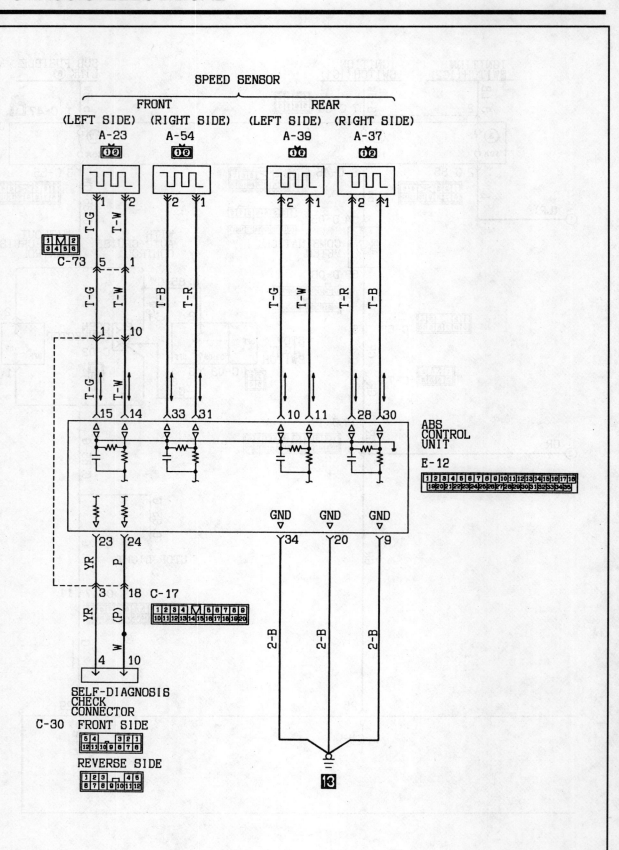

Fig. 149 Anti-lcck Brake System (ABS) circuit diagram — 1991 vehicle with FWD, cont.

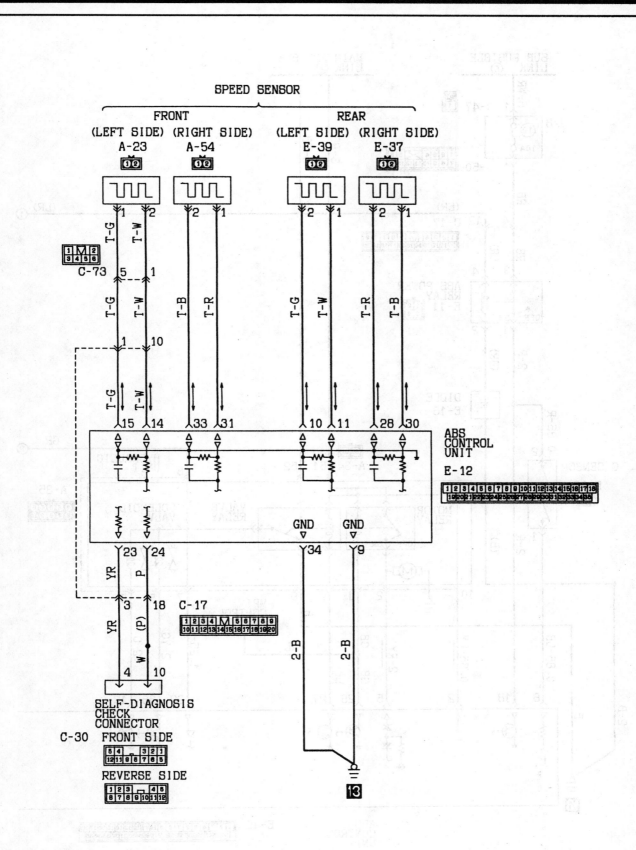

Fig. 150 Anti-lock Brake System (ABS) circuit diagram — 1991 vehicle with AWD

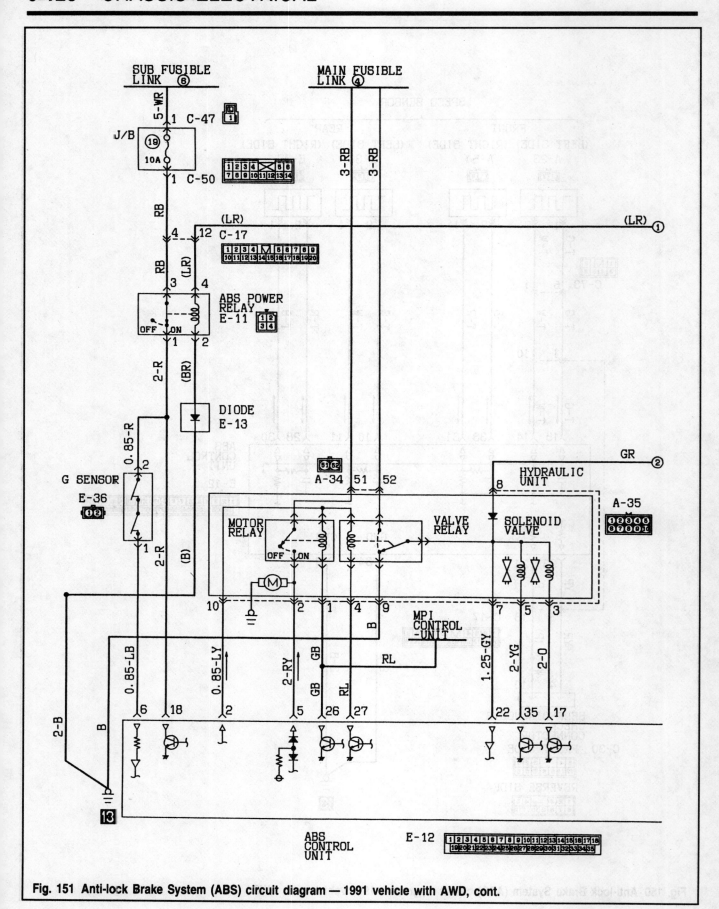

Fig. 151 Anti-lock Brake System (ABS) circuit diagram — 1991 vehicle with AWD, cont.

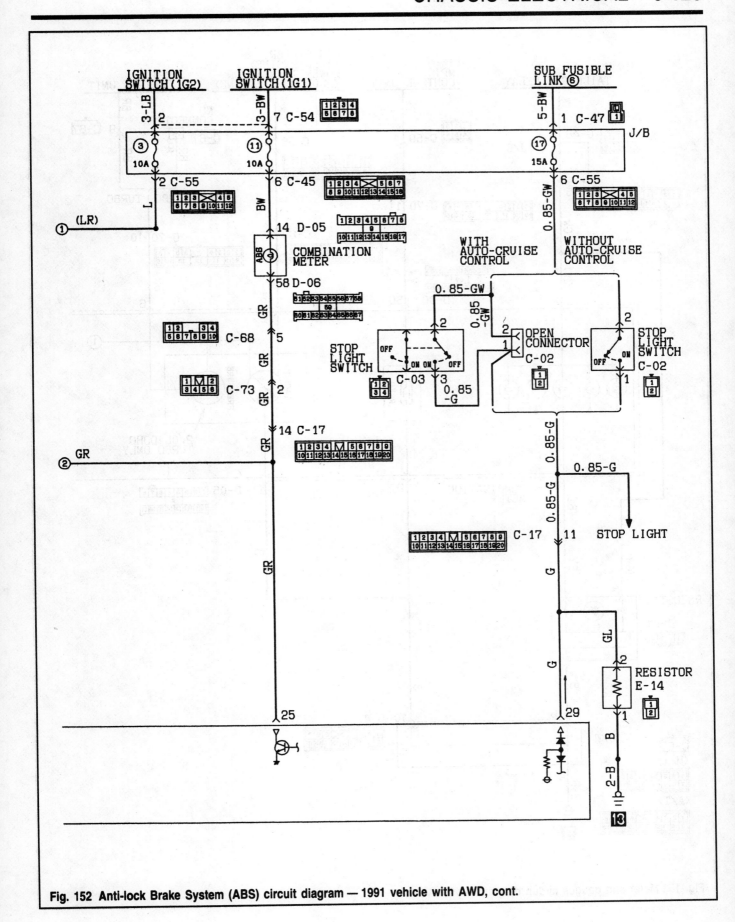

Fig. 152 Anti-lock Brake System (ABS) circuit diagram — 1991 vehicle with AWD, cont.

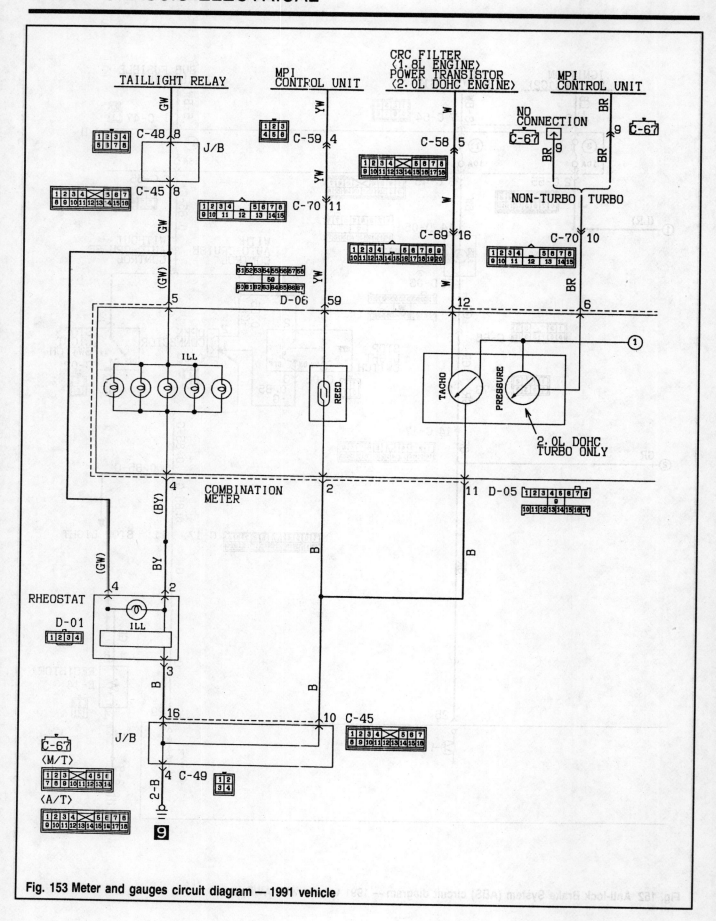

Fig. 153 Meter and gauges circuit diagram — 1991 vehicle

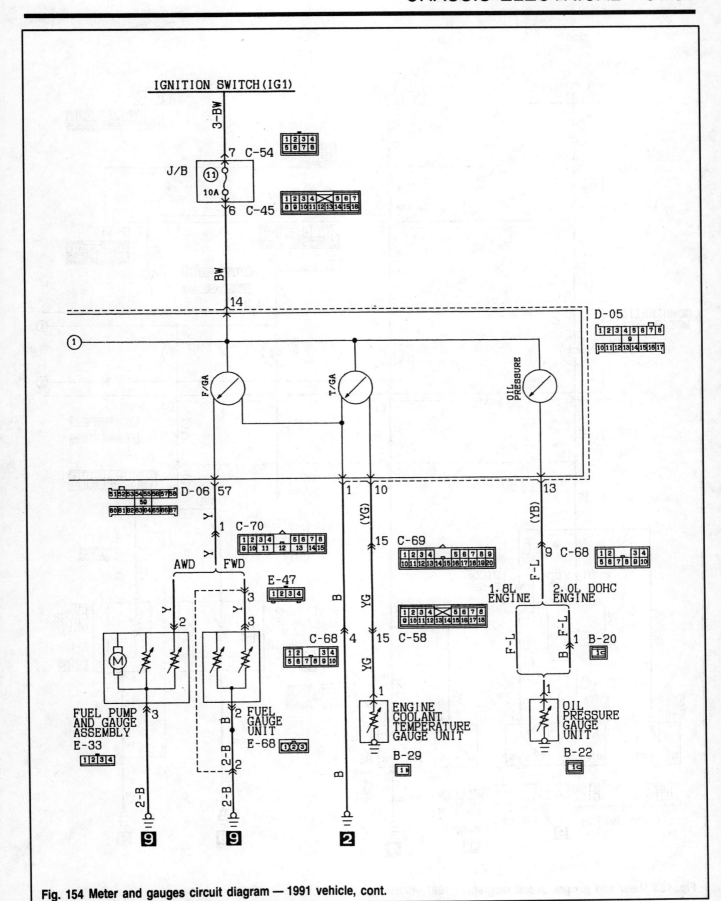

Fig. 154 Meter and gauges circuit diagram — 1991 vehicle, cont.

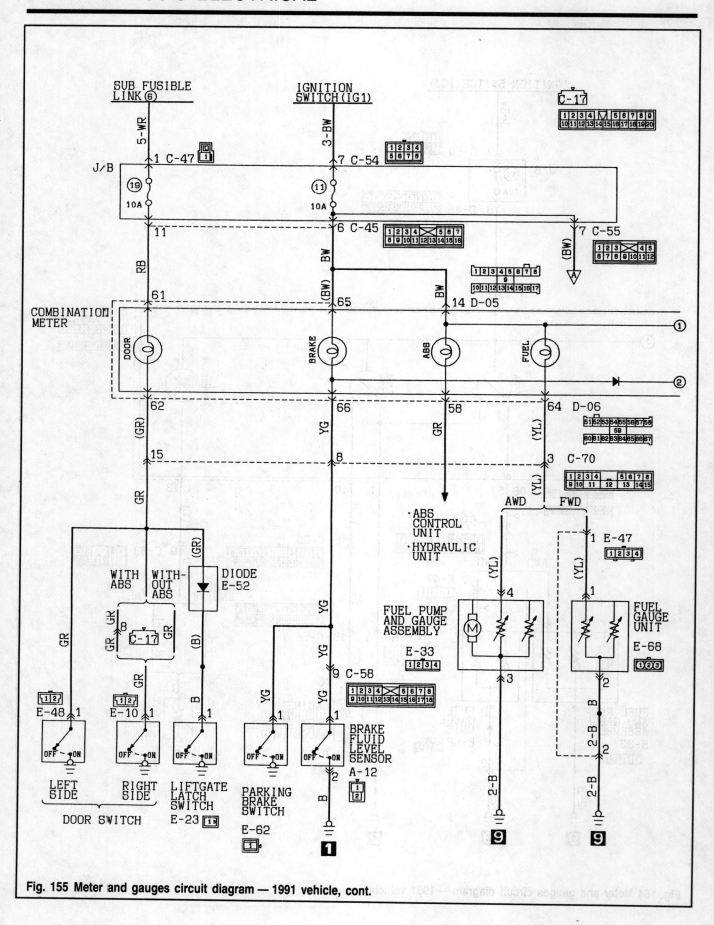

Fig. 155 Meter and gauges circuit diagram — 1991 vehicle, cont.

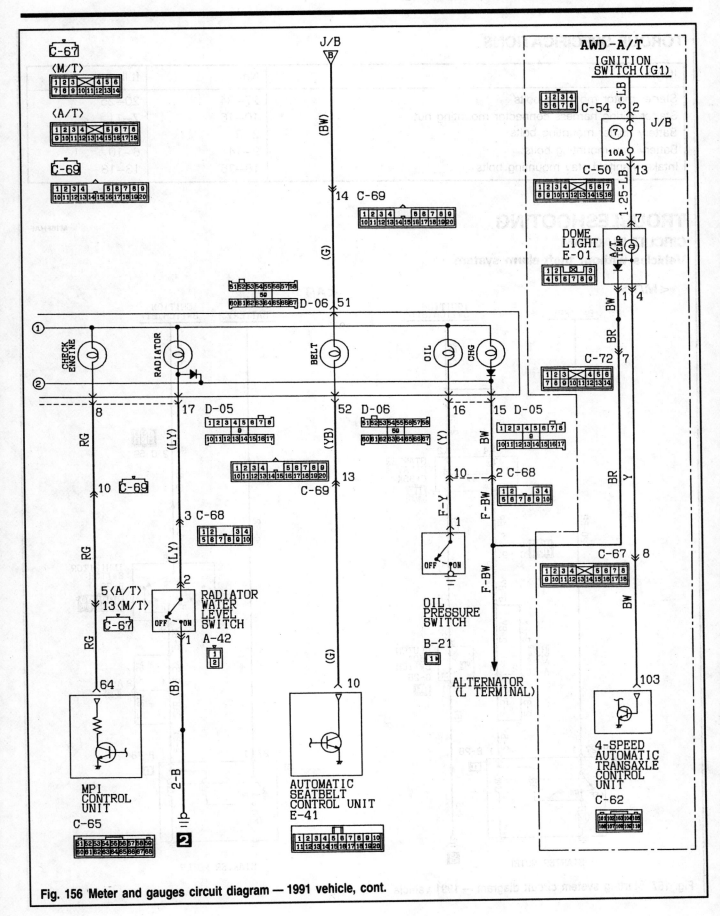

Fig. 156 Meter and gauges circuit diagram — 1991 vehicle, cont.

TORQUE SPECIFICATIONS

Items	Nm	ft.lbs.
Starter motor mounting bolts	27—34	20—25
Starter wiring harness connector mounting nut	10—16	7—11
Battery holder mounting bolts	2—3	1.5—2
Battery tray mounting bolts	9—14	6—10
Intake manifold stay mounting bolts	18—25	13—18

TROUBLESHOOTING

CIRCUIT DIAGRAM
Vehicles without theft-alarm system

M16FHAF

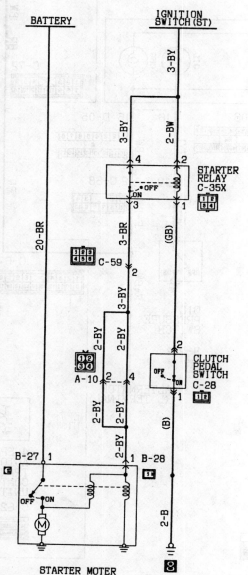

Fig. 157 Starting system circuit diagram — 1991 vehicle

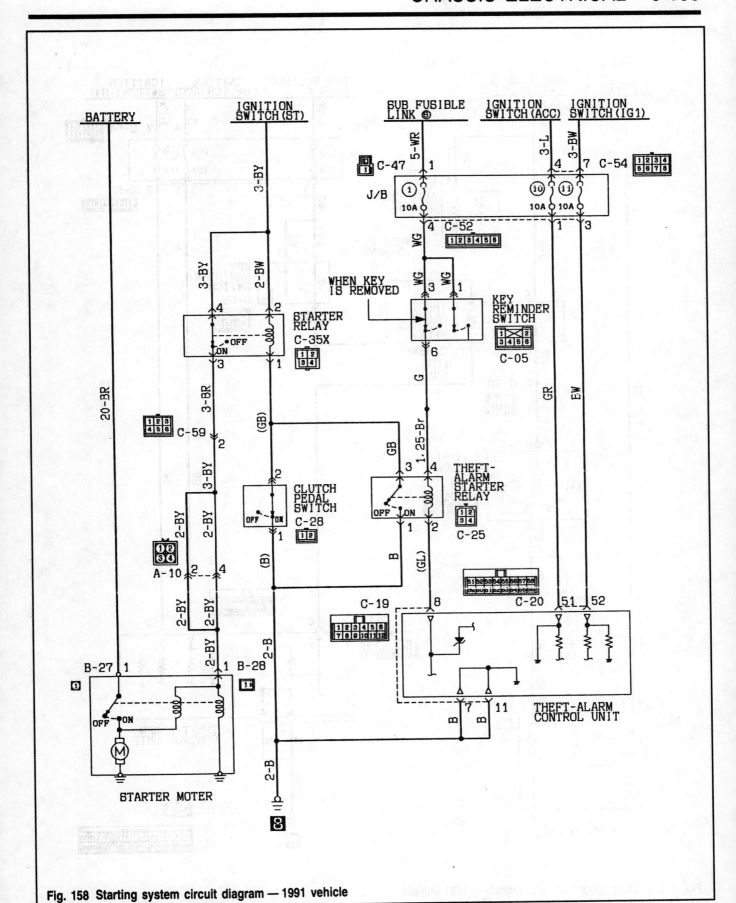

Fig. 158 Starting system circuit diagram — 1991 vehicle

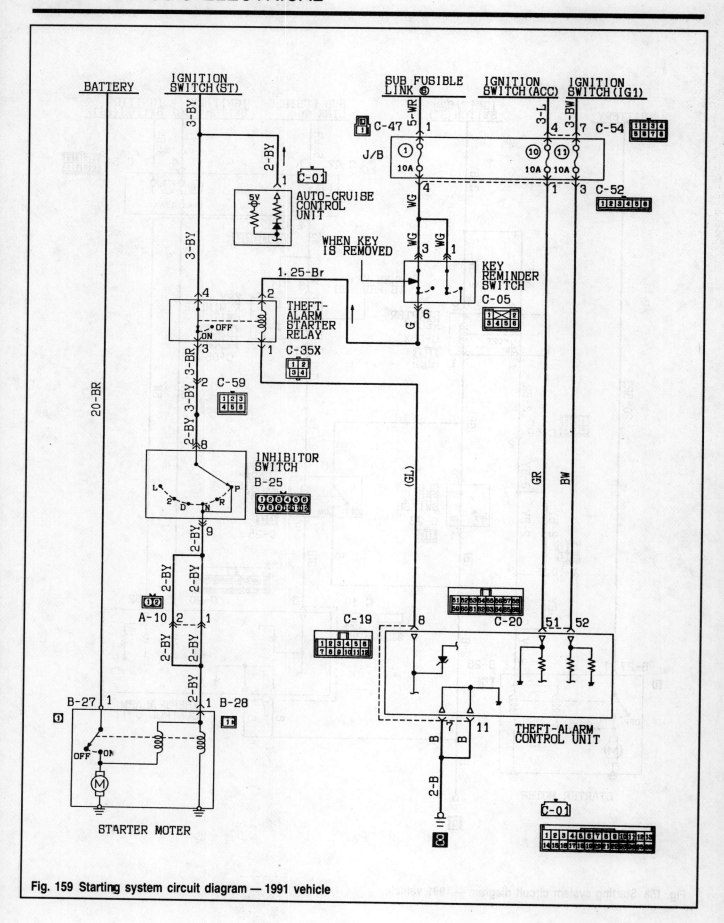

Fig. 159 Starting system circuit diagram — 1991 vehicle

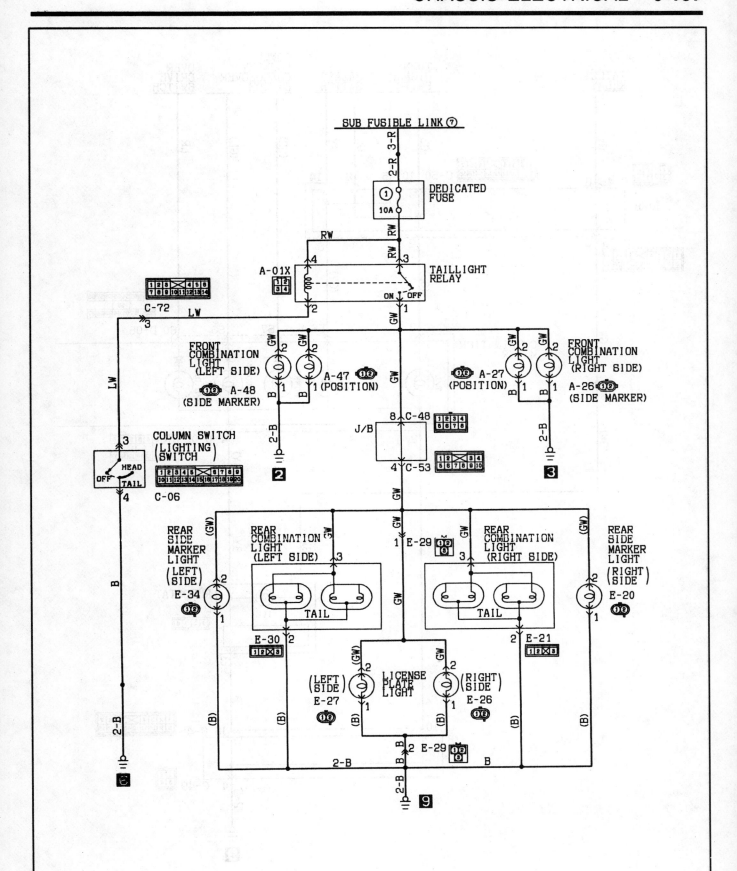

Fig. 160 Tail light, position light, side marker and license plate light circuit diagram — 1991 vehicle

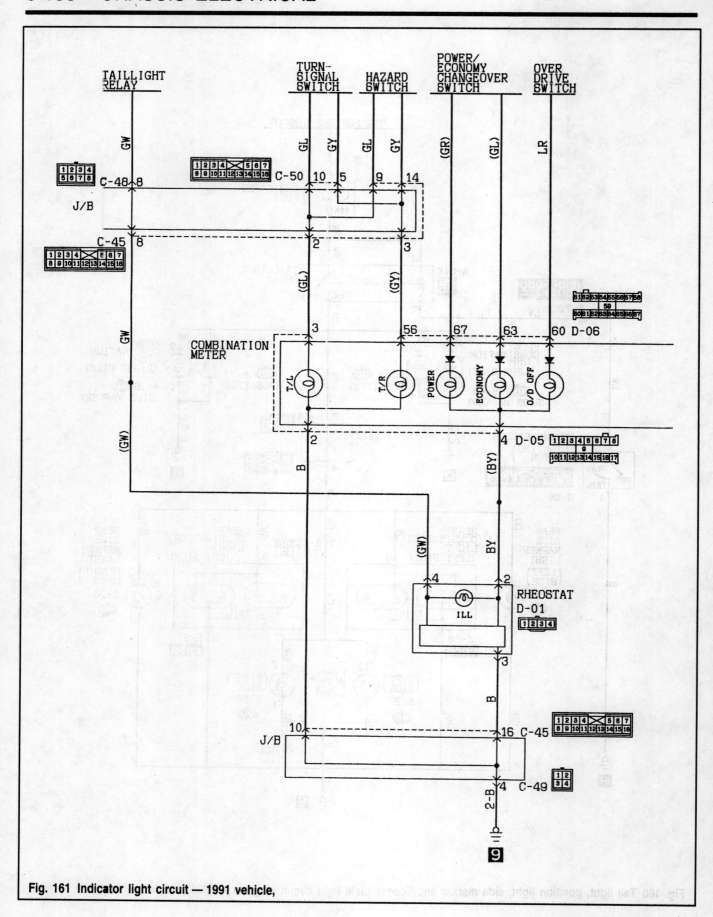

Fig. 161 Indicator light circuit — 1991 vehicle,

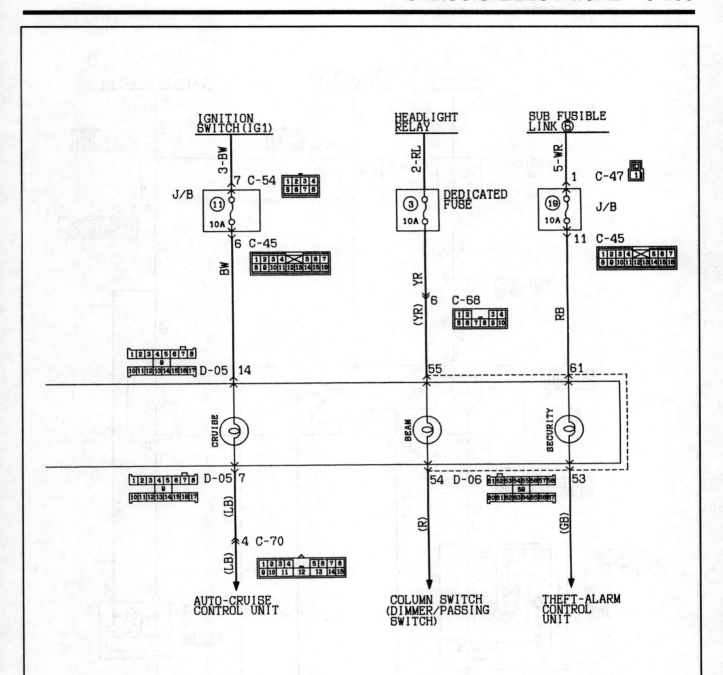

Fig. 162 Indicator light circuit — 1991 vehicle, cont.

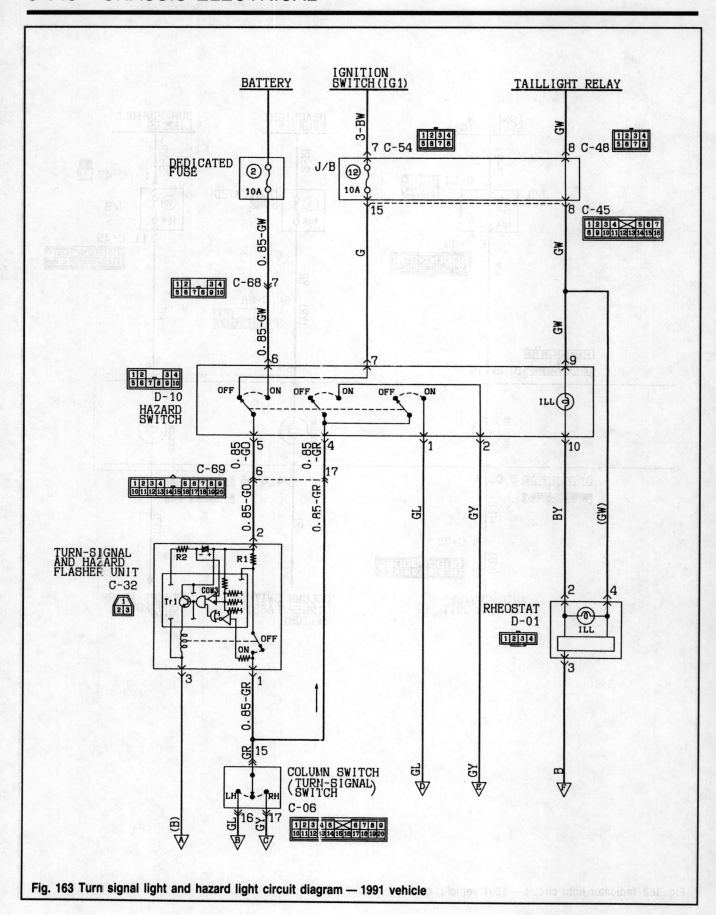

Fig. 163 Turn signal light and hazard light circuit diagram — 1991 vehicle

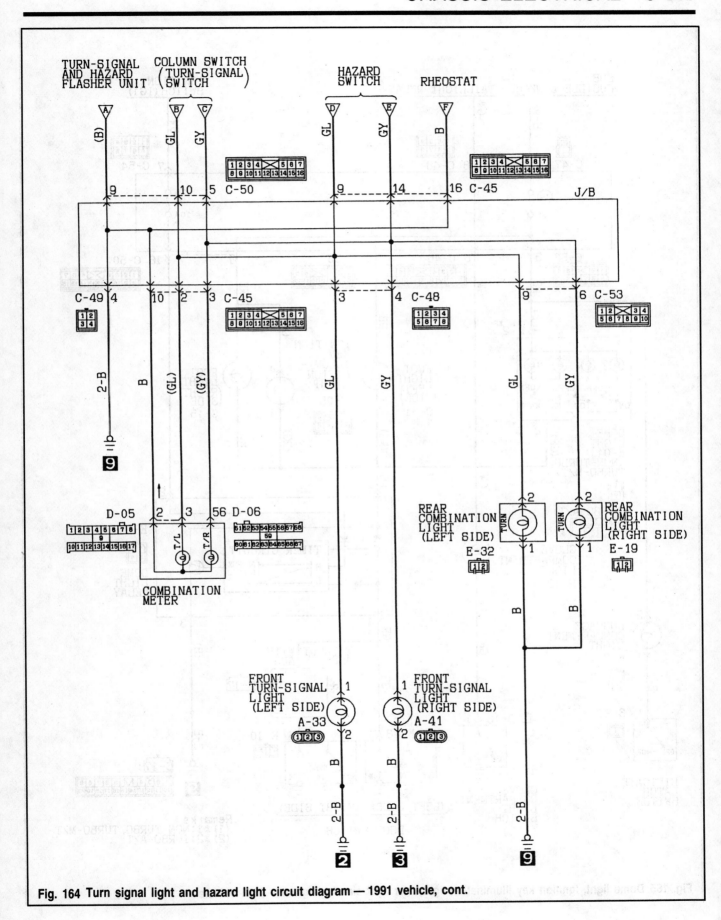

Fig. 164 Turn signal light and hazard light circuit diagram — 1991 vehicle, cont.

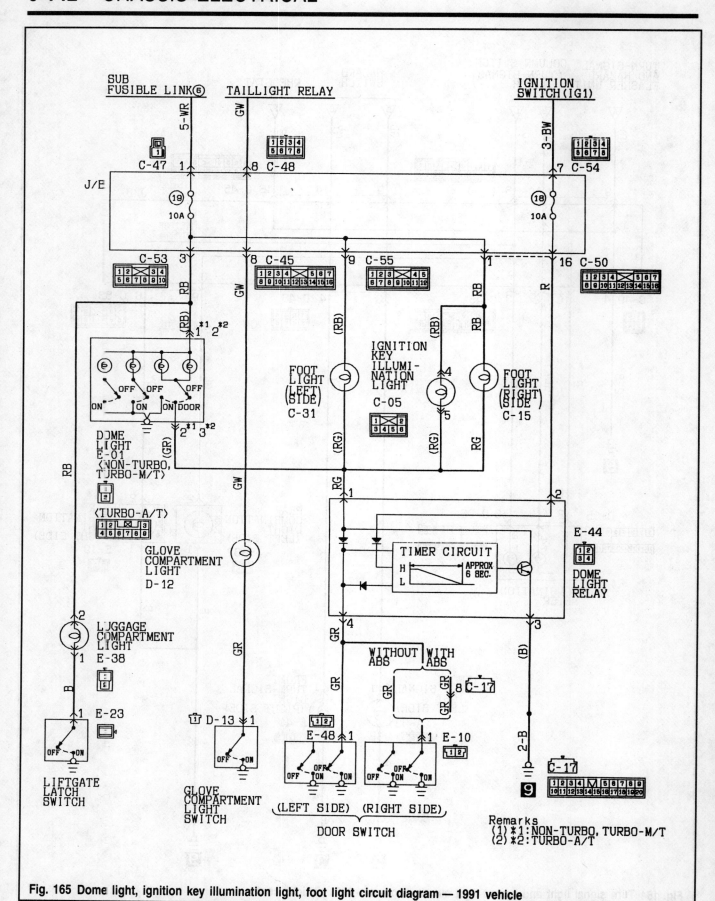

Fig. 165 Dome light, ignition key illumination light, foot light circuit diagram — 1991 vehicle

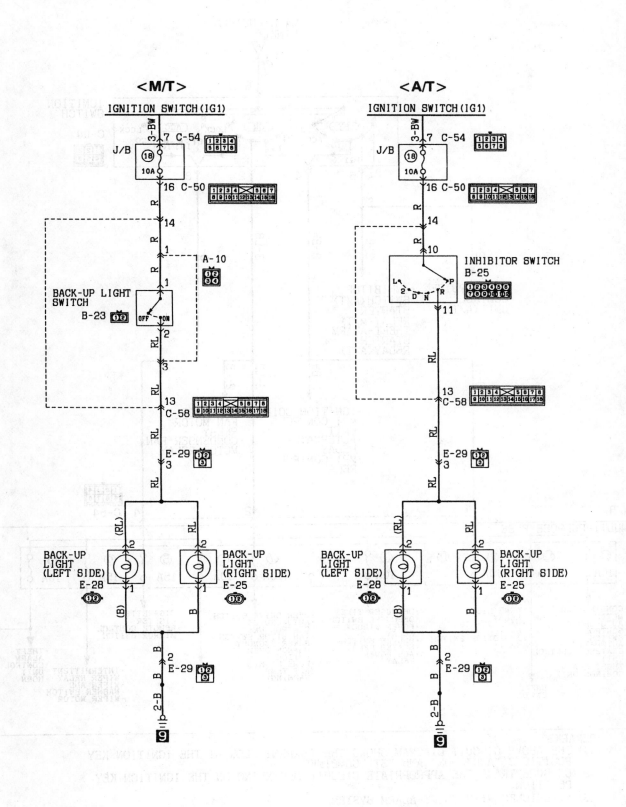

Fig. 166 Back-up light circuit diagram — 1991 vehicle

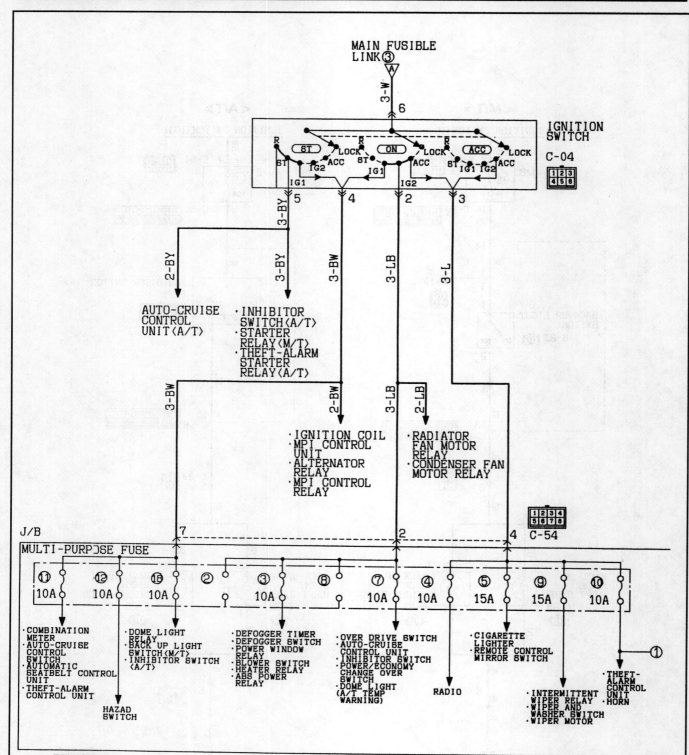

Fig. 167 **Power distribution circuit diagram — 1991 vehicle**

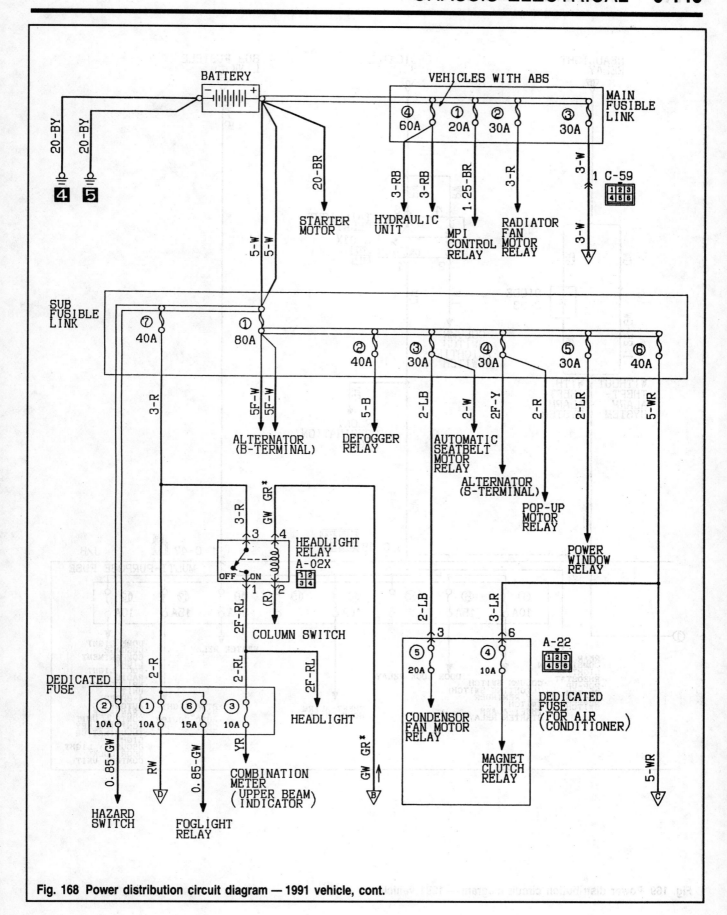

Fig. 168 Power distribution circuit diagram — 1991 vehicle, cont.

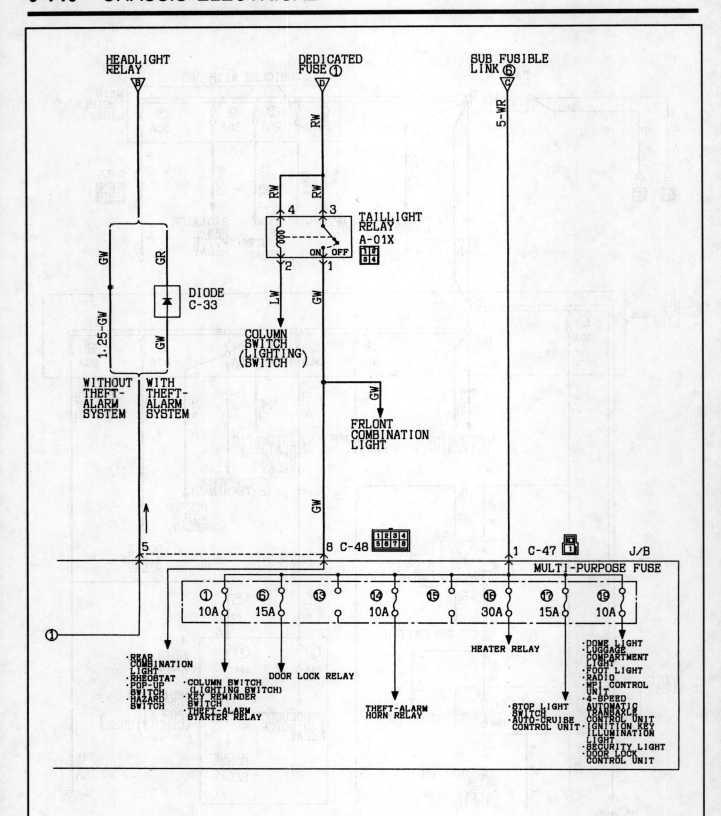

Fig. 169 Power distribution circuit diagram — 1991 vehicle

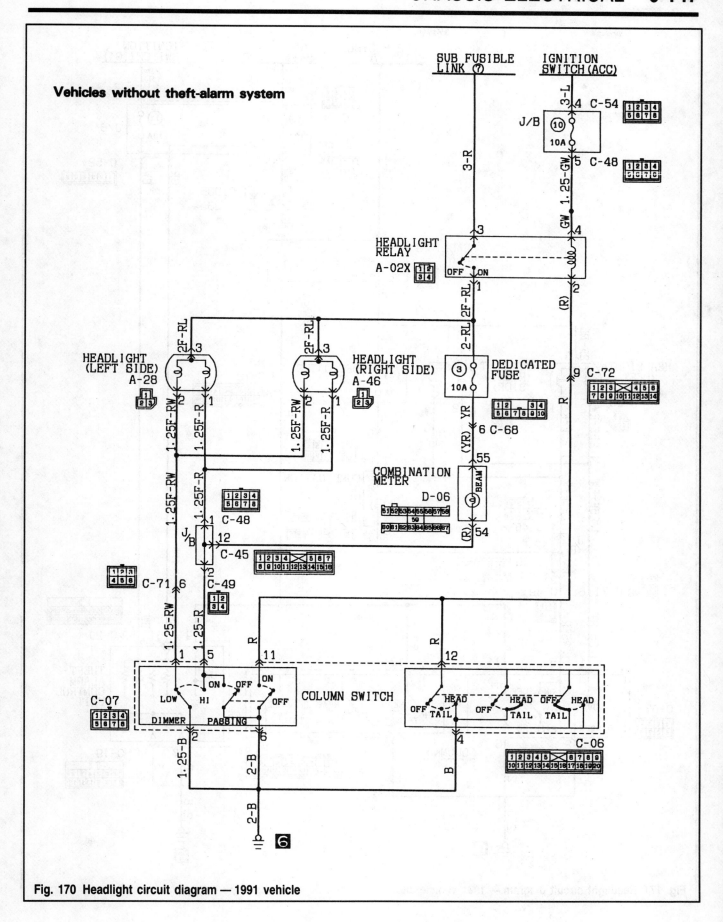

Fig. 170 Headlight circuit diagram — 1991 vehicle

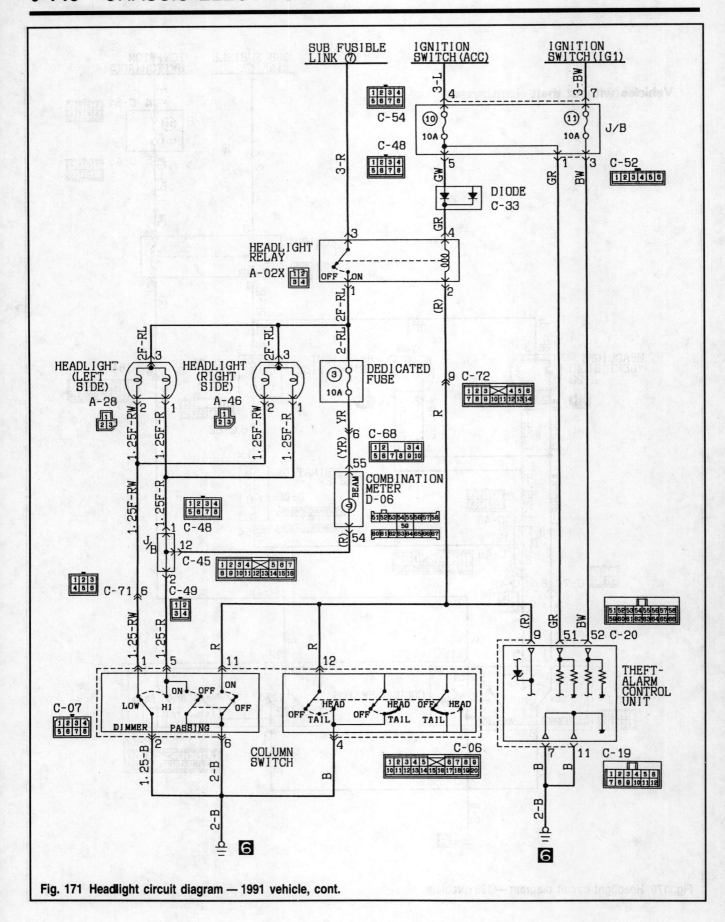

Fig. 171 Headlight circuit diagram — 1991 vehicle, cont.

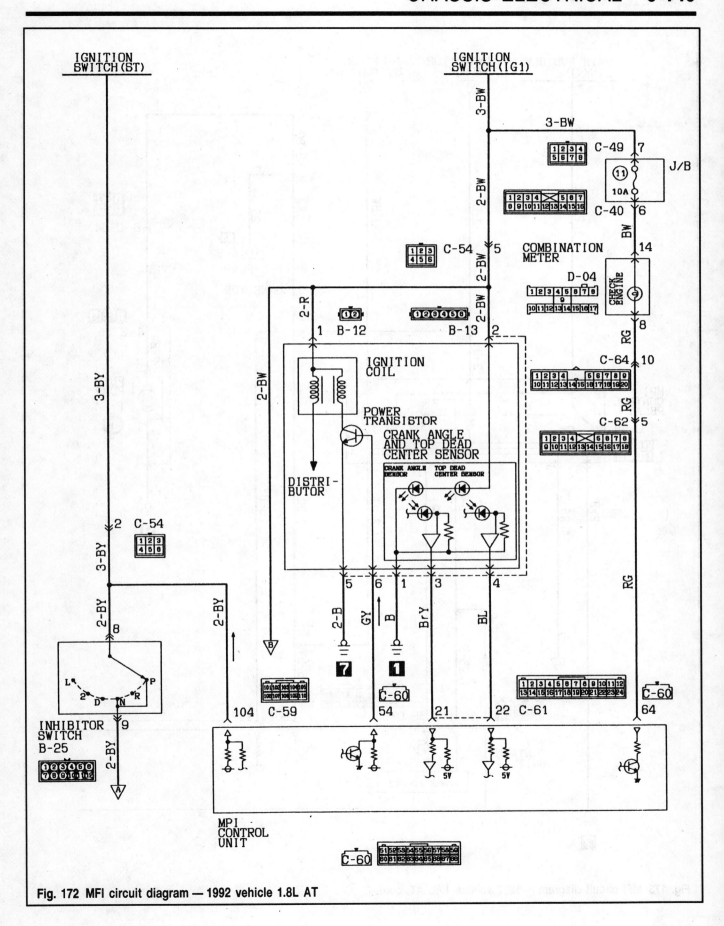

Fig. 172 MFI circuit diagram — 1992 vehicle 1.8L AT

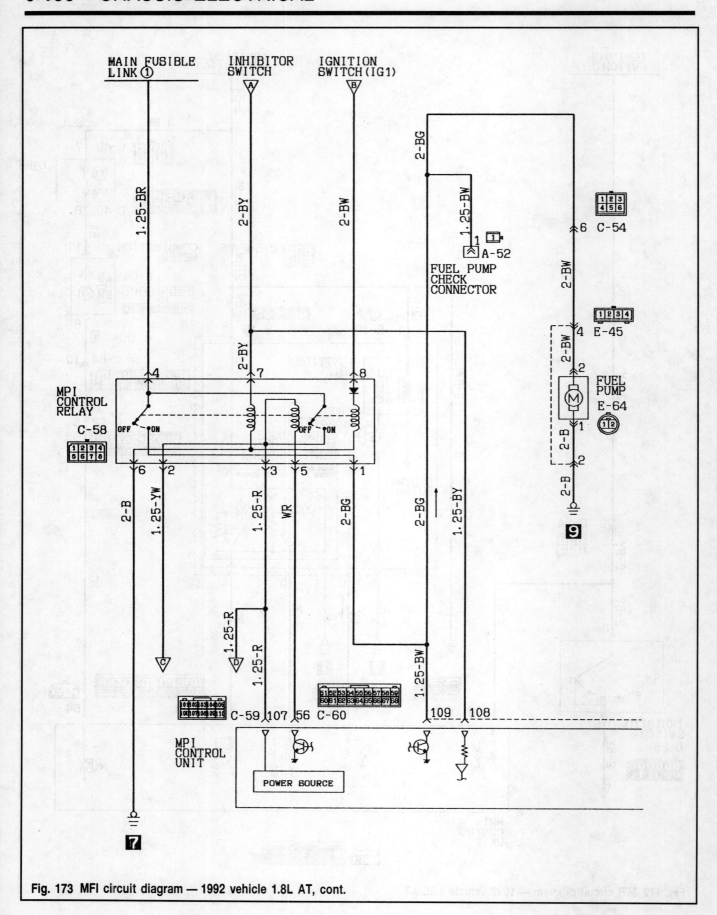

Fig. 173 MFI circuit diagram — 1992 vehicle 1.8L AT, cont.

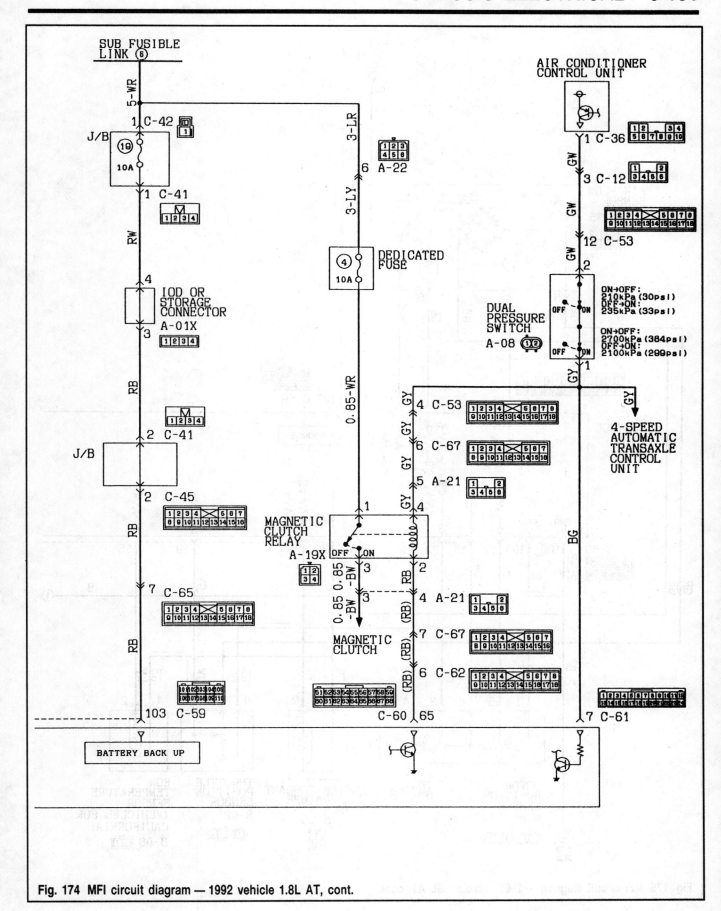

Fig. 174 MFI circuit diagram — 1992 vehicle 1.8L AT, cont.

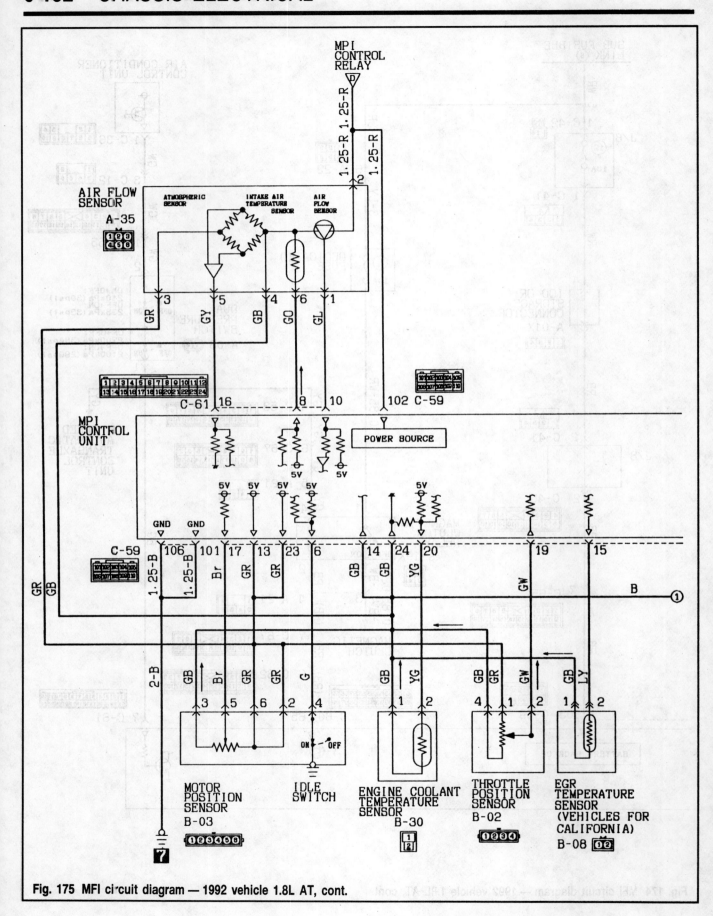

Fig. 175 MFI circuit diagram — 1992 vehicle 1.8L AT, cont.

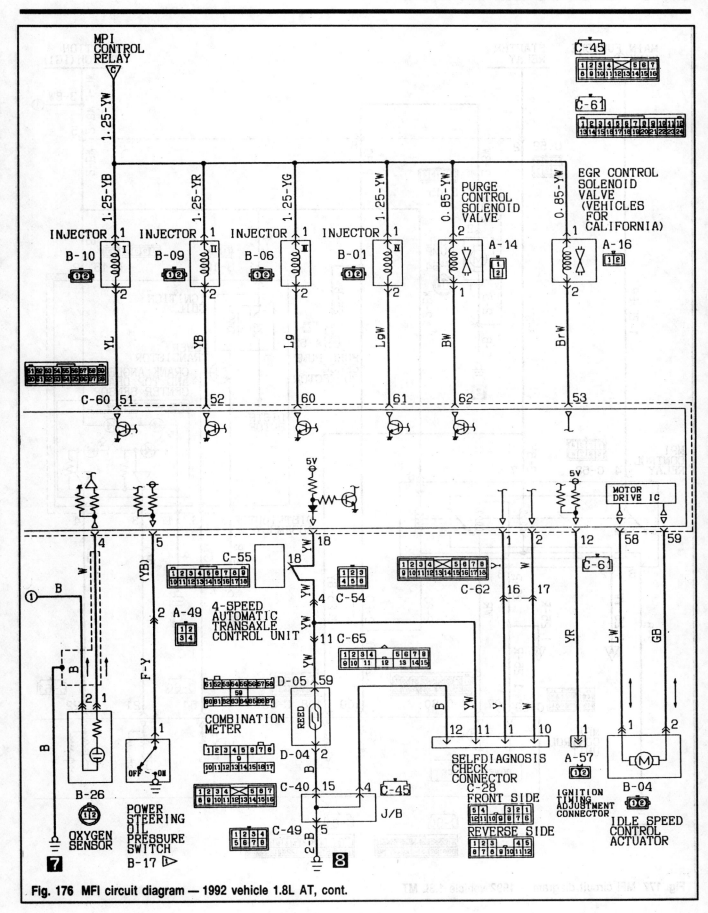

Fig. 176 MFI circuit diagram — 1992 vehicle 1.8L AT, cont.

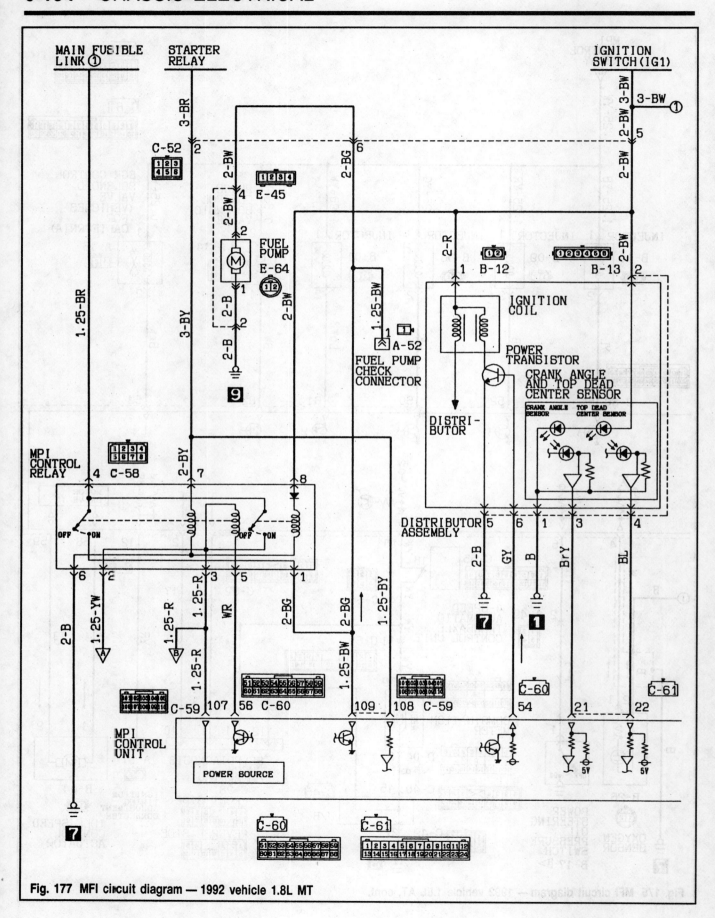

Fig. 177 MFI circuit diagram — 1992 vehicle 1.8L MT

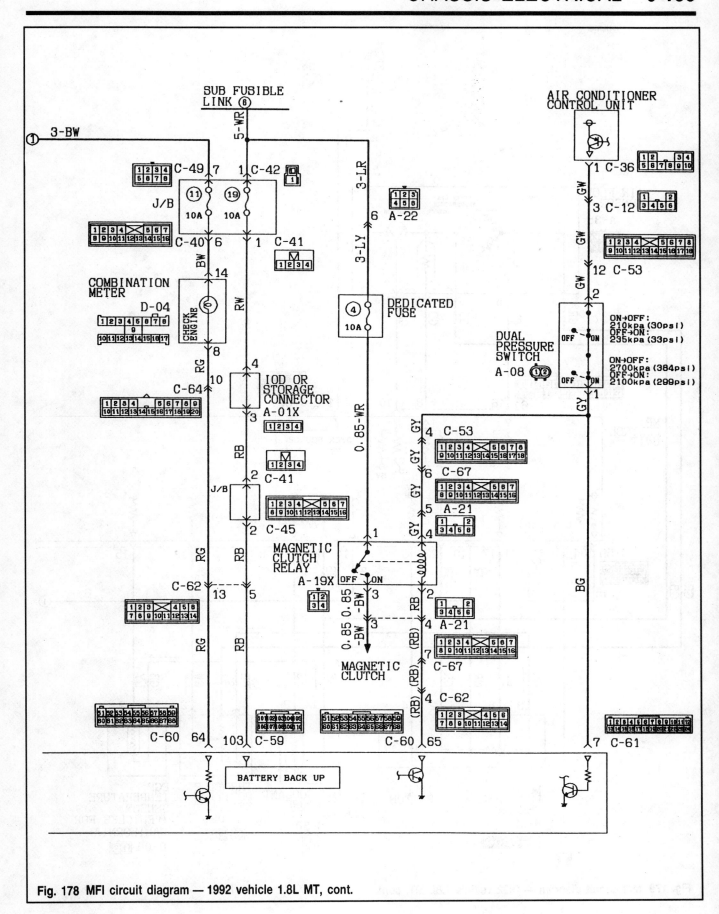

Fig. 178 MFI circuit diagram — 1992 vehicle 1.8L MT, cont.

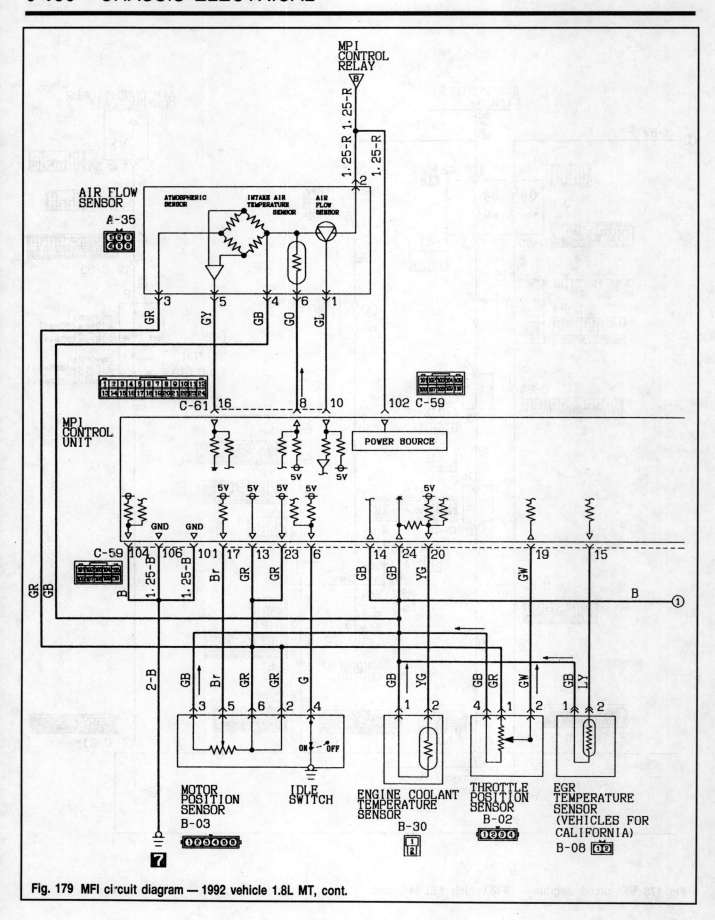

Fig. 179 MFI circuit diagram — 1992 vehicle 1.8L MT, cont.

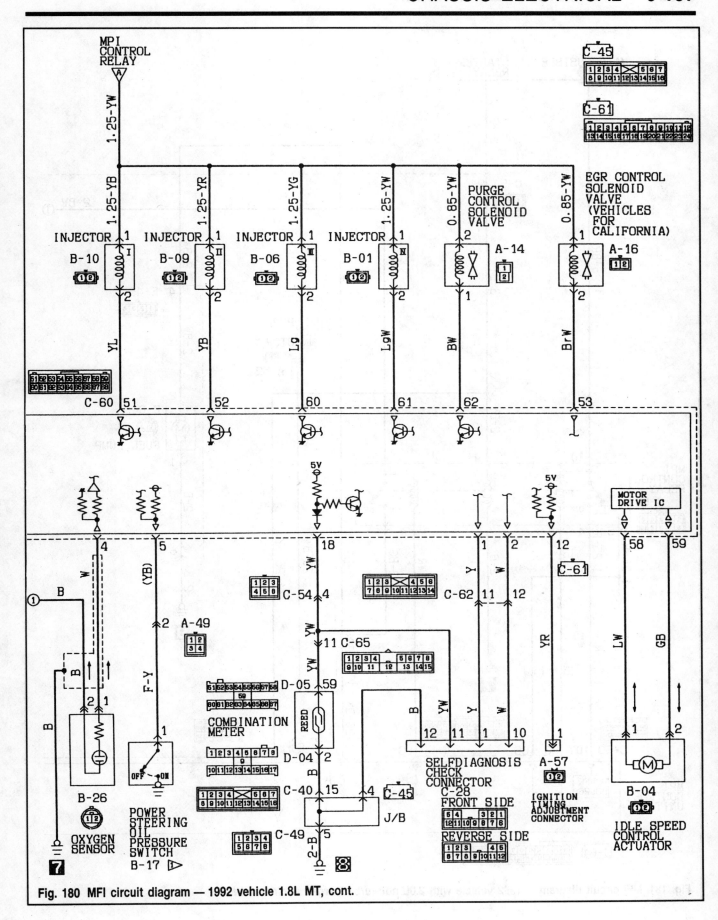

Fig. 180 MFI circuit diagram — 1992 vehicle 1.8L MT, cont.

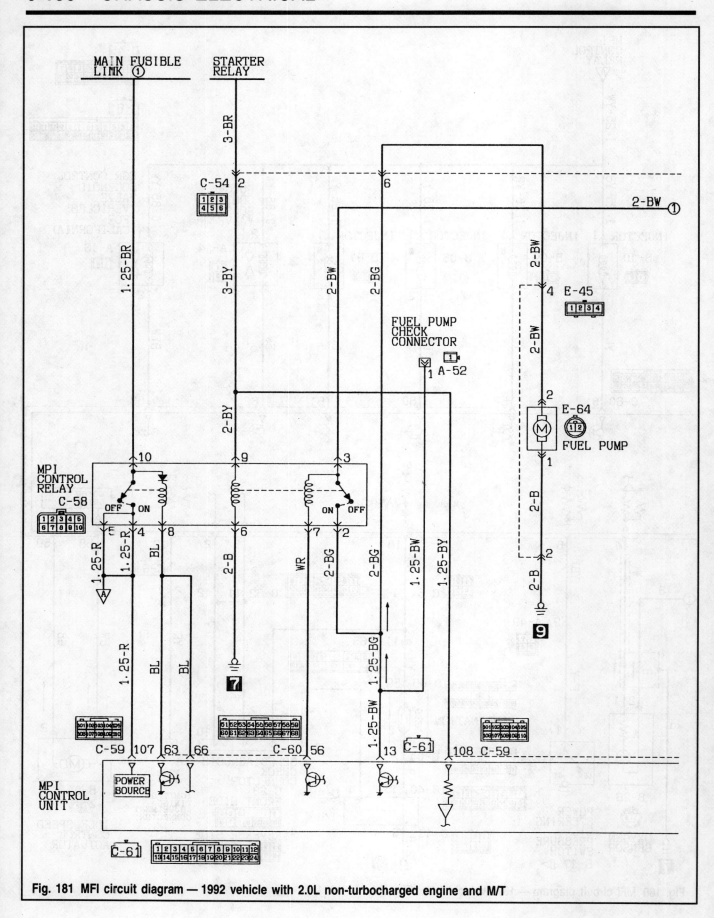

Fig. 181 MFI circuit diagram — 1992 vehicle with 2.0L non-turbocharged engine and M/T

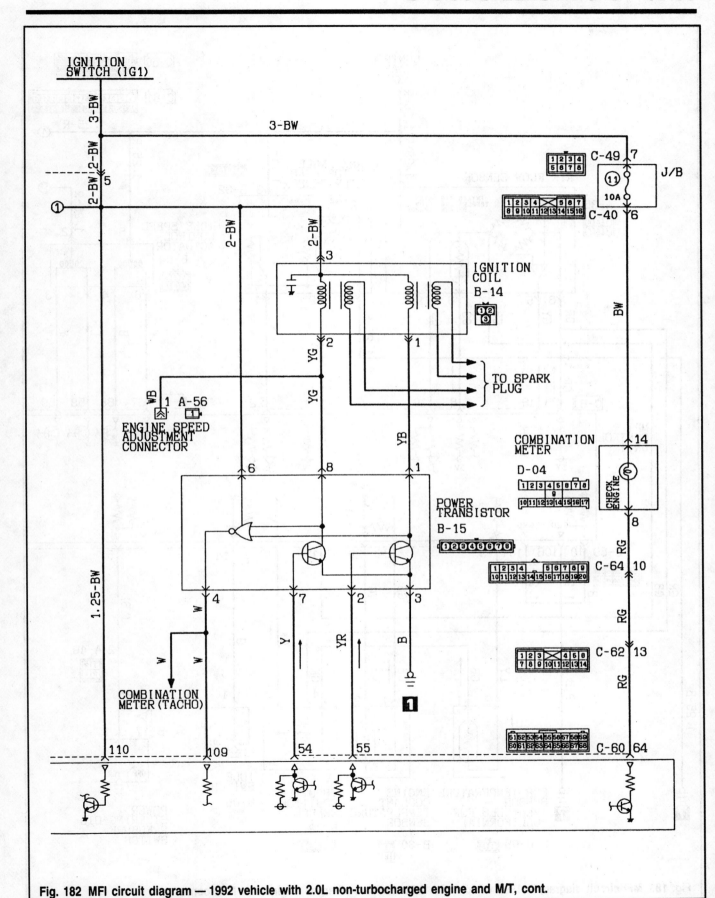

Fig. 182 MFI circuit diagram — 1992 vehicle with 2.0L non-turbocharged engine and M/T, cont.

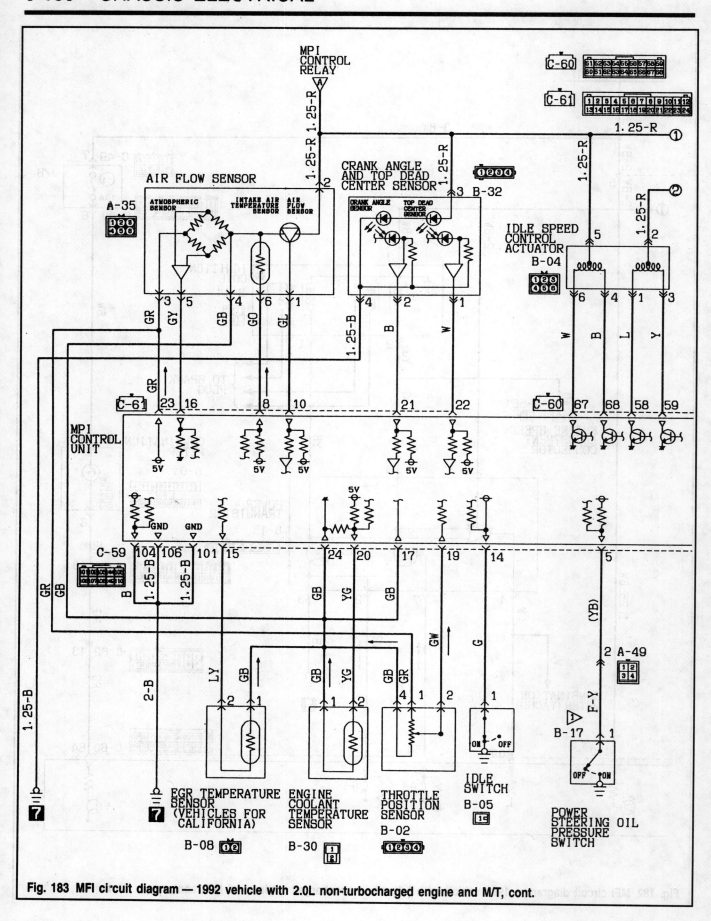

Fig. 183 MFI circuit diagram — 1992 vehicle with 2.0L non-turbocharged engine and M/T, cont.

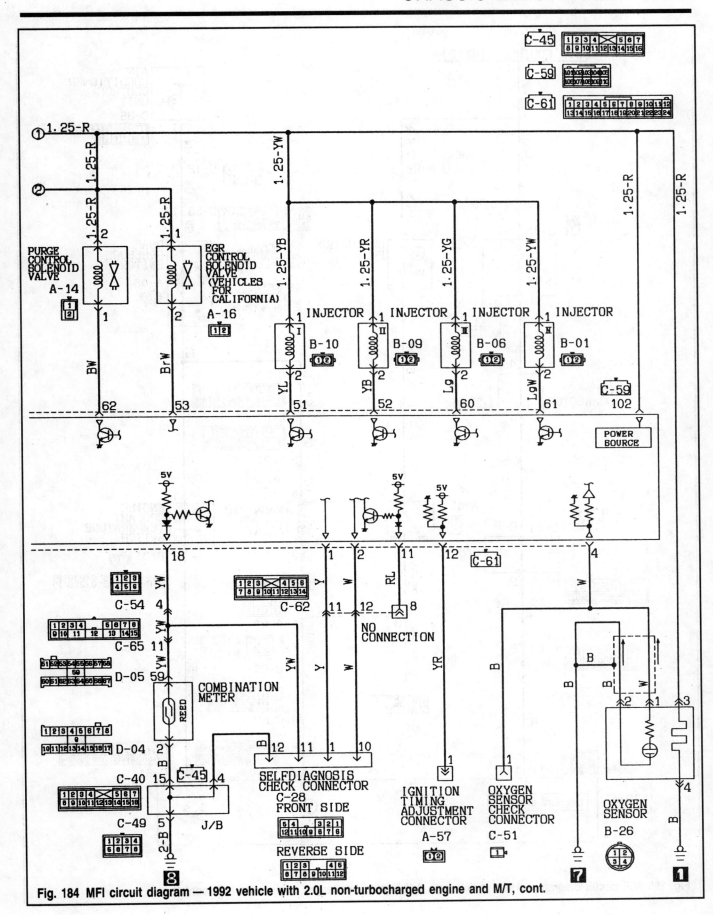

Fig. 184 MFI circuit diagram — 1992 vehicle with 2.0L non-turbocharged engine and M/T, cont.

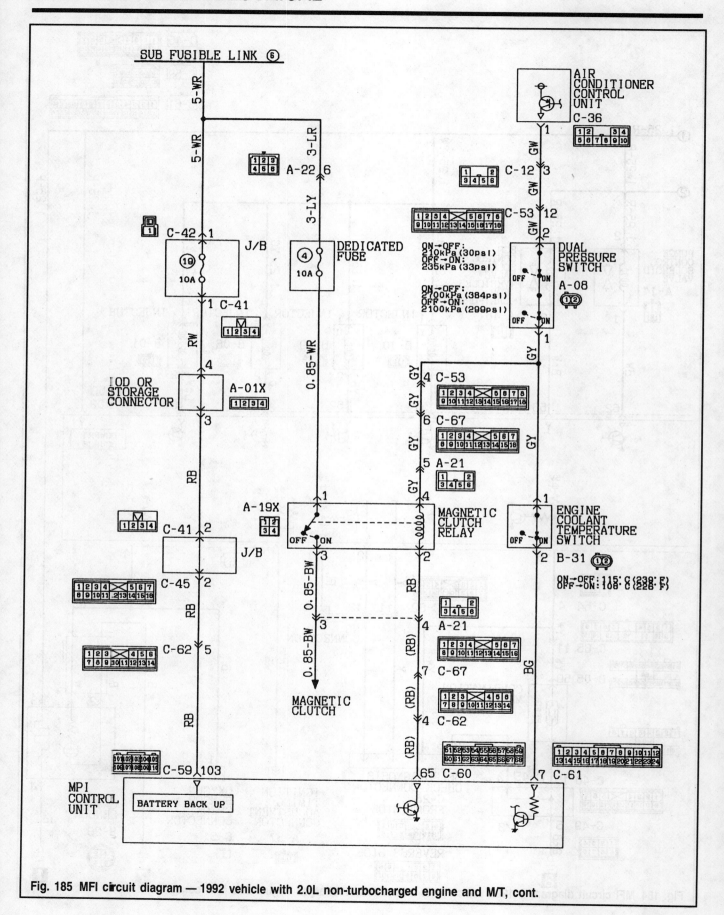

Fig. 185 MFI circuit diagram — 1992 vehicle with 2.0L non-turbocharged engine and M/T, cont.

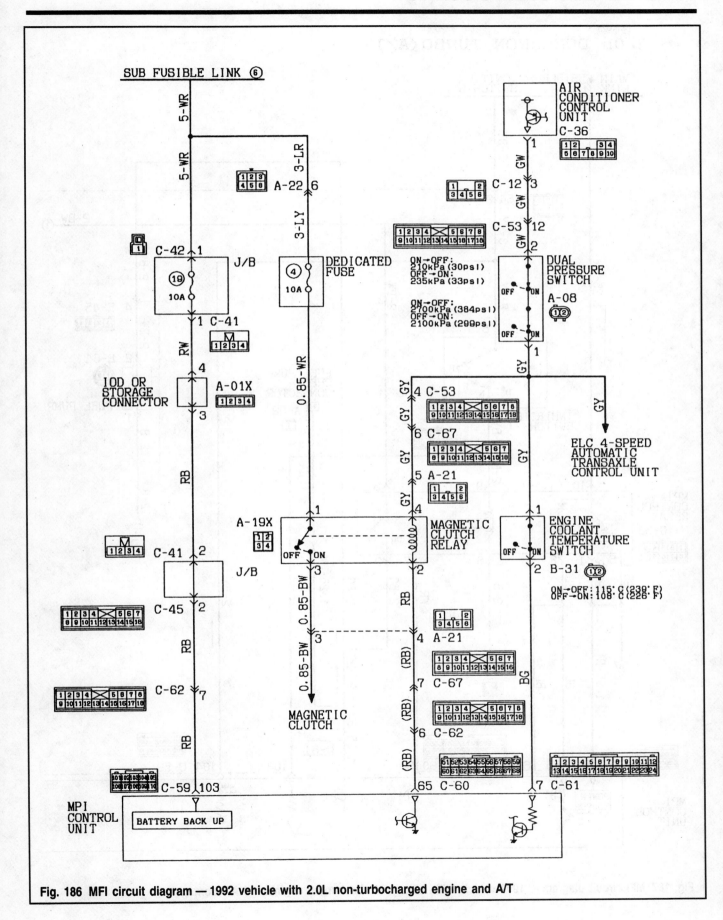

Fig. 186 MFI circuit diagram — 1992 vehicle with 2.0L non-turbocharged engine and A/T

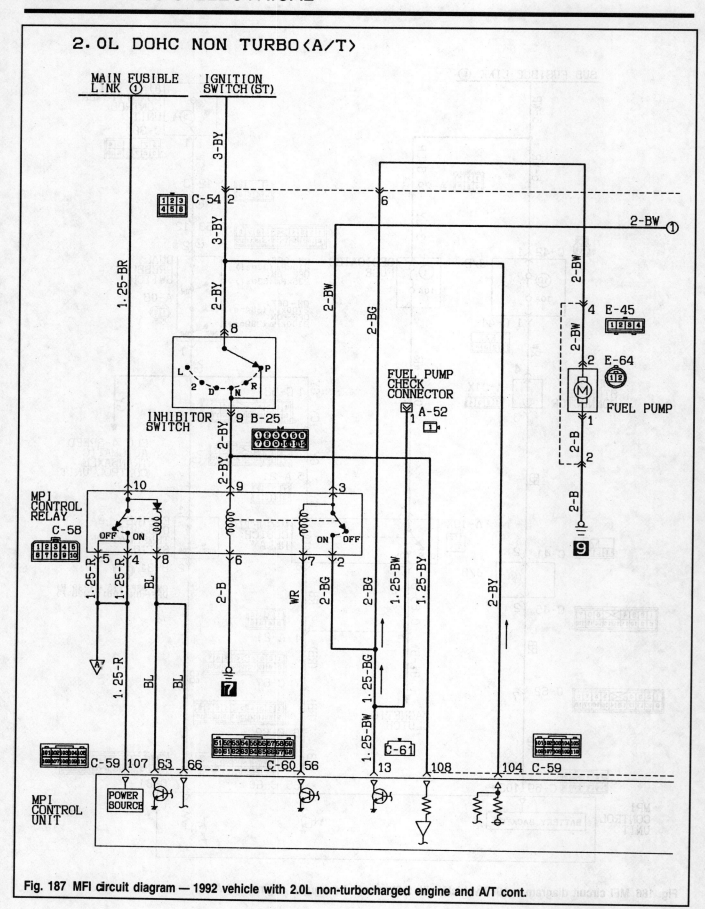

Fig. 187 MFI circuit diagram — 1992 vehicle with 2.0L non-turbocharged engine and A/T cont.

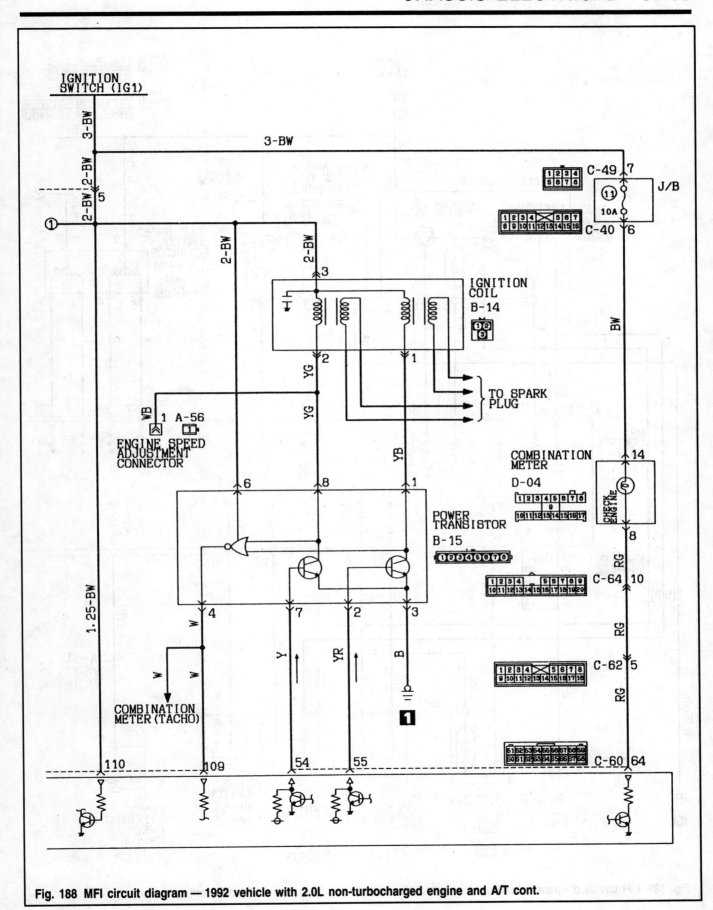

Fig. 188 MFI circuit diagram — 1992 vehicle with 2.0L non-turbocharged engine and A/T cont.

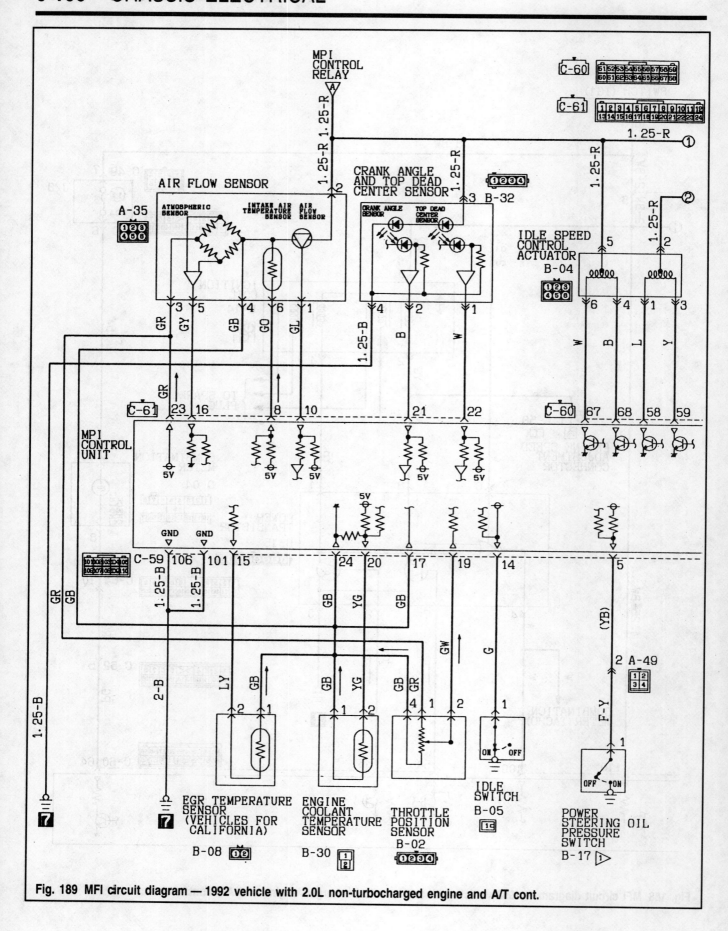

Fig. 189 MFI circuit diagram — 1992 vehicle with 2.0L non-turbocharged engine and A/T cont.

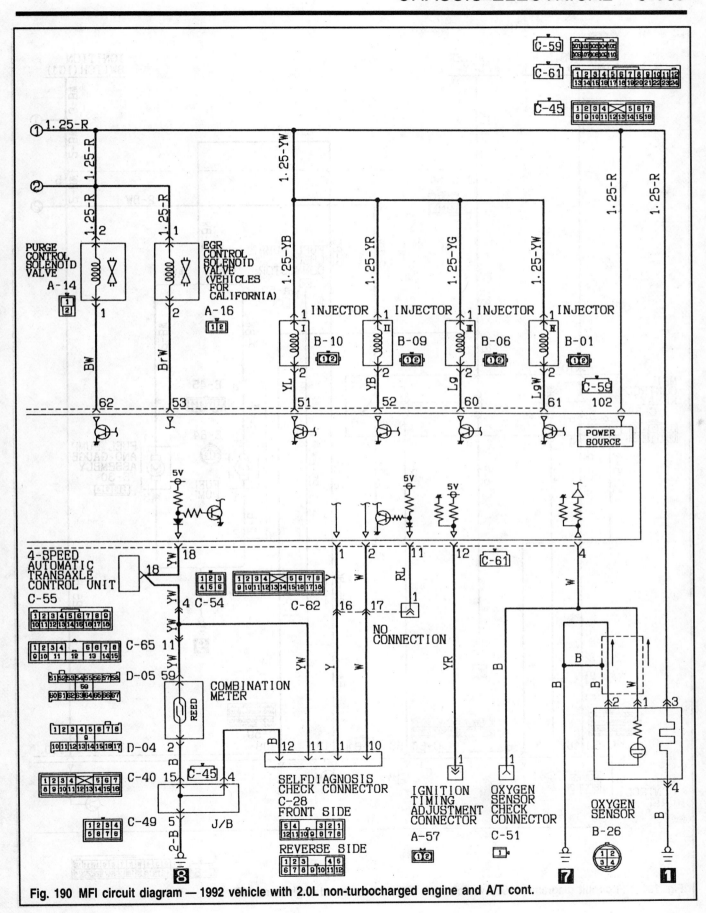

Fig. 190 MFI circuit diagram — 1992 vehicle with 2.0L non-turbocharged engine and A/T cont.

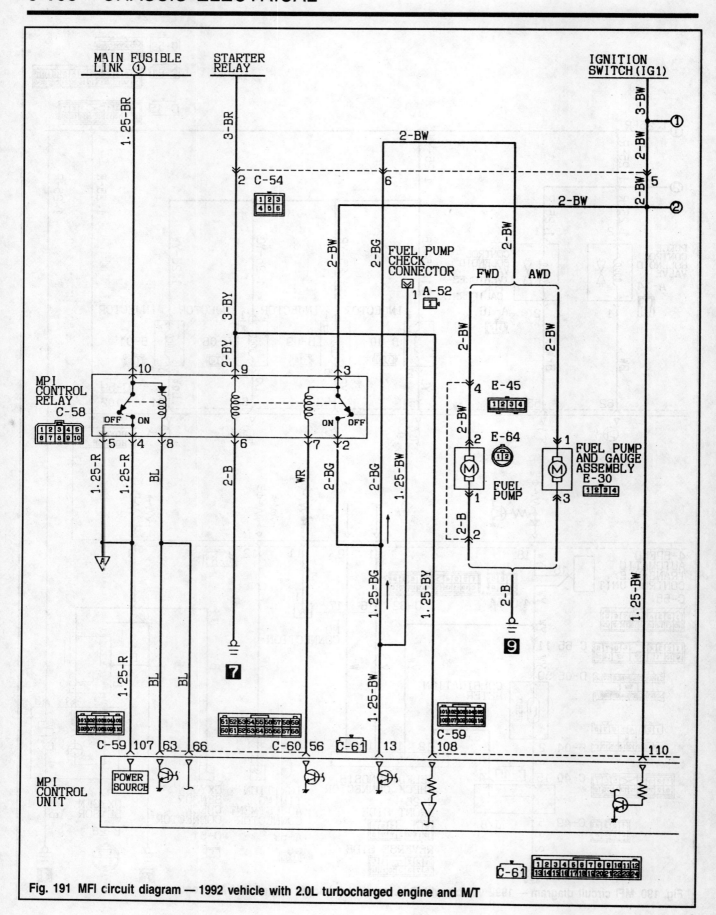

Fig. 191 MFI circuit diagram — 1992 vehicle with 2.0L turbocharged engine and M/T

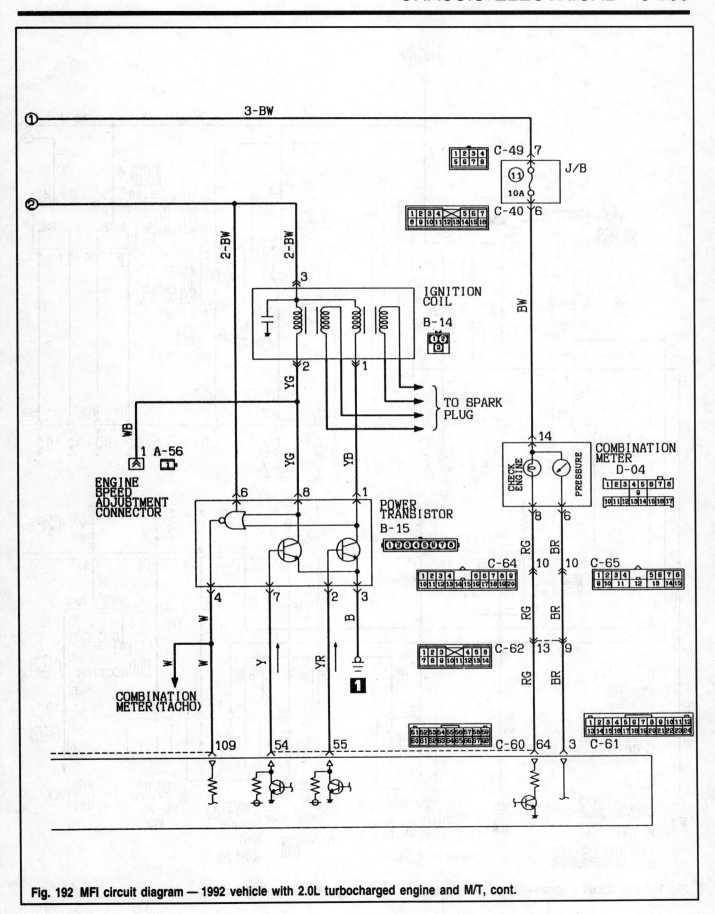

Fig. 192 MFI circuit diagram — 1992 vehicle with 2.0L turbocharged engine and M/T, cont.

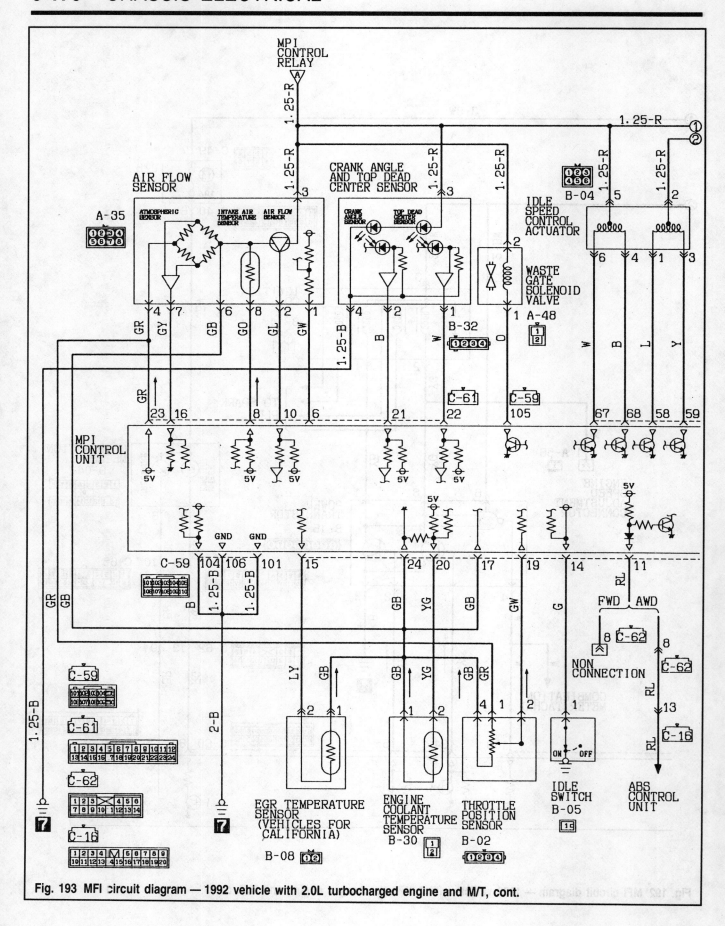

Fig. 193 MFI circuit diagram — 1992 vehicle with 2.0L turbocharged engine and M/T, cont.

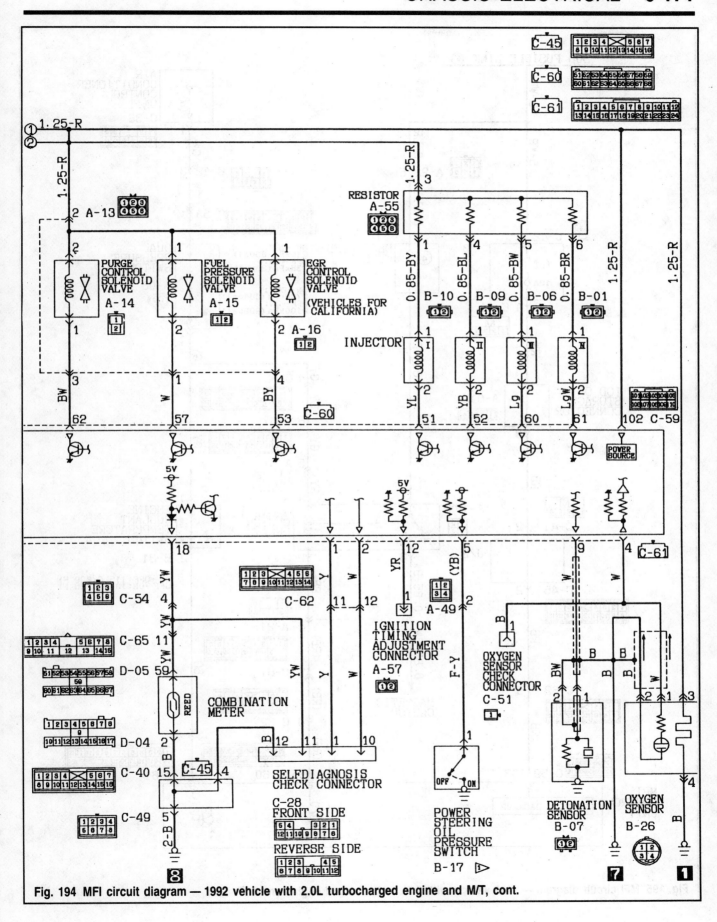

Fig. 194 MFI circuit diagram — 1992 vehicle with 2.0L turbocharged engine and M/T, cont.

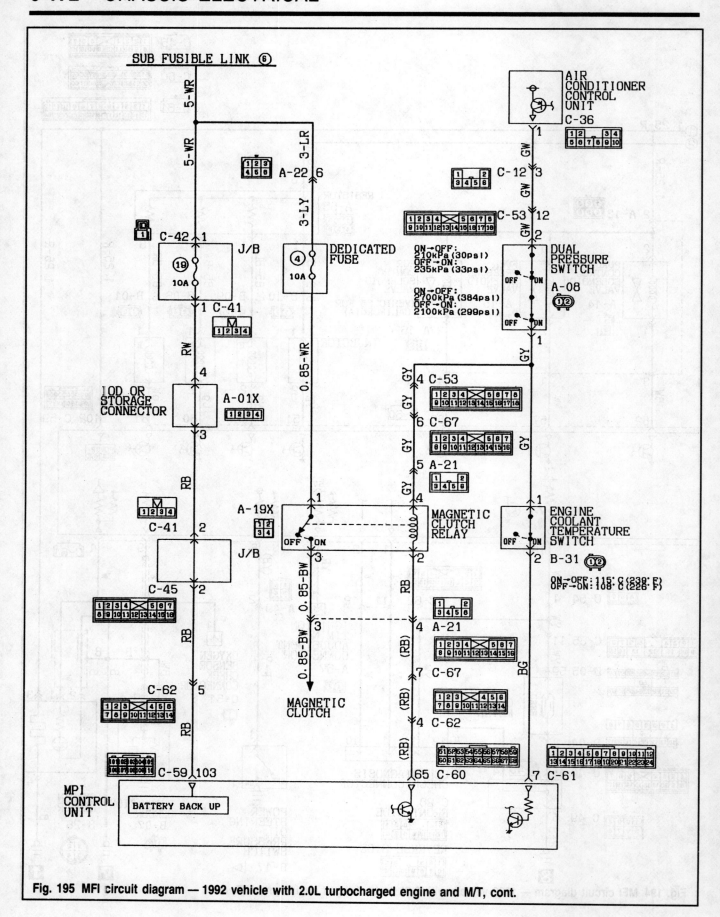

Fig. 195 MFI circuit diagram — 1992 vehicle with 2.0L turbocharged engine and M/T, cont.

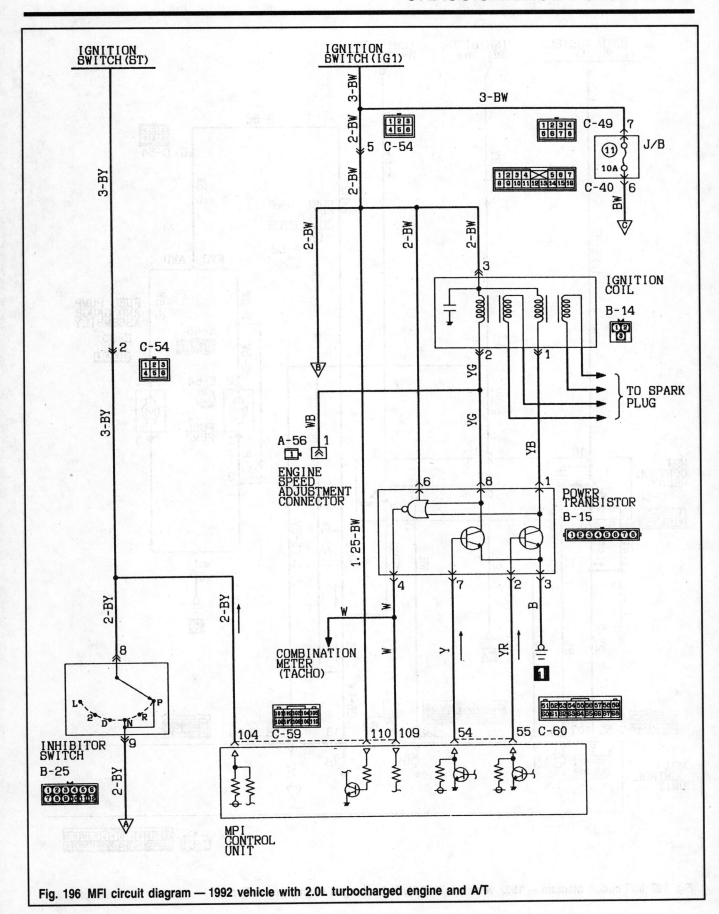

Fig. 196 MFI circuit diagram — 1992 vehicle with 2.0L turbocharged engine and A/T

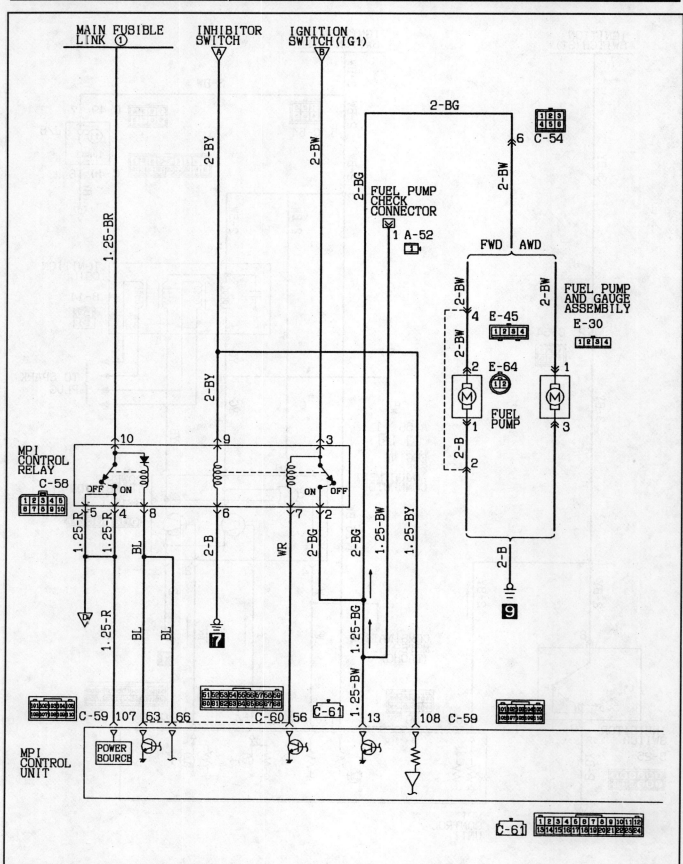

Fig. 197 MFI circuit diagram — 1992 vehicle with 2.0L turbocharged engine and A/T, cont.

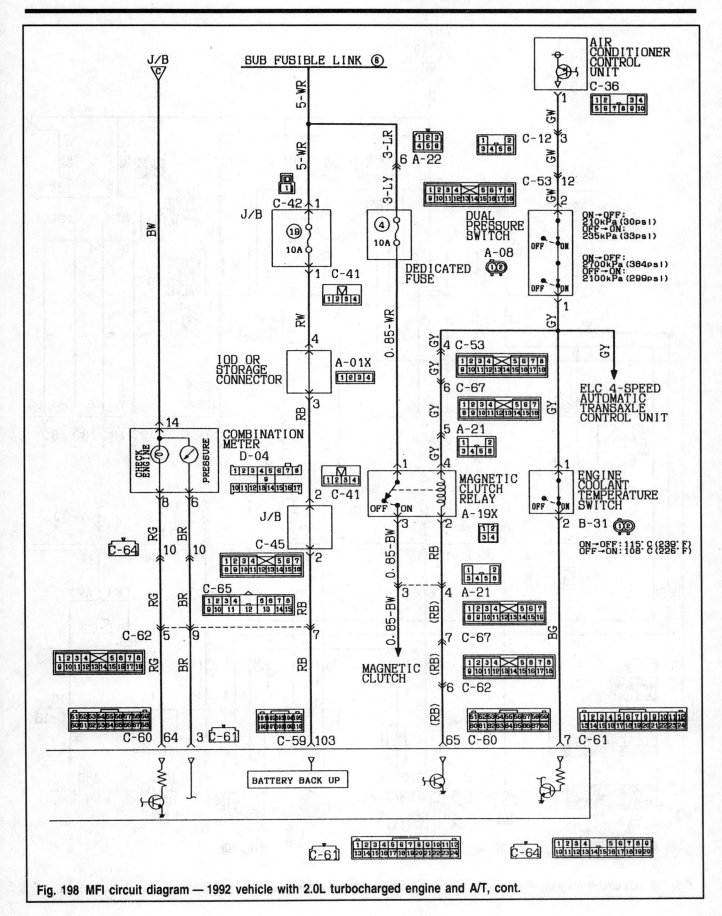

Fig. 198 MFI circuit diagram — 1992 vehicle with 2.0L turbocharged engine and A/T, cont.

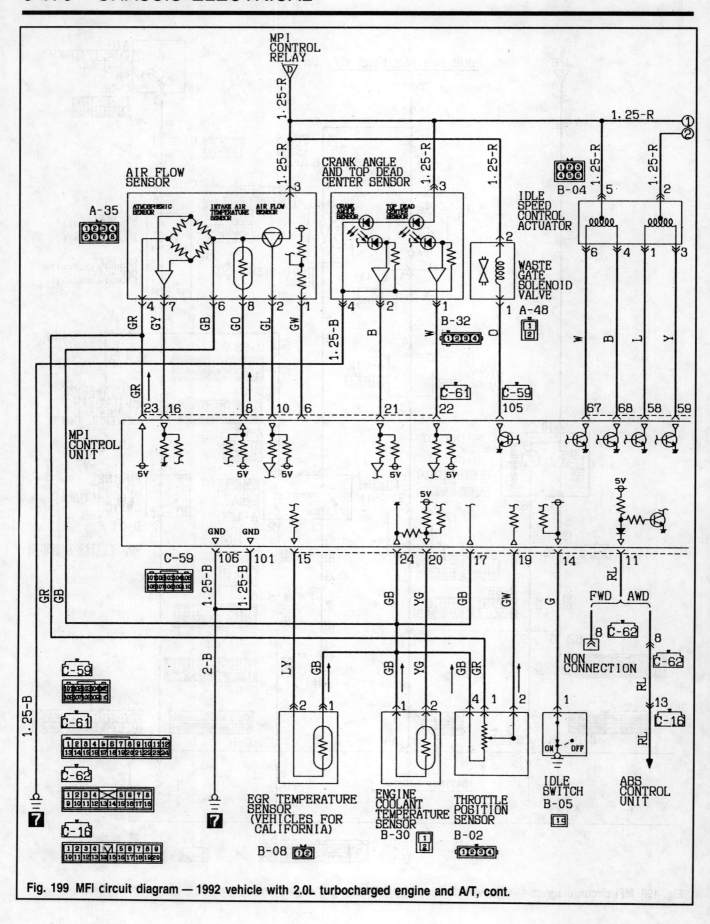

Fig. 199 MFI circuit diagram — 1992 vehicle with 2.0L turbocharged engine and A/T, cont.

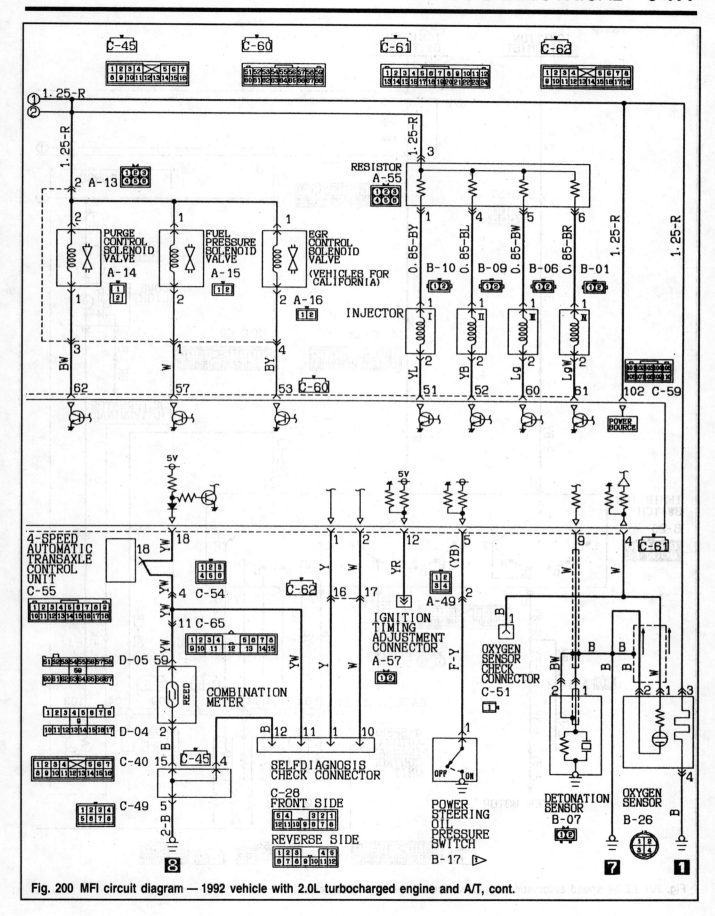

Fig. 200 MFI circuit diagram — 1992 vehicle with 2.0L turbocharged engine and A/T, cont.

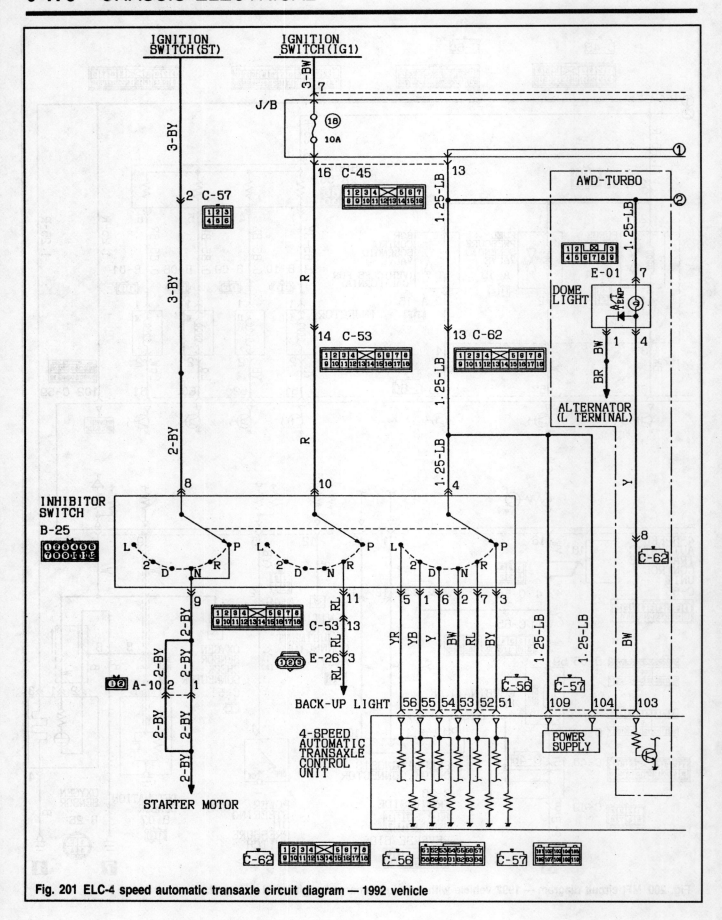

Fig. 201 ELC-4 speed automatic transaxle circuit diagram — 1992 vehicle

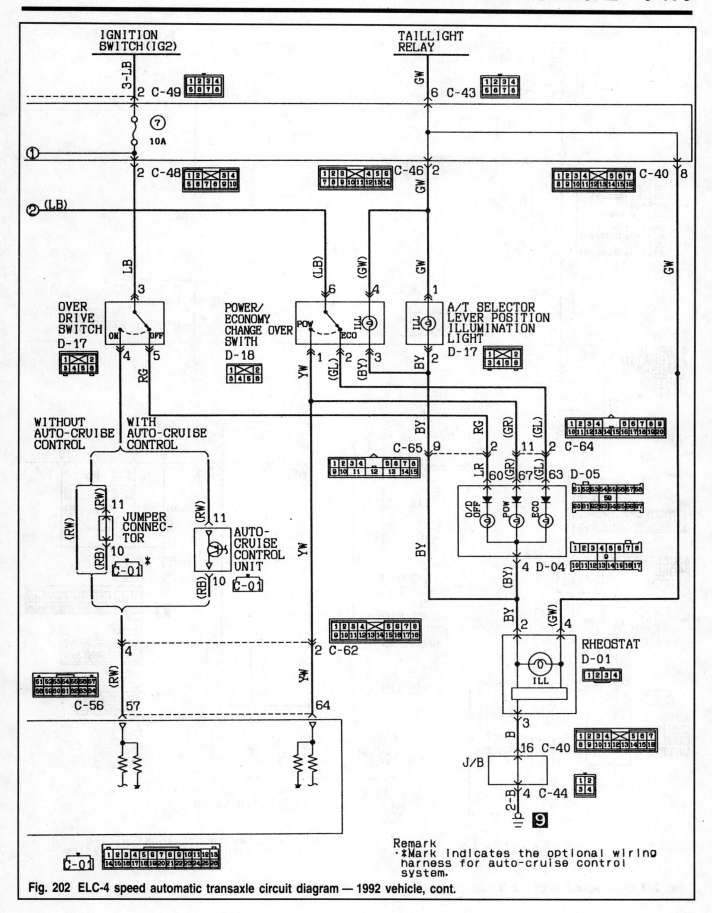

Fig. 202 ELC-4 speed automatic transaxle circuit diagram — 1992 vehicle, cont.

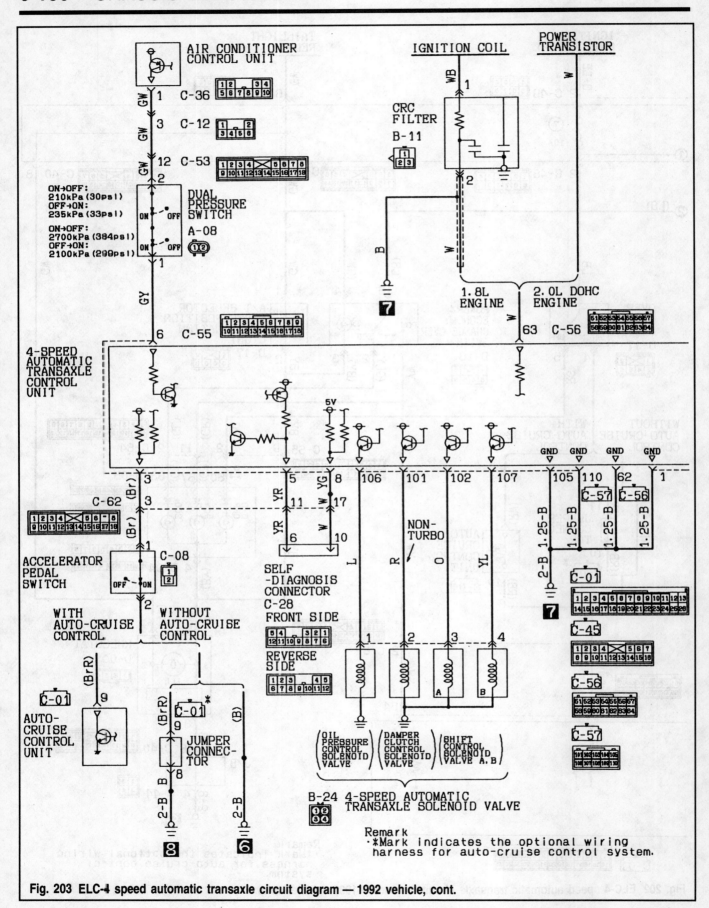

Fig. 203 ELC-4 speed automatic transaxle circuit diagram — 1992 vehicle, cont.

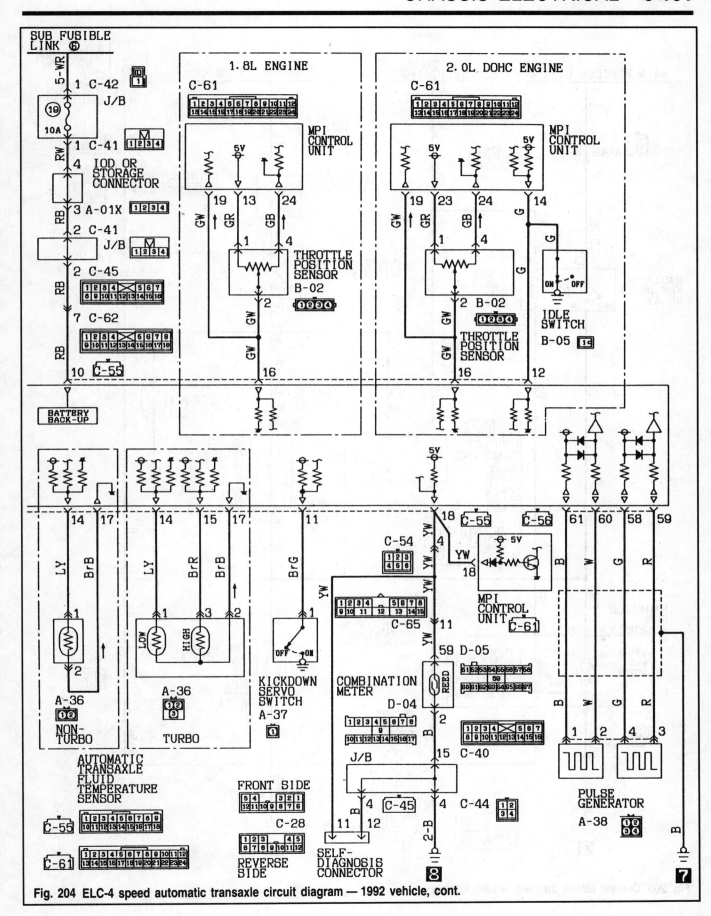

Fig. 204 ELC-4 speed automatic transaxle circuit diagram — 1992 vehicle, cont.

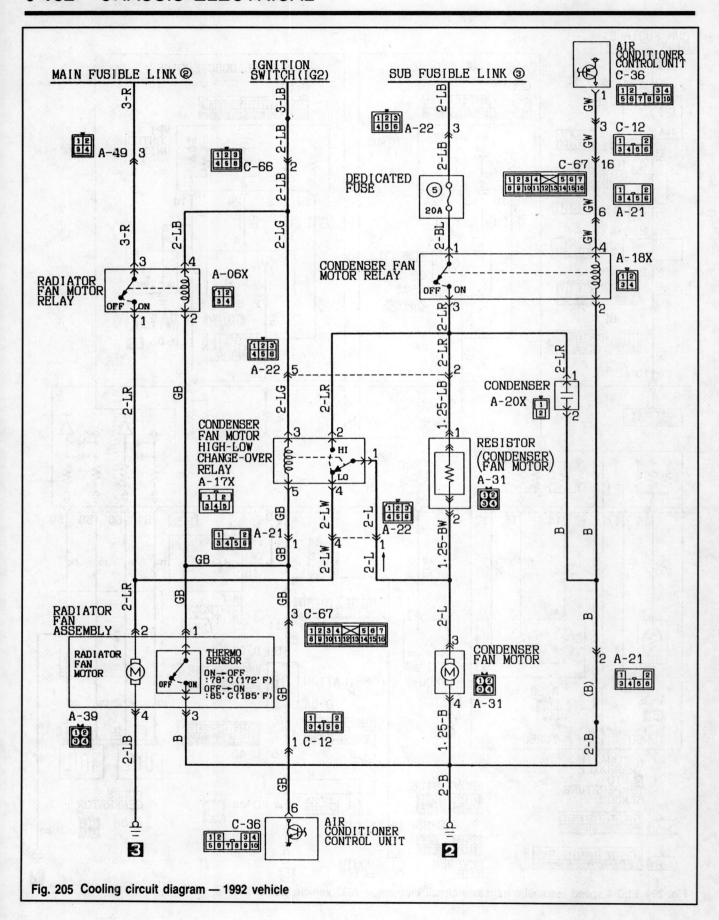

Fig. 205 Cooling circuit diagram — 1992 vehicle

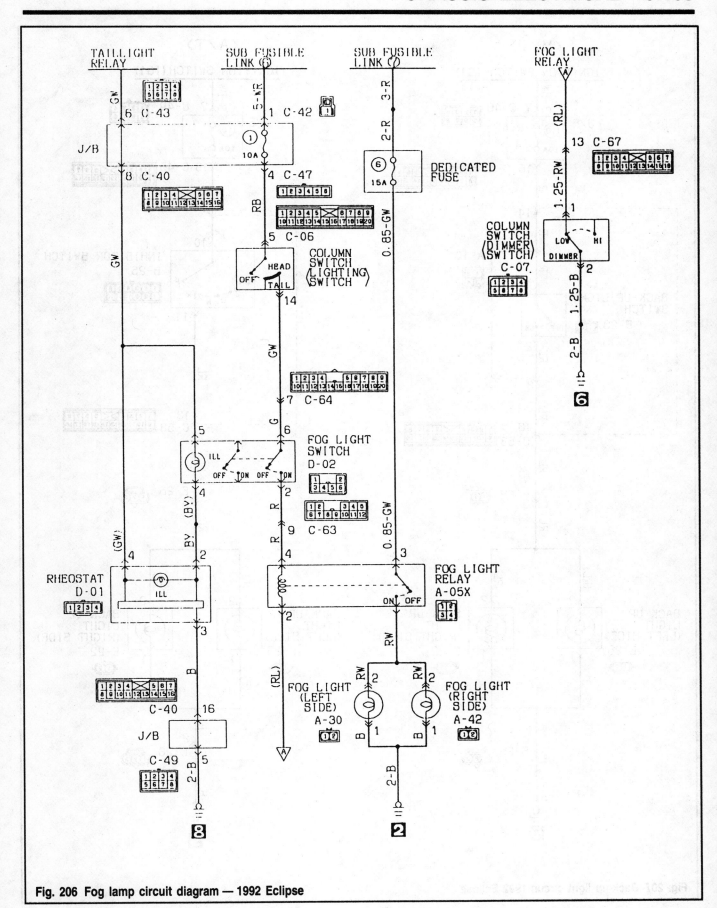

Fig. 206 Fog lamp circuit diagram — 1992 Eclipse

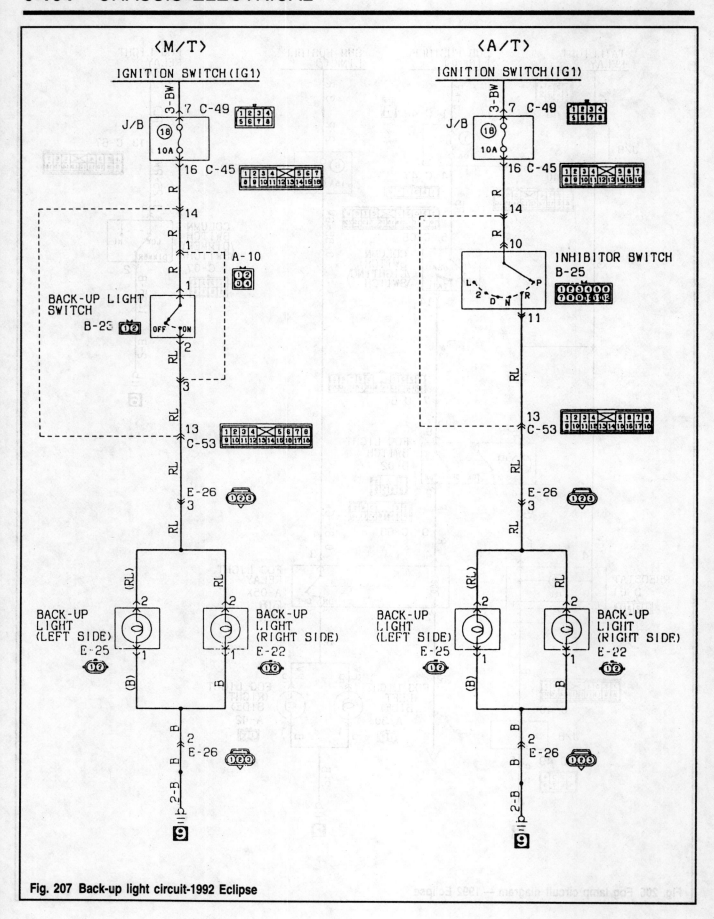

Fig. 207 Back-up light circuit-1992 Eclipse

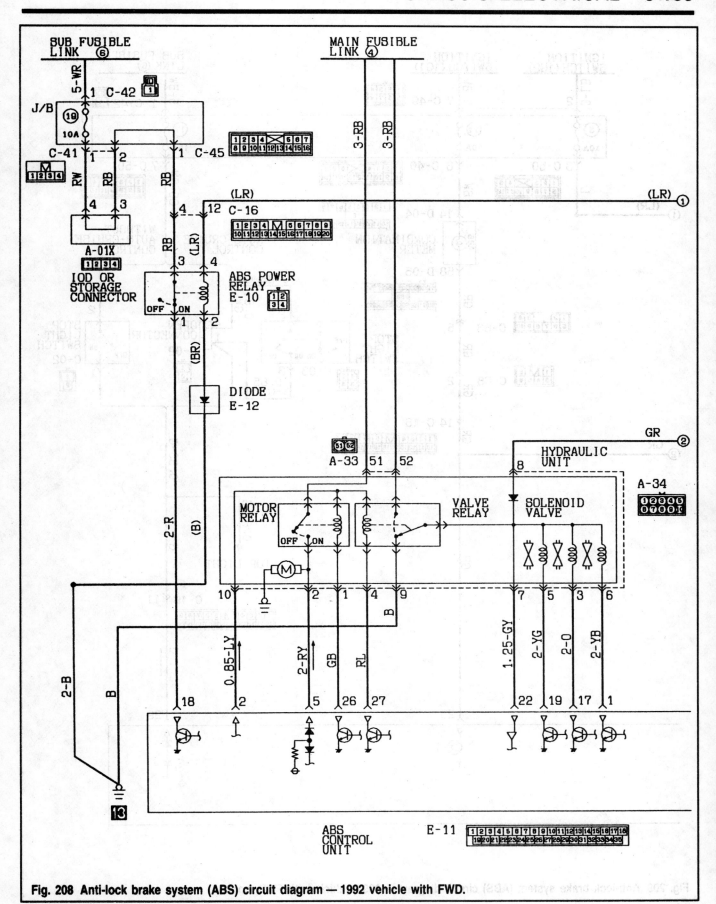

Fig. 208 Anti-lock brake system (ABS) circuit diagram — 1992 vehicle with FWD.

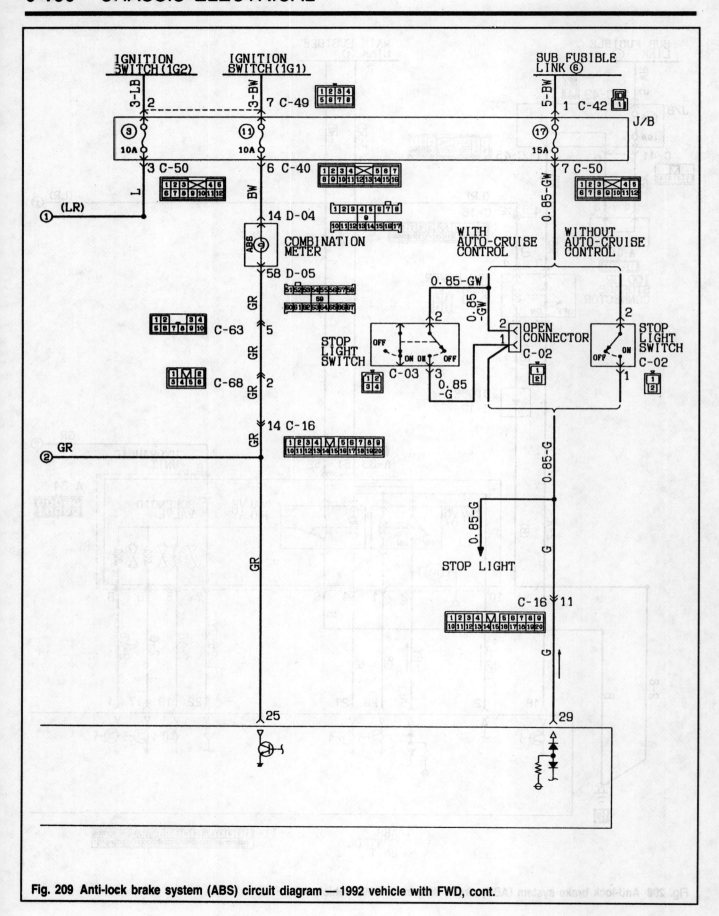

Fig. 209 Anti-lock brake system (ABS) circuit diagram — 1992 vehicle with FWD, cont.

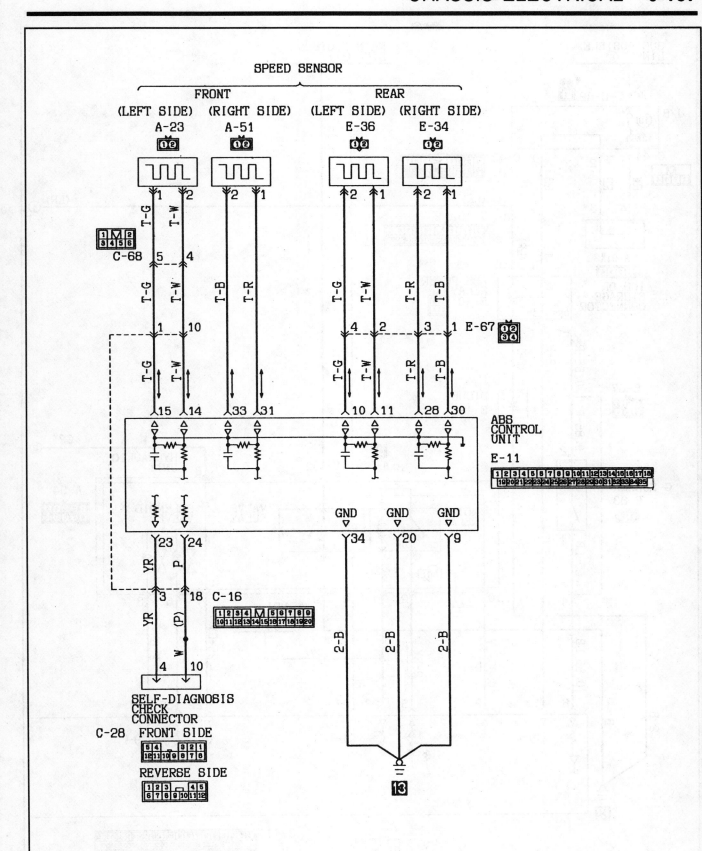

Fig. 210 Anti-lock brake system (ABS) circuit diagram — 1992 vehicle with FWD, cont.

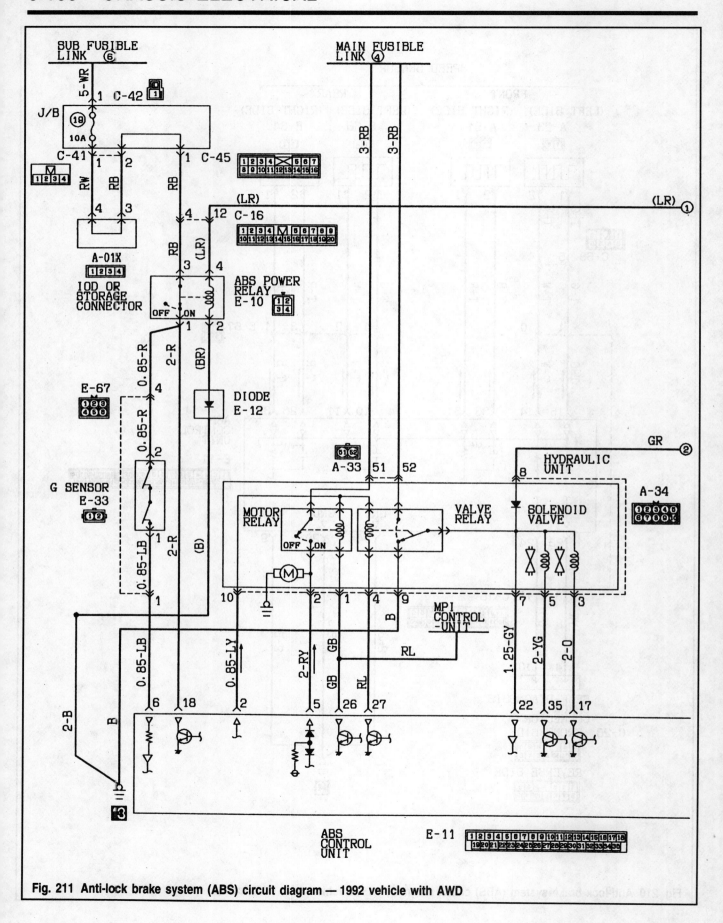

Fig. 211 Anti-lock brake system (ABS) circuit diagram — 1992 vehicle with AWD

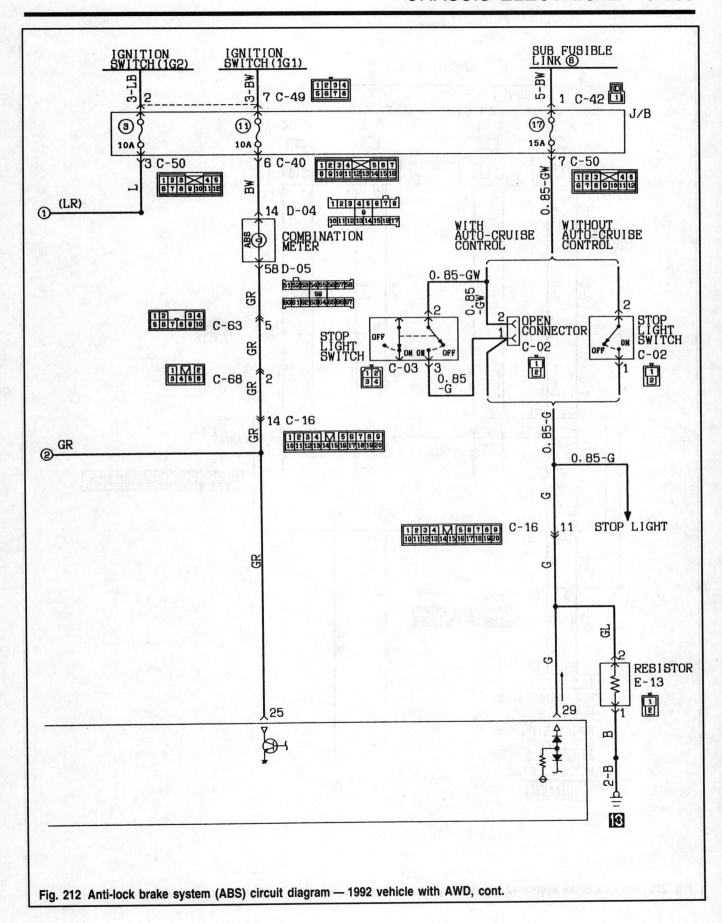

Fig. 212 Anti-lock brake system (ABS) circuit diagram — 1992 vehicle with AWD, cont.

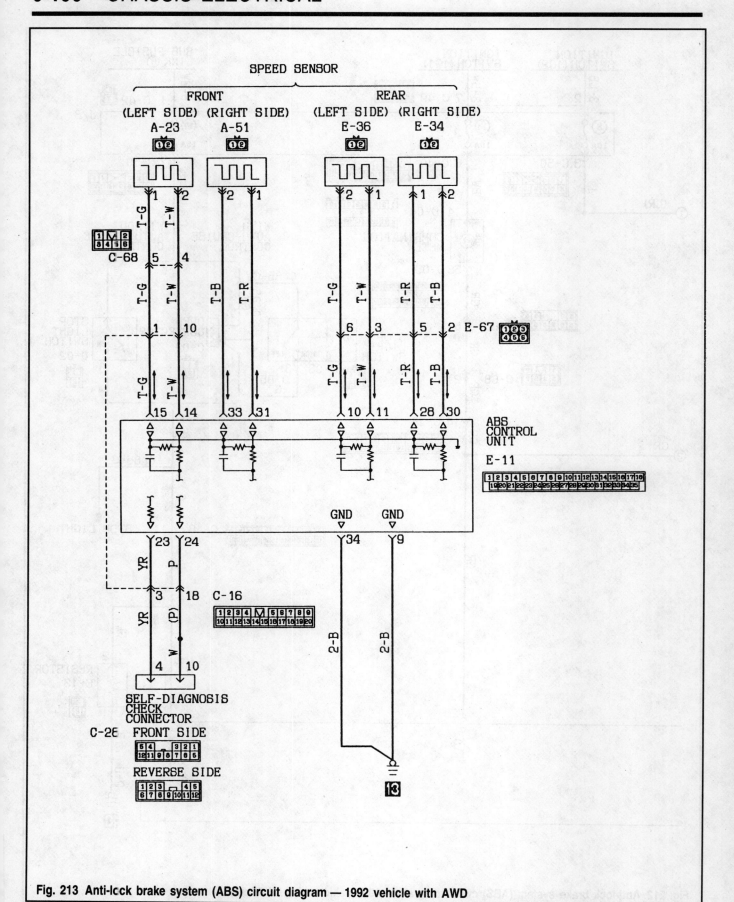

Fig. 213 Anti-lock brake system (ABS) circuit diagram — 1992 vehicle with AWD

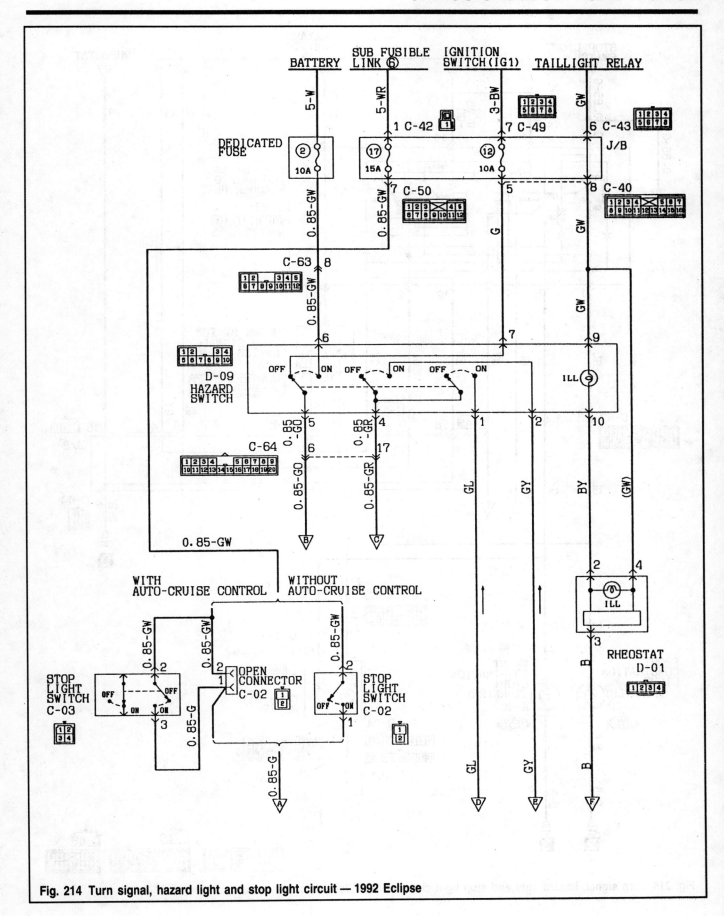

Fig. 214 Turn signal, hazard light and stop light circuit — 1992 Eclipse

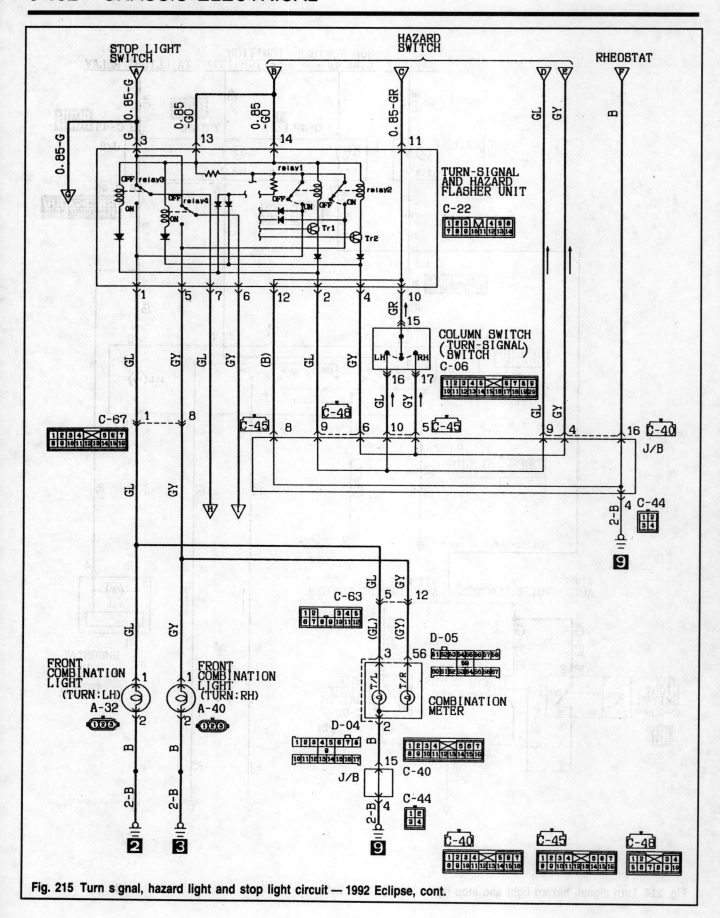

Fig. 215 Turn signal, hazard light and stop light circuit — 1992 Eclipse, cont.

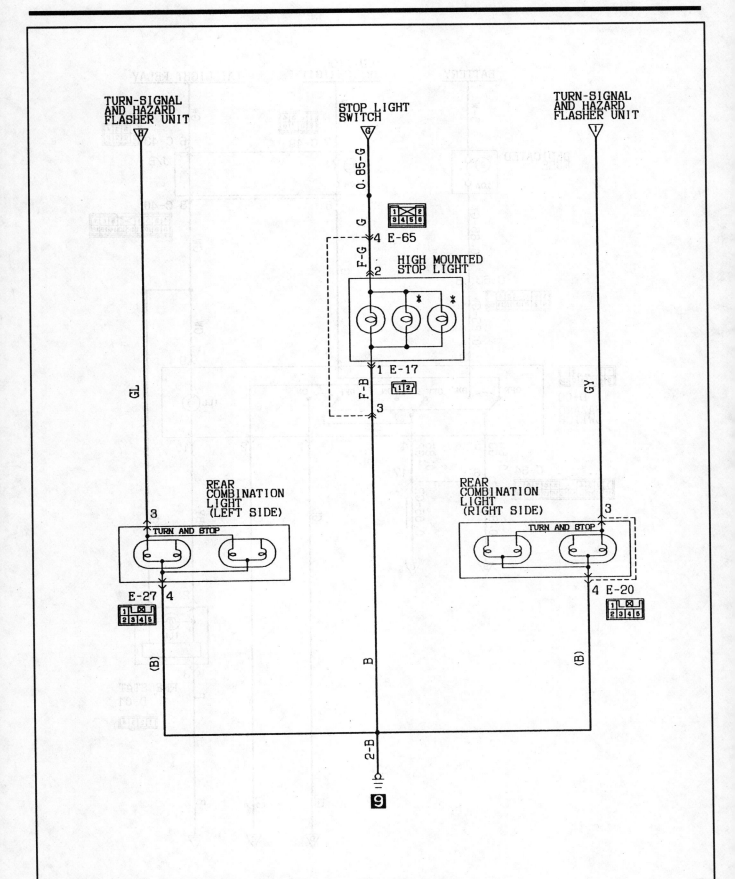

Fig. 216 Turn signal, hazard light and stop light circuit — 1992 Eclipse, cont.

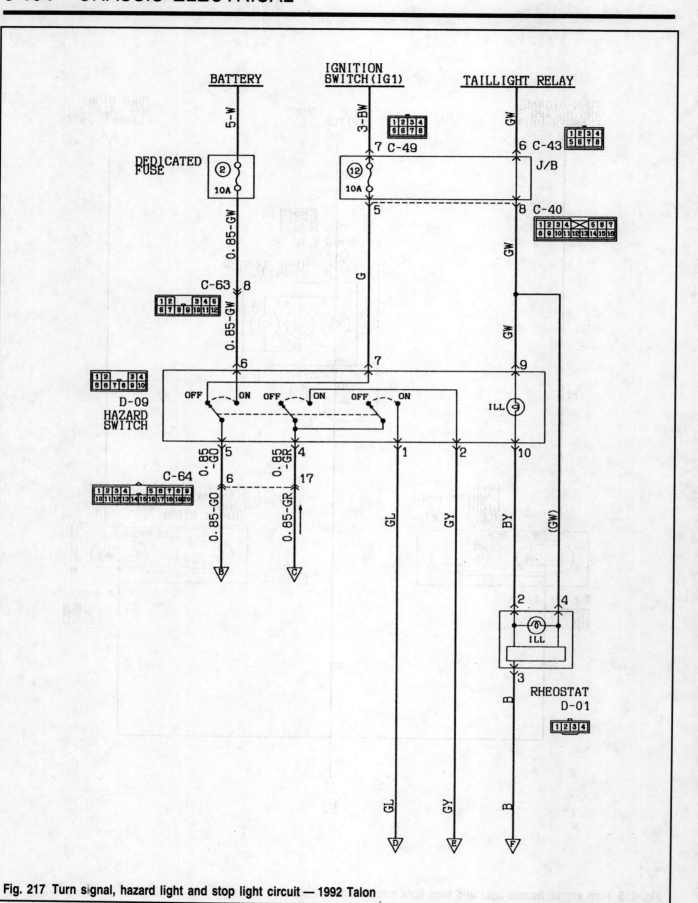

Fig. 217 Turn signal, hazard light and stop light circuit — 1992 Talon

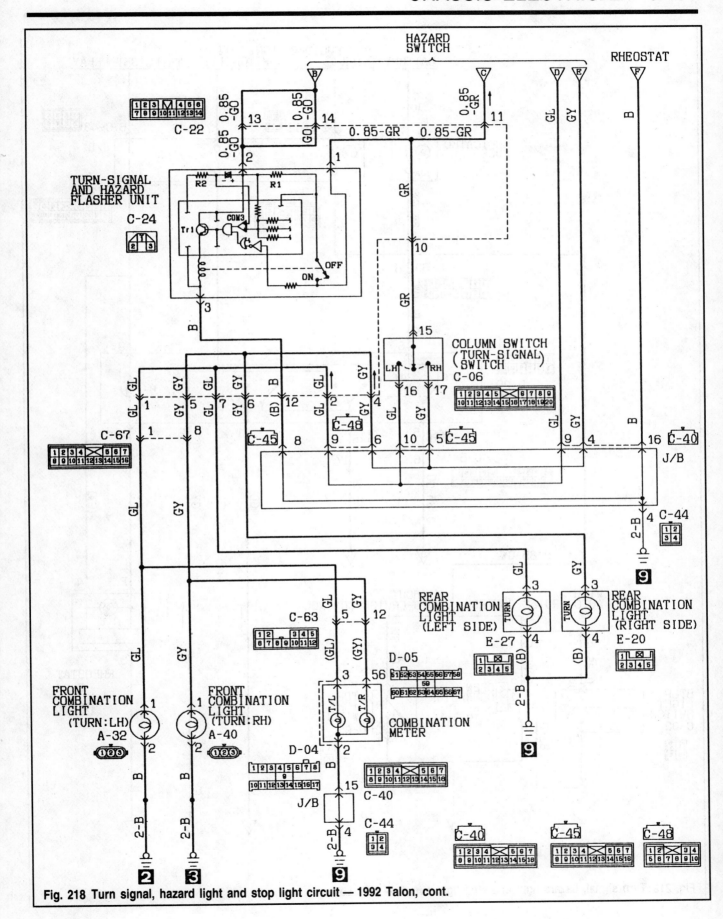

Fig. 218 Turn signal, hazard light and stop light circuit — 1992 Talon, cont.

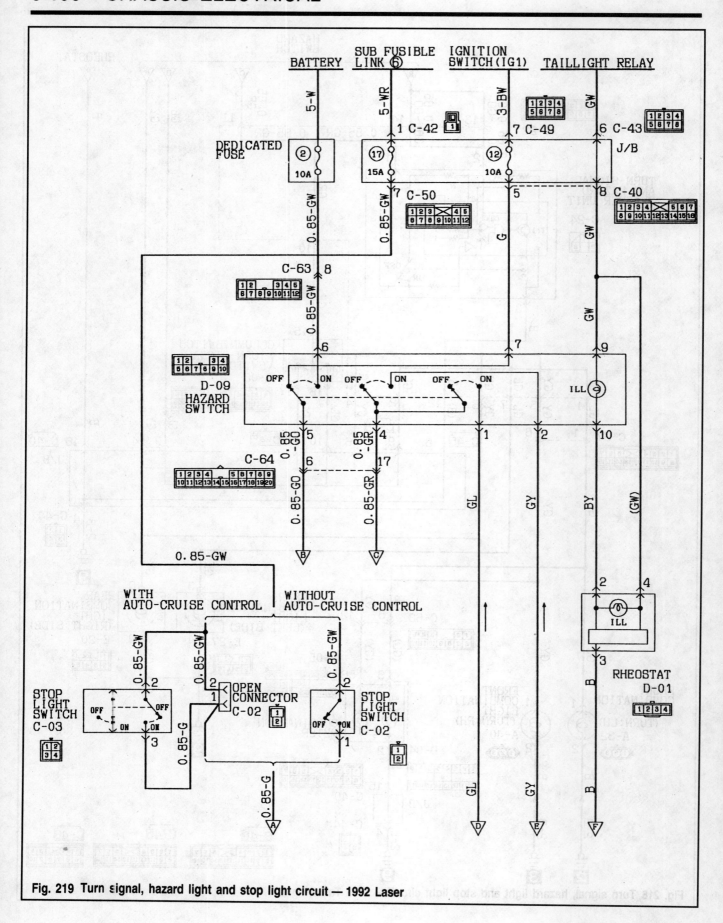

Fig. 219 Turn signal, hazard light and stop light circuit — 1992 Laser

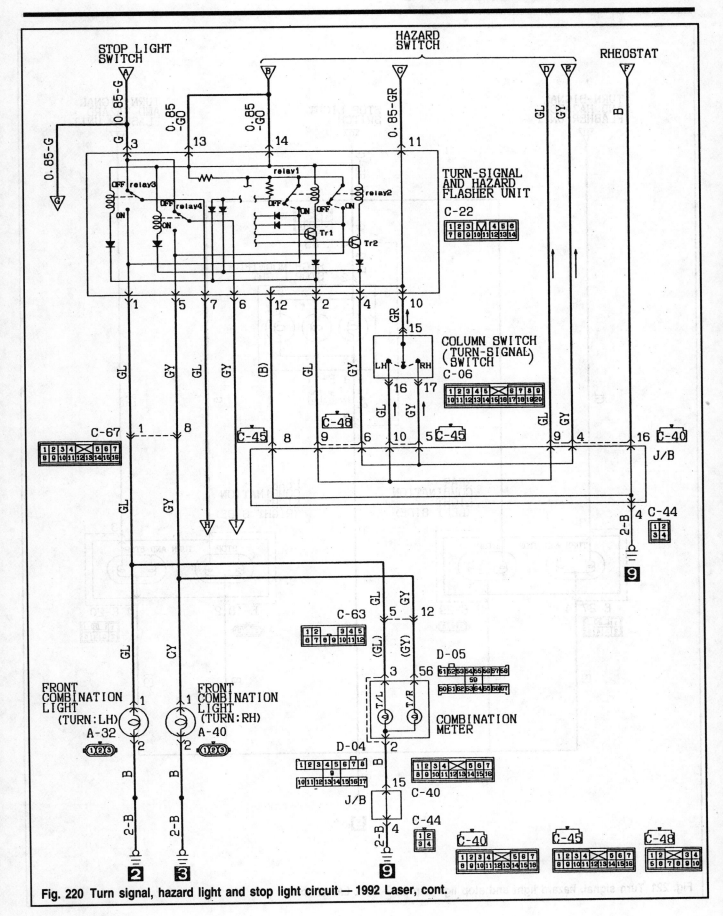

Fig. 220 Turn signal, hazard light and stop light circuit — 1992 Laser, cont.

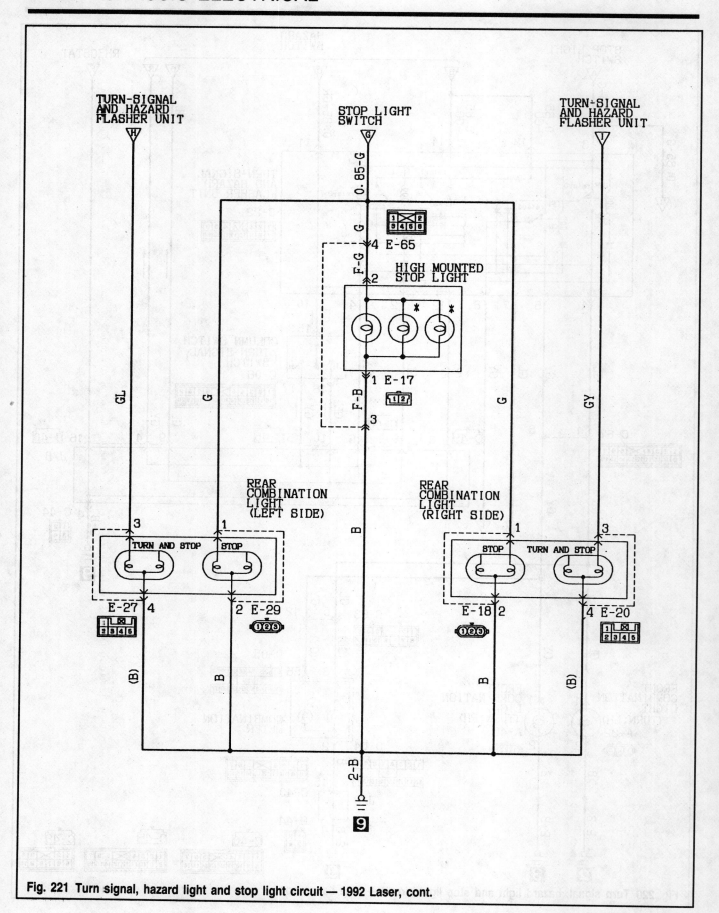

Fig. 221 Turn signal, hazard light and stop light circuit — 1992 Laser, cont.

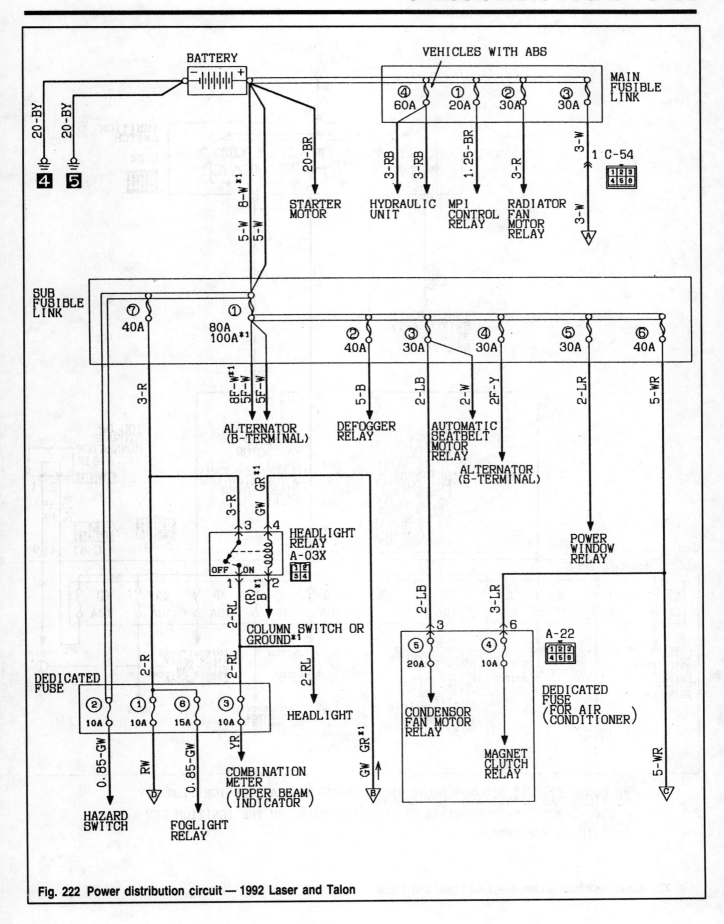

Fig. 222 Power distribution circuit — 1992 Laser and Talon

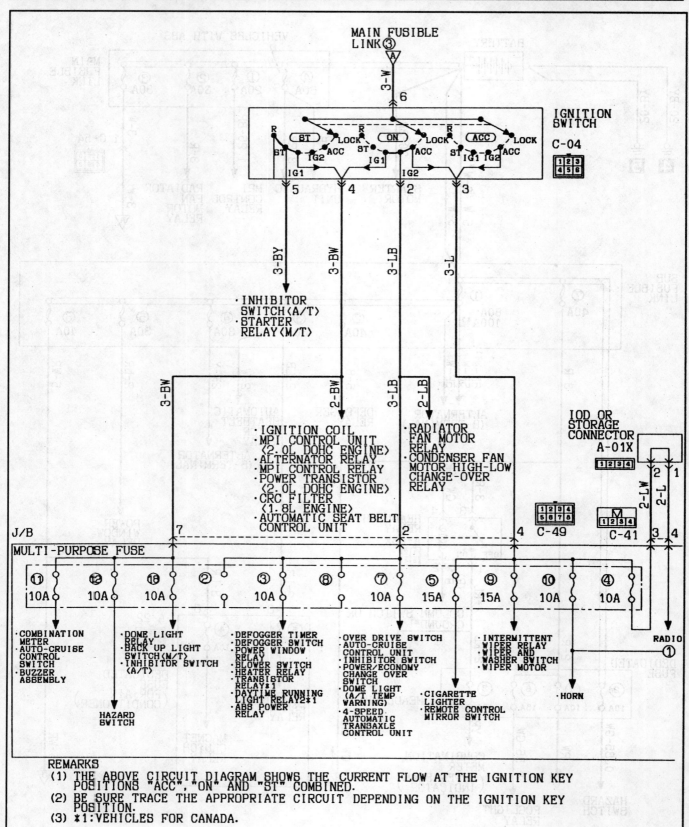

Fig. 223 Power distribution circuit — 1992 Laser and Talon

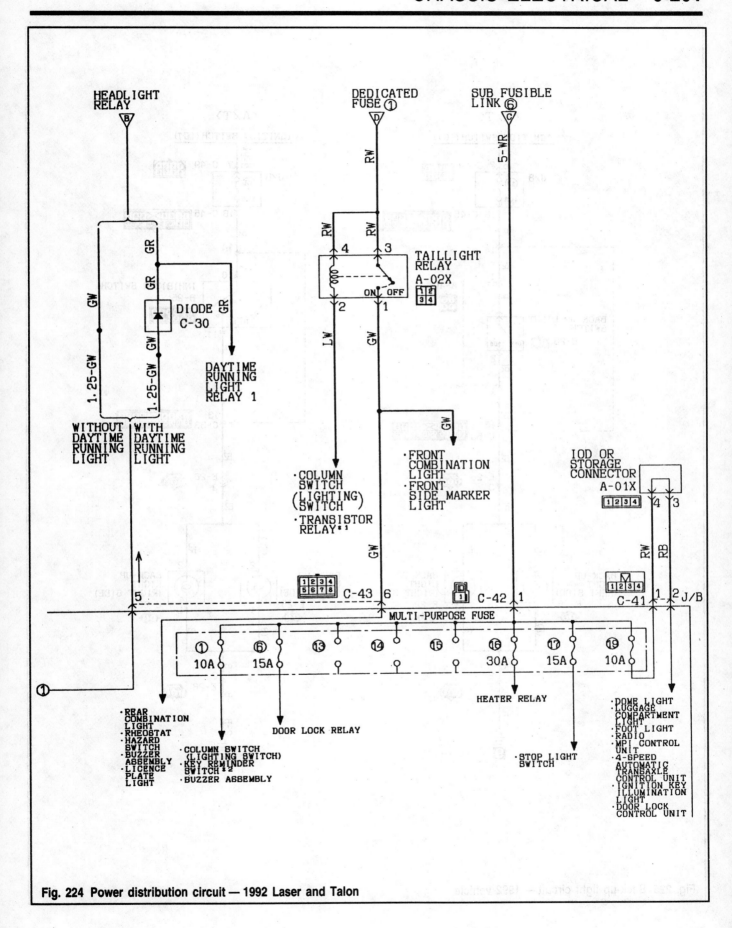

Fig. 224 Power distribution circuit — 1992 Laser and Talon

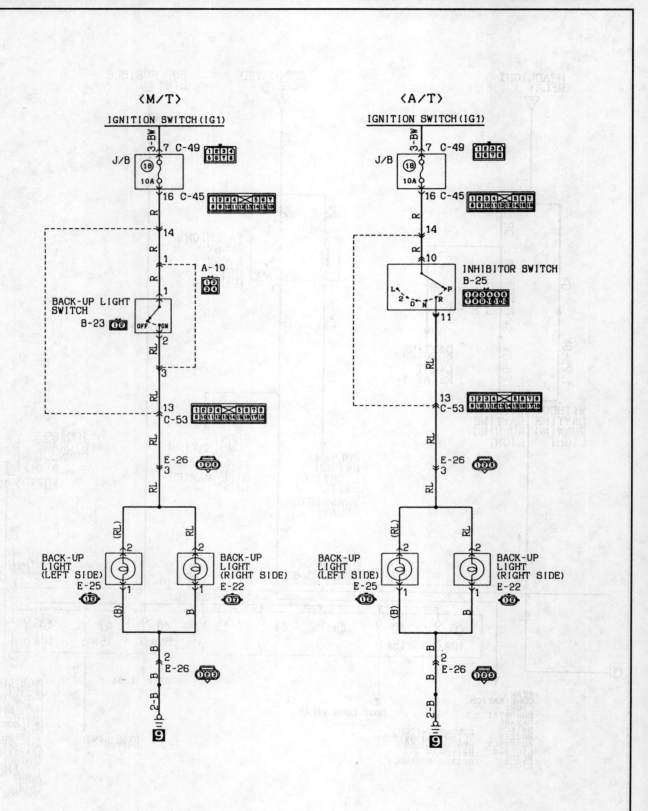

Fig. 225 Back-up light circuit — 1992 vehicle

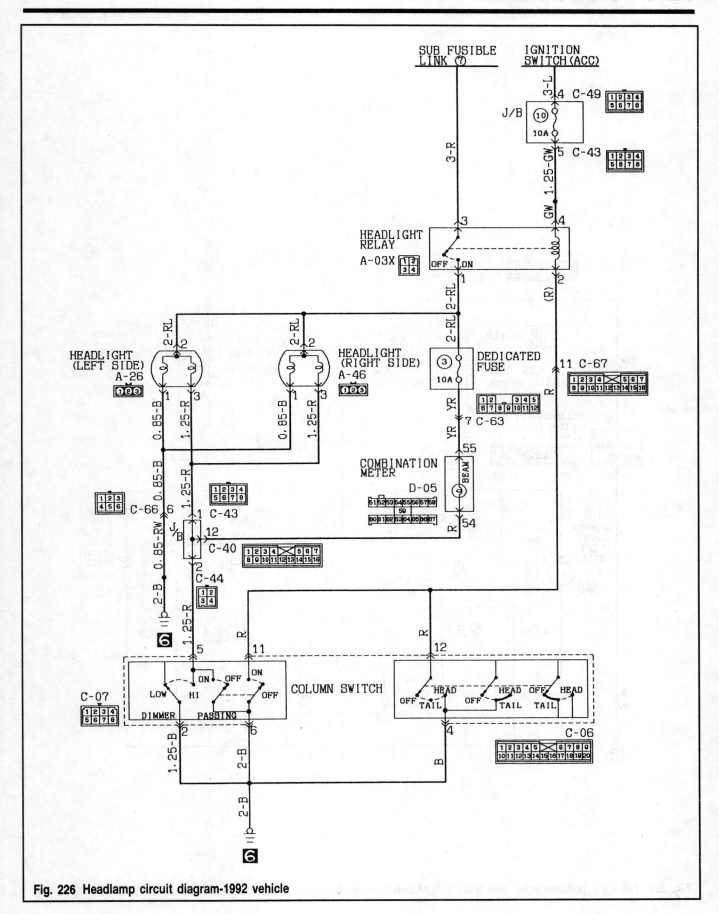

Fig. 226 Headlamp circuit diagram-1992 vehicle

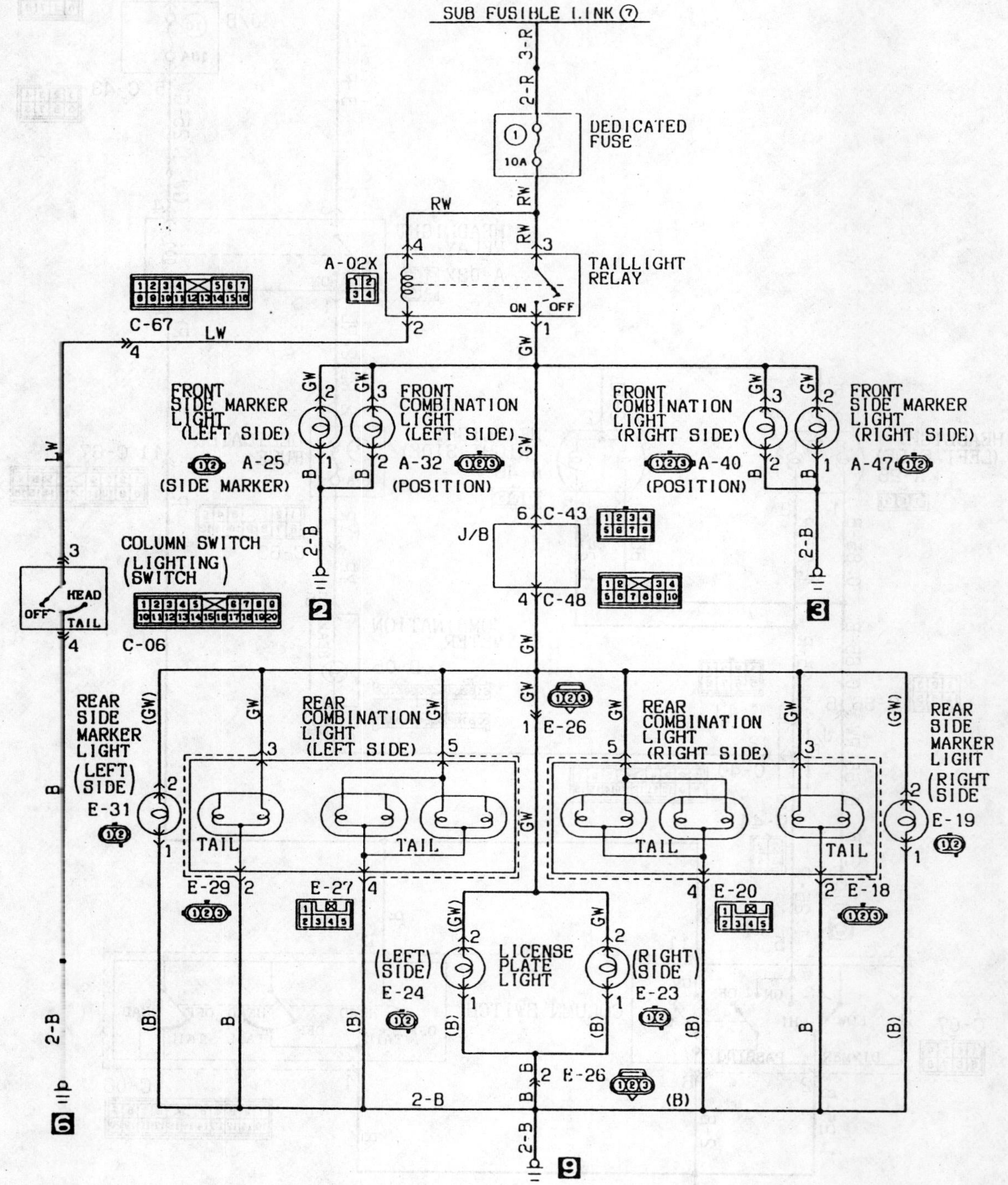

SUB FUSIBLE LINK ⑦

DEDICATED FUSE

1
10A

A-02X

TAILLIGHT RELAY

ON   OFF

C-67   LW

FRONT SIDE MARKER LIGHT (LEFT SIDE)

A-25 (SIDE MARKER)

FRONT COMBINATION LIGHT (LEFT SIDE)

A-32 (POSITION)

FRONT COMBINATION LIGHT (RIGHT SIDE)

A-40 (POSITION)

FRONT SIDE MARKER LIGHT (RIGHT SIDE)

A-47

6 C-43

J/B

4 C-48

COLUMN SWITCH (LIGHTING SWITCH)

OFF HEAD TAIL

C-06

2

3

REAR SIDE MARKER LIGHT (LEFT SIDE)

E-31

REAR COMBINATION LIGHT (LEFT SIDE)

TAIL   TAIL

E-29   E-27

REAR COMBINATION LIGHT (RIGHT SIDE)

TAIL   TAIL

E-20   E-18

REAR SIDE MARKER LIGHT (RIGHT SIDE)

E-19

E-26

LICENSE PLATE LIGHT

(LEFT SIDE)   (RIGHT SIDE)
E-24   E-23

2 E-26

2-B

9

6

Fig. 227 Tail light, position light, side marker light and license plate light circuit diagram — 1992 vehicle

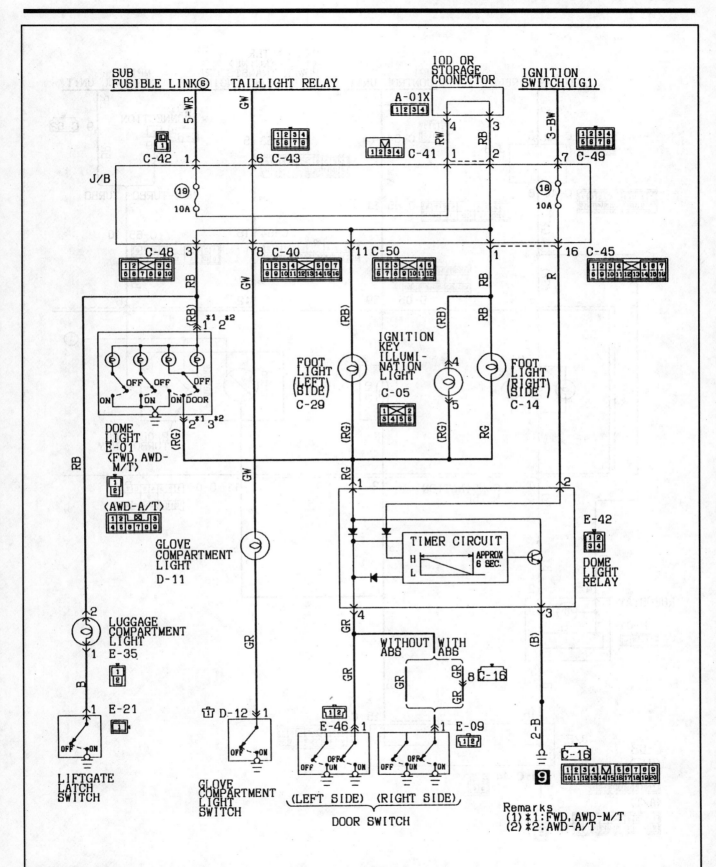

Fig. 228 Dome light, ignition key illumination light, foot light, glove compartment light and luggage compartment light circuit diagram — 1992 vehicle

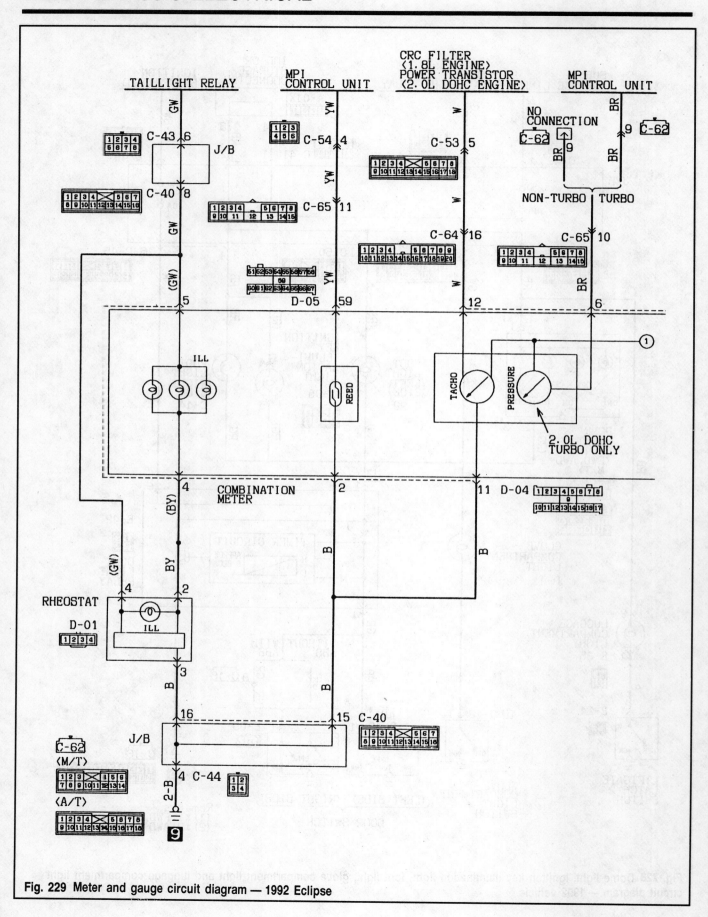

Fig. 229 Meter and gauge circuit diagram — 1992 Eclipse

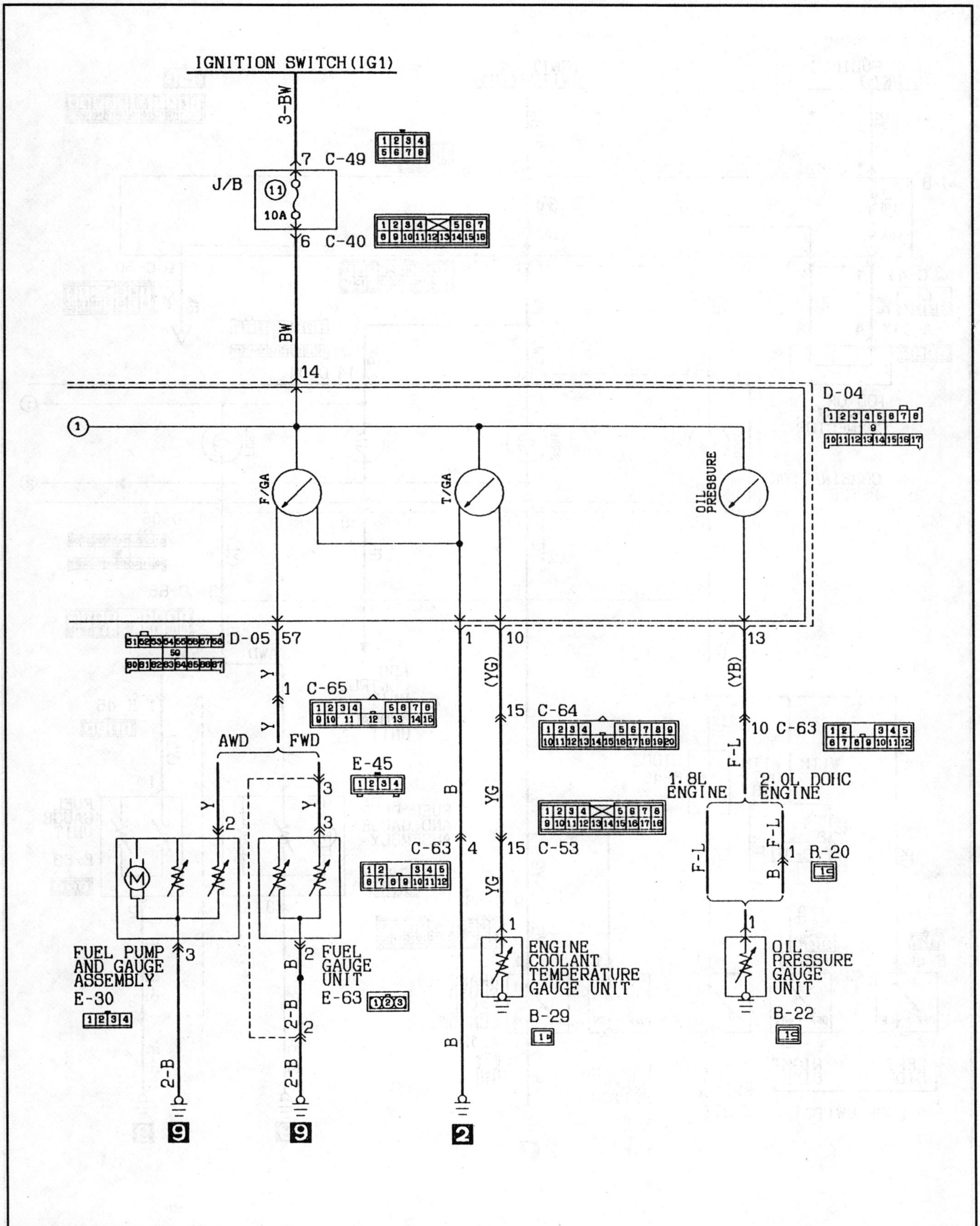

Fig. 230 Meter and gauge circuit diagram — 1992 Eclipse cont.

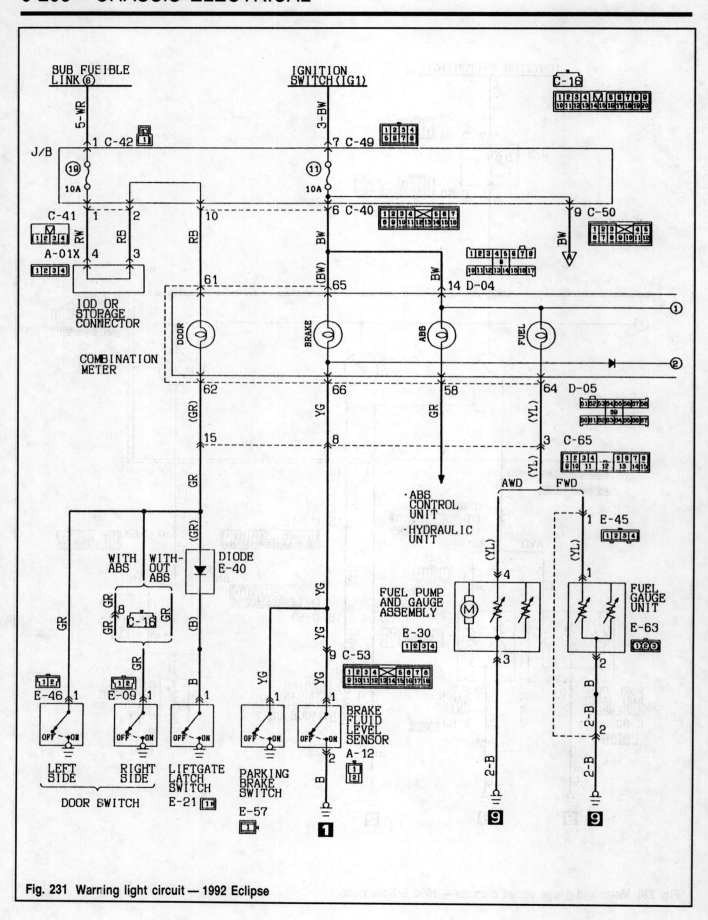

Fig. 231 Warning light circuit — 1992 Eclipse

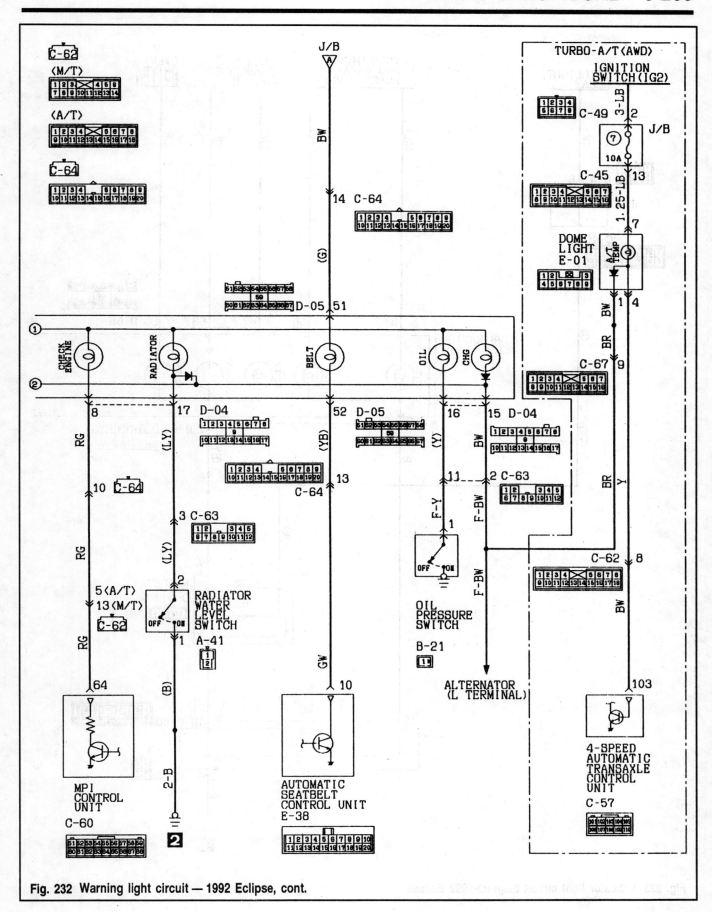

Fig. 232 Warning light circuit — 1992 Eclipse, cont.

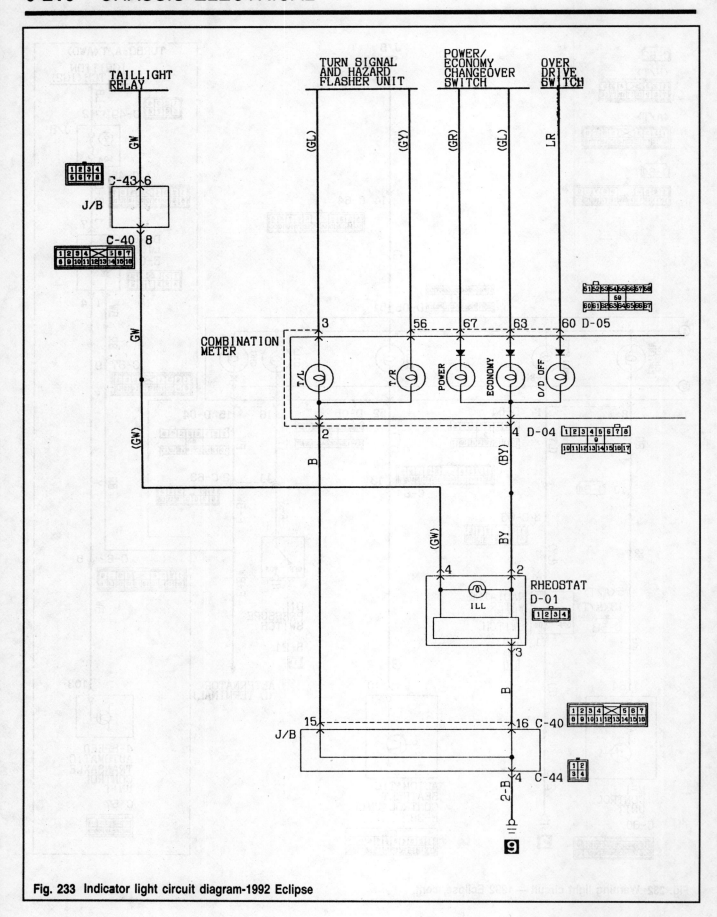

Fig. 233 Indicator light circuit diagram-1992 Eclipse

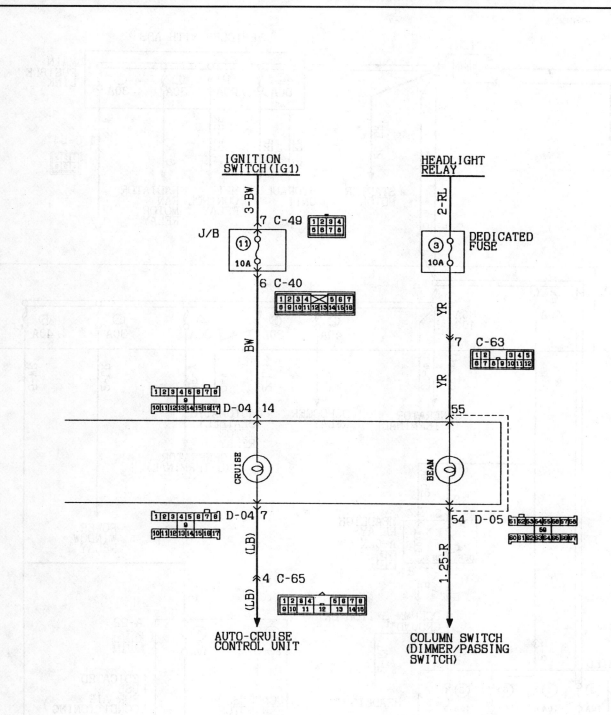

Fig. 234 Indicator light circuit diagram-1992 Eclipse, cont.

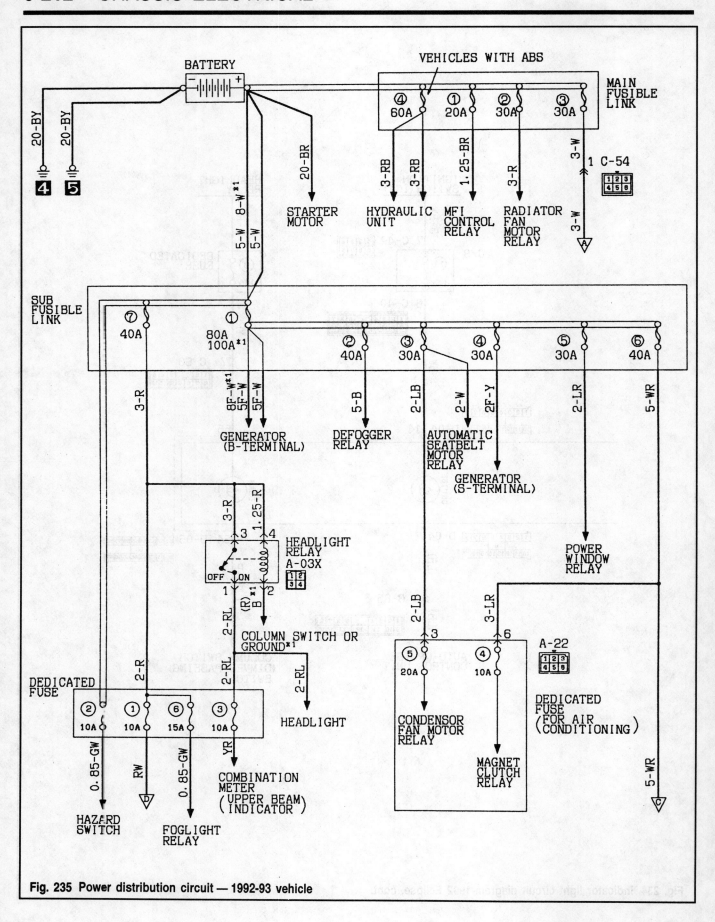

Fig. 235 Power distribution circuit — 1992-93 vehicle

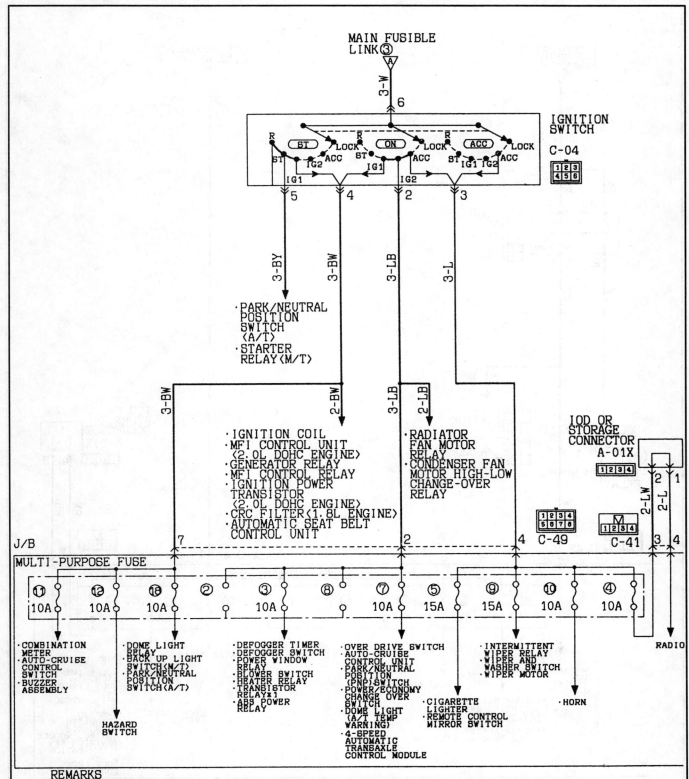

Fig. 236 Power distribution circuit — 1992-93 vehicle, cont.

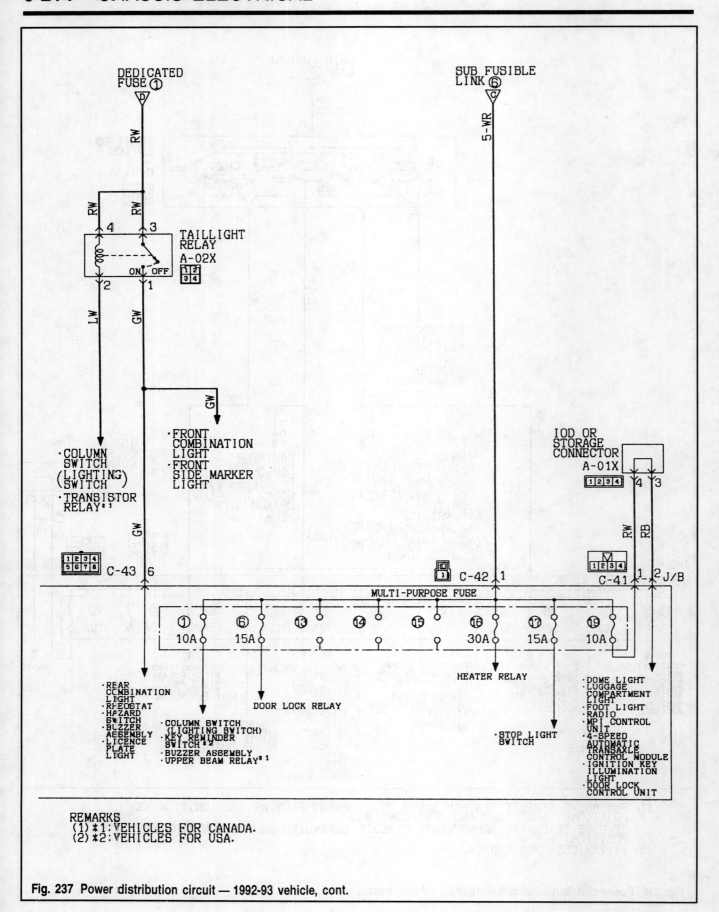

Fig. 237 Power distribution circuit — 1992-93 vehicle, cont.

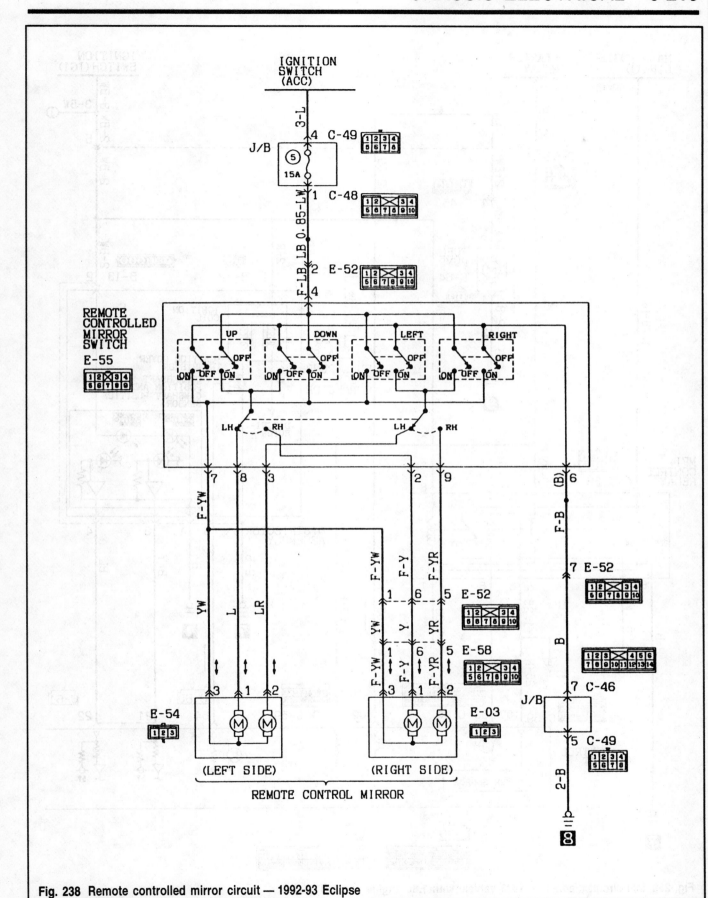

Fig. 238 Remote controlled mirror circuit — 1992-93 Eclipse

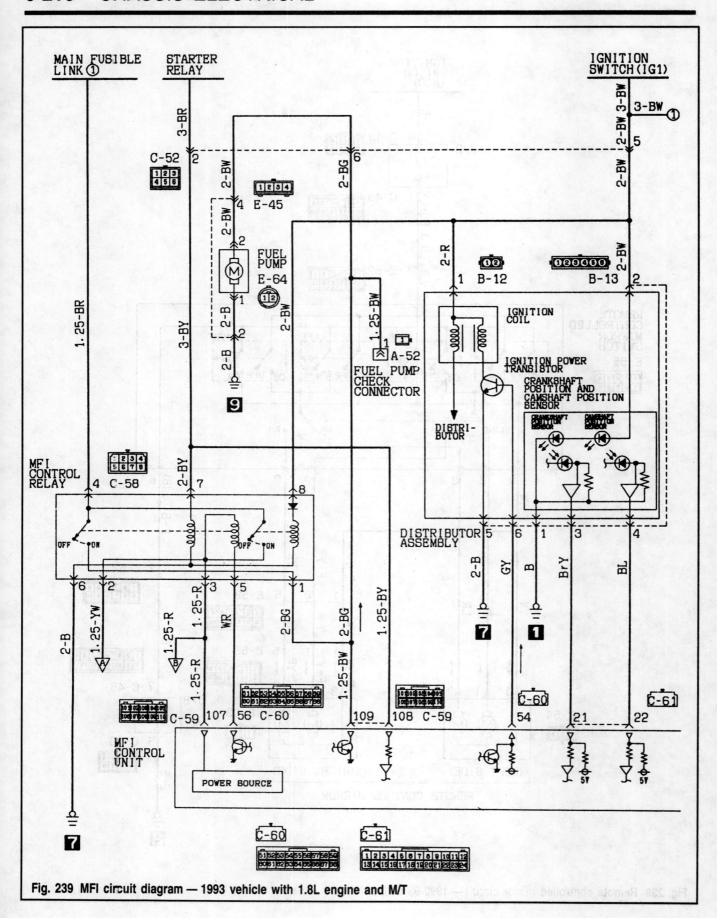

Fig. 239 MFI circuit diagram — 1993 vehicle with 1.8L engine and M/T

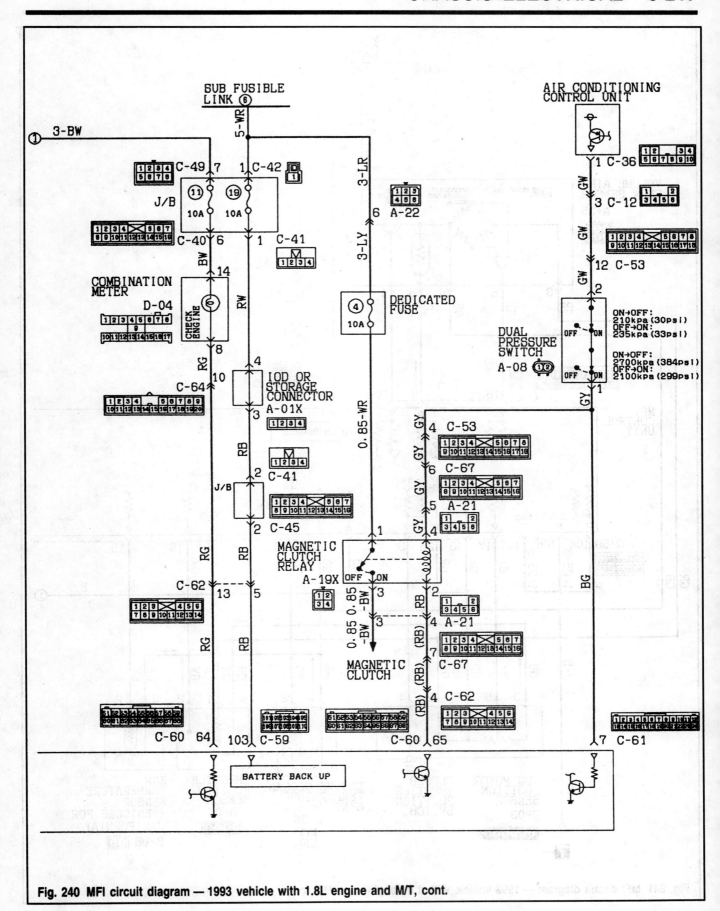

Fig. 240 MFI circuit diagram — 1993 vehicle with 1.8L engine and M/T, cont.

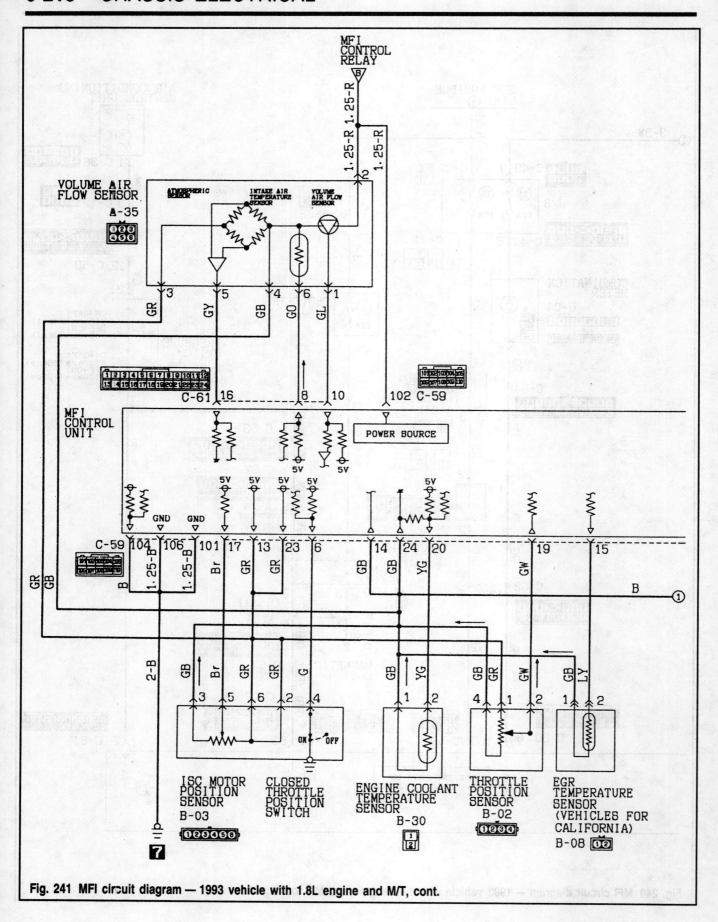

Fig. 241 MFI circuit diagram — 1993 vehicle with 1.8L engine and M/T, cont.

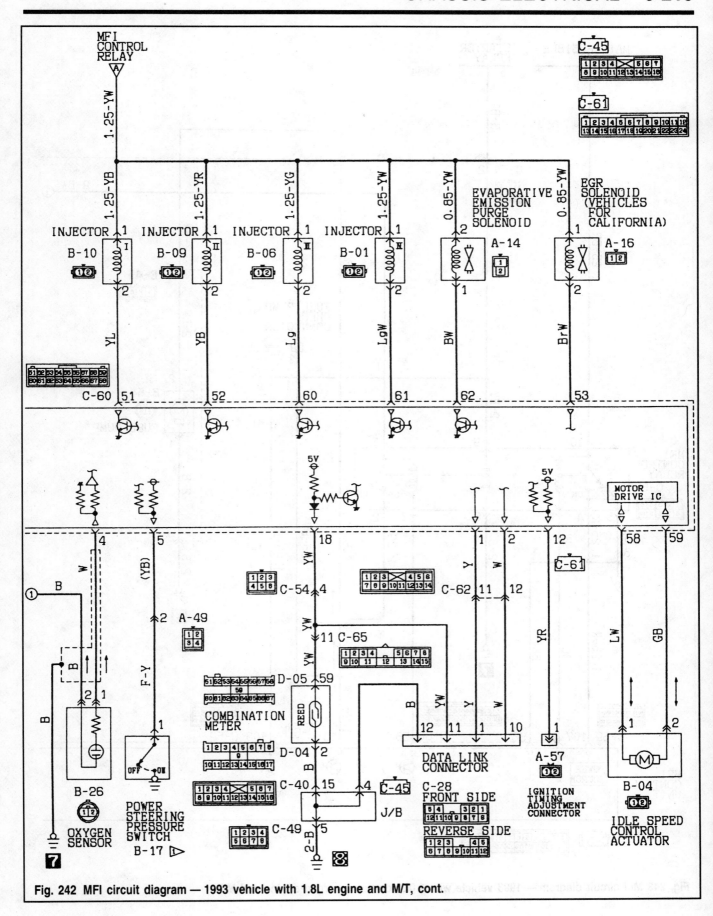

Fig. 242 MFI circuit diagram — 1993 vehicle with 1.8L engine and M/T, cont.

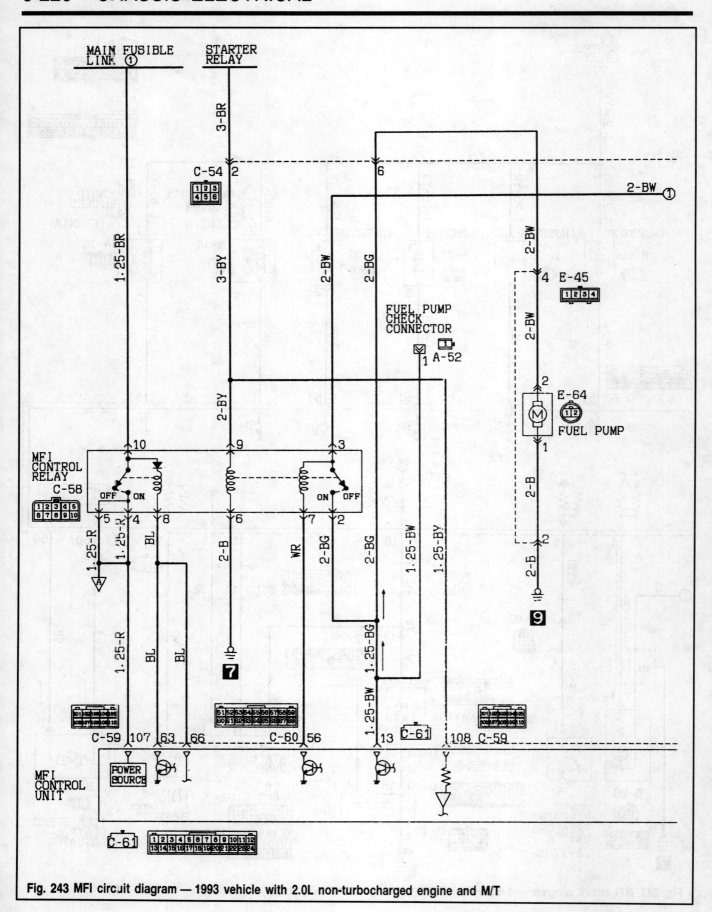

Fig. 243 MFI circuit diagram — 1993 vehicle with 2.0L non-turbocharged engine and M/T

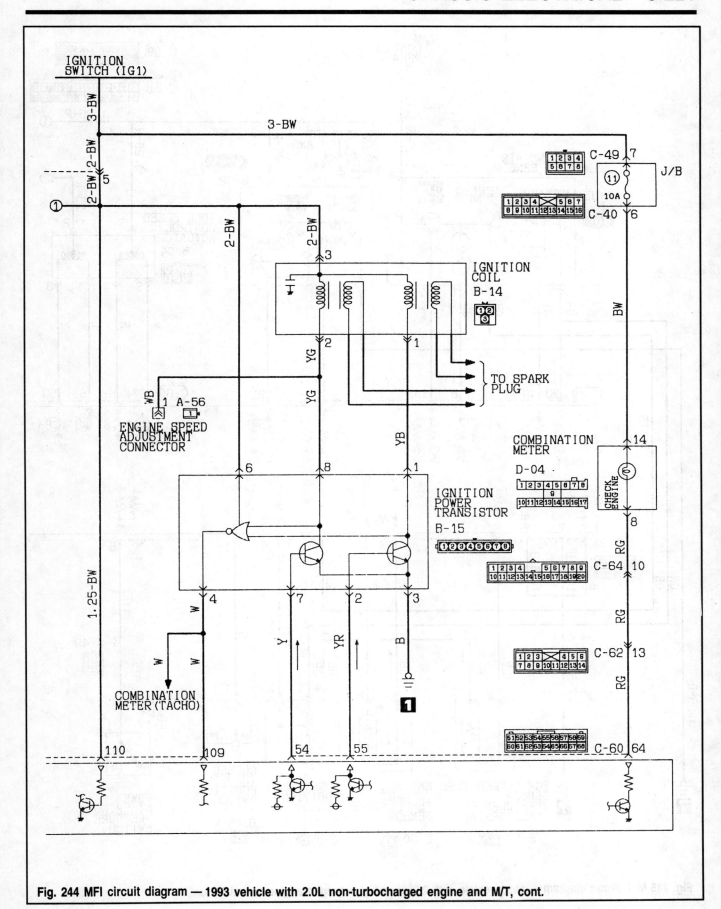

Fig. 244 MFI circuit diagram — 1993 vehicle with 2.0L non-turbocharged engine and M/T, cont.

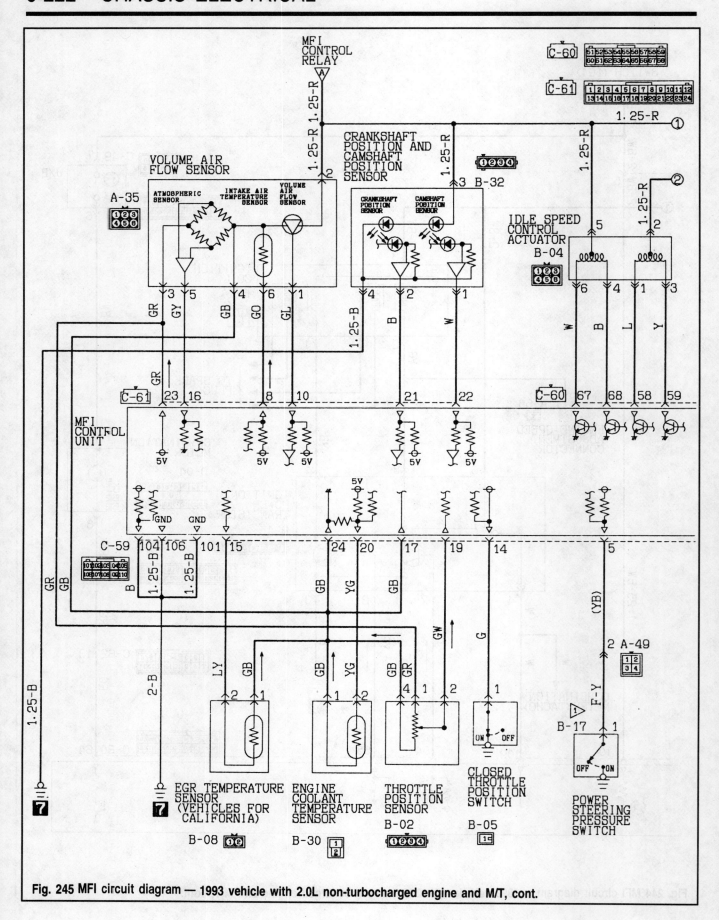

Fig. 245 MFI circuit diagram — 1993 vehicle with 2.0L non-turbocharged engine and M/T, cont.

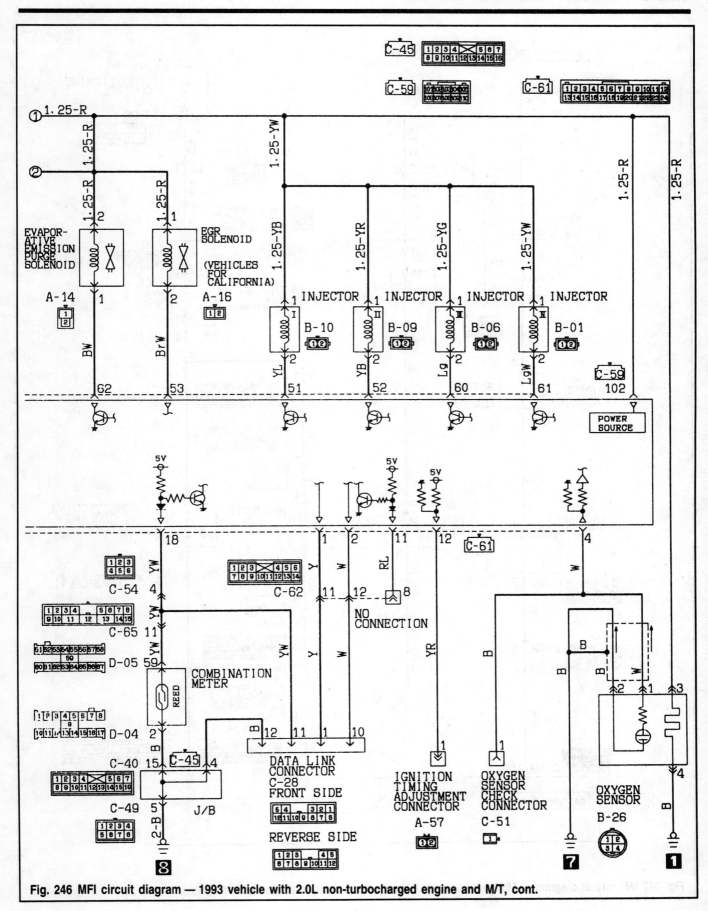

Fig. 246 MFI circuit diagram — 1993 vehicle with 2.0L non-turbocharged engine and M/T, cont.

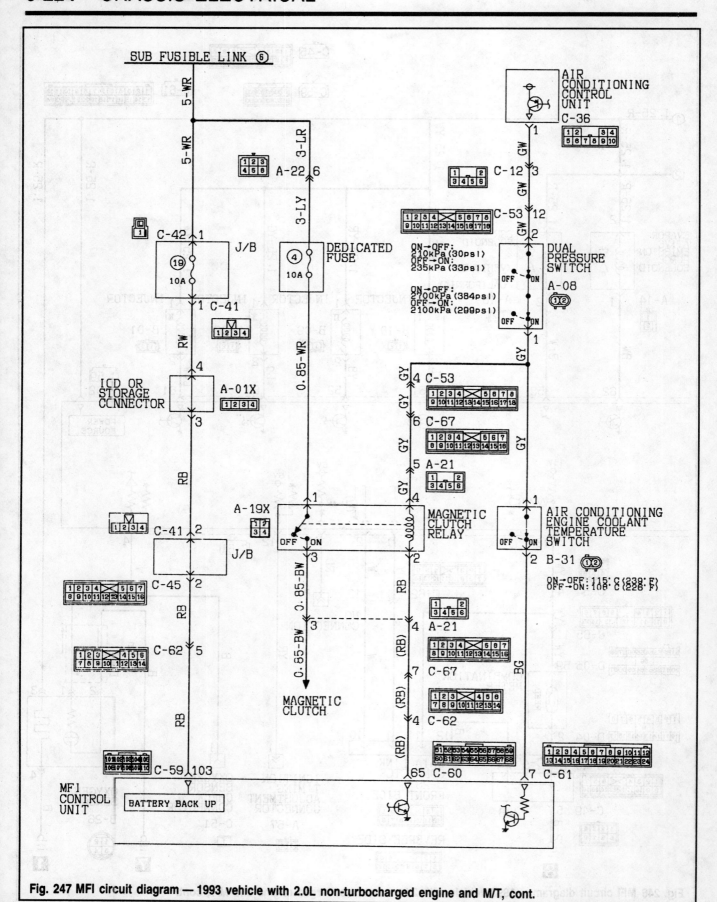

Fig. 247 MFI circuit diagram — 1993 vehicle with 2.0L non-turbocharged engine and M/T, cont.

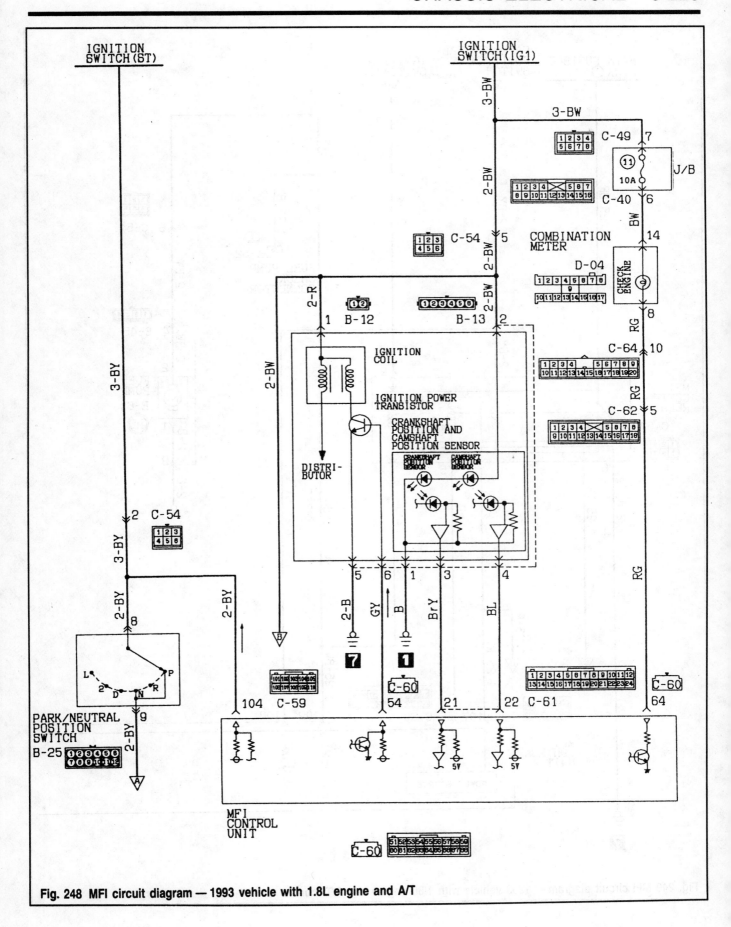

Fig. 248 MFI circuit diagram — 1993 vehicle with 1.8L engine and A/T

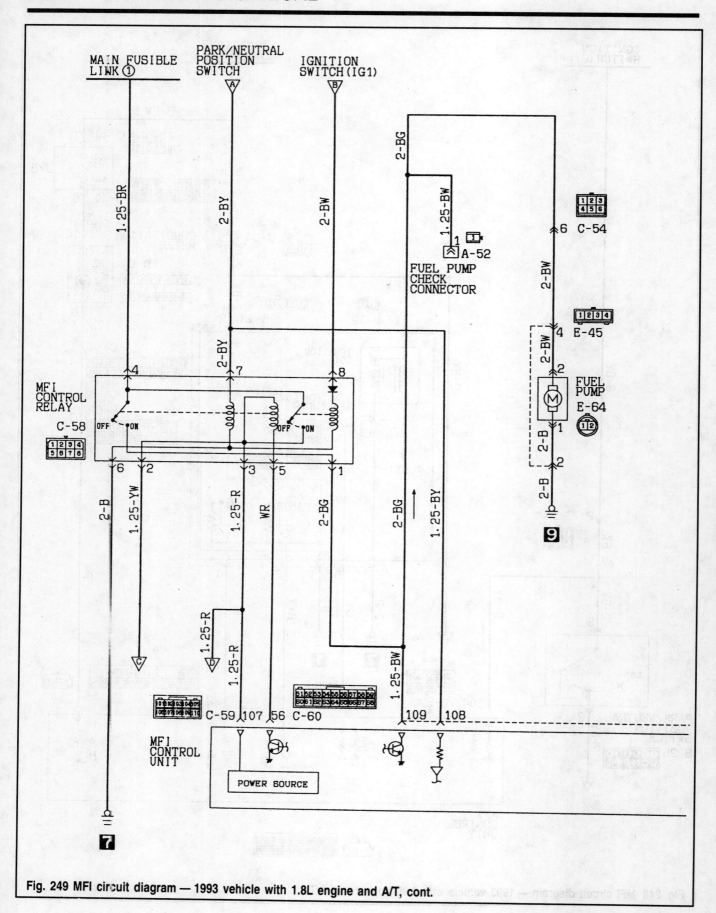

Fig. 249 MFI circuit diagram — 1993 vehicle with 1.8L engine and A/T, cont.

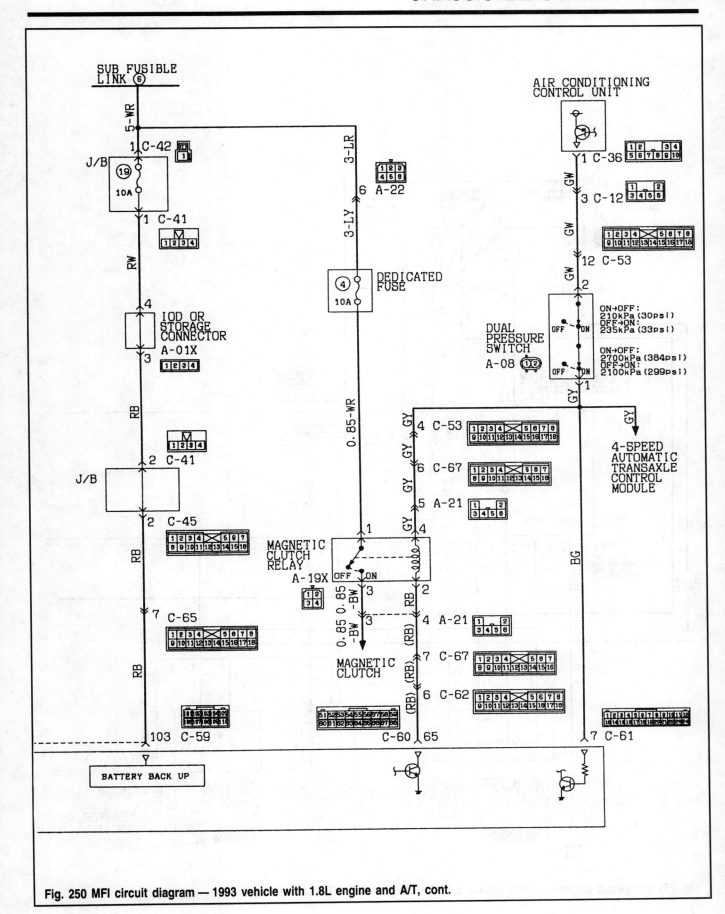

Fig. 250 MFI circuit diagram — 1993 vehicle with 1.8L engine and A/T, cont.

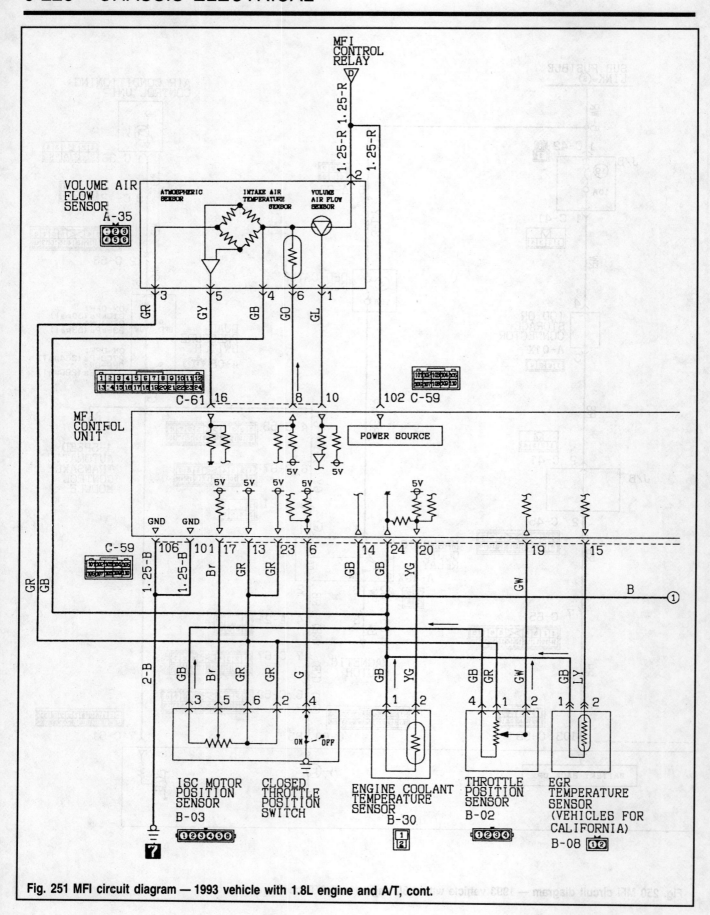

Fig. 251 MFI circuit diagram — 1993 vehicle with 1.8L engine and A/T, cont.

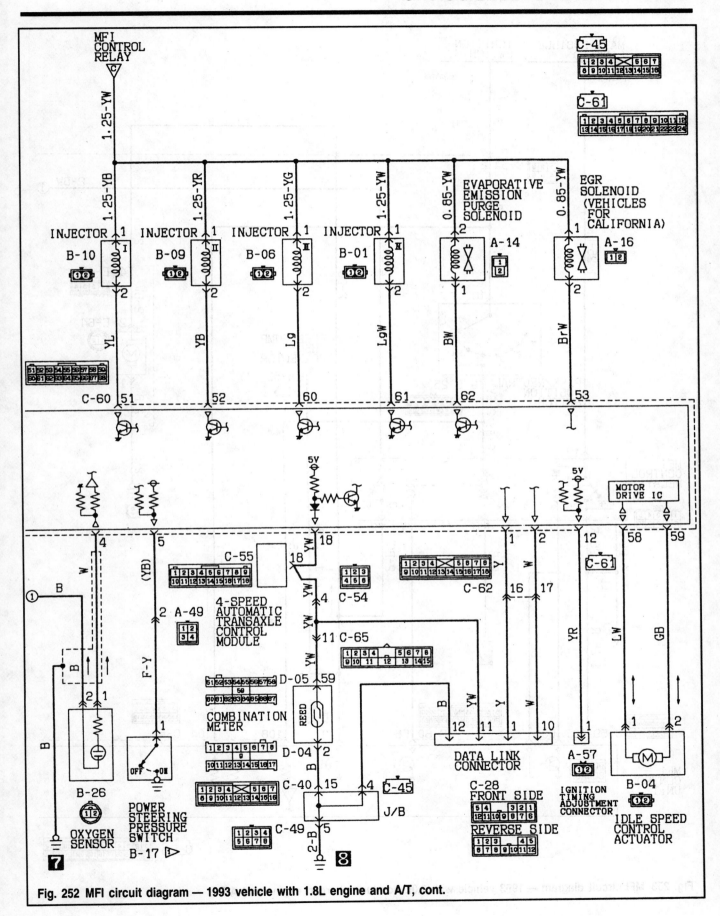

Fig. 252 MFI circuit diagram — 1993 vehicle with 1.8L engine and A/T, cont.

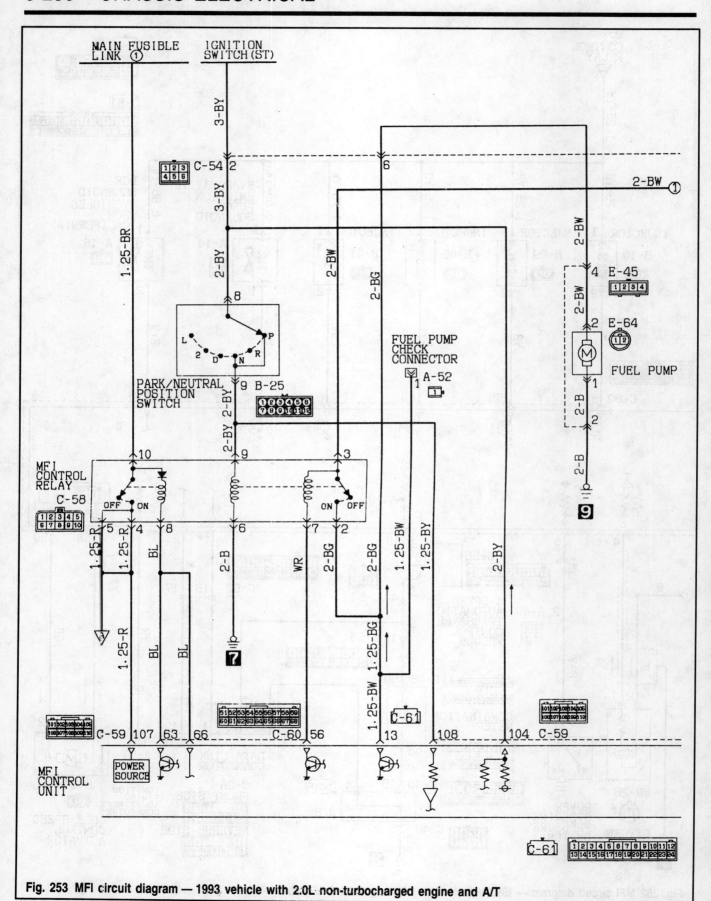

Fig. 253 MFI circuit diagram — 1993 vehicle with 2.0L non-turbocharged engine and A/T

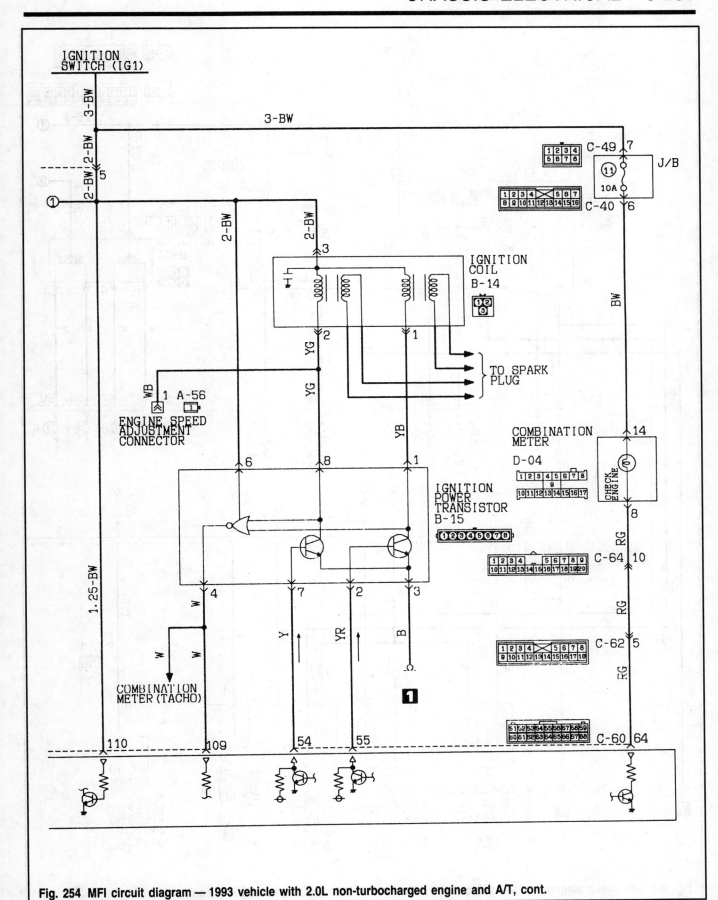

Fig. 254 MFI circuit diagram — 1993 vehicle with 2.0L non-turbocharged engine and A/T, cont.

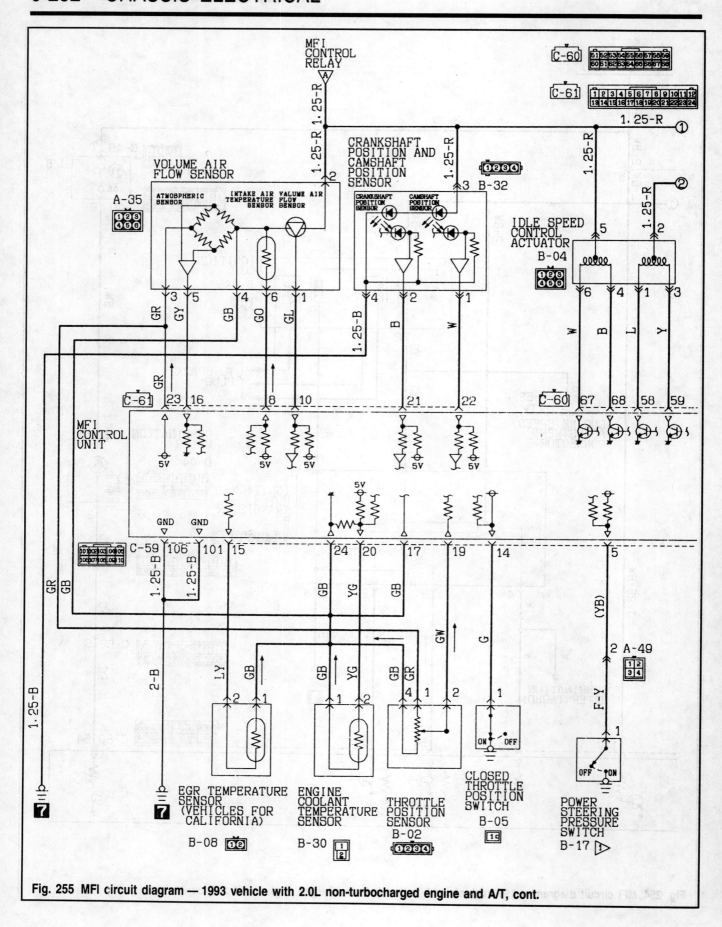

Fig. 255 MFI circuit diagram — 1993 vehicle with 2.0L non-turbocharged engine and A/T, cont.

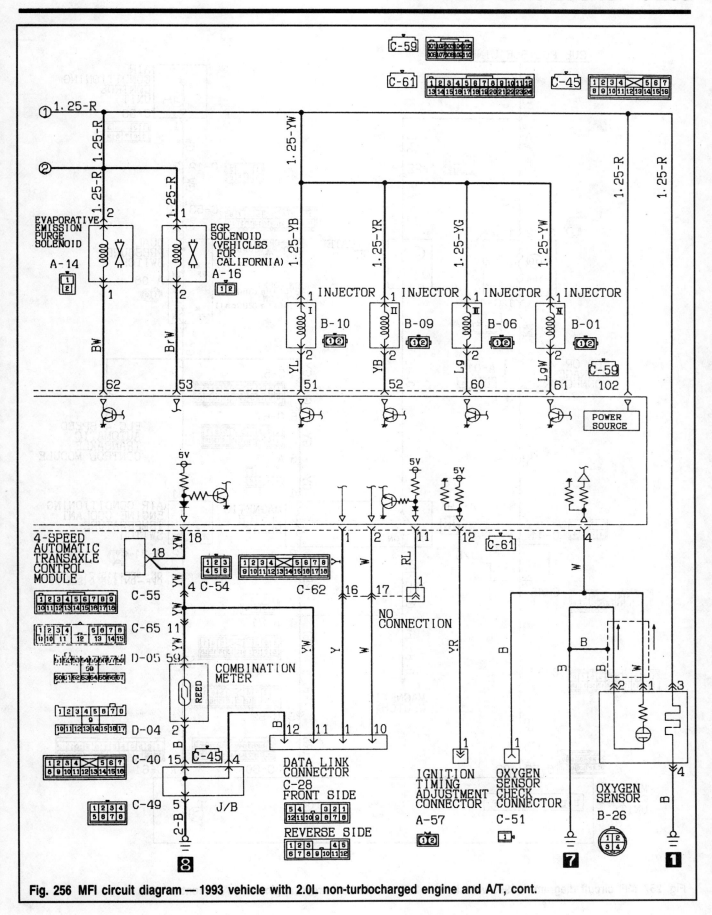

Fig. 256 MFI circuit diagram — 1993 vehicle with 2.0L non-turbocharged engine and A/T, cont.

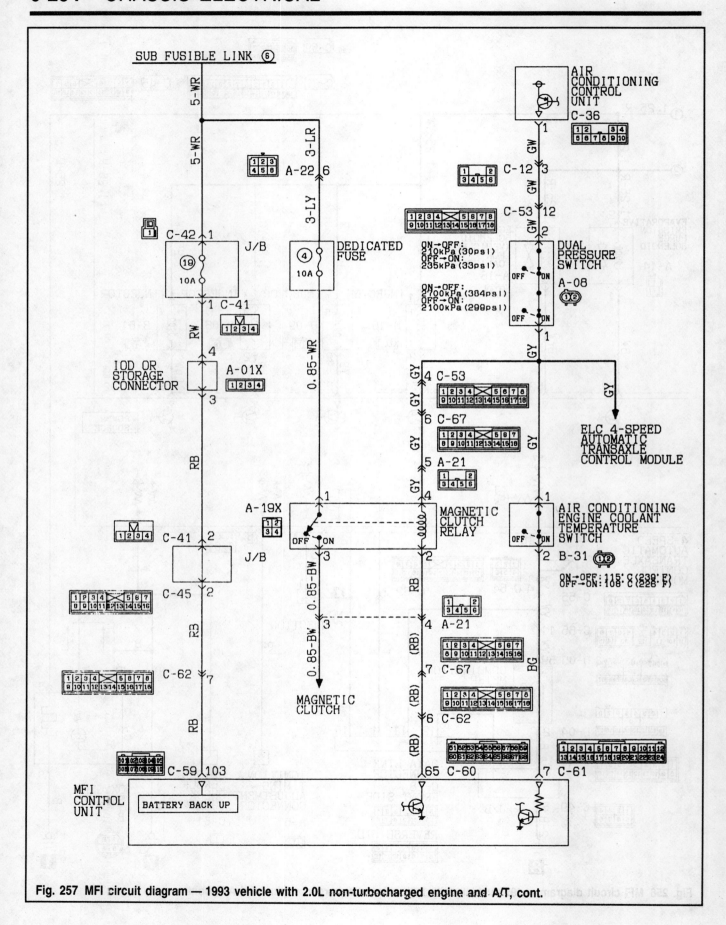

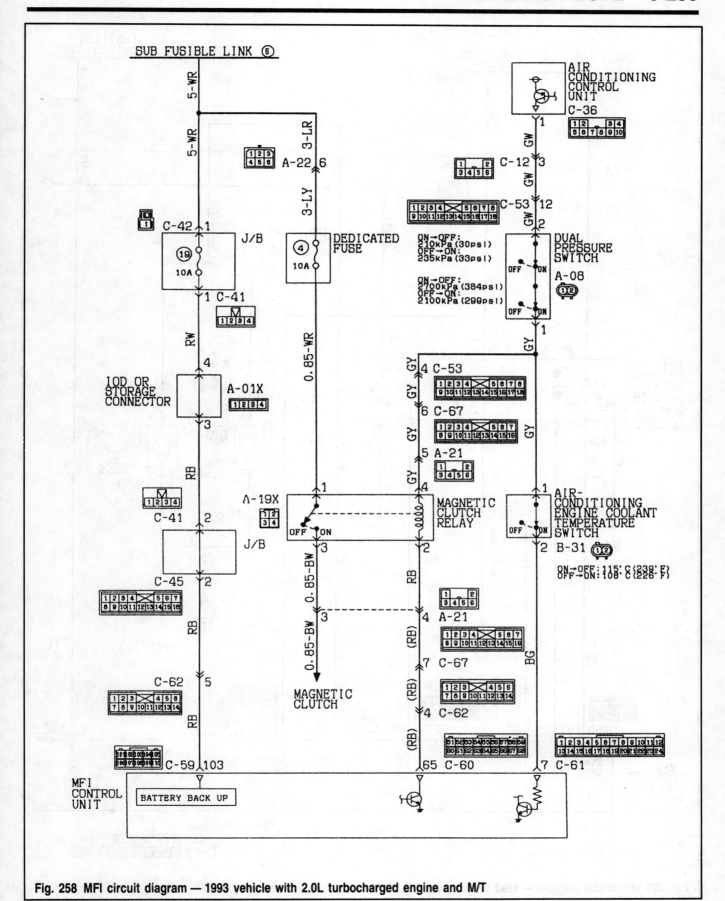

Fig. 258 MFI circuit diagram — 1993 vehicle with 2.0L turbocharged engine and M/T

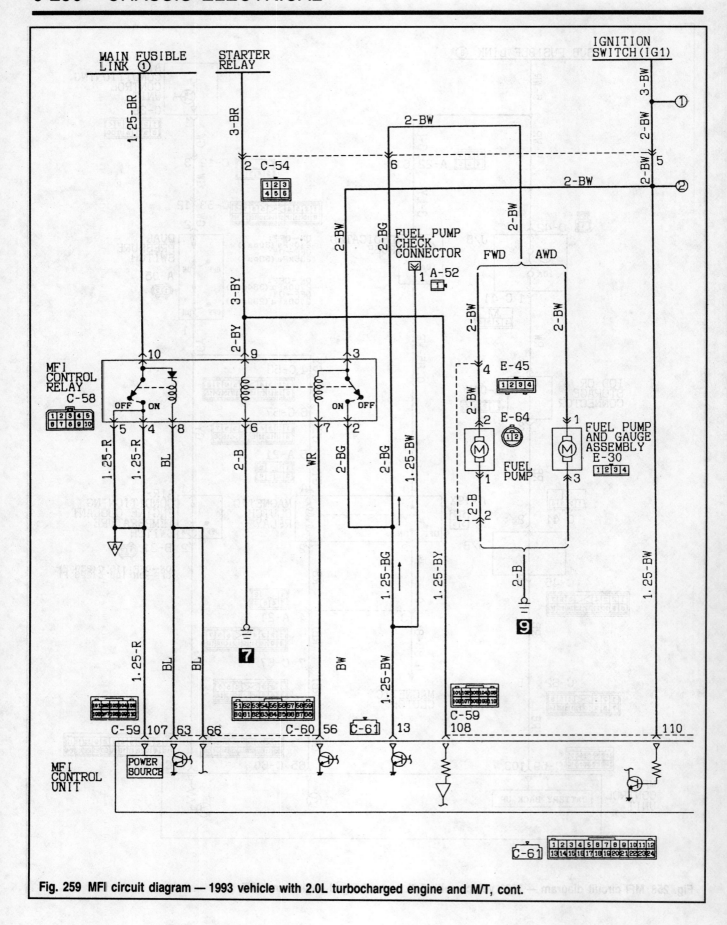

Fig. 259 MFI circuit diagram — 1993 vehicle with 2.0L turbocharged engine and M/T, cont.

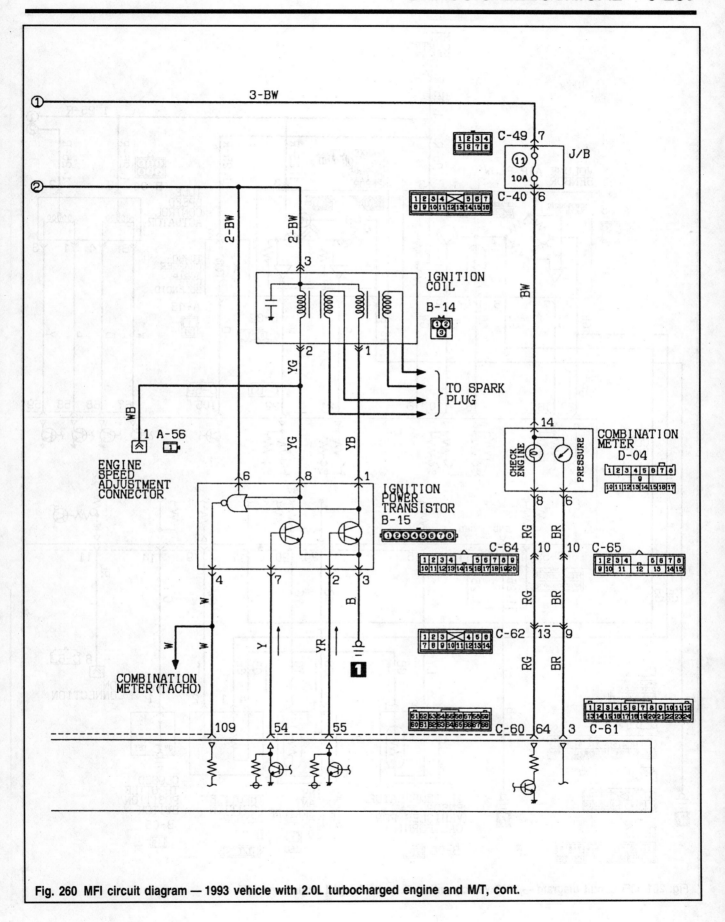

Fig. 260 MFI circuit diagram — 1993 vehicle with 2.0L turbocharged engine and M/T, cont.

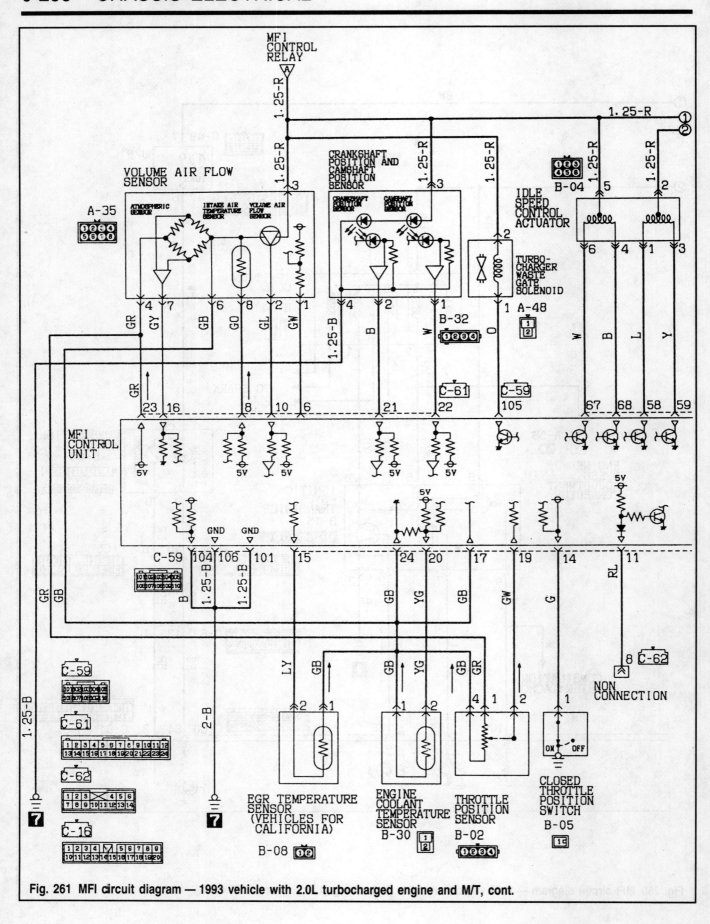

Fig. 261 MFI circuit diagram — 1993 vehicle with 2.0L turbocharged engine and M/T, cont.

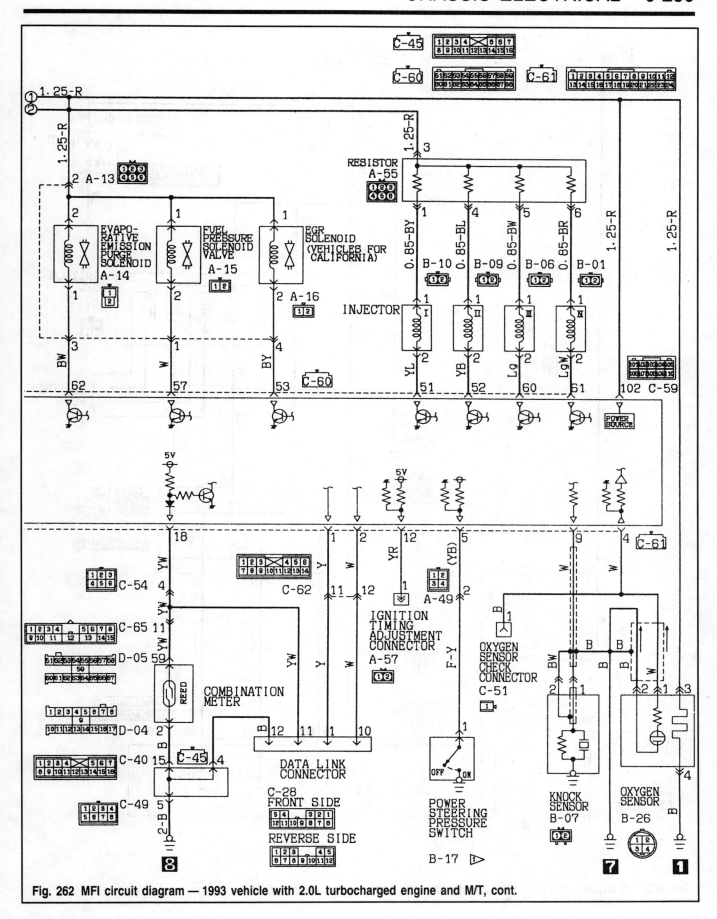

Fig. 262 MFI circuit diagram — 1993 vehicle with 2.0L turbocharged engine and M/T, cont.

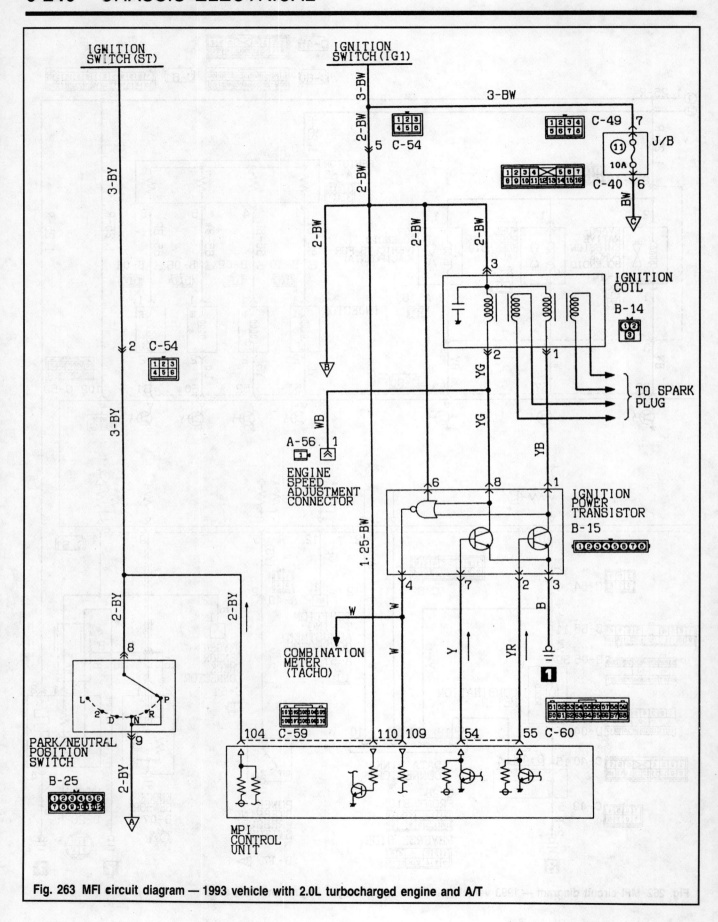

Fig. 263 MFI circuit diagram — 1993 vehicle with 2.0L turbocharged engine and A/T

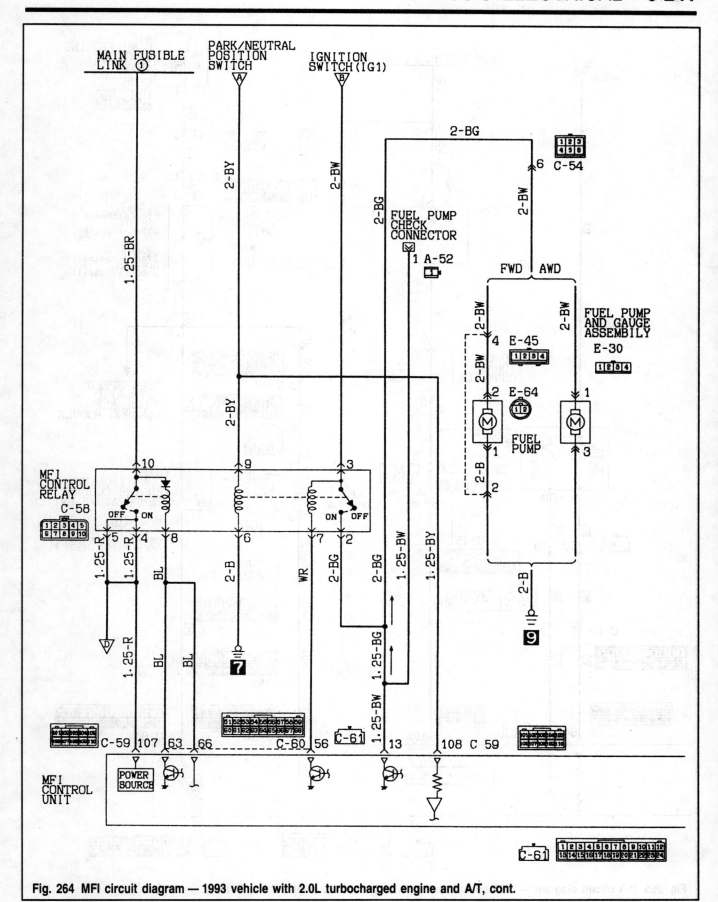

Fig. 264 MFI circuit diagram — 1993 vehicle with 2.0L turbocharged engine and A/T, cont.

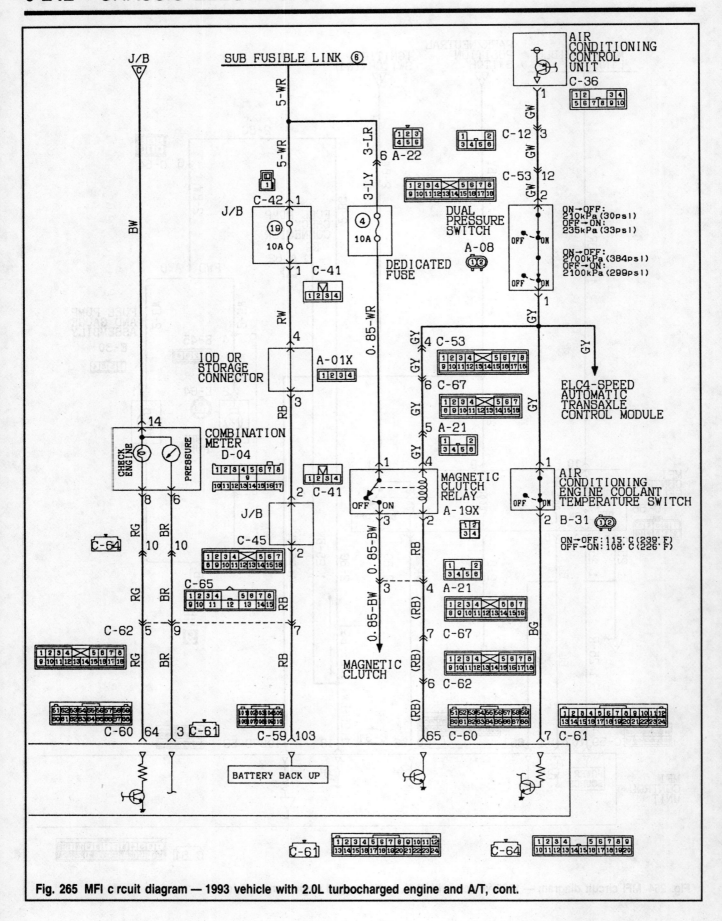

Fig. 265 MFI circuit diagram — 1993 vehicle with 2.0L turbocharged engine and A/T, cont.

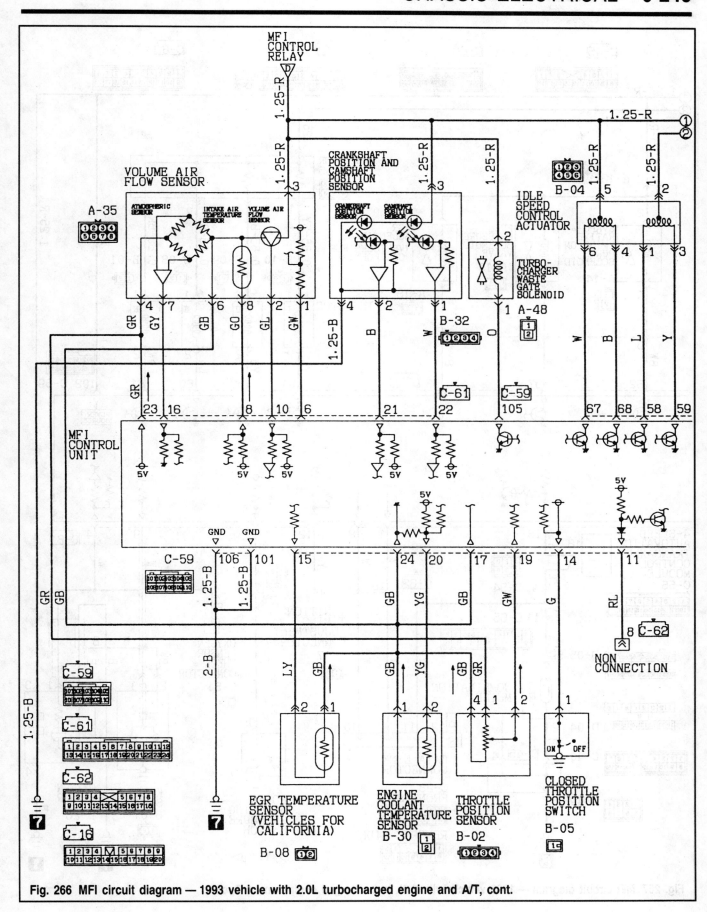

Fig. 266 MFI circuit diagram — 1993 vehicle with 2.0L turbocharged engine and A/T, cont.

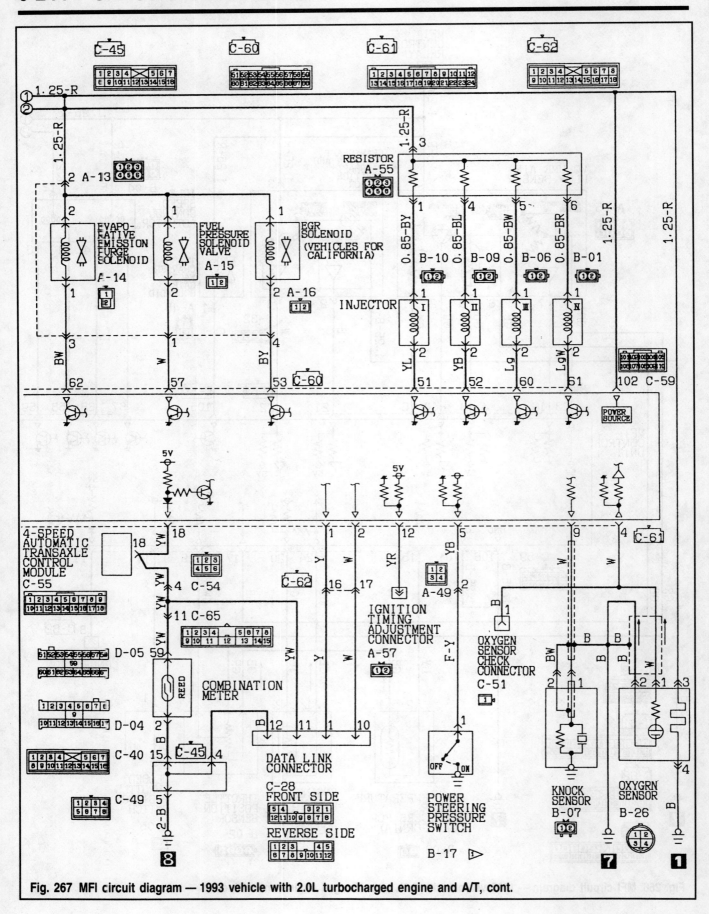

Fig. 267 MFI circuit diagram — 1993 vehicle with 2.0L turbocharged engine and A/T, cont.

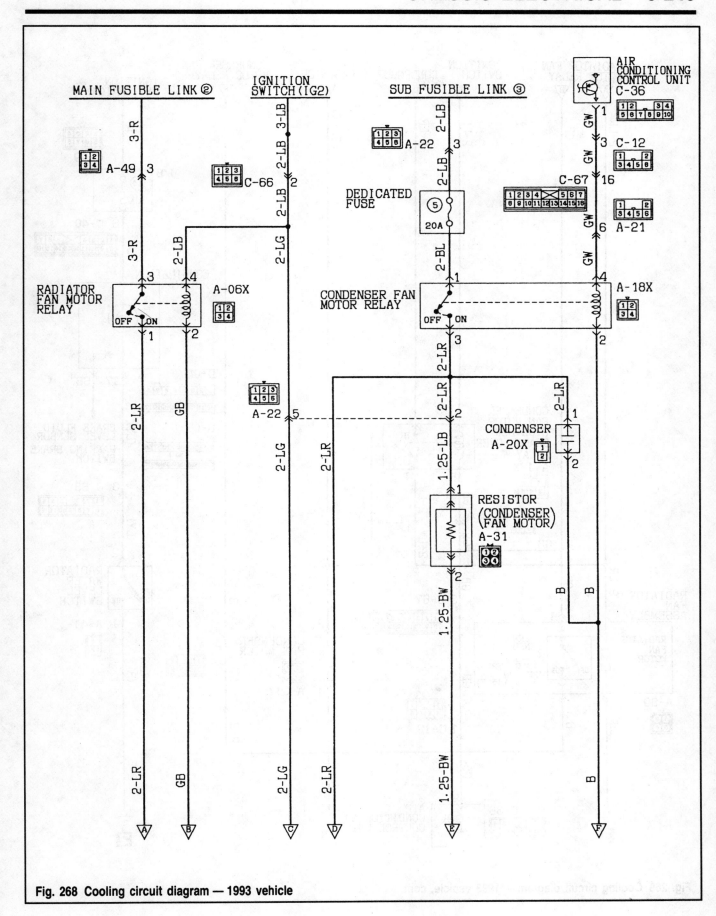

Fig. 268 Cooling circuit diagram — 1993 vehicle

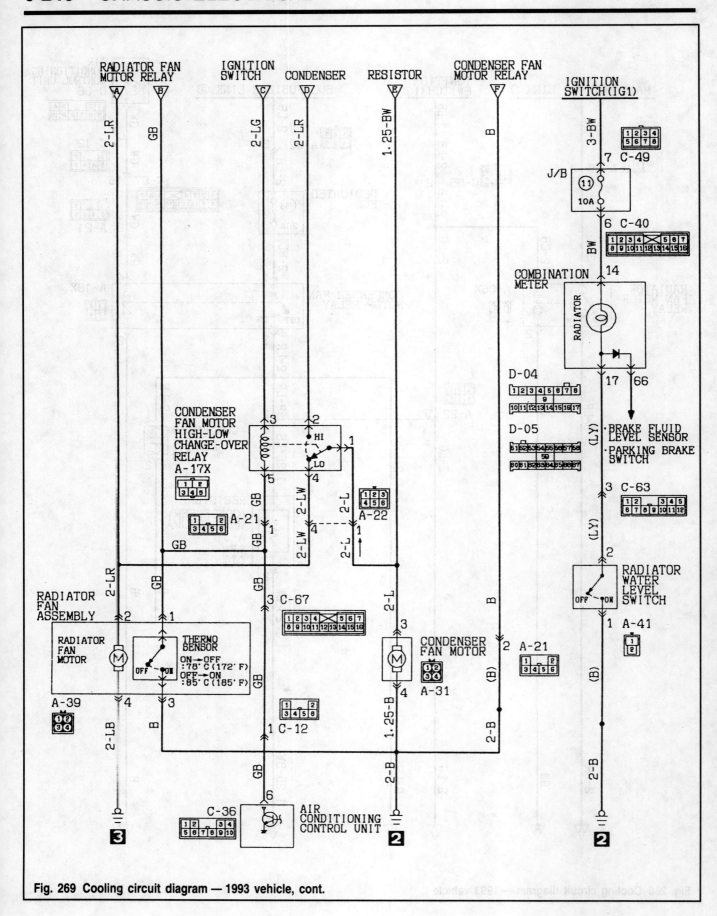

Fig. 269 Cooling circuit diagram — 1993 vehicle, cont.

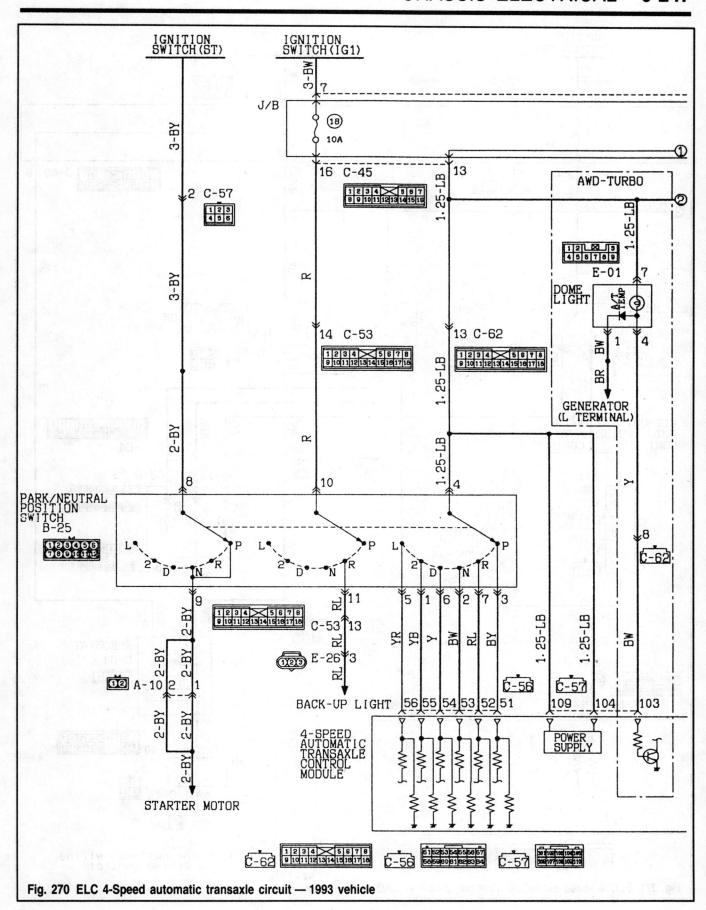

Fig. 270 ELC 4-Speed automatic transaxle circuit — 1993 vehicle

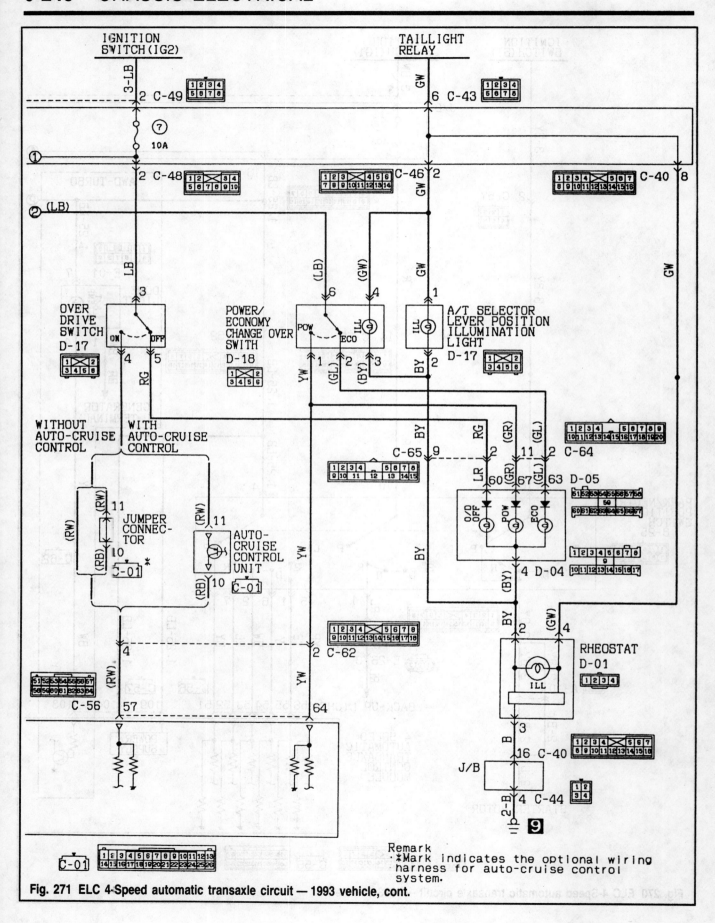

Fig. 271 ELC 4-Speed automatic transaxle circuit — 1993 vehicle, cont.

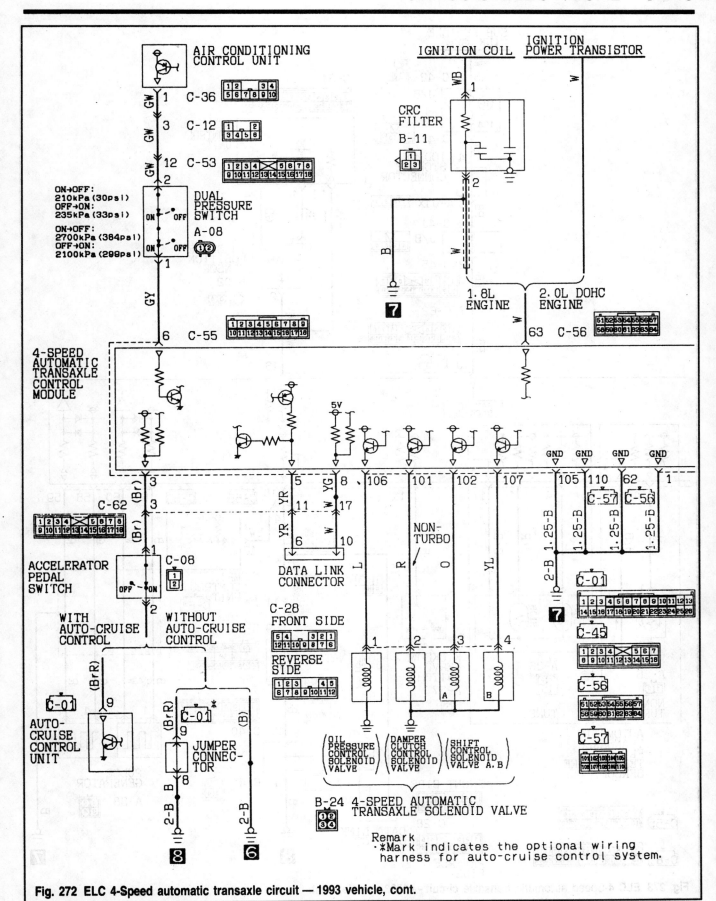

Fig. 272 ELC 4-Speed automatic transaxle circuit — 1993 vehicle, cont.

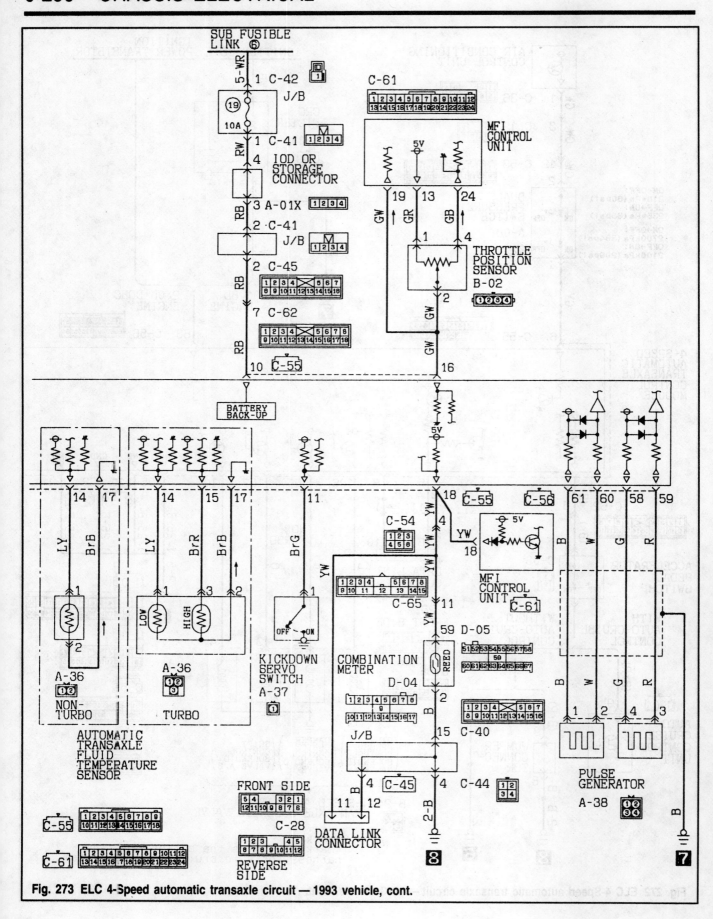

Fig. 273 ELC 4-Speed automatic transaxle circuit — 1993 vehicle, cont.

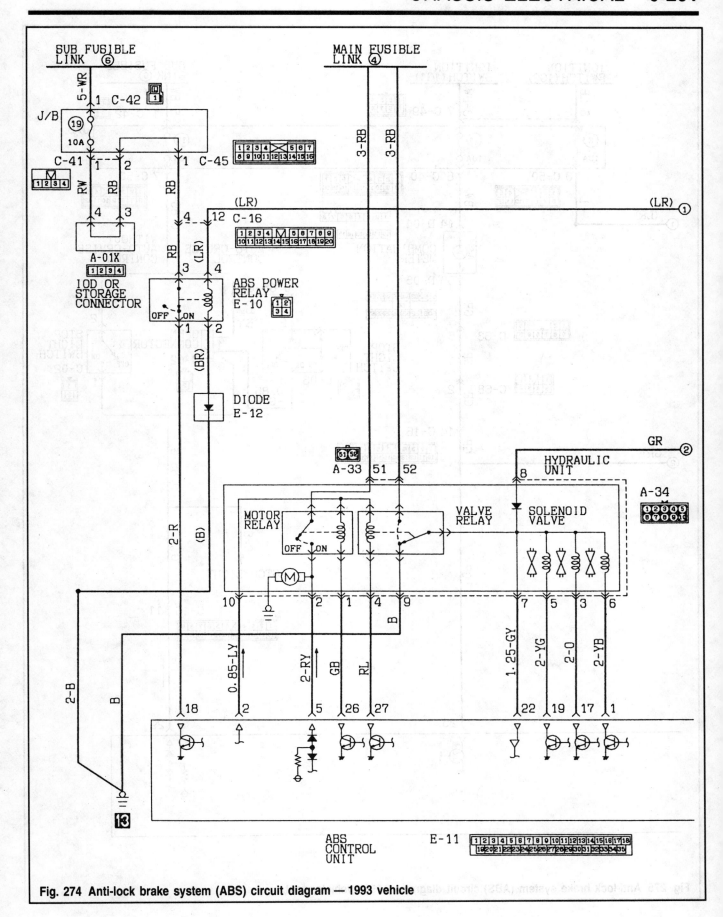

Fig. 274 Anti-lock brake system (ABS) circuit diagram — 1993 vehicle

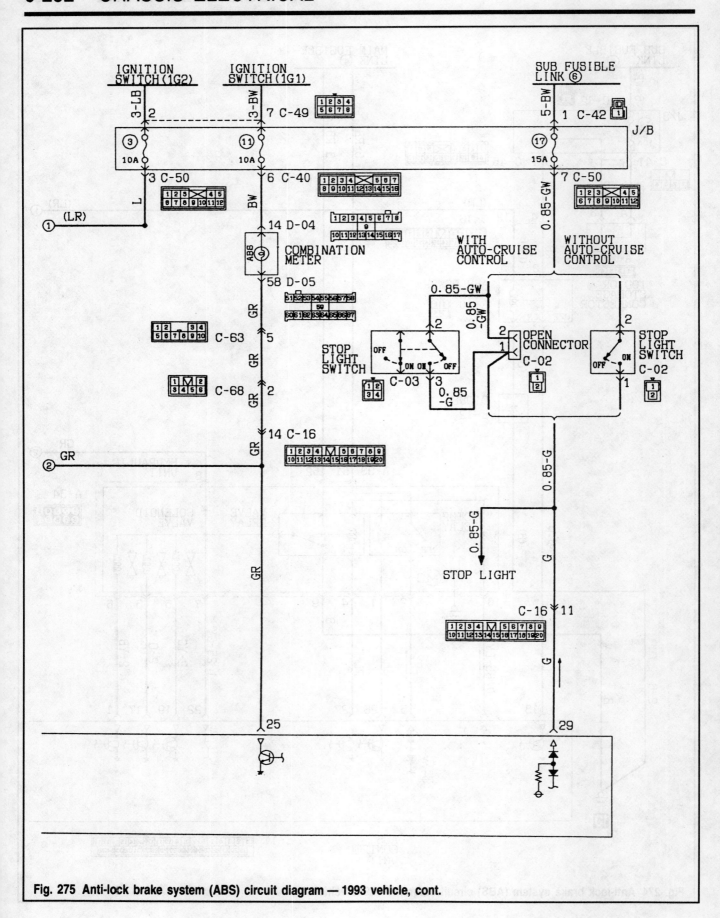

Fig. 275 Anti-lock brake system (ABS) circuit diagram — 1993 vehicle, cont.

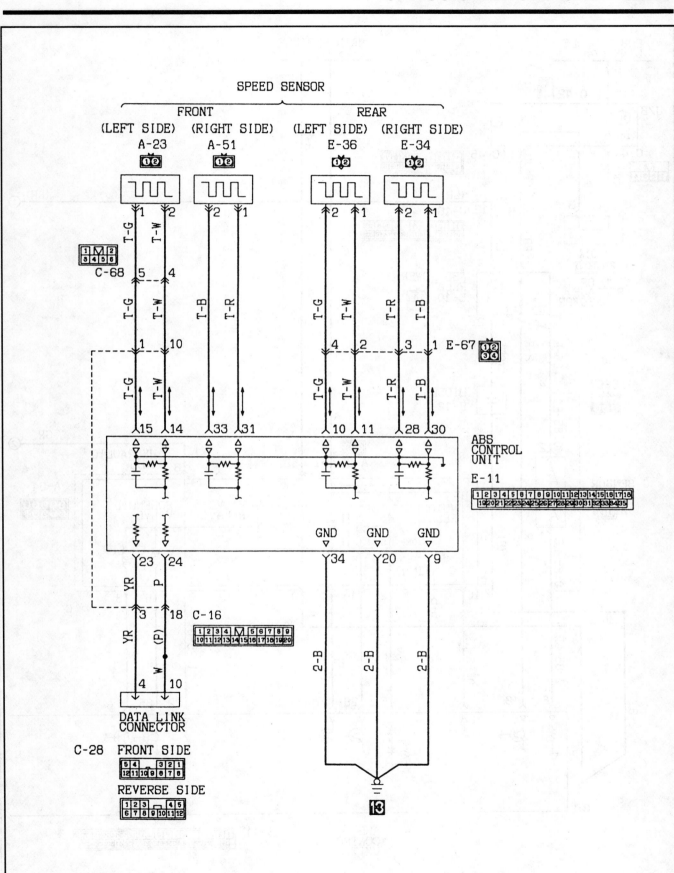

Fig. 276 Anti-lock brake system (ABS) circuit diagram — 1993 vehicle, cont.

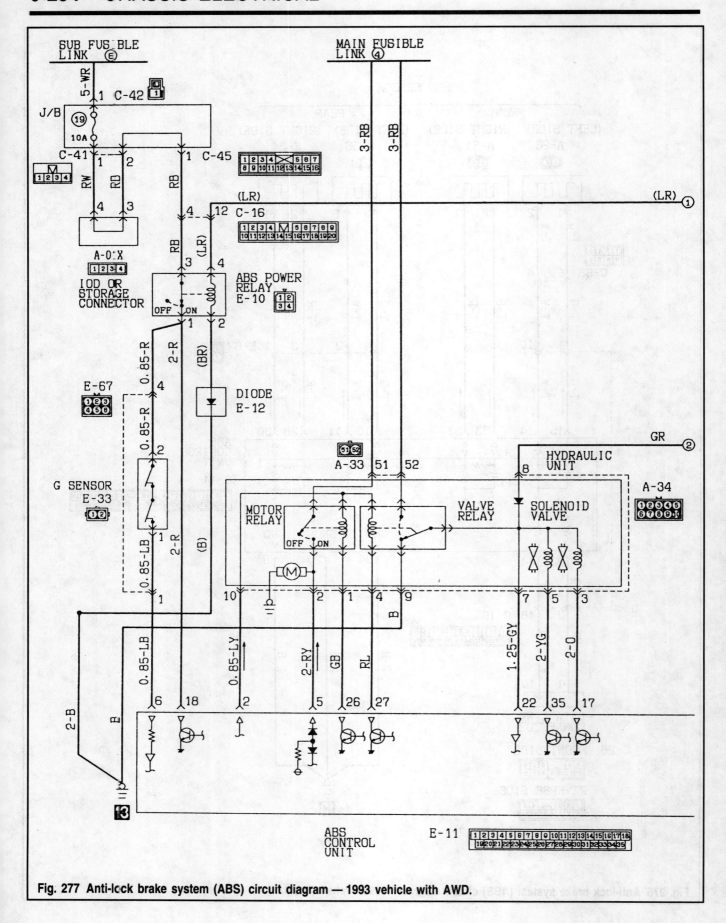

Fig. 277 Anti-lock brake system (ABS) circuit diagram — 1993 vehicle with AWD.

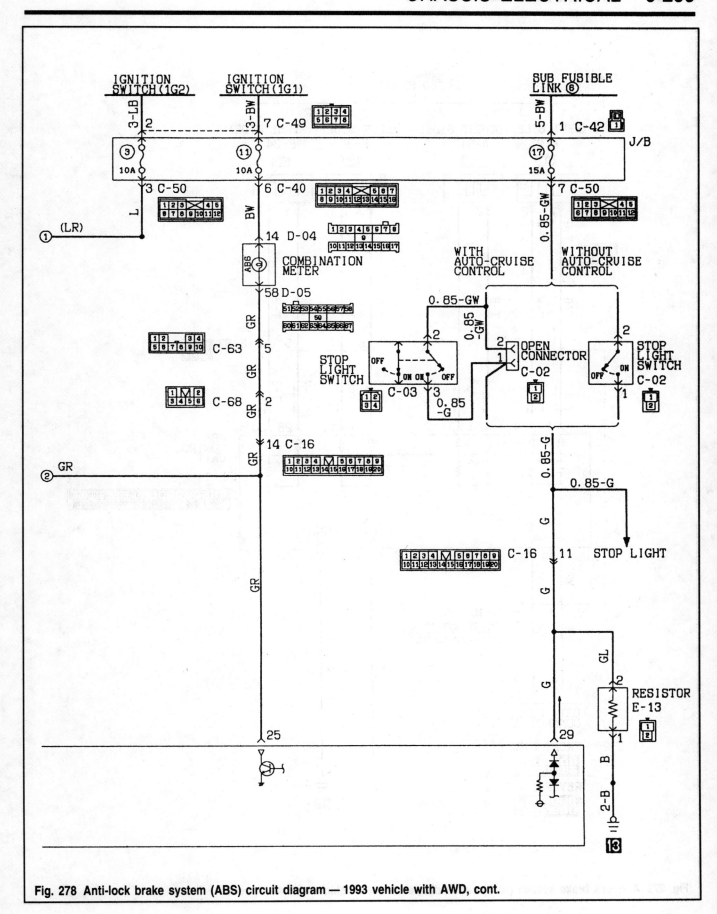

Fig. 278 Anti-lock brake system (ABS) circuit diagram — 1993 vehicle with AWD, cont.

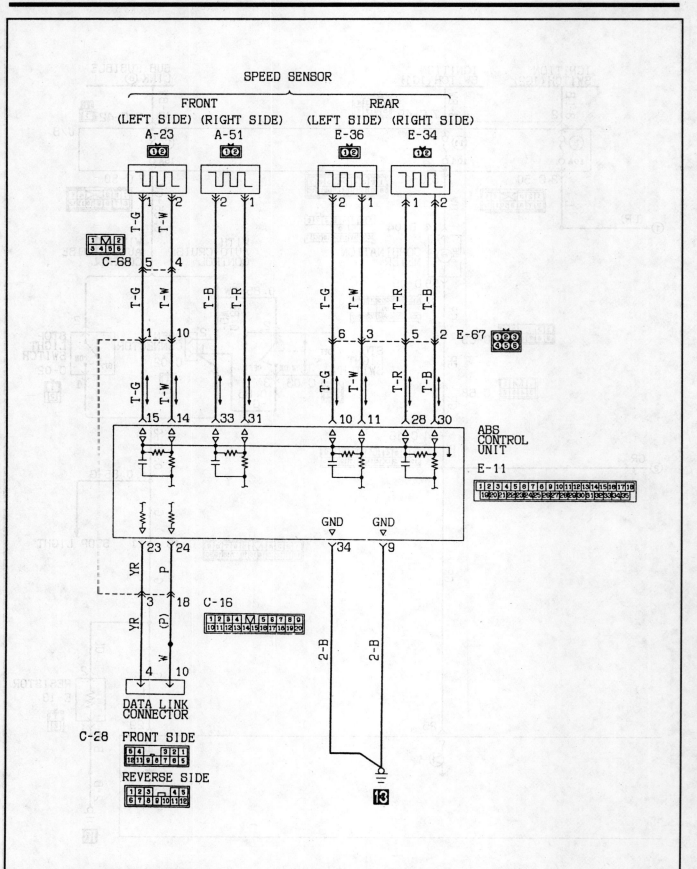

Fig. 279 Anti-lock brake system (ABS) circuit diagram — 1993 vehicle with AWD, cont.

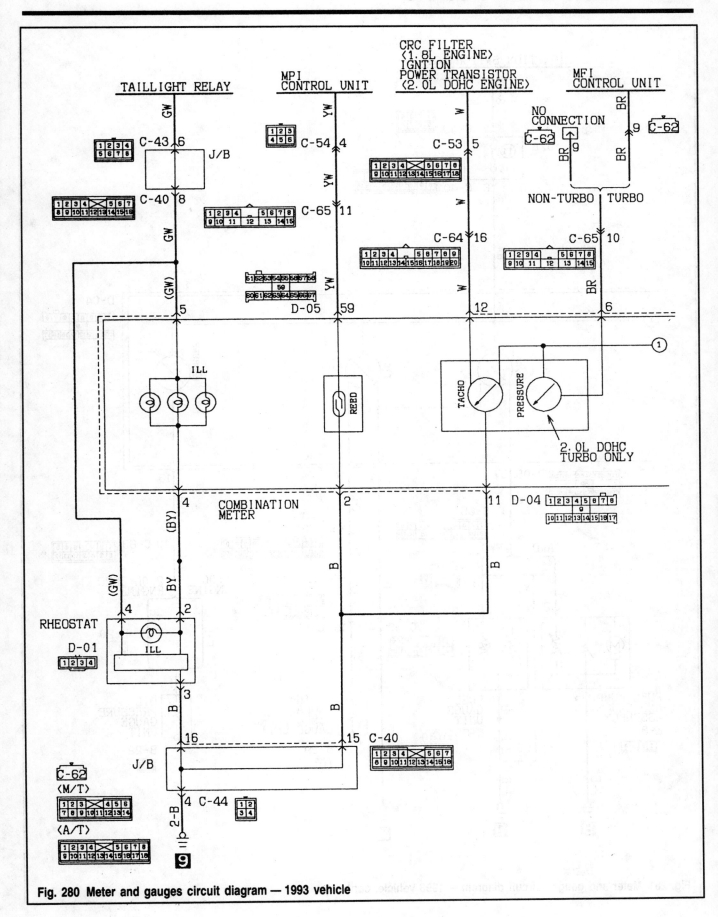

Fig. 280 Meter and gauges circuit diagram — 1993 vehicle

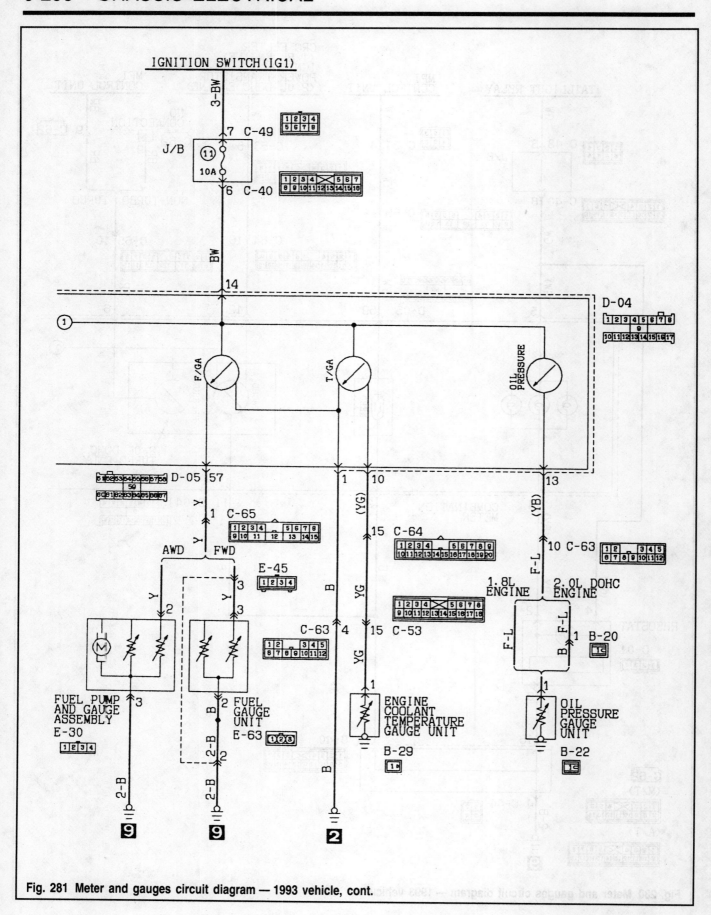

Fig. 281 Meter and gauges circuit diagram — 1993 vehicle, cont.

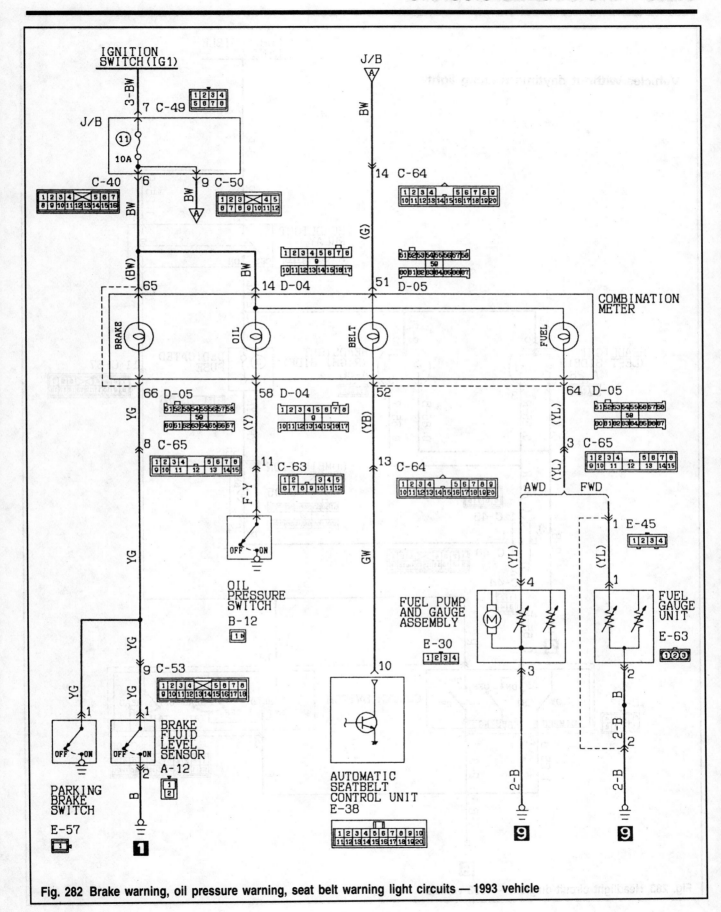

Fig. 282 Brake warning, oil pressure warning, seat belt warning light circuits — 1993 vehicle

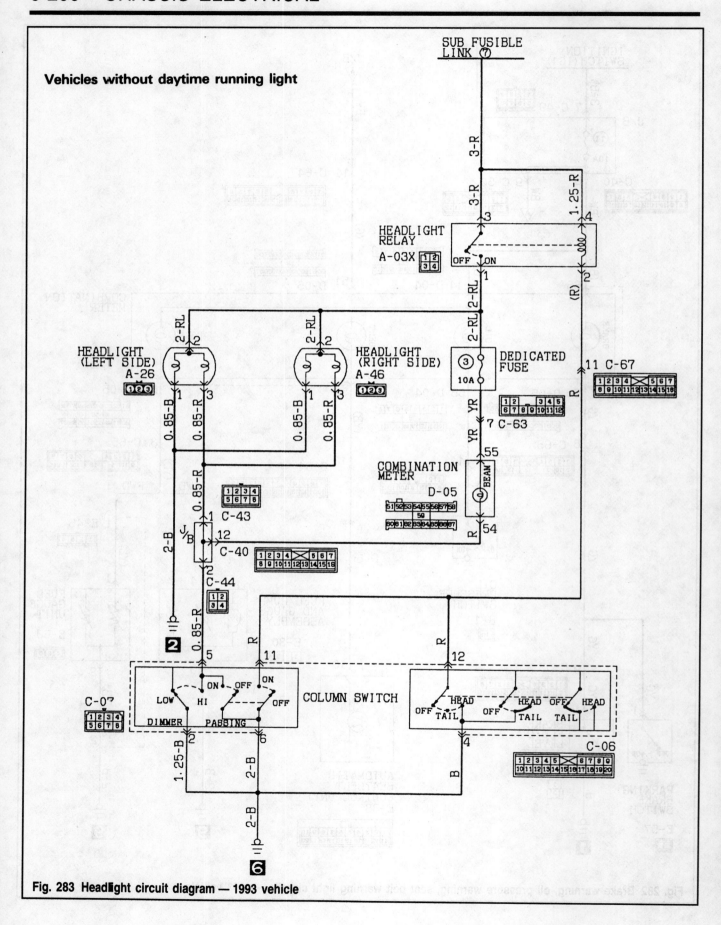

Fig. 283 Headlight circuit diagram — 1993 vehicle

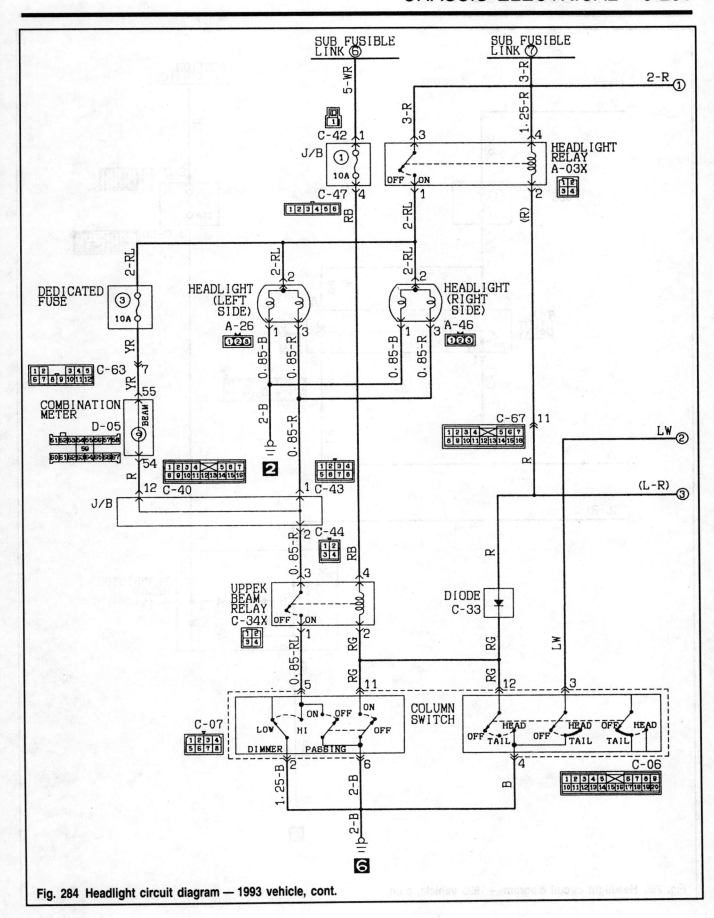

Fig. 284 Headlight circuit diagram — 1993 vehicle, cont.

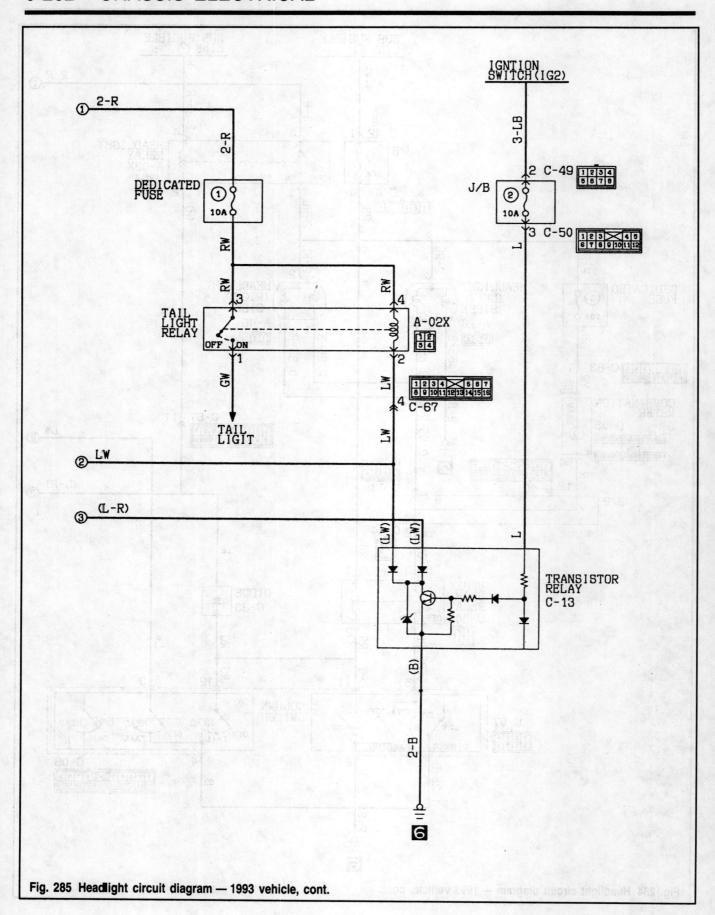

Fig. 285 Headlight circuit diagram — 1993 vehicle, cont.

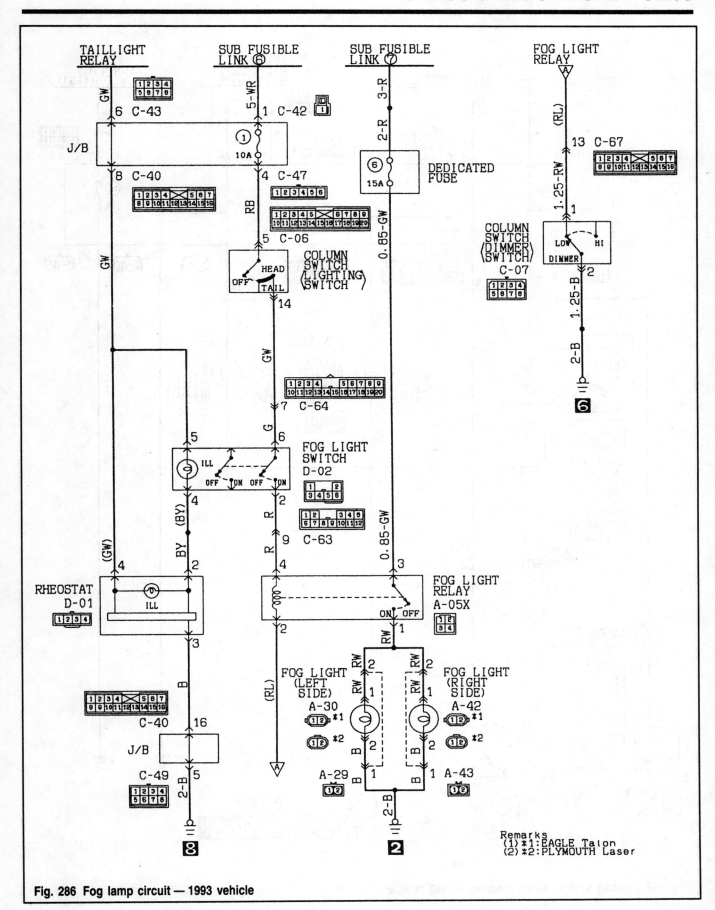

Fig. 286 Fog lamp circuit — 1993 vehicle

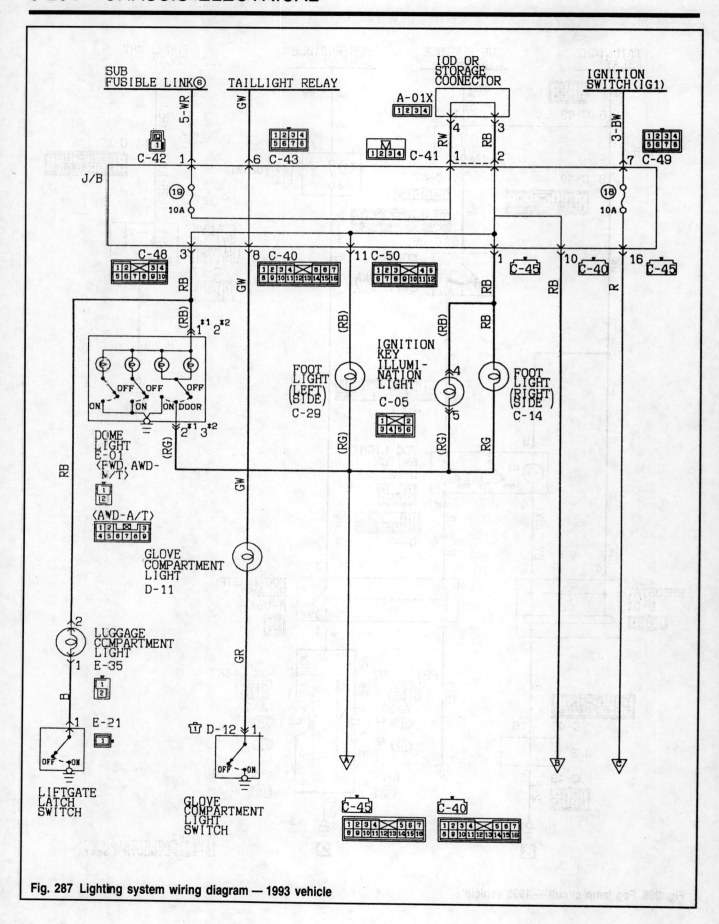

Fig. 287 Lighting system wiring diagram — 1993 vehicle

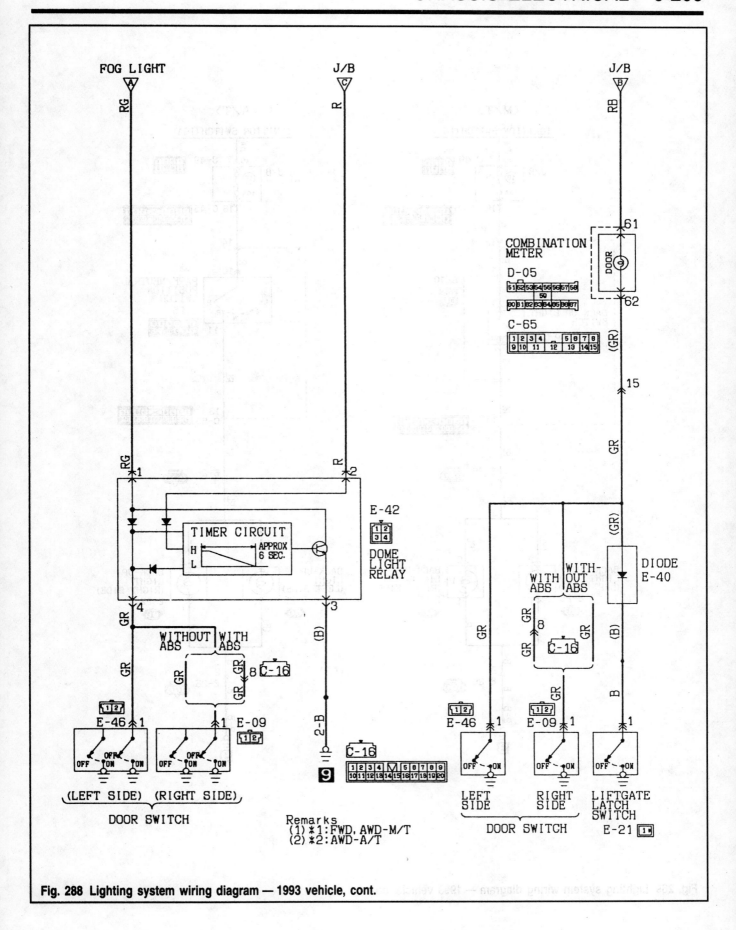

Fig. 288 Lighting system wiring diagram — 1993 vehicle, cont.

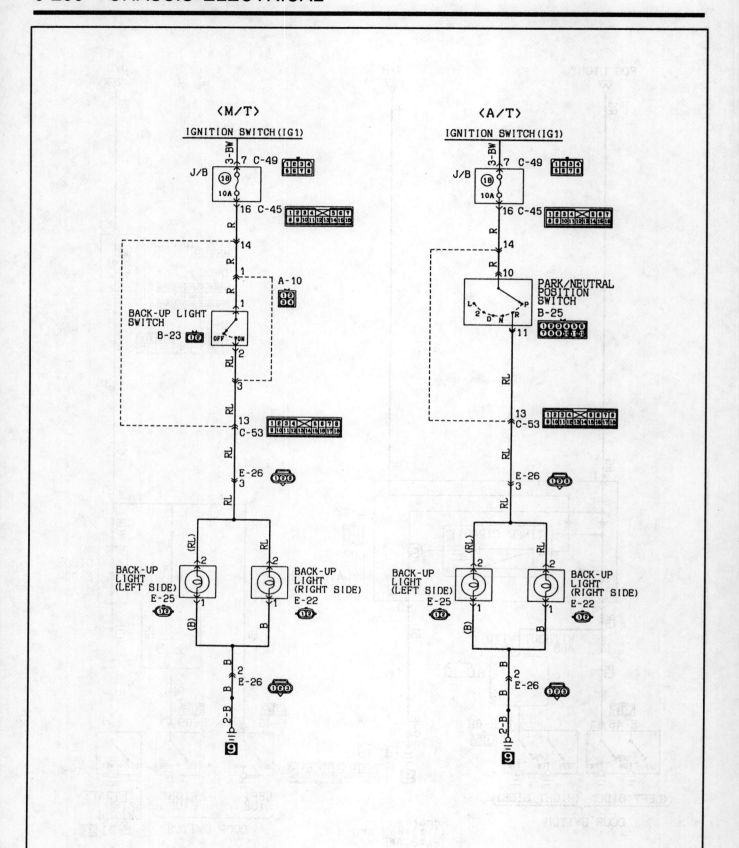

Fig. 289 Lighting system wiring diagram — 1993 vehicle, cont.

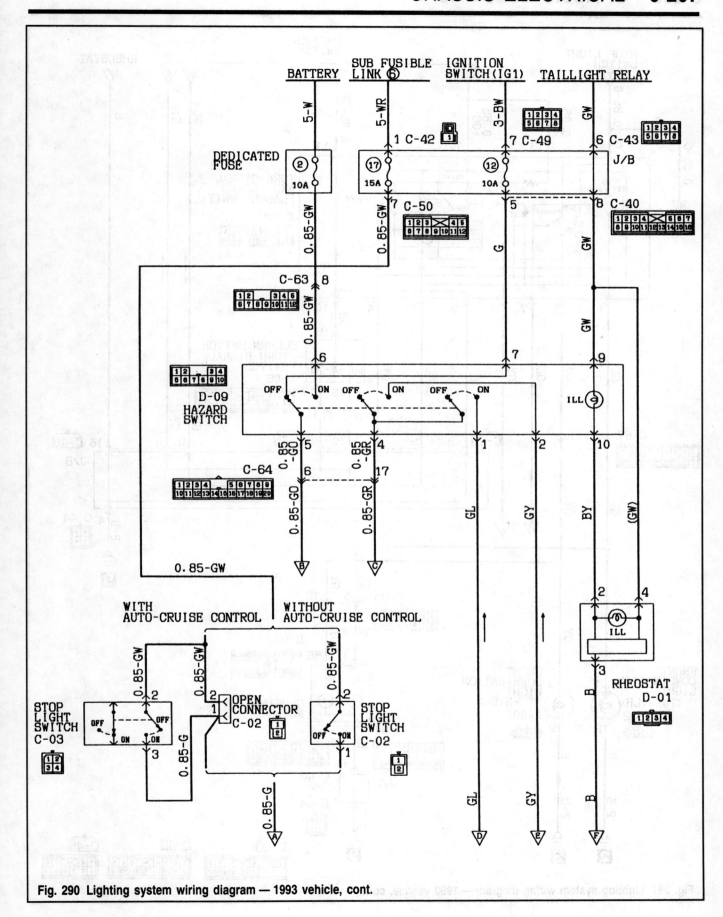

Fig. 290 Lighting system wiring diagram — 1993 vehicle, cont.

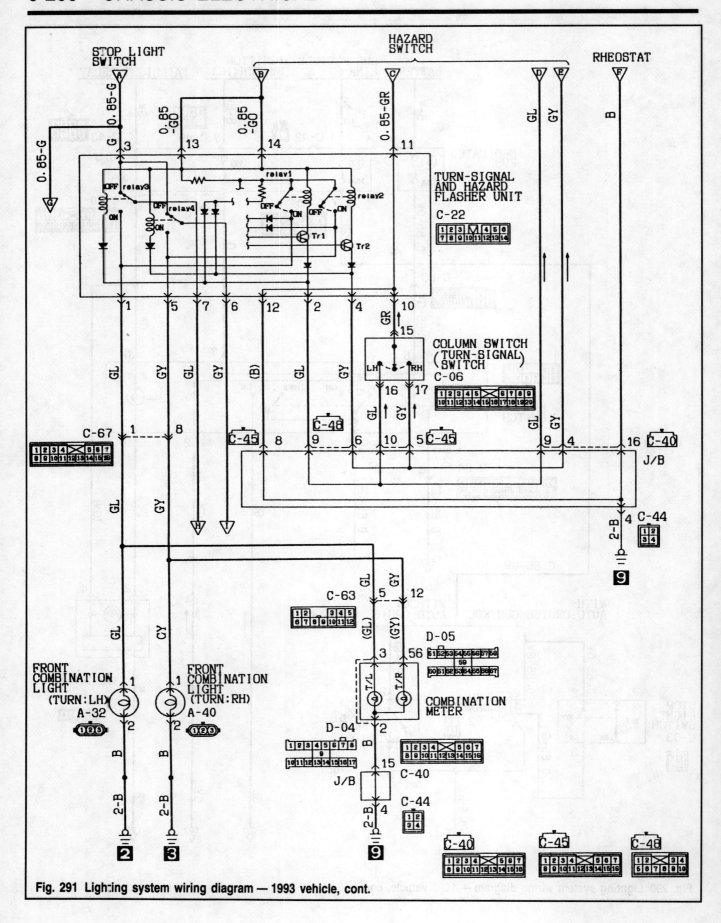

Fig. 291 Lighting system wiring diagram — 1993 vehicle, cont.

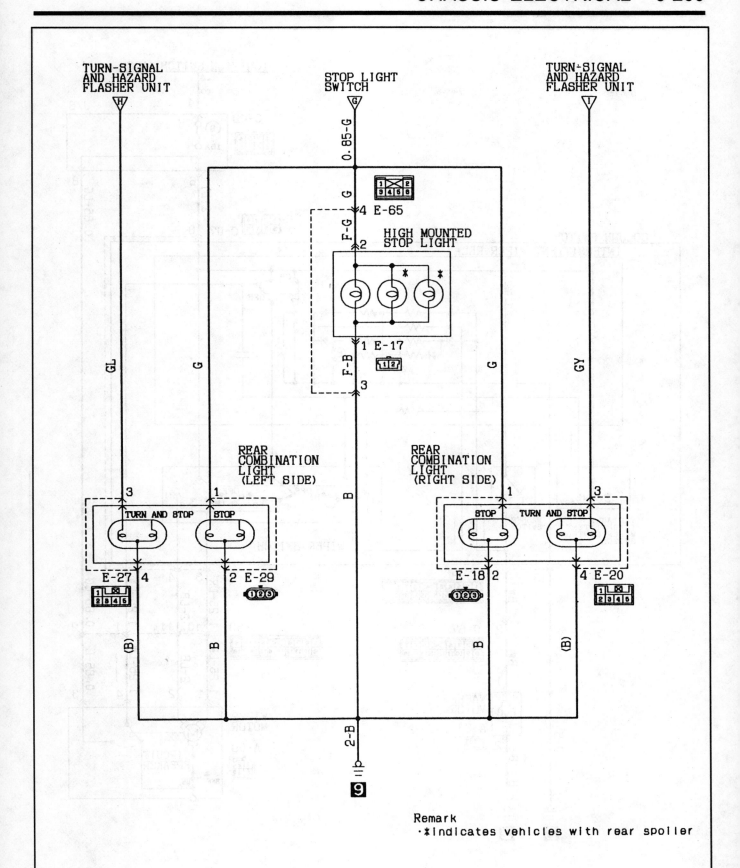

Fig. 292 Lighting system wiring diagram — 1993 vehicle, cont.

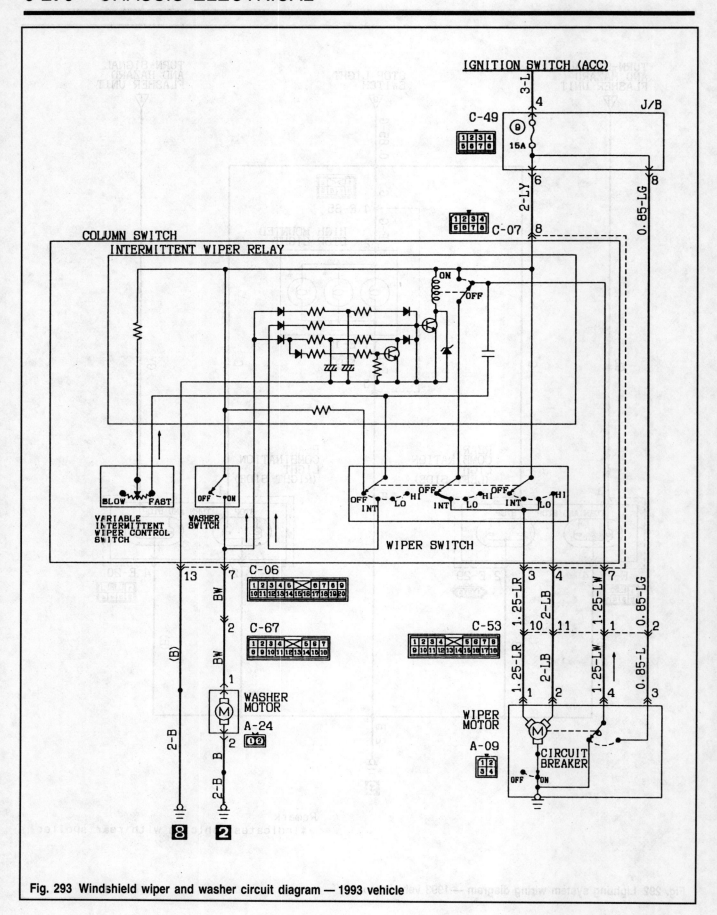

Fig. 293 Windshield wiper and washer circuit diagram — 1993 vehicle

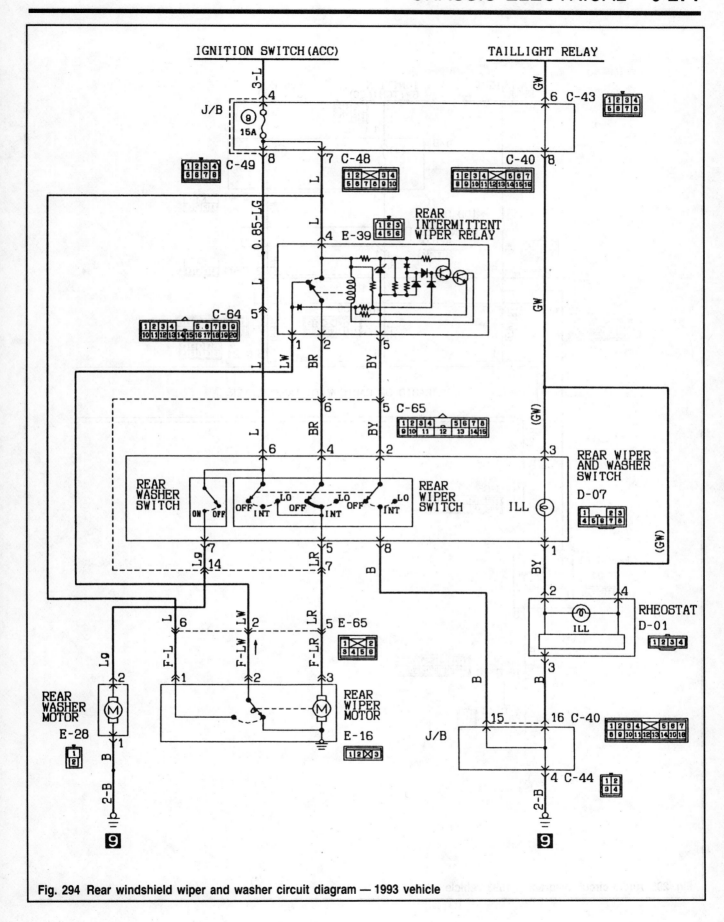

Fig. 294 Rear windshield wiper and washer circuit diagram — 1993 vehicle

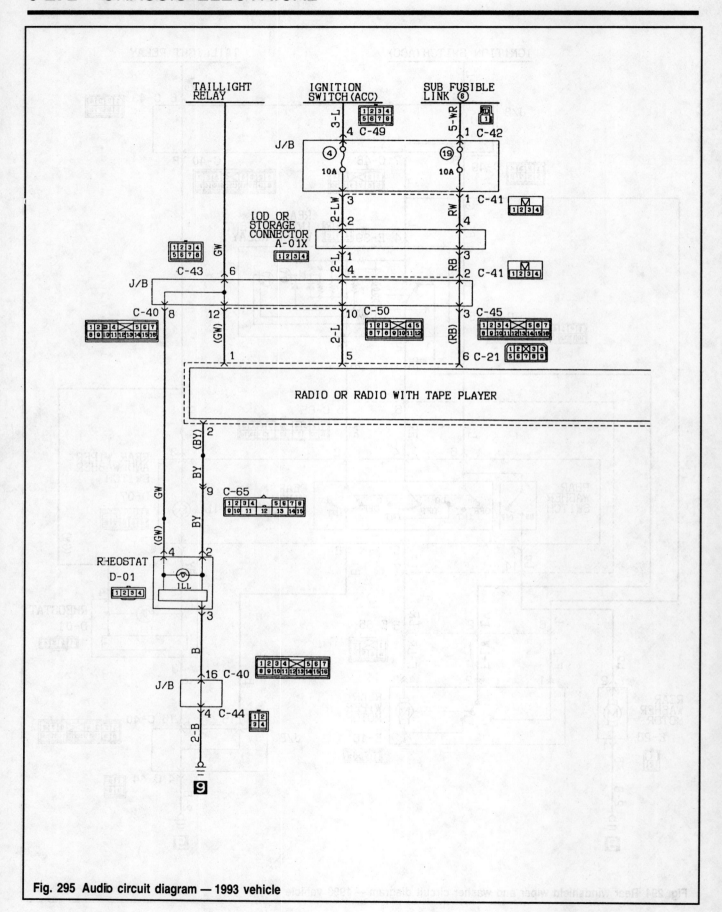

Fig. 295 Audio circuit diagram — 1993 vehicle

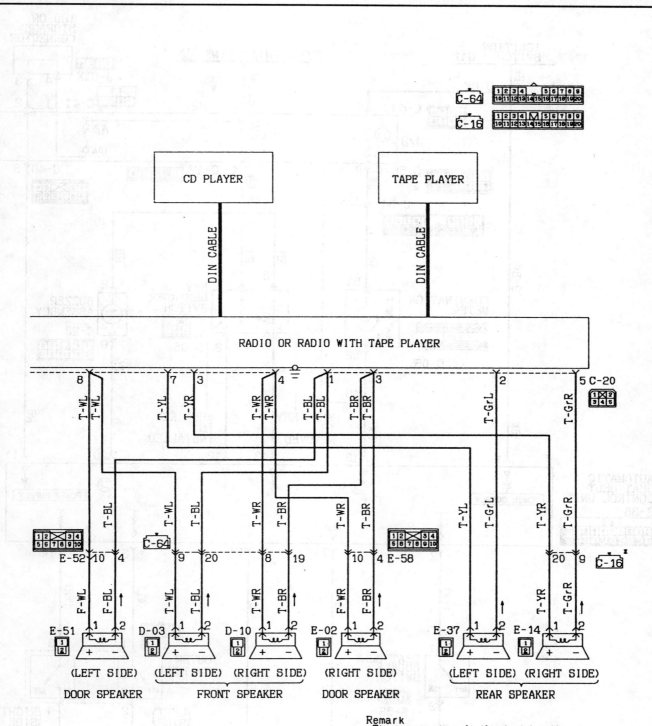

Fig. 296 Audio circuit diagram — 1993 vehicle, cont.

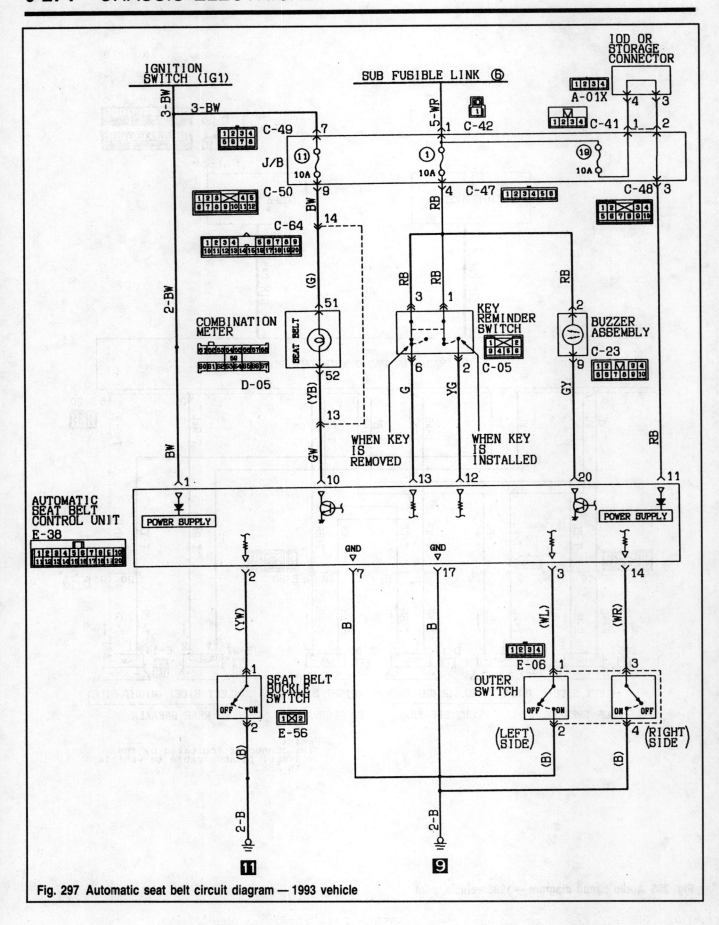

Fig. 297 Automatic seat belt circuit diagram — 1993 vehicle

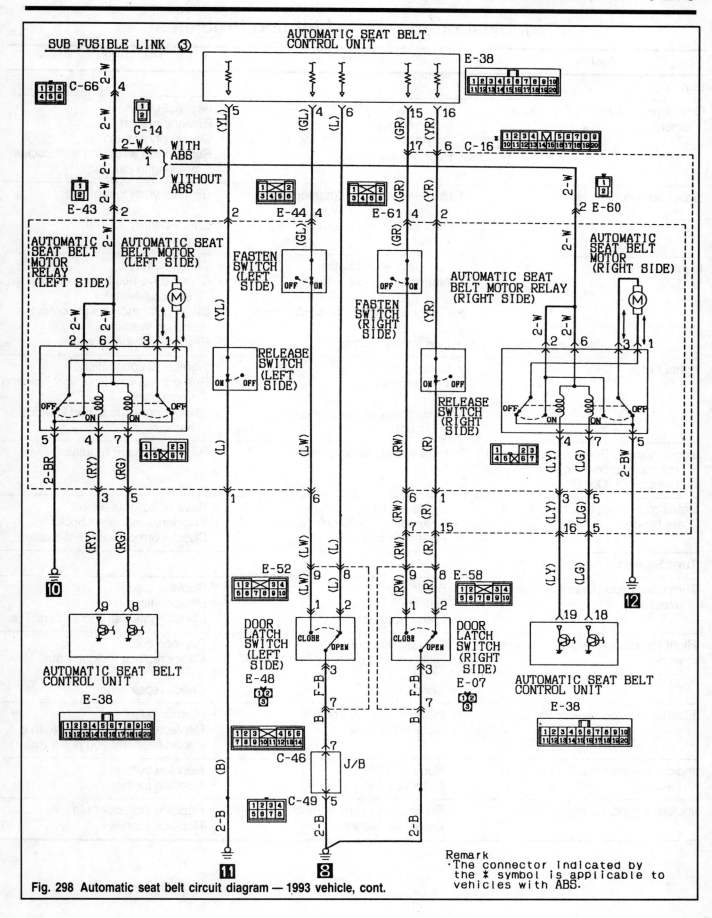

Fig. 298 Automatic seat belt circuit diagram — 1993 vehicle, cont.

Troubleshooting Basic Lighting Problems

Problem	Cause	Solution
Lights		
One or more lights don't work, but others do	• Defective bulb(s) • Blown fuse(s) • Dirty fuse clips or light sockets • Poor ground circuit	• Replace bulb(s) • Replace fuse(s) • Clean connections • Run ground wire from light socket housing to car frame
Lights burn out quickly	• Incorrect voltage regulator setting or defective regulator • Poor battery/alternator connections	• Replace voltage regulator • Check battery/alternator connections
Lights go dim	• Low/discharged battery • Alternator not charging • Corroded sockets or connections • Low voltage output	• Check battery • Check drive belt tension; repair or replace alternator • Clean bulb and socket contacts and connections • Replace voltage regulator
Lights flicker	• Loose connection • Poor ground • Circuit breaker operating (short circuit)	• Tighten all connections • Run ground wire from light housing to car frame • Check connections and look for bare wires
Lights "flare"—Some flare is normal on acceleration—if excessive, see "Lights Burn Out Quickly"	• High voltage setting	• Replace voltage regulator
Lights glare—approaching drivers are blinded	• Lights adjusted too high • Rear springs or shocks sagging • Rear tires soft	• Have headlights aimed • Check rear springs/shocks • Check/correct rear tire pressure
Turn Signals		
Turn signals don't work in either direction	• Blown fuse • Defective flasher • Loose connection	• Replace fuse • Replace flasher • Check/tighten all connections
Right (or left) turn signal only won't work	• Bulb burned out • Right (or left) indicator bulb burned out • Short circuit	• Replace bulb • Check/replace indicator bulb • Check/repair wiring
Flasher rate too slow or too fast	• Incorrect wattage bulb • Incorrect flasher	• Flasher bulb • Replace flasher (use a variable load flasher if you pull a trailer)
Indicator lights do not flash (burn steadily)	• Burned out bulb • Defective flasher	• Replace bulb • Replace flasher
Indicator lights do not light at all	• Burned out indicator bulb • Defective flasher	• Replace indicator bulb • Replace flasher

Troubleshooting Basic Dash Gauge Problems

Problem	Cause	Solution
Coolant Temperature Gauge		
Gauge reads erratically or not at all	• Loose or dirty connections • Defective sending unit • Defective gauge	• Clean/tighten connections • Bi-metal gauge: remove the wire from the sending unit. Ground the wire for an instant. If the gauge registers, replace the sending unit. • Magnetic gauge: disconnect the wire at the sending unit. With ignition ON gauge should register COLD. Ground the wire; gauge should register HOT.
Ammeter Gauge—Turn Headlights ON (do not start engine). Note reaction		
Ammeter shows charge Ammeter shows discharge Ammeter does not move	• Connections reversed on gauge • Ammeter is OK • Loose connections or faulty wiring • Defective gauge	• Reinstall connections • Nothing • Check/correct wiring • Replace gauge
Oil Pressure Gauge		
Gauge does not register or is inaccurate	• On mechanical gauge, Bourdon tube may be bent or kinked • Low oil pressure • Defective gauge • Defective wiring • Defective sending unit	• Check tube for kinks or bends preventing oil from reaching the gauge • Remove sending unit. Idle the engine briefly. If no oil flows from sending unit hole, problem is in engine. • Remove the wire from the sending unit and ground it for an instant with the ignition ON. A good gauge will go to the top of the scale. • Check the wiring to the gauge. If it's OK and the gauge doesn't register when grounded, replace the gauge. • If the wiring is OK and the gauge functions when grounded, replace the sending unit
All Gauges		
All gauges do not operate All gauges read low or erratically All gauges pegged	• Blown fuse • Defective instrument regulator • Defective or dirty instrument voltage regulator • Loss of ground between instrument voltage regulator and car • Defective instrument regulator	• Replace fuse • Replace instrument voltage regulator • Clean contacts or replace • Check ground • Replace regulator

Troubleshooting Basic Dash Gauge Problems

Problem	Cause	Solution
Warning Lights		
Light(s) do not come on when ignition is ON, but engine is not started	• Defective bulb • Defective wire • Defective sending unit	• Replace bulb • Check wire from light to sending unit • Disconnect the wire from the sending unit and ground it. Replace the sending unit if the light comes on with the ignition ON.
Light comes on with engine running	• Problem in individual system • Defective sending unit	• Check system • Check sending unit (see above)

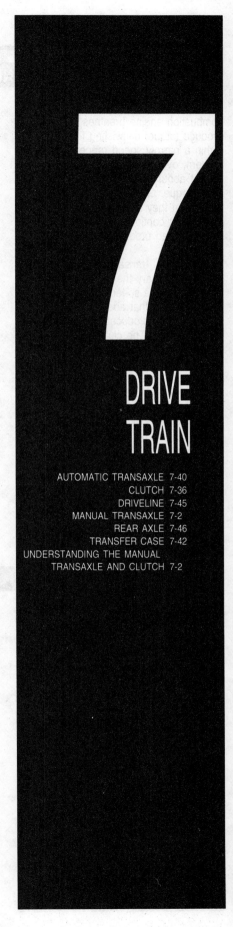

7

DRIVE
TRAIN

UNDERSTANDING THE MANUAL TRANSAXLE AND CLUTCH

Because of the way an internal combustion engine breathes, it can produce torque, or twisting force, only within a narrow speed range. Most modern, overhead valve engines must turn at about 3,000 rpm to produce their peak torque. Generally speaking, by 5,000 rpm they are producing so little torque that continued increases in engine speed produce no power increases.

The manual transaxle and clutch are employed to vary the relationship between engine speed and the speed of the wheels so that adequate engine power can be produced under all circumstances. The clutch allows engine torque to be applied to the transaxle input shaft gradually, due to mechanical slippage. The car can, consequently, be started smoothly from a full stop. The transaxle changes the ratio between the rotating speeds of the engine and the wheels by the use of gears. The lower gears allow full engine power to be applied to the rear wheels during acceleration at low speeds.

The clutch drive plate is a thin disc, the center of which is splined to the transaxle input shaft. Both sides of the disc are covered with a layer of material which is similar to brake lining and

which is capable of allowing slippage without roughness or excessive noise. The clutch cover is bolted to the engine flywheel and incorporates a diaphragm spring which provides the pressure to engage the clutch. The cover also houses the pressure plate. The driven disc is sandwiched between the pressure plate and the smooth surface of the flywheel when the clutch pedal is released, thus forcing it to turn at the same speed as the engine crankshaft. The transaxle contains a mainshaft which passes all the way through the transaxle, from the clutch to the driveshaft. This shaft is separated at one point, so that front and rear portions can turn at different speeds. Power is transmitted by a countershaft in the lower gears and reverse. The gears of the countershaft mesh with gears on the mainshaft, allowing power to be carried from one to the other. All the countershaft gears are integral with that shaft, while several of the mainshaft gears can either rotate independently of the shaft or be locked to it. Shifting from one gear to the next causes one of the gears to be freed from rotating with the shaft and locks another to it. Gears are locked and unlocked by internal dog clutches which slide between the center

of the gear and the shaft. The forward gears usually employ synchronizers; friction members which smoothly bring gear and shaft to the same speed before the toothed dog clutches are engaged.

The clutch is operating properly if:
1. It will stall the engine when released with the vehicle held stationary.
2. The shift lever can be moved freely between first and reverse gears when the vehicle is stationary and the clutch disengaged. A clutch pedal free-play adjustment is incorporated in the linkage. If there is about 1-2 inch motion before the pedal begins to release the clutch, it is adjusted properly. Inadequate free-play wears all parts of the clutch releasing mechanisms and may cause slippage. Excessive free-play may cause inadequate release and hard shifting of gears. Some clutches use a hydraulic system in place of mechanical linkage. If the clutch fails to release, fill the clutch master cylinder with fluid to the proper level and pump the clutch pedal to fill the system with fluid. Bleed the system in the same way as a brake system. If leaks are located, tighten loose connections or overhaul the master or slave cylinder as necessary.

MANUAL TRANSAXLE

Identification

The transaxle can be identified by the vehicle information code plate which is riveted onto the bulkhead in the engine compartment. The plate shows model code, engine model, transaxle model and body color code.

The Front Wheel Drive (FWD) vehicles covered in this manual use either the F5M22 or the F5M33 transaxles. The All Wheel Drive (AWD) vehicles covered by this manual use the W5M33 transaxle. All of these transaxles are 5-speed transaxles.

Shift Linkage

ADJUSTMENT

1. On the transaxle, put select lever

in **N** and move the transaxle shift lever to put it in **4th** gear. Depress the clutch, if necessary, to shift.
2. Move the shift lever in the vehicle to the **4th** gear position until it contacts the stop.
3. Turn the adjuster turn buckle so the shift cable eye aligns with the eye in the gear shift lever. When installing the cable eye, make sure the flange side of the plastic bushing at the shift cable end is on the cotter pin side.
4. The cables should be adjusted so the clearance between the shift lever and the 2 stoppers are equal when the shift lever is moved to 3rd and 4th gear. Move the shift lever to each position and check that the shifting is smooth.

Back-Up Light Switch

REMOVAL & INSTALLATION

The switch is screwed into the right side of the transaxle case and is replaceable, but not adjustable. To remove the switch, disconnect the wiring harness and unscrew the switch from the transaxle case. Do not remove the steel ball from the switch mounting bore. If it falls out, make sure that you retrieve it and put it back. Install the new switch in reverse of the removal procedure using a new gasket. Torque the switch to 22-25 ft. lbs. (29-33 Nm). Check the new switch for proper operation.

Transaxle

REMOVAL & INSTALLATION

➡️**If the vehicle is going to be rolled while the halfshafts are out of the vehicle, obtain 2 outer CV-joints or proper equivalent tools and install to the hubs. If the vehicle is rolled without the proper torque applied to the front wheel bearings, the bearings will no longer be usable.**

1. Remove the battery and the air intake hoses.

2. Remove the auto-cruise actuator and underhood bracket, located on the passenger side inner fender wall.

3. Drain the transaxle and transfer case fluid, if equipped, into a suitable waste container.

4. Remove the retainer bolt and pull the speedometer cable from the transaxle assembly.

5. Remove the cotter pin securing the select and shift cables and remove the cable ends from the transaxle.

6. Remove the connection for the clutch release cylinder and without disconnecting the hydraulic line, secure aside.

7. Disconnect the backup light switch harness and position aside.

8. Disconnect the starter electrical connections, if necessary, remove the starter motor and position aside.

9. Remove the transaxle mount bracket. Remove the upper transaxle mounting bolts.

10. Raise the vehicle and support safely on jackstands. Remove the undercover and the front wheels.

11. Remove the cotter pin and disconnect the tie rod end from the steering knuckle.

12. Remove the self-locking nut from the halfshafts. Disconnect the lower arm ball joint from the steering knuckle.

13. Remove the halfshafts from the transaxle.

14. On AWD vehicle, disconnect the front exhaust pipe.

15. On AWD vehicle, remove the transfer case by removing the attaching bolts, moving the transfer case to the left and lowering the front side. Remove it from the rear driveshaft. Be careful of the oil seal. Do not allow the driveshaft to hang; once the front is removed from the transfer, tie it up. Cover the transfer case openings to keep out dirt.

16. Remove the cover from the transaxle bellhousing. On AWD, also remove the crossmember and the triangular gusset.

17. Remove the transaxle lower coupling bolt. It is just above the halfshaft opening on 2WD or transfer case opening on AWD.

18. Support the weight of the engine from above (chain hoist). Support the transaxle using a transmission jack and remove the remaining lower mounting bolts.

19. On turbocharged vehicle, be careful not to damage the lower radiator hose with the transaxle housing during

removal. Wrap tape on both the lower hose and the transaxle housing to prevent damage. Move the transaxle assembly to the right and carefully lower it from the vehicle.

To install:

20. Install the transaxle to the engine and install the mounting bolts. Install the transaxle lower coupling bolt.

21. Install the underpan, crossmember and the triangular gusset.

22. Install the transfer case on AWD vehicles and connect the exhaust pipe.

23. Install the halfshafts, using new circlips on the axle ends. Try to keep the inboard joint straight in relation to the axle. Be careful not to damage the oil seal lip of the transaxle with the serrated part of the halfshaft.

24. Connect the tie rod and ball joint to the steering knuckle.

25. Install the transaxle mount bracket.

26. Install the starter motor.

27. Connect the backup light switch and the speedometer cable.

28. Install the clutch release cylinder.

29. Connect the select and shift cables and install new cotter pins.

30. Install the air intake hose.

31. Install the auto-cruise actuator and bracket.

32. Install the battery.

33. Make sure the vehicle is level when refilling the transaxle. Use Hypoid gear oil or equivalent, GL-4 or higher.

34. Check the transaxle and transfer case for proper operation. Make sure the reverse lights come on when in reverse.

MANUAL TRANSAXLE—SPECIAL TOOLS

Tools	Number	Name	Use
	MD998325-01	Differential oil seal installer	Installation of differential oil seal
	MD998321-01	Oil seal installer	Installation of input shaft oil seal
	MD998802-01	Input shaft holder	Installation and removal of input shaft and intermediate gear lock nut
	MD998348-01	Bearing remover	Removal of gears and bearings of input shaft, intermediate gear and output shaft
	MD998323-01	Bearing installer	Installation of input shaft bearing
	GENERAL SERVICE TOOL	Installer cap	Use with installer and adapter
	GENERAL SERVICE TOOL	Installer handle	Use with installer cap and adapter
	MIT304180	Installer handle	Use with installer cap and adapter
	GENERAL SERVICE TOOL (30) (34) MD998818 (38) MD998819 (40) MIT215013 (42) MD998822-01 (46) GENERAL SERVICE TOOL (50) (52) MD998827 (56)	Installer adapter	Installation of each bearing

MANUAL TRANSAXLE—SPECIAL TOOLS, CONT.

Tools	Number	Name	Use
	MD998304-01	Oil seal installer	Installation of transfer extension housing oil seal
	MD998917	Bearing remover	Removal of intermediate gear bearing
	GENERAL SERVICE TOOL	Differential oil seal installer	Installation of differential oil seal
	MD998806-01	Wrench adapter	Adjustment of tooth contact and inspection of turning drive torque
	MD998808-01	Snap ring installer	Installation of input shaft rear snap ring
	GENERAL SERVICE TOOL	Claw	Removal of bearing outer race
	GENERAL SERVICE TOOL	Preload socket	Measurement of drive bevel gear shaft rotating torque
	MB991144	Side gear holding Tool	Measurement of drive bevel gear shaft rotating torque
	MIT307098	Special spanner	Installation and removal of driven bevel gear lock nut

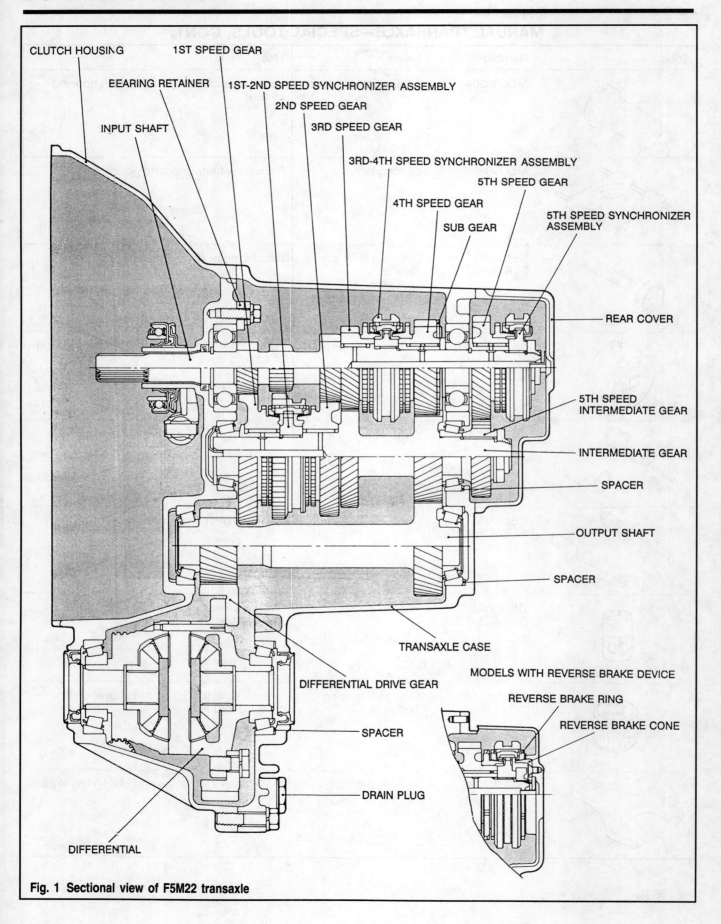

Fig. 1 Sectional view of F5M22 transaxle

➡Metric tools will be required to service this transaxle. Due to the large number of alloy parts used in this transaxle, torque specifications should be strictly observed. Before installing capscrews into aluminum parts, dip the bolts into clean transmission fluid as this will prevent the screws from galling the aluminum threads, thus causing damage. Do not attempt to interchange metric fasteners for inch system fasteners. Mismatched or incorrect fasteners can cause damage to the automatic transmission unit and possible personal injury. Care should be taken to reuse fasteners in their original position.

OVERHAUL

◆ See Figure 1

Before Disassembly

Cleanliness is an important factor in the overhaul of the manual transaxle. Before opening up this unit, the entire outside of the transaxle assembly should be cleaned, preferable with a high pressure washer such as a car wash spray unit. Dirt entering the transaxle internal parts will negate all the time and effort spent on the overhaul. During inspection and reassembly all parts should be thoroughly cleaned with solvent then dried with compressed air. Wiping cloths and rags should not be used to dry parts.

Wheel bearing grease, long used to hold thrust washers and lube parts, should not be used. Lube seals with clean transaxle oil and use ordinary unmedicated petroleum jelly to hold the thrust washers and to ease the assembly of seals, since it will not leave a harmful residue as grease often will. Do not use solvent on neoprene seals, if they are to be reused, or thrust washers. Before installing bolts into aluminum parts, always dip the threads into clean transaxle oil. Anti-seize compound can also be used to prevent bolts from galling the aluminum and seizing. Always use a torque wrench to keep from stripping the threads. The internal snaprings should be compressed and the external rings should be expanded if they are to be reused. This will help insure proper seating when installed.

TRANSAXLE CASE DISASSEMBLY

◆ See Figures 2, 3 and 4

1. Remove the transaxle from the vehicle and position the assembly on a suitable holding fixture.
2. Unbolt and remove the transaxle mounting bracket.
3. Remove the back-up light switch and the steel ball.
4. Remove the speedometer gear locking bolt.
5. Withdraw the speedometer gear assembly from the clutch housing.
6. Remove the bolts and the rear cover from the transaxle case.
7. Remove the reverse brake cone and wave spring.
8. Remove the three poppet plugs and gasket, then the three poppet springs and steel balls.
9. Remove the air breather.
10. Remove the spring pin using a pin punch.
11. Unstake the locknuts of the input shaft and intermediate shaft.
12. Shift the transaxle in reverse using the control lever and select lever.
13. Install tool MD998802 onto the input shaft. Screw a bolt 10mm into the bolt hole on the surface of the clutch housing and attach a spinner handle to the tool to remove the locknut.
14. Remove the 5th speed synchronizer assembly and shift fork.
15. Remove the synchronizer ring and the 5th gear.
16. Remove the needle bearing and the 5th intermediate gear.
17. Remove the reverse idler gear shaft bolt and gasket.
18. Remove the 13 bolts and separate the transaxle case from the clutch housing.
19. Remove the differential oil seal and guide.
20. Remove the bolt, spring washer and stopper bracket.
21. Remove the restriction ball assembly and gasket.
22. Remove the oil seal.
23. Remove the 3 bearing outer races and spacers.
24. Remove the bolt and the reverse shift lever assembly, shift lever shoe, idler gear shaft and idler gear.
25. Remove the shift pins and the shift rails and forks.

26. Remove the 2 bolts and the bearing retainer.
27. Lift up the input shaft assembly and remove the intermediate shaft assembly.
28. Remove the input shaft assembly.
29. Remove the output shaft assembly.
30. Remove the differential gear assembly.
31. Remove the 3 bearing outer races and the oil guide.
32. Remove the 2 oil seals.
33. Remove the magnet and the magnet holder.

5th SPEED SYNCHRONIZER

◆ See Figures 5 and 6

Disassembly

1. Remove the stop plate.
2. Remove the synchronizer spring.
3. Remove the synchronizer k.
4. Remove the synchronizer hub.

Inspection

1. Combine the synchronizer sleeve and hub and check that they slide smoothly.
2. Check that the sleeve is free from damage at its inside front and rear ends.
3. Check for wear of the hub end (the surface in contact with the 5th speed gear).

➡When replacing, replace the synchronizer hub and sleeve as a set.

4. Check for wear of the synchronizer key center protrusion.
5. Check the spring for weakness, deformation and breakage.
6. Replace and worn or damaged components.

Assembly

1. Assemble the synchronizer hub, sleeve and key noting their direction. Assemble so that the projections of the synchronizer springs fit into the grooves of the synchronizer keys.

➡Be sure to assemble so that the front and rear spring projections are not fitted to the same key.

2. Install the synchronizer spring and the stop plate.

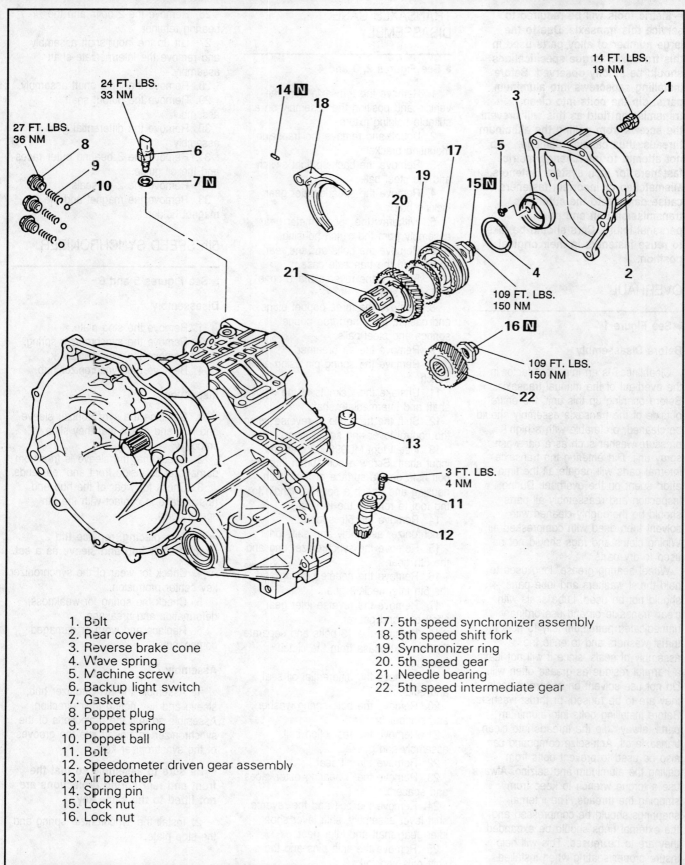

14 FT. LBS.
19 NM

24 FT. LBS.
33 NM

14 N

27 FT. LBS.
36 NM

109 FT. LBS.
150 NM

16 N

109 FT. LBS.
150 NM

3 FT. LBS.
4 NM

1. Bolt
2. Rear cover
3. Reverse brake cone
4. Wave spring
5. Machine screw
6. Backup light switch
7. Gasket
8. Poppet plug
9. Poppet spring
10. Poppet ball
11. Bolt
12. Speedometer driven gear assembly
13. Air breather
14. Spring pin
15. Lock nut
16. Lock nut

17. 5th speed synchronizer assembly
18. 5th speed shift fork
19. Synchronizer ring
20. 5th speed gear
21. Needle bearing
22. 5th speed intermediate gear

Fig. 2 Disassembly and Assembly — F5M22 transaxle

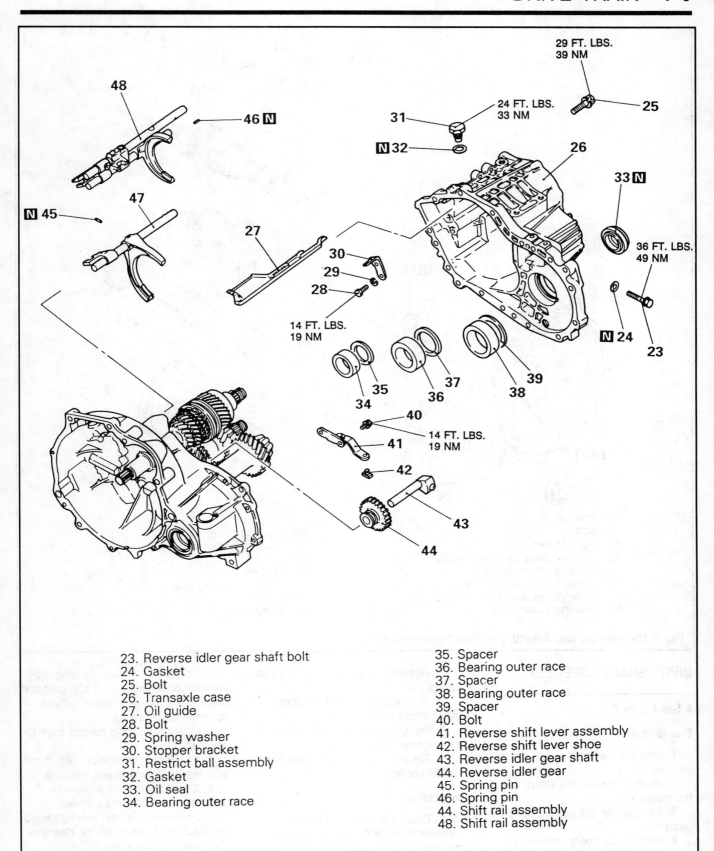

Fig. 3 Disassembly and Assembly — F5M22 transaxle, cont.

23. Reverse idler gear shaft bolt
24. Gasket
25. Bolt
26. Transaxle case
27. Oil guide
28. Bolt
29. Spring washer
30. Stopper bracket
31. Restrict ball assembly
32. Gasket
33. Oil seal
34. Bearing outer race

35. Spacer
36. Bearing outer race
37. Spacer
38. Bearing outer race
39. Spacer
40. Bolt
41. Reverse shift lever assembly
42. Reverse shift lever shoe
43. Reverse idler gear shaft
44. Reverse idler gear
45. Spring pin
46. Spring pin
44. Shift rail assembly
48. Shift rail assembly

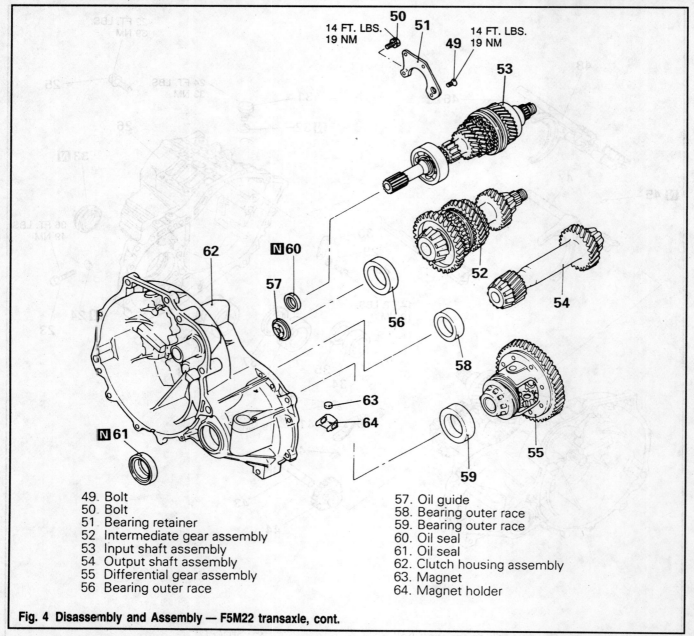

Fig. 4 Disassembly and Assembly — F5M22 transaxle, cont.

49. Bolt
50. Bolt
51. Bearing retainer
52. Intermediate gear assembly
53. Input shaft assembly
54. Output shaft assembly
55. Differential gear assembly
56. Bearing outer race
57. Oil guide
58. Bearing outer race
59. Bearing outer race
60. Oil seal
61. Oil seal
62. Clutch housing assembly
63. Magnet
64. Magnet holder

INPUT SHAFT ASSEMBLY

▶ **See Figure 7**

Disassembly

1. Remove the snapring and the front ball bearing.
2. Remove the bearing sleeve using the proper tool.
3. Remove the ball bearing using the proper tool.
4. Remove the spacer, snapring, spring and the sub gear.
5. Remove the 4th speed gear.
6. Remove the needle bearing and sleeve.

7. Remove the synchronizer ring and spring.
8. Remove the 3rd and 4th speed synchronizer sleeve and key.
9. Remove the 3rd and 4th speed synchronizer hub and ring.
10. Remove the 3rd speed gear and needle bearing.

Inspection

1. Check the outer surface of the input shaft where the needle bearing is mounted for damage, abnormal wear and seizure.
2. Check the input shaft splines for damage and wear.

3. Combine the needle bearing with the shaft or bearing sleeve and gear and check that it rotates smoothly without abnormal noise or play.
4. Check the needle bearing cage for deformation.
5. Check the synchronizer ring clutch gear teeth for damage and breakage.
6. Check the internal surface for damage, wear and broken threads.
7. Force the synchronizer ring toward the clutch gear and check the clearance. The clearance should be 0.02 in. (0.5mm) maximum.
8. Check the bevel gear and clutch gear teeth for damage and wear.

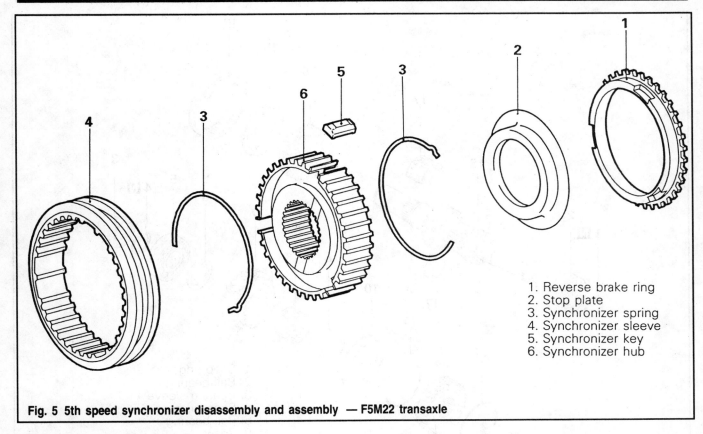

1. Reverse brake ring
2. Stop plate
3. Synchronizer spring
4. Synchronizer sleeve
5. Synchronizer key
6. Synchronizer hub

Fig. 5 5th speed synchronizer disassembly and assembly — F5M22 transaxle

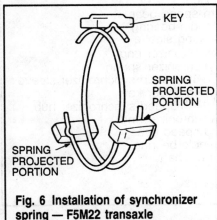

Fig. 6 Installation of synchronizer spring — F5M22 transaxle

9. Check the synchronizer cone for rough surface, damage and wear.

Assembly

1. Install the needle bearing and 3rd speed gear.
2. Install the 3rd and 4th speed synchronizer hub and ring.
3. Install the 3rd and 4th speed synchronizer sleeve and key.
4. Install the synchronizer ring and spring.
5. Install the sleeve needle bearing.
6. Install the 4th gear.

7. Install the sub gear, cone spring, snapring and the spacer.
8. Install the ball bearing using the proper tool.
9. Install the bearing sleeve using the proper tool.
10. Install the front ball bearing and the snapring.

INTERMEDIATE SHAFT ASSEMBLY

▶ See Figure 8

Disassembly

1. Remove the snapring, taper roller bearing, 1st speed gear and bearing sleeve using the proper tool.
2. Remove the synchronizer ring.
3. Remove the synchronizer spring and the 1st and 2nd speed synchronizer sleeve.
4. Remove the 1st and 2nd speed synchronizer hub.
5. Remove the synchronizer ring.
6. Remove the 2nd speed gear.
7. Remove the needle bearing.
8. Remove the taper roller bearing using the proper tool.
9. Remove the intermediate gear.

Inspection

1. Check the outer surface of the intermediate shaft where the needle bearing is mounted for damage, abnormal wear and seizure.
2. Check the intermediate shaft splines for damage and wear.
3. Combine the needle bearing with the shaft or bearing sleeve and gear and check that it rotates smoothly without abnormal noise or play.
4. Check the needle bearing cage for deformation.
5. Check the synchronizer ring clutch gear teeth for damage and breakage.
6. Check the internal surface for damage, wear and broken threads.

Assembly

1. Install the intermediate gear. Install the taper roller bearing using the proper tool.
2. Install the needle bearing.
3. Install the 2nd speed gear.
4. Install the synchronizer ring.
5. Install the 1st and 2nd speed synchronizer hub.
6. Install the 1st and 2nd speed synchronizer sleeve and the synchronizer spring.
7. Install the synchronizer ring.

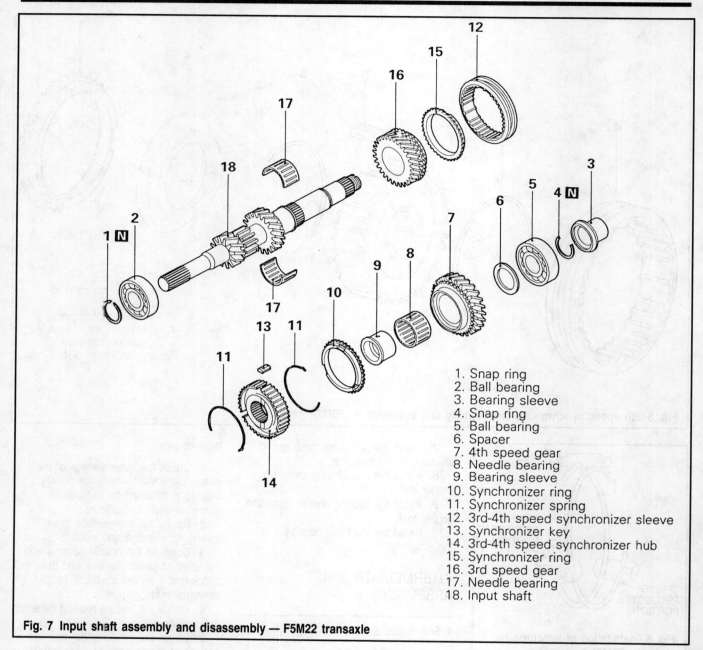

1. Snap ring
2. Ball bearing
3. Bearing sleeve
4. Snap ring
5. Ball bearing
6. Spacer
7. 4th speed gear
8. Needle bearing
9. Bearing sleeve
10. Synchronizer ring
11. Synchronizer spring
12. 3rd-4th speed synchronizer sleeve
13. Synchronizer key
14. 3rd-4th speed synchronizer hub
15. Synchronizer ring
16. 3rd speed gear
17. Needle bearing
18. Input shaft

Fig. 7 Input shaft assembly and disassembly — F5M22 transaxle

8. Install the snapring, taper roller bearing, 1st speed gear and bearing sleeve using the proper tool.

DIFFERENTIAL

▶ See Figure 9

Disassembly

1. Remove the bolts and the differential drive gear from the case.
2. Remove the taper roller bearings and discard. Do not reuse the taper roller bearing or races.
3. Remove the lockpin, pinion shaft and pinion gears and washer.

4. Remove the side gears and spacers.

Inspection

Check both the gears and the bearings for wear or damage. Make sure they are clean and dry before reassembly. Coat them with gear oil before reassembly.

Assembly

1. Install the side gears and spacers.
2. Install new lockpin, pinion shaft and pinion gears and washers.
3. Measure the backlash between the side gears and pinions. The standard

backlash is 0.001-0.006 in. (0.025-0.150mm). If the backlash is out of specification, disassemble and use a different spacer.

4. Install the taper roller bearings.

➡**When press fitting the bearings, push on the inner race only.**

5. Install the differential drive gear and the retaining bolts. Apply stud locking sealant to the threads of the bolts and quickly tighten in the a alternating cross pattern. This will assure even tightening of the bolts. Tighten bolts to 98 ft. lbs. (135 Nm).

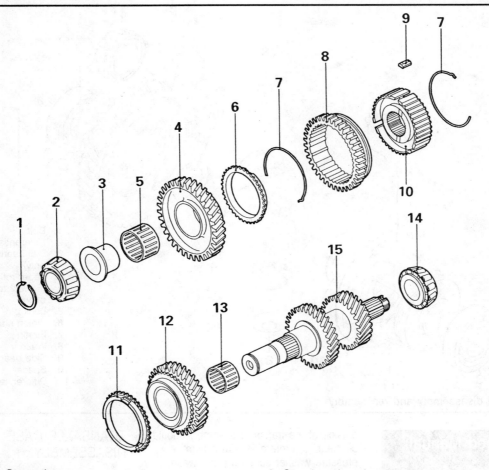

1. Snap ring
2. Taper roller bearing
3. Bearing sleeve
4. 1st speed gear
5. Needle bearing
6. Synchronizer ring
7. Synchronizer spring
8. 1st-2nd speed synchronizer sleeve
9. Synchronizer key
10. 1st-2nd speed synchronizer hub
11. Synchronizer ring
12. 2nd speed gear
13. Needle bearing
14. Taper roller bearing
15. Intermediate gear

Fig. 8 Intermediate shaft disassembly and assembly — F5M22 transaxles

TRANSAXLE CASE ASSEMBLY

1. Install the magnet and the magnet holder.

2. Install the 2 oil seals.

3. Install the 3 bearing outer races and the oil guide.

4. Install the differential gear assembly.

5. Install the output shaft assembly.

6. Install the input shaft assembly.

7. Lift up the input shaft assembly and install the intermediate shaft assembly.

8. Install the 2 bolts and the bearing retainer.

9. Install the shift pins and the shift rails and forks.

10. Install the bolt and the reverse shift lever assembly, shift lever shoe, idler gear shaft and idler gear.

11. Install the 3 bearing outer races and spacers.

12. Install the oil seal.

13. Install the restriction ball assembly and gasket.

14. Install the bolt, spring washer and stopper bracket.

15. Install the oil guide.

16. Assemble the transaxle case and the clutch housing. Install the bolts.

17. Install the reverse idler gear shaft bolt and gasket.

18. Install the needle bearing and the 5th intermediate gear.

19. Install the synchronizer ring and the 5th gear.

20. Install the 5th speed synchronizer assembly and shift fork.

21. Install the spring pin.

22. Install the air breather.

23. Install the speedometer driven gear assembly.

24. Install the poppet plug, spring and steel ball.

25. Install the backup light switch, gasket and steel ball.

26. Install the rear cover with new gasket in place.

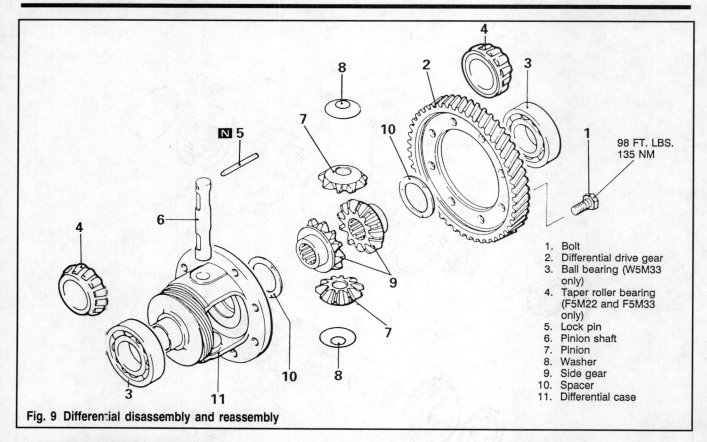

Fig. 9 Differential disassembly and reassembly

98 FT. LBS.
135 NM

1. Bolt
2. Differential drive gear
3. Ball bearing (W5M33 only)
4. Taper roller bearing (F5M22 and F5M33 only)
5. Lock pin
6. Pinion shaft
7. Pinion
8. Washer
9. Side gear
10. Spacer
11. Differential case

F5M33 5-Speed Manual Transaxle

▶ See Figure 10

➡Metric tools will be required to service this transmission. Due to the large number of alloy parts used in this transmission, torque specifications should be strictly observed. Before installing capscrews into aluminum parts, dip the bolts into clean transmission fluid as this will prevent the screws from galling the aluminum threads, thus causing damage. Do not attempt to interchange metric fasteners for inch system fasteners. Mismatched or incorrect fasteners can cause damage to the automatic transmission unit and possible personal injury. Care should be taken to reuse fasteners in their original position.

OVERHAUL

Before Disassembly

Cleanliness is an important factor in the overhaul of the manual transaxle. Before opening up this unit, the entire outside of the transaxle assembly should be cleaned, preferable with a high pressure washer such as a car wash spray unit. Dirt entering the transaxle internal parts will negate all the time and effort spent on the overhaul. During inspection and reassembly all parts should be thoroughly cleaned with solvent then dried with compressed air. Wiping cloths and rags should not be used to dry parts.

Wheel bearing grease, long used to hold thrust washers and lube parts, should not be used. Lube seals with clean transaxle oil and use ordinary unmedicated petroleum jelly to hold the thrust washers and to ease the assembly of seals, since it will not leave a harmful residue as grease often will. Do not use solvent on neoprene seals, if they are to be reused, or thrust washers. Before installing bolts into aluminum parts, always dip the threads into clean transaxle oil. Anti-seize compound can also be used to prevent bolts from galling the aluminum and seizing. Always use a torque wrench to keep from stripping the threads. The internal snaprings should be compressed and the external rings should be expanded if they are to be reused. This will help insure proper seating when installed.

TRANSAXLE CASE DISASSEMBLY

▶ See Figures 11, 12 and 13

✳✳WARNING

The use of the correct special tools or their equivalent is REQUIRED for some of these procedures involving disassembly. Additionally, some tools are required which may not be in your tool box. Be prepared to use snapring pliers, a micrometer, small drifts or punches and assorted gear and bearing pullers.

1. Remove the transaxle from the vehicle and position the assembly on a suitable holding fixture.
2. Unbolt and remove the transaxle mounting bracket.
3. Remove the back-up light switch and the steel ball.
4. Remove the speedometer gear locking bolt.
5. Withdraw the speedometer gear assembly from the clutch housing.
6. Remove the bolts and the rear cover from the transaxle case.

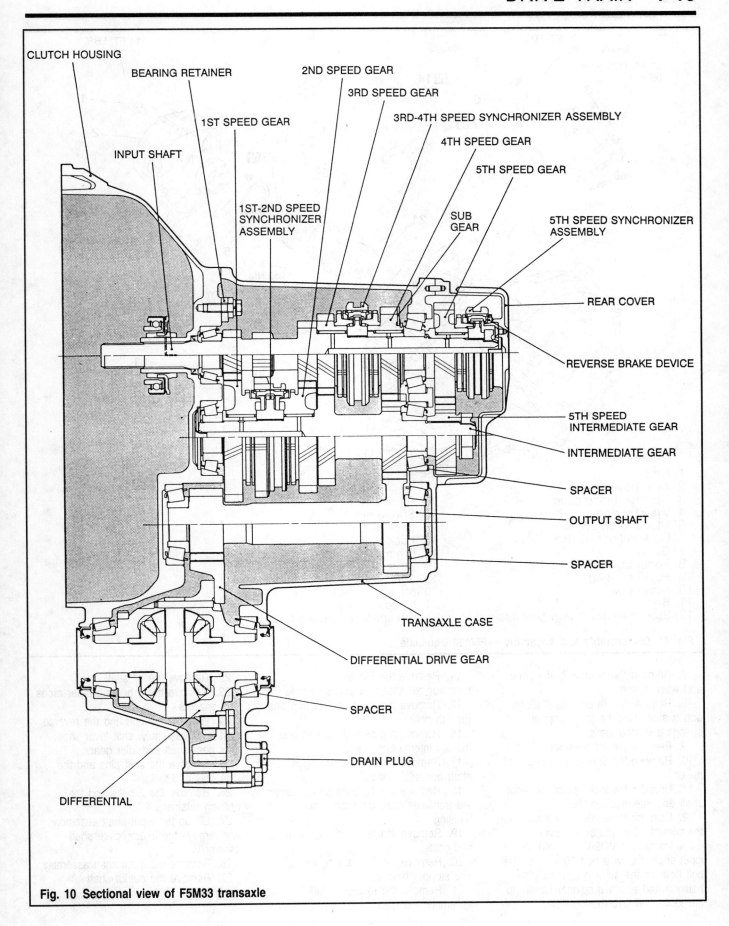

Fig. 10 Sectional view of F5M33 transaxle

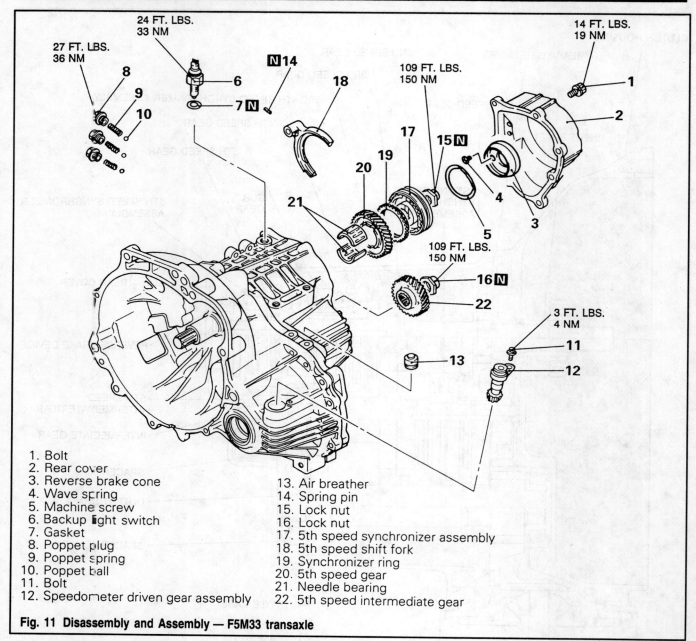

Fig. 11 Disassembly and Assembly — F5M33 transaxle

1. Bolt
2. Rear cover
3. Reverse brake cone
4. Wave spring
5. Machine screw
6. Backup light switch
7. Gasket
8. Poppet plug
9. Poppet spring
10. Poppet ball
11. Bolt
12. Speedometer driven gear assembly

13. Air breather
14. Spring pin
15. Lock nut
16. Lock nut
17. 5th speed synchronizer assembly
18. 5th speed shift fork
19. Synchronizer ring
20. 5th speed gear
21. Needle bearing
22. 5th speed intermediate gear

7. Remove the reverse brake cone and wave spring.

8. Remove the three poppet plugs and gasket, then the three poppet springs and steel balls.

9. Remove the air breather.

10. Remove the spring pin using a pin punch.

11. Unstake the locknuts of the input shaft and intermediate shaft.

12. Shift the transaxle in reverse using the control lever and select lever.

13. Install tool MD998802 onto the input shaft. Screw a bolt 10mm into the bolt hole on the surface of the clutch housing and attach a spinner handle to the tool to remove the locknut.

14. Remove the 5th speed synchronizer assembly and shift fork.

15. Remove the synchronizer ring and the 5th gear.

16. Remove the needle bearing and the 5th intermediate gear.

17. Remove the reverse idler gear shaft bolt and gasket.

18. Remove the 13 bolts and separate the transaxle case from the clutch housing.

19. Remove the differential oil seal and guide.

20. Remove the bolt, spring washer and stopper bracket.

21. Remove the restriction ball assembly and gasket.

22. Remove the oil seal.

23. Remove the 3 bearing outer races and spacers.

24. Remove the bolt and the reverse shift lever assembly, shift lever shoe, idler gear shaft and idler gear.

25. Remove the shift pins and the shift rails and forks.

26. Remove the 2 bolts and the bearing retainer.

27. Lift up the input shaft assembly and remove the intermediate shaft assembly.

28. Remove the input shaft assembly.

29. Remove the output shaft assembly.

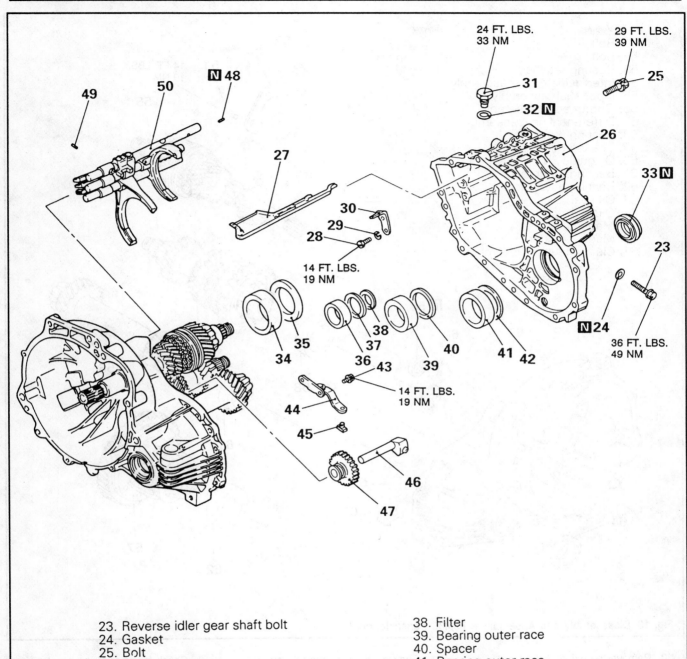

Fig. 12 Disassembly and Assembly — F5M33 transaxle, cont.

23. Reverse idler gear shaft bolt
24. Gasket
25. Bolt
26. Transaxle case
27. Oil guide
28. Bolt
29. Spring washer
30. Stopper bracket
31. Restrict ball assembly
32. Gasket
33. Oil seal
34. Bearing outer race
35. Spacer
36. Bearing outer race
37. Spacer

38. Filter
39. Bearing outer race
40. Spacer
41. Bearing outer race
42. Spacer
43. Bolt
44. Reverse shift lever assembly
45. Reverse shift lever shoe
46. Reverse idler gear shaft
47. Reverse idler gear
48. Spring pin
49. Spring pin
50. Shift rail assembly

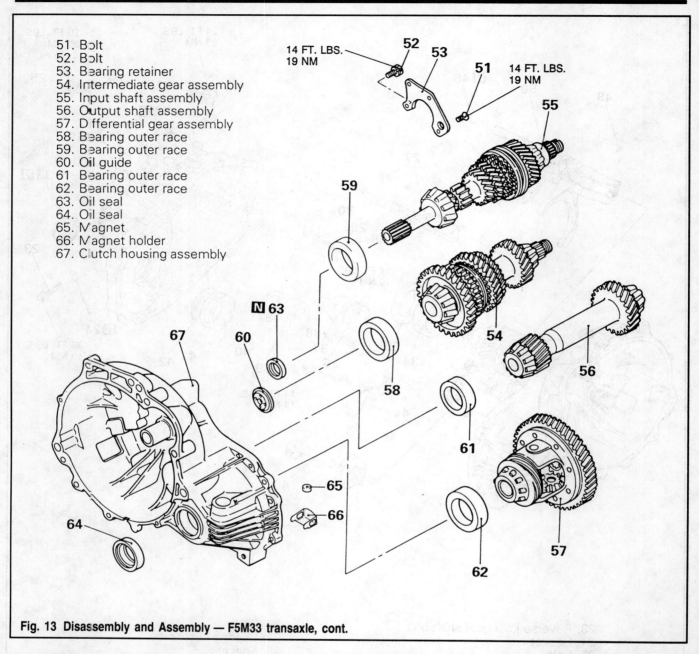

51. Bolt
52. Bolt
53. Bearing retainer
54. Intermediate gear assembly
55. Input shaft assembly
56. Output shaft assembly
57. Differential gear assembly
58. Bearing outer race
59. Bearing outer race
60. Oil guide
61. Bearing outer race
62. Bearing outer race
63. Oil seal
64. Oil seal
65. Magnet
66. Magnet holder
67. Clutch housing assembly

Fig. 13 Disassembly and Assembly — F5M33 transaxle, cont.

30. Remove the differential gear assembly.
31. Remove the 3 bearing outer races and the oil guide.
32. Remove the 2 oil seals.
33. Remove the magnet and the magnet holder.

5th SPEED SYNCHRONIZER

▶ **See Figures 14 and 15**

Disassembly

1. Remove the stop plate.
2. Remove the synchronizer spring.
3. Remove the synchronizer sleeve.
4. Remove the synchronizer key.
5. Remove the synchronizer hub.

Inspection

1. Combine the synchronizer sleeve and hub and check that they slide smoothly.
2. Check that the sleeve is free from damage at its inside front and rear ends.
3. Check for wear of the hub end (the surface in contact with the 5th speed gear).

➡**When replacing, replace the synchronizer hub and sleeve as a set.**

4. Check for wear of the synchronizer key center protrusion.
5. Check the spring for weakness, deformation and breakage.
6. Replace and worn or damaged components.

Assembly

1. Assemble the synchronizer hub, sleeve and key noting their direction. Assemble so that the projections of the synchronizer springs fit into the grooves of the synchronizer keys.

➡**Be sure to assemble so that the front and rear spring projections are not fitted to the same key.**

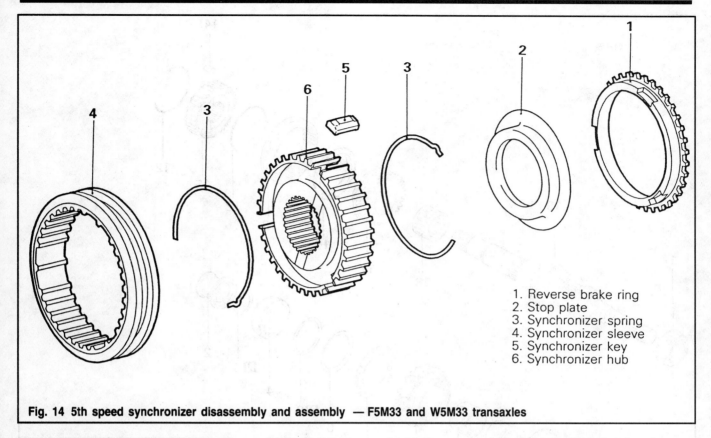

Fig. 14 5th speed synchronizer disassembly and assembly — F5M33 and W5M33 transaxles

1. Reverse brake ring
2. Stop plate
3. Synchronizer spring
4. Synchronizer sleeve
5. Synchronizer key
6. Synchronizer hub

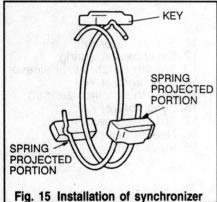

Fig. 15 Installation of synchronizer spring — F5M33 and W5M33 transaxles

2. Install the synchronizer spring and the stop plate.

INPUT SHAFT ASSEMBLY

◆ **See Figure 16**

Disassembly

1. Remove the front taper roller bearing using the proper tool.
2. Remove the bearing sleeve using the proper tool.
3. Remove the taper roller bearing using the proper tool.
4. Remove the spacer, snapring, cone spring and the sub gear.
5. Remove the 4th speed gear.
6. Remove the needle bearing and sleeve.
7. Remove the synchronizer ring and spring.
8. Remove the 3rd and 4th speed synchronizer sleeve.
9. Remove the 3rd and 4th speed synchronizer hub and ring.
10. Remove the 3rd speed gear and needle bearing.

Inspection

1. Check the outer surface of the input shaft where the needle bearing is mounted for damage, abnormal wear and seizure.
2. Check the input shaft splines for damage and wear.
3. Combine the needle bearing with the shaft or bearing sleeve and gear and check that it rotates smoothly without abnormal noise or play.
4. Check the needle bearing cage for deformation.
5. Check the synchronizer ring clutch gear teeth for damage and breakage.
6. Check the internal surface for damage, wear and broken threads.
7. Force the synchronizer ring toward the clutch gear and check the clearance. The clearance should be 0.02 in. (0.5mm) maximum.
8. Check the bevel gear and clutch gear teeth for damage and wear.
9. Check the synchronizer cone for rough surface, damage and wear.

Assembly

1. Install the needle bearing and 3rd speed gear.
2. Install the 3rd and 4th speed synchronizer hub and ring.
3. Install the 3rd and 4th speed synchronizer sleeve and key.
4. Install the synchronizer ring and spring.
5. Install the sleeve needle bearing.
6. Install the 4th gear.
7. Install the sub gear, cone spring, snapring and the spacer.
8. Install the taper roller bearing and snapring. Install the bearing sleeve and taper roller bearing using the proper equipment.

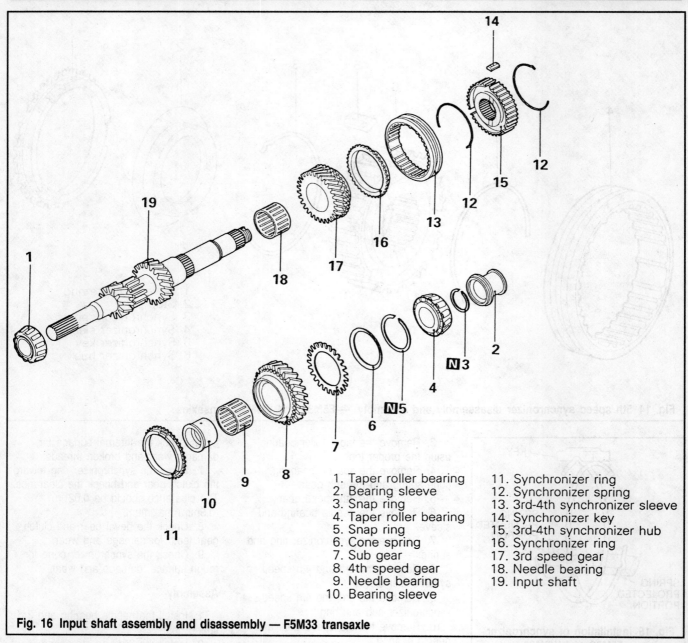

1. Taper roller bearing
2. Bearing sleeve
3. Snap ring
4. Taper roller bearing
5. Snap ring
6. Cone spring
7. Sub gear
8. 4th speed gear
9. Needle bearing
10. Bearing sleeve
11. Synchronizer ring
12. Synchronizer spring
13. 3rd-4th synchronizer sleeve
14. Synchronizer key
15. 3rd-4th synchronizer hub
16. Synchronizer ring
17. 3rd speed gear
18. Needle bearing
19. Input shaft

Fig. 16 Input shaft assembly and disassembly — F5M33 transaxle

INTERMEDIATE SHAFT ASSEMBLY

♦ **See Figure 17**

Disassembly

1. Remove the snapring, taper roller bearing, 1st speed gear and bearing sleeve using the proper tool.

2. Remove the needle bearing and synchronizer ring.

3. Remove the synchronizer spring and the 1st and 2nd speed synchronizer sleeve.

4. Remove the 1st and 2nd speed synchronizer hub.

5. Remove the synchronizer ring.
6. Remove the 2nd speed gear.
7. Remove the needle bearing.
8. Remove the taper roller bearing using the proper tool.
9. Remove the intermediate gear.

Inspection

1. Check the outer surface of the intermediate shaft where the needle bearing is mounted for damage, abnormal wear and seizure.

2. Check the intermediate shaft splines for damage and wear.

3. Combine the needle bearing with the shaft or bearing sleeve and gear and

check that it rotates smoothly without abnormal noise or play.

4. Check the needle bearing cage for deformation.

5. Check the synchronizer ring clutch gear teeth for damage and breakage.

6. Check the internal surface for damage, wear and broken threads.

Assembly

1. Install the intermediate gear. Install the taper roller bearing using the proper tool.

2. Install the needle bearing.
3. Install the 2nd speed gear.
4. Install the synchronizer ring.

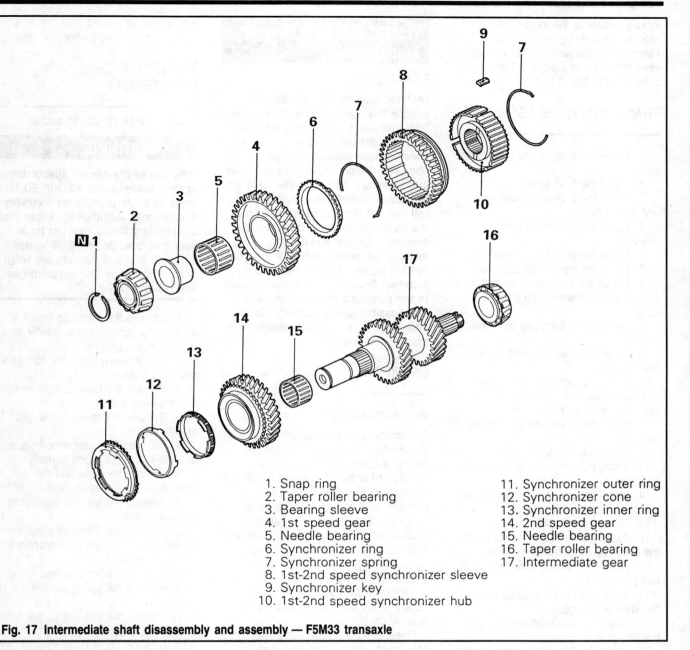

1. Snap ring
2. Taper roller bearing
3. Bearing sleeve
4. 1st speed gear
5. Needle bearing
6. Synchronizer ring
7. Synchronizer spring
8. 1st-2nd speed synchronizer sleeve
9. Synchronizer key
10. 1st-2nd speed synchronizer hub
11. Synchronizer outer ring
12. Synchronizer cone
13. Synchronizer inner ring
14. 2nd speed gear
15. Needle bearing
16. Taper roller bearing
17. Intermediate gear

Fig. 17 Intermediate shaft disassembly and assembly — F5M33 transaxle

5. Install the 1st and 2nd speed synchronizer hub.

6. Install the 1st and 2nd speed synchronizer sleeve and the synchronizer spring.

7. Install the synchronizer ring.

8. Install the snapring, taper roller bearing, 1st speed gear and bearing sleeve using the proper tool.

DIFFERENTIAL

Disassembly

1. Remove the bolts and the differential drive gear from the case.

2. Remove the taper roller bearings and discard. Do not reuse the taper roller bearing or races.

3. Remove the lockpin, pinion shaft and pinion gears and washers.

4. Remove the side gears and spacers.

Inspection

Check both the gears and the bearings for wear or damage. Make sure they are clean and dry before reassembly. Coat them with gear oil before reassembly.

Assembly

1. Install the side gears and spacers.

2. Install new lockpin, pinion shaft and pinion gears and washers.

3. Measure the backlash between the side gears and pinions. The standard backlash is 0.001-0.006 in. (0.025-0.150mm). If the backlash is out of specification, disassemble and use a different spacer.

4. Install the taper roller bearings.

➡**When press fitting the bearings, push on the inner race only.**

5. Install the differential drive gear and the retaining bolts. Apply stud

locking sealant to the threads of the bolts and quickly tighten in the a alternating cross pattern. This will assure even tightening of the bolts. Tighten bolts to 98 ft. bs. (135 Nm).

TRANSAXLE CASE ASSEMBLY

1. Install the magnet and the magnet holder.
2. Install the 2 oil seals.
3. Install the 3 bearing outer races and the oil guide.
4. Install the differential gear assembly.
5. Install the output shaft assembly.
6. Install the input shaft assembly.
7. Lift up the input shaft assembly and install the intermediate shaft assembly.
8. Install the 2 bolts and the bearing retainer.
9. Install the shift pins and the shift rails and forks.
10. Install the bolt and the reverse shift lever assembly, shift lever shoe, idler gear shaft and idler gear.
11. Install the 3 bearing outer races and spacers.
12. Install the oil seal.
13. Install the restriction ball assembly and gasket.
14. Install the bolt, spring washer and stopper bracket.
15. Install the oil guide.
16. Assemble the transaxle case and the clutch housing. Install the bolts.
17. Install the reverse idler gear shaft bolt and gasket.
18. Install the needle bearing and the 5th intermediate gear.
19. Install the synchronizer ring and the 5th gear.
20. Install the 5th speed synchronizer assembly and shift fork.
21. Install the spring pin.
22. Install the air breather.
23. Install the speedometer driven gear assembly.
24. Install the poppet plug, spring and steel ball.
25. Install the backup light switch, gasket and steel ball.
26. Install the rear cover with new gasket in place.

W5M33 5-Speed Manual Transaxle

▶ See Figure 18

➡Metric tools will be required to service this transmission. Due to the large number of alloy parts used in this transmission, torque specifications should be strictly observed. Before installing capscrews into aluminum parts, dip the bolts into clean transmission fluid as this will prevent the screws from galling the aluminum threads, thus causing damage. Do not attempt to interchange metric fasteners for inch system fasteners. Mismatched or incorrect fasteners can cause damage to the automatic transmission unit and possible personal injury. Care should be taken to reuse fasteners in their original position.

OVERHAUL

Before Disassembly

Cleanliness is an important factor in the overhaul of the manual transaxle. Before opening up this unit, the entire outside of the transaxle assembly should be cleaned, preferable with a high pressure washer such as a car wash spray unit. Dirt entering the transaxle internal parts will negate all the time and effort spent on the overhaul. During inspection and reassembly all parts should be thoroughly cleaned with solvent then dried with compressed air. Wiping cloths and rags should not be used to dry parts.

Wheel bearing grease, long used to hold thrust washers and lube parts, should not be used. Lube seals with clean transaxle oil and use ordinary unmedicated petroleum jelly to hold the thrust washers and to ease the assembly of seals, since it will not leave a harmful residue as grease often will. Do not use solvent on neoprene seals, if they are to be reused, or thrust washers. Before installing bolts into aluminum parts, always dip the threads into clean transaxle oil. Anti-seize compound can also be used to prevent bolts from galling the aluminum and seizing. Always use a torque wrench to keep from stripping the threads. The internal snaprings should be compressed and the external rings should be expanded if

they are to be reused. This will help insure proper seating when installed.

TRANSAXLE CASE DISASSEMBLY

▶ See Figures 19, 20, 21 and 22

✳✳WARNING

The use of the correct special tools or their equivalent is REQUIRED for some of these procedures involving disassembly. Additionally, some tools are required which may not be in your tool box. Be prepared to use snaping pliers, a micrometer, small drifts or punches and assorted gear and bearing pullers.

1. Remove the transaxle from the vehicle and position the assembly on a suitable holding fixture.
2. Unbolt and remove the transaxle mounting bracket.
3. Remove the back-up light switch and the steel ball.
4. Remove the speedometer gear locking bolt.
5. Withdraw the speedometer gear assembly from the clutch housing.
6. Remove the bolts and the rear cover from the transaxle case.
7. Remove the reverse brake cone and wave spring.
8. Remove the 3 poppet plugs and gasket, then the three poppet springs and steel balls.
9. Remove the air breather.
10. Remove the spring pin using a pin punch.
11. Unstake the locknuts of the input shaft and intermediate shaft.
12. Shift the transaxle in reverse using the control lever and select lever.
13. Install tool MD998802 onto the input shaft. Screw a bolt 10mm into the bolt hole on the surface of the clutch housing and attach a spinner handle to the tool to remove the locknut.
14. Remove the 5th speed synchronizer assembly and shift fork.
15. Remove the synchronizer ring and the 5th gear.
16. Remove the needle bearing and the 5th speed intermediate gear.
17. Remove the reverse idler gear shaft bolt and gasket.
18. Remove the snapring and the viscous coupling. Remove the steel ball.

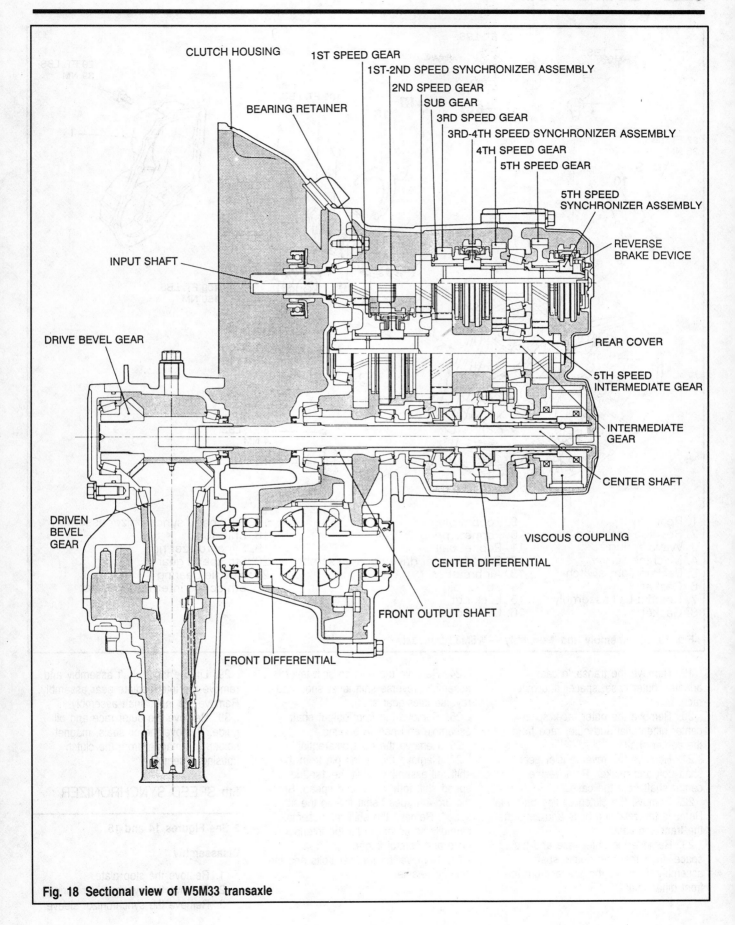

Fig. 18 Sectional view of W5M33 transaxle

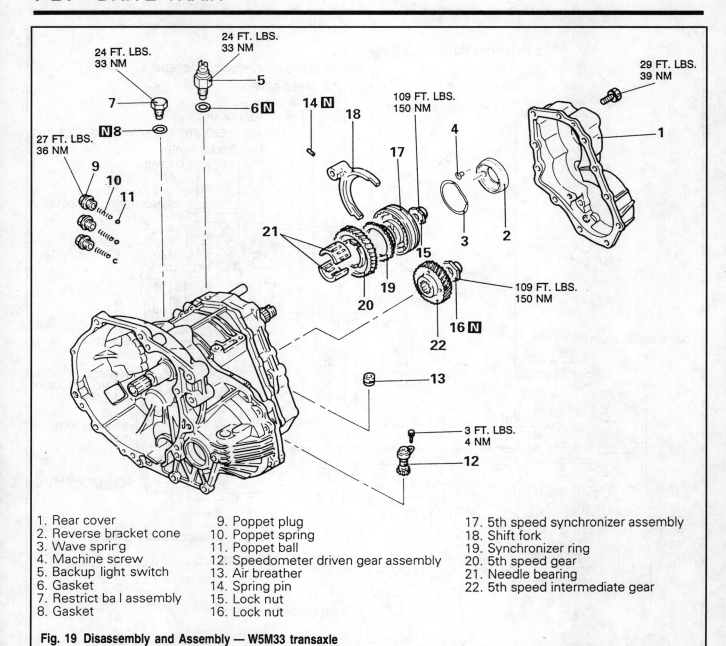

1. Rear cover
2. Reverse bracket cone
3. Wave spring
4. Machine screw
5. Backup light switch
6. Gasket
7. Restrict ball assembly
8. Gasket
9. Poppet plug
10. Poppet spring
11. Poppet ball
12. Speedometer driven gear assembly
13. Air breather
14. Spring pin
15. Lock nut
16. Lock nut
17. 5th speed synchronizer assembly
18. Shift fork
19. Synchronizer ring
20. 5th speed gear
21. Needle bearing
22. 5th speed intermediate gear

Fig. 19 Disassembly and Assembly — W5M33 transaxle

19. Remove the transaxle case adapter, outer case, spacer and outer race.

20. Remove the outer race, spacer center differential and outer race from the center shaft.

21. Remove the reverse idler gear shaft bolt and gasket. Remove the center shaft from the case.

22. Remove the clutch oil line bracket. Remove the retaining bolts and separate the transaxle case.

23. Remove the outer race and the spacer from the front output shaft assembly. Remove the spacer from the front differential.

24. Remove the reverse shift lever assembly, reverse shift lever shoe and reverse idler gear shaft.

25. Remove the front output shaft assemble and needle bearing.

26. Remove the front differential.

27. Remove the spring pin from the shift rail assembly. Shift the 1st-2nd speed shift fork to the 2nd speed. Shift the 3rd-4th speed shift fork to the 4th speed. Remove the shift rail assembly carefully so as not to hit the interlock plate and control finger.

28. Remove the retainer bolts and the bearing retainer.

29. Lift the input shaft assembly and remove the intermediate gear assembly. Remove the input shaft assembly.

30. Remove the outer race and oil guide. Remove the oil seals, magnet holder and magnet from the clutch housing assembly.

5th SPEED SYNCHRONIZER

▶ **See Figures 14 and 15**

Disassembly

1. Remove the stop plate.
2. Remove the synchronizer spring.
3. Remove the synchronizer sleeve.

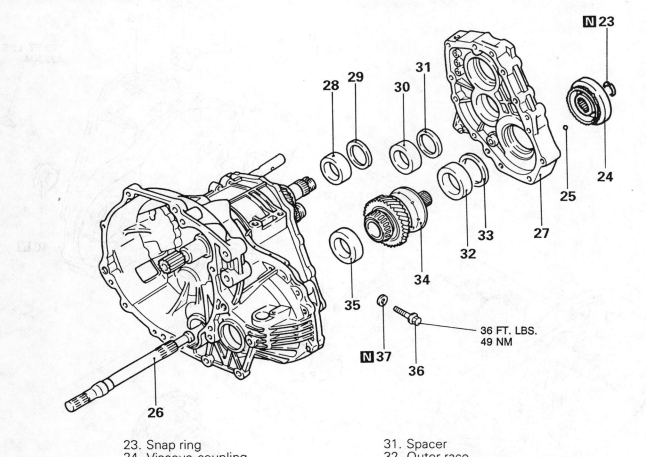

23. Snap ring
24. Viscous coupling
25. Steel ball
26. Center shaft
27. Transaxle case adapter
28. Outer case
29. Spacer
30. Outer race
31. Spacer
32. Outer race
33. Spacer
34. Center differential
35. Outer race
36. Reverse idler gear shaft bolt
37. Gasket

Fig. 20 Disassembly and Assembly — W5M33 transaxle, cont.

4. Remove the synchronizer key.
5. Remove the synchronizer hub.

Inspection

1. Combine the synchronizer sleeve and hub and check that they slide smoothly.
2. Check that the sleeve is free from damage at its inside front and rear ends.
3. Check for wear of the hub end (the surface in contact with the 5th speed gear).

➡**When replacing, replace the synchronizer hub and sleeve as a set.**

4. Check for wear of the synchronizer key center protrusion.
5. Check the spring for weakness, deformation and breakage.
6. Replace and worn or damaged components.

Assembly

1. Assemble the synchronizer hub, sleeve and key noting their direction. Assemble so that the projections of the synchronizer springs fit into the grooves of the synchronizer keys.

➡**Be sure to assemble so that the front and rear spring projections are not fitted to the same key.**

2. Install the synchronizer spring and the stop plate.

INPUT SHAFT ASSEMBLY

▶ **See Figure 23**

Disassembly

1. Remove the front taper roller bearing using the proper tool.
2. Remove the bearing sleeve using the proper tool.
3. Remove the snapring and press the taper roller bearing from the shaft using the proper equipment.

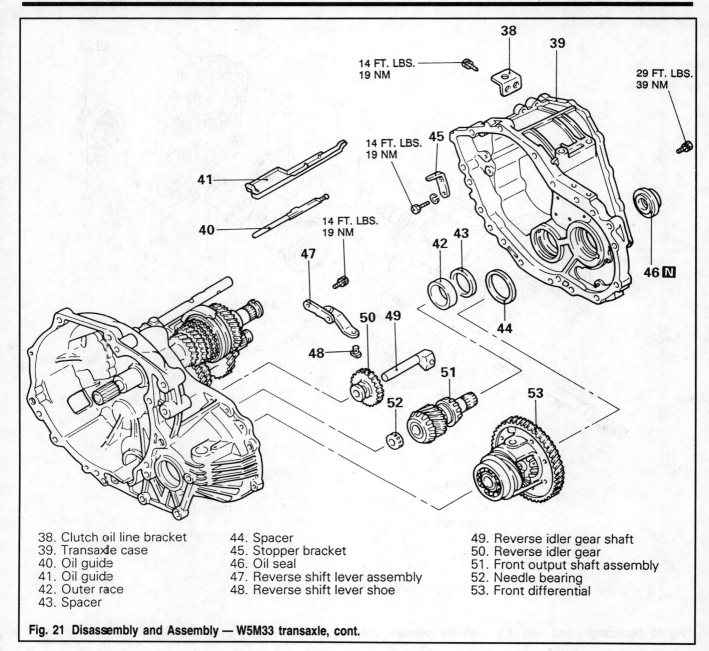

14 FT. LBS.
19 NM

38

39

29 FT. LBS.
39 NM

14 FT. LBS.
19 NM

45

41

40

14 FT. LBS.
19 NM

47

42   43

46 **N**

44

48   50   49

52   51

53

38. Clutch oil line bracket
39. Transaxle case
40. Oil guide
41. Oil guide
42. Outer race
43. Spacer

44. Spacer
45. Stopper bracket
46. Oil seal
47. Reverse shift lever assembly
48. Reverse shift lever shoe

49. Reverse idler gear shaft
50. Reverse idler gear
51. Front output shaft assembly
52. Needle bearing
53. Front differential

Fig. 21 Disassembly and Assembly — W5M33 transaxle, cont.

4. Remove the 4th speed gear, needle bearing and bearing sleeve. Remove the 3rd-4th speed synchronizer hub and ring.

5. Remove the 3rd speed gear using the proper adapter and press. Remove the needle bearing.

6. Remove the cone spring, sub gear and needle bearing.

Inspection

1. Check the outer surface of the input shaft where the needle bearing is mounted for damage, abnormal wear and seizure.

2. Check the input shaft splines for damage and wear.

3. Combine the needle bearing with the shaft or bearing sleeve and gear and check that it rotates smoothly without abnormal noise or play.

4. Check the needle bearing cage for deformation.

5. Check the synchronizer ring clutch gear teeth for damage and breakage.

6. Check the internal surface for damage, wear and broken threads.

7. Force the synchronizer ring toward the clutch gear and check the clearance. The clearance should be 0.02 in. (0.5mm) maximum.

8. Check the bevel gear and clutch gear teeth for damage and wear.

9. Check the synchronizer cone for rough surface, damage and wear.

Assembly

1. Install the needle bearing and the sub gear on the input shaft.

2. Install the cone spring, snapring and 3rd speed gear. Install the 3rd-4th speed synchronizer sleeve and hub assembly.

3. Install the synchronizer spring. When installing springs, be sure to position each spring with respect to the

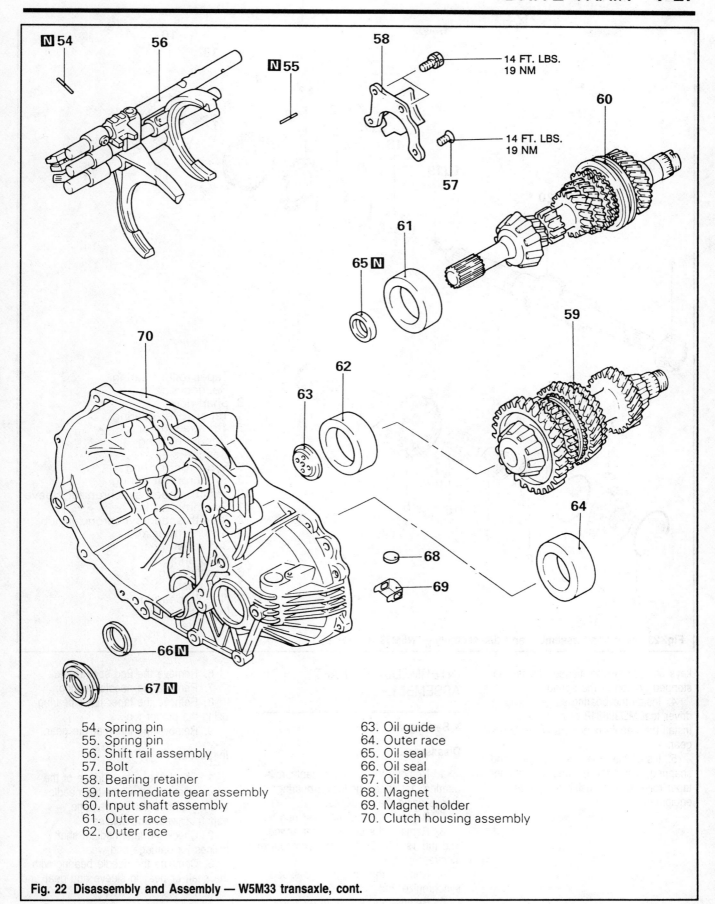

54. Spring pin
55. Spring pin
56. Shift rail assembly
57. Bolt
58. Bearing retainer
59. Intermediate gear assembly
60. Input shaft assembly
61. Outer race
62. Outer race

63. Oil guide
64. Outer race
65. Oil seal
66. Oil seal
67. Oil seal
68. Magnet
69. Magnet holder
70. Clutch housing assembly

Fig. 22 Disassembly and Assembly — W5M33 transaxle, cont.

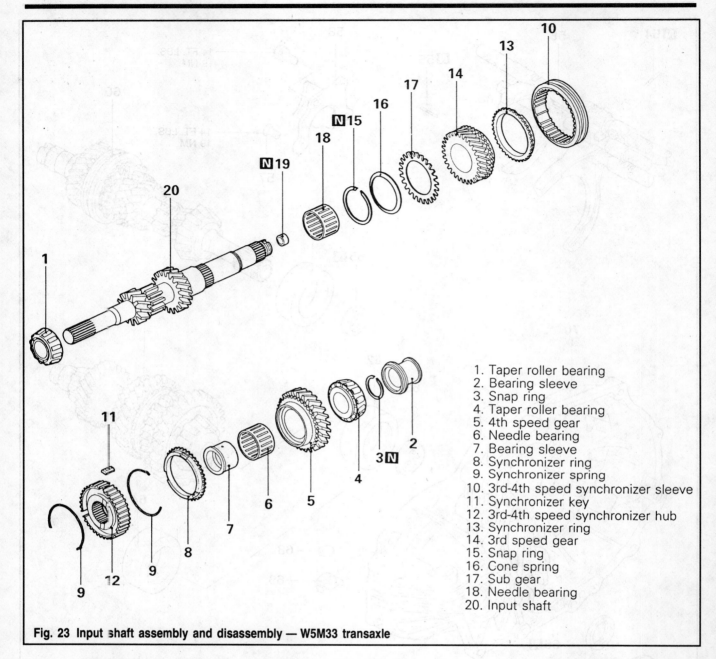

1. Taper roller bearing
2. Bearing sleeve
3. Snap ring
4. Taper roller bearing
5. 4th speed gear
6. Needle bearing
7. Bearing sleeve
8. Synchronizer ring
9. Synchronizer spring
10. 3rd-4th speed synchronizer sleeve
11. Synchronizer key
12. 3rd-4th speed synchronizer hub
13. Synchronizer ring
14. 3rd speed gear
15. Snap ring
16. Cone spring
17. Sub gear
18. Needle bearing
20. Input shaft

Fig. 23 Input shaft assembly and disassembly — W5M33 transaxle

keys so they are positioned against the stepped portion of the spring.

4. Install the bearing sleeve using driver tool MD998818 or equivalent. Install the needle bearing and 4th speed gear.

5. Install the taper roller bearing and snapring. Install the bearing sleeve and taper roller bearing using the proper equipment.

INTERMEDIATE SHAFT ASSEMBLY

▶ **See Figure 24**

Disassembly

1. Remove the snapring, taper roller bearing, 1st speed gear and bearing sleeve using the proper tool.

2. Remove the synchronizer ring.

3. Remove the synchronizer spring and the 1st and 2nd speed synchronizer sleeve.

4. Remove the 1st and 2nd speed synchronizer hub.

5. Remove the synchronizer ring.

6. Remove the 2nd speed gear.

7. Remove the needle bearing.

8. Remove the taper roller bearing using the proper tool.

9. Remove the intermediate gear.

Inspection

1. Check the outer surface of the intermediate shaft where the needle bearing is mounted for damage, abnormal wear and seizure.

2. Check the intermediate shaft splines for damage and wear.

3. Combine the needle bearing with the shaft or bearing sleeve and gear and

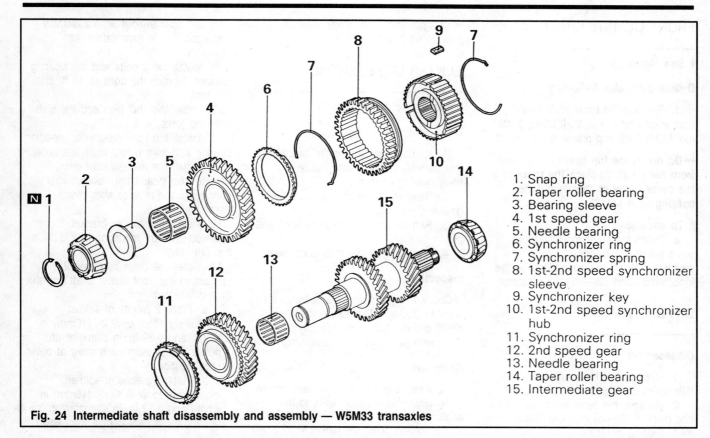

1. Snap ring
2. Taper roller bearing
3. Bearing sleeve
4. 1st speed gear
5. Needle bearing
6. Synchronizer ring
7. Synchronizer spring
8. 1st-2nd speed synchronizer sleeve
9. Synchronizer key
10. 1st-2nd speed synchronizer hub
11. Synchronizer ring
12. 2nd speed gear
13. Needle bearing
14. Taper roller bearing
15. Intermediate gear

Fig. 24 Intermediate shaft disassembly and assembly — W5M33 transaxles

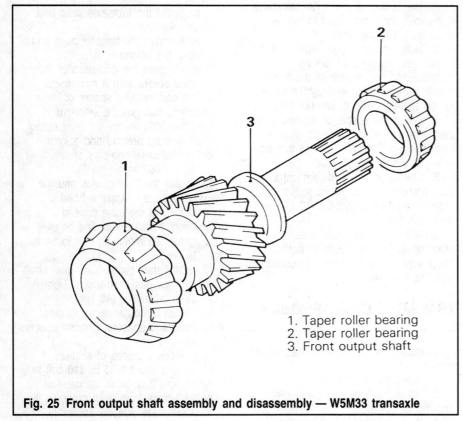

1. Taper roller bearing
2. Taper roller bearing
3. Front output shaft

Fig. 25 Front output shaft assembly and disassembly — W5M33 transaxle

check that it rotates smoothly without abnormal noise or play.

4. Check the needle bearing cage for deformation.

5. Check the synchronizer ring clutch gear teeth for damage and breakage.

6. Check the internal surface for damage, wear and broken threads.

Assembly

1. Install the intermediate gear. Install the taper roller bearing using the proper tool.

2. Install the needle bearing.

3. Install the 2nd speed gear.

4. Install the synchronizer ring.

5. Install the 1st and 2nd speed synchronizer hub.

6. Install the 1st and 2nd speed synchronizer sleeve and the synchronizer spring.

7. Install the synchronizer ring.

8. Install the snapring, taper roller bearing, 1st speed gear and bearing sleeve using the proper tool.

FRONT OUTPUT SHAFT

▶ See Figure 25

Disassembly and Assembly

1. Remove the taper roller bearings from each end of the shaft using guide tool MD998348 and pressing from shaft.

➡ Do not reuse the bearing removed from the shaft. Replace the inner and the outer races of the taper roller bearing as a set.

To assemble:

2. Install the taper roller bearings using driver tool MD998818 or equivalent. Apply the special tool to the inner race when installing the bearings.

DIFFERENTIAL

Disassembly

1. Remove the bolts and the differential drive gear from the case.
2. Remove the taper roller bearings and discard. Do not reuse the taper roller bearing or races.
3. Remove the lockpin, pinion shaft and pinion gears and washers.
4. Remove the side gears and spacers.

Inspection

Check both the gears and the bearings for wear or damage. Make sure they are clean and dry before reassembly. Coat them with gear oil before reassembly.

Assembly

1. Install the side gears and spacers.
2. Install new lockpin, pinion shaft and pinion gears and washers.
3. Measure the backlash between the side gears and pinions. The standard backlash is 0.001-0.006 in. (0.025-0.150mm). If the backlash is out of specification, disassemble and use a different spacer.
4. Install the taper roller bearings.

➡ When press fitting the bearings, push on the inner race only.

5. Install the differential drive gear and the retaining bolts. Apply stud locking sealant to the threads of the bolts and quickly tighten in the a alternating cross pattern. This will assure

even tightening of the bolts. Tighten bolts to 98 ft. lbs. (135 Nm).

CENTER DIFFERENTIAL

▶ See Figure 26

Disassembly

1. Remove the taper roller bearings and discard. Do not reuse the taper roller bearing or races.
2. Remove the retainer bolts, output gear and spacer.
3. Remove the side gear, pinions and pinion washer.
4. Remove the side gear and spacer.

Inspection

Check the gears and the bearings for wear or damage. Make sure they are clean and dry before reassembly. Coat them with gear oil before reassembly.

Assembly

1. Install the spacer, side gear, pinion gear, washer and pinion shaft to the center differential case.
2. Holding down the pinion shaft, select the spacer of maximum thickness that allows the pinion gear to turn lightly and install it to the shaft.
3. Install the side gear, spacer, and output gear. Apply stud locking compound to the threads of the bolts. Install in output gear and tighten evenly in star pattern to 55 ft. lbs. (75 Nm).
4. Select the spacer of maximum thickness that allows the side gear to turn lightly and install it. Check that both side gears turn lightly.
5. Check the center differential side gear end-play and compare to the desired value of 0.0020-0.0010 in. (0.05-0.25mm).
6. Install the taper roller bearings using driver MD998822-01 or equivalent. Make sure the driver comes in contact with the inner race only.

TRANSAXLE CASE ASSEMBLY

1. Install the magnet and the magnet holder.
2. Install the 3 oil seals.
3. Install the 3 bearing outer races and the oil guide.
4. Install the input shaft assembly.

5. Lift up the input shaft assembly and install the intermediate shaft assembly.
6. Install the 2 bolts and the bearing retainer. Tighten the bolts to 14 ft. lbs. (19 Nm).
7. Install the shift pins and the shift rails and forks.
8. Install the front differential, needle bearing and front output shaft assembly.
9. Install the reverse idler gear, reverse idler gear shaft, reverse idler gear shoe and reverse shift lever assembly.
10. Install the stopper bracket, if removed. Tighten mounting bolt to 14 ft. lbs. (19 Nm).
11. Install the outer race and 2 spacers to the front output shaft and the front differential:
    a. Place 2 pieces of soldier measuring about 0.39 in. (10mm) in length and 0.12 in. in diameter at points across from each other at outer race location.
    b. Place 2 pieces of soldier measuring about 0.39 in. (10mm) in length and 0.12 in. in diameter on the bearing outer race across from each other.
    c. Install the transaxle case and tighten the bolts.
    d. Remove the transfer case and remove the soldier.
    e. Measure the thickness of the crushed soldier with a micrometer. Select and install a spacer of thickness that gives a standard 0.0031-0.0051 in. for the front output shaft bearing preload and a front differential case end-play of 0.0020-0.0067 in.
12. Install the 2 oil guides and the transaxle oil case. Apply a bead of sealant around the case prior to installation. Install the oil line bracket and tighten the mounting bolt to 14 ft. lbs. (19 Nm).
13. Install the reverse idler gear shaft bolt with new gasket in place. Tighten the bolt to 36 ft. lbs. (49 Nm).
14. Install the outer races, center differential and spacers. Choose spacers as follows:
    a. Place 2 pieces of soldier measuring about 0.39 in. (10mm) in length and 0.12 in. in diameter at points across from each other on the transaxle case adapter assembly. Install each outer race.

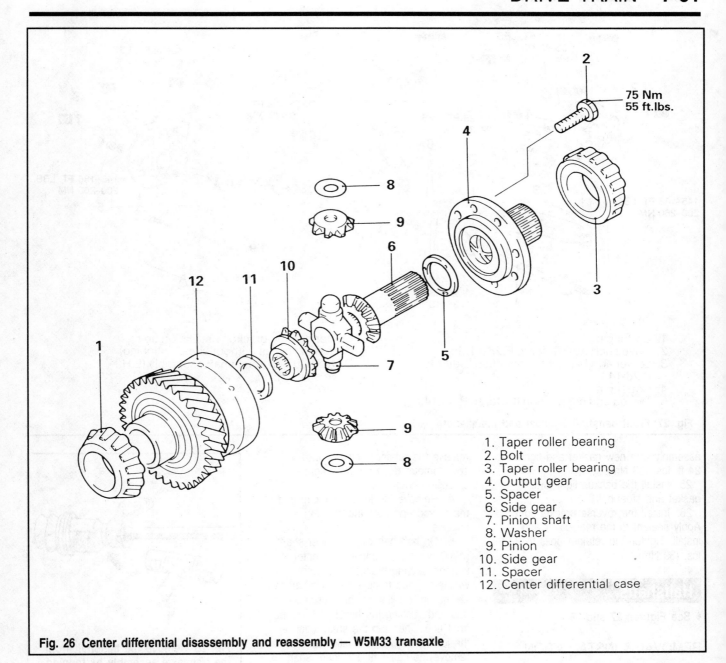

2 75 Nm 55 ft.lbs.

1. Taper roller bearing
2. Bolt
3. Taper roller bearing
4. Output gear
5. Spacer
6. Side gear
7. Pinion shaft
8. Washer
9. Pinion
10. Side gear
11. Spacer
12. Center differential case

Fig. 26 Center differential disassembly and reassembly — W5M33 transaxle

b. Install the transaxle case adapter assembly and rear cover. Tighten the bolts.

c. Remove the transaxle case adapter assembly and rear cover. Remove each outer race and the the soldier.

d. Measure the thickness of the crushed soldier and select the spacer that gives the correct end-play and preload. The desired intermediate gear preload is 0.0031-0.0051 inch. The desired center differential case preload is 0.0031-0.0051 inch. The input shaft end-play should be 0-0.0020 inch.

15. Apply liquid gasket to the transaxle case side of the transaxle case adapter assembly and install.

16. Install the viscous coupling, steel ball and new snapring. Move the center shaft so the steel balls are securely seated in the grooves. Choose a snapring that gives the viscous coupling an end-play of 0.0039-0.102 in. (0.10-0.26mm).

17. Install the needle bearing and the 5th intermediate gear.

18. Install the synchronizer ring and the 5th gear.

19. Install the 5th speed synchronizer assembly and shift fork.

20. Install socket tool MD998802-01 or equivalent onto the input shaft. Screw a bolt 10mm in size into the hole in the periphery of the clutch housing and attach a spinner handle to the tool. Shift the transaxle in reverse using the control lever and select lever. Tighten the locknut to 109 ft. lbs. (150 Nm) while using the bolt attached to the clutch housing as a spinner handle stopper. Stake the locknut in this position.

21. Install the spring pin.

22. Install the air breather.

23. Install the speedometer driven gear assembly.

24. Install the poppet plug, spring and steel ball. Install the restrict ball

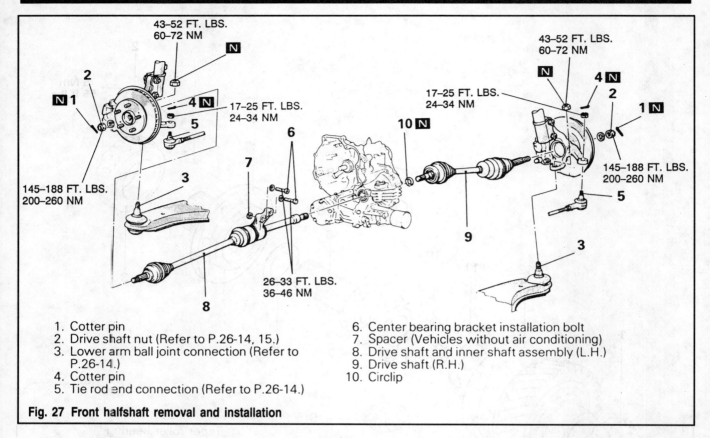

43–52 FT. LBS.
60–72 NM

17–25 FT. LBS.
24–34 NM

145–188 FT. LBS.
200–260 NM

26–33 FT. LBS.
36–46 NM

43–52 FT. LBS.
60–72 NM

17–25 FT. LBS.
24–34 NM

145–188 FT. LBS.
200–260 NM

1. Cotter pin
2. Drive shaft nut (Refer to P.26-14, 15.)
3. Lower arm ball joint connection (Refer to P.26-14.)
4. Cotter pin
5. Tie rod end connection (Refer to P.26-14.)
6. Center bearing bracket installation bolt
7. Spacer (Vehicles without air conditioning)
8. Drive shaft and inner shaft assembly (L.H.)
9. Drive shaft (R.H.)
10. Circlip

Fig. 27 Front halfshaft removal and installation

assembly with new gasket and tighten to 24 ft. lbs. (33 Nm).

25. Install the backup light switch, gasket and steel ball.

26. Install the reverse bracket cone. Apply sealant to the rear cover and install. Tighten the retainer bolts to 29 ft. lbs. (39 Nm).

Halfshafts

▶ **See Figures 27 and 28**

REMOVAL & INSTALLATION

➡If the vehicle is going to be rolled while the halfshafts are out of the vehicle, obtain 2 outer CV-joints or proper equivalent tools and install to the hubs. If the vehicle is rolled without the proper torque applied to the front wheel bearings, the bearings will no longer be usable.

1. Disconnect the negative battery cable.

2. Remove the cotter pin, halfshaft nut and washer.

3. Raise the vehicle and support safely. If removing the right halfshaft,

remove the retainer bolt and the speedometer drive from the right extension housing.

4. Remove the lower ball joint and the tie rod end from the steering knuckle.

5. On halfshaft with an inner shaft (AWD vehicles), remove the center support bearing bracket bolts and washers. Then remove the halfshaft by setting up a puller on the outside wheel hub and pushing the halfshaft from the front hub. Then tap the shaft union at the joint case with a plastic hammer to remove the halfshaft and inner shaft from the transaxle.

6. On one piece halfshafts (FWD vehicles), remove the halfshaft from the hub/knuckle by setting up a puller on the outside wheel hub and pushing the halfshaft from the front hub. After pressing the outer shaft, insert a prybar between the transaxle case and the halfshaft and pry the shaft from the transaxle.

➡**Do not pull on the shaft; doing so damages the inboard joint. Do not insert the prybar too far or the oil seal in the case may be damaged.**

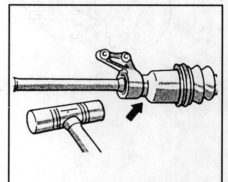

Fig. 28 Remove the halfshaft from the transaxle assembly by tapping the T.J. case with plastic hammer — AWD vehicle

To install:

7. Inspect the halfshaft boot for damage or deterioration. Check the ball joints and splines for wear.

8. Replace the circlips on the ends of the halfshafts.

9. Insert the halfshaft into the transaxle. Make sure it is fully seated.

10. Pull the strut assembly outward and install the other end of the halfshaft into the hub.

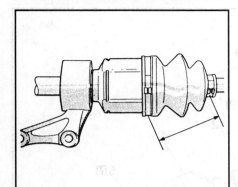

Fig. 29 Set the boot so the bands are at correct distance

11. Install the center bearing bracket bolts and tighten to 33 ft. lbs. (45 Nm), if equipped.

12. Install the washer so the chamfered edge faces outward. Install the halfshaft nut and tighten temporarily.

13. Install the tie rod end and ball joint to the steering knuckle.

14. Install the wheel and lower the vehicle to the floor. Tighten the axle nut with the brakes applied. Tighten the nut to torque of 145-188 ft. lbs. (200-260 Nm).

15. Install a new cotter pin and bend to secure.

CV-JOINTS

These vehicles use different types of CV-joints. Engine size, transaxle type, whether the joint is an inboard or outboard joint, even which side of the vehicle is being serviced could make a difference in joint type. Be sure to properly identify the joint before attempting joint or boot replacement. Look for identification numbers at the large end of the boots and/or on the end of the metal retainer bands.

The 2 types of joints used are the Birfield Joint, (B.J.) and the Tripod Joint (T.J.). Special grease and clamps are used with these joints and is normally supplied with the replacement joint and/or boot kit. Do not use regular chassis grease.

Correct installation of the CV-boot is essential for its longevity. A specification is given for the distance between the large and small boot bands. This is so the boot will not be installed either too loose or too tight, which could cause early wear and cracking, allowing the grease to get out and water and dirt in, leading to early joint failure.

BOOT AND JOINT REPLACEMENT

FWD Vehicle
▶ **See Figures 29 and 30**

Both of the halfshaft boots are going to be removed from the T.J. case side of the halfshaft.

1. Disconnect the negative battery cable. Remove the halfshaft from the vehicle.

2. Remove the T.J. boot bands from the boot. Side cutter pliers can be used to cut off the metal retaining bands. Remove the T.J. case from the halfshaft.

3. Remove the snapring next to the tripod joint spider assembly from the halfshaft with snapring pliers. Remove the spider assembly from the shaft.

➡**Do not disassemble the spider and use care in handling.**

4. If the boot is be reused, wrap vinyl tape around the spline part of the shaft so the boot will not be damaged when removed. Remove the dynamic damper, if used, and boots from the shaft.

To install:

5. Double check that the correct replacement parts are being installed. Wrap vinyl tape around the splines to protect the boot and install the boots and damper, if used, in the correct order.

6. Fill the inside of the boot with the specified grease. Often the grease supplied in the replacement parts kit is meant to be divided in half, with half being used to lubricate the joint and half being used inside the boot. Keep grease off the rubber part of the dynamic damper (if used).

7. Secure the boot bands with the halfshaft in a horizontal position. Make sure the boot span on the halfshaft is 3.35 ± 0.12 in. (85 ± 3mm) in length.

8. Install halfshaft into vehicle.

AWD Vehicle
▶ **See Figures 31 and 32**

1. Disconnect the negative battery cable. Remove the halfshaft from the vehicle.

2. Remove the T.J. large and small boot bands. Remove the T.J. case from inner shaft assembly.

3. Remove the snapring next to the tripod joint spider assembly from the halfshaft with snapring pliers. Remove the spider assembly from the shaft.

➡**Do not disassemble the spider and use care in handling.**

4. If the boot is be reused, wrap vinyl tape around the spline part of the shaft so the boot will not be damaged when removed.

5. Remove the inner and the outer dust seals from the center support bearing assembly. Remove the center bearing from the shaft.

6. Remove the inner shaft assembly, together with the seal plate, from the T.J. case. Using puller tool, remove the inner shaft from the center bearing bracket.

To install:

7. Apply multi-purpose grease to the center bearing and inside the the center bearing bracket. Using proper size driver, press fit the center bearing into the center bearing bracket.

8. Apply multi-purpose grease to the rear surfaces of both dust seals and install. Use a pipe to hold the inner race of the center bearing and force the inner shaft into place.

9. Install the boots in place. Apply grease to the inner shaft splines, then press fit it into the T.J. case. Press the seal plate into the T.J. case.

10. Fill the join and the boot with the specified grease, enclosed in the repair kit. Divide the grease in half between the joint and the boot. Keep grease off the rubber part of the dynamic damper (if used).

11. Secure the boot bands with the halfshaft in a horizontal position. Make sure the boot span on the halfshaft is 3.35 ± 0.12 in. (85 ± 3mm) in length.

12. Install halfshaft into vehicle.

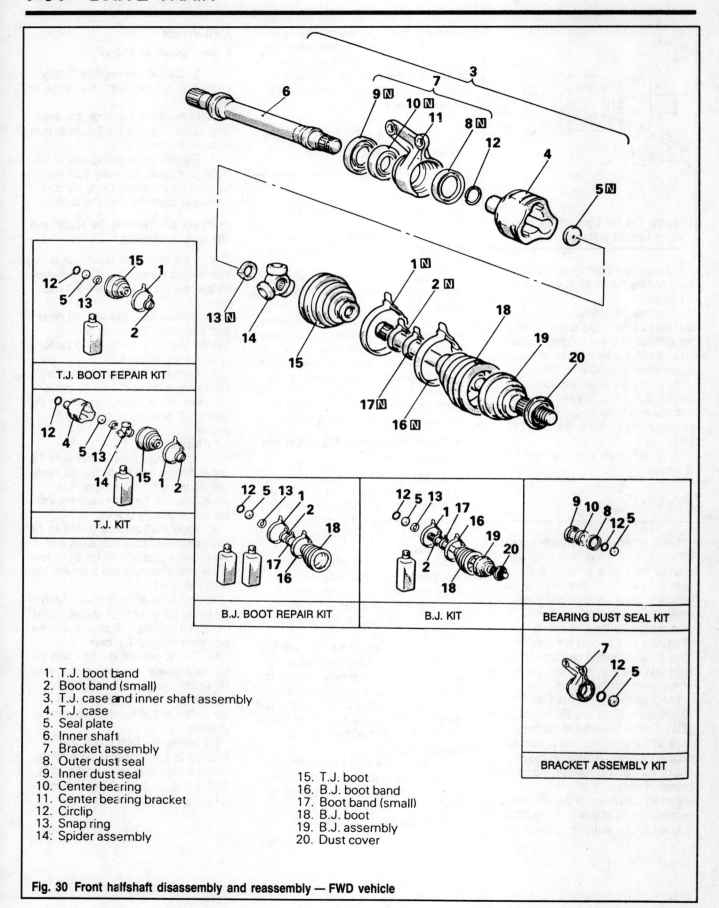

T.J. BOOT FEPAIR KIT

T.J. KIT

B.J. BOOT REPAIR KIT

B.J. KIT

BEARING DUST SEAL KIT

BRACKET ASSEMBLY KIT

1. T.J. boot band
2. Boot band (small)
3. T.J. case and inner shaft assembly
4. T.J. case
5. Seal plate
6. Inner shaft
7. Bracket assembly
8. Outer dust seal
9. Inner dust seal
10. Center bearing
11. Center bearing bracket
12. Circlip
13. Snap ring
14. Spider assembly
15. T.J. boot
16. B.J. boot band
17. Boot band (small)
18. B.J. boot
19. B.J. assembly
20. Dust cover

Fig. 30 Front halfshaft disassembly and reassembly — FWD vehicle

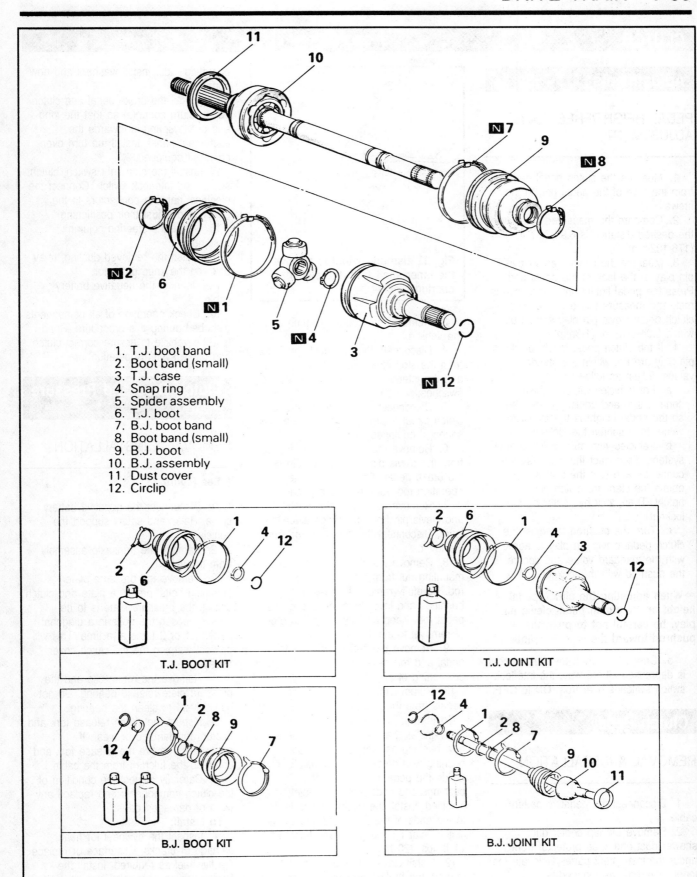

1. T.J. boot band
2. Boot band (small)
3. T.J. case
4. Snap ring
5. Spider assembly
6. T.J. boot
7. B.J. boot band
8. Boot band (small)
9. B.J. boot
10. B.J. assembly
11. Dust cover
12. Circlip

T.J. BOOT KIT

T.J. JOINT KIT

B.J. BOOT KIT

B.J. JOINT KIT

Fig. 31 Front halfshaft disassembly and reassembly — FWD vehicle

CLUTCH

Adjustments

PEDAL HEIGHT/FREE-PLAY ADJUSTMENT

1. Measure the clutch pedal height from the face of the pedal pad to the firewall.

2. Compare the measured value with the desired distance is 6.93-7.17 in. (176-182mm).

3. Measure the clutch pedal clevis pin play at the face of the pedal pad. Press the pedal lightly until resistance is met, and measure this distance. The clutch pedal clevis pin play should be within 0.04-0.12 in (1-3mm).

4. If the clutch pedal height or clevis pin play are not within the standard values, adjust as follows:

   a. For vehicles without cruise control, turn and adjust the stop bolt so the pedal height is the standard value, then tighten the locknut.

   b. Vehicles with auto-cruise control system, disconnect the clutch switch connector and turn the switch to obtain the standard clutch pedal height. Then, lock by tightening the locknut.

   c. Turn the pushrod to adjust the clutch pedal clevis pin play to agree with the standard value and secure the pushrod with the locknut.

➡**When adjusting the clutch pedal height or the clutch pedal clevis pin play, be careful not to push the pushrod toward the master cylinder.**

   d. Check that when the clutch pedal is depressed all the way, the interlock switch switches over from **ON** to **OFF**.

Clutch Pedal

REMOVAL & INSTALLATION

1. Disconnect the negative battery cable.

2. Remove the lap cooler duct, shower duct and knee protector from under the instrument panel. Relocate the inner relay box, as required.

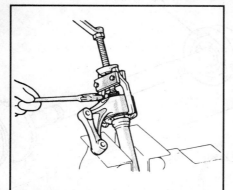

Fig. 32 Use puller tool to remove the inner shaft from the center bearing bracket

3. Remove the steering column from the vehicle.

4. Disconnect the electrical connector from the stop light switch, clutch switch and the interlock switch and remove the switches.

5. Disconnect the brake pedal and clutch pedal return springs and turn over spring, if equipped.

6. Remove the cotter pin and washer from the clevis pin and remove pin from the brake pedal. This will disconnect the operating rod from the brake pedal.

7. Remove the cotter pin, washer and clevis pin from the clutch actuation lever. Disconnect the actuating rod from the lever.

8. Remove the clutch actuator lever mounting nut from the end of the pedal rod. Carefully remove the washers, bushings and lever from the end of the pedal rod. Keep all components in order of removal to aid in installation.

9. Remove the brake pedal, clutch pedal and rod assembly. Note positioning of all bushings.

10. Remove the brake and clutch pedals from the vehicle.

To install:

11. Install the pedal(s) in position with new bushing as required. Lubricate bushing with grease prior to installation. Guide the pedal rod through the bushings and pedal until completely installed. Install the bushings and lever to the ends of the pedal rod. Install the clutch lever mounting nut and tighten to 14 ft. lbs. (20 Nm).

12. Install clevis pin and washers through the brake and clutch pedal operating rods. Install washers and new cotter pins.

13. Install the brake pedal and clutch pedal return spring(s) so that the long end of the spring is towards the instrument panel. Install the turn over spring, if equipped.

14. Install the stop light switch, clutch switch and interlock switch. Connect the electrical harness connectors to the switch and adjust their positioning.

15. Install the steering column assembly.

16. Install the removed ducting, relay box and the knee protector.

17. Connect the negative battery cable.

18. Check operation of all components disturbed during this procedure. Road test the vehicle to assure correct clutch and brake system operation.

Clutch Disc and Pressure Plate

REMOVAL & INSTALLATION

▶ **See Figure 33**

1. Disconnect the negative battery cable. Raise and safely support the vehicle.

2. Remove the transaxle assembly from the vehicle.

3. Remove the pressure plate attaching bolts, pressure plate and clutch disc. If the pressure plate is to be reused, loosen the bolts in a diagonal pattern, 1 or 2 turns at a time. This will prevent warping the the clutch cover assembly.

4. Remove the return clip and the pressure plate release bearing. Do not use solvent to clean the bearing.

5. Inspect the clutch release fork and fulcrum for damage or wear. If necessary, remove the release fork and unthread the fulcrum from the cable.

6. Carefully inspect the condition of the clutch components and replace any worn or damaged parts.

To install:

7. Inspect the flywheel for heat damage or cracks. Resurface or replace the flywheel as required. Install the flywheel using new bolts.

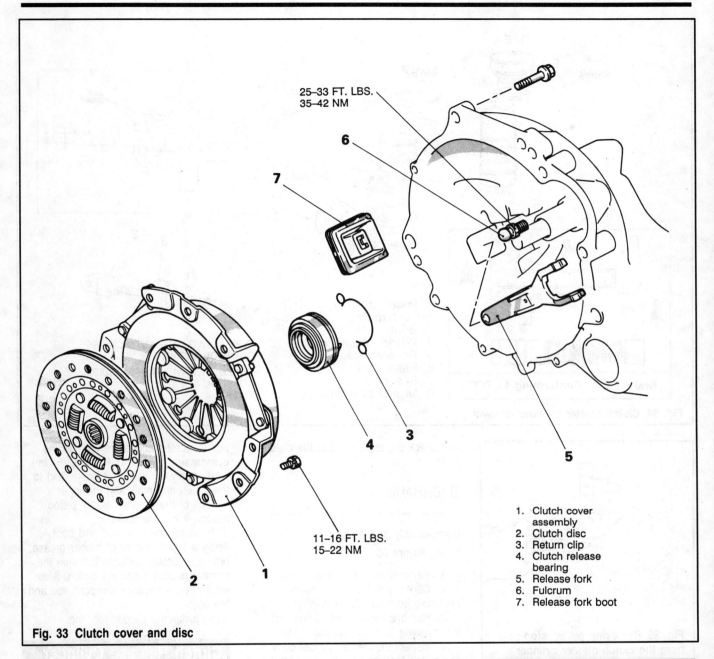

25–33 FT. LBS.
35–42 NM

6

7

11–16 FT. LBS.
15–22 NM

5

4   3

2   1

1. Clutch cover
   assembly
2. Clutch disc
3. Return clip
4. Clutch release
   bearing
5. Release fork
6. Fulcrum
7. Release fork boot

Fig. 33 Clutch cover and disc

8. Install the fulcrum and tighten to 25 ft. lbs. (35 Nm). Install the release fork. Apply a coating of multi-purpose grease to the point of contact with the fulcrum and the point of contact with the release bearing. Apply a coating of multi-purpose grease to the end of the release cylinder's push rod and the push rod hole in the release fork.

➡When installing the clutch, apply grease to each part, but be careful not to apply excessive grease; excessive grease will cause clutch slippage and shudder.

9. Apply multi-purpose grease to the clutch release bearing. Pack the bearing inner surface and the groove with grease. Do not apply grease to the resin portion of the bearing. Place the bearing in position and install return clip.

10. Apply a coating of grease to the clutch disc splines and then use a brush to rub it in the grooves. Using a universal clutch disc aligner, position the clutch disc on the flywheel. Install the retainer bolts and tighten a little at a time, in a diagonal sequence. Tighten them to a final torque of 16 ft. lbs. (22 Nm). Remove the aligning tool.

11. Install the transaxle assembly and check for proper clutch operation.

Clutch Master Cylinder

REMOVAL & INSTALLATION

▶ See Figure 34

1. Disconnect the negative battery cable.

2. Remove necessary underhood components in order to gain access to the clutch master cylinder.

3. Loosen the clutch fluid line at the cylinder and allow the fluid to drain. Use care; brake fluid damages paint.

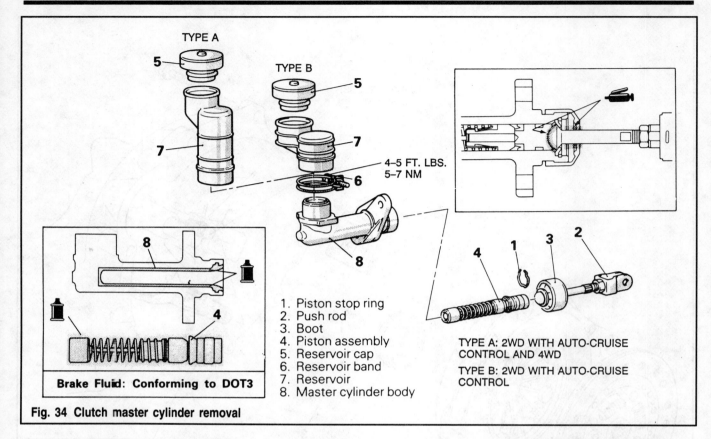

1. Piston stop ring
2. Push rod
3. Boot
4. Piston assembly
5. Reservoir cap
6. Reservoir band
7. Reservoir
8. Master cylinder body

TYPE A: 2WD WITH AUTO-CRUISE CONTROL AND 4WD

TYPE B: 2WD WITH AUTO-CRUISE CONTROL

4–5 FT. LBS.
5–7 NM

Brake Fluid: Conforming to DOT3

Fig. 34 Clutch master cylinder removal

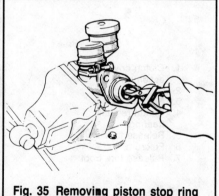

Fig. 35 Removing piston stop ring from the clutch master cylinder

4. Remove the clevis pin retainer at the clutch pedal and remove the washer and clevis pin.

5. Remove the 2 nuts and pull the cylinder from the firewall. A seal should be between the mounting flange and firewall. This seal should be replace.

6. The installation is the reverse of the removal procedure.

7. Lubricate all pivot points with grease.

8. Bleed the system at the slave cylinder using fresh DOT 3 brake fluid

and check the adjustment of the clutch pedal.

OVERHAUL

Disassembly and Assembly
♦ See Figure 35

1. Remove the piston stop ring.
2. Carefully remove the piston assembly front the clutch cylinder.
3. Remove the reservoir band and the reservoir from the cylinder body.
4. Inspect the inside cylinder body and piston for rust or scars. Inspect the piston cup for wear of deformation.

➡Do not disassemble the piston assembly.

5. Measure and record the inside of the cylinder bore with a bore gauge or inside micrometer. Take measurements at the top, middle and bottom of the cylinder bore. Compare reading with specification of ⅝ in. (15.87mm).

6. If the bore diameter exceeds specifications, replace the release cylinder assembly.

To assemble:

7. Install the reservoir to the cylinder body. Position the band so the tightening

bolt is located above the body of the cylinder and tighten to 4 ft. lbs. (5 Nm).

8. Apply fresh DOT 3 brake fluid to the piston assembly and the inner surface of the cylinder. Install piston assembly into the cylinder.

9. Install the push rod and boot. Apply a small amount of rubber grease onto the contact surfaces between the damper push rod and the piston. Also apply grease between the push rod and the boot.

10. Install the piston stop ring.

Clutch Release Cylinder

REMOVAL & INSTALLATION

♦ See Figure 36

1. Disconnect the negative battery cable. Remove necessary underhood components in order to gain access to the clutch release cylinder.

2. Remove the hydraulic line and allow the system to drain.

3. Remove the bolts and pull the cylinder from the transaxle housing.

4. The installation is the reverse of the removal procedure.

5. Lubricate all pivot points with grease.

6. Fill the system with clean brake fluid meeting DOT 3 specifications.

7. Bleed the system and adjust the clutch pedal height and the clevis pin play.

OVERHAUL

Disassembly

1. Remove the clutch release cylinder from the vehicle and mount on a vise.

2. Remove the valve plate, spring, pushrod and boot.

3. Cover the piston bore opening with a rag and slowly introduce compressed air through the inlet opening to blow the piston and cup from the bore.

Inspection

1. Wipe the inside of the release cylinder body with a clean, lint-free rag.

2. Check the inside of the cylinder body for scratches or irregular wear. Replace if the piston cup outer circumference is scratched or shows signs of fatigue.

3. Measure and record the inside of the cylinder bore with a bore gauge or inside micrometer. Take measurements at the top, middle and bottom of the cylinder bore. Compare reading with the following specifications:

   a. 1.8L or 2.0L non-turbocharged engine — $^{13}/_{16}$ in. (20.64mm)

   b. 2.0L turbocharged engine — $^{3}/_{4}$ in. (19.05mm).

4. If the bore diameter exceeds specifications, replace the release cylinder assembly.

Assembly

1. Wipe all parts with a clean, lint-free rag.

2. Coat the inside surface of the release cylinder bore and the entire piston outer surface with clean brake fluid meeting DOT 3 specifications.

3. Insert the piston assembly into the bore.

4. Install the valve plate, spring and boot.

5. Install the release cylinder.

6. Fill the system with clean brake fluid meeting DOT 3 specifications. Bleed the system.

7. Adjust the clutch pedal height and the clevis pin play as required.

HYDRAULIC CLUTCH SYSTEM BLEEDING

1. Fill the reservoir with clean brake fluid meeting DOT 3 specifications.

2. Loosen the bleed screw, have the clutch pedal pressed to the floor.

3. Tighten the bleed screw and release the clutch pedal.

4. Repeat the procedure until the fluid is free of air bubbles.

➡**It is suggested to attach a hose to the bleeder and place the other end into a container at least ½ full of brake fluid during the bleeding operation. Do not allow the reservoir to run out of fluid during bleeding.**

CLUTCH SYSTEM TROUBLESHOOTING

Symptom	Probable cause	Remedy
Clutch chatter	Wear or damage of clutch disc	Replace
	Grease or oil on disc facings	Replace
	Break or looseness of engine mounting	Replace or tighten mounting
Clutch slips	Clutch pedal free play insufficient	Adjust pedal free play
	Clogged hydraulic system	Repair or replace parts
	Clutch disc lining oily or worn out	Inspect clutch disc
	Pressure plate faulty	Replace clutch cover
	Release fork binding	Inspect release fork
Clutch grabs/chatters	Clutch disc lining oily or worn out	Inspect clutch disc
	Pressure plate faulty	Replace clutch cover
	Clutch diaphragm spring bent	Replace clutch cover
	Worn or broken torsion spring	Replace clutch disc
Clutch noisy	Damaged clutch pedal bushing	Replace clutch pedal bushing
	Loose part inside housing	Repair as necessary
	Release bearing worn or dirty	Replace release bearing
	Release fork or linkage sticks	Repair as necessary

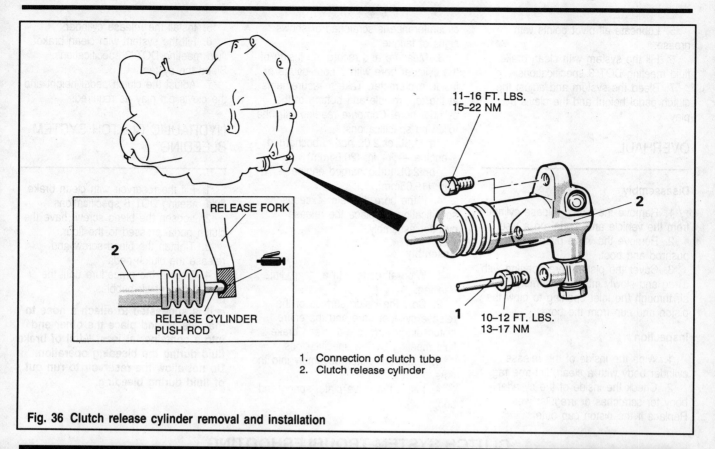

1. Connection of clutch tube
2. Clutch release cylinder

Fig. 36 Clutch release cylinder removal and installation

RELEASE FORK

RELEASE CYLINDER
PUSH ROD

11–16 FT. LBS.
15–22 NM

10–12 FT. LBS.
13–17 NM

AUTOMATIC TRANSAXLE

Identification

On most models, the Vehicle Information Code Plate mounted on the firewall in the engine compartment. On the third line, labeled 'Transaxle" the transaxle model is listed, preceding a space. The serial number is stamped following the space.

Fluid Pan

REMOVAL & INSTALLATION

▶ See Figures 37 and 38

1. Disconnect the negative battery cable.
2. Inspect the fluid and check for contamination. If found, replacement of the fluid and the filter is required.
3. Remove the drain plugs and allow the fluid from the transaxle to drain.
4. Remove the retainer bolts and the pan.

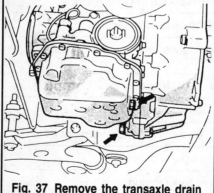

Fig. 37 Remove the transaxle drain plugs

5. Installation is the reverse of the removal procedure. Tighten the pan mounting bolts to 8 ft. lbs. (12 Nm).
6. Fill the transaxle with the proper level using Dexron II, Mopar ATF Plus type 7176, Mitsubishi Plus ATF or equivalent, automatic transaxle fluid.

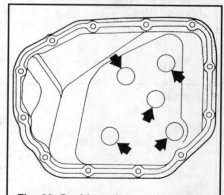

Fig. 38 Position of the magnets in the oil pan

FILTER SERVICE

1. Disconnect the negative battery cable. Raise and safely support the vehicle.
2. Remove the drain plugs and allow the fluid from the transaxle to drain.
3. Remove the retainer bolts and the pan.

4. Remove the retainers and the filter.

5. Clean the inside of the oil pan and the 5 magnets. Attach the magnets to the concave party cf the oil pan. Clean the mating surfaces on the oil pan and the transaxle.

To install:

6. Install the oil filter and retainers.

7. Install the oil pan with new gasket in place. Tighten the retaining bolts to 8 ft. lbs. (12 Nm).

8. Fill the transaxle with the proper level using Dexron II, Mopar ATF Plus type 7176, Mitsubishi Plus ATF or equivalent, automatic transaxle fluid.

9. Connect the negative battery cable.

Neutral Safety Switch

REMOVAL & INSTALLATION

1. Disconnect the negative battery cable.

2. Disconnect the selector cable from the lever.

3. Remove the 2 retaining screws and lift off the switch.

4. The installation is the reverse of the removal procedure. Do not tighten the bolts until the switch is adjusted.

5. Make sure the engine only starts in **P** and **N**. Also make sure the reverse lights turn ON in **R**.

ADJUSTMENT

✳✳WARNING

If the switch is faulty, the engine may start with the vehicle 'in gear''. If this happens the vehicle will accelerate when the engine starts. Keep the brakes on and be ready to switch the key off instantly.

1. Locate the neutral safety switch on the top of the transaxle.

2. Place the selector lever in **N**.

3. Loosen the 2 adjusting nuts to free up the cable and lever.

4. Place the safety switch manual control lever in **N**.

5. Note that 1 end of the safety switch manual control lever and the switch body have a hole in the ends. These holes, if properly adjusted, should be aligned.

6. If adjustment is required, loosen the mounting bolt and rotate the switch body until they are in alignment. Tighten the locknut to 9 ft. lbs. (12 Nm).

7. Loosen the adjuster nuts and gently pull the transaxle control cable to remove any slack. Tighten the nut to 10 ft. lbs. (14 Nm).

8. Verify that the switch lever moves to positions corresponding to each position of the selector lever.

9. Make sure the engine only starts in **P** and **N**. Also make sure the reverse lights turn ON in **R**.

Automatic Transaxle

REMOVAL & INSTALLATION

1. Remove the battery and battery tray.

2. On 1990-91 vehicles equipped with auto-cruise, remove the control actuator and bracket.

3. Drain the transaxle fluid.

4. Remove the air cleaner assembly, intercooler and air hose, as required.

5. Remove the adjusting nut and disconnect the shift cable.

6. Disconnect and tag the electrical connectors for the solenoid, neutral safety switch (inhibitor switch), the pulse generator kickdown servo switch and oil temperature sensor.

7. Disconnect the speedometer cable and oil cooler lines.

8. Disconnect the wires to the starter motor and remove the starter.

9. Remove the upper transaxle to engine bolts.

10. Support the transaxle and remove the transaxle mounting bracket.

11. Raise the vehicle and support safely. Remove the sheet metal under guard.

12. Remove the tie rod ends and the ball joints from the steering knuckle.

13. Remove the halfshafts by inserting a prybar between the transaxle case and

the driveshaft and prying the shaft from the transaxle. Do not pull on the driveshaft. Doing so damages the inboard joint. Use the prybar. Do not insert the prybar so far the oil seal in the case is damaged. Tie the halfshafts aside.

14. On AWD vehicles, disconnect the exhaust pipe and remove the transfer case.

15. Remove the lower bellhousing cover and remove the special bolts holding the flexplate to the torque converter. To remove, turn the engine crankshaft with a box wrench and bring the bolts into a position appropriate for removal, one at a time. After removing the bolts, push the torque converter toward the transaxle so it doesn't stay on the engine allowing oil to pour out the converter hub or cause damage to the converter.

16. Remove the lower transaxle to engine bolts and remove the transaxle assembly.

To install:

17. After the torque converter has been mounted on the transaxle, install the transaxle assembly on the engine. Tighten the driveplate bolts to 34-38 ft. lbs. (46-53 Nm). Install the bellhousing cover.

18. On AWD, install the transfer case and frame pieces. Connect the exhaust pipe using a new gasket.

19. Replace the circlips and install the halfshafts to the transaxle.

20. Install the tie rods and ball joint to the steering arm.

21. Install the transaxle mounting bracket.

22. Install the under guard.

23. Install the starter.

24. Connect the speedometer cable and oil cooler lines.

25. Connect the solenoid, neutral safety switch (inhibitor switch), the pulse generator kickdown servo switch and oil temperature sensor.

26. Install the shift control cable.

27. Install the air hose, intercooler and air cleaner assembly.

28. If equipped with auto-cruise, install the control actuator and bracket.

29. Refill with Dexron II, Mopar ATF Plus type 7176, Mitsubishi Plus ATF or equivalent, automatic transaxle fluid.

30. Start the engine and allow to idle for 2 minutes. Apply parking brake and move selector through each gear position, ending in **N**. Recheck fluid level and add if necessary. Fluid level should be between the marks in the **HOT** range.

TRANSFER CASE

Rear Output Shaft Seal

▶ **See Figures 39 and 40**

REMOVAL & INSTALLATION

1. Raise and support the vehicle safely.
2. Remove the propeller shaft from the transfer assembly. Place a drain pan under the rear of the transfer assembly to catch any fluid that leaks out.
3. Using a flat-tiped prying tool, remove the oil seal from the transfer dust seal guard.

Halfshaft

REMOVAL & INSTALLATION

For halfshaft removal, installation and component replacement, refer to

Halfshaft under Manual Transaxles, in this section.

MD998304-01

Fig. 40 Installation of oil seal using proper driver

To install:

4. Using proper size seal driver tool, install the seal into the dust seal guard and the transfer assembly.
5. Install the rear propeller shaft.
6. Lower the vehicle and inspect the transfer assembly fluid lever.

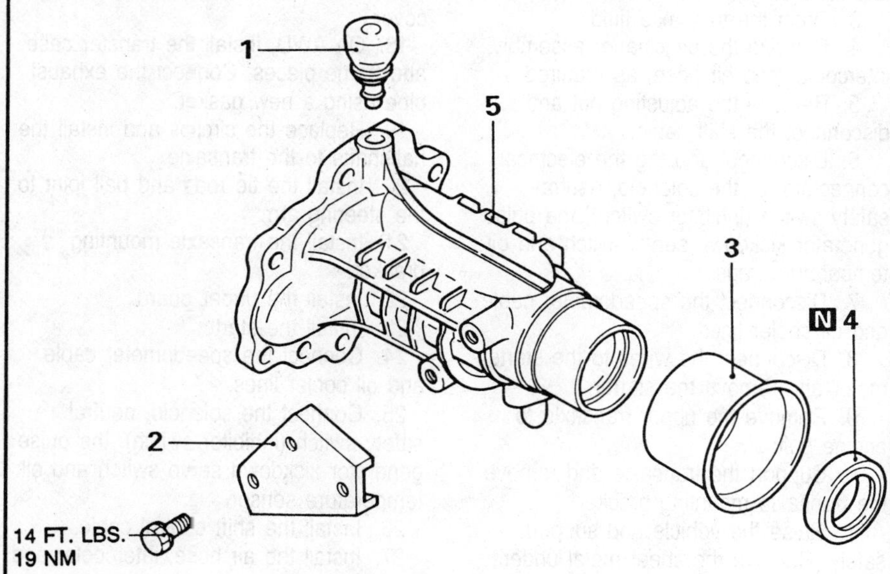

Transfer Assembly

REMOVAL & INSTALLATION

▶ **See Figures 41 and 42**

1. Disconnect the battery negative cable.
2. Raise the vehicle and support safely. Drain the transfer oil.
3. Disconnect the front exhaust pipe.
4. Unbolt the transfer case assembly and remove by sliding it off the rear propeller shaft. Be careful not to damage the oil seal in the transfer case output housing. Do not let the rear propeller shaft hang; suspend it from a frame piece. Cover the opening in the transaxle and transfer case to keep oil from dripping and to keep dirt out.

To install:

5. Lubricate the driveshaft sleeve yoke and oil seal lip on the transfer extension housing. Install the transfer case assembly to the transaxle. Use care when installing the rear propeller shaft to the transfer case output shaft.
6. Tighten the transfer case to transaxle bolts to 40-43 ft. lbs. (55-60 Nm) on manual transaxle vehicles or 43-58 ft. lbs. (60-80 Nm) on automatic transaxle vehicles.
7. Install the exhaust pipe using a new gasket.
8. Refill the transfer case with gear oil of classification GL-4 or higher, SAE 75W-85W or 75W-90. Check fluid level in transaxle and add as required.

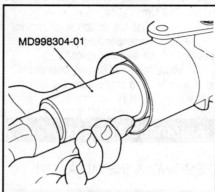

14 FT. LBS.
19 NM

1. Air bleeder
2. Hanger bracket (with differential lock only)
3. Dust seal guard
4. Oil seal
5. Extention housing

Fig. 39 Transfer assembly extension housing and related components

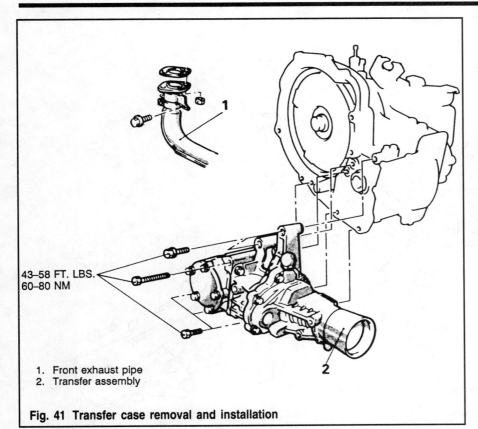

43–58 FT. LBS.
60–80 NM

1. Front exhaust pipe
2. Transfer assembly

Fig. 41 Transfer case removal and installation

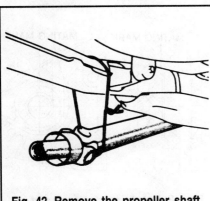

Fig. 42 Remove the propeller shaft and position aside

Transfer

▶ **See Figures 43, 44, 45, 46, 47, 48, 49 and 50**

DISASSEMBLY

1. Remove the transfer assembly from the vehicle.
2. Remove the retainer bolts and the upper cover. Remove the cover gasket.

3. Remove the extension housing retainer bolts and slide the extension housing back off of the transfer case adapter.
4. Remove the retainers, adapter sub assembly, spacer and O-ring from the transfer case sub assembly.
5. Inspect all components for excess wear or damage. Replace components as required.

ASSEMBLY

1. Apply a light an uniform coating of machine blue or red lead to the driven bevel gear teeth, both sides, using a brush. Install the used spacer and new O-ring.
2. Install the transfer case and transfer adapter with the mating marks aligned. Tighten the transfer case adapter sub assembly to the transfer case sub assembly bolts to 14 ft. lbs. (19 Nm).
3. Using socket tool MB991144 or equivalent, turn the bevel drive gear shaft one turn in the normal direction and one turn in the reverse direction.
4. Check to be sure that the mating marks of the driven bevel gear and transfer case are aligned. Check to see

if the driven gear tooth contact is normal.
5. Check to see if the drive bevel gear and the driven bevel gear backlash are 0.0031-0.0051 in.
6. Apply sealant to the adapter flange and install the extension housing. Install the retainer bolts and tighten to 14 ft. lbs. (19 Nm).
7. Install cover with new gasket in place. Apply a light coat of sealant to the cover prior to installation. Tighten the retainer bolts to 7 ft. lbs. (9 Nm).

Transfer Case

▶ **See Figures 51, 52 and 53**

DISASSEMBLY

1. Remove the transfer assembly from the vehicle.
2. Disassemble the transfer assembly.
3. Remove the mounting bolts, transfer case cover, O-ring and spacer.
4. Remove the outer races, drive bevel gear assembly and spacer.
5. Remove the oil seal from the transfer case.
6. Inspect all components for excess wear or damage. Replace components as required.

ASSEMBLY

1. Using the proper size driver, install the oil seal into the transfer case.
2. Use the existing spacer during assembly the transfer case. Install the old spacer, outer race, drive bevel gear assembly, outer race, old spacer, new O-ring and cover to the transfer case. Tighten the retainer bolts to 29 ft. lbs. (39 Nm).
3. Once assembled, use torque wrench and turning tool MB991144 or equivalent to check the bevel gear turning torque. Compare measured value to desired reading of 1.23-1.81 ft. lbs. (1.7-2.5 NM) torque.
4. If the rotating torque is outside the desired range, adjust using adjusting spacers.

➡**For adjustment, use 2 spacers of which thickness is as close as possible to each other.**

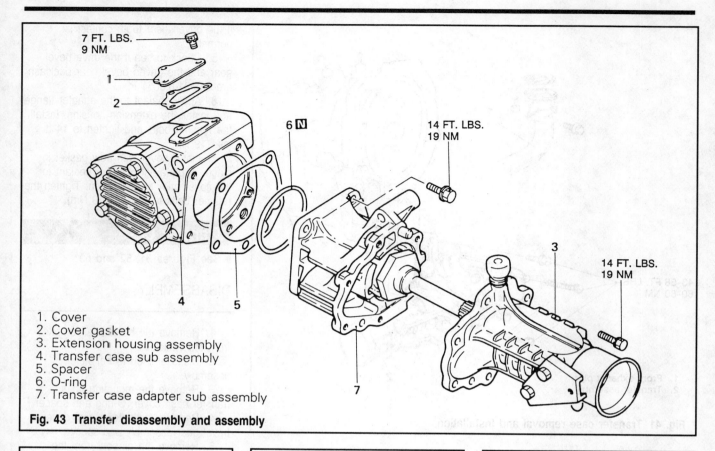

1. Cover
2. Cover gasket
3. Extension housing assembly
4. Transfer case sub assembly
5. Spacer
6. O-ring
7. Transfer case adapter sub assembly

Fig. 43 Transfer disassembly and assembly

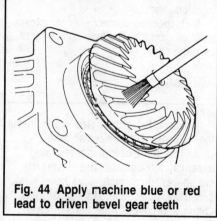

Fig. 44 Apply machine blue or red lead to driven bevel gear teeth

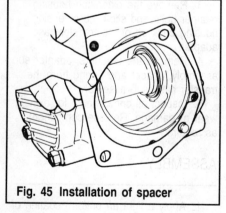

Fig. 45 Installation of spacer

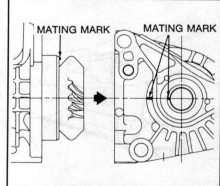

Fig. 46 Install transfer case and transfer adapter with mating marks aligned

5. Assemble the remaining portions of the transfer.

Transfer Case Adapter

▶ See Figures 54 and 55

DISASSEMBLY

1. Remove the transfer assembly from the vehicle.
2. Remove the transfer case adapter from the transfer case.

3. Unlock the locknut by straightening the bent flange on the nut. Holding the driven bevel gear in a vice and using wrench adapter MB991013 or equivalent, remove the locknut.
4. Using a press, remove the driven bevel gear assembly.
5. Remove the taper roller bearing, spacer and collar.
6. Remove the outer race using a hammer and flat-tiped tool.
7. Inspect all components for damage or excess wear and replace as required.

ASSEMBLY

1. Using the proper driver, press fit the bearing outer races into the adapter.
2. Install the collar, original spacer and driven bevel gear.
3. Install the taper roller bearing using the proper equipment.
4. Install the locknut. While holding the drive bevel gear in a vice and using wrench tool MB991013 or equivalent, tighten the locknut to 109 ft. lbs. (150

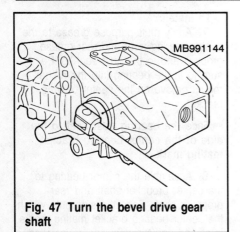

Fig. 47 Turn the bevel drive gear shaft

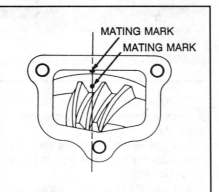

Fig. 48 Be sure mating marks on driven bevel gear and transfer case are aligned

Nm). Lock in position by crimping flange on nut in 2 places.

5. Using torque wrench and socket tool MD998806-01 or equivalent, rotate the bevel gear and measure the rotating torque. Compare measured torque to desired reading of 0.72-1.23 ft. lbs. (1.0-1.7 Nm) torque.

6. If the rotating torque is outside of the desired range, adjust using different spacers.

DRIVELINE

Propeller Shaft

▶ **See Figures 56 and 57**

REMOVAL & INSTALLATION

1. Disconnect the negative battery cable. Raise the vehicle and support safely.

2. The rear driveshaft is a 3-piece unit, with a front, center and rear propeller shaft. Remove the nuts and insulators from the center support bearing. Work carefully. There will be a number of spacers which will differ from vehicle to vehicle. Check the number of spacers and write down their locations for reference during reassembly.

3. Matchmark the rear differential companion flange and the rear driveshaft flange yoke. Remove the companion shaft bolts and remove the driveshaft, keeping it as straight as possible so as to ensure that the boot is not damaged or pinched. Use care to keep from damaging the oil seal in the output housing of the transfer case.

➡**Damage to the boot can be avoided and work will be easier if a piece of cloth or similar material is inserted in the boot.**

4. Do not lower the rear of the vehicle or oil will flow from the transfer case. Cover the opening to keep dirt out.

To install:

5. Install the driveshaft to the vehicle and align the matchmarks at the rear yoke.

6. Install the bolts at the rear differential flange and torque to 22-25 ft. lbs. (30-35 Nm).

7. Install the center support bearing with all spacers in place. Torque the retaining nuts to 22-25 ft. lbs. (30-35 Nm).

8. Check the fluid levels in the transfer case and rear differential case.

U-joints

REMOVAL & INSTALLATION

1. Make mating marks on the yoke and the universal joint that is to be disassembled. Remove the snaprings from the yoke with snapring pliers.

2. Force out the bearing journals from the yoke using a large C-clamp. Install a collar on the fixed side of the C-clamp. Press the journal bearing into the collar by applying pressure with the C-clamp, on the opposite side.

3. Pull the journal bearing from the yoke.

➡**If the journal bearing is hard to remove, strike the yoke with a plastic hammer.**

4. Press the journal shaft using C-clamp or similar tool, to remove the remaining bearings.

5. Once all bearings are removed, remove the journal.

To install:

6. Apply multi-purpose grease to the shafts, grease sumps, dust seal lips and needle roller bearings of the replacement U-joint. Do not apply excessive grease. Otherwise, faulty fitting of bearing caps and errors in selection of snaprings may result.

7. Press fit the journal bearings to the yoke using a C-clamp as follows:

a. Install a solid base onto the bottom of the C-clamp.

b. Insert both bearings into the yoke. Hold and press fit them by tightening the C-clamp.

c. Install snaprings of the same thickness onto both sides of each yoke.

d. Press the bearing and journal into 1 side by using a brass bar with diameter of 0.59 in. (15mm).

8. Measure the clearance between the snapring and the groove wall of the yoke with a feeler gauge. If the clearance exceeds 0.0008-0.0024 in. (0.02-0.06mm), the snap rings should be replaced.

Center Bearing

REMOVAL & INSTALLATION

1. Place mating marks on the companion flange and the Lobro joint assembly.

2. Remove the Lobro joint installation bolts. Separate the Lobro joint from the companion flange.

3. Place mating marks on the center yoke and center propeller shaft, and the companion flange and the rear propeller shaft.

4. Remove the self-locking nuts. Remove the center yoke and companion flange.

5. Place mating marks on the center bearing assembly front bracket and the center propeller shaft, and the center bearing assembly rear bracket and the rear propeller shaft. Remove the center bearing bracket.

➡**The mounting rubber can not be removed from the center bearing bracket.**

6. Pull out the front and rear center bearings with a commercially available puller.

To install:

7. Apply multi-purpose grease to the center bearing front and rear grease grooves and to the dust seal lip. Be sure to fit the bearing into the rubber mount groove on the center bearing bracket.

➡**Face the bearing dust seal to the side of the center bearing bracket mating mark.**

8. Assemble the center bearing to the center propeller shaft and rear propeller shaft. Face the side onto which the center bearing bracket mating marks is placed and the dust seal is installed toward the side of the center propeller shaft and rear propeller shaft.

9. Apply a thin and even coat of the grease, enclosed with the repair kit, to the rubber packing on the companion flange. Align the mating marks on the center propeller shaft and the companion flange, then press fit the center bearing with self-locking nuts.

10. Install the Lobro joint assembly installation bolts. Secure the companion flange and Lobro joint assembly with the installation bolts. Check for grease leakage from the Lobro joint boot and companion flange installation parts.

REAR AXLE

Rear Driveshaft and Seal

▶ **See Figures 58 and 59**

REMOVAL & INSTALLATION

1. Disconnect the negative battery cable. Raise the vehicle and support safely.

2. Remove the bolts that attach the rear halfshaft to the companion flange.

3. Use a prybar to pry the inner shaft out of the differential case. Don't insert the prybar too far or the seal could be damaged.

4. Remove the rear driveshaft from the vehicle.

5. If necessary, pry the oil seal from the rear differential using a flat tipped prying tool.

To install:

6. Install a new oil seal into the rear differential housing using proper size driver.

7. Replace the circlip and install the rear driveshaft to the differential case. Make sure it snaps in place.

8. Install the companion flange bolts and tighten to 40-47 ft. lbs. (55-65 Nm).

9. Check the fluid level in the rear differential.

Differential Carrier

REMOVAL & INSTALLATION

1. Raise the vehicle and support safely.

2. Drain the differential gear oil and remove the center exhaust pipe.

3. Matchmark and remove the rear driveshaft.

4. Remove the rear halfshafts.

5. Remove the center exhaust pipe and muffler assembly, as required.

6. The large mounting bolts that hold the differential carrier support plate to the underbody may use self-locking nuts. Before removing them, support the rear axle assembly in the middle with a transaxle jack. Remove the nuts, then

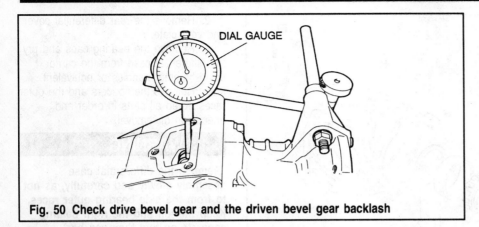

Fig. 50 Check drive bevel gear and the driven bevel gear backlash

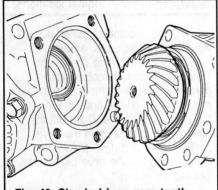

Fig. 49 Check driven gear tooth contact

remove the support plate(s) and the square dynamic damper from the rear of the carrier.

7. Lower the differential carrier and remove from the vehicle.

To install:

8. Raise the rear differential carrier into position and install support member bolts. Install new locknuts on all support bolts.

9. Install new circlips on both rear driveshafts and install.

10. Install the propeller shaft.

11. Install the center exhaust pipe and muffler.

12. Lower the vehicle. With the vehicle level, fill the rear differential.

Stub Axle Shaft, Bearing and Seal

REMOVAL & INSTALLATION

1. Disconnect the negative battery cable.

2. Raise and support the vehicle safely.

3. Remove the tire and wheel assembly from the vehicle.

4. If equipped with ABS, remove the rear wheel speed sensor.

➡**Be cautious to ensure that the tip of the pole piece on the rear speed sensor does not come in contact with other parts during removal. Sensor damage could occur.**

5. Remove the rear caliper and support assembly out of the way. Remove the brake disc.

6. Remove the driveshaft and companion flange installation bolts, nuts and washers. Move the end of shaft slightly to access the self-locking nut.

7. Using axle holding tool MB990211-01 or equivalent, secure the rear axle shaft in position, then remove the self-locking nut.

8. Using puller and adapter MB990211-01 and MB990241-01 or equivalents, remove the rear axle shaft from the trailing arm.

9. If equipped with ABS, remove the rear rotor from the axle assembly using collar and press. The rotor is a press fit.

10. Remove the outer bearing and dust cover concurrently from the axle shaft using a press.

11. Using puller, remove the oil seal and inner bearing from the trailing arm.

12. Inspect the companion flange and axle shaft for wear or damage. Inspect the dust cover for deformation or damage. Inspect the bearings for burning or declaration. Replace components as required.

To install:

13. Using the proper driver, press fit the inner bearing onto the trailing arm. Press fit the oil seal onto the trailing arm with the depression in the oil seal facing

upward, and until it contacts the shoulder on the inner arm.

➡**When tapping the oil seal in, use a plastic hammer to lightly tap the top and circumference of the seal installation tool, press fitting gradually and evenly.**

14. Press fit the dust covers onto the axle until it contacts the axle shaft shoulder. Install the innermost cover so the depression is facing upward.

➡**When tapping the oil seal in, use a plastic hammer to lightly tap the top and circumference of the seal installation tool, press fitting gradually and evenly.**

15. Apply multi-purpose grease around the entire circumference of the inner side of the outer bearing seal lip. Press fit the outer bearing to the axle shaft so that the bearing seal lip surface is facing towards the axle shaft flange.

16. Press fit the rear rotor to the axle shaft with the rear rotor groove surface towards the axle shaft flange.

17. Install the rear axle shaft to the trailing arm temporarily. Install the companion flange to the rear axle shaft, then install a new self-locking nut.

18. While holding the rear axle shaft in position using holding fixture tool MB990767-01 or equivalent, tighten a new self-locking nut to 159 ft. lbs. (220 Nm).

19. Install the drive shaft nuts, washers and bolts. Tighten to 47 ft. lbs. (65 Nm).

20. Install the rear brake disc, caliper assembly and parking brake.

21. Install the tire and wheel assembly and lower the vehicle. Check the parking brake stroke and adjust as required.

22. Before moving the vehicle, pump the brakes until a firm pedal is achieved.

Rear Pinion Seal

REMOVAL & INSTALLATION

1. Raise the vehicle and support safely.

2. Matchmark the rear propeller shaft and companion flange and remove the shaft. Don't let it hang from the transaxle. Tie it up to the underbody.

3. Hold the companion flange stationary and remove the large self-

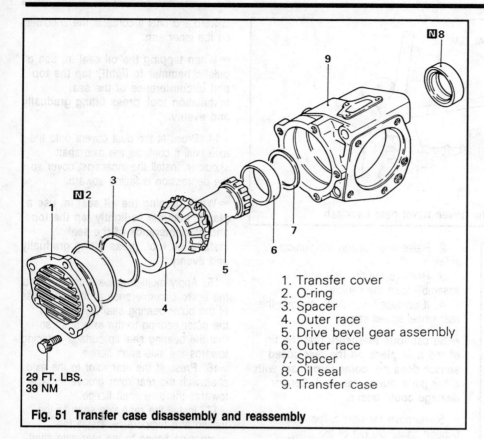

1. Transfer cover
2. O-ring
3. Spacer
4. Outer race
5. Drive bevel gear assembly
6. Outer race
7. Spacer
8. Oil seal
9. Transfer case

29 FT. LBS.
39 NM

Fig. 51 Transfer case disassembly and reassembly

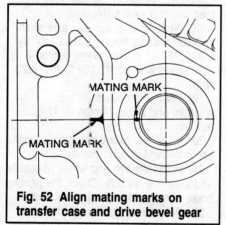

Fig. 52 Align mating marks on transfer case and drive bevel gear

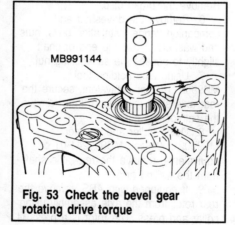

MB991144

Fig. 53 Check the bevel gear rotating drive torque

locking nut in the center of the companion flange.

4. Using a puller, remove the flange. Pry the old seal out.

To install:

5. Apply a thin coat of multi-purpose grease to the seal lip and the companion flange seal contacting surface. Install the new seal with an appropriate driver.

6. Install the companion flange. Install a new locknut and torque to 116-159 ft. lbs. (157-220). The rotation torque of the

drive pinion should be about 4 inch lbs. for new or reused, oiled bearings.

7. Install the propeller shaft.

Rear Differential

▶ See Figures 60, 61, 62 and 63

DISASSEMBLY

1. Remove the rear differential carrier from the vehicle.

2. Remove the rear differential cover and vent plate.

3. Remove the bearing caps and pry the differential case from the carrier using hammer handles or equivalent tools. Remove the spacers and the outer races. Keep all parts in order and orientation of removal.

✳✳WARNING

Remove the differential case assembly slowly and carefully, as not to from the side bearing outer races. Keep the left and the right bearing separate so that they can be reinstalled in their original locations.

4. Using appropriate puller, pull out the bearing inner races, as required.

5. While holding the companion flange using special tool, remove the companion flange self-locking nut. Make mating marks on the drive pinion and the companion flange. Drive out the drive pinion together with the drive pinion spacer and drive pinion front shims. Drive out the drive pinion front bearing, as required.

INSPECTION

1. Check the companion flange for damage or wear.
2. Check the oil seal for wear or deterioration.
3. Check the bearing for wear or discoloration.
4. Check the gear carrier for cracks.
5. Check the drive pinion and drive gear for wear or cracks.
6. Replace components as required.

ASSEMBLY

1. Using the proper driver, press fit the oil seals into the gear carrier until they are flush with the end of the gear carrier.

2. Press fit the drive pinion rear bearing outer races into the gear carrier using the proper tools.

3. Adjust the drive pinion height as follows:

  a. Using the proper driver tools, install the front and rear bearing inner races on the gear carrier.

  b. Tighten the handle of socket tool MB990905-01 until the drive pinion

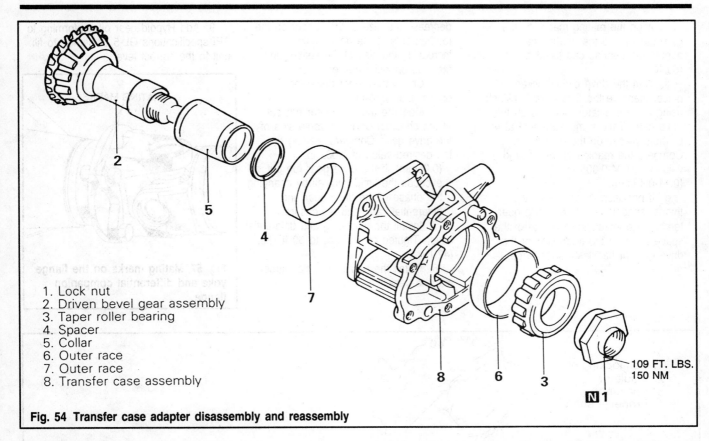

1. Lock nut
2. Driven bevel gear assembly
3. Taper roller bearing
4. Spacer
5. Collar
6. Outer race
7. Outer race
8. Transfer case assembly

109 FT. LBS.
150 NM

Fig. 54 Transfer case adapter disassembly and reassembly

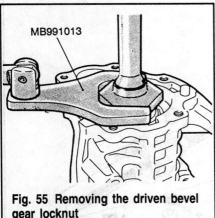

MB991013

Fig. 55 Removing the driven bevel gear locknut

turning torque of 3-4 inch lbs. (0.4-0.5 Nm) is obtained.

c. Position spacer tool MB990392-01 or equivalent in the side bearing seat of the rear carrier. Select a drive pinion rear shim that corresponds to the gap between the special tool. Keep the number of shims to a minimum.

d. Fit the selected drive pinion rear shims to the drive pinion and press fit the drive pinion rear bearing inner race using driver tool MIT215013 or equivalent.

4. Adjust the drive pinion turning torque as follows:

a. Fit the drive pinion front shims between the drive pinion spacer and the drive pinion front bearing inner race.

b. Tighten the companion flange to 116-159 ft. lbs. (160-220 Nm). Do not install the oil seal.

c. Measure the drive pinion turning torque using special socket and torque wrench.

d. Compare measured turning torque to specifications of 3-4 inch lbs. (0.4-0.5 Nm).

e. If the drive pinion turning torque is not within range, adjust by replacing the drive pinion front shims or the front spacer.

f. Remove the drive pinion and companion flange. Drive the oil seal into the gear carrier front lip using proper driver.

g. Install the drive pinion assembly and companion flange with the mating marks aligned. Tighten the the companion flange self-locking nut to 116-159 ft. lbs. (160-220 Nm).

h. Measure the drive pinion turning torque to verify that the drive pinion turning torque agrees with the

specification of 4-5 inch lbs. (0.5-0.6 Nm).

i. If there is a deviation from the standard value, check whether or not there is incorrect tightening torque of the companion flange self-locking nut or incorrect fitting of the oil seal.

5. Press fit the side bearing inner races to the differential case if removed.

6. Adjust the final drive gear backlash as follows:

a. Install the side bearing spacers, which are thinner than those removed, to the side bearing outer races and install the differential case assembly into the gear carrier.

b. Push the differential case to one side and measure the clearance between the gear carrier and the side bearing.

c. Measure the thickness of the side bearing spacers on one side, select the 2 pairs of spacers which correspond to that thickness plus 0.002 in. (0.05mm). Install one pair of each to the drive pinion side and the drive gear side.

d. Install the side bearing spacers and differential case to the gear carrier. Tap the side bearing spacers with a brass bar to fit them to the side bearing outer race.

e. Align the mating marks on the gear carrier and the bearing cap. tighten the bearing cap to 43 ft. lbs. (60 Nm).

f. With the drive pinion locked in place, measure the final drive backlash using a dial indicator mounted on the drive gear. Take a measurement at 4 or more places on the drive gear. Compare the reading to the desired value of 0.004-0.006 in. (0.11-0.16mm).

g. If backlash is too small, add thinner spacer on side opposite gear teeth, while increasing the size of the space used on the tooth side of the drive gear. If backlash is too large,

decrease the size of the spacer on the toothed side of the gear, while increasing the size of the spacer on the side opposite the teeth.

h. Check for proper drive and driven gear contact.

i. Measure the drive gear run out at the shoulder on the reverse side of the drive gear. Compare reading to the desired value of 0.002 in. (0.05mm). If the drive gear run out exceeds the limit, reinstall by changing the phase of the drive gear and differential case and remeasure.

7. Install the vent plug and differential cover. Tighten cover bolts to 30 ft. lbs. (42 Nm).

8. Install the rear differential carrier into the vehicle.

9. Add Hypoid gear oil conforming to API specifications GL-5 or higher, to fill unit to the correct level.

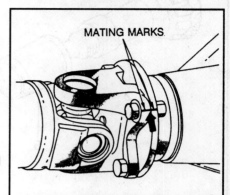

Fig. 57 Mating marks on the flange yoke and differential companion flange

1. Self locking nut
2. Insulator
3. Spacer
4. Propeller shaft

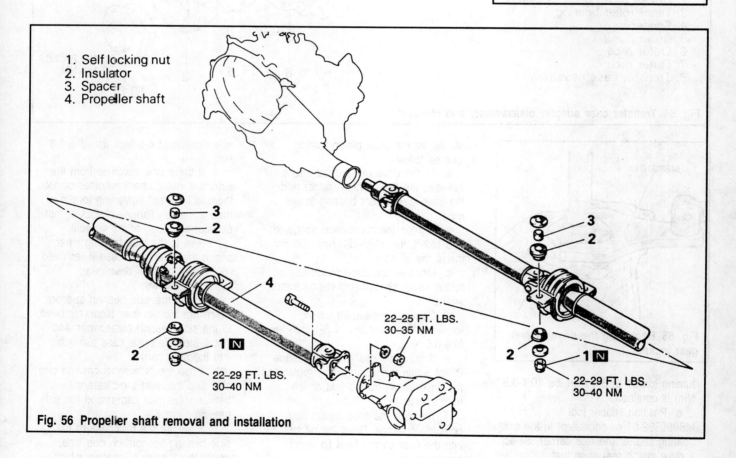

22–25 FT. LBS.
30–35 NM

22–29 FT. LBS.
30–40 NM

22–29 FT. LBS.
30–40 NM

Fig. 56 Propeller shaft removal and installation

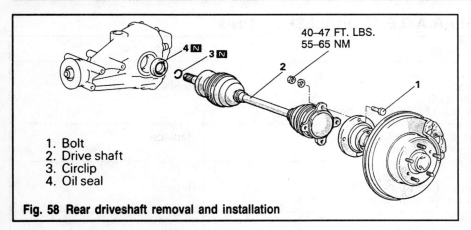

1. Bolt
2. Drive shaft
3. Circlip
4. Oil seal

Fig. 58 Rear driveshaft removal and installation

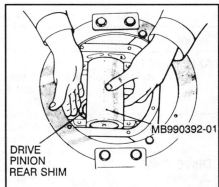

Fig. 61 Selecting a drive pinion rear shim while using tool MB990392-01

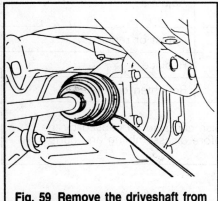

Fig. 59 Remove the driveshaft from the differential carrier

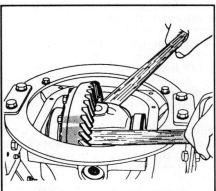

Fig. 60 Removal of differential case assembly

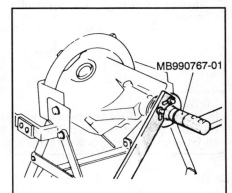

Fig. 62 Adjusting drive pinion preload

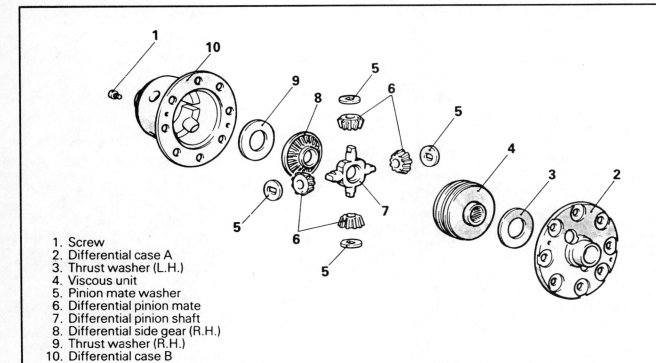

1. Screw
2. Differential case A
3. Thrust washer (L.H.)
4. Viscous unit
5. Pinion mate washer
6. Differential pinion mate
7. Differential pinion shaft
8. Differential side gear (R.H.)
9. Thrust washer (R.H.)
10. Differential case B

Fig. 63 Limited slip differential assembly

REAR AXLE TROUBLESHOOTING

Symptom	Probable cause	Remedy
AXLE SHAFT Noise while wheels are rotating	Brake drag Bent axle shaft Worn or scarred axle shaft bearing	Replace
Grease leakage	Worn or damaged oil seal Malfunction of bearing seal	Replace
DRIVE SHAFT Noise	Wear, play or seizure of ball joint Excessive drive shaft spline looseness	Replace
DIFFERENTIAL Abnormal noise during driving or gear changing	Excessive final drive gear backlash Insufficient drive pinion preload	Adjust
	Excessive differential gear backlash	Adjust or replace
	Worn spline of a side gear	Replace
	Loose companion flange self-locking nut	Retighten or replace

NOTE
In addition to a malfunction of the differential carrier components, abnormal noise can also be caused by the universal joint of the propeller shaft, the axle shafts, the wheel bearings, etc. Before disassembling any parts, take all possibilities into consideration and confirm the source of the noise.

REAR AXLE TROUBLESHOOTING

Symptom	Probable cause	Remedy
Abnormal noise when cornering	Damaged differential gears	Replace
	Insufficient gear oil quantity	Replenish
Gear noise	Improper final drive gear tooth contact adjustment	Adjust or replace
	Incorrect final drive gear backlash Improper drive pinion preload adjustment	Adjust
	Damaged, broken, and/or seized tooth surfaces of the drive gear and drive pinion Damaged, broken, and/or seized drive pinion bearings Damaged, broken, and/or seized side bearings Damaged differential case Inferior gear oil	Replace
	Insufficient gear oil quantity	Replenish

NOTE
Noise from the engine, muffler vibration, transaxle, propeller shaft, wheel bearings, tires, body, etc., is easily mistaken as being caused by malfunctions in the differential carrier components. Be extremely careful and attentive when performing the driving test, etc.
Test methods to confirm the source of the abnormal noise include: coasting acceleration, constant speed driving, raising the rear wheels on a jack, etc. Use the method most appropriate to the circumstances.

Gear oil leakage	Worn or damaged front oil seal, or an improperly installed oil seal Damaged gasket	Replace
	Loose companion flange self-locking nut	Retighten or replace
	Loose filler or drain plug	Retighten or apply adhesive
	Clogged or damaged vent plug	Clean or replace
Seizure	Insufficient final drive gear backlash Excessive drive pinion preload Excessive side bearing preload Insufficient differential gear backlash Excessive clutch plate preload	Adjust
	Inferior gear oil	Replace
	Insufficient gear oil quantity	Replenish

NOTE
In the event of seizure, disassemble and replace the parts involved, and also be sure to check all components for any irregularities and repair or replace as necessary.

Breakdown	Incorrect final drive gear backlash Insufficient drive pinion preload Insufficient side bearing preload Excessive differential gear backlash	Adjust
	Loose drive gear clamping bolts	Retighten

NOTE
In addition to disassembling and replacing the failed parts, be sure to check all components for irregularities and repair or replace as necessary.

The limited slip differential does not function (on snow, mud, ice, etc.)	The limited slip device is damaged	Disassemble, check the functioning, and replace the damaged parts

TORQUE SPECIFICATIONS

Component	US	Metric
Automatic transaxle oil pan bolts	8 ft. lbs.	12 Nm
Axle nut (front)	145–188 ft. lbs.	200–260 Nm
Back-up light switch	22–25 ft. lbs.	29–33 Nm
Clutch cover assembly mounting bolts	16 ft. lbs.	22 Nm
Clutch lever mounting nut	14 ft. lbs.	20 Nm
Differential drive gear retainer bolts	98 ft. lbs.	135 Nm
Front halfshaft center support bracket retainer bolts	33 ft. lbs.	45 Nm
Fulcrum (clutch)	25 ft. lbs.	35 Nm
Rear driveshaft companion flange bolts	40–47 ft. lbs.	55–65 Nm
Transfer case to transaxle bolts	40–43 ft. lbs.	55–60 Nm

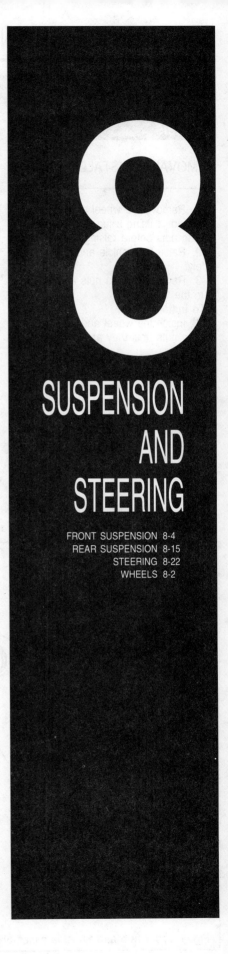

8

SUSPENSION AND STEERING

WHEELS

Wheels

REMOVAL & INSTALLATION

1. Remove the wheel covers, if equipped. If using a lug wrench, loosen the lug nuts before raising the vehicle.

2. Raise the vehicle and support safely.

3. Remove the lug nuts and wheel from the vehicle.

To install:

4. Install the wheel and hand tighten the lug nuts until they are snug.

➡Some vehicles are equipped with **directional tires, meaning that they are to rotate in a specific direction. These tires are marked on the wheels and tires. They are to be installed as indicated by the markings. If positioned so they rotate is the opposite direction, vehicle handling and passenger safety may be affected.**

5. Lower the vehicle until the wheel just touches the ground. In this position,

the wheel lug nuts should be able to be tightened without the wheel spinning.

6. Tighten the wheel lug nut to 87-101 ft. lbs. (120-140 Nm) torque.

7. Install the wheel covers, if equipped.

Wheel Lug Studs

REPLACEMENT

Front Wheel

▶ **See Figures 1 and 2**

➡If the vehicle is equipped with ABS, **removal and disassembly of the front hub is required. Refer to the appropriate procedure in this Section for the removal and installation procedure.**

1. Raise and support the vehicle safely. Remove the front wheel.

2. Remove the caliper and support assembly and position out of the way. It is not necessary to disconnect the brake hose from the caliper. Do not allow the

brake hose to stretch or twist or damage to the hose may occur.

3. Remove the brake disc.

4. Position the wheel stud to be replaced towards the cut out area of the dust shield.

5. Carefully press the stud out of the front hub and remove from the vehicle.

To install:

6. Install the new stud into the hub. Draw the into place using the nut and a stack of washers. Make sure the stud is fully seated.

➡If there is not enough clearance **between the hub and the dust shield to install the new stud, removal of the front hub from the steering knuckle will be required. Refer to the appropriate procedure in this Section for the procedure.**

7. Install the brake disc.

8. Install the caliper support and caliper assembly onto the vehicle.

9. Install the tire and wheel assembly. Lower the vehicle.

10. Before moving the vehicle, pump the brakes until a firm pedal is achieved.

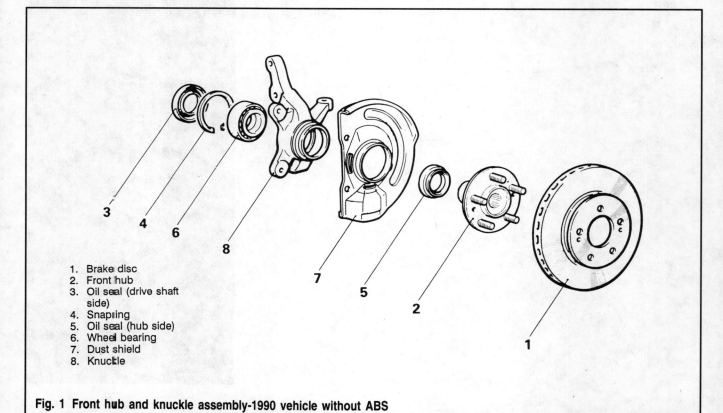

1. Brake disc
2. Front hub
3. Oil seal (drive shaft side)
4. Snapring
5. Oil seal (hub side)
6. Wheel bearing
7. Dust shield
8. Knuckle

Fig. 1 Front hub and knuckle assembly-1990 vehicle without ABS

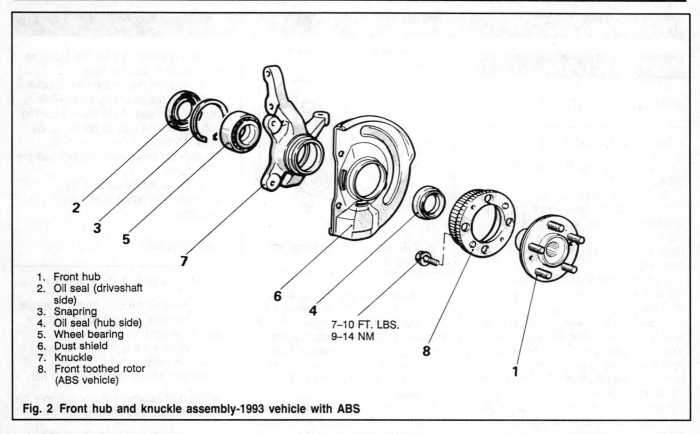

1. Front hub
2. Oil seal (driveshaft side)
3. Snapring
4. Oil seal (hub side)
5. Wheel bearing
6. Dust shield
7. Knuckle
8. Front toothed rotor (ABS vehicle)

7–10 FT. LBS.
9–14 NM

Fig. 2 Front hub and knuckle assembly-1993 vehicle with ABS

Rear Wheel

2WD VEHICLE

1. Disconnect the negative battery cable.
2. Raise and support the vehicle safely.
3. Remove the tire and wheel assembly from the vehicle.
4. Remove the rear caliper. Remove the brake disc.
5. Remove the hub cap and the wheel bearing nut. Remove the tounged washer from the rear axle hub.
6. Remove the rear hub bearing unit.
7. Using the appropriate equipment, drive the lug nut stud from the hub bearing unit.

To install:

8. Install the new stud into the hub. Draw the into place using the nut and a stack of washers. Make sure the stud is fully seated.
9. Install the rear bearing hub assembly onto the vehicle and install the tounged washer. Install a new wheel bearing nut. Tighten new bearing nut to 188 ft. lbs. (260 Nm) torque.
10. Once the bearing nut is torqued, align with the indentation in the spindle

and crimp. This will lock the nut in place.
11. Install the hub (bearing nut) cap.
12. Install the brake disc, caliper and parking brake cable.
13. Install the tire and wheel assembly.
14. Lower the vehicle. Before moving the vehicle, pump the brakes until a firm pedal is achieved.

AWD VEHICLE

1. Disconnect the negative battery cable.
2. Raise and support the vehicle safely.
3. Remove the tire and wheel assembly from the vehicle.
4. Remove the rear caliper and support assembly out of the way. Remove the brake disc.
5. Remove the driveshaft and companion flange installation bolts, nuts and washers. Move end of shaft slightly to access the self-locking nut.
6. Using axle holding tool MB990211-01 or equivalent, secure the

rear axle shaft in position, then remove the self-locking nut.
7. Using puller and adapter MB990211-01 and MB990241-01 or equivalents, remove the rear axle shaft.
8. Press the broken lug nut stud from the axle shaft.

To install:

9. Install the rear axle shaft to the trailing arm temporarily. Install the companion flange to the rear axle shaft, then install the self-locking nut.
10. While holding the rear axle shaft in position using tool MB990767-01 or equivalent, tighten a new self-locking nut to 159 ft. lbs. (220 Nm).
11. Install the drive shaft nuts, washers and bolts. Tighten to 47 ft. lbs. (65 Nm).
12. Install the rear brake disc, caliper assembly and parking brake.
13. Install the tire and wheel assembly and lower the vehicle. Check the parking brake stroke and adjust as required.
14. Before moving the vehicle, pump the brakes until a firm pedal is achieved.

FRONT SUSPENSION

MacPherson Strut

REMOVAL & INSTALLATION

♦ **See Figures 3 and 4**

1. Disconnect the negative battery cable.
2. Raise and safely support vehicle.
3. Remove the brake hose and tube bracket retainer bolt and bracket from the front strut. Do not pry the brake hose and tube clamp away when removing.
4. If equipped with ABS, disconnect the front speed sensor mounting clamp from the strut.
5. Support the lower arm using floor jack or equivalent. Remove the lower strut to knuckle bolts. Once the mounting bolts have been removed, jack up the lower arm. Use a piece of wire to attach the brake hose, tube and driveshaft to the knuckle and to help keep the weight off. These components are not to be pulled.

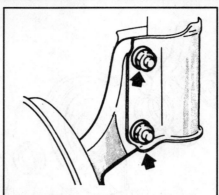

Fig. 4 Removal of front strut lower mounting bolts

6. Before removing the top bolts, make matchmarks on the body and the strut insulator for proper reassembly. If this plate is installed improperly, the wheel alignment will be wrong. Remove the strut upper mounting bolts. Remove the strut assembly from the vehicle.

To install:

7. Install the strut to the vehicle and install the top mounting bolts. Make sure the insulator is installed so the matchmarks made during disassembly are in alignment. Tighten the mounting bolts to 36 ft. lbs. (50 Nm).

8. Position the strut on the knuckle and install the mounting bolts. While holding the head of the lower mounting bolt, tighten the nuts to 80-101 ft. lbs. (110-140 Nm)9.

Install the brake hose bracket and the ABS clamp.

9. Install the wheel and tire assembly. Perform a front end alignment.

DISASSEMBLY

♦ **See Figures 5, 6, 7, 8, 9 and 10**

1. Remove the strut from the vehicle.
2. Remove the dust cover from the top of the strut.
3. While holding the spring upper seat with retainer tool CT-1112 or equivalent, loosen the self-locking nut.

➡**The self-locking nut should only be loosened. Do not remove the nut.**

4. Compress the coil spring using spring compressor tool C-4838 or equivalent. Install the compressor tools so they are of equal distance apart. Tighten the compressor tools slowly and evenly. Do not use air tools on the spring compressors.
5. Once the spring is compressed, remove the locknut.
6. Remove the strut insulator, spring seat spring pad, bumper rubber and dust cover. Remove the coil spring.
7. Inspect all parts for rust, damage or corrosion and replace as required.

ASSEMBLY

♦ **See Figures 11 and 12**

1. Install the spring onto the strut assembly.
2. Join the dust cover and the bump rubber and install onto the strut.
3. Assembly the spring upper seat tot he piston rod, fitting the notch in the rod to the shaped hole in the spring seat. Line up the holes in the strut assembly spring lower seat with the hole in the spring upper seat. This is made easier using a long piece of stock to act as a guide.
4. Install the strut insulator.

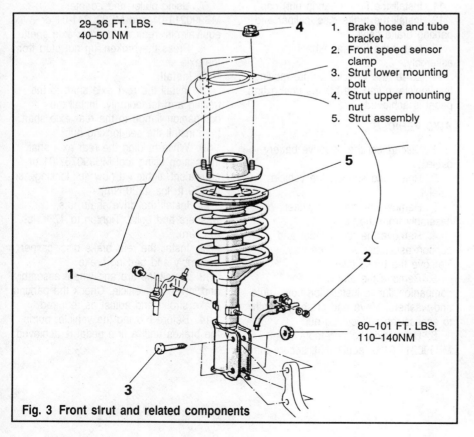

29–36 FT. LBS.
40–50 NM

1. Brake hose and tube bracket
2. Front speed sensor clamp
3. Strut lower mounting bolt
4. Strut upper mounting nut
5. Strut assembly

80–101 FT. LBS.
110–140NM

Fig. 3 Front strut and related components

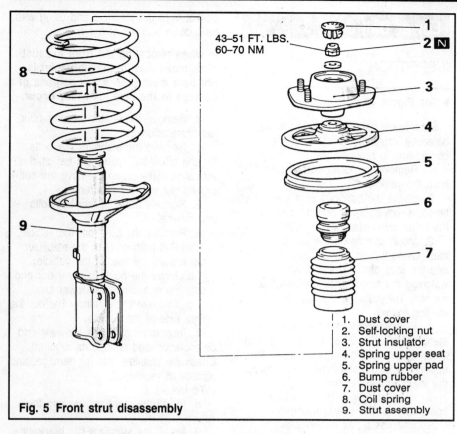

43–51 FT. LBS. 60–70 NM

1. Dust cover
2. Self-locking nut
3. Strut insulator
4. Spring upper seat
5. Spring upper pad
6. Bump rubber
7. Dust cover
8. Coil spring
9. Strut assembly

Fig. 5 Front strut disassembly

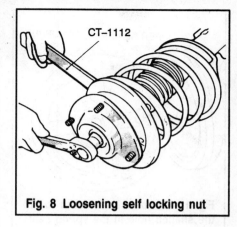

Fig. 8 Loosening self locking nut

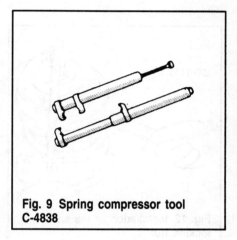

Fig. 9 Spring compressor tool C-4838

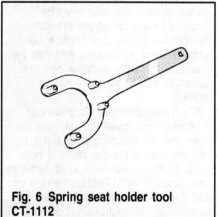

Fig. 6 Spring seat holder tool CT-1112

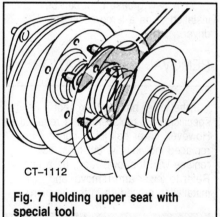

Fig. 7 Holding upper seat with special tool

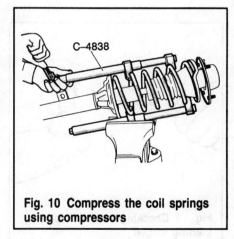

Fig. 10 Compress the coil springs using compressors

5. Install the self-locking nut. With the coil spring held compressed, provisionally tighten the self-locking nut. Correctly align both ends of the coil spring with the grooves in the spring seat, and loosen the spring compressor tool.

6. While holding the spring upper seat with tool CT-1112 or equivalent,

tighten the new locknut to 43-51 ft. lbs. (60-70 Nm).

7. Apply multi-purpose grease to the bearing part of the strut and insulator.

➡**When applying grease to the strut and insulator, make sure the grease does not adhere to the rubber portion of the insulator.**

8. Install the dust cover onto the strut. Install the strut onto the vehicle.

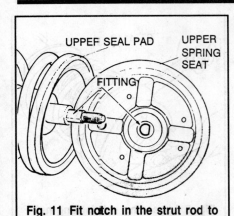

Fig. 11 Fit notch in the strut rod to the shaped hole in the spring seat

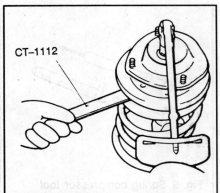

Fig. 12 Installation of the self-locking nut

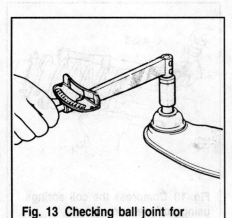

Fig. 13 Checking ball joint for starting torque

Lower Ball Joint

INSPECTION

▶ See Figure 13

If the lower ball joint is found to be defective, replacement of the lower control arm is required.

1. Remove the lower control arm from the vehicle.
2. Shake the ball joint stud several times. If movement is noticed, replace the lower arm assembly.
3. Mount 2 nuts on the ball joint stud. Using torque wrench, turn the nuts and the stud of the ball joint while watching the torque measured on the wrench. This value is the ball joint starting torque.
4. Compare obtained torque to the standard value of 26-87 in. lbs. (10 Nm).
5. If the starting torque exceeds the upper limit of the standard value, replace the lower arm assembly.
6. If the starting torque is within specifications, the ball joint can be reused, unless it has excess play.
7. A new grease boot can be installed using a large socket for a driver.

REMOVAL & INSTALLATION

If the lower ball joint exceeds specifications, replacement is required. However, lower ball joints can not be replaced individually, but must be replace with the lower arm, as a unit. Refer to lower arm removal and installation procedure in this section.

Stabilizer Bar

REMOVAL & INSTALLATION

FWD Vehicle

▶ See Figures 14, 15, 16, 17 and 18

1. Disconnect the negative battery cable.
2. Raise and safely support vehicle. Remove the front exhaust pipe and gasket from the manifold and using wire, tie it down and out of the way.

➡When relocating the front exhaust pipe, make sure the flexible joint is not bent more than a few degrees of damage to the pipe joint may occur.

3. Remove the center crossmember rear installation bolts.
4. Remove the stabilizer link bolts. On the pillow-ball type, hold ball stud with a hex wrench and remove the self-locking nut with a box wrench.
5. Remove the stabilizer bar bolts and mounts.
6. Remove the bar from the vehicle.
   a. Pull both ends of the stabilizer bar toward the rear of the vehicle.
   b. Move the right stabilizer bar end until the end clears the lower arm.
   c. Remove the stabilizer bar out the right side of the vehicle.
7. Inspect all bushings for wear and deterioration and replace as required. Check the stabilizer bar for damage, and replace as required.

To install:

8. Install the stabilizer bar into the vehicle.
9. Install the stabilizer bar brackets on the vehicle, following any side locational markings on the brackets. Temporarily tighten the stabilizer bar bracket. Align the bushing end with the marked part of the stabilizer bar and then fully tighten the stabilizer bar bracket.
10. If equipped with the pillow-ball type mounting, install the stabilizer bar links and link mounting nuts. Using a wrench, secure the ball studs at both ends of the stabilizer link while tightening the mounting nuts. Tighten the nuts on the stabilizer bar bolt so that the distance of bolt protrusion above the top of the nut is 0.63-0.70 in. (16-18mm).
11. Install the front exhaust pipe with new gasket in place. Tighten new self-locking nuts to 29 ft. lbs. (40 Nm).
12. Connect the negative battery cable.

AWD Vehicle

▶ See Figure 19

1. Disconnect the negative battery cable.
2. Remove the front exhaust pipe.
3. Remove the center gusset and transfer assembly.
4. Using a wrench to secure the ball studs at both ends of the stabilizer link,

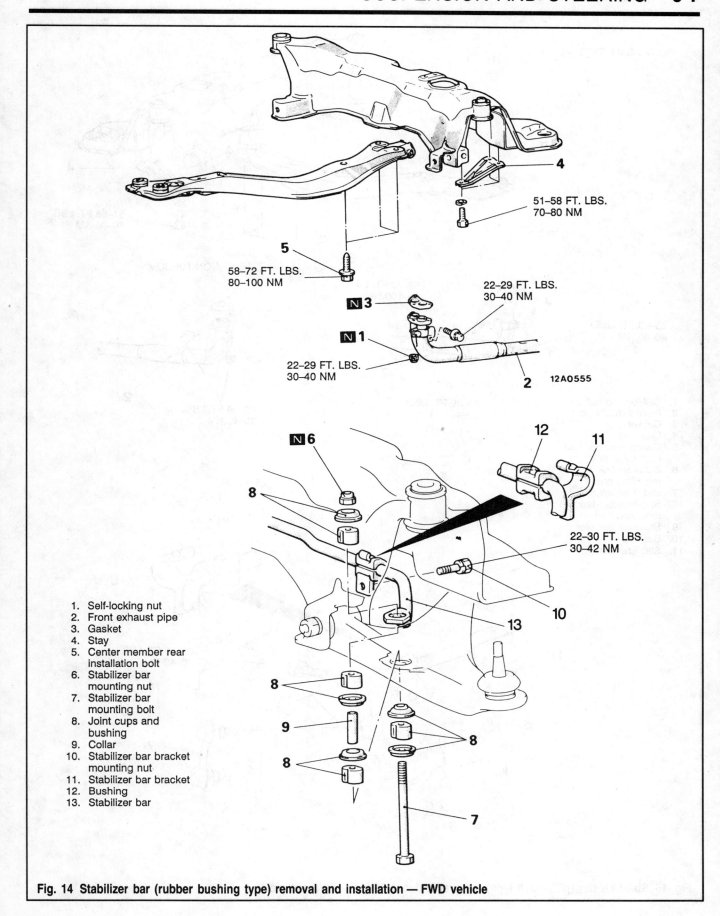

51–58 FT. LBS.
70–80 NM

58–72 FT. LBS.
80–100 NM

22–29 FT. LBS.
30–40 NM

22–29 FT. LBS.
30–40 NM

12A0555

22–30 FT. LBS.
30–42 NM

1. Self-locking nut
2. Front exhaust pipe
3. Gasket
4. Stay
5. Center member rear installation bolt
6. Stabilizer bar mounting nut
7. Stabilizer bar mounting bolt
8. Joint cups and bushing
9. Collar
10. Stabilizer bar bracket mounting nut
11. Stabilizer bar bracket
12. Bushing
13. Stabilizer bar

Fig. 14 Stabilizer bar (rubber bushing type) removal and installation — FWD vehicle

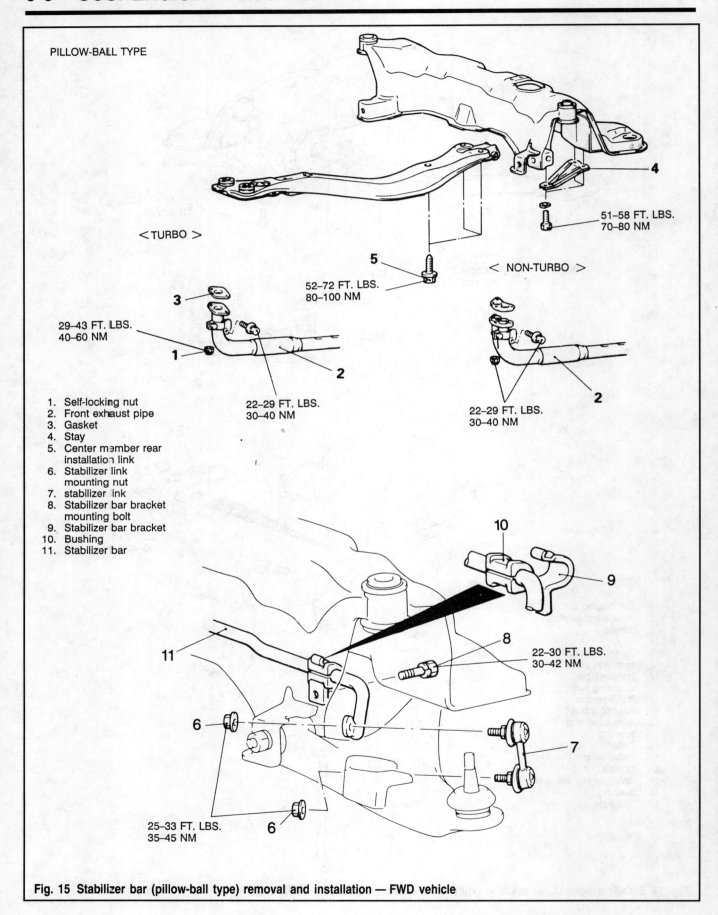

PILLOW-BALL TYPE

< TURBO >

< NON-TURBO >

4

51–58 FT. LBS.
70–80 NM

5

52–72 FT. LBS.
80–100 NM

3

29–43 FT. LBS.
40–60 NM

1

2

22–29 FT. LBS.
30–40 NM

2

22–29 FT. LBS.
30–40 NM

1. Self-locking nut
2. Front exhaust pipe
3. Gasket
4. Stay
5. Center member rear
   installation link
6. Stabilizer link
   mounting nut
7. stabilizer ink
8. Stabilizer bar bracket
   mounting bolt
9. Stabilizer bar bracket
10. Bushing
11. Stabilizer bar

10

9

11

8

22–30 FT. LBS.
30–42 NM

6

7

6

25–33 FT. LBS.
35–45 NM

Fig. 15 Stabilizer bar (pillow-ball type) removal and installation — FWD vehicle

Fig. 16 Use wrench to secure ball stud of stabilizer link while removing the mounting nuts

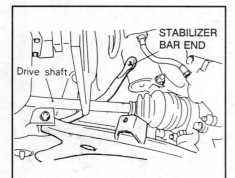

Fig. 17 Pull both ends of the stabilizer bar to the rear of the drive shaft

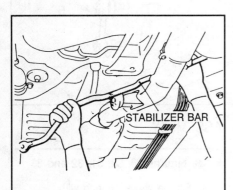

Fig. 18 Pull the stabilizer bar out diagonally

remove the stabilizer link mounting nuts. Remove the stabilizer link.

5. Remove the stabilizer bar bracket installation bolt and the stabilizer bar bracket and bushing.

6. Disconnect the stabilizer bar coupling at the right lower control arm.

Pull out the left side stabilizer edge, pulling it out between the drive shaft and the lower arm. Pull out the right side bar below the lower arm.

To install:

7. Install the bar into the vehicle in the same manner as removal.

8. Temporarily tighten the stabilizer bar bracket. Align the bushing end with the marked part of the stabilizer bar and then fully tighten the stabilizer bar bracket.

9. Install and tighten the stabilizer bar bracket bolt.

10. Install the stabilizer bar links and link mounting nuts. Using a wrench, secure the ball studs at both ends of the stabilizer link while tightening the mounting nuts. Tighten the nuts on the stabilizer bar bolt so that the distance of bolt protrusion above the top of the nut is 0.63-0.70 in. (16-18mm).

11. Install the transfer assembly and gusset.

12. Install the left crossmember. Tighten the rear mounting bolts to 58 ft. lbs. (80 Nm) and the front mounting bolts to 72 ft. lbs. (100 Nm).

Lower Control Arm

REMOVAL & INSTALLATION

▶ **See Figures 20, 21 and 22**

1. Disconnect the negative battery cable.

2. Raise the vehicle and support safely.

3. Remove sway bar links or mounting nuts and bolts from lower control arm. Remove the joint cups and bushings, if equipped.

4. Disconnect the ball joint stud from the steering knuckle.

5. Remove the inner lower arm mounting bolts and nut.

6. Remove the rear mount bolts. Remove the rear retainer clamp if equipped.

7. Remove the arm from the vehicle.

8. Remove the rear rod bushing, if service is required.

To install:

9. Assemble the control arm and bushing. Install the control arm to the vehicle and install the inner mounting bolts. Install new nut and snug temporarily.

10. Install the rear mount clamp, bolts and replacement nuts. Torque the clamp mounting nuts to 34 ft. lbs. (47 Nm). Temporarily tighten the clamp mounting bolt. Once the weight of the vehicle is on the suspension, the bolt will be tightened to 72 ft. lbs. (100 Nm).

11. Connect the ball joint stud to the knuckle. Install a new nut and torque to 43-52 ft. lbs. (60-72 Nm).

12. Install the sway bar and links.

13. Lower the vehicle to the floor for the final torquing of the inner frame mount bolt.

14. Once the full weight of the vehicle is on the suspension, torque the inner lower arm mounting bolt nuts to 87 ft. lbs. (120 Nm). Tighten the inner clamp mounting bolt to 72 ft. lbs. (100 Nm).

15. Inspect all suspension bolts, making sure they all have been fully tightened.

16. Connect the negative battery cable.

Front Wheel Hub, Knuckle and Bearing

REMOVAL

▶ **See Figures 23, 24, 25 and 26**

1. Disconnect the negative battery cable.

2. Remove the cotter pin from the driveshaft nut. With the brakes applied, loosen the halfshaft nut.

3. Raise the vehicle and support safely. Remove the halfshaft nut. If equipped with ABS, remove the front wheel speed sensor.

4. Remove the caliper assembly and brake pads. Suspend the caliper with a wire.

5. Remove the ball joint and tie rod end from the steering knuckle.

6. Remove the halfshaft by setting up a puller on the outside wheel hub and pushing the halfshaft from the front hub. After pressing the outer shaft, insert a prybar between the transaxle case and the halfshaft and pry the shaft from the transaxle.

7. Unbolt the lower end of the strut and remove the hub and steering knuckle assembly from the vehicle.

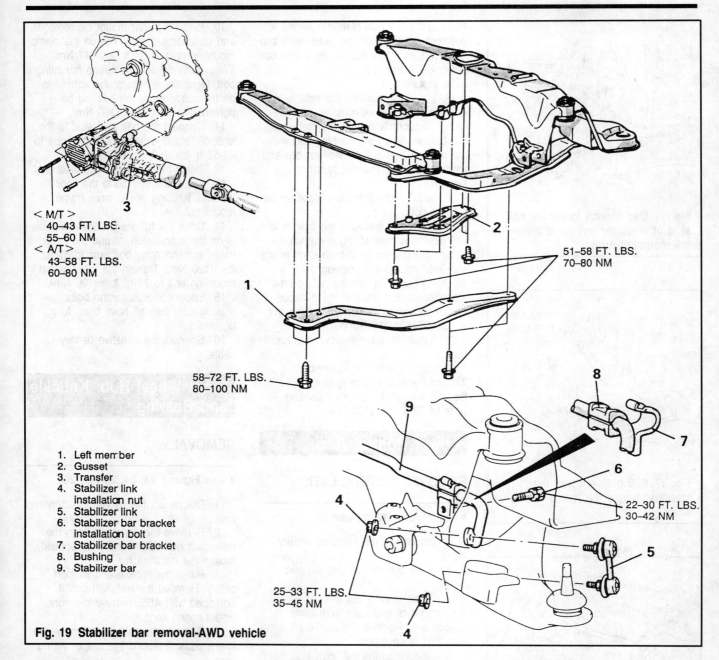

< M/T >
40–43 FT. LBS.
55–60 NM
< A/T >
43–58 FT. LBS.
60–80 NM

51–58 FT. LBS.
70–80 NM

58–72 FT. LBS.
80–100 NM

22–30 FT. LBS.
30–42 NM

25–33 FT. LBS.
35–45 NM

1. Left member
2. Gusset
3. Transfer
4. Stabilizer link installation nut
5. Stabilizer link
6. Stabilizer bar bracket installation bolt
7. Stabilizer bar bracket
8. Bushing
9. Stabilizer bar

Fig. 19 Stabilizer bar removal-AWD vehicle

DISASSEMBLY

▶ See Figures 27 and 28

1. Install the hub/knuckle assembly in a vise. Using a puller, remove the hub from the knuckle.

➡**Do not use a hammer to accomplish this or the bearing will be damaged.**

2. Remove the oil seal from the axle side of the knuckle using a small prying tool.

3. Remove the wheel bearing inner race from the front hub using a puller.

➡**Be careful that the front hub does not fall when the inner race is removed.**

4. Remove the snapring from the axle side of the knuckle. Remove the bearing from the knuckle using a puller.

5. Once the bearing is removed, the bearing outer race can be removed by tapping out with a brass drift pin and a hammer.

ASSEMBLY

▶ See Figures 29, 30, 31, 32 and 33

1. Fill the wheel bearing with multipurpose grease. Apply a thin coating of multipurpose grease to the knuckle and bearing contact surfaces.

2. Press the wheel bearing into the knuckle using the appropriate pressing tool. Once the bearing is installed, install the inner race using the proper driving tool.

3. Drive the oil seal into the knuckle by using the proper size driver. Drive

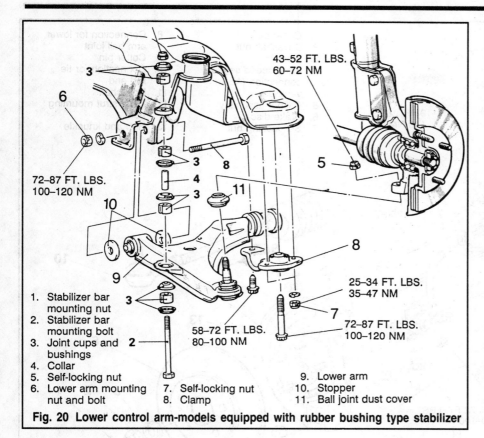

1. Stabilizer bar mounting nut
2. Stabilizer bar mounting bolt
3. Joint cups and bushings
4. Collar
5. Self-locking nut
6. Lower arm mounting nut and bolt
7. Self-locking nut
8. Clamp
9. Lower arm
10. Stopper
11. Ball joint dust cover

72–87 FT. LBS. 100–120 NM

43–52 FT. LBS. 60–72 NM

25–34 FT. LBS. 35–47 NM

58–72 FT. LBS. 80–100 NM

72–87 FT. LBS. 100–120 NM

Fig. 20 Lower control arm-models equipped with rubber bushing type stabilizer

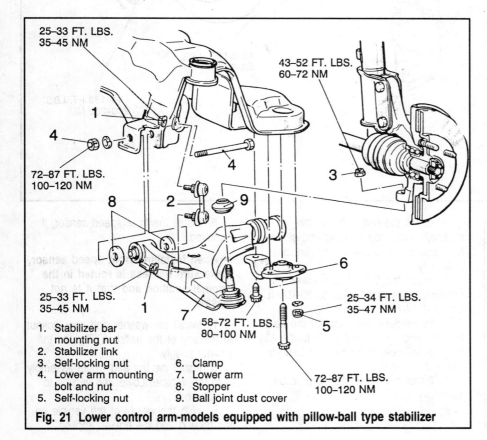

1. Stabilizer bar mounting nut
2. Stabilizer link
3. Self-locking nut
4. Lower arm mounting bolt and nut
5. Self-locking nut
6. Clamp
7. Lower arm
8. Stopper
9. Ball joint dust cover

25–33 FT. LBS. 35–45 NM

72–87 FT. LBS. 100–120 NM

43–52 FT. LBS. 60–72 NM

25–33 FT. LBS. 35–45 NM

58–72 FT. LBS. 80–100 NM

25–34 FT. LBS. 35–47 NM

72–87 FT. LBS. 100–120 NM

Fig. 21 Lower control arm-models equipped with pillow-ball type stabilizer

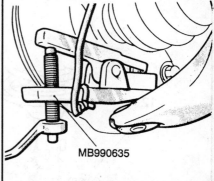

Fig. 22 Disconnecting the lower arm ball joint from the knuckle using separating tool MB990635

seal into knuckle until it is flush with the knuckle end surface.

4. Using pressing tool MB990998-01 or equivalent, mount the front hub assembly into the knuckle. Tighten the nut of the pressing tool to 144-188 ft. lbs. (200-260 Nm). Rotate the hub to seat the bearing.

5. Mount the knuckle assembly in a vise. Check the hub assembly turning torque and end-play as follows:

    a. Using a torque wrench and socket MB990998-01 or equivalent, turn the hub in the knuckle assembly. Note the reading on the torque wrench and compare to the desired reading of 16 inch lbs. (1.8 Nm) or less.

    b. Check for roughness when turning the bearing.

    c. Mount a dial indicator on the hub so the pointer contacts the machined surface on the hub.

    d. Rotate the hub and read the movement of the needle.

    e. Compare the reading to the limit of 0.008 in. (0.2mm).

6. If the starting torque or the hub end-play are not within specifications while the nut is tightened to 144-188 ft. lbs. (200-260 Nm), the bearing, hub or knuckle have probably not been installed correctly. Repeat the disassembly and assembly procedure and recheck starting torque and end-play.

INSTALLATION

1. Install the hub and knuckle assembly onto the vehicle. Install the lower ball joint stud into the steering knuckle and install new nut. Tighten to 52 ft. lbs. (72 Nm).

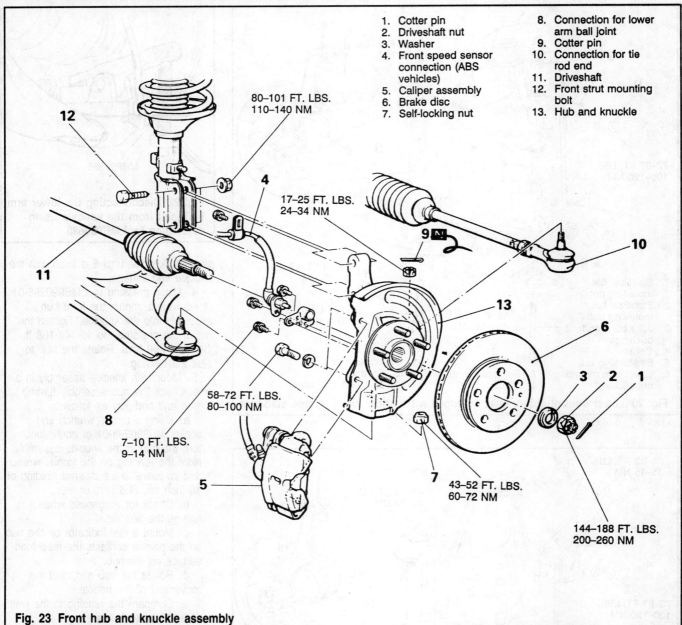

1. Cotter pin
2. Driveshaft nut
3. Washer
4. Front speed sensor connection (ABS vehicles)
5. Caliper assembly
6. Brake disc
7. Self-locking nut
8. Connection for lower arm ball joint
9. Cotter pin
10. Connection for tie rod end
11. Driveshaft
12. Front strut mounting bolt
13. Hub and knuckle

80–101 FT. LBS. 110–140 NM

17–25 FT. LBS. 24–34 NM

58–72 FT. LBS. 80–100 NM

7–10 FT. LBS. 9–14 NM

43–52 FT. LBS. 60–72 NM

144–188 FT. LBS. 200–260 NM

Fig. 23 Front hub and knuckle assembly

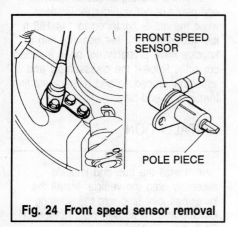

FRONT SPEED SENSOR

POLE PIECE

Fig. 24 Front speed sensor removal

2. Install the halfshaft into the transaxle extension housing and guide the outer end through the hub/knuckle assembly.

3. Install the 2 front strut lower mounting bolts and tighten to 80-101 ft. lbs. (110-140 Nm).

4. Install the connection for the tie rod end and tighten nut to 25 ft. lbs. (34 Nm). Install new cotter pin and bend to lock nut in position.

5. Install the brake disc and caliper assembly.

6. Install the front speed sensor, if removed.

➡When installing front speed sensor, make sure harness is routed in the original position and that it is not twisted.

7. Install the washer and new locknut to the end of the halfshaft. Tighten the locknut snugly.

8. Install the tire and wheel assembly onto the vehicle. Lower the vehicle to the ground.

9. With the weight of the vehicle on the ground and the brakes applied,

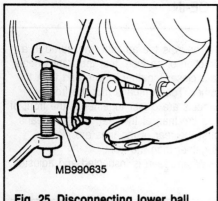

Fig. 25 Disconnecting lower ball joint from control arm using tool MB990635

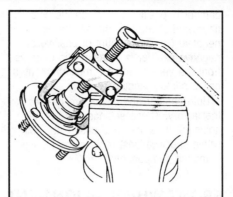

Fig. 28 Removing the wheel bearing inner race from the front hub using puller

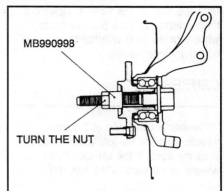

Fig. 31 Mount the front hub assembly to the knuckle and tighten nut to specified torque

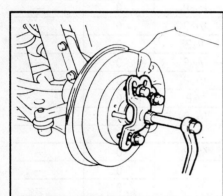

Fig. 26 Removing the front drive shaft from the hub using suitable tool

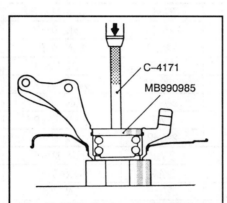

Fig. 29 With the inner wheel bearing race removed, press in the bearing using tool C-4171 and MB990985

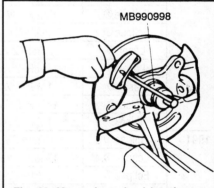

Fig. 32 Measuring wheel bearing starting torque

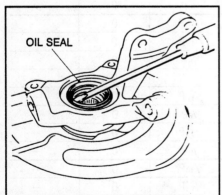

Fig. 27 Removing oil seal from knuckle

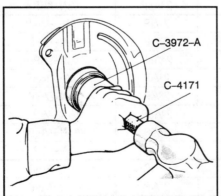

Fig. 30 Drive the oil seal into the hub side of the knuckle using proper driver

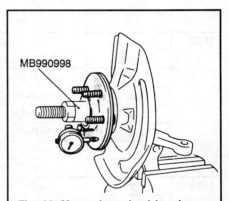

Fig. 33 Measuring wheel bearing end-play

tighten the locknut to 144-188 ft. lbs. (260 Nm).

10. Install the cotter pin in the first matching holes and bend it securely.

Front End Alignment

Front end alignment measurements require the use of special equipment. We recommend that you leave front end alignment adjustments to professional alignment technicians.

CASTER

Caster is the tilt of the front steering axis either forward or backward away from the front of the vehicle. Rearward tilt is referred to as a positive caster, while forward tilt is referred to as negative caster.

CAMBER

Camber is the slope of the front wheels from the vertical when viewed from the front of the vehicle. When the wheels tilt outward at the top, the camber is positive (+). When the wheels tilt inward at the top, the camber is negative (-). The amount of positive and negative camber is measured in degrees from the vertical and the measurement is called camber angle.

Both camber and caster are pre-set at the factory. These settings can not be adjusted. If camber and caster are not within specifications, replace the bent or damaged components.

TOE-IN

Toe-in is the amount, measured in a fraction of an inch, that the front wheels are closer together at one end than the other. Toe-in means that the front wheels are closer together at the front of the tire than at the rear; toe-out means that the rear of the tires are closer together than the front.

When toe-in is inspected and adjusted,

FRONT WHEEL ALIGNMENT

		Caster		Camber			Steering Axis
Year	Model	Range (deg.)	Preferred Setting (deg.)	Range (deg.)	Preferred Setting (deg.)	Toe-in (in.)	Inclination (deg.)
1990	Eclipse ①	$1\frac{5}{16}$P–$2\frac{5}{6}$P	$2\frac{1}{3}$P	$\frac{4}{15}$N–$\frac{11}{15}$P	$\frac{7}{30}$P	0	—
	Eclipse ②	$1\frac{9}{10}$P–$2\frac{9}{10}$P	$2\frac{2}{5}$P	$\frac{5}{12}$N–$\frac{7}{12}$P	$\frac{1}{12}$P	0	—
	Eclipse ③	$1\frac{4}{5}$P–$2\frac{4}{5}$P	$2\frac{3}{10}$P	$\frac{1}{3}$N–$\frac{2}{3}$P	$\frac{1}{6}$P	0	—
	Laser ①	$1\frac{5}{16}$P–$2\frac{5}{6}$P	$2\frac{1}{3}$P	$\frac{4}{15}$N–$\frac{11}{15}$P	$\frac{7}{30}$P	0	—
	Laser ②	$1\frac{9}{10}$P–$2\frac{9}{10}$P	$2\frac{2}{5}$P	$\frac{5}{12}$N–$\frac{7}{12}$P	$\frac{1}{12}$P	0	—
	Laser ③	$1\frac{4}{5}$P–$2\frac{4}{5}$P	$2\frac{3}{10}$P	$\frac{1}{3}$N–$\frac{2}{3}$P	$\frac{1}{6}$P	0	—
	Talon ②	$1\frac{9}{10}$P–$2\frac{9}{10}$P	$2\frac{2}{5}$P	$\frac{5}{12}$N–$\frac{7}{12}$P	$\frac{1}{12}$P	0	—
	Talon ③	$1\frac{4}{5}$P–$2\frac{4}{5}$P	$2\frac{3}{10}$P	$\frac{1}{3}$N–$\frac{2}{3}$P	$\frac{1}{6}$P	0	—
1991	Eclipse ①	$1\frac{5}{16}$P–$2\frac{5}{6}$P	$2\frac{1}{3}$P	$\frac{4}{15}$N–$\frac{11}{15}$P	$\frac{7}{30}$P	0	—
	Eclipse ②	$1\frac{9}{10}$P–$2\frac{9}{10}$P	$2\frac{2}{5}$P	$\frac{5}{12}$N–$\frac{7}{12}$P	$\frac{1}{12}$P	0	—
	Eclipse ③	$1\frac{4}{5}$P–$2\frac{4}{5}$P	$2\frac{3}{10}$P	$\frac{1}{3}$N–$\frac{2}{3}$P	$\frac{1}{6}$P	0	—
	Laser ①	$1\frac{5}{16}$P–$2\frac{5}{6}$P	$2\frac{1}{3}$P	$\frac{4}{15}$N–$\frac{11}{15}$P	$\frac{7}{30}$P	0	—
	Laser ②	$1\frac{9}{10}$P–$2\frac{9}{10}$P	$2\frac{2}{5}$P	$\frac{5}{12}$N–$\frac{7}{12}$P	$\frac{1}{12}$P	0	—
	Laser ③	$1\frac{4}{5}$P–$2\frac{4}{5}$P	$2\frac{3}{10}$P	$\frac{1}{3}$N–$\frac{2}{3}$P	$\frac{1}{6}$P	0	—
	Talon ②	$1\frac{9}{10}$P–$2\frac{9}{10}$P	$2\frac{2}{5}$P	$\frac{5}{12}$N–$\frac{7}{12}$P	$\frac{1}{12}$P	0	—
	Talon ③	$1\frac{4}{5}$P–$2\frac{4}{5}$P	$2\frac{3}{10}$P	$\frac{1}{3}$N–$\frac{2}{3}$P	$\frac{1}{6}$P	0	—
1992	Eclipse ①	$1\frac{5}{16}$P–$2\frac{5}{6}$P	$2\frac{1}{3}$P	$\frac{4}{15}$N–$\frac{11}{15}$P	$\frac{7}{30}$P	0	—
	Eclipse ②	$1\frac{9}{10}$P–$2\frac{9}{10}$P	$2\frac{2}{5}$P	$\frac{5}{12}$N–$\frac{7}{12}$P	$\frac{1}{12}$P	0	—
	Eclipse ③	$1\frac{4}{5}$P–$2\frac{4}{5}$P	$2\frac{3}{10}$P	$\frac{1}{3}$N–$\frac{2}{3}$P	$\frac{1}{6}$P	0	—
	Laser ①	$1\frac{5}{16}$P–$2\frac{5}{6}$P	$2\frac{1}{3}$P	$\frac{4}{15}$N–$\frac{11}{15}$P	$\frac{7}{30}$P	0	—
	Laser ②	$1\frac{9}{10}$P–$2\frac{9}{10}$P	$2\frac{2}{5}$P	$\frac{5}{12}$N–$\frac{7}{12}$P	$\frac{1}{12}$P	0	—
	Laser ③	$1\frac{4}{5}$P–$2\frac{4}{5}$P	$2\frac{3}{10}$P	$\frac{1}{3}$N–$\frac{2}{3}$P	$\frac{1}{6}$P	0	—
	Talon ②	$1\frac{9}{10}$P–$2\frac{9}{10}$P	$2\frac{2}{5}$P	$\frac{5}{12}$N–$\frac{7}{12}$P	$\frac{1}{12}$P	0	—
	Talon ③	$1\frac{4}{5}$P–$2\frac{4}{5}$P	$2\frac{3}{10}$P	$\frac{1}{3}$N–$\frac{2}{3}$P	$\frac{1}{6}$P	0	—
1993	Eclipse ①	$1\frac{5}{16}$P–$2\frac{5}{6}$P	$2\frac{1}{3}$P	$\frac{4}{15}$N–$\frac{11}{15}$P	$\frac{7}{30}$P	0	—
	Eclipse ②	$1\frac{9}{10}$P–$2\frac{9}{10}$P	$2\frac{2}{5}$P	$\frac{5}{12}$N–$\frac{7}{12}$P	$\frac{1}{12}$P	0	—
	Eclipse ③	$1\frac{4}{5}$P–$2\frac{4}{5}$P	$2\frac{3}{10}$P	$\frac{1}{3}$N–$\frac{2}{3}$P	$\frac{1}{6}$P	0	—
	Laser ①	$1\frac{5}{16}$P–$2\frac{5}{6}$P	$2\frac{1}{3}$P	$\frac{4}{15}$N–$\frac{11}{15}$P	$\frac{7}{30}$P	0	—
	Laser ②	$1\frac{9}{10}$P–$2\frac{9}{10}$P	$2\frac{2}{5}$P	$\frac{5}{12}$N–$\frac{7}{12}$P	$\frac{1}{12}$P	0	—
	Laser ③	$1\frac{4}{5}$P–$2\frac{4}{5}$P	$2\frac{3}{10}$P	$\frac{1}{3}$N–$\frac{2}{3}$P	$\frac{1}{6}$P	0	—
	Talon ①	$1\frac{5}{16}$P–$2\frac{5}{6}$P	$2\frac{1}{3}$P	$\frac{4}{15}$N–$\frac{11}{15}$P	$\frac{7}{30}$P	0	—
	Talon ②	$1\frac{9}{10}$P–$2\frac{9}{10}$P	$2\frac{2}{5}$P	$\frac{5}{12}$N–$\frac{7}{12}$P	$\frac{1}{12}$P	0	—
	Talon ③	$1\frac{4}{5}$P–$2\frac{4}{5}$P	$2\frac{3}{10}$P	$\frac{1}{3}$N–$\frac{2}{3}$P	$\frac{1}{6}$P	0	—

N—Negative
P—Positive

① 1.8L engine
② 2.0L engine with Front Wheel Drive (FWD)

③ All Wheel Drive (AWD) vehicle

the wheels must be dead straight ahead, all fluids must be at their proper levels, all other suspension and steering adjustments must be correct and the tires must be properly inflated to their cold specifications.

The toe-in is adjusted by undoing the clips and turning the left and right tie rod turnbuckles by the same amount in opposite directions. The toe will move out as the left turnbuckle is turned toward the front of the vehicle and the right turnbuckle is turned toward the rear of the vehicle.

After making adjustments, a turning radius gauge should be used to confirm that the steering wheel turning angle is within specifications.

REAR SUSPENSION

Shock Absorber

REMOVAL & INSTALLATION

▶ **See Figures 34, 35 and 36**

1. Disconnect the negative battery cable. Remove the trunk interior trim to gain access to the top mounting nuts.
2. Remove the top cap and upper shock mounting nuts.
3. Remove the brake tube bracket bolt.
4. mc>Raise and safely support torsion axle and arm assembly slightly. Make sure the jack does not contact the lateral rod.

➡**Always use a wooden block between the jack receptacle and the axle beam. Place the jack at the center of the axle beam.**

5. Remove the shock lower mounting bolt and remove the assembly from the vehicle.

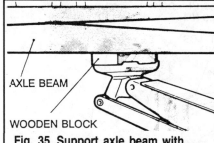

AXLE BEAM

WOODEN BLOCK

Fig. 35 Support axle beam with wooden block between jack and center of axle beam

6. Installation is the reverse of the removal procedure. Tighten the upper shock mounting nuts to 29 ft. lbs. (40 Nm) and the lower mounting bolt to 72 ft. lbs. (100 Nm).

DISASSEMBLY & ASSEMBLY

▶ **See Figures 37, 38 and 39**

1. Compress the spring on the shock absorber.

➡**Do not use an air tool to tighten the bolt on the spring compressor tool.**

2. Once the spring is compressed, remove the piston rod tightening nut.
3. Remove the upper bushings, washer, bracket, collar and spring pad from the shock taking note of their exact position and orientation.
4. Remove the compressed spring from the shock absorber.

To assemble:

5. Install the compressed spring onto the shock absorber, making sure the edge of the coil spring is aligned against the edge of the shock absorber spring seat.
6. Install the dust cover, cup assembly, spring pad, bracket assembly and washer to the shock piston in their original locations.
7. Align the top bracket assembly with the end of the spring and install a new piston rod tightening nut. While holding the piston rod, tighten the nut to 18 ft. lbs. (25 Nm).
8. Align the spring so the lower edge fits into the indent in the spring seat and the upper edge fits into the spring pad groove, then slowly release the spring compressor tool.
9. Fill the shock cap with grease and install onto the shock absorber assembly.

Upper Control Arm

REMOVAL & INSTALLATION

AWD Vehicle
▶ **See Figure 40**

1. Disconnect the negative battery cable. Raise and safely support vehicle. Remove the tire and wheel assembly.
2. Support the rear lower control arm. Remove the brake line clamp bolt.
3. Remove the nut and separate the upper ball joint from the rear trailing arm/steering knuckle.

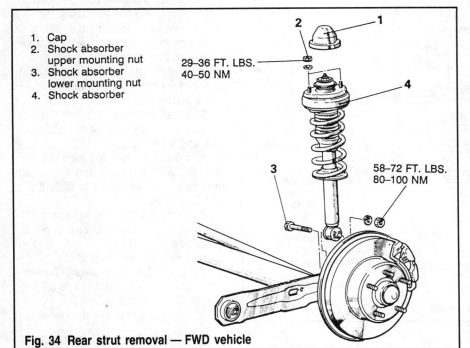

1. Cap
2. Shock absorber upper mounting nut
3. Shock absorber lower mounting nut
4. Shock absorber

29–36 FT. LBS.
40–50 NM

58–72 FT. LBS.
80–100 NM

Fig. 34 Rear strut removal — FWD vehicle

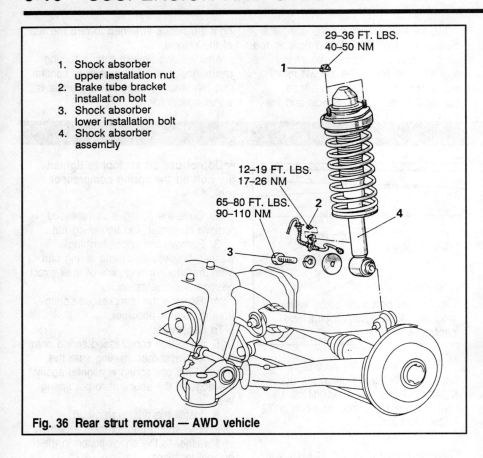

29–36 FT. LBS.
40–50 NM

1. Shock absorber upper installation nut
2. Brake tube bracket installation bolt
3. Shock absorber lower installation bolt
4. Shock absorber assembly

12–19 FT. LBS.
17–26 NM

65–80 FT. LBS.
90–110 NM

Fig. 36 Rear strut removal — AWD vehicle

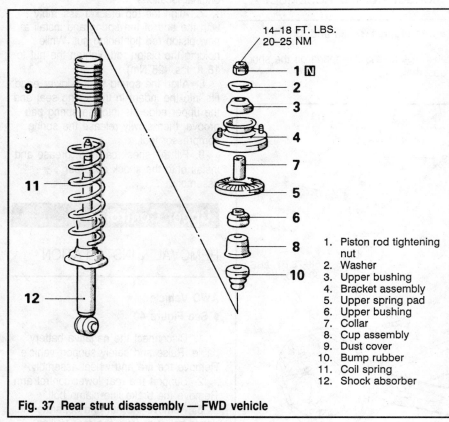

14–18 FT. LBS.
20–25 NM

1. Piston rod tightening nut
2. Washer
3. Upper bushing
4. Bracket assembly
5. Upper spring pad
6. Upper bushing
7. Collar
8. Cup assembly
9. Dust cover
10. Bump rubber
11. Coil spring
12. Shock absorber

Fig. 37 Rear strut disassembly — FWD vehicle

4. Matchmark the eccentric on the upper installation bolt and remove remove from the control arm.

5. Remove the upper arm from the vehicle.

To install:

6. Install the arm to the vehicle and install the upper arm installation bolt. Align the matchmarks and tighten the nut snugly only.

7. Install the upper arm ball joint to the rear spindle assembly and install new nut. Tighten to 52 ft. lbs. (72 Nm) torque.

8. Install the tire and wheel assembly.

9. Lower the vehicle until the suspension supports its weight. Tighten the upper arm installation bolt to 116 ft. lbs. (160 Nm).

10. Check the rear wheel alignment.

Lower Control Arm

REMOVAL & INSTALLATION

AWD Vehicle

▶ See Figure 40

1. Disconnect the negative battery cable. Raise and safely support vehicle. Remove the tire and wheel assembly.

2. Remove the stabilizer link installation nut. Remove the spacers, bushings and washers.

3. Loosen the lower arm ball joint nut and separate the ball joint using the appropriate tool. Once the ball joint stud is broken loose, remove the nut and the stud from the lower arm.

4. Remove the lower arm installation bolt and the arm from the vehicle.

To install:

5. Install the lower arm onto the vehicle. Install the lower arm installation bolt and tighten snugly only.

6. Install the lower ball joint into the hole in the lower control arm and secure with new nut. Tighten nut to 52 ft. lbs. (72 Nm) torque.

7. Install the stabilizer link spacers, bushings and washers. Secure in place using new nut.

8. Lower the vehicle until the suspension supports its weight. Tighten the lower arm installation bolt to 80 ft. lbs. (110 Nm) torque.

9. Check the rear wheel alignment.

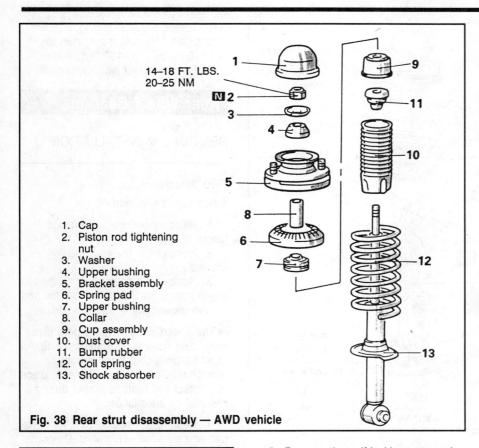

14–18 FT. LBS.
20–25 NM

1. Cap
2. Piston rod tightening nut
3. Washer
4. Upper bushing
5. Bracket assembly
6. Spring pad
7. Upper bushing
8. Collar
9. Cup assembly
10. Dust cover
11. Bump rubber
12. Coil spring
13. Shock absorber

Fig. 38 Rear strut disassembly — AWD vehicle

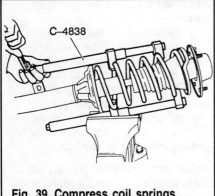

C–4838

Fig. 39 Compress coil springs before removing piston rod tightening nut

Stabilizer Bar

REMOVAL & INSTALLATION

AWD Vehicle

▶ See Figure 41

1. Raise and support the vehicle safely.

2. Place a jack under the rear axle and suspension assembly.

3. Remove the self-locking nuts and crossmember bracket.

4. Remove the retainer bolts and the stabilizer bar brackets. Remove the bushing.

5. Hold the stabilizer bar with a wrench. Remove the self-locking nut.

6. Once the stabilizer bar nut is removed, remove the joint cups and stabilizer rubber bushing.

7. Hold the stabilizer link with a wrench and remove the self-locking nuts. Remove the stabilizer link.

8. Lower the jack supporting the rear axle slightly. Maintain a slight gap between the rear suspension and the body of the vehicle.

9. Remove the stabilizer bar.

10. Inspect the bar for damage, wear and deterioration and replace as required.

To install:

11. Install the stabilizer bar into the vehicle. Raise the rear axle and suspension into place.

12. Install the stabilizer link into the stabilizer bar and install a new self-locking nut. Tighten the nut to 33 ft. lbs. (45 Nm).

13. Install the joint cups and stabilizer rubber to the link. Install a new self-locking nut onto the link. While holding the stabilizer link ball studs with a wrench, tighten the self-locking nut so the protrusion of the stabilizer link is within 0.354-0.433 in. (9-11mm).

14. Install the center stabilizer bar bushings, brackets and bolts. Tighten the bolts to 10 ft. lbs. (14 Nm).

15. Install the parking brake cable and rear speed sensor installation bolt.

16. Install the crossmember bracket and tighten the bolt to 61 ft. lbs. (85 Nm). Tighten the crossmember bracket mounting nut to 94 ft. lbs. (130 Nm).

17. Install the rubber insulators and new self-locking nuts onto the crossmember brackets. Tighten the nuts to 80-94 ft. lbs. (110-130 Nm).

18. Lower the vehicle.

Rear Trailing Arm

REMOVAL & INSTALLATION

AWD Vehicle

▶ See Figures 42, 43 and 44

1. Disconnect the negative battery cable. Raise and safely support the vehicle.

2. Disconnect the parking brake cable from the caliper and trailing arm. Remove the rear caliper from the brake disc and suspend with a wire. Remove the brake disc.

3. Remove the bolt(s) holding the speed sensor bracket to the knuckle and remove the assembly from the vehicle.

➡**The speed sensor has a pole piece projecting from it. This exposed tip must be protected from impact or scratches. Do not allow the pole piece to contact the toothed wheel during removal or installation.**

4. Remove the driveshaft to companion flange bolts and nuts and separate the axle from the companion flange.

5. While holding the axle stationary using tool MB990767-01 or equivalent, remove the self-locking nut and remove the axle hub companion flange. Remove the dust shield.

6. Remove the upper arm and lower arm mounting bolts, and using a special splitter tool, separate the ball joints from the arms.

7. Remove the rear shock absorber lower mounting bolt.

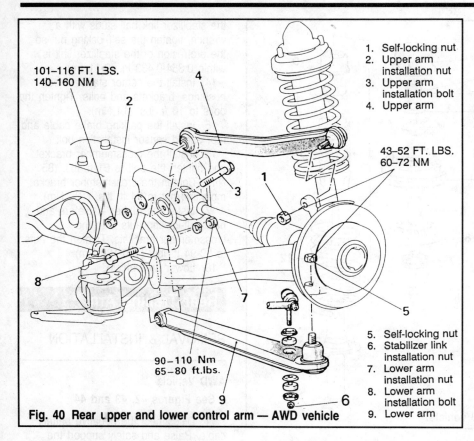

101–116 FT. LBS.
140–160 NM

43–52 FT. LBS.
60–72 NM

90–110 Nm
65–80 ft.lbs.

1. Self-locking nut
2. Upper arm installation nut
3. Upper arm installation bolt
4. Upper arm

5. Self-locking nut
6. Stabilizer link installation nut
7. Lower arm installation nut
8. Lower arm installation bolt
9. Lower arm

Fig. 40 Rear upper and lower control arm — AWD vehicle

8. Remove the trailing arm front mounting nuts and bolts and remove the trailing arm from the vehicle.

9. Remove the connecting rod from the front of the trailing arm using puller tool MB991254 or equivalent.

To install:

10. Assemble the trailing arm and connecting rod. Install the trailing arm to the vehicle and secure in place with the front mounting nuts and bolts. Tighten the nuts snugly at this time. Complete the final tightening of the trailing arm installation bolt when the full weight of the vehicle is on the suspension.

11. Install both control arms to the trailing arm, using new self-locking nuts. Tighten the stud nuts to 52 ft. lbs. (72 Nm).

12. Install the lower shock bolt and tighten to 80 ft. lbs. (110 Nm).

13. On FWD Stealth, install the sway bar link. Install the parking brake parts and axle hub unit.

14. On AWD, install the dust shield, axle hub and companion flange with a new self-locking nut. Connect the rear axle to the companion flange.

15. Temporarily install the speed sensor to the knuckle; tighten the bolts only finger-tight.

16. Route the cable correctly and loosely install the clips and retainers. All clips must be in their original position and the sensor cable must not be twisted. Improper installation may cause cable damage and possibly system failure.

➡**The wiring in the harness is easily damaged by twisting and flexing. Use the white stripe on the outer insulation to keep the sensor harness properly placed.**

17. Use a brass or other non-magnetic feeler gauge to check the air gap between the tip of the pole piece and the toothed wheel. Correct gap is 0.012-0.035 inch (0.3-0.9mm). Tighten the 2 sensor bracket bolts to 10 ft. lbs. (14 Nm) with the sensor located so the gap is the same at several points on the toothed wheel. If the gap is incorrect, it is likely that the toothed wheel is worn or improperly installed.

18. Install the brake disc, caliper and connect the parking brake cable, if not already done. Install the mounting clamps bolts.

19. Double check everything for correct routing and installation. Lower the vehicle so its full weight is on the floor.

20. Torque the front trailing arm/spindle assembly mount nuts to 101-116 ft. lbs. (140-160 Nm).

21. Perform a rear wheel alignment.

Rear Wheel Bearings

REMOVAL & INSTALLATION

FWD Vehicle
▶ **See Figures 45 and 46**

1. Raise the vehicle and support safely.

2. Remove the tire and wheel assembly.

3. Remove the bolt(s) holding the speed sensor bracket to the knuckle and remove the assembly from the vehicle.

➡**The speed sensor has a pole piece projecting from it. This exposed tip must be protected from impact or scratches. Do not allow the pole piece to contact the toothed wheel during removal or installation.**

4. Remove the caliper from the brake disc and suspend with a wire. Do not disconnect the brake hose from the caliper.

5. Remove the brake disc.

6. Remove the grease cap, self-locking nut and tounged washer.

7. Remove the rear hub assembly.

8. If equipped with ABS, remove the retainer bolts and the rear rotor assembly from the back side of the bearing assembly. Replace the rear hub bearing assembly as required.

➡**The rear hub assembly can not be disassembled. If bearing replacement is required, replace the assembly as a unit.**

To install:

9. Install the rear rotor to the back side of the bearing assembly and secure with the retainer bolts. Install the hub assembly.

10. Install the tounged washer and a new self-locking nut. Torque the nut to 144-188 ft. lbs. (200-260 Nm), align with the indentation in the spindle, and crimp the wall of the nut. This will lock the nut in position.

11. Set up a dial indicator and measure the end-play while moving the hub in and out. If the end-play exceeds 0.004 in. (0.01mm), retorque the nut. If

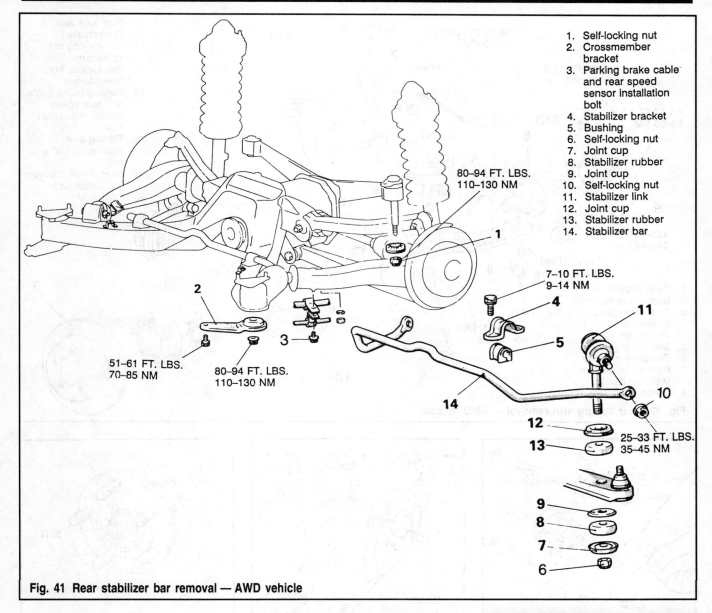

1. Self-locking nut
2. Crossmember bracket
3. Parking brake cable and rear speed sensor installation bolt
4. Stabilizer bracket
5. Bushing
6. Self-locking nut
7. Joint cup
8. Stabilizer rubber
9. Joint cup
10. Self-locking nut
11. Stabilizer link
12. Joint cup
13. Stabilizer rubber
14. Stabilizer bar

80–94 FT. LBS.
110–130 NM

7–10 FT. LBS.
9–14 NM

51–61 FT. LBS.
70–85 NM

80–94 FT. LBS.
110–130 NM

25–33 FT. LBS.
35–45 NM

Fig. 41 Rear stabilizer bar removal — AWD vehicle

still beyond the limit, replace the hub unit.

12. Install the grease cap and brake components removed during this procedure.

13. Temporarily install the speed sensor to the knuckle; tighten the bolts only finger-tight.

14. Route the cable correctly and loosely install the clips and retainers. All clips must be in their original position and the sensor cable must not be twisted. Improper installation may cause cable damage and possibly system failure.

→**The wiring in the harness is easily damaged by twisting and flexing. Use the white stripe on the outer**

insulation to keep the sensor harness properly placed.

15. Use a brass or other non-magnetic feeler gauge to check the air gap between the tip of the pole piece and the toothed wheel. Correct gap is 0.012-0.035 in. (0.3-0.9mm). Tighten the 2 sensor bracket bolts to 10 ft. lbs. (14 Nm) with the sensor located so the gap is the same at several points on the toothed wheel. If the gap is incorrect, it is likely that the toothed wheel is worn or improperly installed.

16. Install the tire and wheel assembly.

17. Prior to moving the vehicle, pump the brakes unit a firm pedal is obtained.

Rear End Alignment

On FWD vehicles, camber and toe-in are pre-set at the factory. These settings can not be adjusted. If not within specifications, replace the bent or damaged components. On AWD vehicles, Camber and toe-in can be adjusted. However, proper measurement and adjustment required special tool and equipment. Again, we wish to express that rear alignment inspection and adjustments should be left to those who have the proper equipment and experience.

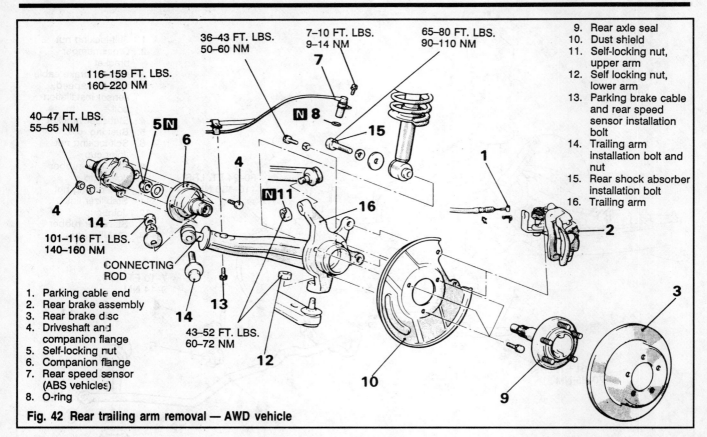

36–43 FT. LBS.
50–60 NM

7–10 FT. LBS.
9–14 NM

65–80 FT. LBS.
90–110 NM

116–159 FT. LBS.
160–220 NM

40–47 FT. LBS.
55–65 NM

101–116 FT. LBS.
140–160 NM

CONNECTING ROD

43–52 FT. LBS.
60–72 NM

9. Rear axle seal
10. Dust shield
11. Self-locking nut, upper arm
12. Self locking nut, lower arm
13. Parking brake cable and rear speed sensor installation bolt
14. Trailing arm installation bolt and nut
15. Rear shock absorber installation bolt
16. Trailing arm

1. Parking cable end
2. Rear brake assembly
3. Rear brake disc
4. Driveshaft and companion flange
5. Self-locking nut
6. Companion flange
7. Rear speed sensor (ABS vehicles)
8. O-ring

Fig. 42 Rear trailing arm removal — AWD vehicle

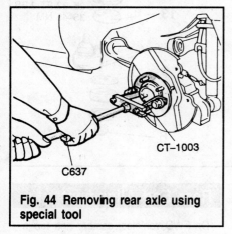

Fig. 44 Removing rear axle using special tool

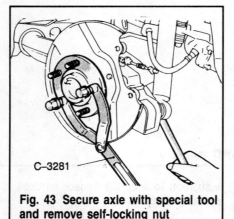

Fig. 43 Secure axle with special tool and remove self-locking nut

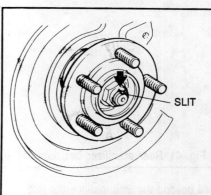

Fig. 46 Crimp wheel bearing nut at spindle indentation-FWD vehicle

REAR WHEEL ALIGNMENT

Year	Model	Caster Range (deg.)	Caster Preferred Setting (deg.)	Camber Range (deg.)	Camber Preferred Setting (deg.)	Toe-in (in.)	Steering Axis Inclination (deg.)
1990	Eclipse ①	—	—	$1\frac{1}{4}N$–$\frac{1}{4}N$	$\frac{3}{4}N$	0	—
	Eclipse ②	—	—	$1\frac{1}{4}N$–$\frac{1}{4}N$	$\frac{3}{4}N$	0	—
	Eclipse ③	—	—	$2\frac{1}{20}N$–$1\frac{1}{20}N$	$1\frac{11}{20}N$	0.14	—
	Laser ①	—	—	$1\frac{1}{4}N$–$\frac{1}{4}N$	$\frac{3}{4}N$	0	—
	Laser ②	—	—	$1\frac{1}{4}N$–$\frac{1}{4}N$	$\frac{3}{4}N$	0	—
	Laser ③	—	—	$2\frac{1}{20}N$–$1\frac{1}{20}N$	$1\frac{11}{20}N$	0.14	—
	Talon ②	—	—	$1\frac{1}{4}N$–$\frac{1}{4}N$	$\frac{3}{4}N$	0	—
	Talon ③	—	—	$2\frac{1}{20}N$–$1\frac{1}{20}N$	$1\frac{11}{20}N$	0.14	—
1991	Eclipse ①	—	—	$1\frac{1}{4}N$–$\frac{1}{4}N$	$\frac{3}{4}N$	0	—
	Eclipse ②	—	—	$1\frac{1}{4}N$–$\frac{1}{4}N$	$\frac{3}{4}N$	0	—
	Eclipse ③	—	—	$2\frac{1}{20}N$–$1\frac{1}{20}N$	$1\frac{11}{20}N$	0.14	—
	Laser ①	—	—	$1\frac{1}{4}N$–$\frac{1}{4}N$	$\frac{3}{4}N$	0	—
	Laser ②	—	—	$1\frac{1}{4}N$–$\frac{1}{4}N$	$\frac{3}{4}N$	0	—
	Laser ③	—	—	$2\frac{1}{20}N$–$1\frac{1}{20}N$	$1\frac{11}{20}N$	0.14	—
	Talon ②	—	—	$1\frac{1}{4}N$–$\frac{1}{4}N$	$\frac{3}{4}N$	0	—
	Talon ③	—	—	$2\frac{1}{20}N$–$1\frac{1}{20}N$	$1\frac{11}{20}N$	0.14	—
1992	Eclipse ①	—	—	$1\frac{1}{4}N$–$\frac{1}{4}N$	$\frac{3}{4}N$	0	—
	Eclipse ②	—	—	$1\frac{1}{4}N$–$\frac{1}{4}N$	$\frac{3}{4}N$	0	—
	Eclipse ③	—	—	$2\frac{1}{20}N$–$1\frac{1}{20}N$	$1\frac{11}{20}N$	0.14	—
	Laser ①	—	—	$1\frac{1}{4}N$–$\frac{1}{4}N$	$\frac{3}{4}N$	0	—
	Laser ②	—	—	$1\frac{1}{4}N$–$\frac{1}{4}N$	$\frac{3}{4}N$	0	—
	Laser ③	—	—	$2\frac{1}{20}N$–$1\frac{1}{20}N$	$1\frac{11}{20}N$	0.14	—
	Talon ②	—	—	$1\frac{1}{4}N$–$\frac{1}{4}N$	$\frac{3}{4}N$	0	—
	Talon ③	—	—	$2\frac{1}{20}N$–$1\frac{1}{20}N$	$1\frac{11}{20}N$	0.14	—
1993	Eclipse ①	—	—	$1\frac{1}{4}N$–$\frac{1}{4}N$	$\frac{3}{4}N$	0	—
	Eclipse ②	—	—	$1\frac{1}{4}N$–$\frac{1}{4}N$	$\frac{3}{4}N$	0	—
	Eclipse ③	—	—	$2\frac{1}{20}N$–$1\frac{1}{20}N$	$1\frac{11}{20}N$	0.14	—
	Laser ①	—	—	$1\frac{1}{4}N$–$\frac{1}{4}N$	$\frac{3}{4}N$	0	—
	Laser ②	—	—	$1\frac{1}{4}N$–$\frac{1}{4}N$	$\frac{3}{4}N$	0	—
	Laser ③	—	—	$2\frac{1}{20}N$–$1\frac{1}{20}N$	$1\frac{11}{20}N$	0.14	—
	Talon ①	—	—	$1\frac{1}{4}N$–$\frac{1}{4}N$	$\frac{3}{4}N$	0	—
	Talon ②	—	—	$1\frac{1}{4}N$–$\frac{1}{4}N$	$\frac{3}{4}N$	0	—
	Talon ③	—	—	$2\frac{1}{20}N$–$1\frac{1}{20}N$	$1\frac{11}{20}N$	0.14	—

N—Negative
P—Positive
① 1.8L engine
② 2.0L engine with Front Wheel Drive (FWD)
③ All Wheel Drive (AWD) vehicle

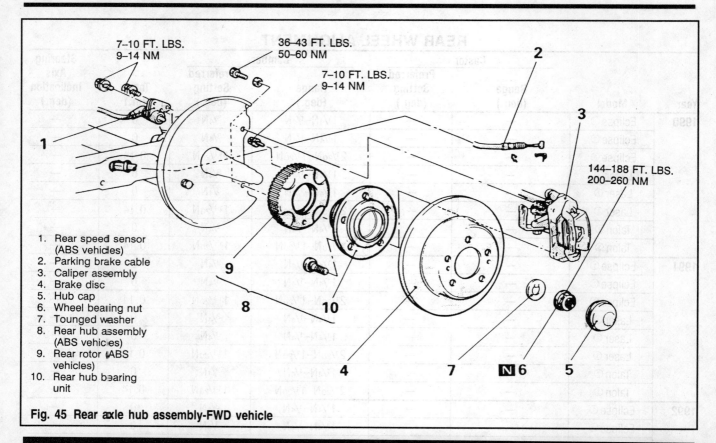

7–10 FT. LBS.
9–14 NM

36–43 FT. LBS.
50–60 NM

7–10 FT. LBS.
9–14 NM

144–188 FT. LBS.
200–260 NM

1. Rear speed sensor
   (ABS vehicles)
2. Parking brake cable
3. Caliper assembly
4. Brake disc
5. Hub cap
6. Wheel bearing nut
7. Tounged washer
8. Rear hub assembly
   (ABS vehicles)
9. Rear rotor (ABS
   vehicles)
10. Rear hub bearing
    unit

Fig. 45 Rear axle hub assembly-FWD vehicle

STEERING

Steering Wheel

REMOVAL & INSTALLATION

▶ **See Figures 47 and 48**

1. Disconnect the negative battery cable.
2. Remove the horn pad from the steering wheel. Remove the retainers in the horn pad. Push pad upward to remove. Disconnect horn button connector.
3. Remove steering wheel retaining nut.

4. Matchmark the steering wheel to the shaft.
5. Use a steering wheel puller to remove the steering wheel. Do not hammer on steering wheel to remove it. The collapsible column mechanism may be damaged.

To install:

6. Line up the matchmarks and install the steering wheel to the shaft.
7. Torque the steering wheel attaching nut to 33 ft. lbs. (45 Nm).
8. Reconnect the horn connector and install the horn pad.
9. Connect the negative battery cable.

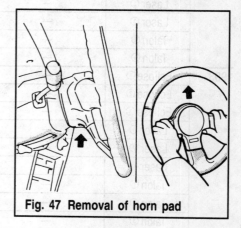

Fig. 47 Removal of horn pad

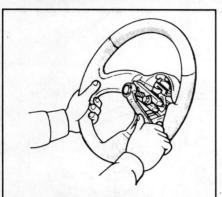

Fig. 48 Using steering wheel puller, remove steering wheel from shaft

Combination Switch

REMOVAL & INSTALLATION

▶ **See Figure 49**

➡ **The headlights, turn signals, dimmer switch, windshield/washer and, on some models, the cruise control function are all built into 1 multi-function combination switch that is mounted on the steering column.**

1. Disconnect the negative battery cable.

2. Remove the knee protector panel under the steering column, then the upper and lower column cover.

3. Remove the horn pad attaching screw on the under side of the steering wheel and remove the horn pad by pushing the pad upward.

4. Matchmark and remove the steering wheel with a steering wheel puller. Do not hammer on the steering wheel to remove it or the collapsible mechanism may be damaged.

5. Locate the rectangular plugs in the knee protector on either side of the steering column. Pry these plugs out and remove the screws. Remove the screws from the hood lock release lever and position lever aside. Remove the knee protector.

6. Remove the upper and lower column cover.

7. Remove the lap cooler duct.

8. Remove the band retaining the switch wiring.

9. Disconnect all connectors, remove the wiring clip. Remove the retainers and the column switch assembly.

To install:

10. Install the switch assembly and secure the clip. Make sure no wires are pinched or out of place.

11. Install the lap cooler ducts.

12. Install the column covers and knee protector.

13. Install the steering wheel. Torque the steering wheel-to-column nut to 33 ft. lbs. (45 Nm).

14. Connect the negative battery cable and check all functions of the combination switch for proper operation.

Ignition Lock/Switch

REMOVAL & INSTALLATION

▶ **See Figures 50, 51 and 52**

1. Disconnect the negative battery cable. Remove the hood lock release lever from the lower panel.

2. Remove the lower instrument panel knee protector.

3. Remove the upper and lower steering column cover.

4. Remove the clip that holds the wiring against the steering column.

5. Remove the key reminder switch, if equipped. Unplug the ignition switch from the steering lock cylinder and remove.

6. Insert the key into the steering lock cylinder and turn to the **ACC** position.

7. With a small cross-tip screwdriver, push the lock pin of the steering lock cylinder inward and then pull the lock cylinder out towards you.

➡ **When equipped with automatic transaxle, the vehicles have safety-lock systems and will have a key interlock cable installed in a slide lever on the side of the key cylinder. Carefully unhook the interlock cable from the lock cylinder while withdrawing cylinder from lock housing.**

To install:

8. With the ignition key removed, install the slide lever to the steering lock cylinder. Connect the interlock cable to the slide lever and the steering lock cylinder. Apply grease to the interlock cable and install cylinder into the lock housing.

9. If equipped with key interlock system, place the gearshift selector in **P** with the engine **OFF**. Check the system operation as follows:

a. Check that the gear select lever can not be moved and the button on the lever can not be pushed under with the ignition key in the **LOCK** or **OFF** position, and the brake pedal not depresses.

b. Turn the ignition key to the **ACC** position. Depress the brake pedal. Press the button on the select lever. Check to be sure that under this conditions, the select lever can be moved from the from the **P** position to any other position. Press the button a few times to be sure that the select lever moves smoothly.

c. Check to be sure that at all other positions of the select lever other than **P**, the ignition key can not be turned to the **OFF** position. Check to be sure the ignition key turns smoothly to the **OFF** position when the selector lever is in the **P** position.

10. If a malfunction is discovered with the shift lock mechanism, adjust or check the key interlock system.

11. Install the ignition switch plug carefully and make sure no wires in the harness are pinched.

12. Install the wiring clip. Align the matchmarks and install the steering wheel to the steering shaft. Install the steering wheel retainer nut and tighten to 33 ft. lbs. (45 Nm).

13. Install the steering column covers.

14. Install the knee protector.

15. Connect the negative battery cable and check the ignition switch and lock for proper operation.

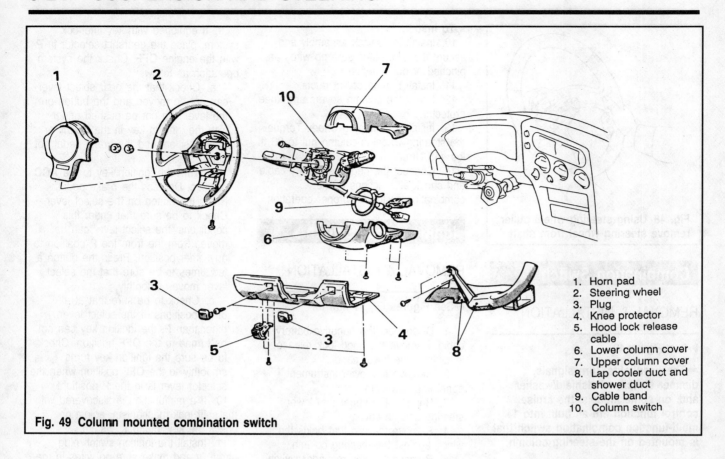

1. Horn pad
2. Steering wheel
3. Plug
4. Knee protector
5. Hood lock release cable
6. Lower column cover
7. Upper column cover
8. Lap cooler duct and shower duct
9. Cable band
10. Column switch

Fig. 49 Column mounted combination switch

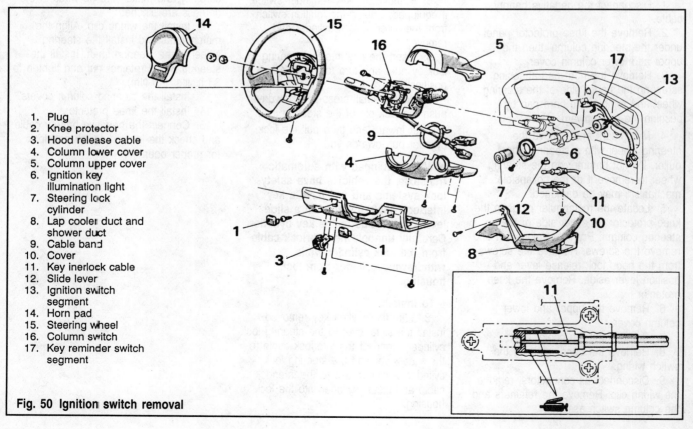

1. Plug
2. Knee protector
3. Hood release cable
4. Column lower cover
5. Column upper cover
6. Ignition key illumination light
7. Steering lock cylinder
8. Lap cooler duct and shower duct
9. Cable band
10. Cover
11. Key inerlock cable
12. Slide lever
13. Ignition switch segment
14. Horn pad
15. Steering wheel
16. Column switch
17. Key reminder switch segment

Fig. 50 Ignition switch removal

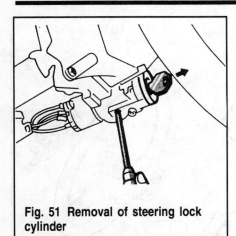

Fig. 51 Removal of steering lock cylinder

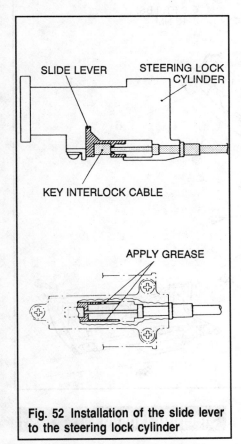

Fig. 52 Installation of the slide lever to the steering lock cylinder

Steering Column

REMOVAL & INSTALLATION

▶ **See Figure 53**

1. Disconnect the negative battery cable.
2. Remove the instrument panel undercover or knee protector.

3. Remove the trim clip, foot shower duct and lap shower duct.
4. Remove the steering wheel. Remove the column upper and lower covers. Disconnect the key interlock cable, if equipped.
5. Disconnect all connector to column-mounted items.
6. Remove the band from the steering joint cover. Remove the joint assembly and gear box connecting body.
7. Remove the screws that attach the rubber seal to the firewall.
8. Remove the lower and upper column mounting bolt.
9. Remove the steering column assembly from the vehicle.

To install:

10. Install the column so the splines are inserted around the rack input shaft. Install the pinch bolt and tighten to 14 ft. lbs. (20 Nm).
11. Install the column mounting bolts.
12. Install the rubber seal screws.
13. Connect the connectors and interlock cable. Check key interlock system operation as follows:

   a. Check that the gear select lever can not be moved and the button on the lever can not be pushed under with the ignition key in the **LOCK** or **OFF** position, and the brake pedal not depresses.

   b. Turn the ignition key to the **ACC** position. Depress the brake pedal. Press the button on the select lever. Check to be sure that under this conditions, the select lever can be moved from the from the **P** position to any other position. Press the button a few times to be sure that the select lever moves smoothly.

   c. Check to be sure that at all other positions of the select lever other than **P**, the ignition key can not be turned to the **OFF** position. Check to be sure the ignition key turns smoothly to the **OFF** position when the selector lever is in the **P** position.

14. If a malfunction is discovered with the shift lock mechanism, adjust or check the key interlock system.
15. Install the column covers.
16. Install the remaining interior pieces.
17. Connect the negative battery cable and check all column-mounted switches for proper operation.

DISASSEMBLY

▶ **See Figures 54, 55 and 56**

1. Disconnect the negative battery cable. Remove the steering column from the vehicle.
2. Remove the combination switch from the steering shaft. If necessary to remove the steering lock, use a haxsaw to cut the special bolt at the steering lock bracket side.
3. Remove the boot and cover assembly from the lower portion of the steering shaft assembly.
4. Carefully remove the lower bearing assembly.
5. Remove the retainer bolt at the upper portion of the lower joint assembly. Remove the lower joint assembly from the shaft.
6. Using a socket wrench or similar tool, remove the bearing from above the lower joint assembly. Remove the snap rings above the bearing.
7. At the top of the shaft, remove the snapring, stopper and bearing spacer.
8. Remove the centrally located column tube clamp. Remove the steering shaft from the upper column tube. Replace the column tube spacer and bushing as required.

To install:

9. Install the steering shaft into the upper column tube. Install the column tube clamp. Slide the column tube so the protrusion out the lower tube protrusion out of the clamp measures 0.73-0.81 in. (24.8-25.2mm). Lock clamp in this position.
10. Install the steering lock bracket and special bolt. When installing the steering lock bracket to the column tube, temporarily install the steering lock in alignment with the column boss. After checking that the locks works properly, tighten the special bolt until the head twists off.

✳✳CAUTION

The steering lock bracket and bolt must be replaced with new ones when the steering lock is installed.

11. Install the snapring, stopper and bearing spacer to the top of the steering shaft assembly.
12. Using a socket wrench or similar tool, install the bearing at the bottom of the shaft. Install the snapring above the bearings.

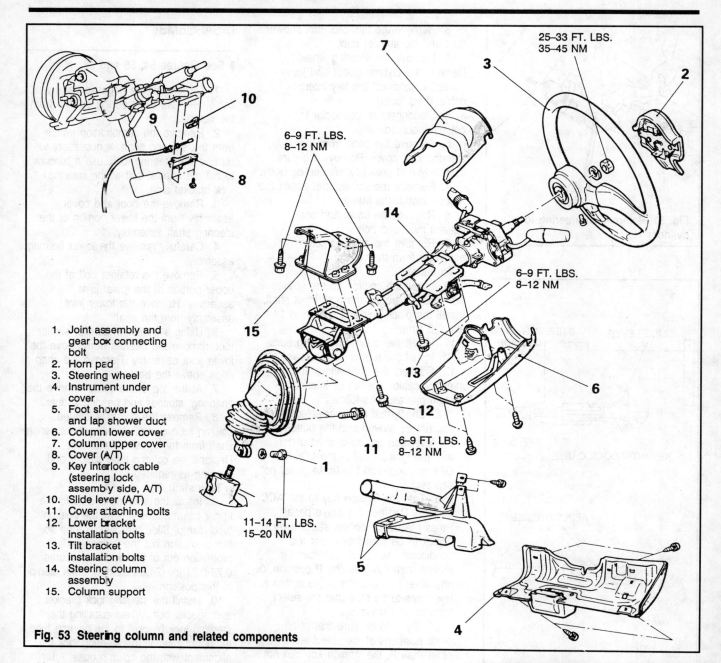

1. Joint assembly and gear box connecting bolt
2. Horn pad
3. Steering wheel
4. Instrument under cover
5. Foot shower duct and lap shower duct
6. Column lower cover
7. Column upper cover
8. Cover (A/T)
9. Key interlock cable (steering lock assembly side, A/T)
10. Slide lever (A/T)
11. Cover attaching bolts
12. Lower bracket installation bolts
13. Tilt bracket installation bolts
14. Steering column assembly
15. Column support

Fig. 53 Steering column and related components

13. Install the lower joint assembly and tighten pinch bolt to 14 ft. lbs. (20 Nm).

14. Install the lower bearing assembly as follows:

a. Install the cover on the joint assembly.

b. Fill the inside of the bearing with multi-purpose grease.

c. Install the bearings to the shaft on the joint assembly.

d. Wrap vinyl tape approximately 1½ times around the concave circumference of the bearings. Press fit the bearings into the cover assembly.

e. Apply multi-purpose grease to the mating surfaces of the joint and cover assemblies.

15. Install the boot assembly.

16. Install the combination switch to the steering shaft.

17. Install the steering column into the vehicle.

Steering Linkage

REMOVAL & INSTALLATION

Tie Rod Ends

1. Disconnect the battery negative cable. Raise the vehicle and support safely.

2. Wire brush the threads on the tie rod shaft and lubricate with penetrating oil. Loosen the locknut.

3. Remove the cotter pin and nut and press the tie rod end from the steering knuckle.

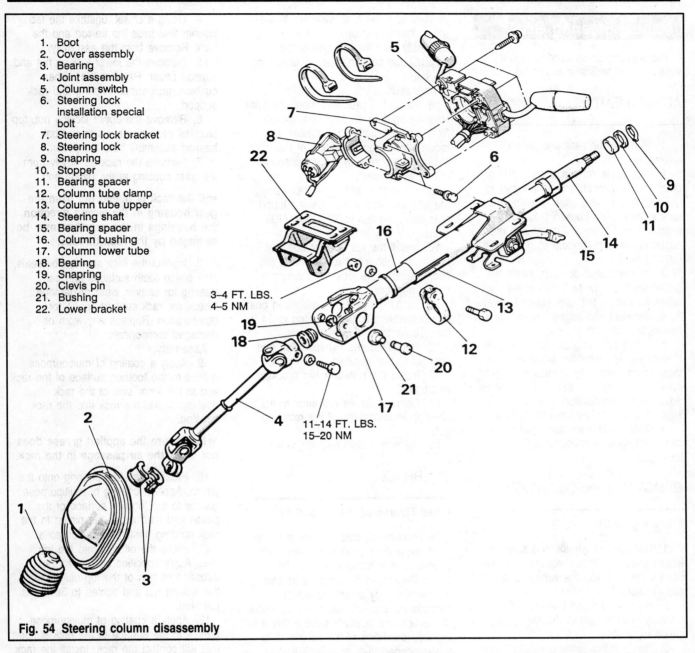

1. Boot
2. Cover assembly
3. Bearing
4. Joint assembly
5. Column switch
6. Steering lock installation special bolt
7. Steering lock bracket
8. Steering lock
9. Snapring
10. Stopper
11. Bearing spacer
12. Column tube clamp
13. Column tube upper
14. Steering shaft
15. Bearing spacer
16. Column bushing
17. Column lower tube
18. Bearing
19. Snapring
20. Clevis pin
21. Bushing
22. Lower bracket

3–4 FT. LBS.
4–5 NM

11–14 FT. LBS.
15–20 NM

Fig. 54 Steering column disassembly

Fig. 55 Removal of steering lock installation bolt

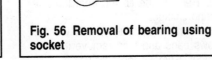

Socket

Fig. 56 Removal of bearing using socket

4. Hold the tie rod shaft with locking pliers and turn the tie rod end off, counting the number of turns for installation.

To install:

5. Install the tie rod end the same number of turns that it took to remove the old one. Tighten the locknut to secure tie rod in place.

6. Install the tie rod stud into the steering knuckle and install nut. Tighten the nut to 25 ft. lbs. (34 Nm) and install new cotter pin.

7. Perform front end alignment.

Manual Steering Gear

The steering gears used on these vehicles is of rack and pinion type.

ADJUSTMENT

1. Remove the rack and pinion assembly.

2. Mount the rack in a vise. With a small torque wrench and an adapter to connect to the input shaft, position the rack at its center. Loosen the rack support cover locknut. Tighten the rack support cover, the bottom plug, to 11 ft. lbs. (15 Nm).

3. In the neutral position, rotate the pinion shaft clockwise 1 turn in 4-6 seconds. Return the rack support cover 30-60 degrees and adjust the total pinion torque to 5-11 inch lbs.

4. When adjusting, set to the higher side of the specification. Make sure there is no ratcheting or catching when operating the rack. If the rack cannot be adjusted to specification, check the rack support cover components or replace.

5. After adjustment has been made, lock the rack support cover with the locking nut.

REMOVAL & INSTALLATION

▶ **See Figure 57**

1. Position the wheels in a straight ahead position. Disconnect the negative battery cable. Raise the vehicle and support safely.

2. Remove the bolt holding lower steering column joint to the rack and pinion input shaft.

3. Remove the cotter pins and, using the proper separating tools, disconnect the tie rod ends from the knuckle.

4. Locate the triangular brace near the stabilizer bar brackets on the crossmember and remove both the brace and the stabilizer bar bracket.

5. Place a jack under the center member. Remove the through bolt from the round roll stopper. Remove the rear bolts from the center crossmember.

6. Disconnect the front exhaust pipe and tie out of the way. Lower the center member slightly.

7. Remove the rack and pinion steering assembly and its rubber mounts. Move the rack to the right to remove from the crossmember. While tilting downward, remove the rack assembly from the left side of the vehicle. Use caution to avoid damaging the boots.

To install:

8. Install the rack and mounting bolts, torquing bolts to 43-58 ft. lbs. (60-80 Nm). When installing the rubber rack mounts, align the projection of the mounting rubber with the indentation in the crossmember.

9. Raise the center member using the jack and install the center support rear bolts. Tighten to 72 ft. lbs. (100 Nm).

10. Install the roll stopper bolt and new nut. Tighten nut to 47 ft. lbs. (65 Nm). Remove the jack supporting the center member.

11. Install the joint assembly and gear box connecting bolt and tighten to 14 ft. lbs. (20 Nm).

12. Reposition the exhaust pipe and connect to the manifold.

13. Install the stabilizer bar brackets and brace.

14. Connect the tie rod ends to the steering knuckles. Install the retaining nuts.

15. Perform a front end alignment.

OVERHAUL

▶ **See Figures 58, 59, 60 and 61**

To successfully overhaul the steering gear, experience at rack overhaul and special care to tolerances must be taken. Installation of components that are worn slightly or are not seated properly will prevent successful overhaul of the unit and possibly pose a threat to the safe operation of the vehicle. It is recommended that the overhaul of a steering gear be performed by qualified steering gear rebuilders. Furthermore, the cost for specific components and overhaul kits, the time invested in the repair, and an unsuccessful result may not be as cost effective as purchasing a rebuilt unit. It is recommended that all possibilities be researched before taking on such a repair.

1. Remove the steering gear from the vehicle. Loosen the tire rod locking nuts.

2. Remove the tie rod from the steering linkage.

3. Remove the bellows clip. Cut the bellows band and then remove it from the boot. Remove the bellows.

4. Using a chisel, unstake the tab washer that fixes the tie rod and the rack. Remove from the assembly.

5. Remove the lower locking nut and support cover. Remove the rubber cushion, rack support spring and rack support.

6. Remove the upper locking nut, top plug, oil seal, pinion collar and ball bearing assembly from the pinion.

7. Remove the rack assembly from the gear housing at the pinion end.

➡**If the rack is pulled out from the gear housing in the wrong direction, the bushings in the gear box may be damaged by the rack threads.**

8. Inspect the rack support, oil seals, rack pinion tooth surfaces and ball bearing for uneven wear or damage. Check the rack support spring for deterioration. Replace any worn or damaged components.

Assembly:

9. Apply a coating of multipurpose grease to the toothed surface of the rack and to the inner side of the rack bushing. Install the rack into the rack housing.

➡**Make sure the applied grease does not block the air passage in the rack.**

10. Press fit the ball bearing onto the pinion. Apply a coating of multipurpose grease to the toothed surface of the pinion and then install the pinion to the rack housing. Install the pinion collar.

11. Press the oil seal into the top plug. Apply specified sealant to the threaded portion of the top plug. Install the locking nut and tighten to 36 ft. lbs. (50 Nm).

12. Apply a coating of multipurpose grease to the surface of the rack support that will contact the rack. Install the rack support into the rack housing.

13. Fill the inner side of the rack support spring with multi-purpose grease and then install into the rack housing. Install the rubber cushion to the rack support cover to the rack housing. Install the lock nut and tighten to 36 ft. lbs. (50 Nm).

14. Adjust the total pinion torque at this time. Once adjusted, install and tighten the locknut to specifications.

15. Install the tie rod assembly. After installing the tie rod to the rack, fold the tab washer end to tie rod notch at 2 locations. Pack the tie rod bellows lock

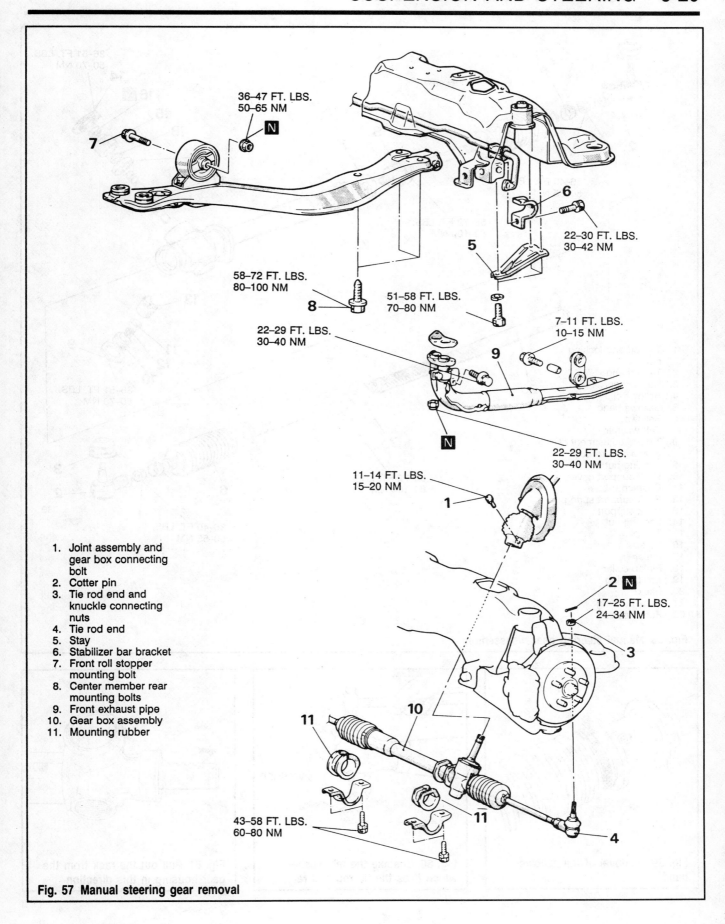

36–47 FT. LBS.
50–65 NM

7

22–30 FT. LBS.
30–42 NM

6

5

58–72 FT. LBS.
80–100 NM

8

51–58 FT. LBS.
70–80 NM

22–29 FT. LBS.
30–40 NM

9

7–11 FT. LBS.
10–15 NM

22–29 FT. LBS.
30–40 NM

11–14 FT. LBS.
15–20 NM

1

2

17–25 FT. LBS.
24–34 NM

3

10

11

11

43–58 FT. LBS.
60–80 NM

4

1. Joint assembly and
   gear box connecting
   bolt
2. Cotter pin
3. Tie rod end and
   knuckle connecting
   nuts
4. Tie rod end
5. Stay
6. Stabilizer bar bracket
7. Front roll stopper
   mounting bolt
8. Center member rear
   mounting bolts
9. Front exhaust pipe
10. Gear box assembly
11. Mounting rubber

Fig. 57 Manual steering gear removal

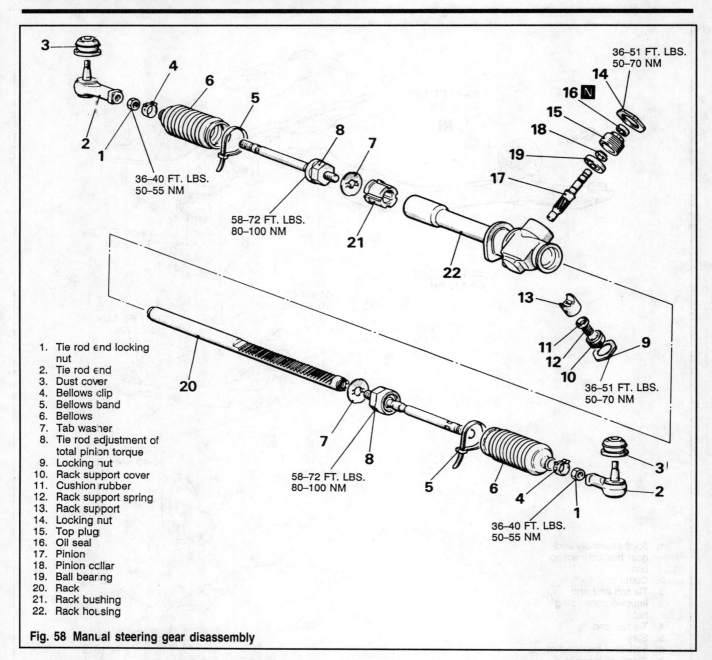

36–51 FT. LBS.
50–70 NM

36–40 FT. LBS.
50–55 NM

58–72 FT. LBS.
80–100 NM

36–51 FT. LBS.
50–70 NM

58–72 FT. LBS.
80–100 NM

36–40 FT. LBS.
50–55 NM

1. Tie rod end locking nut
2. Tie rod end
3. Dust cover
4. Bellows clip
5. Bellows band
6. Bellows
7. Tab washer
8. Tie rod adjustment of total pinion torque
9. Locking nut
10. Rack support cover
11. Cushion rubber
12. Rack support spring
13. Rack support
14. Locking nut
15. Top plug
16. Oil seal
17. Pinion
18. Pinion collar
19. Ball bearing
20. Rack
21. Rack bushing
22. Rack housing

Fig. 58 Manual steering gear disassembly

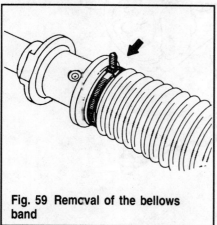

Fig. 59 Removal of the bellows band

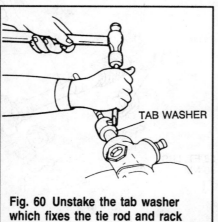

TAB WASHER

Fig. 60 Unstake the tab washer which fixes the tie rod and rack

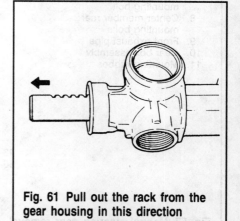

Fig. 61 Pull out the rack from the gear housing in this direction

groove with silicone grease. Install the bellows band, bellows and bellows clip.

16. Install the tie rod end to the gear assembly and tighten the locknut.

17. Install the gear assembly into the vehicle.

18. Align the front suspension.

Power Steering Gear

The steering gears used on these vehicles is of rack and pinion type.

ADJUSTMENT

1. Disconnect the negative battery cable.

2. Raise the vehicle and support safely.

3. Remove the steering rack assembly from the vehicle.

4. Secure the steering rack assembly in a vise. Do not clamp the vise jaws on the steering housing tubes. Clamp the vise jaws only on the housing cast metal.

5. With rack at center position, check torque on the rack support cover to 11 ft. lbs. (15 Nm).

6. With rack at center position, rotate the shaft clockwise 1 turn in 4-6 seconds. Return the rack support cover 30-60 degrees and adjust the total pinion torque to 5-11 inch lbs. Set the standard value at its highest value when adjusting. Assure no ratcheting or catching when operating the rack towards the shaft direction.

REMOVAL & INSTALLATION

♦ **See Figure 62**

1. Disconnect the negative battery cable. Drain the power steering fluid. Raise the vehicle and support safely.

2. Remove the bolt holding lower steering column joint to the rack and pinion input shaft.

3. Remove the transfer case, if equipped.

4. Remove the cotter pins and using the proper tools, separate the tie rod ends from the steering knuckle.

5. Locate the triangular brace near the stabilizer bar brackets on the crossmember and remove both the brace and the stabilizer bar bracket.

6. Support the center crossmember. Remove the through bolt from the round

roll stopper and remove the rear bolts from the center crossmember.

7. Disconnect the front exhaust pipe, if equipped with FWD.

8. Disconnect the power steering fluid pressure pipe and return hose from the rack fittings. Plug the fittings to prevent excess fluid leakage.

9. Lower the crossmember slightly. Remove the rack and pinion steering assembly and its rubber mounts. Move the rack to the right to remove from the crossmember. Tilt the assembly downward and remove from the left side of the vehicle. Use caution to avoid damaging the boots.

To install:

10. Install the rack and install the mounting bolts. Torque the mounting bolts to 43-58 ft. lbs. (60-80 Nm). When installing the rubber rack mounts, align the projection of the mounting rubber with the indentation in the crossmember.

11. Connect the power steering fluid lines to the rack.

12. Connect the exhaust pipe, if removed.

13. Raise the crossmember into position. Install the center member mounting bolts and tighten to 72 ft. lbs. (100 Nm). Install the roll stopper bolt and new nut. Tighten nut to 47 ft. lbs. (65 Nm).

14. Install the stabilizer bar brackets and brace.

15. Connect the tie rod ends and tighten nuts to 25 ft. lbs. (34 Nm).

16. Install the transfer case, if removed.

17. Refill the reservoir with power steering fluid and bleed the system.

18. Perform a front end alignment.

OVERHAUL

♦ **See Figure 63**

To successfully overhaul the steering gear, special tools, experience at gear overhaul and special care to tolerances must be taken. Incorrect component installation or installation of components that are worn slightly will prevent successful overhaul of the unit and possibly pose a threat to the safe operation of the vehicle. It is recommended that overhaul of a steering gear be performed by qualified steering gear rebuilders. Furthermore, the cost for worn components, overhaul kits, required tools, time (time invested in the repair

procedure) and an unsuccessful result may not be as cost effective as purchasing an already rebuilt unit. It is recommended that all possibilities be researched before taking on such a repair.

1. Remove the steering gear from the vehicle. Loosen the tire rod locking nuts.

2. Remove the tie rod from the steering linkage.

3. Remove the bellows clip. Cut the bellows band and then remove it from the boot. Remove the bellows.

4. Using a chisel or equivalent tool, unstake the tab washer which attach the tie rod end and the rack.

5. Remove the tie rods from the assembly. Disconnect and remove the oil feed tubes as required.

6. Remove the end plug caulking. Tap the end plug to loosen and remove from the lower portion of the rack housing.

7. Remove the support cover locknut. Using Snap-on tool S6161 or equivalent, remove the rack support cover from the gear box. Remove the rack support spring and rack support.

8. Remove the retaining bolts and the valve housing from the top end of the gear housing.

9. Using a plastic hammer, gently tap the pinion to remove it. Remove the oil seal. Using a socket, remove the oil seal and the ball bearing from the valve housing simultaneously.

10. Turn the rack stopper clockwise until the end of the circlip comes out of the slot in the rack housing. Turn the rack stopper counterclockwise to remove the circlip.

11. Pull the rack from the housing slowly, in the direction away from the pinion and valve assembly. At this time, take out the rack stopper and the rack bushing simultaneously.

12. Partially bend the rack bushing oil seal and remove from the bushing. Be careful not to damage the oil seal press fitting surface during seal removal.

13. Use a brass bar, remove the ball bearing from the gear housing. Remove the roller bearing from the rack housing.

14. Using a piece of pipe or similar tool, remove the back-up washer and oil seal from the gear housing. Be careful not to damage the inner surface of the rack cylinder of the gear housing.

15. Inspect the rack as follows:
    a. Check the rack toothed surfaces for damage or wear.

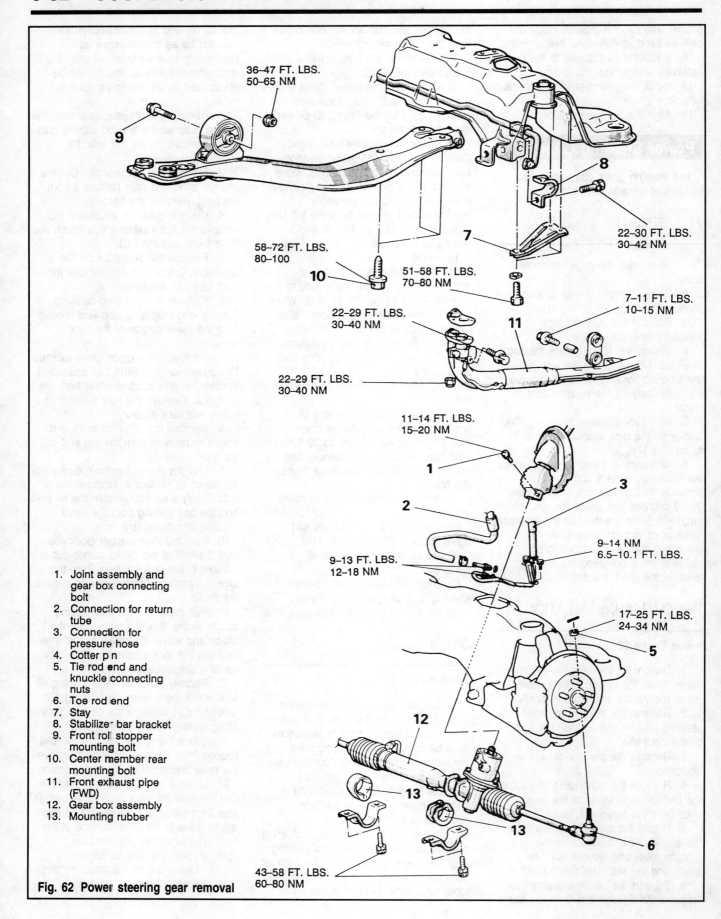

36–47 FT. LBS.
50–65 NM

9

8

22–30 FT. LBS.
30–42 NM

7

58–72 FT. LBS.
80–100

10

51–58 FT. LBS.
70–80 NM

7–11 FT. LBS.
10–15 NM

22–29 FT. LBS.
30–40 NM

11

22–29 FT. LBS.
30–40 NM

11–14 FT. LBS.
15–20 NM

1

2

3

9–13 FT. LBS.
12–18 NM

9–14 NM
6.5–10.1 FT. LBS.

17–25 FT. LBS.
24–34 NM

5

12

12

13

13

6

43–58 FT. LBS.
60–80 NM

1. Joint assembly and
   gear box connecting
   bolt
2. Connection for return
   tube
3. Connection for
   pressure hose
4. Cotter pin
5. Tie rod end and
   knuckle connecting
   nuts
6. Toe rod end
7. Stay
8. Stabilizer bar bracket
9. Front roll stopper
   mounting bolt
10. Center member rear
    mounting bolt
11. Front exhaust pipe
    (FWD)
12. Gear box assembly
13. Mounting rubber

Fig. 62 Power steering gear removal

1. Tie rod end locking nut
2. Tie rod end
3. Dust cover
4. Bellows clip
5. Bellows band
6. Bellows
7. Tab washer
8. Tie rods
9. Feed tubes
10. O-rings
11. End plug
12. Self-locking nut
13. Locking nut
14. Rack support cover
15. Rack support spring
16. Rack support
17. Valve housing
18. Oil seal
19. Pinion and valve assembly
20. Seal ring
21. Ball bearing
22. Oil seal
23. Circlip
24. Rack stopper
25. Rack bushing
26. Rack
27. O-ring
28. Oil seal
29. Seal ring
30. O-ring
31. Ball bearing
32. Needle roller bearing
33. Oil seal
34. Back-up washer
35. Rack housing

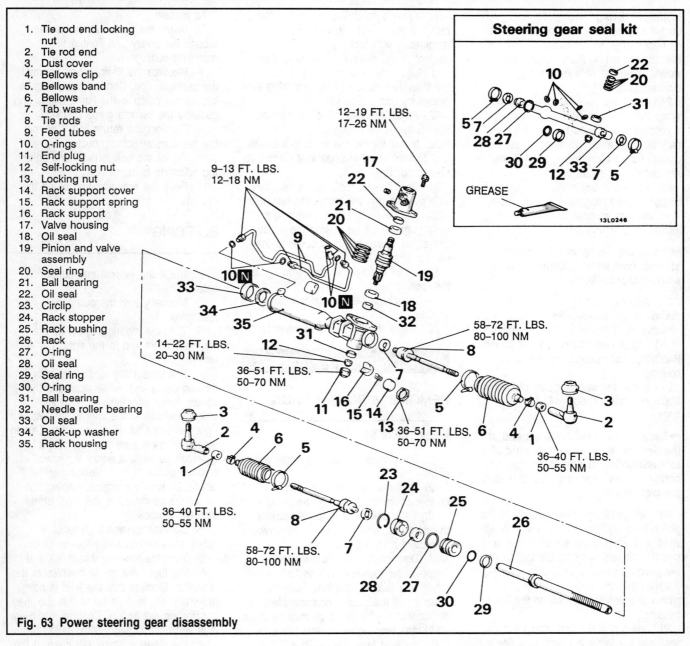

Steering gear seal kit

GREASE

Fig. 63 Power steering gear disassembly

b. Check the oil seal contact surface for uneven wear.

c. Check the rack for bends.

16. Inspect the pinion and valve assembly as follows:

a. Check the pinion gear toothed surfaces for damage and wear.

b. Check for worn or defective seal ring.

17. Inspect other related components as follows:

a. Check the cylinder inner surface of the rack housing for damage.

b. Check the boots for damage, cracking or deterioration.

c. Check the rack support for uneven wear or dents.

d. Check the rack bushing for uneven wear or damage.

18. Replace all components that show wear or damage.

To assemble:

19. Apply a coating of MOPAR ATF PLUS (automatic transmission fluid type 7176), Dextron or Dextron II type fluid to the outside of the oil seal. Using a tool, press the back-up washer and the oil seal into the rack housing to where the upper end of the press in guide coincides with the stepped part of the press in tool.

20. Install the roller needle roller bearing. Apply MOPAR ATF PLUS (automatic transmission fluid type 7176),

Dextron or Dextron II type fluid to the housing, bearing and oil seal press fitting surface. press in the needle roller bearing. Make sure to install bearing straight, the housing is only aluminum and will enlarge if installed incorrectly.

21. Install the ball bearing, which is also a press fit.

22. Apply a coating of ATF to the outside of the oil seal and O-ring and install. Press the seal in until it touches the rack bush end.

23. Install the rack assembly. Apply a coating of multi-purpose grease to the rack teeth face. Make sure not to close the vent hole in the rack with grease. Cover the serrations with rack installation

tool MB991213 or equivalent. Apply ATF on the rack installation tool. Match the oil seal center with the rack to prevent the retainer spring from slipping and slowly insert the rack from the power cylinder side.

24. Install the rack bushing assembly. Wrap the rack end with vinyl tape, apply a coating of ATF and then install the rack bushing and rack stopper. Be careful not to allow the spring retainer to slip out during installation.

25. Insert the circlip into the rack stopper hole through the cylinder hole. Turn the rack stopper clockwise and insert circlip firmly.

➡️**Insert the circlip and the rack stopper hole while turning the rack stopper clockwise.**

26. Apply a coating of ATF to the outside of the oil seal. Press the oil seal into the valve housing. Apply a coating of ATF to the outside of the ball bearing. Press the ball bearing into the valve housing.

27. Install the seal rings by pressing firmly into groove. Apply AFT to the rings.

➡️**Because the seal rings expand at the time of installation, use special tool MB991317 or equivalent to compress the seal rings so that they are well seated.**

28. Apply multi-purpose grease to the pinion gear and housing bearing. Wrap vinyl tape around the serrated part so that the oil seal wont be damaged when the pinion and valve assembly is installed into the housing. Mount the pinion and valve assembly to the valve housing.

29. Using a press tool, press the oil seal into the valve housing. In order to eliminate a seal malfunction at the valve housing alignment surface, the upper surface of the oil seal should project outward approximately 0.040 in. (1mm) from the housing edge surface.

30. Apply multi-purpose grease to the rack support surface in contact with the rack bar. Install rack support and spring.

31. Apply semi-drying sealant to the rack support cover screw. Lock temporarily with locking nut. Apply sealant to the threaded portion of the end plug. Install the end plug. Secure the threaded portion of the end plug at 2 places by using a punch.

32. Adjust the total pinion torque to specifications as outlined.

33. Install the tie rod ends to the rack and fold tab washer ends in 2 places, to tie rod notch.

34. Pack the tie rod bellows groove with silicone grease. Install the retainers.

35. Fill the tie rod dust covers with multi-purpose grease. Using the proper press tool, press the dust cover to the tie rod end.

36. Install the tie rod ends and tighten the locking nuts.

37. Install the steering gear into the vehicle.

38. Adjust the front end alignment.

Power Steering Pump

REMOVAL & INSTALLATION

◆ **See Figure 64**

1. Disconnect the battery negative cable.

2. Remove the pressure switch connector from the side of the pump.

3. If the alternator is located under the oil pump, cover it with a shop towel to protect it from oil.

4. Disconnect the return fluid line. Remove the reservoir cap and allow the return line to drain the fluid from the reservoir. If the fluid is contaminated, disconnect the ignition high tension cable and crank the engine several times to drain the fluid from the gearbox.

5. Disconnect the pressure line.

6. Remove the pump drive belt and unbolt the pump from its bracket.

To install:

7. Install the pump, wrap the belt around the pulley and tighten the mounting bolts.

8. Replace the O-rings and connect the pressure line. Connect the pressure line so the notch in the fitting aligns and contacts the pump's guide bracket.

9. Connect the return line. Connect the pressure switch connector.

10. Adjust the belt tension and tighten the adjusting bolts.

11. Refill the reservoir and bleed the system.

BLEEDING

1. Raise the vehicle and support safely.

2. Manually turn the pump pulley a few times.

3. Turn the steering wheel all the way to the left and to the right 5 or 6 times.

4. Disconnect the ignition high tension cable and, while operating the starter motor intermittently, turn the steering wheel all the way to the left and right 5-6 times for 15-20 seconds. During bleeding, make sure the fluid in the reservoir never falls below the lower position of the filter. If bleeding is attempted with the engine running, the air will be absorbed in the fluid. Bleed only while cranking.

5. Connect ignition high tension cable, start engine and allow to idle.

6. Turn the steering wheel left and right until there are no air bubbles in the reservoir. Confirm that the fluid is not milky and the level is up to the specified position on the gauge. Confirm that there is is very little change in the fluid level when the steering wheel is turned. If the fluid level changes more than 0.2 in., the air has not been completely bled. Repeat the process.

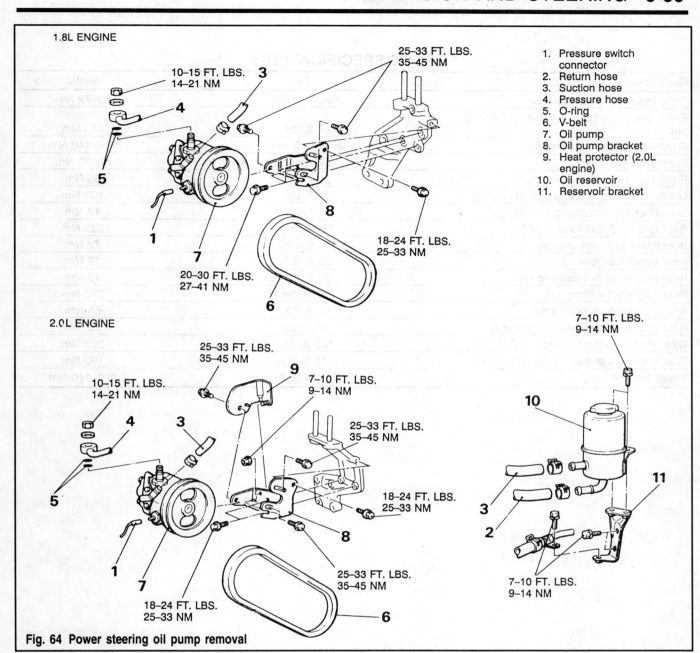

1.8L ENGINE

10–15 FT. LBS.
14–21 NM

25–33 FT. LBS.
35–45 NM

20–30 FT. LBS.
27–41 NM

18–24 FT. LBS.
25–33 NM

2.0L ENGINE

10–15 FT. LBS.
14–21 NM

25–33 FT. LBS.
35–45 NM

7–10 FT. LBS.
9–14 NM

25–33 FT. LBS.
35–45 NM

18–24 FT. LBS.
25–33 NM

25–33 FT. LBS.
35–45 NM

18–24 FT. LBS.
25–33 NM

7–10 FT. LBS.
9–14 NM

7–10 FT. LBS.
9–14 NM

1. Pressure switch
   connector
2. Return hose
3. Suction hose
4. Pressure hose
5. O-ring
6. V-belt
7. Oil pump
8. Oil pump bracket
9. Heat protector (2.0L
   engine)
10. Oil reservoir
11. Reservoir bracket

Fig. 64 Power steering oil pump removal

TORQUE SPECIFICATIONS

Component	US	Metric
Front lower ball joint stud nut	43–52 ft. lbs.	60–72 Nm
Front lower control arm inner mounting nut	87 ft. lbs.	120 Nm
Front strut lower mounting nut	80–101 ft. lbs.	110–140 Nm
Front strut rod locknut	43–51 ft. lbs.	60–70 Nm
Front strut upper mounting nut	36 ft. lbs.	50 Nm
Lower control arm clamp mounting bolt	72 ft. lbs.	100 Nm
Lower control arm clamp mounting nuts	34 ft. lbs.	47 Nm
Rear axle shaft self-locking nut	159 ft. lbs.	220 Nm
Rear lower ball joint stud nut	52 ft. lbs.	72 Nm
Rear upper ball joint stud nut	52 ft. lbs.	72 Nm
Rear shock upper mounting nut	29 ft. lbs.	49 Nm
Rear axle shaft locking nut	159 ft. lbs.	220 Nm
Rear driveshaft mounting nut	47 ft. lbs.	65 Nm
Rear strut piston rod attaching nut	18 ft. lbs.	25 Nm
Rear wheel bearing hub nut (FWD vehicle)	188 ft. lbs.	260 Nm
Rear upper control arm installation bolt	116 ft. lbs.	160 Nm
Wheel lug nut	87–101 ft. lbs.	120–140 Nm

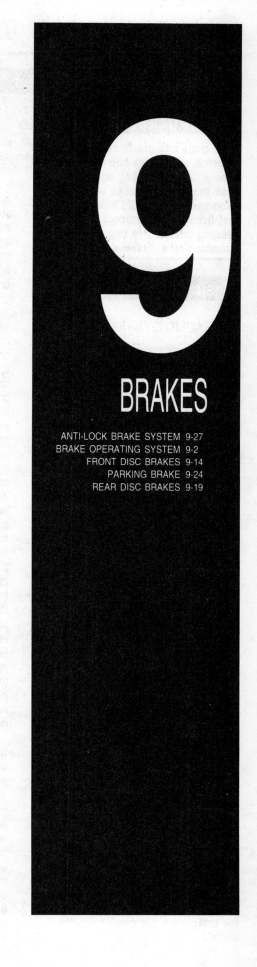

9

BRAKES

BRAKE OPERATING SYSTEM

❊❊CAUTION

Brake pads and shoes contain asbestos, which has been determined to be a cancer causing agent. Never clean the brake surfaces with compressed air! Avoid inhaling any dust from brake surfaces! When cleaning brakes, use commercially available brake cleaning fluids.

Basic System Operation

HYDRAULIC SYSTEM

Hydraulic systems are used to actuate the brakes of all modern automobiles. A hydraulic system rather than a mechanical system is used for two reasons. First, fluid under pressure can be carried to all parts of an automobile by small hoses — some of which are flexible — without taking up a significant amount of room or posing routing problems. Second, hydraulics can offer a great deal of mechanical advantage, producing a great deal of pressure at the wheels generated by little pressure at the pedal.

The master cylinder consists of a fluid reservoir and a single or double cylinder and piston assembly. Double (or dual) master cylinders are designed to separate the front and rear braking systems hydraulically in case of a leak. The master cylinder coverts mechanical motion from the pedal into hydraulic pressure within the lines. This pressure is translated back into mechanical motion at the wheels by either the wheel cylinder (drum brakes) or the caliper (disc brakes).

Steel lines carry the brake fluid to a point on the vehicle's frame near each of the vehicle's wheels. The fluid is then carried to the calipers and wheel cylinders (if equipped with drum brakes) by flexible tubes. Flexible tubes allow for suspension and steering movements.

In drum brake systems, wheel cylinders are used to apply the brakes. Each wheel cylinder contains two pistons, one at either end, which push outward in opposite directions and force the brake shoe into contact with the drum.

In disc brake systems, the cylinders are part of the calipers. One, two or four cylinders are used to force the brake pads against the disc, but all cylinders contain one piston only.

All brake cylinder (master cylinder, wheel cylinder or caliper) pistons employ some type of seal, usually made of rubber, to minimize the leak age of fluid around the piston. A rubber dust boot seals the outer end of the cylinder against dust and dirt. The boot fits around the outer end of either the piston or the brake actuating rod.

When at rest the entire hydraulic system, from the piston(s) in the master cylinder to those in the wheel cylinders or calipers, is full of brake fluid. Upon application of the brake pedal, fluid trapped in front of the master cylinder piston(s) is forced through the lines to the caliper or wheel cylinders. Here it applies pressure on the pistons, which forces the brake linings against the rotors or drums.

Upon release of the brake pedal, a spring located inside the master cylinder immediately returns the master cylinder pistons to the normal position. The pistons contain check valves and the master cylinder has compensating ports drilled in it. These are uncovered as the pistons reach their normal position. The piston check valves allow fluid to flow toward the wheel cylinders or calipers as the pistons withdraw. Then, as the return springs force the brake pads or shoes into the released position, the excess fluid reservoir through the compensating ports. It is during the time the pedal is in the released position that any fluid that has leaked out of the system will be replaced through the compensating ports.

Dual circuit master cylinders employ two pistons, located one behind the other, in the same cylinder. The primary piston is actuated directly by mechanical linkage from the brake pedal through the power booster. The secondary piston is actuated by fluid trapped between the two pistons. If a leak develops in front of the secondary piston, it moves forward until it bottoms against the front of the master cylinder, and the fluid trapped between the pistons will operate the rear brakes. If the rear brakes develop a leak, the primary piston will move forward until direct contact with the

secondary piston takes place, and it will force the secondary piston to actuate the front brakes. In either case, the brake pedal moves farther when the brakes are applied, and less braking power is available.

All dual-circuit systems incorporate a switch which senses either line pressure or fluid level. This system will warn the driver when only half of the brake system is operational.

Disc brake systems also contain a metering valve and a proportioning valve. The metering valve keeps pressure from traveling to the disc brakes on the front wheels until the brake shoes on the rear wheels have contacted the drum, ensuring that the front brakes will never be used alone. The proportioning valve controls the pressure to the rear brakes avoiding rear wheel lock-up during braking.

DISC BRAKES

Instead of the traditional expanding brakes that press out ward against a circular drum, disc brake systems employ a cast iron disc with brake pads positioned on either side of it. Braking is achieved in a way similar to the braking of a bicycle using hand brakes. On the bicycle, the brake pads squeeze onto the rim of the wheel, slowing its motion. Automobile disc brakes use the same principal but apply the braking effort to a separate disc, normally called the rotor.

The disc or rotor is a one-piece casting mounted just inside the wheel. Some rotors are one solid piece while others have cooling fins between the two braking surfaces. These vented rotors enable air to circulate between the braking surfaces cooling them quicker and making them less sensitive to heat buildup, warpage and fade. Disc brakes are only slightly affected by dirt and water, contaminants are thrown off of the braking surface by centrifugal force. Secondly, since disc pads are constantly in contact with the rotors they tend to clean dirt and contaminants from the braking surface during vehicle movement. The equal clamping action of the two brake pads, present in a properly operation system, tend to deliver uniform, straight-line stops. Unequal application of the pads between the left and right wheels can cause a vicious pull during braking. All disc brakes are inherently self-adjusting.

There are three general types of disc brake:

1. A fixed caliper.
2. A floating caliper.
3. A sliding caliper.

The sliding and floating designs are quite similar. In fact, these two types are often lumped together. In both designs, the pad on the inside of the rotor is moved into contact with the rotor by hydraulic force. The caliper, which is not held in a fixed position, moves slightly on its mount, bringing the other pad into contact with the rotor. There are various methods of attaching floating calipers. Some pivot at the bottom or top, and some slide on mounting bolts. Many uneven brake wear problems can be caused by dirty or seized slides and pivots.

DRUM BRAKES

Drum brakes employ two brake shoes mounted on a stationary backing plate. These shoes are positioned inside a circular cast iron drum which rotates with the wheel. The shoes are held in place by springs; this allows them to slide toward the drum (when they are applied) while keeping the linings in alignment.

The shoes are actuated by a wheel cylinder which is mounted at the top of the backing plate. When the brakes are applied, hydraulic pressure forces the wheel cylinder's two actuating links outward. Since these links bear directly against the top of the brake shoe webbing, the tops of the shoes are then forced outward against the inside of the drum. This action forces the bottoms of the two shoes to contact the brake drum by rotating the entire assembly slightly. When the pressure within the wheel cylinder is relaxed, return springs pull the shoes away from the drum.

Most drum brakes are designed to self-adjust during application of the brakes while the vehicle is moving in reverse. This causes both shoes to rotate very slightly with the drum, moving an adjusting lever or cable and thereby causing rotation of the adjusting screw via a star wheel. This automatic adjustment system reduces the need for maintenance adjustments but in most cases, periodic adjustment is still required.

GENERAL BRAKE SYSTEM TROUBLESHOOTING CHART

Symptom	Probable cause	Remedy
Scraping or grinding noise when brakes are applied	Worn brake pad	Replace
	Caliper to wheel interference	Correct or replace
	Cracked brake disc	
Squealing, groaning or chattering noise when brakes are applied	Disc brakes—missing or damaged brake pad anti-squeak shim	Replace
	Brake discs and pads worn or scored	Correct or replace
	Improper lining parts	
	Disc brake—burred or rusted calipers	Clean or deburr
	Dirty, greased, contaminated or glazed pad	Clean or replace
	Incorrect adjustment of brake pedal or booster push-rod	Adjust
Squealing noise when brakes are not applied	Disc brakes—rusted, stuck	Lubricate or replace
	Loose or extra parts in brakes	Retighten
	Improper positioning of pads in caliper	Correct
	Improper installation of support mounting to caliper body	
	Poor return of brake booster or master cylinder	Replace
	Incorrect adjustment of brake pedal or booster push-rod	Adjust

GENERAL BRAKE SYSTEM TROUBLESHOOTING CHART

Groaning, clicking or rattling noise when brakes are not applied	Stones or foreign material trapped inside wheel covers	Remove stones, etc.
	Disc brakes—failure of shim	Replace
	Disc brakes—loose installation bolt	Retighten
	Loose wheel nuts	
	Incorrect adjustment of brake pedal or booster push-rod	Adjust

FRONT DISC BRAKE ROTOR INSPECTION

Inspection items	Remarks
Scratches, rust, saturated lining materials and wear	• If the vehicle is not driven for a certain period, the sections of the discs that are not in contact with lining will become rusty, causing noise and shuddering. • If grooves resulting from excessive disc wear and scratches are not removed prior to installing a new pad assembly, there will momentarily be inappropriate contact between the disc and the lining (pad).
Run-out or drift	Excessive run-out or drift of the discs will increase the pedal depression resistance due to piston knock-back.
Change in thickness (parallelism)	If the thickness of the disc changes, this will cause pedal pulsation, shuddering and surging.
Inset or warping (flatness)	Overheating and improper handling while servicing will cause inset or warping.

Adjustment

DISC BRAKES

Disc brakes are inherently self adjusting. Periodic adjustment of disc brakes is not required. After new disc brake pads have been installed, the brake pedal must be pumped to seat the pads against the rotors, prior to moving the vehicle. If the pads are not seated before putting the vehicle in gear or driving the vehicle, the brake pedal will have to be applied a number of times before any braking action will be achieved.

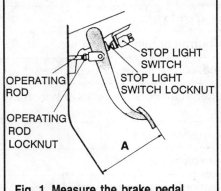

Fig. 1 Measure the brake pedal height at A and compare to specifications

BRAKE PEDAL HEIGHT

▶ See Figures 1, 2 and 3

Measure the brake pedal height from the floor of the vehicle to the upper surface of the brake pedal. The distance should be 6.9-7.1 in (176-181mm). If the brake pedal height is incorrect, adjust as follows:

1. Disconnect the stop lamp switch connector.
2. Loosen the locknut on the base of the stop light switch and move the switch to a position where it does not contact the brake pedal.
3. Loosen the operating rod locknut. Adjust the height of the brake pedal by turning the operating rod using pliers.

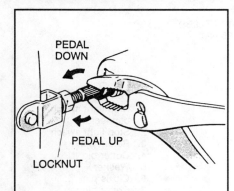

Fig. 2 Adjust the brake pedal height by increasing or decreasing the length of the operating rod

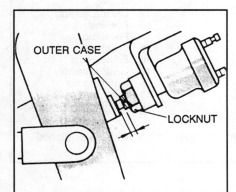

Fig. 3 Inspect the clearance between the stop light switch and the brake pedal stop and compare to specifications

Once the desired pedal height is obtained, tighten the locknut on the operating rod.

4. Screw the stop light switch until the it contacts the brake pedal stopper. Turn switch in until the brake pedal just starts to move. At this point, return (loosen) the stoplight switch ½-1 turn and secure in this position by tightening the locknut. In this position, the distance between the lower stop light switch case and the brake pedal stop should be 0.02-0.04 in. (0.5-1.0mm).

5. Connect the electrical connector to the stop light switch.

6. Check to be sure that the stop lights are not illuminated with no pressure on the brake pedal.

7. Without starting the vehicle, depress the brake pedal. If the brake light switch is properly connected, the brake lights will illuminate.

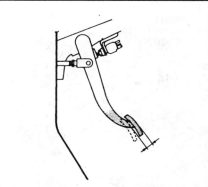

Fig. 4 Inspect brake pedal free-play

BRAKE PEDAL FREE-PLAY

▶ See Figure 4

1. With the engine off, depress the brake pedal fully several times to evacuate the vacuum in the booster.

2. Once all the vacuum assist has been eliminated, press the brake pedal down by hand and confirm that the amount of movement before resistance is felt is within 0.1-0.3 in. (3-8mm).

3. If the free-play is less than desired, confirm that the brake light switch is in proper adjustment.

4. If there is excessive free-play, look for wear or play in the clevis pin and brake pedal arm. Replace worn parts as required and recheck brake pedal free-play.

Brake Light Switch

REMOVAL & INSTALLATION

1. Disconnect the negative battery cable.

2. Disconnect the stop lamp switch electrical harness connector.

3. Loosen the locknut holding the switch to the bracket. Remove the locknut and the switch.

To install:

4. Install the new switch and install the locknut, tightening it just snug.

5. Reposition the brake light switch so that the distance between the outer case of the switch and the pedal is 0.02-0.04 in. (0.5-1.0mm). Note that the switch plunger must press against the pedal to keep the brake lights off. As the pedal moves away from the switch, the plunger extends and closes the switch, which turns on the stop lights.

6. Hold the switch in the correct position and tighten the locknut.

7. Connect the wiring to the switch.

8. Check the operation of the switch. Turn the ignition key to the ON position but do not start the engine. Have an assistant observe the brake lights at the rear of the vehicle while you push on the brake pedal. The lights should come on just as the brake pedal passes the point of free play.

9. Adjust the brake light switch as necessary. The small amount of free play in the pedal should not trigger the brake lights; if the switch is set incorrectly, the brake lights will flicker due to pedal vibration on road bumps.

Brake Pedal

REMOVAL & INSTALLATION

▶ See Figures 5, 6, 7 and 8

1. Disconnect the negative battery cable.

2. Remove the knee protector, on the bottom portion of the instrument panel.

3. Remove the lower lap cooler and shower ducting.

4. If equipped with manual transaxle, remove the steering column from the vehicle.

5. Disconnect the electrical connector from the stop light switch. Remove the stop lamp switch. Disconnect the harness at the clutch switch and remove switch, if equipped.

6. Disconnect the brake pedal and clutch pedal (if equipped) return springs. Remove the cotter pin and washer from the clevis pin and remove pin from the brake pedal. This will disconnect the operating rod from the brake pedal.

7. Disconnect the shift-lock cable connection from the lever assembly. Remove the lever assembly mounting nut from the end of the pedal rod.

8. Carefully remove the washers, bushings and lever from the end of the pedal rod. Keep all components in order of removal to aid in installation.

9. Using plastic hammer, tap on the end of the brake pedal rod and remove from the support bracket. Note positioning of all bushings.

10. Remove the pedal(s) from the vehicle.

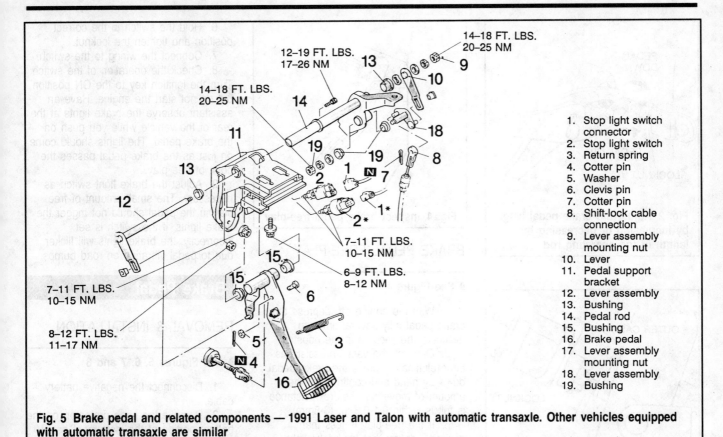

14–18 FT. LBS.
20–25 NM

12–19 FT. LBS.
17–26 NM

14–18 FT. LBS.
20–25 NM

7–11 FT. LBS.
10–15 NM

7–11 FT. LBS.
10–15 NM

6–9 FT. LBS.
8–12 NM

8–12 FT. LBS.
11–17 NM

1. Stop light switch connector
2. Stop light switch
3. Return spring
4. Cotter pin
5. Washer
6. Clevis pin
7. Cotter pin
8. Shift-lock cable connection
9. Lever assembly mounting nut
10. Lever
11. Pedal support bracket
12. Lever assembly
13. Bushing
14. Pedal rod
15. Bushing
16. Brake pedal
17. Lever assembly mounting nut
18. Lever assembly
19. Bushing

Fig. 5 Brake pedal and related components — 1991 Laser and Talon with automatic transaxle. Other vehicles equipped with automatic transaxle are similar

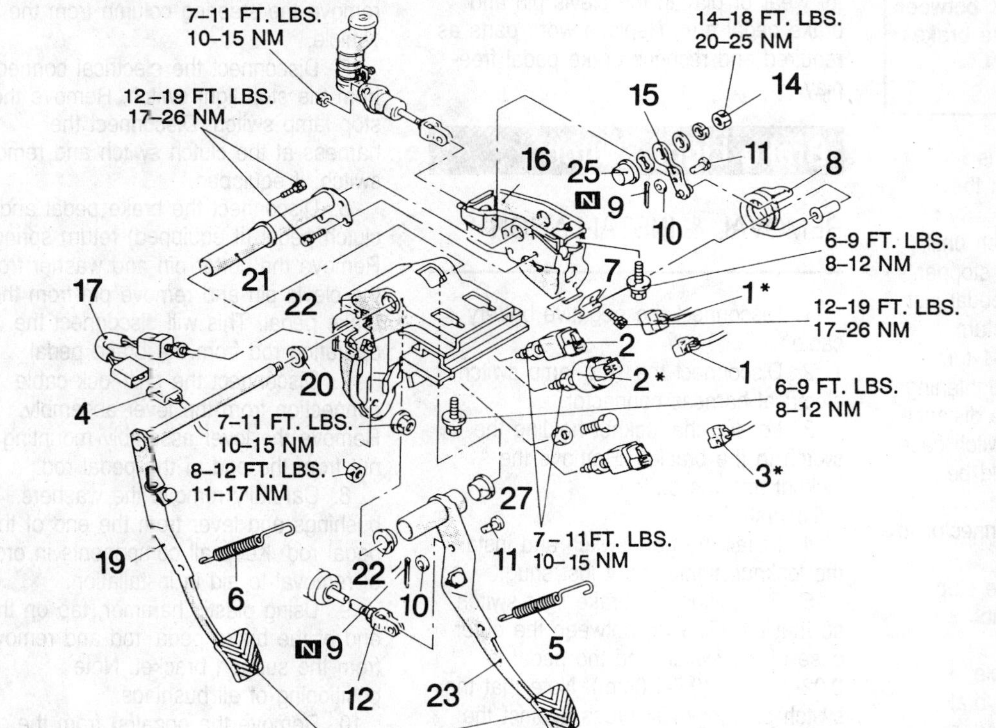

7–11 FT. LBS.
10–15 NM

12–19 FT. LBS.
17–26 NM

14–18 FT. LBS.
20–25 NM

6–9 FT. LBS.
8–12 NM

12–19 FT. LBS.
17–26 NM

6–9 FT. LBS.
8–12 NM

7–11 FT. LBS.
10–15 NM

8–12 FT. LBS.
11–17 NM

7–11 FT. LBS.
10–15 NM

1. Stop light switch connector
2. Stop light switch
3. Clutch switch connector and clutch switch
4. Inter lock switch connector
5. Return spring
6. Return spring (non-turbo)
7. Clip (turbo)
8. Turnover spring (turbo)
9. Cotter pin
10. Washer
11. Clevis pin
12. Brake booster push rod
13. Pedal and bracket assembly
14. Clutch pedal mounting nut
15. Lever assembly
16. Clutch pedal mounting bracket
17. Inter lock switch
18. Pedal support bracket
19. clutch pedal
20. Bushing
21. Pedal rod
22. Bushing
23. Brake pedal

Fig. 6 Brake pedal and related components — 1991 Laser and Talon with manual transaxle. Other vehicles equipped with manual transaxle are similar

Fig. 7 Insert the clevis pin into the brake pedal. Install washer and bend new cotter pin tightly.

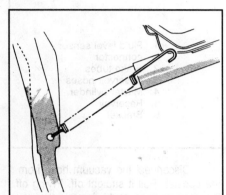

Fig. 8 Install the return spring with the short end towards the brake pedal

To install:

11. Install the pedal(s) in position with new bushing as required. Lubricate bushing with grease prior to installation. Guide the pedal rod through the bushings and the brake pedal until the completely installed. Install the bushings and lever to the ends of the pedal rod and install mounting nut. Tighten mounting nut to 14 ft. lbs. (20 Nm).

12. Install the shift-lock cable connection to the lever assembly. Install clevis pin and washers through operating rod and brake pedal. Install new cotter pin.

13. Install the brake pedal and clutch pedal return spring(s) so that the long end of the spring is towards the instrument panel. Install the stop light switch and reconnect the electrical connector.

14. Install the steering column assembly.

15. Install the removed ducting and the knee protector.

16. Connect the negative battery cable.

17. Adjust the brake pedal and the brake light switch.

Master Cylinder

REMOVAL & INSTALLATION

♦ **See Figure 9**

1. Disconnect the negative battery cable.

2. Disconnect the fluid level sensor connector, if equipped.

3. Disconnect the brake lines from the master cylinder. A separate fluid reservoir is used. Plug the lines to prevent drainage.

4. Remove the 2 nuts securing the master cylinder to the brake booster and remove the master cylinder.

To install:

5. Install master cylinder to the mounting studs and install the mounting nuts. Tighten mounting nuts to 9 ft. lbs. (12 Nm).

6. Fill the reservoir to the proper level with clean DOT 3 brake fluid. Bleed the master cylinder.

7. Install the brake lines to the master cylinder.

8. Apply the brake pedal and check for firmness. If the pedal is spongy,. air is present in the system. If air remains in the system, bleeding the entire system is required.

9. Check the brakes for proper operation and leaks.

OVERHAUL

♦ **See Figures 10, 11, 12 and 13**

1. Drain the brake fluid from the master cylinder reservoir. Remove the master cylinder from the vehicle.

2. Remove the piston stopper bolt and washer from the bottom of the master cylinder body while depressing the piston slightly. On vehicles equipped with ABS, the piston stopper bolt is located on the side of the cylinder body.

3. While lightly depressing the piston, remove the piston stopper ring from the groove in the master cylinder body.

4. Allow the piston to return to position. Extract the piston assembly

from the cylinder in the master cylinder body.

➡**If it is hard to remove the secondary piston from the cylinder, gradually apply compressed air into the outlet port on the secondary end of the master cylinder.**

5. Inspect the inner surface of the master cylinder body for rust or pitting. Check the primary and secondary pistons for rust, scoring, wear or damage. Replace any damaged or worn components.

➡**The primary and the secondary piston assemblies are not to be disassembled. If a component on the assembly is damaged or worn, the entire piston assembly is to be replaced.**

6. Remove the retainers and the reservoir nipples from the master cylinder body. Remove the reservoir seals and inspect for wear or damage. Replace the reservoir seals as required.

To install:

7. Lubricate the piston assemblies and the master cylinder bore with clean brake fluid. Install the secondary and primary piston assemblies into the master cylinder body.

8. While pushing inward on the piston assembly, install the piston stopper ring. Make sure the ring is fully seated.

9. Install the piston stopper bolt into the master cylinder bore with a new gasket in place. Tighten to 1-2 ft. lbs. (1.5-3 Nm).

10. Install the reservoir seals into the cylinder body bores. Install the primary and secondary nipple over the reservoir seals and install the retainers, tightened to 2 ft. lbs. (3 Nm) torque.

➡**When installing the primary and the secondary nipples to the master cylinder body, make sure not to confuse one for the other. The primary nipple is the most rearward nipple, and will extend at 90 degrees from the master cylinder body, once installed. The secondary nipple is located in the front of the master cylinder body. It extends at a 40 degree angle to the cylinder body once it is installed.**

11. Install the brake master cylinder onto the vehicle. Attach the brake fluid lines from the reservoir to the nipples on

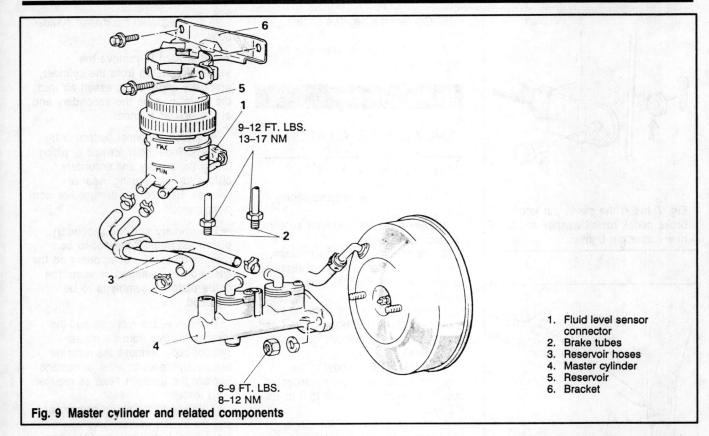

1. Fluid level sensor connector
2. Brake tubes
3. Reservoir hoses
4. Master cylinder
5. Reservoir
6. Bracket

9–12 FT. LBS.
13–17 NM

6–9 FT. LBS.
8–12 NM

Fig. 9 Master cylinder and related components

the master cylinder. Install the hose retainers.

12. Fill the brake fluid reservoir with clean brake fluid. Bleed the master cylinder. Install the brake tubes to the cylinder and tighten to 9 ft. lbs. (13 Nm) torque.

13. Inspect the brake pedal for proper adjustment.

14. Test the brake system for proper operation.

Power Brake Booster

TESTING BOOSTER OPERATION

1. Start the engine and run at idle for 1-2 minutes. Shut the engine OFF.

2. Step on the brake pedal several times using normal pressure.

3. If the brake pedal depressed fully the first time but gradually becomes higher when depressed succeeding times, the booster is operating properly.

4. If the pedal height remains unchanged, the booster is faulty.

5. With the engine stopped, step on the brake pedal several times with the

same pressure and make sure the pedal height does not change.

6. Step on the brake pedal and start the engine. If the pedal moves downward slightly, the booster is in good condition. If there is no change in the position of the pedal, the booster is faulty.

7. With the engine running, step on the brake pedal and then stop the engine. Hold the pedal depressed for 30 seconds. If the pedal height does not change, the booster is in good condition. If the pedal rises, the booster is faulty.

8. If the above test results are okay, the booster performance can be determined as good. If one of the above 3 tests is not okay, the check valve, vacuum hose or booster is faulty.

REMOVAL & INSTALLATION

▶ See Figure 14

1. Disconnect the negative battery cable. Siphon the brake fluid from the master cylinder reservoir.

2. Remove and relocate the air conditioning relay box and the solenoid valve located at the power brake unit.

3. Disconnect the vacuum hose from the booster. Pull it straight off. Prying off the vacuum hose could damage the check valve installed in the brake booster vacuum hose.

4. Disconnect the electrical harness connector at the brake level sensor.

5. Remove the nuts attaching the master cylinder to the booster and remove the master cylinder and position aside. If necessary, disconnect and plug the brake fluid lines at the master cylinder.

6. From inside the passenger compartment, remove the cotter pin and clevis pin that secures the booster pushrod to the brake pedal.

7. From inside the vehicle, remove the nuts that attach the booster to the dash panel. Remove the brake booster from the engine compartment.

To install:

8. Install the brake booster to the dash panel. From inside the vehicle, install the attaching nuts and tighten to 12 ft. lbs. (17 Nm).

9. Apply grease to the clevis pin and install with washers in place. Install new cotter pin and bend to secure in place.

10. Install the vacuum hose to the booster fitting.

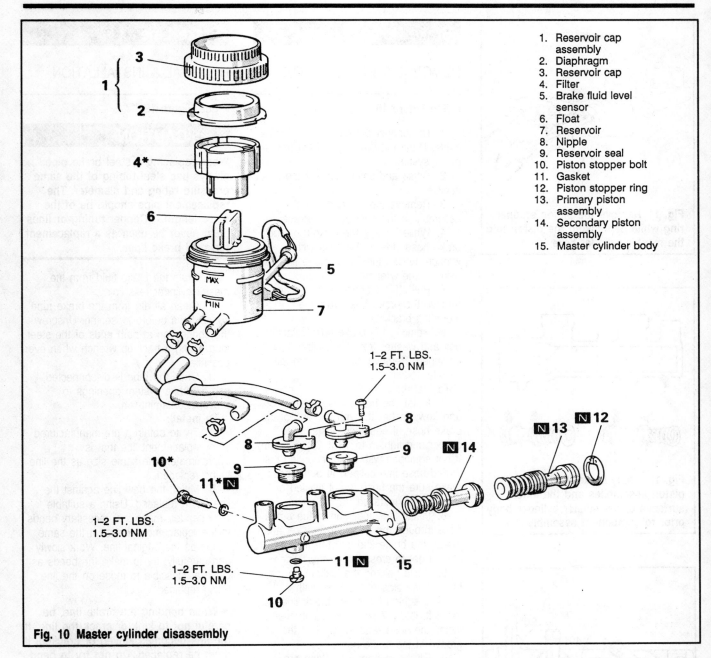

1. Reservoir cap assembly
2. Diaphragm
3. Reservoir cap
4. Filter
5. Brake fluid level sensor
6. Float
7. Reservoir
8. Nipple
9. Reservoir seal
10. Piston stopper bolt
11. Gasket
12. Piston stopper ring
13. Primary piston assembly
14. Secondary piston assembly
15. Master cylinder body

Fig. 10 Master cylinder disassembly

11. Install the master cylinder assembly to the mounting studs on the brake booster. Install the master cylinder mounting nuts and tighten to 9 ft. lbs. (12 Nm).

12. Reconnect the brake fluid reservoir to the master cylinder, if disconnected. Reconnect the electrical connector to the brake fluid level sensor.

13. Install the solenoid valve assembly and the relay box, if removed. Connect the negative battery cable.

14. Add fluid to the brake fluid reservoir as required. Bleed the master cylinder. If after bleeding the master cylinder the brake pedal feels soft, bleed the brake system at all wheels.

15. Check the brake system for proper operation.

Proportioning Valve

REMOVAL & INSTALLATION

1. Disconnect the negative battery cable. Drain the brake fluid from the brake system.

2. Label and disconnect the brake lines at the proportioning valve.

3. Remove the proportioning valve mounting bolt and the valve from the engine compartment.

→Do not disassemble the proportioning valve because its performance depends on the set load of the spring inside the valve. If defective, replace the proportioning valve.

4. Installation is the reverse of the removal procedure. Bleed the brake system once the valve is installed.

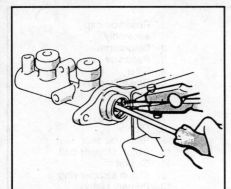

Fig. 11 Remove the piston stopper ring while depressing the piston into the master cylinder bore

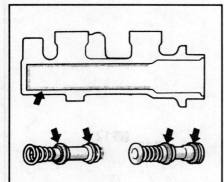

Fig. 12 Apply brake fluid to the piston assemblies and the inner surfaces of the master cylinder body prior to component assembly

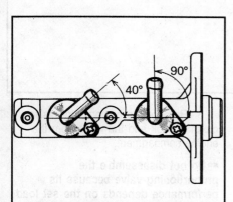

Fig. 13 Correct positioning of the primary and the secondary side nipples

Brake Hose

REMOVAL & INSTALLATION

♦ **See Figure 15**

1. Disconnect the negative battery cable. Drain the brake fluid from the brake system.
2. Raise and safely support the vehicle.
3. Remove the tire and wheel assembly of the hose to be replaced.
4. While holding the locknut on the brake hose side of the hose with a wrench, loosen the flared brake line nut using a line wrench.
5. Pull off the brake hose clip and remove the brake hose from the mounting bracket.
6. Remove the brake hose retainer bolt and washer from the caliper, if equipped. Remove the hose from the vehicle.

To install:

7. Install the brake hose retainer bolt and new washer through the end of the brake hose, if equipped. Install onto caliper and tighten to 18 ft. lbs. (25 Nm). If not equipped with hose mounting bolt, thread hose into caliper. In both cases, make sure the hose is not twisted or kinked once installed.
8. Install the other end of the brake hose through the mounting bracket and thread the brake line into the fitting. Be careful not to cross-thread the fittings.
9. While holding the locknut on the brake hose side of the hose with a wrench, tighten the flared brake line nut to 12 ft. lbs. (17 Nm). Once tightened, install the hose retainer clip into the groove in the brake hose.
10. Fill the system with clean brake fluid. Bleed the brake system.
11. Install the tire and wheel assembly. Check the brake system for proper operation.

Brake Line

REMOVAL & INSTALLATION

♦ **See Figure 16**

❊❊CAUTION

When replacing a steel brake pipe, always use steel tubing of the same pressure rating and diameter. The replacement pipe should be of the same length. Copper tubing or lines must never be used as a replacement for steel brake lines.

1. Drain the brake fluid from the master cylinder reservoir.
2. Clean all dirt from the brake tube connections before loosening. Unscrew the connection at both ends of the steel pipe, using a back-up wrench when ever possible.
3. Once the tube is disconnected, cap all brake system openings to prevent contamination.

To install:

4. Try to obtain a pre-manufactured steel replacement line that is approximately the same size as the line being replaced.
5. Match the new line against the line being replaced. Using a suitable tube bender, make the necessary bends in the replacement line so it the same shape as the original line. Work slowly and carefully; try to make the bends as close as possible to those on the line being replaced.

➡**When bending the brake line, be careful not to kink or crack the line. If the line becomes kinked or cracked, it must be replaced. Do not try to bend the steel lines too sharply.**

6. Before installing the replacement brake line, flush it with clean brake fluid. This will remove any foreign material in the line, preventing brake system contamination.
7. Install the replacement brake line onto the vehicle and connect all fittings. Install new sealing washers where used. Make sure to attach the brake lines to the retainer clips, as provided.
8. Inspect the positioning of the replacement brake line. Make sure the brake line will not contact any components that could rub a hole in the line and cause a leak.

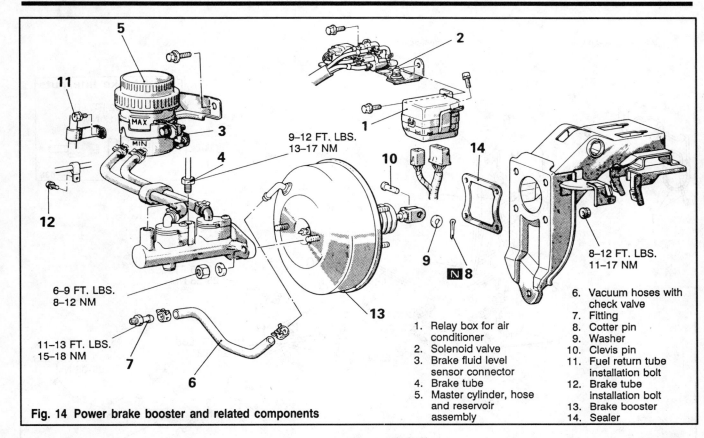

Fig. 14 Power brake booster and related components

9–12 FT. LBS.
13–17 NM

8–12 FT. LBS.
11–17 NM

6–9 FT. LBS.
8–12 NM

11–13 FT. LBS.
15–18 NM

1. Relay box for air conditioner
2. Solenoid valve
3. Brake fluid level sensor connector
4. Brake tube
5. Master cylinder, hose and reservoir assembly
6. Vacuum hoses with check valve
7. Fitting
8. Cotter pin
9. Washer
10. Clevis pin
11. Fuel return tube installation bolt
12. Brake tube installation bolt
13. Brake booster
14. Sealer

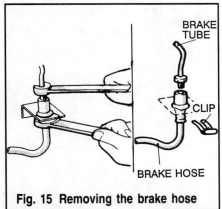

Fig. 15 Removing the brake hose

9. Bleed the brake system.
10. Inspect the system for leaks and proper operation.

BRAKE LINE FLARING

▶ **See Figures 17 and 18**

Use only brake line tubing approved for automotive use; never use copper tubing. Whenever possible, try to work with manufactured brake lines. These lines are available at most parts stores and have machine made flares, the quality of which is hard to duplicate with most of the available inexpensive flaring kits.

When the brakes are applied, there is a great amount of pressure developed in the hydraulic system. An improperly formed or cracked flare can leak with resultant loss of braking power. If you have never made a double-flare, take the time to familiarize yourself with the flaring kit components and there operation. Practice forming double-flares on scrap tubing until you are satisfied with the results. The flare should be uniform in thickness around the total circumference of the tubing. Carefully inspect the side walls of the tubing for cracks. Discard the tubing if deformed or cracked.

The following procedure applies to most commercially available double-flaring kits. If these instructions differ from those provided with your kit, follow the instructions in the kit.

1. Cut the brake line to the necessary length using a tube cutter.
2. Square the end of the tubing with a file and chamfer the edge.
3. Insert the tubing in the proper size hole in the bar until the end of the tube sticks out the thickness of a single flare adapter. Tighten the bar wing nuts tightly so the tube can not move.

4. Place the single flare adapter into the tube and slide the bar into the yoke.
5. Position the yoke screw over the single flare adapter and tighten it until the bar is locked in the yoke. Continue tightening the yoke screw until the adapter bottoms on the bar. This should form a single flare.

➡**Make sure the tube is not forced out of the hole in the bar during the single flare operation. If it is, the single flare will not be formed properly and the procedure must be repeated from Step 1.**

6. Loosen the yoke screw and remove the single flare adapter.
7. Position the yoke screw over the tube and tighten until the taper contacts the single flare and the bar is locked in the yoke. Continue tightening to form the double-flare.

➡**Make sure the tube is not forced out of the hole in the bar during the single flare operation. If it is, the single flare will not be formed properly and the procedure must be repeated from Step 1.**

8. Loosen the screw and remove the bar from the yoke. Remove the tubing from the bar.

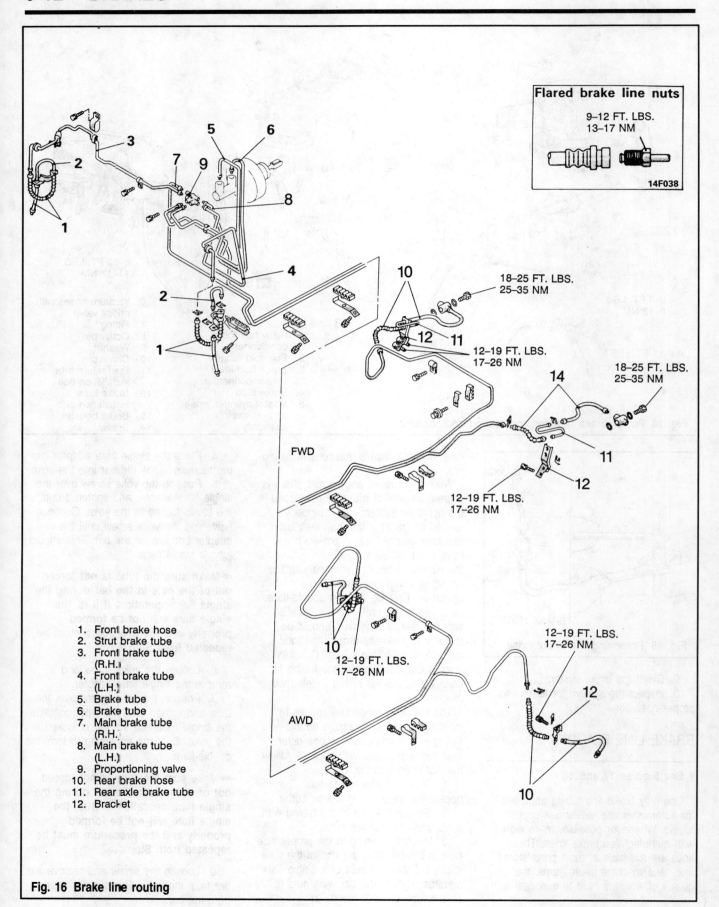

Flared brake line nuts

9–12 FT. LBS.
13–17 NM

14F038

18–25 FT. LBS.
25–35 NM

12–19 FT. LBS.
17–26 NM

18–25 FT. LBS.
25–35 NM

FWD

12–19 FT. LBS.
17–26 NM

12–19 FT. LBS.
17–26 NM

AWD

12–19 FT. LBS.
17–26 NM

1. Front brake hose
2. Strut brake tube
3. Front brake tube
   (R.H.)
4. Front brake tube
   (L.H.)
5. Brake tube
6. Brake tube
7. Main brake tube
   (R.H.)
8. Main brake tube
   (L.H.)
9. Proportioning valve
10. Rear brake hose
11. Rear axle brake tube
12. Bracket

Fig. 16 Brake line routing

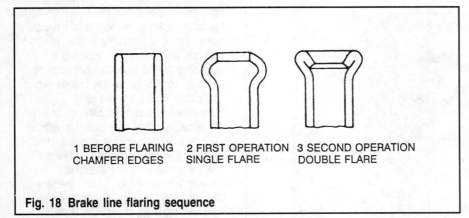

1 BEFORE FLARING CHAMFER EDGES   2 FIRST OPERATION SINGLE FLARE   3 SECOND OPERATION DOUBLE FLARE

Fig. 18 Brake line flaring sequence

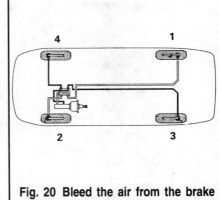

Fig. 20 Bleed the air from the brake system in the sequence shown

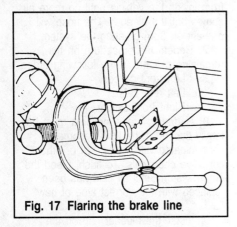

Fig. 17 Flaring the brake line

9. Check the flare for cracks or uneven flaring. If the flare is not perfect, cut it off and begin again at Step 1.

Brake System Bleeding

Bleeding the brake system is required anytime the normally closed system has been opened to the atmosphere. When bleeding the system, keep the brake fluid level in the master cylinder reservoir above ½full. If the reservoir is empty, air will be pushed through the system. If equipped with ABS, refer to the ABS portion of Section 5 for bleeding procedure.

PROCEDURE

➡️If using a pressure bleeder, follow the instructions furnished with the unit and choose the correct adaptor for the application. Do not substitute an adapter that 'almost fits" as it will not work and could be dangerous.

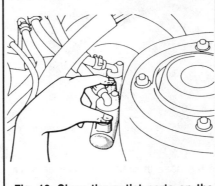

Fig. 19 Close the outlet ports on the master cylinder using fingers

Master Cylinder
➧ See Figure 19

Due to the location of the fluid reservoir, bench bleeding of the master cylinder is not recommended. The master cylinder is to be bled while mounted on the brake booster. If the fluid reservoir runs dry, bleeding of the entire system will be necessary. Two people will be required to bleed the brake system.

1. Fill the brake fluid reservoir with clean brake fluid. Disconnect the brake tube from the master cylinder.
2. Have a helper slowly depress the brake pedal. Once depressed, hold it in that position. Brake fluid will be expelled from the master cylinder.

✳️✳️CAUTION

When bleeding the brakes, keep face away from area. Spewing fluid may cause facial and/or visual damage. Do not allow brake fluid to spill on the car's finish; it will remove the paint.

3. While the pedal is held down, use a finger to close the outlet port of the master cylinder. While the port is closed, have the helper release the brake pedal.
4. Repeat this procedure until all air is bled from the master cylinder. Check the brake fluid in the reservoir every 4-5 times, making sure the reservoir does not run dry. Add clean DOT 3 brake fluid to the reservoir as needed. All air is bled from the master cylinder when the fluid expelled from the port is free of bubbles.
5. Connect the brake tube to the port on the master cylinder. Add clean fluid to fill the reservoir to the appropriate level.

Calipers
➧ See Figures 20 and 21

1. Fill the master cylinder with fresh brake fluid. Check the level often during this procedure. Raise and safely support the vehicle.
2. Starting with the wheel farthest from the master cylinder, remove the protective cap from the bleeder and place where it will not be lost. Clean the bleeder screw.
3. Start the engine and run at idle.

✳️✳️CAUTION

When bleeding the brakes, keep face away from the brake area. Spewing fluid may cause physical and/or visual damage. Do not allow brake fluid to spill on the car's finish; it will remove the paint.

4. If the system is empty, the most efficient way to get fluid down to the wheel is to loosen the bleeder about ½-¾turn, place a finger firmly over the bleeder and have a helper pump the

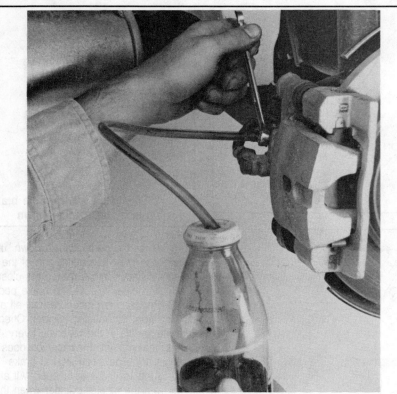

Fig. 21 Bleeding brake caliper. The end of the hose in brake fluid prevents air from entering the system

brakes slowly until fluid comes out the bleeder. Once fluid is at the bleeder, close it before the pedal is released inside the vehicle.

➡ **If the pedal is pumped rapidly, the fluid will churn and create small air bubbles, which are almost impossible to remove from the system. These air bubbles will accumulate and a spongy pedal will result.**

5. Once fluid has been pumped to the caliper, open the bleed screw again,

have the helper press the brake pedal to the floor, lock the bleeder and have the helper slowly release the pedal. Wait 15 seconds and repeat the procedure (including the 15 second wait) until no more air comes out of the bleeder upon application of the brake pedal. Remember to close the bleeder before the pedal is released inside the vehicle each time the bleeder is opened. If not, air will be introduced into the system.

6. If a helper is not available, connect a small hose to the bleeder, place the end in a container of brake fluid and proceed to pump the pedal from inside the vehicle until no more air comes out the bleeder. The hose will prevent air from entering the system.

7. Repeat the procedure on the remaining calipers in the following order:
   a. Left front caliper
   b. Left rear caliper
   c. Right front caliper

8. Hydraulic brake systems must be totally flushed if the fluid becomes contaminated with water, dirt or other corrosive chemicals. To flush, bleed the entire system until all fluid has been replaced with the correct type of new fluid.

9. Install the bleeder cap on the bleeder to keep dirt out. Always road test the vehicle after brake work of any kind is done.

FRONT DISC BRAKES

✳✳CAUTION

Brake pads and shoes contain asbestos, which has been determined to be a cancer causing agent. Never clean the brake surfaces with compressed air! Avoid inhaling any dust from brake surfaces! When cleaning brakes, use commercially available brake cleaning fluids.

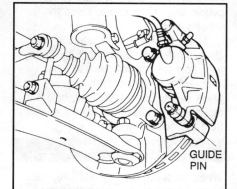

Fig. 22 Front caliper guide pin (bottom) and lock pin (top)

Brake Pads

REMOVAL & INSTALLATION

◆ **See Figures 22, 23, 24, 25, 26, 27, 28, 29, 30 and 31**

1. Disconnect the battery negative cable.

2. Remove some of the brake fluid from the master cylinder reservoir. The reservoir should be no more than ½full. When the pistons are depressed into the calipers, excess fluid will flow up into the reservoir.

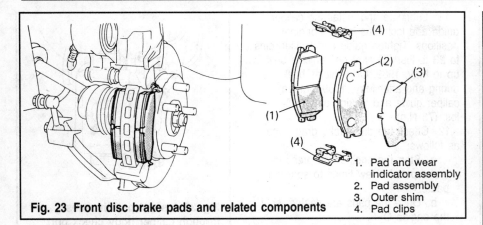

Fig. 23 Front disc brake pads and related components

1. Pad and wear indicator assembly
2. Pad assembly
3. Outer shim
4. Pad clips

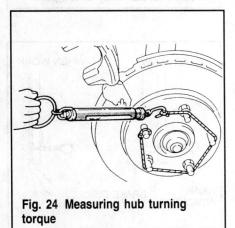

Fig. 24 Measuring hub turning torque

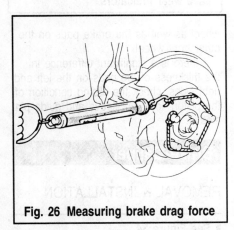

Fig. 26 Measuring brake drag force

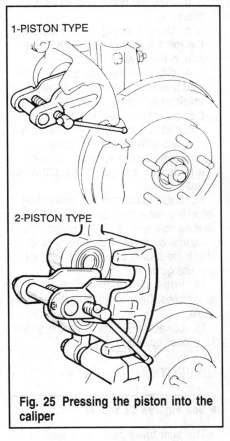

1-PISTON TYPE

2-PISTON TYPE

Fig. 25 Pressing the piston into the caliper

Fig. 27 Front caliper assembly

Fig. 28 Remove the caliper assembly and tie out of the way using wire

3. Raise the vehicle and support safely.

4. Remove the appropriate tire and wheel assemblies.

5. Remove the caliper guide and lock pins and lift the caliper assembly from the caliper support. Tie the caliper out of the way using wire. Do not allow the caliper to hang by the brake line.

6. Remove the brake pads, spring clip and shims. Take note of positioning to aid installation.

7. Install the wheel lug nuts onto the studs and tighten. This is done to hold the disc on the hub.

8. Using a spring scale, turn the disc in a forward direction and measure the rotation sliding resistance of the hub.

9. Clean the piston and using the appropriate tool, compress it into the caliper bore.

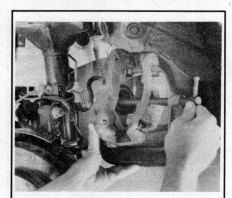

Fig. 29 Press the piston into the caliper bore

Fig. 30 Install the caliper assembly

Fig. 31 Make sure the brake disc and pads are free of dirt and grease

To install:

10. Install the brake pads, shims and spring clip to the caliper support. Install the caliper over the brake pads.

➡Be careful that the piston boot does not become caught when lowering the caliper onto the support. — Do not twist the brake hose during caliper installation.

11. Lubricate and install the caliper guide and lock pins in their original positions. Tighten guide and locking pins to 23 ft. lbs. (32 Nm) on vehicles built up to May, 1989. On vehicles built during and after May, 1989, tighten caliper guide and locking pins to 54 ft. lbs. (75 Nm).

12. Check the disc brake drag force as follows:

   a. Start the engine and press the brake pedal a few times to seat the pads.

   b. Once the pads are seated, shut the engine OFF.

   c. Turn the brake disc forward 10 times.

   d. Using a spring scale, measure the rotation sliding resistance of the hub in the forward direction.

   e. Calculate the drag torque of the disc brake by subtracting the value obtained in Step 7 (force required to turn hub alone) from the value obtained in Step D (force required to turn hub with caliper and pads installed). Compare calculated force with desired force of 15 lbs. (70 N) or less.

13. If the calculated disc brake drag force is greater than specifications, disassemble and clean the piston. Check for corrosion or worn piston seal and check the sliding condition of the lock pin and guide pin.

14. Install the tire and wheel assemblies. Connect the negative battery cable.

15. Lower the vehicle. Test the brakes for proper operation.

INSPECTION

▶ **See Figures 32 and 33**

The front brake pads have wear indicators that contact the brake disc when the brake pad thickness becomes 0.08 in. (2.0mm) and emit a squealing sound to worn the driver.

Inspect the thickness of the brake linings by looking through the brake caliper body check port. The thickness limit of the lining is 0.08 in. (2.0mm).

When the limit is exceeded, replace the pads on both sides of the brake disc and also the brake pads on the wheel on the opposite side of the vehicle. Do not replace 1 pad on a caliper because the wear indicator is hitting, without replacing the other pad on the same

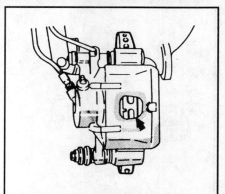

Fig. 32 Inspect pad thickness through caliper body check port

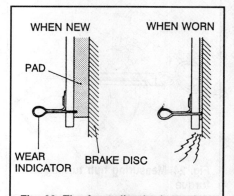

Fig. 33 The front disc brake pads have wear indicators

wheel as well as the brake pads on the other front wheel.

If there is a significant difference in the thickness of the pads on the left and right sides, check the sliding condition of the piston, lock pin sleeve and guide pin sleeve.

Brake Caliper

REMOVAL & INSTALLATION

▶ **See Figure 34**

1. Raise the vehicle and support safely.

2. Remove the appropriate tire and wheel assembly.

3. To disconnect the front brake hose, hold the nut on the brake hose side and loosen the flared brake line nut. Remove the brake hose from the caliper.

4. Remove the caliper guide and lock pins and lift the caliper assembly from the caliper support.

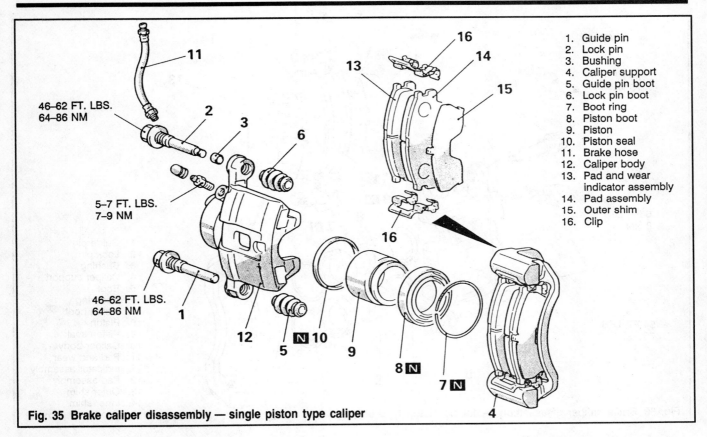

1. Guide pin
2. Lock pin
3. Bushing
4. Caliper support
5. Guide pin boot
6. Lock pin boot
7. Boot ring
8. Piston boot
9. Piston
10. Piston seal
11. Brake hose
12. Caliper body
13. Pad and wear indicator assembly
14. Pad assembly
15. Outer shim
16. Clip

46–62 FT. LBS.
64–86 NM

5–7 FT. LBS.
7–9 NM

46–62 FT. LBS.
64–86 NM

Fig. 35 Brake caliper disassembly — single piston type caliper

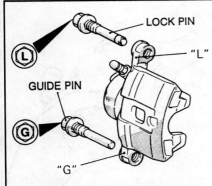

LOCK PIN

"L"

GUIDE PIN

"G"

Fig. 34 Install the guide pin and lock pin so the identification mark on the caliper body and head mark on the pins are aligned

5. Installation is the reverse of the removal procedure. Tighten guide and locking pins to 23 ft. lbs. (32 Nm) on vehicles built up to May, 1989. On vehicles built during and after May, 1989, tighten caliper guide and locking pins to 54 ft. lbs. (75 Nm).

6. Bleed the brake system.

OVERHAUL

▶ **See Figures 35, 36, 37, 38 and 39**

1. Remove the caliper from the vehicle.

2. Drain the remaining fluid from the caliper.

3. Remove the rubber lock pin and guide pin from the caliper.

4. Carefully remove the boot ring from around the piston using a flat tip tool.

5. Position the caliper on table top so the open side is facing down. Protect the caliper body with a cloth. Blow compressed air through the brake hose opening in the caliper. This will force the piston out of the caliper bore. Do not blow compressed air into the caliper with full force. Start out with a little air pressure and gradually increase the air pressure until the piston is forced from the bore.

✳✳CAUTION

Do not place fingers near the open area of the caliper. The piston can be blown out of the caliper bore with great force, causing personal injury.

6. Remove the piston from the caliper. Using finger tip, remove the O-ring from inside the caliper.

7. Check the inside of the caliper for rust, pitting, deterioration or cracking. If deep pitting or corrosion is present, do not recondition the caliper. Replacement will be required.

8. Using crocus cloth, lightly sand the sides of the caliper bore to remove rust or minor pitting.

9. Clean the components to be reused with an aerosol brake cleaner and dry them thoroughly using compressed air.

To assemble:

10. Obtain a caliper kit and new piston to be used during caliper assembly. Lubricate the inside of the caliper bore with clean DOT 3 brake fluid.

11. Install the piston seal into the bore groove. Lubricate the piston with clean brake fluid. Install new piston into the bore without twisting it.

12. Lubricate the edges of the piston with brake fluid. Install the piston boots and boot ring.

13. Install the guide pin boot and lock pin boot, if removed.

14. Position the caliper onto the caliper support. Install the guide pin and lock pin. On some calipers, identification

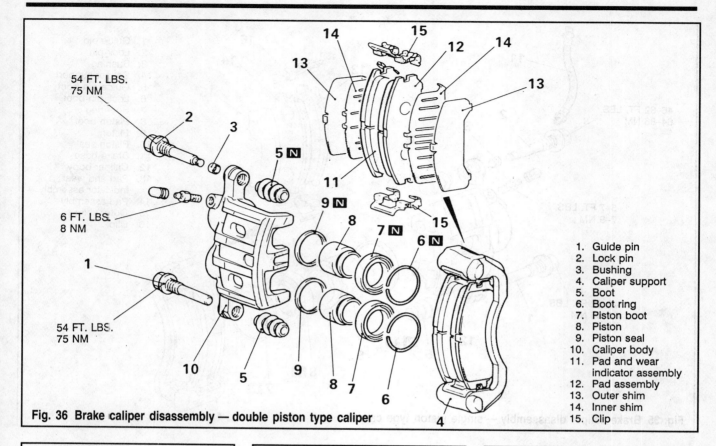

54 FT. LBS.
75 NM

6 FT. LBS.
8 NM

54 FT. LBS.
75 NM

1. Guide pin
2. Lock pin
3. Bushing
4. Caliper support
5. Boot
6. Boot ring
7. Piston boot
8. Piston
9. Piston seal
10. Caliper body
11. Pad and wear
    indicator assembly
12. Pad assembly
13. Outer shim
14. Inner shim
15. Clip

Fig. 36 Brake caliper disassembly — double piston type caliper

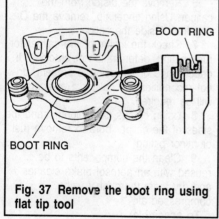

BOOT RING

BOOT RING

Fig. 37 Remove the boot ring using flat tip tool

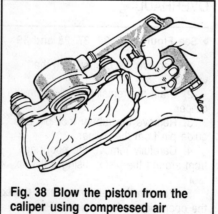

Fig. 38 Blow the piston from the caliper using compressed air

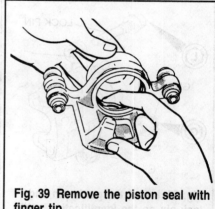

Fig. 39 Remove the piston seal with finger tip

markings are present to locate the locking pin and guide pin. Identifications markings are also present on each pin head. Install each in their original locations and tighten to 54 ft. lbs. (75 Nm).

15. Install the remaining components and bleed the brake system.

Brake Disc (Rotor)

REMOVAL & INSTALLATION

▶ See Figure 40

1. Raise the vehicle and support

safely. Remove appropriate wheel assembly.

2. Remove the caliper and brake pads. Support the caliper out of the way using wire.

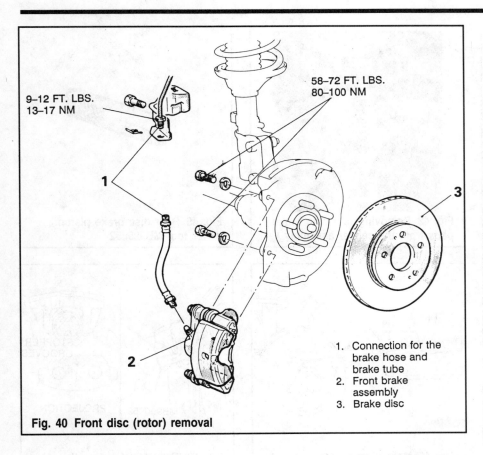

1. Connection for the brake hose and brake tube
2. Front brake assembly
3. Brake disc

9–12 FT. LBS.
13–17 NM

58–72 FT. LBS.
80–100 NM

Fig. 40 Front disc (rotor) removal

3. The rotor on most models is held to the hub by 2 small threaded screws. Remove screws, if equipped, and pull off the rotor.

4. Installation is the reverse of the removal process.

INSPECTION

Using a micrometer, measure the disc thickness at eight positions, approximately 45 degrees apart and 0.39 in. (10mm) in from the outer edge of the disc. The minimum thickness is 0.882 in. (22.4mm), with a maximum thickness variation of 0.0006 in. (0.015mm).

If the disc is beyond limits for thickness, remove it and install a new one. If the thickness variation exceeds the specifications, replace the disc or turn rotor with on the car type brake lathe.

REAR DISC BRAKES

✳✳CAUTION

Brake pads and shoes contain asbestos, which has been determined to be a cancer causing agent. Never clean the brake surfaces with compressed air! Avoid inhaling any dust from brake surfaces! When cleaning brakes, use commercially available brake cleaning fluids.

Brake Pads

REMOVAL & INSTALLATION

▶ See Figures 41, 42, 43, 44, 45, 46, 47 and 48

1. Disconnect the battery negative cable.

2. Remove some of the brake fluid from the master cylinder reservoir. The reservoir should be no more than ½ full. When the pistons are depressed into the calipers, excess fluid will flow up into the reservoir.

3. Raise the vehicle and support safely.

4. Remove the appropriate tire and wheel assemblies. Loosen the parking brake cable adjustment from inside the vehicle.

5. Disconnect the parking brake cable end installed to the rear brake caliper assembly.

6. Remove the caliper lock and guide pins and lift the caliper assembly from the caliper support. Tie the caliper out of the way using wire. Do not allow the caliper to hang by the brake line.

7. Remove the outer shim, brake pads and spring clips from the caliper support. Take note of positioning of each to aid in installation.

8. Install the wheel lug nuts onto the studs and tighten. This is done to hold the disc on the hub.

9. Using a spring scale, turn the disc in a forward direction and measure the rotation sliding resistance of the hub.

10. Clean the caliper piston. Using rear disc brake driver tool MB990652 or equivalent, thread the piston into the caliper bore. Be sure at this point, that the stopper groove of the piston correctly fits into the projection on the replacement brake pads rear surface.

To install:

11. Install the brake pads, shims and spring clip to the caliper support. Install the caliper over the brake pads.

➡**Be careful that the piston boot does not become caught when lowering the caliper onto the support. — Do not twist the brake hose during caliper installation.**

12. Lubricate and install the caliper guide and lock pins. Tighten the pins to 23 ft. lbs. (32 Nm). Attach the parking brake cable to the rear brake assembly.

13. Start the engine and forcefully depress the brake pedal 5-6 times. Apply the parking brake and make sure the adjustment is within specifications.

Fig. 41 Rear brake caliper assembly

Fig. 42 Removing rear caliper assembly

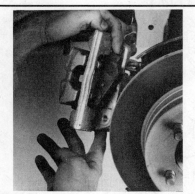

Fig. 43 Removing rear disc brake pads from the caliper assembly

Fig. 46 Rear disc brake piston driver tool MB990652

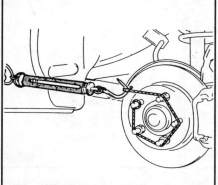

Fig. 45 Measuring hub torque with brake pads removed

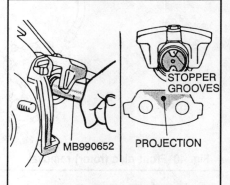

STOPPER GROOVES

MB990652 | PROJECTION

Fig. 47 Thread piston into the caliper bore

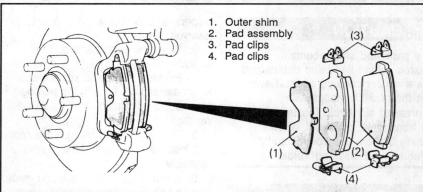

1. Outer shim
2. Pad assembly
3. Pad clips
4. Pad clips

Fig. 44 Removing the brake pads, shims and spring clips from the rear caliper support

Adjust the parking brake cable, as required.

14. Check the disc brake drag force as follows:

a. Start the engine and press the brake pedal a few times.

b. Shut the engine OFF.

c. Turn the brake disc forward 10 times.

d. Using a spring scale, measure the rotation sliding resistance of the hub in the forward direction.

e. Calculate the drag torque of the disc brake by subtracting the value obtained in Step 9 (force required to turn disc alone) from the value obtained in Step D (force required to turn disc with caliper and pads installed). Compare calculated force with desired force of 15 lbs. (70 N) or less.

15. If the calculated disc brake drag force is greater than specifications,

disassemble and clean the piston. Check for corrosion or worn piston seal and check the sliding condition of the lock pin and guide pin.

16. Install the tire and wheel assemblies. Lower the vehicle.

17. Test the brakes for proper operation.

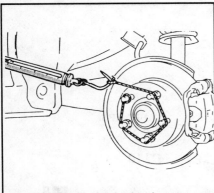

Fig. 48 Check the brake drag torque

Fig. 49 Inspecting rear disc brake pads

INSPECTION

▶ **See Figure 49**

Inspect the thickness of the brake linings by looking through the brake caliper body check port. The minimum allowable thickness of the lining is 0.08 in. (2.0mm).

When the limit is exceeded, replace the pads on both sides of the brake disc and also the brake pads on the wheel on the opposite side of the vehicle. Do not replace 1 pad on a caliper because the lining is below specifications, without replacing the other pad on the same wheel as well as the brake pads on the other rear wheel.

Brake Caliper

REMOVAL & INSTALLATION

1. Disconnect the battery negative cable.
2. Raise the vehicle and support safely.
3. Remove the appropriate tire and wheel assemblies. Loosen the parking brake cable adjustment from inside the vehicle.
4. Disconnect the parking brake cable end installed to the rear brake caliper assembly.
5. Remove the caliper lock and guide pins. Lift the caliper assembly from the caliper support.
6. Remove the rear brake hose from the caliper. Remove the caliper from the vehicle.

To install:
7. Install the rear brake hose onto the caliper with new washers in place. If equipped with brake hose retainer bolt, tighten bolt to 25 ft. lbs. (35 Nm) torque. If no bolt is used, tighten the brake hose fitting to 12 ft. lbs. (17 Nm).

➡**Do not twist the brake hose during installation.**

8. Install the caliper over the brake pads. Lubricate and install the lock pin and tighten to 23 ft. lbs. (32 Nm). Install the guide pin and tighten to 23 ft. lbs. (32 Nm).
9. Bleed the brake system.
10. Inspect the brake system for proper operation.

OVERHAUL

▶ **See Figures 50, 51, 52, 53 and 54**

1. Remove the caliper from the vehicle.
2. Remove the lock and guide pins from the caliper, if still installed. Pull the caliper support from the caliper body.
3. Remove the caliper boot ring and piston boot.
4. Using rear disc brake piston driver tool MB990652 or equivalent, twist the piston out of the caliper bore.

5. Remove the piston seal using your finger. Do not use a metal tool to remove the seal, the side of the bore may be damaged.
6. Clean the caliper inner cylinder wall with alcohol or DOT 3 brake fluid.
7. While using a 0.75 in. (19mm) diameter steel pipe to press the spring case into the caliper body, use snapring pliers to remove the snapring from the caliper body.
8. Remove the spring case, return spring, stopper plate, stopper, auto-adjuster spindle and connecting link. Note positioning of all components to aid in assembly. Remove the parking brake lever return spring, retainer nut, lever boot and spindle lever as required.
9. Check the connection link and the spindle for wear or damage. Check the caliper body for cracks or rust. Check the spindle lever shaft and the piston for rust. Check the piston seal and boot for cracks or deterioration. Replace components as required.
10. Inspect the brake pads, measuring the thickness in the thinnest place. The minimum thickness is 0.08 in. (2.0mm). If worn beyond the limit, replace the brake pads.

To install:
11. Apply the grease furnished with the repair kit to the lever boot, spindle lever, O-ring, connecting link and auto-adjuster spindle. Install the spindle lever, lever boot, retainer nut and washer, if removed. Tighten the retainer nut to 29-40 ft. lbs. (40-55 Nm). Install the brake lever return spring.
12. Install the connecting link and the auto-adjuster with a new O-ring in place. Install the stopper, stopper plate, return spring and spring case.
13. While using a 0.75 in. (19mm) diameter steel pipe to press in the spring case, use snapring pliers to attach the snapring to the caliper body.

➡**Attach the snapring to the caliper body with the opening facing the bleeder.**

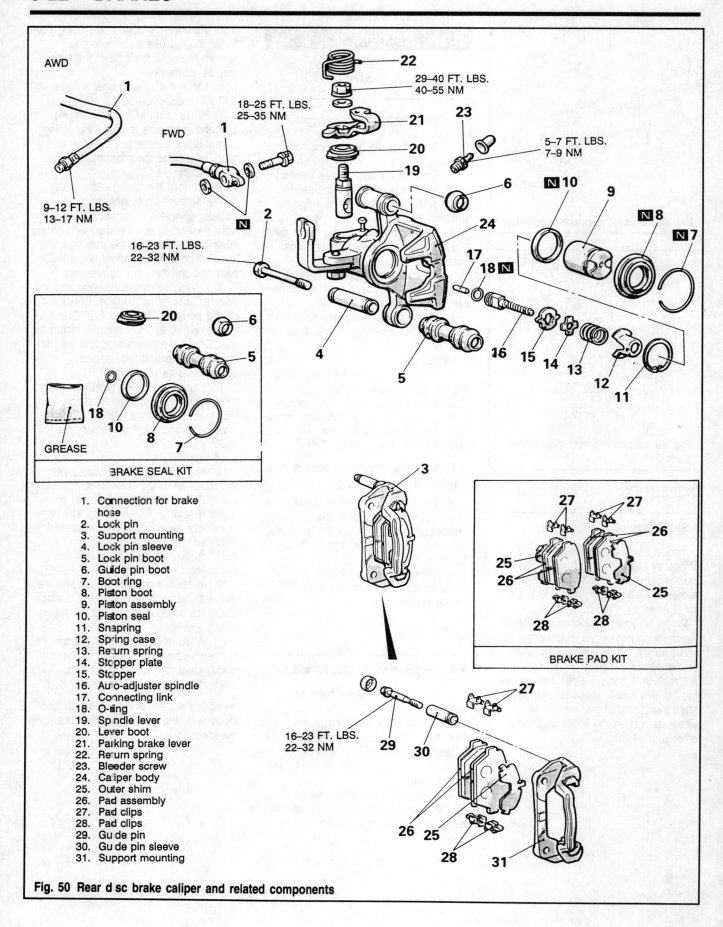

AWD

1

FWD

18–25 FT. LBS.
25–35 NM

1

9–12 FT. LBS.
13–17 NM

16–23 FT. LBS.
22–32 NM

2

2

22

29–40 FT. LBS.
40–55 NM

21

20

19

23

5–7 FT. LBS.
7–9 NM

6

N 10

9

N 8

N 7

24

17

18 N

4

5

5

16

15

14

13

12

11

20

6

5

18

10

8

7

GREASE

BRAKE SEAL KIT

1. Connection for brake hose
2. Lock pin
3. Support mounting
4. Lock pin sleeve
5. Lock pin boot
6. Guide pin boot
7. Boot ring
8. Piston boot
9. Piston assembly
10. Piston seal
11. Snapring
12. Spring case
13. Return spring
14. Stopper plate
15. Stopper
16. Auto-adjuster spindle
17. Connecting link
18. O-ring
19. Spindle lever
20. Lever boot
21. Parking brake lever
22. Return spring
23. Bleeder screw
24. Caliper body
25. Outer shim
26. Pad assembly
27. Pad clips
28. Pad clips
29. Guide pin
30. Guide pin sleeve
31. Support mounting

3

27

27

26

25

26

25

28

28

BRAKE PAD KIT

27

16–23 FT. LBS.
22–32 NM

29

30

26

25

28

31

Fig. 50 Rear disc brake caliper and related components

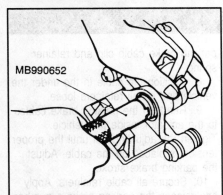

Fig. 51 Twist the piston out of the caliper using rear disc brake piston driver MB990652 or equivalent

Fig. 52 Removing the piston seal

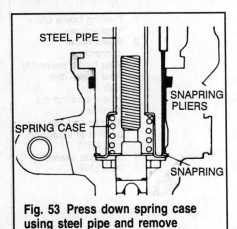

Fig. 53 Press down spring case using steel pipe and remove snapring from caliper body

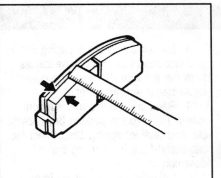

Fig. 54 Measuring rear disc brake pads

14. Apply the supplied grease to the piston seal and the cylinder walls. Insert the piston seal into the body of the caliper.

15. Push the piston into the caliper using the rear disc brake piston driver tool or equivalent. The pins on the back side of the brake pad must be placed in the grooves in the piston.

16. Apply the supplied grease to the piston boot mounting grooves in the caliper body and the piston and then install the piston boot.

17. Apply grease to the guide on the boot inner surface, lock pin boot inner surface and the lock pin sleeve (not on the threads).

18. Install the caliper to the caliper guide. Install the guide and lock pins, tightening to 23 ft. lbs. (32 Nm).

19. Install the brake hose to the caliper. Connect the parking brake cable to the caliper. Bleed the brake system.

20. Check the adjustment of the parking brake as follows:

   a. Free the parking brake cable.

   b. With the engine running, firmly press the brake pedal 5-6 times.

   c. Check whether or not the stroke of the parking brake lever is within the standard value of 5-7 notches. If out of specifications, adjust the parking brake.

21. Inspect the brake system for proper operation.

Brake Disc (Rotor)

REMOVAL & INSTALLATION

▶ **See Figures 55 and 56**

1. Raise the vehicle and support safely. Remove appropriate wheel assembly.

2. Disconnect the parking brake connection at the rear caliper assembly.

3. Remove the caliper and brake pads. Support the caliper out of the way using wire.

4. Remove the brake rotor (disc) from the rear hub assembly.

5. Installation is the reverse of the removal procedure.

INSPECTION

Using a micrometer, measure the disc thickness at eight positions, approximately 45 degrees apart and 0.39 in. (10mm) in from the outer edge of the disc. The minimum thickness is 0.331 in. (8.4mm). If the disc is beyond limits for thickness, remove it and install a new one.

PARKING BRAKE

Cables

REMOVAL & INSTALLATION

♦ **See Figures 57, 58 and 59**

1. Disconnect the negative battery cable.

2. Remove the floor console from the vehicle as follows:

    a. Remove the screw plugs in the side covers. Remove the retainer screws and the side covers from the vehicle.

    b. Remove the front mounting screw cover from the floor console. Remove the manual transaxle shift lever knob.

    c. Remove the cup holder and the carpet inserts from the floor console assembly.

    d. Label and disconnect the electrical wire harness connections for the floor console.

    e. Remove the mounting bolts and the floor console from the vehicle.

3. Loosen the cable adjusting nut and disconnect the rear brake cables from the actuator. Remove the center cable clamp and grommet.

4. Raise the vehicle and support safely. Remove the parking brake cable clip and retainer spring. Disconnect the cable end from the parking brake assembly.

5. Unfasten any remaining frame retainers and remove the cables from the vehicle.

To install:

6. The parking brake cables may be color coded to indicate side. Check the parking brake cables for an identification mark. If present, position the cables as follows:

    a. AWD vehicle — yellow cable goes on left side

    b. AWD vehicle — orange cable goes on right side

    c. FWD vehicle — white cable goes on right side

    d. FWD vehicle — no color marking goes on left side

7. Install the cable to the rear actuator. Secure in place with the parking brake cable clip and retainer spring.

8. Position the cable in the under the vehicle and install retainers loose.

9. Reattach the parking brake cables to the actuator inside the vehicle. Tighten the adjusting nut until the proper tension is placed on the cable. Adjust the parking brake stroke.

10. Secure all cable retainers. Apply and release the parking brake a number of times once all adjustments have been made. With the rear wheels raised, make sure the parking brake is not causing excess drag on the rear wheels.

11. Install the floor console assembly as follows:

    a. Install the floor console in position in the vehicle. Position the seat belts as required. Install the console retainer bolts.

    b. Reconnect the electrical harness connectors to the vehicle body harness.

    c. Install the carpet and the cup holder to the console assembly. Install the manual shift knob.

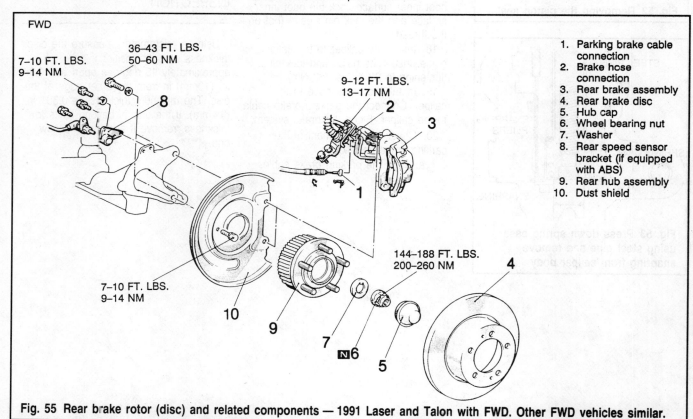

FWD

7–10 FT. LBS.
9–14 NM

36–43 FT. LBS.
50–60 NM

9–12 FT. LBS.
13–17 NM

7–10 FT. LBS.
9–14 NM

144–188 FT. LBS.
200–260 NM

1. Parking brake cable connection
2. Brake hose connection
3. Rear brake assembly
4. Rear brake disc
5. Hub cap
6. Wheel bearing nut
7. Washer
8. Rear speed sensor bracket (if equipped with ABS)
9. Rear hub assembly
10. Dust shield

Fig. 55 Rear brake rotor (disc) and related components — 1991 Laser and Talon with FWD. Other FWD vehicles similar.

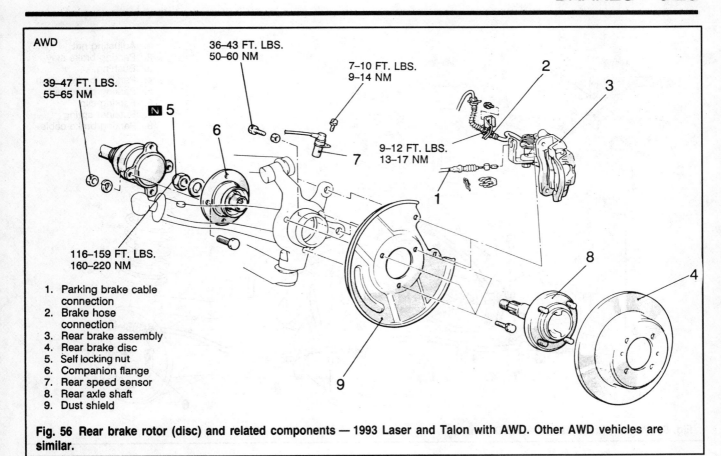

AWD

36–43 FT. LBS.
50–60 NM

7–10 FT. LBS.
9–14 NM

39–47 FT. LBS.
55–65 NM

9–12 FT. LBS.
13–17 NM

116–159 FT. LBS.
160–220 NM

1. Parking brake cable connection
2. Brake hose connection
3. Rear brake assembly
4. Rear brake disc
5. Self locking nut
6. Companion flange
7. Rear speed sensor
8. Rear axle shaft
9. Dust shield

Fig. 56 Rear brake rotor (disc) and related components — 1993 Laser and Talon with AWD. Other AWD vehicles are similar.

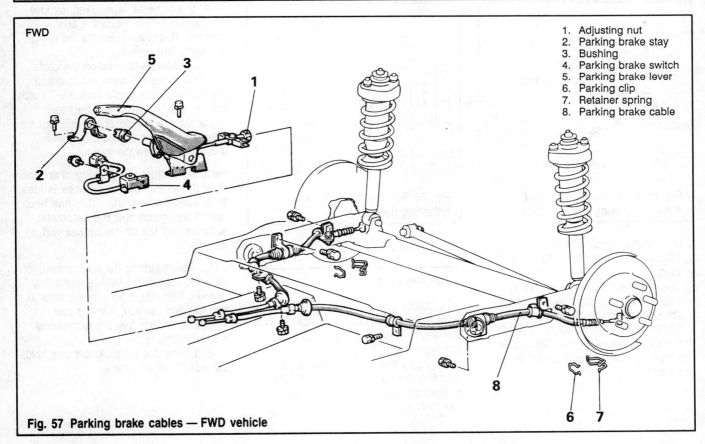

FWD

1. Adjusting nut
2. Parking brake stay
3. Bushing
4. Parking brake switch
5. Parking brake lever
6. Parking clip
7. Retainer spring
8. Parking brake cable

Fig. 57 Parking brake cables — FWD vehicle

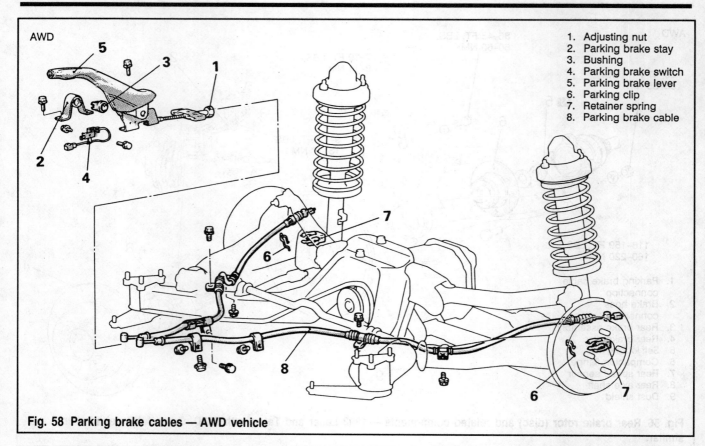

AWD

1. Adjusting nut
2. Parking brake stay
3. Bushing
4. Parking brake switch
5. Parking brake lever
6. Parking clip
7. Retainer spring
8. Parking brake cable

Fig. 58 Parking brake cables — AWD vehicle

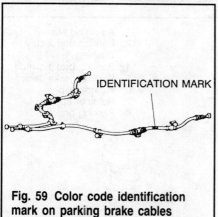

IDENTIFICATION MARK

Fig. 59 Color code identification mark on parking brake cables

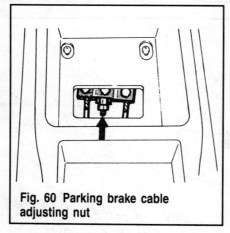

Fig. 60 Parking brake cable adjusting nut

d. Install the side covers and retainers. Cover retainer screws with plugs.

e. Connect the negative battery cable and check console electrical components for proper operation.

12. Road test the vehicle and check for proper brake operation. Check that the parking brake holds the vehicle on an incline.

ADJUSTMENT

▶ See Figure 60

1. Make sure the parking brake cable is free and is not frozen or sticking. With the engine running, forcefully depress the brake pedal 5-6 times.

2. Apply the parking brake while counting the number of notches. Check desired parking brake stroke should be 5-7 notches.

3. If adjustment is required, remove the carpeting from inside the floor console. This will expose the adjusting nut within the console.

4. Loosen the locknut on the cable rod. Rotate the adjusting nut to adjust the parking brake stroke to the 5-7 notch setting. After making the adjustment, check there is no looseness between the adjusting nut and the parking brake lever, then tighten the locknut.

➡**Do not adjust the parking brake too tight. If the number of notches is less than specification, the cable has been pulled too much and the automatic adjuster will fail or the brakes will drag.**

5. After adjusting the lever stroke, raise the rear of the vehicle and safely support. With the parking brake lever in the released position, turn the rear wheels to confirm that the rear brakes are not dragging.

6. Check that the parking brake holds the vehicle on an incline.

Brake Lever

REMOVAL & INSTALLATION

1. Disconnect the negative battery cable.

2. Remove the floor console from the vehicle as follows:

   a. Remove the screw plugs in the side covers. Remove the retainer screws and the side covers from the vehicle.

   b. Remove the front mounting screw cover from the floor console. Remove the manual transaxle shift lever knob.

   c. Remove the cup holder and the carpet inserts from the floor console assembly.

   d. Label and disconnect the electrical wire harness connections for the floor console.

   e. Remove the mounting bolts and the floor console from the vehicle.

3. Loosen the cable adjusting nut and disconnect the rear brake cables from the actuator. Remove the parking brake stay and bushing.

4. Disconnect the harness connector from the parking brake switch. Remove the switch.

5. Remove the mounting bolts and the parking brake lever.

To install:

6. Lubricate the sliding parts of the ratchet plate and the ratchet pawl on the parking brake lever assembly. Install assembly into the vehicle.

7. Install the parking brake switch to the lever assembly and secure in position. Connect the electrical harness connector.

8. Lubricate and install the bushing and the parking brake stay.

9. Install the cables if removed. Install the parking brake cable rod and adjusting nut in position. Tighten the adjusting nut until the proper tension is obtained. Adjust the parking brake stroke, as required.

10. Apply and release the parking brake a number of times once all adjustments have been made. With the rear wheels raised, make sure the parking brake is not causing excess drag on the rear wheels.

11. Install the floor console assembly as follows:

   a. Install the floor console in position in the vehicle. Position the seat belts as required. Install the console retainer bolts.

   b. Reconnect the electrical harness connectors to the vehicle body harness.

   c. Install the carpet and the cup holder to the console assembly. Install the manual shift knob.

   d. Install the side covers and retainers. Cover retainer screws with plugs.

   e. Connect the negative battery cable and check console electrical components for proper operation.

12. Road test the vehicle and check for proper brake operation. Check that the parking brake holds the vehicle on an incline.

ANTI-LOCK BRAKE SYSTEM

General Description

Anti-lock braking systems are designed to prevent locked-wheel skidding during hard braking or during braking on slippery surfaces. The front wheels of a vehicle cannot apply steering force if they are locked and sliding; the vehicle will continue in its previous direction of travel. The four wheel anti-lock brake systems found on these vehicles holds the individual wheels just below the point of locking, thereby allowing some steering response and preventing the rear of the vehicle from sliding sideways.

Electrical signals are sent from the wheel speed sensors to the ABS control unit; when the system detects impending lock-up at any wheel, solenoid valves within the hydraulic unit cycle to control the line pressure as needed. The systems employ normal master cylinder and vacuum booster arrangements; no hydraulic accumulator is used, nor is any high pressure fluid stored within the system. The system employs a conventional master cylinder and vacuum booster arrangements; no hydraulic accumulator is used, nor is any high pressure fluid stored within the system.

The Front Wheel Drive (FWD) vehicle family uses a 3-channel anti-lock system. The 3-channel system uses 3 solenoids in the hydraulic unit to control brake pressure in the left front, right front and rear circuits. A proportioning valve within the rear circuit equalizes pressure to each rear wheel.

The All Wheel Drive (AWD) vehicle family use a 2 channel system. The left front and right rear wheels share a control solenoid as do the right front and left rear wheels. The system contains a select-low valve which reacts to reduced pressure in one circuit and balances the pressure to the opposite rear wheel. In this fashion, the anti-lock function is provided at 3 wheels, rather than just the one originating the lock-up signal.

System Operation

Both FWD and AWD systems monitor and compare wheel speed based on the inputs from the wheel speed sensors. The brake pressure is controlled according to the impending lock-up computations of the ABS control unit.

On the FWD vehicles using a 3-channel system, if either front wheel approaches lock-up, the controller actuates the individual solenoid for that wheel, reducing pressure in the line. Impending lock-up at either rear wheel will engage the rear control solenoid; hydraulic pressure is reduced equally to both rear wheels, reducing the tendency of the rear to skid sideways under braking.

The AWD vehicles incorporate a G-sensor into the system to give the ABS control unit an acceleration signal; this is used in conjunction with engine rpm and wheel speed signals to determine high or low friction road conditions. If the system detects impending wheel lock at, for example, the left front wheel, the solenoid controlling the left front/right rear circuit is activated to reduce line pressures. The select-low valve (common to both channels) reacts to the pressure change and reduces the pressure to the other rear wheel. This system overcomes some of the inherent problems of applying ABS to all wheel drive vehicles.

The system incorporates an idle-up control which raises engine speed to 1800 rpm during braking. This eliminates engine braking during ABS-engaged stops and allows the control system to apply maximum stopping effort.

Also found on the AWD is a delay valve which prevents simultaneous front and rear wheel slip by momentarily delaying rear wheel cylinder pressure when ever front wheel line pressure increases sharply. This insures more even application of braking force to the ground and better vehicle control for the operator.

Three separate relays aid the operation of the ABS system. The ABS motor relay (for the pump motor) and the ABS valve relay are located together immediately adjacent to the hydraulic unit under the hood.

The ABS power relay is mounted on a separate bracket next to the anti-lock control unit in the right rear quarter panel. Each of the relays may be replaced in the usual fashion, although care must be taken to release wiring connector clips before removing the relay.

Flat Battery Remedy

When booster cables are used to start the engine when the battery is completely flat and the vehicle is immediately driven without waiting for the battery to recharge itself, the engine may misfire, and driving may not be possible. This is due to the fact that the ABS consumes a great amount of current for it self-check functions; the remedy is to either allow the battery to recharge sufficiently, or to disconnect the electrical harness connector for the ABS circuit, thus disabling the ABS system. The ABS warning light will illuminate when the connector is disconnected.

After the battery has been sufficiently charged, connect the ABS electrical connector and restart the engine. Check to be sure the ABS warning lamp is not illuminated.

System Precautions

• Certain components within the ABS system are not intended to be serviced or repaired individually. Only those components with removal and installation procedures should be serviced.

• Do not use rubber hoses or other parts not specifically specified for the ABS system. When using repair kits, replace all parts included in the kit. Partial or incorrect repair may lead to functional problems and require the replacement of components.

• Lubricate rubber parts with clean, fresh brake fluid to ease assembly. Do not use lubricated shop air to clean parts; damage to rubber components may result.

• Use only DOT 3 brake fluid from an unopened container.

• If any hydraulic component or line is removed or replaced, it may be necessary to bleed the entire system.

• A clean repair area is essential. Always clean the reservoir and cap thoroughly before removing the cap. The slightest amount of dirt in the fluid may plug an orifice and impair the system function. Perform repairs after components have been thoroughly cleaned; use only denatured alcohol to clean components. Do not allow ABS components to come into contact with any substance containing mineral oil; this includes used shop rags.

• The Anti-Lock control unit is a microprocessor similar to other computer units in the vehicle. Ensure that the ignition switch is **OFF** before removing or installing controller harnesses. Avoid static electricity discharge at or near the controller.

• If any arc welding is to be done on the vehicle, the ALCU connectors should be disconnected before welding operations begin.

DEPRESSURIZING THE SYSTEM

The ABS system requires no special system depressurization prior to the opening of hydraulic lines or bleeding of the system.

System Diagnosis

Diagnosis of the ABS system consists of 3 general steps, performed in order:

1. The visual or preliminary inspection, including inspection of the basic brake system, is always required before any other steps are taken.

2. Initial diagnosis is then made by a careful analysis of the ANTI-LOCK Warning Lamp display during start-up and operation. The warning lamp troubleshooting chart will direct the use of further charts and detailed testing based on initial findings.

3. The ABS system may be further checked with the DRB-II or similar diagnostic scan tool, provided the correct cartridges are used. The DRB-II or equivalent, will allow various components of the system to be operated for testing purposes. Connect the DRB II or equivalent, according to instructions furnished with the tool. The system will enter diagnostic mode and prompt the operator through the assorted system checks and tests.

Visual Inspection

Remember to first determine if the problem is related to the anti-lock system or not. The anti-lock system is made up of 2 basic sub-systems:

1. The hydraulic system, which may be diagnosed and serviced using normal brake system procedures, however, there is a need to determine whether the problem is related to the ABS components or not.

2. The electrical system which may be diagnosed using the charts and diagnostic tools.

Before diagnosing an apparent ABS problem, make absolutely certain that the normal braking system is in correct working order. Many common brake problems (dragging lining, seepage, etc.) will affect the ABS system. A visual check of specific system components may reveal problems creating an apparent ABS malfunction. Performing this inspection may reveal a simple failure, thus eliminating extended diagnostic time.

3. Inspect the brake fluid level in the reservoir.

4. Inspect brake lines, hoses, master cylinder assembly, and brake calipers for leakage.

5. Visually check brake lines and hoses for excessive wear, heat damage, punctures, contact with other parts, missing clips or holders, blockage or crimping.

6. Check the calipers for rust or corrosion. Check for proper sliding action if applicable.

7. Check the caliper pistons for freedom of motion during application and release.

8. Inspect the wheel speed sensors for proper mounting and connections.

9. Inspect the toothed wheels for broken teeth or poor mounting.

ABS WARNING LIGHT

Fig. 61 ABS warning lamp location — FWD and AWD vehicles

10. Inspect the wheels and tires on the vehicle. They must be of the same size and type to generate accurate speed signals. Check also for approximately equal tire pressures.

11. Confirm the fault occurrence with the operator. Certain driver induced faults may cause dash warning lamps to light. Excessive wheel spin on low-traction surfaces or high speed acceleration may also set fault codes and trigger a warning lamp. These induced faults are not system failures but examples of vehicle performance outside the parameters of the controller.

12. The most common cause of intermittent faults is not a failed sensor but a loose, corroded or dirty connector.

Incorrect installation of the wheel speed sensor will cause a loss of wheel speed signal. Check harness and component connectors carefully.

➡**If the battery on the vehicle has been completely drained, always recharge the battery before driving. If the vehicle is driven immediately after jump starting, the ABS self-check may draw enough current to make the engine run improperly. An alternate solution is to disconnect the ABS connector at the hydraulic unit. This will disable the ABS and illuminate the dash warning lamp. Reconnect the ABS when the battery is sufficiently charged.**

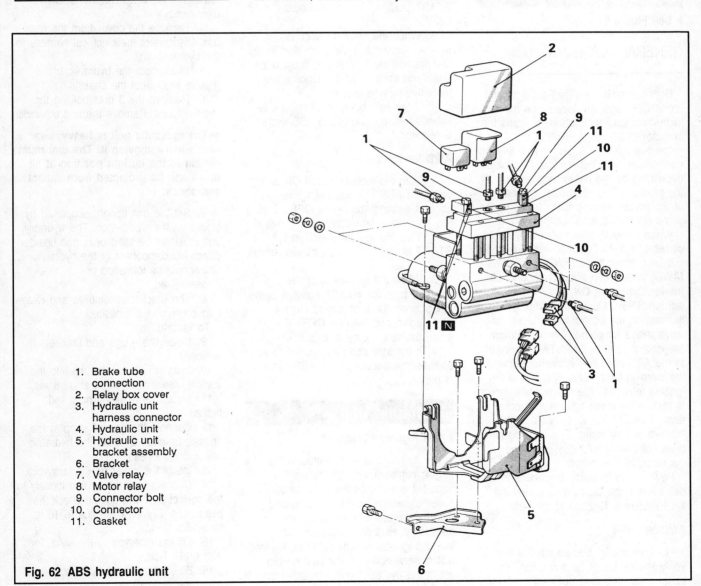

1. Brake tube connection
2. Relay box cover
3. Hydraulic unit harness connector
4. Hydraulic unit
5. Hydraulic unit bracket assembly
6. Bracket
7. Valve relay
8. Motor relay
9. Connector bolt
10. Connector
11. Gasket

Fig. 62 ABS hydraulic unit

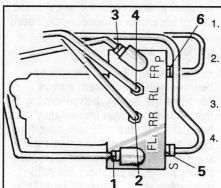

1. From the hydraulic unit to the front brake (L.H.)
2. From the hydraulic unit to the proportioning valve (Rear, R.H.)
3. From the hydraulic unit to the front brake (R.H.)
4. From the hydraulic unit to the proportioning valve (Rear, L.H.)
5. From the master cylinder for the left front and the right rear
6. From the master cylinder for the right front and the left rear

Fig. 63 Connection of brake tubes to ABS hydraulic unit

Anti-Lock Warning Lamp

▶ See Figure 61

GENERAL INFORMATION

Both the amber ANTI-LOCK light and red BRAKE light are located on the instrument cluster. Each lamp warns the operator of a possible fault in the respective system. A fault in one system may cause the other lamp to illuminate depending on the nature and severity of the problem. The operation or behavior of the amber warning lamp is one of the prime diagnostic tools for the system.

When the system is operating correctly, the ANTI-LOCK warning lamp will flash either twice (FWD) or 4 times (AWD) vehicles, in about 1 second with the ignition switch ON, then the lamp will turn OFF. During the lamp illumination the control unit checks the valve relays for proper function. When the ignition is turned to START, power to the ABS controller is interrupted and the warning lamp stays ON. Once the ignition returns to the ON position, power is restored and the system re-checks itself. This self test again yields either the two or four blinks. The warning lamp goes out and should stay off during operation of the vehicle.

Perform the following procedure to determine if the warning lamp and the ABS system is functioning properly:

FWD Vehicles

1. Turn the ignition switch ON and verify the ABS light flashes twice in about 1 second; then goes OFF. During this time the valve relay is being tested.

2. Turn the ignition switch to the START position. The light should remain ON.

3. When the ignition switch is returned from the START position to the ON position the light should flash twice again and then go OFF. Once again the valve relay is tested.

4. If the light does not illuminate as specified, inspection of the ABS system is required.

AWD Vehicles

1. Turn the ignition switch ON and verify the ABS light flashes 4 times in about 1 second; then goes OFF. During this time the valve relay is being tested.

2. Turn the ignition switch to the START position. The light should remain ON.

3. When the ignition switch is returned from the START position to the ON position the light should flash 4 times again and then go OFF. Once again the valve relay is tested.

4. If the light does not illuminate as specified, inspection of the ABS system is required.

Hydraulic Unit

▶ See Figures 62 and 63

The hydraulic unit is located in the engine compartment. It contains the solenoid valves and the pump/motor assembly which provides pressurized fluid for the anti-lock system when necessary. Hydraulic units are not interchangeable on any vehicles. Neither unit is serviceable; if any fault occurs within the hydraulic unit, the entire unit must be replaced.

REMOVAL & INSTALLATION

1. Disconnect the negative battery cable. Use a syringe or similar device to remove as much fluid as possible from the reservoir. Some fluid will be spilled from lines during removal of the hydraulic unit; protect adjacent painted surfaces.

2. On turbocharged engine, remove the center intercooler duct. Loosen the clamps and remove the bolts holding the duct to the air cleaner.

3. Disconnect the brake lines from the hydraulic unit. Correct reassembly is critical. Label or identify the lines before removal. Plug each line immediately after removal.

4. Remove the cover from the relay box. Disconnect the electrical harness to the hydraulic unit.

5. Disconnect the hydraulic unit ground strap from the chassis.

6. Remove the 3 nuts holding the hydraulic unit. Remove the unit upwards.

➡The hydraulic unit is heavy; use care when removing it. The unit must remain in the upright position at all times and be protected from impact and shock.

7. Set the unit upright supported by blocks on the workbench. The hydraulic unit must not be tilted or turned upside down. No component of the hydraulic unit should be loosened or disassembled.

8. The bracket assemblies and relays may be removed if desired.

To install:

9. Install the relays and brackets if removed.

10. Install the hydraulic unit into the vehicle, keeping it upright at all times.

11. Install the retaining nuts and tighten.

12. Connect the ground strap to the chassis bracket. Connect the hydraulic unit wiring harness.

13. Install the cover on the relay box.

14. Connect each brake line loosely to the correct port and double check the placement. Tighten each line to 10 ft. lbs. (13.5 Nm).

15. Fill the reservoir to the MAX line with brake fluid.

16. Bleed the master cylinder, then bleed the brake lines.

17. If equipped, install the intercooler air duct.

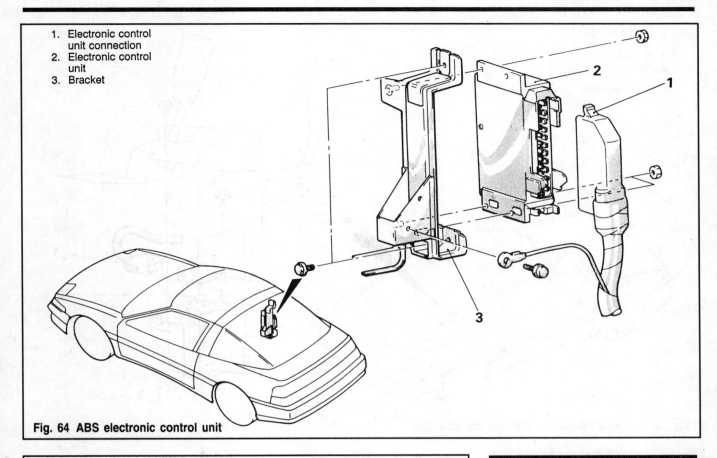

1. Electronic control unit connection
2. Electronic control unit
3. Bracket

Fig. 64 ABS electronic control unit

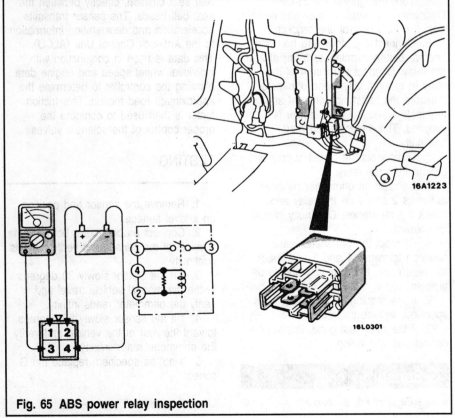

Fig. 65 ABS power relay inspection

Anti-Lock Control Unit (ALCU)

The anti-lock control unit is located behind the right rear quarter trim panel. It is a micro-processor capable of dealing with many inputs simultaneously and controls the function of the solenoid valves within the hydraulic unit.

REMOVAL & INSTALLATION

▶ See Figure 64

1. Turn the key to the **OFF** position. Ensure that the ignition switch remains **OFF** throughout the procedure.
2. Remove the interior right rear quarter trim panel and rear seat back and/or cushion.
3. Release the lock on the bottom of the connector; disconnect the multi-pin connector from the control unit. Access may be easier if the external ground is disconnected from the bracket.
4. Remove the retaining nuts and remove the control unit from its bracket. The bracket may be removed, if desired.

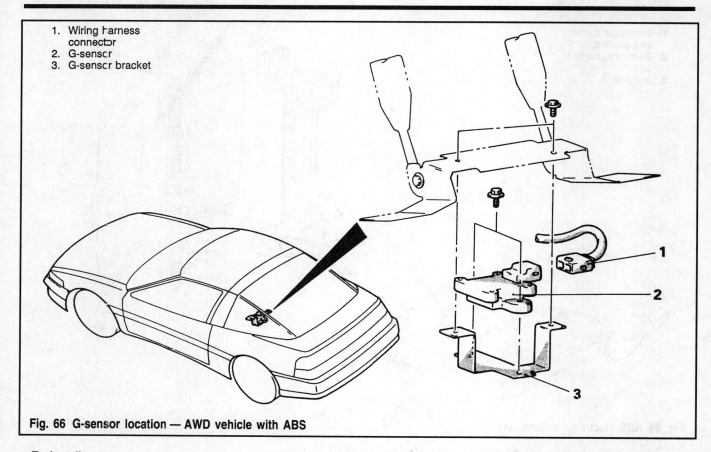

1. Wiring harness connector
2. G-sensor
3. G-sensor bracket

Fig. 66 G-sensor location — AWD vehicle with ABS

To install:

5. Place the bracket in position. Install the controller and tighten the retaining nuts.

6. Connect the ground wire to the bracket if removed. Ensure a proper, tight connection. The ground must be connected before the multi-pin harness is connected.

7. Connect the multi-pin connector and secure the lock.

8. Install the rear quarter trim panel and seat.

ABS Power Relay

TESTING

▶ See Figure 65

1. Disconnect the system relay.

2. Connect the positive lead of a voltmeter to terminal **3** of the relay connector and the other lead to ground.

3. Verify that there is approximately 12 volts with the ignition switch **ON** . If not as specified, inspect the ABS system fuse and wiring to the relay.

4. Turn the ignition switch **OFF** . Disconnect the system relay and remove from the ABS control unit bracket.

5. Connect a jumper wire from the positive battery terminal to terminal **2** of the relay. Connect terminal **4** of the relay to ground. Using and ohmmeter, measure the resistance present at terminals **1** and **3** , while power is supplied. The meter should read continuity.

6. Disconnect the power supply and ground from the relay.

7. Connect an ohmmeter between terminals **2** and **4** on the relay and check the resistance. Continuity should be present.

8. Connect the ohmmeter leads between terminals **1** and **3** and check the resistance. No continuity should be present.

9. If the readings are not as specified, replace the relay.

10. If the relay test good, inspect the control unit and wiring.

G-Sensor

▶ See Figures 66, 67 and 68

The G-sensor is used only on the AWD vehicles. It is mounted under the rear seat cushion, directly between the seat belt heads. The sensor transmits acceleration and deceleration information to the Anti-lock Control Unit (ALCU). This data is used in conjunction with individual wheel speed and engine data, allowing the controller to determine the approximate road friction. This friction factor is then used to compute the proper control of the solenoid valves.

TESTING

1. Remove the sensor and position on a level surface.

2. Connect an ohmmeter between the terminal of the sensor and verify there is continuity.

3. Tilt the sensor slowly 30 degrees in the direction of vehicle travel and verify the ohmmeter reads infinity.

4. Tilt the sensor slowly 30 degrees toward the rear of the vehicle and verify the ohmmeter reads infinity.

5. If not as specified, replace the G sensor.

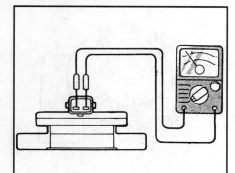

Fig. 67 G-sensor inspection — checking for continuity of sensor while on flat surface

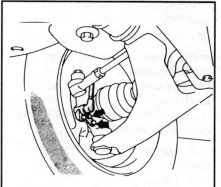

Fig. 69 Front speed sensor and mounting bolt

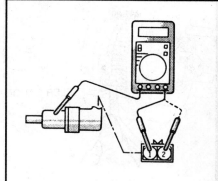

Fig. 72 Measure the resistance between the speed sensor terminals

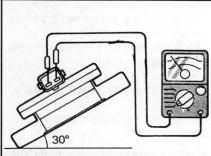

Fig. 68 G-sensor inspection — checking for no continuity of sensor while inclined at angle greater than 30 degrees

Fig. 70 Rear speed sensor and mounting bolt

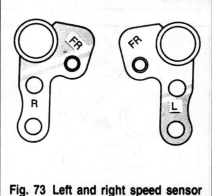

Fig. 73 Left and right speed sensor brackets with identification markings

REMOVAL & INSTALLATION

1. Insure that the ignition switch is **OFF** throughout the procedure.
2. Remove the rear seat cushion.
3. Disconnect the wiring harness to G-sensor.
4. Remove the retaining bolts and remove the sensor.

To install:

5. Position the sensor on the mounting bracket. Tighten the retaining bolts.
6. Connect the electrical harness.
7. Install the rear seat cushion and related components.

Wheel Speed Sensors

⬥ **See Figures 69, 70, 71, 72, 73, 74 and 75**

Each wheel is equipped with a magnetic sensor mounted a fixed

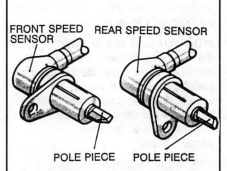

Fig. 71 Inspect removed speed sensor for damaged pole piece. Also check pole piece for foreign material or metal adhesion

distance from a toothed ring which rotates with the wheel. The sensors are replaceable but not interchangeable; each must be fitted to its correct location. The toothed rings are

replaceable although disassembly of the hub or axle shaft is required.

The wheel speed sensor is designed to produce a voltage directly proportional to the speed of the wheel. In other words as the speed of the wheel increases, so does the voltage. The voltage should not fluctuate when the wheel speed (speedometer) reading is steady.

TESTING

1. Raise and safely support the vehicle.
2. Disconnect the suspected malfunctioning sensor wire connector.
3. Connect an ohmmeter to the sensor connector 2 terminals.
4. The resistances should be 0.8-1.2 kilo-ohms.
5. If the resistance is not within specifications, replace the speed sensor.

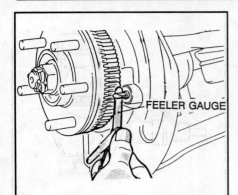

Fig. 74 Checking the clearance between the pole piece of the speed sensor and the rotor's toothed surface

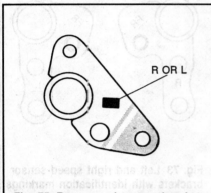

Fig. 75 Rear speed sensor bracket identification marking

REMOVAL & INSTALLATION

1. Elevate and safely support the vehicle.

2. Remove the wheel and tire assembly and the disc brake rotor.

3. Remove the inner fender or splash shield.

4. Beginning at the sensor end, carefully disconnect or release each clip and retainer along the sensor wire. Take careful note of the exact position of each clip; they must be reinstalled in the identical position. Rear wheel sensor harnesses will be held by plastic wire ties; these may be cut away but must be replaced at reassembly.

5. Disconnect the sensor connector at the end of the harness.

6. Remove the two bolts holding the speed sensor bracket to the knuckle and remove the assembly from the vehicle.

➡ **The speed sensor has a pole piece projecting from it. This exposed tip must be protected from impact or scratches. Do not allow the pole piece to contact the toothed wheel during removal or installation.**

7. Remove the sensor from the bracket.

To install:

8. Assemble the sensor onto the bracket. Note that the brackets are different for the left and right front wheels, as well as both side rear wheels. Each bracket has identifying letters stamped on it.

9. Identify the front speed sensor brackets as follows:

   a. FR: Indicates that the bracket is for the front speed sensor.

   b. R: Indicates that the bracket is for the right wheel.

   c. L: Indicates that the bracket is for the left wheel.

10. Identify the rear speed sensor brackets as follows:

   a. R: Indicates that the bracket is for the right wheel.

   b. L: Indicates that the bracket is for the left wheel.

11. Temporarily install the speed sensor to the spindle or knuckle; tighten the bolts only finger tight.

➡ **During speed sensor installation to the mounting bracket, make sure the letters FR are visible. Be careful when installing the speed sensor on the vehicle, that the pole piece at the tip of the sensor does not strike the toothed edge of the rotor, and damage them.**

12. Route the cable correctly and loosely install the clips and retainers. All clips must be in their original position and the sensor cable must not be twisted. Improper installation may cause cable damage and system failure.

13. Use a brass or other non-magnetic feeler gauge to check the air gap between the tip of the pole piece and the toothed wheel. The correct gap is 0.012-0.035 inch (0.3-0.9mm). Tighten the 2 sensor bracket bolts to 10 ft. lbs. (14 Nm) with the sensor located so that the gap is the same at several points on the toothed wheel. If the gap is

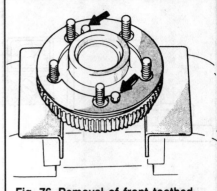

Fig. 76 Removal of front toothed ring

incorrect, it is likely that the toothed wheel is worn or improperly installed.

14. Tighten the screws and bolts for the cable retaining clips.

15. Install the disc brake rotor, inner fender or splash shield.

16. Install the wheel and tire assembly. Lower the vehicle to the ground.

17. Inspect the brake system for proper operation.

Toothed Wheels

REMOVAL & INSTALLATION

Front Wheel Ring
▶ **See Figure 76**

1. Elevate and safely support the vehicle.

2. Remove the wheel and tire assembly.

3. Remove the wheel speed sensor and disconnect sufficient harness clips to allow the sensor and wiring to be moved out of the work area.

➡ **The speed sensor has a pole piece projecting from it. This exposed tip must be protected from impact or scratches. Do not allow the pole piece to contact the toothed wheel during removal or installation.**

4. Remove the front hub and knuckle assembly.

5. Remove the hub from the knuckle.

6. Support the hub in a vise with protected jaws. Remove the retaining bolts from the toothed wheel and remove the toothed wheel.

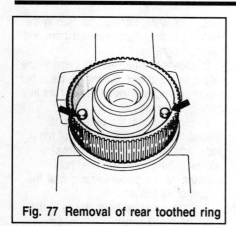

Fig. 77 Removal of rear toothed ring

To install:

7. Fit the new toothed wheel onto the hub and tighten the retaining bolts to 7 ft. lbs. (10 Nm).

8. Assemble the hub to the knuckle

9. Install the hub and knuckle assembly onto the vehicle.

10. Install the wheel speed sensor following procedures given in this section.

11. Install the wheel and tire assembly.

12. Lower the vehicle to the ground. Inspect the brake system for proper operation.

Rear Wheel Ring — FWD Vehicles
▶ See Figure 77

1. Raise and safely support the vehicle.

2. Remove the wheel and tire assembly. Remove the rear brake rotor.

3. Remove the wheel speed sensor and disconnect sufficient harness clips to allow the sensor and wiring to be moved out of the work area.

➡**The speed sensor has a pole piece projecting from it. This exposed tip must be protected from impact or scratches. Do not allow the pole piece to contact the toothed wheel during removal or installation.**

4. Remove the rear hub assembly.

5. Support the hub in a vise with protected jaws. Remove the retaining bolts from the toothed wheel and remove the toothed wheel.

To install:

6. Fit the new toothed wheel onto the hub and tighten the retaining bolts to 7 ft. lbs. (10 Nm).

7. Install the hub assembly to the vehicle. The center hub nut is not reusable. The new nut must be tightened

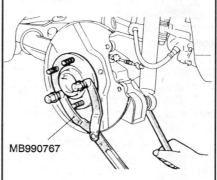

Fig. 78 Holding the rear axle shaft stationary using tool MB990767

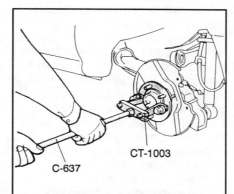

Fig. 79 Removing the rear axle shaft using tool CT-1003 and tool C-637

to 144-188 ft. lbs. (200-260 Nm). After the nut is tightened, align the nut with the spindle indentation and crimp the nut in place.

8. Install the wheel speed sensor following procedures given in this section.

9. Install the brake rotor, caliper and tire assembly.

10. Lower the vehicle to the ground. Check the brakes for proper operation.

Rear Wheel Rings — AWD Vehicles
▶ See Figures 78 and 79

1. Disconnect the negative battery cable. Raise and safely support the vehicle.

2. Remove the wheel and tire assembly.

3. Disconnect the parking brake cable at the caliper.

4. Remove the speed sensor and its O-ring. Disconnect sufficient clamps and

wire ties to allow the sensor to be moved well out of the work area.

➡**The speed sensor has a pole piece projecting from it. This exposed tip must be protected from impact or scratches. Do not allow the pole piece to contact the toothed wheel during removal or installation.**

5. Remove the brake caliper and brake disc.

6. Matchmark the drive shaft to the companion flange. Remove the 3 retaining nuts and bolts holding the outer end of the driveshaft to the companion flange. Swing the axle shaft away and support it with stiff wire. Do not overextend the joint in the axle; do not allow it to hang of its own weight.

7. Remove the retaining nut and washer on the back of the driveshaft. Use special tool MB990767 or equivalent, to counterhold the hub.

8. Remove the companion flange from the knuckle.

9. Using an axle puller which bolts to the wheel lugs, remove the axle shaft assembly.

10. Fit the shaft assembly in a press with the toothed wheel completely supported by a bearing plate such as special tool MB990560 or equivalent.

11. Press the toothed wheel off the axle shaft.

To install:

12. Press the new toothed wheel onto the shaft with the groove facing the axle shaft flange.

13. Install the axle shaft to the knuckle and fit the companion flange in place.

14. Install the lock washer and a new self-locking nut on the axle shaft. Hold the axle shaft stationary and torque nut to 116-159 ft. lbs. (160-220 Nm).

15. Swing the axle assembly into place and install the nuts and bolts. Tighten each to 45 ft. lbs. (61 Nm).

16. Install the brake disc and caliper.

17. Install the wheel speed sensor. Always use a new O-ring.

18. Connect the parking brake cable to the caliper.

19. Install the wheel and tire.

20. Lower the vehicle to the ground and check for proper brake operation.

FILLING THE SYSTEM

The brake fluid reservoir is part of the normal brake system and is filled or checked in the usual manner. Always clean the reservoir cap and surrounding area thoroughly before removing the cap. Fill the reservoir only to the FULL mark; do not overfill. Use only fresh DOT 3 brake fluid from unopened containers. Do not use any fluid containing a petroleum base. Do not use any fluid which has been exposed to water or moisture. Failure to use the correct fluid will affect system function and component life.

BLEEDING THE SYSTEM

Master Cylinder

If the master cylinder has been emptied of fluid or replaced, it must be bled separately from the rest of the system. Since the cylinder has no check valve, air can become trapped within it. To bleed the brake master cylinder after it has been drained, proceed as follows:

1. Disconnect the 2 brake lines from the master cylinder. Plug the lines immediately. The brake fluid reservoir should be in place and connected to the master cylinder. Check the fluid level before beginning.

2. An assistant should slowly depress and hold the brake pedal.

3. With the pedal held down, use 2 fingers to plug each outlet port on the master cylinder and release the brake pedal.

4. Repeat Steps 2 and 3 three or four times. The air will be bled from the cylinder.

5. Connect the brake lines to the master cylinder and tighten the fittings to 10 ft. lbs. (13.5 Nm).

6. Start the engine, allowing the system to pressurize and self-check. Shut the ignition **OFF** and bleed the brake lines.

Lines and Calipers

The brake system must be bled any time a line, hose or component is loosened or removed. Any air trapped within the lines can affect pedal feel and system function. Bleeding the system is performed in the usual manner with an assistant in the car to pump the brake pedal. Make certain the fluid level in the reservoir is maintained at or near correct levels during bleeding operations.

The individual lines may be bled manually at each wheel using the traditional 2 person method.

1. The ignition must remain **OFF** throughout the bleeding procedure.

2. The system should be bled in the following order: Right rear, left front, left rear and right front.

3. Connect a transparent hose to the caliper bleed screw. Submerge the other end of the hose in clean brake fluid in a clear glass container.

4. Slowly pump the brake pedal several times. Use full strokes of the pedal and allow 5 seconds between strokes. After 2 or 3 strokes, hold pressure on the pedal keeping it at the bottom of its travel.

5. With pressure held on the pedal, open the bleed screw 1/2-3/4 turn. Leave the bleed screw open until fluid stops flowing from the hose. Tighten the bleed screw and release the pedal.

6. Repeat Steps 3 and 4 until air-free fluid flows from the hose. Tighten the caliper bleed screw to 7.5 ft. lbs. (10 Nm).

7. Repeat the sequence at each remaining wheel.

➡ **Check the fluid level in the reservoir frequently and maintain it near the full level.**

8. When bleeding is complete, bring fluid level in the reservoir to the correct level. Install the reservoir cap.

BRAKE SPECIFICATIONS
All measurements in inches unless noted.

Year	Model		Master Cylinder Bore	Brake Disc			Brake Drum Diameter			Minimum Lining Thickness	
				Original Thickness	Minimum Thickness	Maximum Runout	Original Inside Diameter	Max. Wear Limit	Maximum Machine Diameter	Front	Rear
1990	Eclipse	front	①	0.940	0.882	0.003	—	—	—	0.080	0.080
		rear disk	—	0.390	0.331	0.003	—	—	—	0.080	0.080
	Laser	front	①	0.940	0.882	0.003	—	—	—	0.080	0.080
		rear disk	—	0.390	0.331	0.003	—	—	—	0.080	0.080
	Talon	front	①	0.940	0.882	0.003	—	—	—	0.080	0.080
		rear disk	—	0.390	0.331	0.003	—	—	—	0.080	0.080

BRAKE SPECIFICATIONS
All measurements in inches unless noted.

Year	Model		Master Cylinder Bore	Brake Disc Original Thickness	Brake Disc Minimum Thickness	Brake Disc Maximum Runout	Brake Drum Diameter Original Inside Diameter	Brake Drum Diameter Max. Wear Limit	Brake Drum Diameter Maximum Machine Diameter	Minimum Lining Thickness Front	Minimum Lining Thickness Rear
1991	Eclipse	front	②	0.940	0.882	0.003	—	—	—	0.080	0.080
		rear disk	—	0.390	0.331	0.003	—	—	—	0.080	0.080
	Laser	front	②	0.940	0.882	0.003	—	—	—	0.080	0.080
		rear disk	—	0.390	0.331	0.003	—	—	—	0.080	0.080
	Talon	front	②	0.940	0.882	0.003	—	—	—	0.080	0.080
		rear disk	—	0.390	0.331	0.003	—	—	—	0.080	0.080
1992	Eclipse	front	②	0.940	0.882	0.003	—	—	—	0.080	0.080
		rear disk	—	0.390	0.331	0.003	—	—	—	0.080	0.080
	Laser	front	②	0.940	0.882	0.003	—	—	—	0.080	0.080
		rear disk	—	0.390	0.331	0.003	—	—	—	0.080	0.080
	Talon	front	②	0.940	0.882	0.003	—	—	—	0.080	0.080
		rear disk	—	0.390	0.331	0.003	—	—	—	0.080	0.080
1993	Eclipse	front	③	0.940	0.882	0.003	—	—	—	0.080	0.080
		rear disk	—	0.390	0.331	0.003	—	—	—	0.080	0.080
	Laser	front	③	0.940	0.882	0.003	—	—	—	0.080	0.080
		rear disk	—	0.390	0.331	0.003	—	—	—	0.080	0.080
	Talon	front	③	0.940	0.882	0.003	—	—	—	0.080	0.080
		rear disk	—	0.390	0.331	0.003	—	—	—	0.080	0.080

① Non-turbocharged engine: 7/8
Turbocharged engine: 15/16

② Non-turbocharged without ABS: 7/8
Non-turbocharged with ABS: 15/16
Turbocharged with FWD: 15/16
Turbocharged with AWD: 1.0

③ Non-turbocharged without ABS: 7/8
Non-turbocharged with ABS: 15/16
Turbocharged with FWD: 1.0
Turbocharged with AWD: 1.0

TORQUE SPECIFICATIONS

Component	US	Metric
Brake booster attaching nuts	12 ft. lbs.	17 Nm
Brake hose to caliper retainer bolt	18 ft. lbs.	25 Nm
Caliper bleed screw	7.5 ft. lbs.	10 Nm
Flared brake line nut	12 ft. lbs.	17 Nm
Front caliper guide and locking pins Vehicles built up to May, 1989: Vehicles built after May, 1989:	23 ft. lbs. 54 ft. lbs.	32 Nm 75 Nm
Front toothed wheel retainer bolts (ABS vehicles)	7 ft. lbs.	10 Nm
Master cylinder mounting nuts	9 ft. lbs.	12 Nm
Master cylinder piston stopper bolt	2 ft. lbs.	3 Nm
Rear caliper guide and locking pins	23 ft. lbs.	32 Nm
Rear toothed wheel retainer bolts (ABS vehicles)	7 ft. lbs.	10 Nm

Troubleshooting the Brake System

Problem	Cause	Solution
Low brake pedal (excessive pedal travel required for braking action.)	• Excessive clearance between rear linings and drums caused by in-operative automatic adjusters	• Make 10 to 15 alternate forward and reverse brake stops to adjust brakes. If brake pedal does not come up, repair or replace adjuster parts as necessary.
	• Worn rear brakelining	• Inspect and replace lining if worn beyond minimum thickness specification
	• Bent, distorted brakeshoes, front or rear	• Replace brakeshoes in axle sets
	• Air in hydraulic system	• Remove air from system. Refer to Brake Bleeding.
Low brake pedal (pedal may go to floor with steady pressure applied.)	• Fluid leak in hydraulic system	• Fill master cylinder to fill line; have helper apply brakes and check calipers, wheel cylinders, differential valve tubes, hoses and fittings for leaks. Repair or replace as necessary.
	• Air in hydraulic system	• Remove air from system. Refer to Brake Bleeding.
	• Incorrect or non-recommended brake fluid (fluid evaporates at below normal temp).	• Flush hydraulic system with clean brake fluid. Refill with correct-type fluid.
	• Master cylinder piston seals worn, or master cylinder bore is scored, worn or corroded	• Repair or replace master cylinder
Low brake pedal (pedal goes to floor on first application—o.k. on subsequent applications.)	• Disc brake pads sticking on abutment surfaces of anchor plate. Caused by a build-up of dirt, rust, or corrosion on abutment surfaces	• Clean abutment surfaces
Fading brake pedal (pedal height decreases with steady pressure applied.)	• Fluid leak in hydraulic system	• Fill master cylinder reservoirs to fill mark, have helper apply brakes, check calipers, wheel cylinders, differential valve, tubes, hoses, and fittings for fluid leaks. Repair or replace parts as necessary.
	• Master cylinder piston seals worn, or master cylinder bore is scored, worn or corroded	• Repair or replace master cylinder
Spongy brake pedal (pedal has abnormally soft, springy, spongy feel when depressed.)	• Air in hydraulic system	• Remove air from system. Refer to Brake Bleeding.
	• Brakeshoes bent or distorted	• Replace brakeshoes
	• Brakelining not yet seated with drums and rotors	• Burnish brakes
	• Rear drum brakes not properly adjusted	• Adjust brakes

Troubleshooting the Brake System (cont.)

Problem	Cause	Solution

Decreasing brake pedal travel (pedal travel required for braking action decreases and may be accompanied by a hard pedal.)	• Caliper or wheel cylinder pistons sticking or seized
	• Master cylinder compensator ports blocked (preventing fluid return to reservoirs) or pistons sticking or seized in master cylinder bore
	• Power brake unit binding internally
		• Repair or replace the calipers, or wheel cylinders
		• Repair or replace the master cylinder
		
		• Test unit according to the following procedure:
		  (a) Shift transmission into neutral and start engine
		  (b) Increase engine speed to 1500 rpm, close throttle and fully depress brake pedal
		  (c) Slow release brake pedal and stop engine
		  (d) Have helper remove vacuum check valve and hose from power unit. Observe for backward movement of brake pedal.
		  (e) If the pedal moves backward, the power unit has an internal bind—replace power unit

Grabbing brakes (severe reaction to brake pedal pressure.)	• Brakelining(s) contaminated by grease or brake fluid • Parking brake cables incorrectly adjusted or seized • Incorrect brakelining or lining loose on brakeshoes • Caliper anchor plate bolts loose • Rear brakeshoes binding on support plate ledges • Incorrect or missing power brake reaction disc • Rear brake support plates loose	• Determine and correct cause of contamination and replace brakeshoes in axle sets • Adjust cables. Replace seized cables. • Replace brakeshoes in axle sets • Tighten bolts • Clean and lubricate ledges. Replace support plate(s) if ledges are deeply grooved. Do not attempt to smooth ledges by grinding. • Install correct disc • Tighten mounting bolts
Chatter or shudder when brakes are applied (pedal pulsation and roughness may also occur.)	• Brakeshoes distorted, bent, contaminated, or worn • Caliper anchor plate or support plate loose • Excessive thickness variation of rotor(s)	• Replace brakeshoes in axle sets • Tighten mounting bolts • Refinish or replace rotors in axle sets
Noisy brakes (squealing, clicking, scraping sound when brakes are applied.)	• Bent, broken, distorted brakeshoes • Excessive rust on outer edge of rotor braking surface	• Replace brakeshoes in axle sets • Remove rust

Troubleshooting the Brake System (cont.)

Problem	Cause	Solution
Hard brake pedal (excessive pedal pressure required to stop vehicle. May be accompanied by brake fade.)	• Loose or leaking power brake unit vacuum hose • Incorrect or poor quality brake-lining • Bent, broken, distorted brakeshoes • Calipers binding or dragging on mounting pins. Rear brakeshoes dragging on support plate.	• Tighten connections or replace leaking hose • Replace with lining in axle sets • Replace brakeshoes • Replace mounting pins and bushings. Clean rust or burrs from rear brake support plate ledges and lubricate ledges with molydisulfide grease. **NOTE:** If ledges are deeply grooved or scored, do not attempt to sand or grind them smooth—replace support plate.
	• Caliper, wheel cylinder, or master cylinder pistons sticking or seized • Power brake unit vacuum check valve malfunction	• Repair or replace parts as necessary • Test valve according to the following procedure: (a) Start engine, increase engine speed to 1500 rpm, close throttle and immediately stop engine (b) Wait at least 90 seconds then depress brake pedal (c) If brakes are not vacuum assisted for 2 or more applications, check valve is faulty
	• Power brake unit has internal bind	• Test unit according to the following procedure: (a) With engine stopped, apply brakes several times to exhaust all vacuum in system (b) Shift transmission into neutral, depress brake pedal and start engine (c) If pedal height decreases with foot pressure and less pressure is required to hold pedal in applied position, power unit vacuum system is operating normally. Test power unit. If power unit exhibits a bind condition, replace the power unit.

Troubleshooting the Brake System (cont.)

Problem	Cause	Solution
Hard brake pedal (excessive pedal pressure required to stop vehicle. May be accompanied by brake fade.)	• Master cylinder compensator ports (at bottom of reservoirs) blocked by dirt, scale, rust, or have small burrs (blocked ports prevent fluid return to reservoirs).	• Repair or replace master cylinder **CAUTION:** Do not attempt to clean blocked ports with wire, pencils, or similar implements. Use compressed air only.
	• Brake hoses, tubes, fittings clogged or restricted	• Use compressed air to check or unclog parts. Replace any damaged parts.
	• Brake fluid contaminated with improper fluids (motor oil, transmission fluid, causing rubber components to swell and stick in bores	• Replace all rubber components, combination valve and hoses. Flush entire brake system with DOT 3 brake fluid or equivalent.
	• Low engine vacuum	• Adjust or repair engine
Dragging brakes (slow or incomplete release of brakes)	• Brake pedal binding at pivot	• Loosen and lubricate
	• Power brake unit has internal bind	• Inspect for internal bind. Replace unit if internal bind exists.
	• Parking brake cables incorrrectly adjusted or seized	• Adjust cables. Replace seized cables.
	• Rear brakeshoe return springs weak or broken	• Replace return springs. Replace brakeshoe if necessary in axle sets.
	• Automatic adjusters malfunctioning	• Repair or replace adjuster parts as required
	• Caliper, wheel cylinder or master cylinder pistons sticking or seized	• Repair or replace parts as necessary
	• Master cylinder compensating ports blocked (fluid does not return to reservoirs).	• Use compressed air to clear ports. Do not use wire, pencils, or similar objects to open blocked ports.
Vehicle moves to one side when brakes are applied	• Incorrect front tire pressure	• Inflate to recommended cold (reduced load) inflation pressure
	• Worn or damaged wheel bearings	• Replace worn or damaged bearings
	• Brakelining on one side contaminated	• Determine and correct cause of contamination and replace brakelining in axle sets
	• Brakeshoes on one side bent, distorted, or lining loose on shoe	• Replace brakeshoes in axle sets
	• Support plate bent or loose on one side	• Tighten or replace support plate
	• Brakelining not yet seated with drums or rotors	• Burnish brakelining
	• Caliper anchor plate loose on one side	• Tighten anchor plate bolts
	• Caliper piston sticking or seized	• Repair or replace caliper
	• Brakelinings water soaked	• Drive vehicle with brakes lightly applied to dry linings
	• Loose suspension component attaching or mounting bolts	• Tighten suspension bolts. Replace worn suspension components.
	• Brake combination valve failure	• Replace combination valve

Troubleshooting the Brake System (cont.)

Problem	Cause	Solution
Noisy brakes (squealing, clicking, scraping sound when brakes are applied.) (cont)	• Brakelining worn out—shoes contacting drum of rotor	• Replace brakeshoes and lining in axle sets. Refinish or replace drums or rotors.
	• Broken or loose holdown or return springs	• Replace parts as necessary
	• Rough or dry drum brake support plate ledges	• Lubricate support plate ledges
	• Cracked, grooved, or scored rotor(s) or drum(s)	• Replace rotor(s) or drum(s). Replace brakeshoes and lining in axle sets if necessary.
	• Incorrect brakelining and/or shoes (front or rear).	• Install specified shoe and lining assemblies
Pulsating brake pedal	• Out of round drums or excessive lateral runout in disc brake rotor(s)	• Refinish or replace drums, re-index rotors or replace

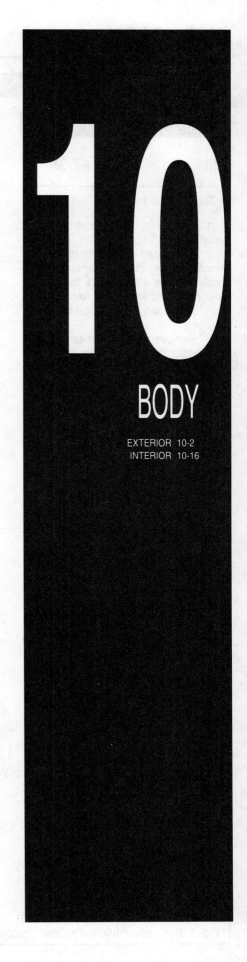

10

BODY

EXTERIOR

Doors

REMOVAL & INSTALLATION

▶ **See Figures 1 and 2**

1. Disconnect the negative battery terminal.

2. If equipped with power door locks, windows or any other power option located on the door, remove the inner door panel and waterproof film from the door. Disconnect the electrical harness connector.

3. Remove the wire harness retainers and extract the harness from the door.

4. Remove the spring pin and disconnect the door check strap. To prevent the strap from falling inside the door, install the retainer into the hole in the end of the check rod.

5. Matchmark the position of both the upper and lower door hinge to aid in alignment during installation. While supporting the door, remove the door-to-hinge bolts and lift the door from the vehicle.

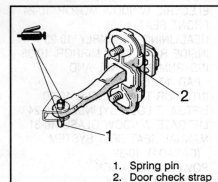

1. Spring pin
2. Door check strap

Fig. 2 Door check strap and spring pin

To install:

6. Position the door on the vehicle and loosely install the hinge bolts.

7. Align each hinge to the matchmarks and secure the hinge mounting bolts.

8. Install and connect the electrical wire harness. Secure harness to the door using retainers.

9. Connect the door check strap and reinstall the interior door trim panel.

10. Close the door slowly and check for proper alignment. Adjust the door as required and reconnect the negative battery cable.

ADJUSTMENT

▶ **See Figures 3, 4 and 5**

Adjustment of the door hinge is made easier with hinge bolt loosening wrench MB990834-01 or equivalent.

1. Apply appropriate tape to the fender and door edges to protect painted surfaces from damage.

2. Use the special wrench to loosen the hinge mounting bolts on the body side of the door slightly.

3. Adjust the door so the clearance around the door is uniform on all sides.

4. Tighten the hinge bolts to 12-19 ft. lbs. (17-26 Nm), once the door is in the desired position.

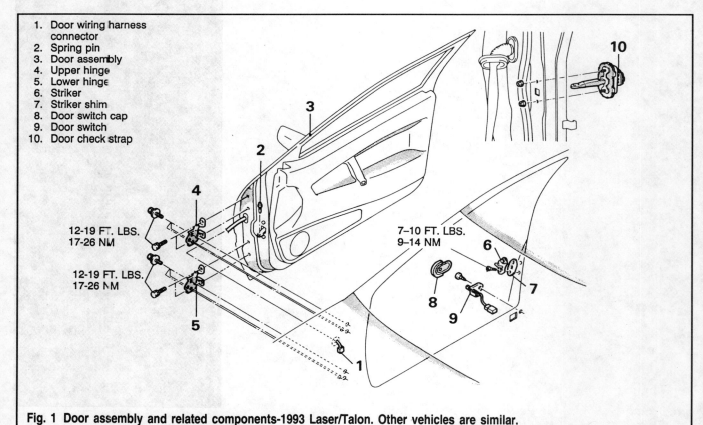

1. Door wiring harness connector
2. Spring pin
3. Door assembly
4. Upper hinge
5. Lower hinge
6. Striker
7. Striker shim
8. Door switch cap
9. Door switch
10. Door check strap

12-19 FT. LBS.
17-26 NM

12-19 FT. LBS.
17-26 NM

7-10 FT. LBS.
9-14 NM

Fig. 1 Door assembly and related components-1993 Laser/Talon. Other vehicles are similar.

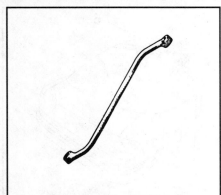

Fig. 3 Door hinge mounting bolt adjustment wrench MB990834-01

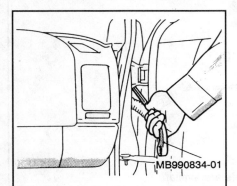

Fig. 4 Using the special tool to loosen the upper hinge mounting bolts

Fig. 5 Adjustment of the door striker

5. When the hinge has been replaced, adjust the alignment as follows:

a. Install the hinge in position on the vehicle aligning the matchmarks made during removal.

b. Loosen the hinge mounting bolts on the door side of the hinge.

c. Adjust the alignment of the fender panel with the front of the door panel.

d. Once the desired positioning is obtained, tighten the hinge mounting bolts in place.

6. To adjust the door striker, loosen the striker mounting bolts and adjust positioning as required. Increase or decease the number of shims behind the striker to adjust the engagement of the striker with the door.

7. Tighten the striker mounting bolts to 10 ft. lbs. (14 Nm) once the adjustment has been made.

Hood

REMOVAL & INSTALLATION

1. Open the hood completely.
2. Protect the cowl panel and hood from scratches during this operation. Apply protection tape or cover body surfaces before starting work.
3. Scribe a mark showing the location of each hinge on the hood to aid in alignment during installation.
4. Have an assistant help hold the hood while you remove the hood-to-hinge bolts. Use care not to damage hood or vehicle during hood removal.
5. Disconnect the connection for the washer tubes and nozzles. Lift the hood off of the vehicle.

To install:

6. Position the hood on hinges and align with the scribe marks made during removal.
7. Install and tighten the mounting bolts with enough torque to hold hood in place.
8. Close the hood slowly to check for proper alignment. Do not slam the hood closed, alignment is normally required.
9. Open the hood and adjust so that all clearances are the same and the hood panel is flush with the body.
10. After all adjustments are complete, torque hood-to-hinge bolts to 10 ft. lbs. (14 Nm) torque.

HOOD ALIGNMENT

▶ See Figures 6 and 7

1. To adjust the hood in a forward or rearward and left or right directions, loosen the hood mounting bolts and

Fig. 6 Hood adjustments can be made by turning hood bumpers or by loosening hinge mounting bolts

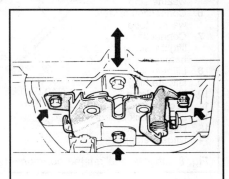

Fig. 7 Loosen the mounting bolts and adjust the hood latch assembly as required

position so the clearance is uniform on all sides.

2. To adjust the front edges of the hood in a vertical direction, turn the hood bumper cushions as required.
3. To adjust the hood lock, remove the retainer clips and the front fascia bracket. Loosen the hood latch attaching bolts slightly. Move the hood latch assembly to adjust the attachment between the hood and latch assembly.

Liftgate

REMOVAL & INSTALLATION

▶ See Figure 8

1. Open the tailgate completely.
2. Remove the inner liftgate trim panels and water deflector.
3. On models equipped with electrical options in the tailgate, disconnect the

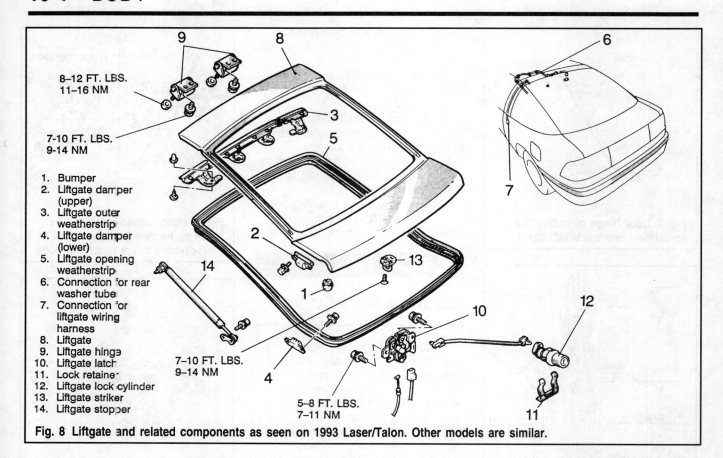

8–12 FT. LBS.
11–16 NM

7-10 FT. LBS.
9-14 NM

1. Bumper
2. Liftgate damper (upper)
3. Liftgate outer weatherstrip
4. Liftgate damper (lower)
5. Liftgate opening weatherstrip
6. Connection for rear washer tube
7. Connection for liftgate wiring harness
8. Liftgate
9. Liftgate hinge
10. Liftgate latch
11. Lock retainer
12. Liftgate lock cylinder
13. Liftgate striker
14. Liftgate stopper

7–10 FT. LBS.
9-14 NM

5–8 FT. LBS.
7–11 NM

Fig. 8 Liftgate and related components as seen on 1993 Laser/Talon. Other models are similar.

wiring harness connectors and pull the wiring from the liftgate.

4. Disconnect the rear washer tube at the liftgate.

5. Scribe hinge location marks on the liftgate to aid in installation.

6. Safely support the liftgate. Disconnect the liftgate stopper from the tailgate and position out of the way. Disconnect the rear defroster connector, if equipped.

7. Remove the liftgate-to-hinge bolts and remove the tailgate from the vehicle.

To install:

8. Position the liftgate on the vehicle and align the hinge scribe marks.

9. Install the tailgate-to-hinge bolts and tighten to 12 ft. lbs. (14 Nm).

10. Install the liftgate stopper and tighten mounting bolt, if removed.

11. Install and reconnect the electrical harness to all connectors at the liftgate.

12. Install all interior trim panels removed.

13. Close the liftgate slowly to check for proper alignment. Adjust liftgate positioning, if necessary.

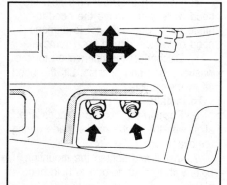

Fig. 9 Vertical and horizontal movement of the liftgate can be obtained by loosening the liftgate hinge mounting bolts and adjusting liftgate

LIFTGATE ALIGNMENT

♦ **See Figures 9 and 10**

To adjust the door in forward/rearward and left/right directions, loosen the hinge

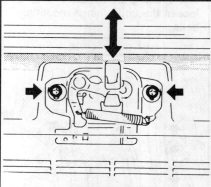

Fig. 10 To adjust the distance between the liftgate striker and the liftgate latch, loosen the latch mounting bolts

bolts and position the tailgate as required.

To adjust the tailgate lock striker, loosen the mounting bolts and using a plastic hammer, tap the striker to the desired position. Removing of the lower trim panel is normally required to access the striker.

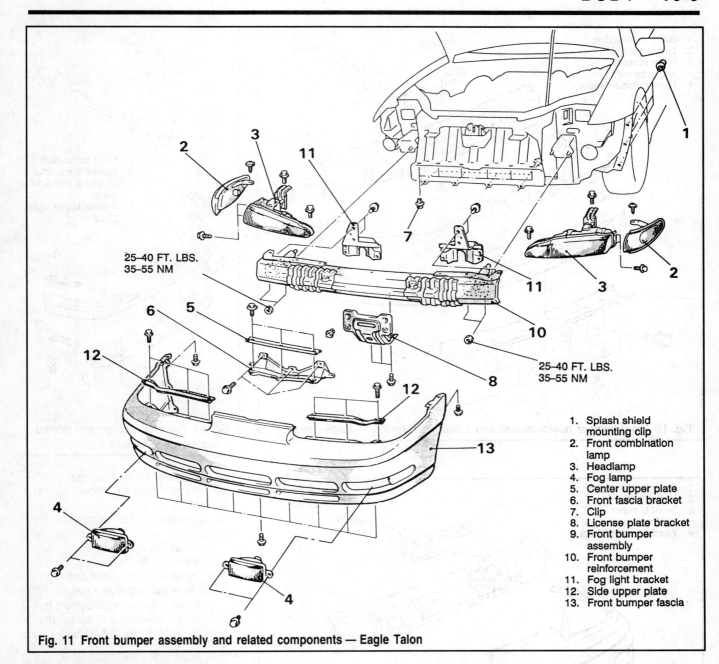

25–40 FT. LBS.
35–55 NM

25–40 FT. LBS.
35–55 NM

1. Splash shield
   mounting clip
2. Front combination
   lamp
3. Headlamp
4. Fog lamp
5. Center upper plate
6. Front fascia bracket
7. Clip
8. License plate bracket
9. Front bumper
   assembly
10. Front bumper
   reinforcement
11. Fog light bracket
12. Side upper plate
13. Front bumper fascia

Fig. 11 Front bumper assembly and related components — Eagle Talon

Front Bumper

REMOVAL & INSTALLATION

Laser and Talon

◗ **See Figures 11, 12 and 13**

1. Disconnect the negative battery cable.

2. Remove the splash shield mounting clips from the forward surface of both front wheel openings.

3. Remove both front combination lamps from the vehicle. Label and disconnect the electrical connectors from each prior to removal.

4. Label and disconnect the electrical harness from the headlamps. Remove the headlamps from the vehicle.

5. Remove the mounting screws and the fog lamps from the vehicle. Label and disconnect the electrical harness from each lamp.

6. Remove the 3 mounting bolts and the center upper plate. Remove the retainers and the front fascia bracket.

7. Remove the license plate bracket.

8. Remove the retainers and the front bumper assembly from the vehicle.

9. Disassemble the front bumper assembly as follows:

a. Remove the front bumper absorbers, if still on bumper assembly.

➡**Do not attempt to reuse a bumper absorber that has been compressed in an accident; always replace it with a new one. Before discarding the bumper absorber, drill a 0.13 in. (3mm) diameter hole in the side of the absorber to discharge the gas contained in the unit.**

b. Remove the retainers and the lower plate from the underside of the bumper fascia.

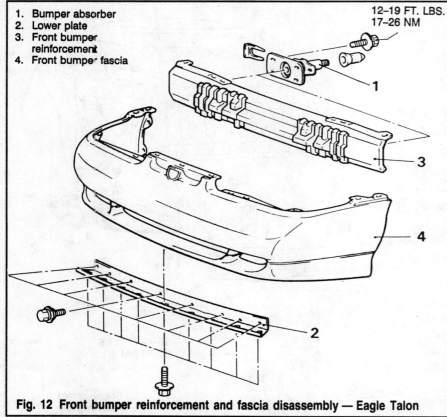

1. Bumper absorber
2. Lower plate
3. Front bumper reinforcement
4. Front bumper fascia

12–19 FT. LBS.
17–26 NM

Fig. 12 Front bumper reinforcement and fascia disassembly — Eagle Talon

1. Front air spoiler
2. Spoiler front plate
3. Spoiler lower side plate
4. Spoiler upper side plate

Fig. 14 Front air spoiler and related components — Eclipse

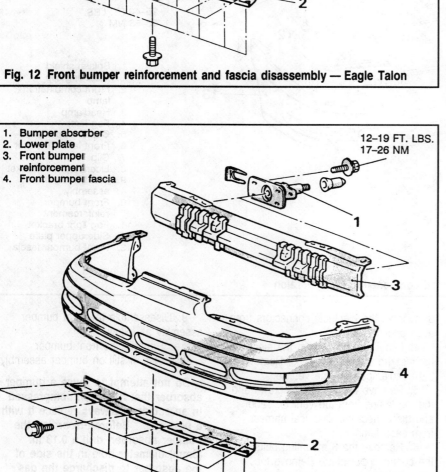

1. Bumper absorber
2. Lower plate
3. Front bumper reinforcement
4. Front bumper fascia

12–19 FT. LBS.
17–26 NM

Fig. 13 Front bumper reinforcement and fascia disassembly — Plymouth Laser

c. Remove the front bumper reinforcement from the fascia.

To install:

10. Assemble the front bumper as follows:

a. Install the front bumper reinforcement into the fascia.

b. Install the lower plate and retainers to the fascia assembly.

c. Install the bumper absorbers to the bumper assembly and tighten the bolts to 14-19 ft. lbs. (17-26 Nm).

➡It is important that the bumper absorbers be installed to the bumper assembly squarely. If the squareness between the front bumper reinforcement and the bumper assembly is improper, adjust it by putting a spacer between them.

11. Install the front bumper assembly and retainers on the vehicle.

12. Install the license plate bracket and secure in position.

13. Install the center upper plate and secure in place using the 3 mounting bolts. Install the front fascia bracket and retainers.

14. Install the fog lamps into the recesses in the bumper and secure in

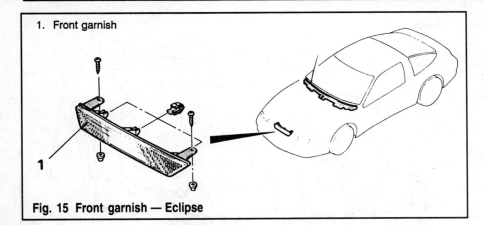

1. Front garnish

Fig. 15 Front garnish — Eclipse

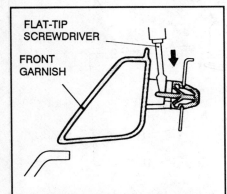

FLAT-TIP
SCREWDRIVER

FRONT
GARNISH

Fig. 16 To remove garnish, press downward on the claws of the retaining clips and pull garnish outward

place using the mounting screws. Connect the electrical harness to each lamp prior to installation.

15. Connect the electrical harness to the headlamps. Install the headlamps to the vehicle.

16. Install both front combination lamps to the vehicle. Connect the electrical connectors to each lamp prior to installation.

17. Install the splash shield mounting clips in both front wheel openings.

Eclipse

▶ See Figures 14, 15, 16, 17, 18, 19, 20 and 21

1. Disconnect the negative battery cable.

2. If equipped with front lower spoiler, remove as follows:

   a. Remove the retaining bolts from the under side of the spoiler, which go into the spoiler front plate. To make removal easier, leave 2 bolts loosely installed to retain the spoiler to the front fascia assembly.

   b. Remove the bolts on both sides of the front spoiler, which go through the spoiler side lower plate. These bolts also retain an upper spoiler side plate.

   c. Remove the remaining front plate bolts and the front lower air spoiler from the vehicle.

3. Remove the splash shield mounting clips from the forward surface of both front wheel openings.

4. Remove the front garnish from the front of the fascia as follows:

   a. Remove the mounting screws at either end of the front garnish.

   b. While pressing downward on the claws of the retainer clips of the garnish, remove the front garnish.

5. Remove both front turn signal lamps from the vehicle. Label and disconnect the electrical connectors from each prior to removal.

6. Remove both front combination lamps from the vehicle. Label and disconnect the electrical connectors from each prior to removal.

7. Remove the air intake cover and/or front fog lamps. Label and disconnect the wiring harness connectors from each lamp prior to removal. Remove the fog lamp brackets.

8. Remove the retainers and the front bumper assembly from the vehicle.

9. Disassemble the front bumper from the fascia as follows:

   a. Remove the front bumper absorbers.

➡**Do not attempt to reuse a bumper absorber that has been compressed in an accident; always replace it with a new one. Before discarding the bumper absorber, drill a 0.13 in. (3mm) diameter hole in the side of the absorber to discharge the gas contained in the unit.**

   b. Remove the retainers and the fascia upper plate.

   c. Remove the retainers and the fascia lower plate.

   d. Remove the front license plate bracket.

   e. Remove the front bumper reinforcement from the fascia.

To install:

10. Assemble the front bumper reinforcement and fascia as follows:

   a. Install the front bumper reinforcement to the fascia.

   b. Install the front license plate bracket and secure with the retainers.

   c. Install the fascia lower plate and retainers.

   d. Install the fascia upper plate and retainers.

   e. Install the bumper absorbers to the bumper assembly and install bolts. Tighten the bolts to 12-19 ft. lbs. (17-26 Nm).

➡**It is important that the bumper absorbers be installed to the bumper assembly squarely. If the squareness between the front bumper reinforcement and the bumper assembly is improper, adjust it by putting a spacer between them.**

11. Install the front bumper assembly and retainers on the vehicle.

12. Install fog lamp brackets and secure.

13. Connect electrical harness connectors to the fog lamps and install lamps into the brackets. Secure in place using retainers. Install the air intake cover, if removed.

14. Install both front combination lamps to the vehicle. Connect the electrical connectors to each lamp prior to installation.

15. Connect the electrical harness to both turn signal lamps and install onto the vehicle.

16. Install the front garnish in place using new spring retainers as required. Secure in place with mounting screws.

17. Install splash shield mounting clips in both front wheel openings.

18. If equipped with lower spoiler, install as follows:

   a. With the aid of a helper, install the lower air spoiler to the vehicle with upper side plates in position. Temporarily hold in place with 2 mounting bolts on both sides of the spoiler.

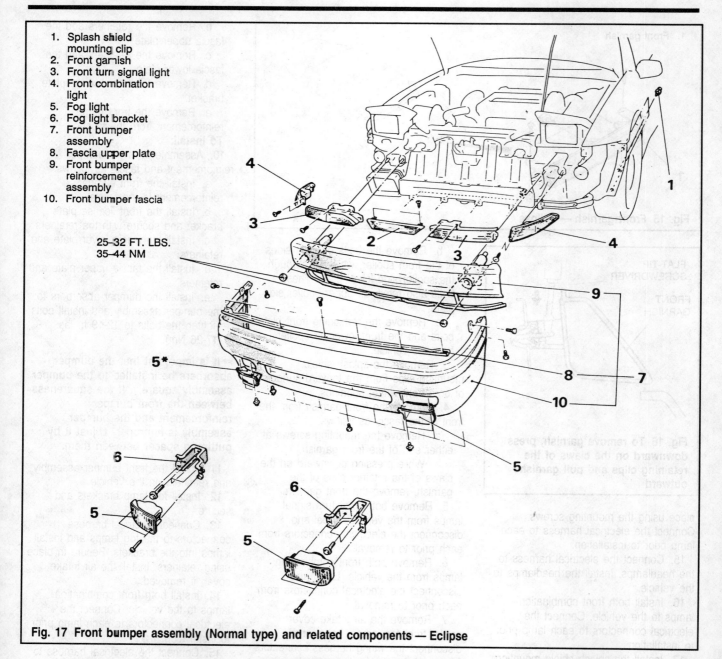

1. Splash shield mounting clip
2. Front garnish
3. Front turn signal light
4. Front combination light
5. Fog light
6. Fog light bracket
7. Front bumper assembly
8. Fascia upper plate
9. Front bumper reinforcement assembly
10. Front bumper fascia

25–32 FT. LBS.
35–44 NM

Fig. 17 Front bumper assembly (Normal type) and related components — Eclipse

b. Install the spoiler front plate and retainers tightening just enough to hold spoiler against fascia assembly.

c. Remove the retainers that were temporarily installed. Position side lower and upper plates on one side of the spoiler in position and secure in place with the retainers. Repeat procedure on remaining side.

d. Once all retainers are installed and the spoiler is in correct alignment, tighten all spoiler retainers securely.

19. Connect the negative battery cable.

Rear Bumper

REMOVAL & INSTALLATION

Eclipse
▶ **See Figures 22, 23, 24, 25, 26 and 27**

1. Disconnect the negative battery cable.

2. Remove the splash shield retainers located in both rear wheel opening.

3. Remove the rear panel garnish as follows:

a. Remove the high mounted stop light cover.

b. Remove the liftgate interior trim. Using a plastic trim tool, remove the clip mounting areas on the back of the liftgate and remove the trim from the liftgate.

c. Remove the retainers and the rear panel garnish from the vehicle.

4. Disconnect and remove the rear combination lights.

5. Disconnect the wiring harness connection for the license plate light and the body wiring harness.

1. Splash shield
   mounting clip
2. Front garnish
3. Front turn signal light
4. Front combination
   light
5. Fog light
6. Fog light bracket
7. Front bumper
   assembly
8. Fascia upper plate
9. Front bumper
   reinforcement
   assembly
10. Front bumper fascia

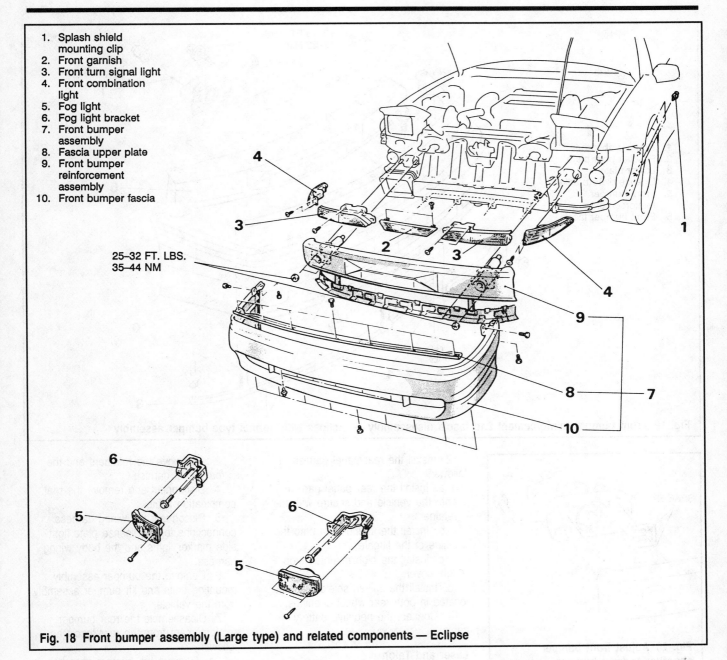

25–32 FT. LBS.
35–44 NM

Fig. 18 Front bumper assembly (Large type) and related components — Eclipse

6. Remove the rear fascia upper plate retainers. Remove the bumper assembly mounting bolts and lift bumper assembly from the vehicle.

7. Disassemble the rear bumper reinforcement from the rear fascia as follows:

   a. Remove the retainers and the rear license plate light.

   b. Remove the back-up light.

   c. Remove the license plate wiring harness from the rear bumper assembly if still in place.

   d. Remove the mounting bolts and the bumper absorbers from the bumper assembly.

➡**Do not attempt to reuse a bumper absorber that has been compressed in an accident; always replace it with a new one. Before discarding the bumper absorber, drill a 0.13 in. (3mm) diameter hole in the side of the absorber to discharge the gas contained in the unit.**

   e. Remove the rear bumper reinforcement from the rear fascia.

To install:

8. Assemble the rear bumper reinforcement to the rear fascia as follows:

   a. Install the rear bumper reinforcement into the rear bumper fascia. Install the rear bumper side plates.

   b. Install the bumper absorbers to the bumper reinforcement and tighten retainer bolts to 19 ft. lbs. (26 Nm).

   c. Install the license plate wiring harness to the rear bumper assembly.

   d. Install the back-up light.

   e. Install the rear license plate light in position and secure using retainers.

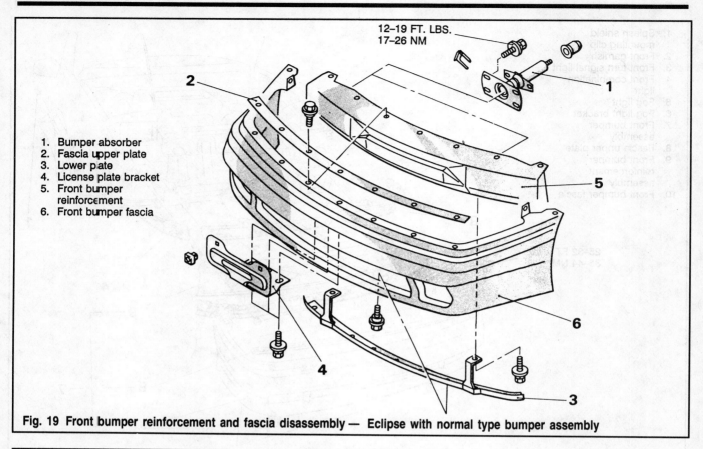

12–19 FT. LBS.
17–26 NM

1. Bumper absorber
2. Fascia upper plate
3. Lower plate
4. License plate bracket
5. Front bumper reinforcement
6. Front bumper fascia

Fig. 19 Front bumper reinforcement and fascia disassembly — Eclipse with normal type bumper assembly

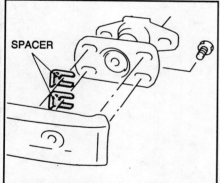

SPACER

Fig. 20 Adjust front bumper absorber to front bumper reinforcement squareness using spacers

9. Install the bumper assembly onto the vehicle and tighten mounting nuts to 39 ft. lbs. (55 Nm).

10. Install the fascia upper plate and retainers. Connect the electrical connector for the license plate and the rear body harness.

11. Install the rear combination lights, making sure to connect the electrical harness connectors prior to installation.

12. Install the rear panel garnish as follows:

   a. Install the rear panel garnish onto the vehicle and secure with the retainers.

   b. Install the interior trim onto the inside of the liftgate.

   c. Install the high mounted stop light cover.

13. Install the splash shield retainers located in both rear wheel openings.

14. Connect the negative battery cable.

Laser and Talon
▶ **See Figures 28, 29, 30, 31 and 32**

1. Disconnect the negative battery cable.

2. Remove the splash shield retainers located in both rear wheel opening.

3. Remove the rear panel garnish as follows:

   a. Remove the liftgate interior trim. Using a plastic trim tool, remove the clip mounting areas on the back of the liftgate and remove the trim from the liftgate.

   b. Remove the retainers and the back-up light.

   c. Remove the retainers and the rear panel garnish.

4. Disconnect and remove the rear combination lights.

5. Disconnect the wiring harness connections for the license plate light, side marker lights and the body wiring harness.

6. Remove the bumper assembly mounting bolts and lift bumper assembly from the vehicle.

7. Disassemble the rear bumper reinforcement from the rear fascia as follows:

   a. Remove the bumper absorber mounting bolts and absorber.

➡ **Do not attempt to reuse a bumper absorber that has been compressed in an accident; always replace it with a new one. Before discarding the bumper absorber, drill a 0.13 in. (3mm) diameter hole in the side of the absorber to discharge the gas contained in the unit.**

   b. Remove the rear bumper reinforcement from the rear fascia.

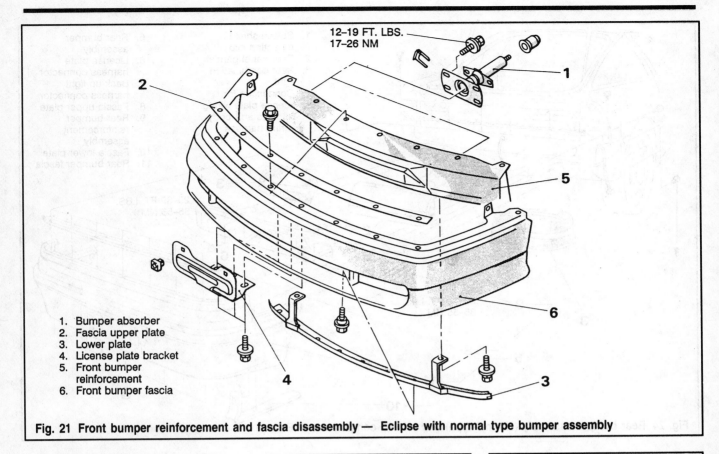

1. Bumper absorber
2. Fascia upper plate
3. Lower plate
4. License plate bracket
5. Front bumper reinforcement
6. Front bumper fascia

12–19 FT. LBS.
17–26 NM

Fig. 21 Front bumper reinforcement and fascia disassembly — Eclipse with normal type bumper assembly

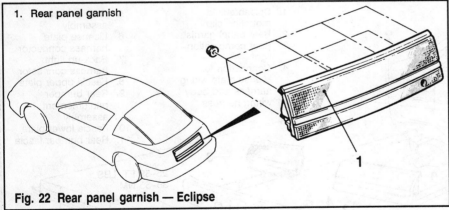

1. Rear panel garnish

Fig. 22 Rear panel garnish — Eclipse

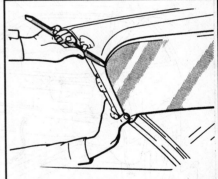

Fig. 23 When removing the liftgate interior trim, use a plastic trim tool to release the retainer clips

To install:

8. Assemble the rear bumper reinforcement to the rear fascia as follows:

a. Install the rear bumper reinforcement into the rear bumper fascia.

b. Install the bumper absorbers to the bumper reinforcement and tighten mounting bolts to 19 ft. lbs. (26 Nm).

9. Install the bumper onto the vehicle and install the mounting nuts. Tighten nuts to 39 ft. lbs. (55 Nm).

10. Reconnect all electrical connections disconnected during the removal procedure.

11. Install the combination lamps. Making sure to connect the wiring harness prior to installation.

12. Install the rear garnish panel as follows:

a. Install the rear garnish onto the vehicle and tighten the retaining nuts.

b. Install the back-up light.

c. Install the rear interior liftgate trim.

13. Install the splash shield retainers located in both rear wheel openings.

14. Connect the negative battery cable.

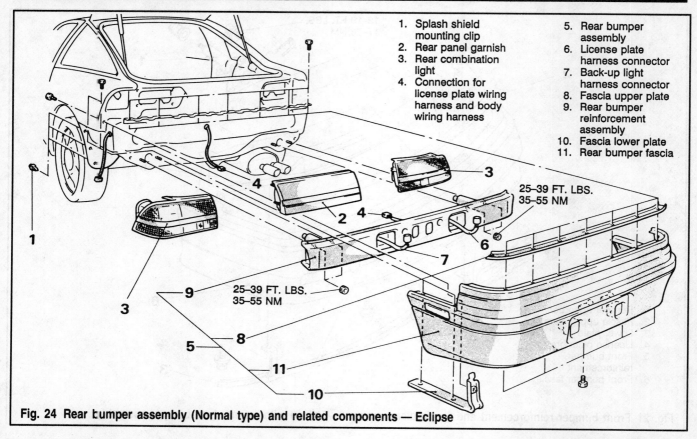

1. Splash shield mounting clip
2. Rear panel garnish
3. Rear combination light
4. Connection for license plate wiring harness and body wiring harness
5. Rear bumper assembly
6. License plate harness connector
7. Back-up light harness connector
8. Fascia upper plate
9. Rear bumper reinforcement assembly
10. Fascia lower plate
11. Rear bumper fascia

25–39 FT. LBS.
35–55 NM

25–39 FT. LBS.
35–55 NM

Fig. 24 Rear bumper assembly (Normal type) and related components — Eclipse

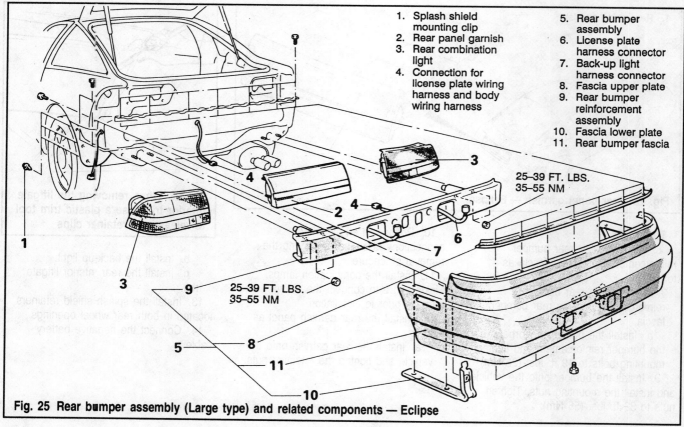

1. Splash shield mounting clip
2. Rear panel garnish
3. Rear combination light
4. Connection for license plate wiring harness and body wiring harness
5. Rear bumper assembly
6. License plate harness connector
7. Back-up light harness connector
8. Fascia upper plate
9. Rear bumper reinforcement assembly
10. Fascia lower plate
11. Rear bumper fascia

25–39 FT. LBS.
35–55 NM

25–39 FT. LBS.
35–55 NM

Fig. 25 Rear bumper assembly (Large type) and related components — Eclipse

1. License plate light
2. Back-up light
3. Rear side marker light
4. License plate light wiring harness
5. Bumper absorber
6. Rear bumper reinforcement
7. License plate bracket
8. Rear bumper fascia

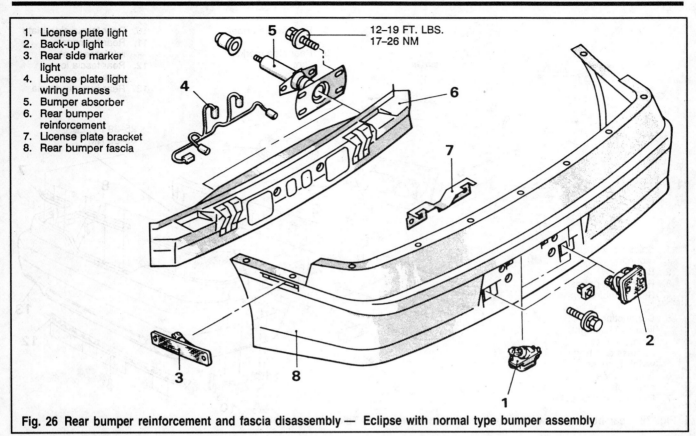

12–19 FT. LBS.
17–26 NM

Fig. 26 Rear bumper reinforcement and fascia disassembly — Eclipse with normal type bumper assembly

1. License plate bracket
2. Back-up light
3. Rear side marker light
4. License plate light wiring harness
5. Bumper absorber
6. Rear bumper reinforcement
7. License plate bracket
8. Rear bumper fascia

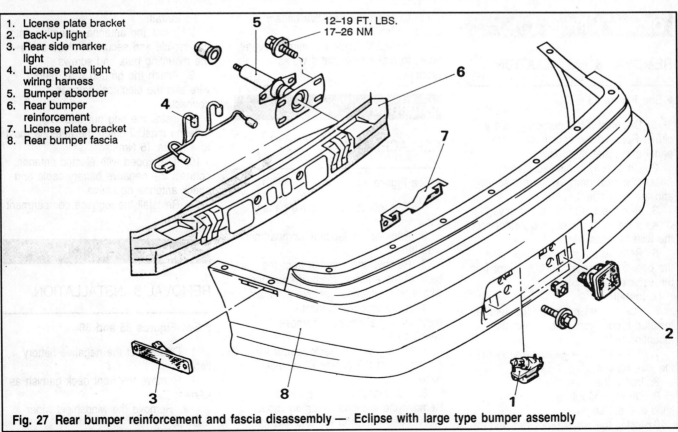

12–19 FT. LBS.
17–26 NM

Fig. 27 Rear bumper reinforcement and fascia disassembly — Eclipse with large type bumper assembly

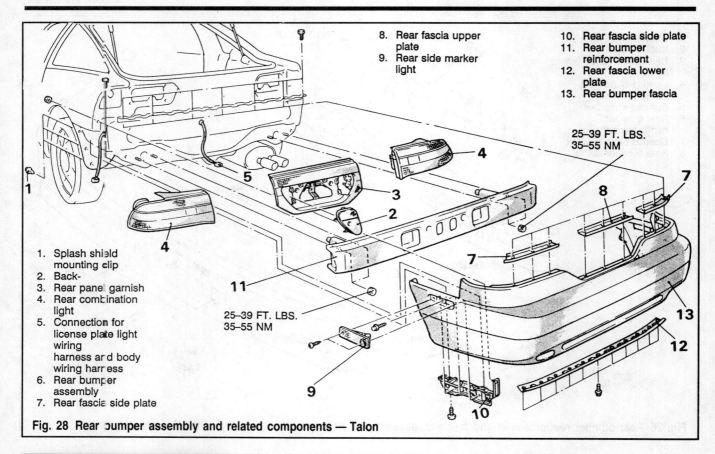

8. Rear fascia upper plate
9. Rear side marker light
10. Rear fascia side plate
11. Rear bumper reinforcement
12. Rear fascia lower plate
13. Rear bumper fascia

25–39 FT. LBS.
35–55 NM

25–39 FT. LBS.
35–55 NM

1. Splash shield mounting clip
2. Back-
3. Rear panel garnish
4. Rear combination light
5. Connection for license plate light wiring harness and body wiring harness
6. Rear bumper assembly
7. Rear fascia side plate

Fig. 28 Rear bumper assembly and related components — Talon

Outside Mirror

REMOVAL & INSTALLATION

▶ See Figure 33

1. Disconnect the negative battery cable. Remove the door interior trim panel and water proof film.
2. On directly controlled mirror (manual mirror), remove the set screw and the adjustment knob.
3. Remove the inner mirror cover.
4. If the mirror is electric, disconnect the wire harness.
5. Remove the mounting nuts and the door trim bracket. Lift the mirror from the vehicle.

To install:

6. Position the mirror on the vehicle. Install the door trim bracket and the mounting nuts.
7. If the mirror is electric, reconnect the wire harness.
8. Install the inner cover.
9. On manual mirrors, install the knob and set screw.
10. Install the water proof film and the door interior trim panel.

11. Connect the negative battery cable.
12. If electric, cycle the mirror several times to make sure that it works properly.

Antenna

REPLACEMENT

▶ See Figure 34

1. Disconnect the negative battery cable.
2. Remove the luggage compartment side trim.
3. Remove the ring nut from the base of the antenna mast. If equipped with a whip antenna, remove the mast.
4. If the antenna is electric, disconnect the electrical harness connector.
5. Disconnect the radio feeder wire connection at the antenna assembly base.
6. Disconnect the ground wire. Remove the antenna assembly mounting screws and nuts and the antenna from the vehicle.

To install:

7. Install the antenna assembly onto the vehicle and secure in place using the mounting nuts and screws.
8. Attach the ground wire, feeder wire and the electrical harness connector.
9. Install the ring nut and the antenna mast. Tighten the antenna mast to 4 ft. lbs. (5 Nm).
10. If equipped with electric antenna, connect the negative battery cable and check antenna operation.
11. Reinstall the luggage compartment side trim.

Fenders

REMOVAL & INSTALLATION

▶ See Figures 35 and 36

1. Disconnect the negative battery cable.
2. Remove the front deck garnish as follows:
   a. Remove the windshield wiper arms.

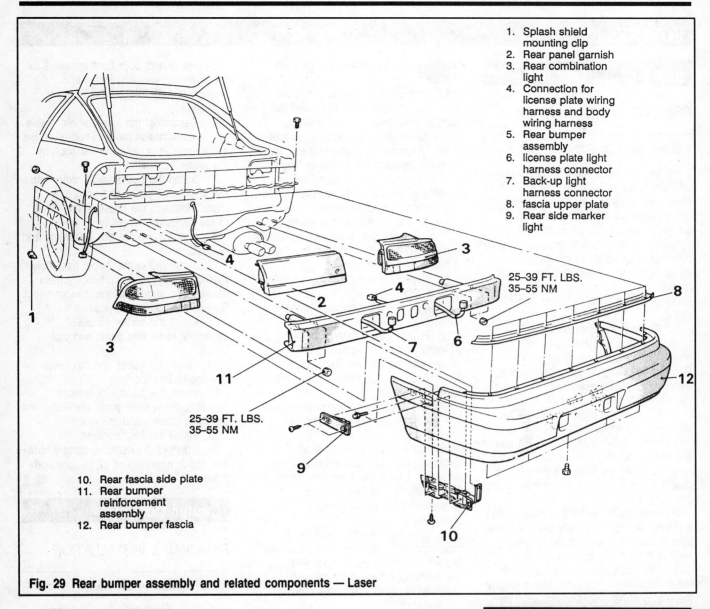

1. Splash shield mounting clip
2. Rear panel garnish
3. Rear combination light
4. Connection for license plate wiring harness and body wiring harness
5. Rear bumper assembly
6. license plate light harness connector
7. Back-up light harness connector
8. fascia upper plate
9. Rear side marker light

25–39 FT. LBS.
35–55 NM

25–39 FT. LBS.
35–55 NM

10. Rear fascia side plate
11. Rear bumper reinforcement assembly
12. Rear bumper fascia

Fig. 29 Rear bumper assembly and related components — Laser

b. Remove the mounting bolts and the front deck garnish from the vehicle.

3. Remove the turn signal lamp and the combination lamp.

4. Remove the headlamp lower bezel.

5. Remove the side air dam and air outlet garnish.

6. Remove the retainer bolts and the front splash shield extension, if equipped.

7. Remove the retainers and the front splash shield.

8. Remove the retainer bolts along the upper, lower and rear side of the fender.

9. Remove the fender from the vehicle.

To install:

10. Apply sealant between the fender and the body panels. This will assure that there are no gaps when the fender is mounted. Install the fender to the vehicle and tighten all retainer bolts to 4 ft. lbs. (6 Nm).

11. Install the fender splash shield and secure in place using retainers.

12. Install front splash shield extension, if removed.

13. Install the side air dam, headlamp bezel, front turn signal lamp and the combination lamp.

14. Install the front deck garnish and wiper arms.

15. Connect the negative battery cable.Sunroof

Pop-Up Sunroof

▶ **See Figures 37 and 38**

The sunroof used on these vehicles is manually operated. Removal of the base plate assembly or hinge female assembly should be done from the interior of the vehicle, after removing the sunshade, roof lid and weather-strip from outside of the vehicle. If a water leak from the sunroof should develop, inspect the drains for restriction and repair as required. The sunroof drains run through the front pillar post and in the body behind the front door. They exit the body of the vehicle at the back side of the front left wheel opening and the front side of the left rear wheel opening.

Instrument Panel and Pad

REMOVAL & INSTALLATION

♦ See Figures 39, 40, 41, 42 and 43

For installation of the instrument panel, different types of fasteners were used. During installation, it is important that these fasteners are installed in their original locations. To aid in this, the specific fasteners and their positions are referenced by letters in the exploded views of the instrument panel which follow.

1. Disconnect the negative battery cable.

2. Remove the floor console.

3. The fasteners for the knee protectors are covered by plugs. Remove the plugs and the screws from the knee protector assembly. Remove the assembly from the vehicle.

4. Remove the retainer and the hood lock release handle from the vehicle.

5. Remove the steering column upper and lower covers.

6. Remove the retainers from the cluster panel assembly and remove from the vehicle.

7. Remove the radio trim plate. Using a plastic trim tool, carefully pry the lower part of the radio panel outward and remove it from the console.

8. Remove the radio retainers and the radio from the console. Disconnect the electrical harness and the antenna lead from the radio and remove it from the vehicle.

9. Remove the center air outlet assembly. While depressing the locking paws of the center air outlet assembly with a flat-tip screwdriver, remove the center air outlet assembly by prying outward with plastic trim tool.

10. Remove the screw retainers from each dial knob on the center cluster. Gently pull the dial knobs from the center cluster panel assembly.

11. Remove the fasteners and the cluster panel assembly from the vehicle. Disconnect the harness connectors as required.

12. Remove the stoppers from inside the glove box assembly. Remove the

retainers and the glove box from the vehicle.

13. Remove the retainers from the combination meter. Pull the meter out slightly. Disconnect the electrical connections from the combination meter.

14. Disconnect the speedometer cable from the combination meter as follows:

  a. Disconnect the speedometer cable at the transaxle assembly.

  b. Pull the speedometer cable slightly toward the vehicle interior.

  c. Release the cable lock by turning the adapter to the left or right, and then remove the adapter.

15. Remove the right and left speaker covers. Disconnect and remove the speakers from the instrument panel assembly.

16. Remove the heater control assembly retaining screws and disconnect the harness connectors. Remove the heater control assembly from the vehicle.

17. Remove the lap cooler duct and the shower duct taking note of their orientation.

18. Remove the steering shaft mounting bolts and allow the steering wheel to rest on the front seat cushion. Make sure no harness wires or connections are being pulled or stretched.

19. Remove the instrument panel retaining bolts, label and disconnect all electrical harness connectors. Remove the instrument panel from the vehicle. Disassemble components as required.

To install:

20. Install the instrument panel into the vehicle, reconnect all harness connections and install retaining bolts. Before installing bolts, make sure the electrical harness wires were not pinched during instrument panel installation.

21. Raise the steering shaft into position and install the mounting bolts. Tighten the bolts to 9 ft. lbs. (12 Nm).

22. Install the lap cooler duct and the shower duct in the same position as removed. Install the mounting screw to hold in position.

23. Install the heater control assembly, making sure to reconnect all connections prior to installation.

24. Reconnect both front speakers to the radio harness and install into instrument panel. Install the speaker garnishes.

25. Reconnect the speedometer cable to the combination meter. Make sure the cable locks in position on the back of the meter assembly.

26. Install the combination meter into the instrument panel and secure in place. Reattach the speedometer cable to the transaxle assembly.

27. Install the glove box assembly and door stops.

28. Install the center panel assembly. Connect the radio equipment harness connections and install into cluster panel. Secure components in position.

29. Install the center air outlet assembly, radio trim panel and dial knobs.

30. Install the upper and the lower steering column covers.

31. Install the hood lock release handle and the knee protector. Install the mounting screws and the plugs.

32. Install the floor console.

33. Connect the negative battery cable and check operation of all gauges and meters.

Center Console

REMOVAL & INSTALLATION

♦ See Figures 44 and 45

1. Disconnect the negative battery cable.

2. Remove the screw plugs in the side covers. Remove the retainer screws and the side covers from the vehicle.

3. Remove the front mounting screw cover from the floor console. Remove the manual transaxle shift lever knob.

4. Remove the cup holder and the carpet inserts from the floor console assembly.

5. Label and disconnect the electrical wire harness connections for the floor console.

6. Remove the mounting bolts and the floor console from the vehicle. Disassemble components as required.

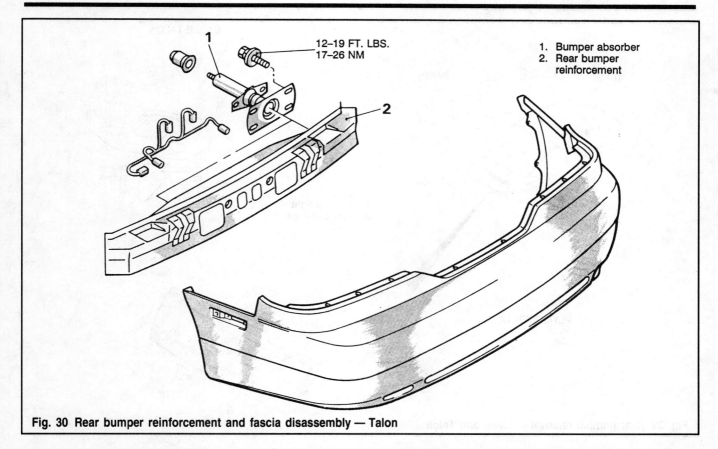

12–19 FT. LBS.
17–26 NM

1. Bumper absorber
2. Rear bumper
   reinforcement

Fig. 30 Rear bumper reinforcement and fascia disassembly — Talon

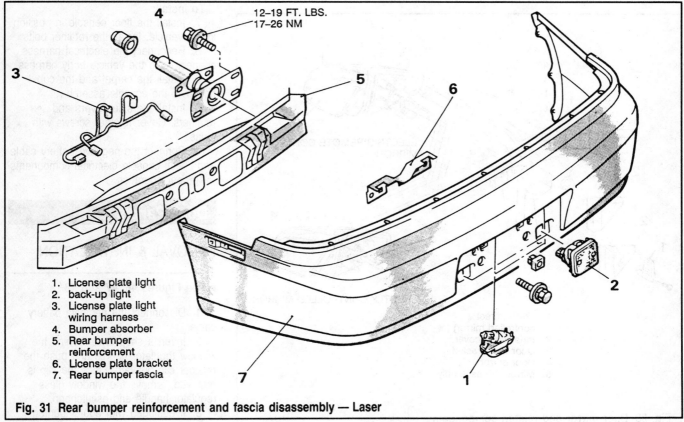

12–19 FT. LBS.
17–26 NM

1. License plate light
2. back-up light
3. License plate light
   wiring harness
4. Bumper absorber
5. Rear bumper
   reinforcement
6. License plate bracket
7. Rear bumper fascia

Fig. 31 Rear bumper reinforcement and fascia disassembly — Laser

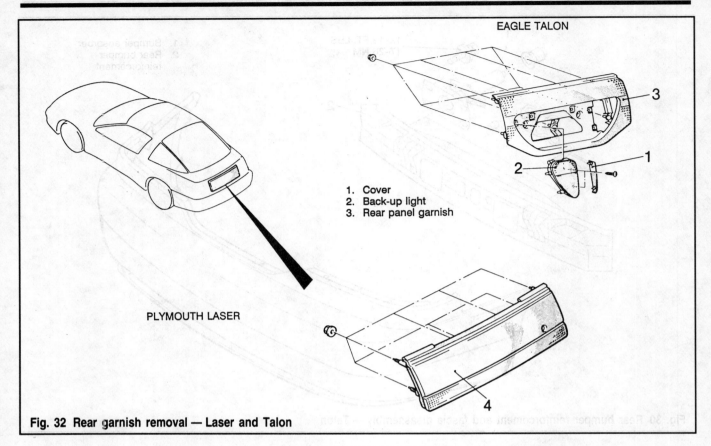

1. Cover
2. Back-up light
3. Rear panel garnish

EAGLE TALON

PLYMOUTH LASER

Fig. 32 Rear garnish removal — Laser and Talon

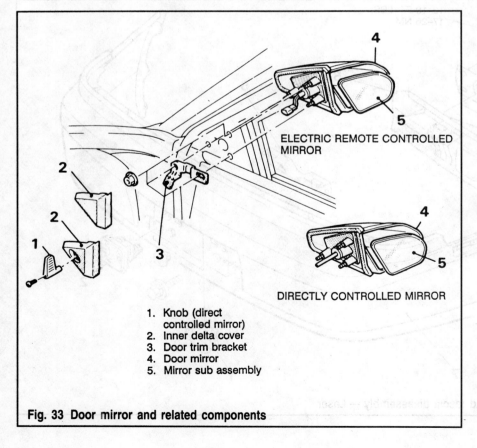

ELECTRIC REMOTE CONTROLLED MIRROR

DIRECTLY CONTROLLED MIRROR

1. Knob (direct controlled mirror)
2. Inner delta cover
3. Door trim bracket
4. Door mirror
5. Mirror sub assembly

Fig. 33 Door mirror and related components

To install:

7. Install the floor console in position in the vehicle. Install the retainer bolts.

8. Reconnect the electrical harness connectors to the vehicle body harness.

9. Install the carpet and the cup holder to the console assembly.

10. Install the side covers and retainers. Cover retainer screws with plugs.

11. Connect the negative battery cable and check console electrical components for proper operation.

Door Panels

REMOVAL & INSTALLATION

▶ See Figures 46, 47 and 48

1. Disconnect the negative battery cable.

2. Insert a shop towel behind the window regulator handle and push the retainer clip outward. Once the clip is removed, remove the window glass regulator handle and escutcheon.

3. If equipped with door mounted speakers, remove the speaker covers

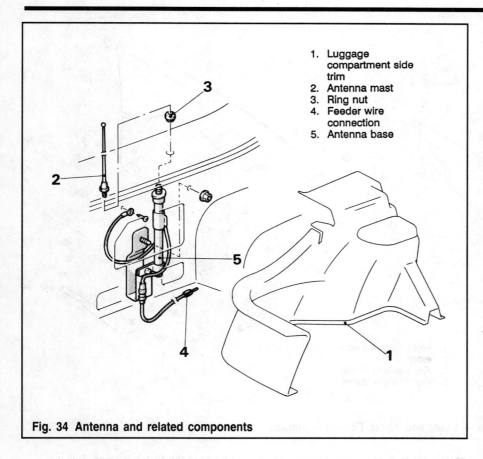

1. Luggage compartment side trim
2. Antenna mast
3. Ring nut
4. Feeder wire connection
5. Antenna base

Fig. 34 Antenna and related components

and the speakers from the door assembly.

4. Remove the door grip mounting screws and the grip from the door panel. remove the pull handle, if equipped.

5. Pull the door panel gently from the door. The panel is retained by spring clips. To separate the clips from the door, slide a small prying tool behind the clip and carefully pull the clip, with the door panel, outward. Try not to bends the door panel or damage may occur.

6. Once all retainer clips are removed from the door, remove the door panel assembly. If equipped with power windows or mirror, remember to disconnect the electrical connector from the switch prior to removal.

7. Carefully remove any retainer clips that remained in the door during panel removal. If damaged, replace the retainer clips.

To install:

8. Install and missing or damaged retainer clips into the door panel. Connect the electrical harness connector to the electric switches on the door panel and install the door panel onto the vehicle. Push retainers into the holes in the door until they lock into position. If any clips do not lock, replace with new.

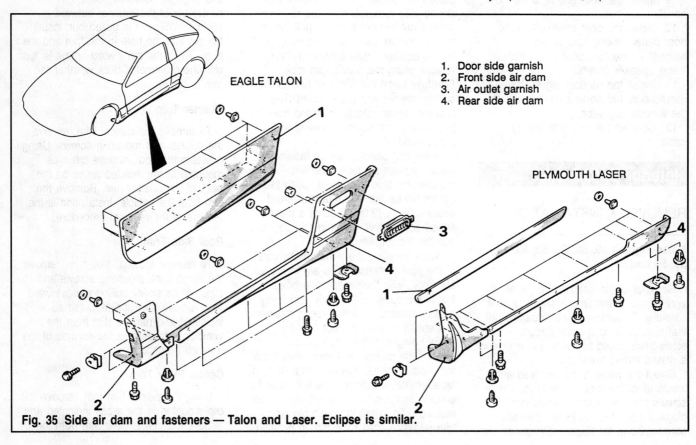

EAGLE TALON

PLYMOUTH LASER

1. Door side garnish
2. Front side air dam
3. Air outlet garnish
4. Rear side air dam

Fig. 35 Side air dam and fasteners — Talon and Laser. Eclipse is similar.

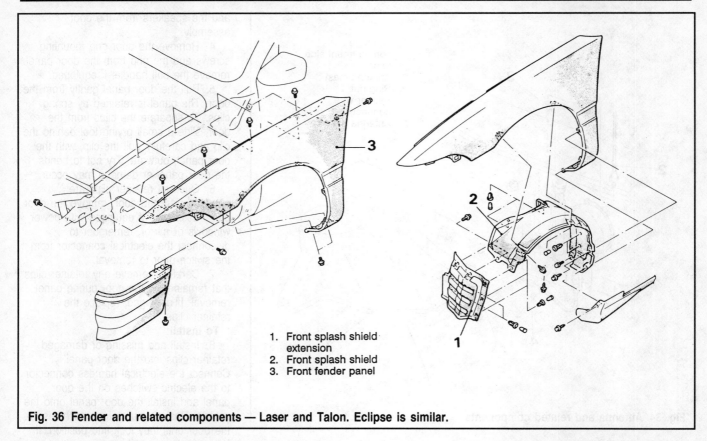

1. Front splash shield extension
2. Front splash shield
3. Front fender panel

Fig. 36 Fender and related components — Laser and Talon. Eclipse is similar.

9. Install the door grip onto the door panel.

10. Install the door speakers to the door panel making sure to attach harness connectors prior to installation. Install speaker covers.

11. Install the window regulator escutcheon, handle and retainer clip to the window regulator.

12. Connect the negative battery cable.

Interior Trim Panels

REMOVAL & INSTALLATION

▶ **See Figures 49, 50, 51, 52, 53, 54, 55, 56 and 57**

Removal of the interior trim panels is very simple, if the time is taken to do it carefully. To remove the trim panels, all that is required is a small tipped screwdriver, cross tipped screwdriver and a plastic trim removal tool.

Select the panel to be removed and locate all of the retaining screws. These screws may be located behind trim plugs. To remove the plugs, carefully pry outward using flat tipped screwdriver. Be

careful not to damage the interior trim during plug removal. After the retainer screws are removed, gently pull the panel outward away from the body. If trim is equipped with spring retainers, release using the plastic trim tool. Carefully insert the trim tool behind the spring retainer and pry outward. The retainer should release from the hole in the body of the vehicle allowing for panel removal.

Some trim panels may be fastened in place by push pin type retainers. To remove this type of retainer, push the center pin on the trim retainer inward about 0.08 in. (2.0mm) using a cross tipped screwdriver. Once in this position, pull the trim outward.

When installing the panel, make sure in the spring retainer clips are in proper alignment with their respective holes. Install the panel to the body and firmly press inward to seat the spring retainers. If a spring retainer will not lock the panel being installed into position, replace the retainer with new. Install the trim panel retainer screws and tighten in an alternate sequence. This will allow for uniform tightness and prevent trim panel warpage. To install the push pin type trim retainers, install the trim in position

and align the fastener holes. Pull the center pin of the retainer outward slightly. While pin is pulled out, install the pin into the hole in the trim and the body. Push the pin inward until it is flush with the grommet. Check whether the trim is secure.

Quarter Trim

To remove the quarter trim, remove the quarter trim mounting screws. Using a plastic trim tool, remove the clips mounted at the indicted areas on the back of the quarter trim. Remove the trim from the vehicle. Installation is the reverse of the removal procedure.

Rear Side Trim

To remove the rear side trim, remove the cargo light, mounting screws and clips. While slowly pulling panel toward you, remove the clips mounted as indicated and remove trim from the vehicle. Installation is the reverse of the removal procedure.

Center Pillar Trim

Using the plastic trim tool, remove the clip mounted in the areas indicated and remove the center pillar trim from the

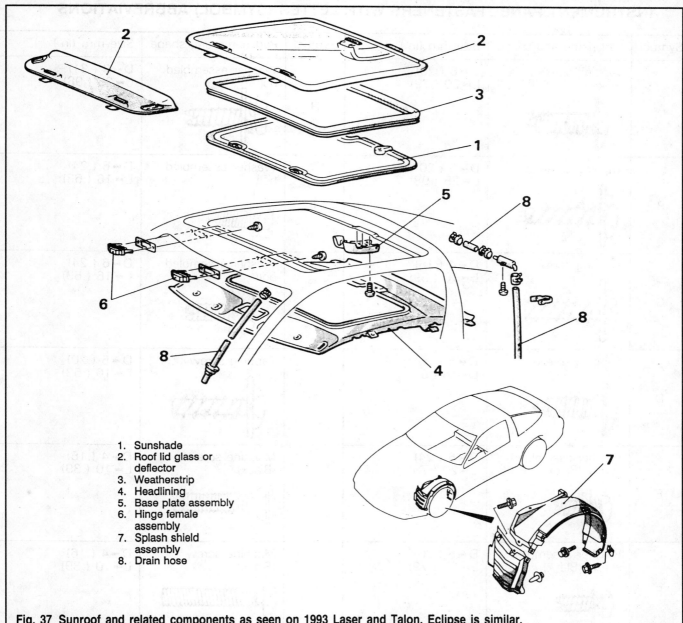

1. Sunshade
2. Roof lid glass or
   deflector
3. Weatherstrip
4. Headlining
5. Base plate assembly
6. Hinge female
   assembly
7. Splash shield
   assembly
8. Drain hose

Fig. 37 Sunroof and related components as seen on 1993 Laser and Talon. Eclipse is similar.

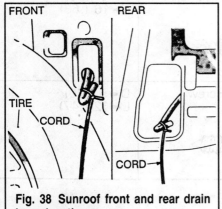

Fig. 38 Sunroof front and rear drain hose locations

vehicle. Installation is the reverse of the removal procedure.

Front Pillar Trim

To remove the front pillar trim, use the plastic trim tool to remove the mounting clips located in the areas indicated. Separate the velcro areas (U.S. vehicles), on the back of the front pillar trim and remove the trim from the vehicle. Installation is the reverse of the removal procedure.

Rear Pillar Trim

To remove the rear pillar trim, use the plastic trim tool to release the retainers located in the indicated areas. Remove the rear pillar post from the vehicle. Installation is the reverse of the removal procedure.

When removing and installing any trim panel, always take the time to do the job right. When properly performed, it should be impossible to tell that a trim panel has been disturbed.

INSTRUMENT PANEL FASTENERS WITH LETTER (SYMBOL) ABBREVIATIONS

Symbol	Part name and shape	Size mm (in.)	Symbol	Part name and shape	Size mm (in.)
A	Tapping screw (Black)	D – 5 (.20) L – 20 (.79)	H	Washer assembled bolt	D – 8 (.31) L – 25 (.98)
B	Tapping screw	D – 5 (.20) L – 25 (.98)	I	Washer assembled bolt	D – 6 (.24) L – 16 (.63)
C	Tapping screw	D – 4 (.16) L – 10 (.39)	J	Washer assembled bolt	D – 6 (.24) L – 16 (.63)
D	Tapping screw	D – 5 (.20) L – 16 (.63)	K	Tapping screw	D – 5 (.20) L – 16 (.63)
E	Washer assembled screw (Black)	D – 6 (.24) L – 20 (.79)	L	Machine screw (Black)	D – 4 (.16) L – 10 (.39)
F	Washer assembled screw (Black)	D – 5 (.20) L – 20 (.79)	M	Machine screw (Black)	D – 4 (.16) L – 10 (.39)
G	Washer assembled bolt	D – 6 (.24) L – 16 (.63)	N	Washer assembled screw	D – 5 (.20) L – 16 (.63)
			O	Tapping screw	D – 5 (.20) L – 12 (.47)

D – Thread diameter
L – Effective thread length

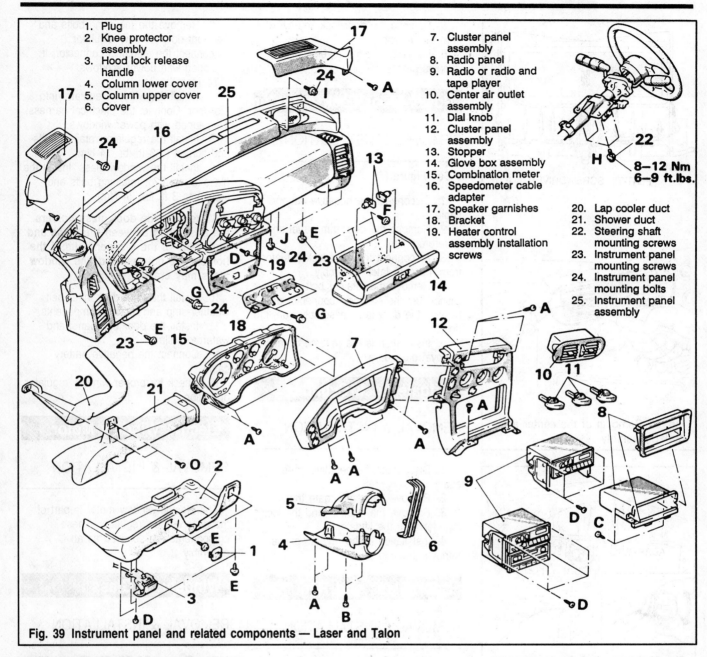

1. Plug
2. Knee protector assembly
3. Hood lock release handle
4. Column lower cover
5. Column upper cover
6. Cover
7. Cluster panel assembly
8. Radio panel
9. Radio or radio and tape player
10. Center air outlet assembly
11. Dial knob
12. Cluster panel assembly
13. Stopper
14. Glove box assembly
15. Combination meter
16. Speedometer cable adapter
17. Speaker garnishes
18. Bracket
19. Heater control assembly installation screws
20. Lap cooler duct
21. Shower duct
22. Steering shaft mounting screws
23. Instrument panel mounting screws
24. Instrument panel mounting bolts
25. Instrument panel assembly

8–12 Nm
6–9 ft.lbs.

Fig. 39 Instrument panel and related components — Laser and Talon

Headlining Assembly

REMOVAL & INSTALLATION

♦ See Figure 58

1. Disconnect the negative battery cable.
2. Remove the center pillar trim, front pillar trim and guide rail.
3. Remove the dome light.
4. Remove the sun visor.
5. If equipped with sunroof, remove the sunshade, roof lid glass and sunroof headlining trim.

6. Remove the headlining trim by turning outward and pulling downward.
7. Remove the retaining clips and the headlining from the vehicle. Be careful not to bend the headlining during removal.

To install:

8. Install the headlining into the vehicle and position in approximate location.
9. Install the headliner retainers. Inspect headliner for proper positioning. Install the headlining trim.
10. Install the sunroof headlining trim, roof lid glass and sunshade, if removed.

11. Install the sunvisor and dome light. Install the center pillar trim, front pillar trim and guide rail.
12. Connect the negative battery cable.

Door Lock Cylinder

REMOVAL & INSTALLATION

♦ See Figure 59

1. Disconnect the negative battery cable.

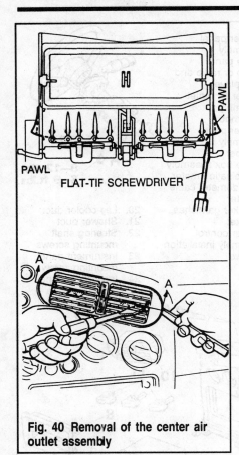

Fig. 40 Removal of the center air outlet assembly

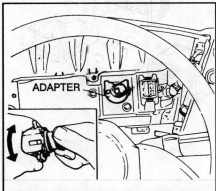

Fig. 41 Removal of the speedometer cable adapter

2. Remove the door interior trim panel and weatherproof film. Position the door glass so the key cylinder and related components are accessible.

3. Disconnect the actuator rod(s) from the lock cylinder. If equipped with a theft alarm system, disconnect the electrical connector at the door key cylinder unlock switch.

4. Remove the retainer ring from the base of the door lock cylinder.

5. Remove the door lock key cylinder from the vehicle.

6. Installation is the reverse of the removal procedure.

Power Door Lock Actuator

REMOVAL & INSTALLATION

▶ See Figure 60

1. Disconnect the negative battery cable.

2. Remove the door trim panel and waterproof film.

3. Disconnect the door lock actuator from the door latch assembly.

4. Remove the retainer bolt and nut. Disconnect the electrical connector and remove the door latch assembly from the door.

5. Installation is the reverse of the removal procedure.

Liftgate Lock Cylinder

REMOVAL & INSTALLATION

1. Disconnect the negative battery cable.

2. Remove the rear liftgate trim.

3. Remove the retainer and the lock cylinder from the liftgate.

4. Installation is the reverse of the removal procedure.

Door Glass and Regulator

REMOVAL & INSTALLATION

▶ See Figures 61, 62 and 63

1. Disconnect the negative battery cable.

2. Remove the interior door trim panel and waterproof film.

3. Remove the door belt line inner weather-strip as follows:

   a. Remove the door mounted mirror.

   b. Pry upward and remove the front door inner belt weather-strip from the top of the door panel.

4. Remove the bolts holding the glass to the window regulator. Remove the door window glass.

5. Remove the door glass holder.

6. Remove the mounting bolts and the front door window regulator. Disconnect the electrical connector, if equipped with power windows.

To install:

7. Install the window regulator into the door. Connect the electrical harness, if equipped with power windows.

8. Install the regulator retainer bolts and tighten to 3 ft. lbs. (4 Nm).

9. Install the window glass. Install the glass-to-regulator retaining bolts and tighten to 3 ft. lbs. (4 Nm).

➡Make sure the door glass holders are positioned between the glass and the heads of the retainer bolts. If the holders are not installed, the window glass may crack.

10. Install the front door inner belt weather-strip and related components.

11. Install the door trim panel and waterproof film.

12. Connect the negative battery cable.

13. Check for proper window regulator operation.

Electric Window Motor

REMOVAL & INSTALLATION

The electric window motor is part of the door window regulator. These components are serviced as an assembly.

Windshield Glass

REMOVAL & INSTALLATION

▶ See Figures 64, 65, 66, 67 and 68

Extreme care must be taken when removing, installing or resealing a windshield. The windshield will crack if a stress in the wrong direction is exerted, even if the pressure is very slight. Undesirable stress on the windshield could take place during a number of operations. Damage could occur days after the repair has been completed, caused by the body of the vehicle flexing during normal vehicle operation, or during the removal of the outer moldings before any repair work has been done. This should be realized

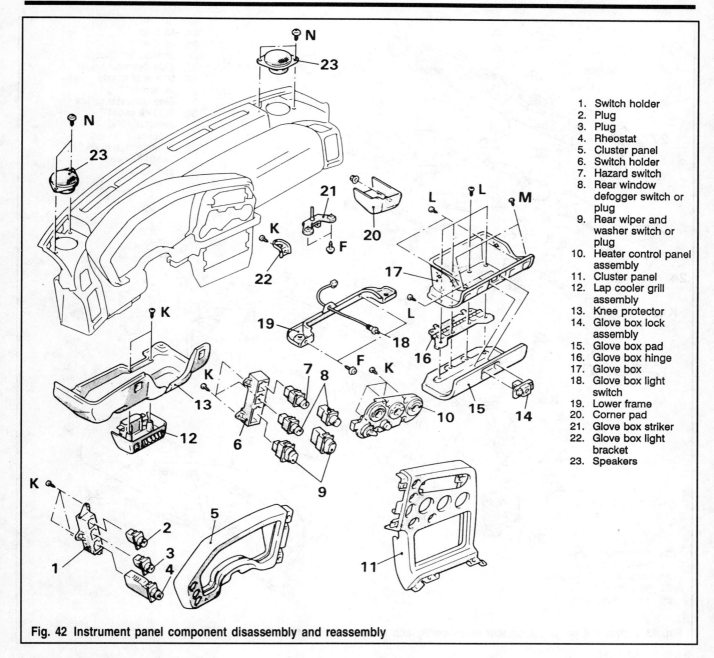

Fig. 42 Instrument panel component disassembly and reassembly

1. Switch holder
2. Plug
3. Plug
4. Rheostat
5. Cluster panel
6. Switch holder
7. Hazard switch
8. Rear window defogger switch or plug
9. Rear wiper and washer switch or plug
10. Heater control panel assembly
11. Cluster panel
12. Lap cooler grill assembly
13. Knee protector
14. Glove box lock assembly
15. Glove box pad
16. Glove box hinge
17. Glove box
18. Glove box light switch
19. Lower frame
20. Corner pad
21. Glove box striker
22. Glove box light bracket
23. Speakers

before the job of removing a windshield is undertaken.

The windshield glass is attached by an urethane based adhesive to the window frame. The adhesive provides improved glass holding and sealing.

1. Disconnect the negative battery cable.

2. Remove the inside rear view mirror from the windshield.

3. Remove the hood, wiper arms and the front deck garnish.

4. Remove the front headlining trim and the drip molding.

5. Tape the end of a scraper and insert between the body and the upper windshield molding. Pry the molding

upward and out of the adhesive. Remove the molding from the vehicle.

6. Using a sharp pointed drill, make a hole in the adhesive area of the windshield.

7. From inside the vehicle, insert 1 end of wire through the adhesive material applied to the upper edge of the windshield.

8. Insert the other end of the wire from inside the vehicle through the adhesive material applied to the lower edge of the windshield.

9. Secure each end of the wire to a piece of wood or similar object.

10. Carefully cut through the adhesive pulling the wire along the windshield

from outside of the vehicle, in a sawing motion.

➡In order to prevent the body of the vehicle from damage, apply cloth tape to all body areas around the installed glass before cutting the adhesive.

11. Using chalk, make mating marks on the glass and the body of the vehicle if the glass is to be reinstalled. Remove the glass from the vehicle using glass holders.

12. Thoroughly remove the adhesive deposited on the windshield glass and clean by applying isopropyl alcohol to

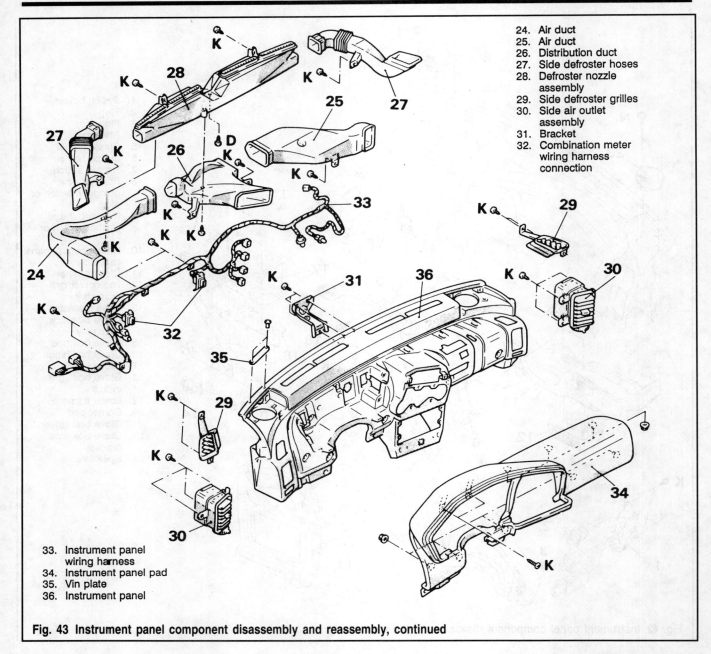

24. Air duct
25. Air duct
26. Distribution duct
27. Side defroster hoses
28. Defroster nozzle assembly
29. Side defroster grilles
30. Side air outlet assembly
31. Bracket
32. Combination meter wiring harness connection

33. Instrument panel wiring harness
34. Instrument panel pad
35. Vin plate
36. Instrument panel

Fig. 43 Instrument panel component disassembly and reassembly, continued

the sealing surfaces. Allow to dry for at least 3 minutes.

To install:

13. Apply 3M Super Fast Urethane Primer or equivalent to the entire bonding surface of the glass, both to the inside surface and the edge of the glass. Allow to dry for 5 minutes.

➡**Use care to avoid spilling primer solution on the trim or a painted surface. Wiper away any spilled solution immediately, because the primer will damage trim or painted surfaces. Do not touch the primer coated surface.**

14. Use a cutter knife to cut away adhesive so that the thickness is within 0.08 in. (2mm) around the entire circumference of the body flange. Finish the flange surfaces so they are smooth.

➡**Be careful not to remove more adhesive than necessary, and also not to damage the paint on the body surface with a knife. If the paint is damaged, repair the damaged area with touch-up paint or Tectyl.**

15. Degrease the bonding surface of the body with isopropyl alcohol and allow to dry for at least 3 minutes.

16. Attach the windshield upper molding to the windshield glass.

17. Using an adhesive gun, coat the Urethane adhesive on the windshield glass mounting surface of the body flange.

➡**Urethane adhesive will normally start to cure after 15 minutes when exposed to air.**

18. Using glass holders, position the windshield glass onto the body opening. Gently press the glass so that no adhesive appears. Make sure the glass in in the marked position. Do not move the glass once in position against the body. Wipe off any adhesive that overflows.

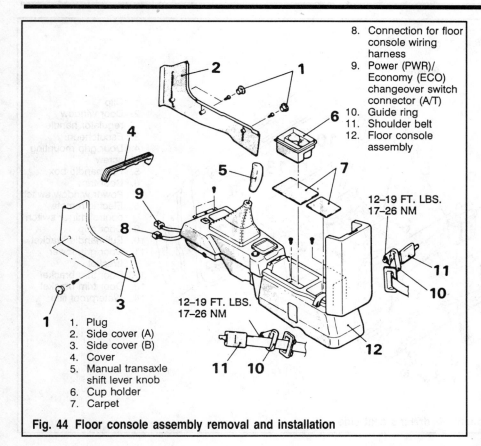

8. Connection for floor console wiring harness
9. Power (PWR)/ Economy (ECO) changeover switch connector (A/T)
10. Guide ring
11. Shoulder belt
12. Floor console assembly

12–19 FT. LBS.
17–26 NM

12–19 FT. LBS.
17–26 NM

1. Plug
2. Side cover (A)
3. Side cover (B)
4. Cover
5. Manual transaxle shift lever knob
6. Cup holder
7. Carpet

Fig. 44 Floor console assembly removal and installation

1. Plug
2. Stoppers
3. Lid inner
4. Lock lever
5. Lock lever spring
6. Accessory box lid
7. Floor console garnish assembly
8. Floor console wiring harness
9. Coin box
10. Power (PWR)/ Economy (ECO) changeover switch (A/T)
11. Coin holder (M/T)
12. Shift lever boot (M/T)
13. Panel (A/T)
14. Ashtray
15. Ashtray spring
16. Cigarette lighter
17. Plug
18. Side cover
19. Console cover (Vehicles for Canada)
20. Floor console

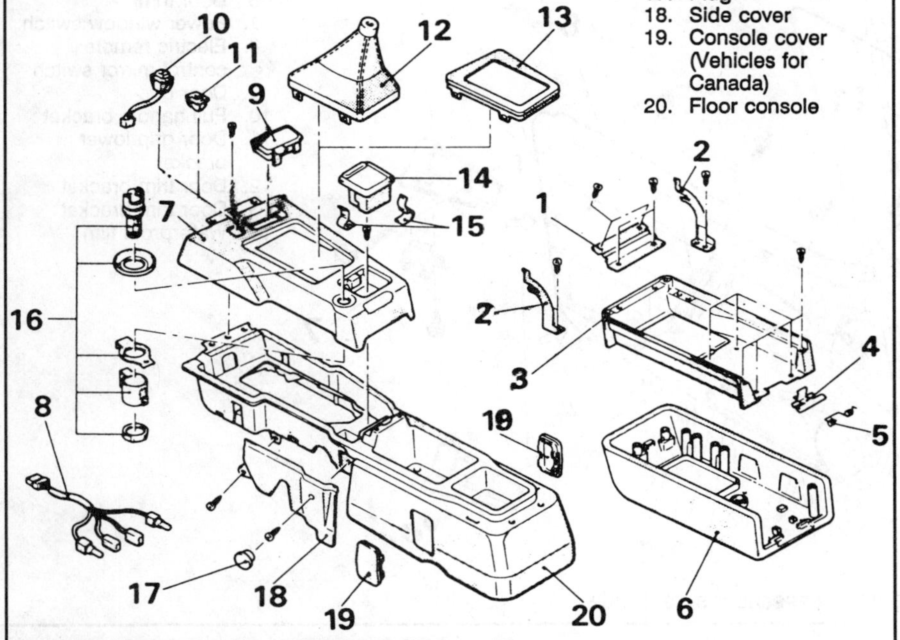

Fig. 45 Floor console component disassembly

➡During installation, use care not to close the water groove in the lower corner of the pinch weld flange with adhesive.

19. Perform a water test on the windshield. Use a cold stream of water, being careful not to direct a powerful stream of water on the new adhesive material. Allow water to spill over the edges of the glass. If there are any leaks, apply sealant at the leak point.

➡If moving the vehicle is necessary, make sure to move the vehicle very slowly. Damage to the windshield or the glass seal may occur.

20. Using a clean, lint-free cloth liberally dampened with isopropyl alcohol, wipe away dirt from the glass perimeter and the body.
21. Install the inside rear view mirror.
22. Install the drip molding, front headliner trim and front deck garnish.
23. Install the front wiper arms and the hood.
24. Reconnect the negative battery cable.

Quarter Window Glass

REMOVAL & INSTALLATION

▶ See Figure 69

1. Disconnect the negative battery cable.
2. Remove the rear side trim, quarter trim, center trim and rear pillar trim.
3. Remove the side rear air spoiler, if equipped, as follows:
   a. If equipped with stationary antenna mast, remove if it impedes spoiler removal.
   b. Remove the antenna mounting nut and the mounting insulator.
   c. Remove the rear wiper arm and liftgate trim.
   d. Remove the high mounted stop light.
   e. Remove the wiper arm grommet.
   f. From the under side of the center spoiler, remove the mounting bolts and lift the center spoiler off of the vehicle.
   g. Remove the side spoiler mounting screws and the bolts from the side air spoiler cover.

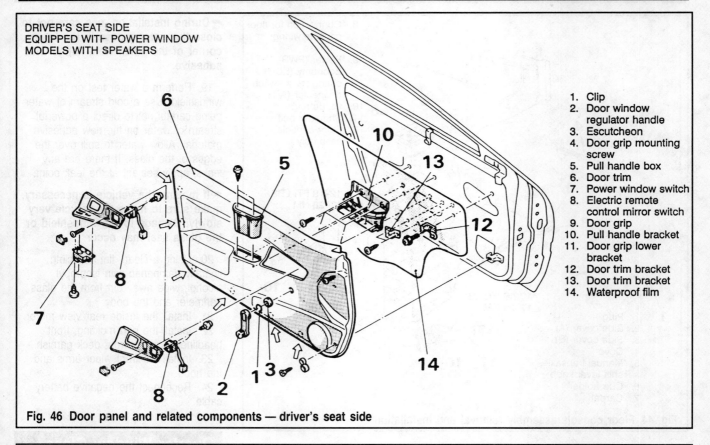

DRIVER'S SEAT SIDE
EQUIPPED WITH POWER WINDOW
MODELS WITH SPEAKERS

1. Clip
2. Door window
   regulator handle
3. Escutcheon
4. Door grip mounting
   screw
5. Pull handle box
6. Door trim
7. Power window switch
8. Electric remote
   control mirror switch
9. Door grip
10. Pull handle bracket
11. Door grip lower
    bracket
12. Door trim bracket
13. Door trim bracket
14. Waterproof film

Fig. 46 Door panel and related components — driver's seat side

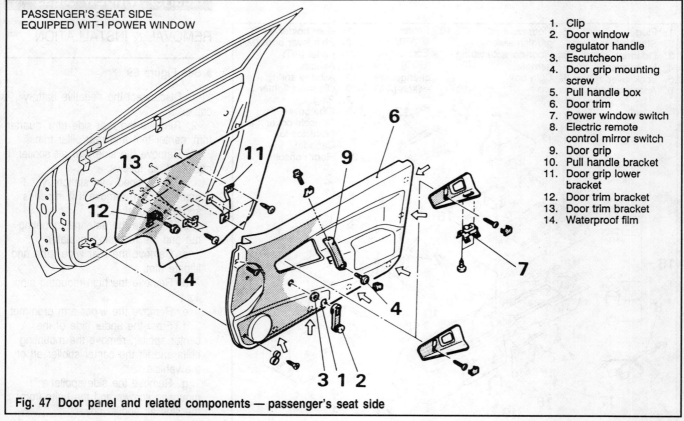

PASSENGER'S SEAT SIDE
EQUIPPED WITH POWER WINDOW

1. Clip
2. Door window
   regulator handle
3. Escutcheon
4. Door grip mounting
   screw
5. Pull handle box
6. Door trim
7. Power window switch
8. Electric remote
   control mirror switch
9. Door grip
10. Pull handle bracket
11. Door grip lower
    bracket
12. Door trim bracket
13. Door trim bracket
14. Waterproof film

Fig. 47 Door panel and related components — passenger's seat side

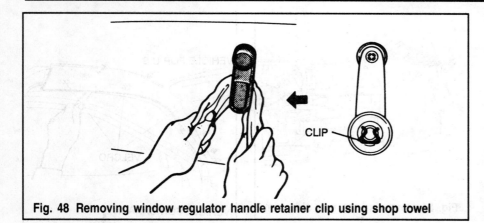

Fig. 48 Removing window regulator handle retainer clip using shop towel

Fig. 50 Push pin type interior trim retainer

1. Scuff plate
2. Cowl side trim
3. Center shelf
4. Shelf holder
5. Quarter trim
6. Quarter trim lower bracket
7. Lid
8. Lid
9. Rear end trim
10. Bracket
11. FWD bracket (FWD)
12. AWD trim bracket (AWD)
13. Rear seat belt anchor plate
14. Retractor cover
15. Rear seat belt protector
16. Rear side trim
17. Shelf catcher
18. Rear speaker bracket assembly
19. Quarter trim upper bracket

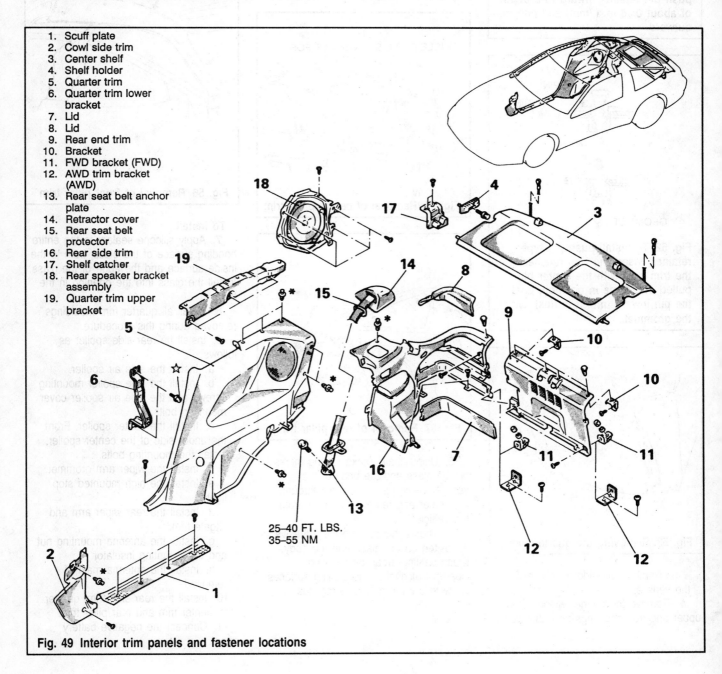

25–40 FT. LBS.
35–55 NM

Fig. 49 Interior trim panels and fastener locations

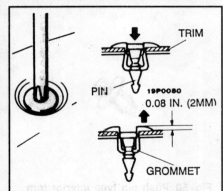

Fig. 51 To remove, push center of push pin retainer inward to a depth of about 0.08 in. (2mm) and pull trim outward

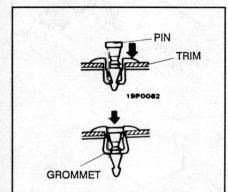

Fig. 52 To install push pin type retainer, insert the trim retainer in the trim hole with the center pin pulled out. Once in position, push the pin inward until it is flush with the grommet.

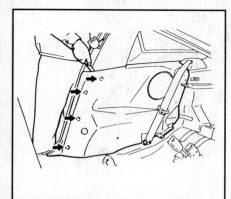

Fig. 53 Removing the quarter trim

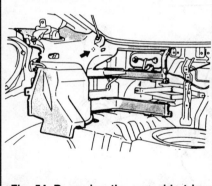

Fig. 54 Removing the rear side trim

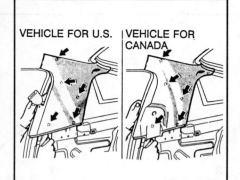

Fig. 55 Removal of center pillar trim

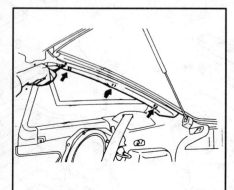

Fig. 57 Removal of rear pillar trim

5. Using chalk, make mating marks on the glass and the body of the vehicle. Using piano wire, cut through the sealer and remove the glass from the vehicle.

6. Thoroughly remove the adhesive deposited on the glass and the body. Clean sealing areas by applying isopropyl alcohol to the sealing surfaces. Allow to dry for at least 3 minutes.

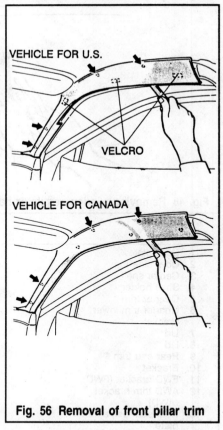

Fig. 56 Removal of front pillar trim

To install:

7. Apply silicone sealant to the entire bonding surface of the glass, both to the inside surface and the edge of the glass. Install the glass into the opening in the body.

8. Install all quarter trim mouldings removed during this procedure.

9. Install he rear side spoiler as follows:

 a. Install the side air spoiler.

 b. Install the side spoiler mounting screws and the side air spoiler cover mounting bolts.

 c. Install the center spoiler. From the under side of the center spoiler, install the mounting bolts.

 d. Install the wiper arm grommet.

 e. Install the high mounted stop light.

 f. Install the rear wiper arm and liftgate trim.

 g. Install the antenna mounting nut and the mounting insulator.

 h. Install the antenna mast, if removed.

10. Install the rear side trim, quarter trim, center trim and rear pillar trim.

11. Connect the negative battery cable.

h. Remove the side air spoiler from the vehicle.

4. Remove the quarter window lower, upper and front moldings from the glass.

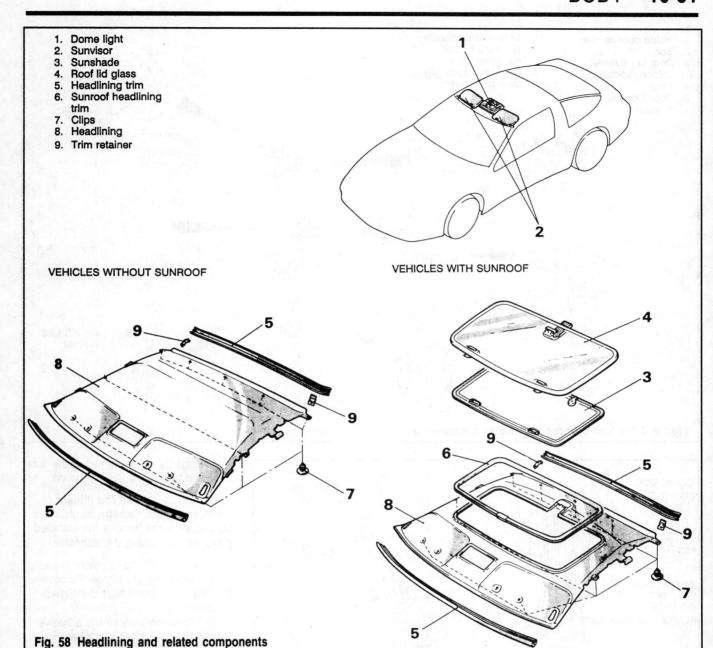

1. Dome light
2. Sunvisor
3. Sunshade
4. Roof lid glass
5. Headlining trim
6. Sunroof headlining trim
7. Clips
8. Headlining
9. Trim retainer

VEHICLES WITHOUT SUNROOF

VEHICLES WITH SUNROOF

Fig. 58 Headlining and related components

Liftgate Window Glass

REMOVAL & INSTALLATION

▶ **See Figure 70**

1. Disconnect the negative battery cable.
2. Remove the rear liftgate trim panel.
3. If equipped with a rear air spoiler, remove as follows:
   a. If equipped with stationary antenna mast, remove it from the vehicle.
   b. Remove the antenna mounting nut and the mounting insulator.
   c. Remove the rear wiper arm.
   d. Remove the high mounted stop light.
   e. Remove the wiper arm grommet.
   f. From the under side of the center spoiler, remove the mounting bolts and lift the center spoiler off of the vehicle.
   g. Remove the side spoiler mounting screws and the bolts from the side air spoiler cover.
   h. Remove both of the side air spoilers from the vehicle.

4. Remove the upper, lower and side liftgate moldings. Remove the molding clips and fasteners, if needed.
5. Disconnect the rear defroster harness connector, if equipped.
6. Using a sharp pointed drill, make a hole in the adhesive area of the liftgate glass.
7. From the rear of the glass, insert 1 end of wire through the adhesive material applied to the upper edge of the glass.
8. Insert the other end of the wire through the adhesive material applied to the lower edge of the glass.
9. Secure each end of the wire to a piece of wood or similar object. Carefully

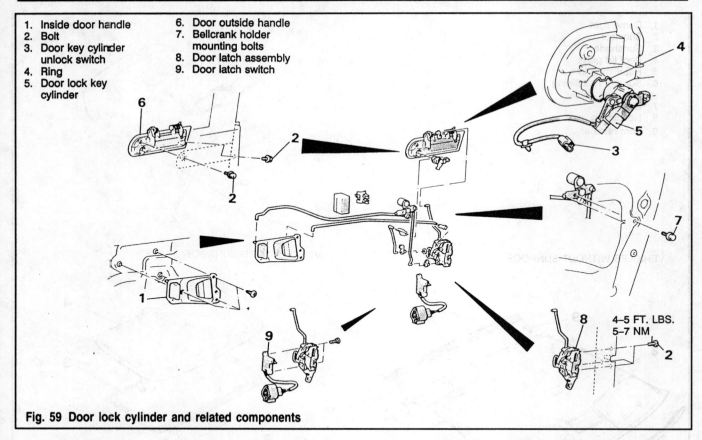

1. Inside door handle
2. Bolt
3. Door key cylinder unlock switch
4. Ring
5. Door lock key cylinder
6. Door outside handle
7. Bellcrank holder mounting bolts
8. Door latch assembly
9. Door latch switch

4–5 FT. LBS.
5–7 NM

Fig. 59 Door lock cylinder and related components

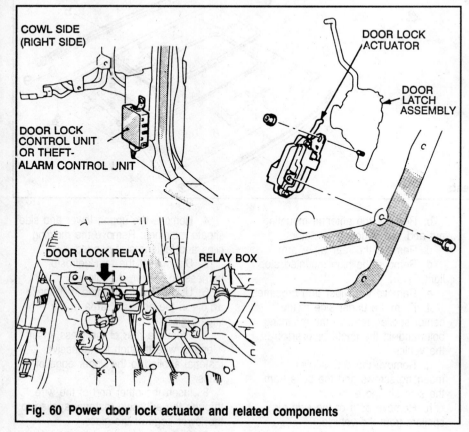

Fig. 60 Power door lock actuator and related components

cut through the adhesive pulling the wire along the glass, in a sawing motion.

➡In order to prevent the liftgate of the vehicle from damage, apply cloth tape to all areas around the installed glass before cutting the adhesive.

10. Using chalk, make mating marks on the glass and the liftgate. Remove the glass from the vehicle using glass holders.

11. Thoroughly remove the adhesive deposited on the glass and clean by applying isopropyl alcohol to the sealing surfaces. Allow to dry for at least 3 minutes.

To install:

12. Apply 3M Super Fast Urethane Primer or equivalent to the bonding surface of the glass, both to the inside surface and the edge of the glass. Allow to dry for 5 minutes.

➡Use care to avoid spilling primer solution on the trim or a painted surface. Wiper away any spilled solution immediately, because the primer will damage trim or painted surfaces. Do not touch the primer coated surface.

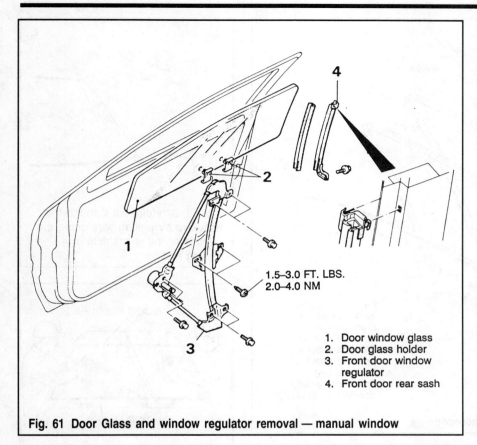

1.5–3.0 FT. LBS.
2.0–4.0 NM

1. Door window glass
2. Door glass holder
3. Front door window regulator
4. Front door rear sash

Fig. 61 Door Glass and window regulator removal — manual window

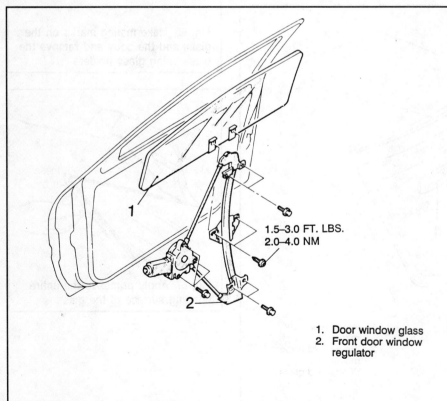

1.5–3.0 FT. LBS.
2.0–4.0 NM

1. Door window glass
2. Front door window regulator

Fig. 62 Door Glass and window regulator removal — power window

13. Use a knife to cut away adhesive so that the thickness is within 0.08 in. (2mm) around the entire circumference of the body flange. Finish the flange surfaces so they are smooth.

➡ **Be careful not to remove more adhesive than necessary and also not to damage the paint on the body surface with a knife. If the paint is damaged, repair the damaged area with touch-up paint.**

14. Degrease the bonding surface of the body with isopropyl alcohol and allow to dry for at least 3 minutes.

15. Using an adhesive gun, coat the Urethane adhesive on the liftgate glass mounting surface.

➡ **Urethane adhesive will normally start to cure after 15 minutes when exposed to air.**

16. Using glass holders, position the liftgate glass onto the body opening. Gently press the glass in place. Make sure the glass is in the marked position. Do not move the glass once in position against the body. Wipe off any adhesive that overflows.

17. Install the liftgate moldings and retainer clips.

18. Attach the harness connector to the rear window defogger, if equipped.

19. Install the rear spoiler as follows:
  a. Position the side air spoilers on the vehicle.
  b. Install the side spoiler mounting screws and the side air spoiler cover mounting bolts.
  c. Install the center spoiler. From the under side of the center spoiler, install the mounting bolts.
  d. Install the wiper arm grommet and wiper arm.
  e. Install the high mounted stop light.
  f. Install the rear liftgate trim.
  g. Install the antenna mounting nut and the mounting insulator.
  h. Install the antenna mast.

20. Connect the negative battery cable.

21. Water test the liftgate glass and inspect for leaks.

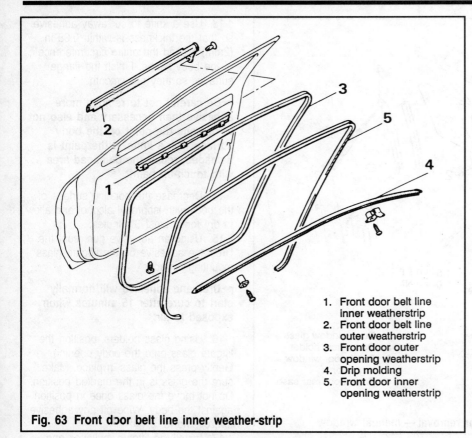

1. Front door belt line inner weatherstrip
2. Front door belt line outer weatherstrip
3. Front door outer opening weatherstrip
4. Drip molding
5. Front door inner opening weatherstrip

Fig. 63 Front door belt line inner weather-strip

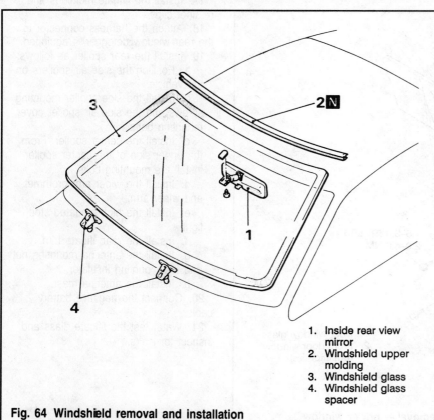

1. Inside rear view mirror
2. Windshield upper molding
3. Windshield glass
4. Windshield glass spacer

Fig. 64 Windshield removal and installation

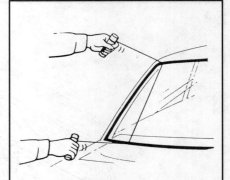

Fig. 65 Carefully cut through the adhesive by pulling wire along the outside of the windshield in a sawing motion

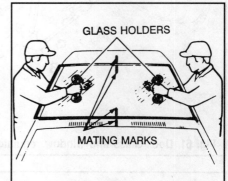

GLASS HOLDERS

MATING MARKS

Fig. 66 Make mating marks on the glass and the body and remove the glass using glass holders

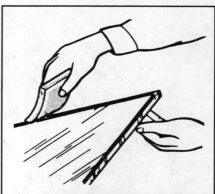

Fig. 67 Apply primer to the entire bonding surface of the glass

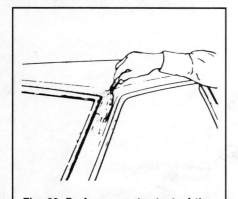

Fig. 68 Perform a water test of the windshield to assure proper sealing

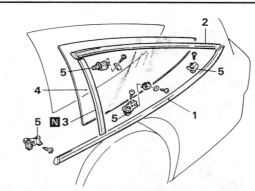

Fig. 69 Quarter window glass and related components

1. Quarter window lower molding
2. Quarter window upper molding
3. Quarter window front molding
4. Quarter window molding clip

Inside Rear View Mirror

REMOVAL & INSTALLATION

Remove the inner rear view mirror by loosening the set screw on the mirror stem and lifting mirror off of the base, which is glued onto the windshield. The installation is the reverse of the removal procedure. If the mirror base falls off of the windshield it can be installed as follows:

1. Scrape the base mounting area with a razor blade to remove the old adhesive.

2. Thoroughly clean the base mounting area with glass cleaner.

3. Obtain a mirror adhesive kit. Apply the cleaning compound to the windshield in the area that the base is to be mounted. Allow to dry.

4. Apply the adhesive to both the base plate and the windshield. Install the base to the glass. Hold in position until the adhesive has a chance to set. Make sure the correct side of the base is installed against the glass.

5. Allow around 24 hours for the adhesive to dry completely before installing the mirror to the base plate. This will assure proper adhesion.

Front Seat

REMOVAL & INSTALLATION

▶ See Figures 71, 72 and 73

1. Disconnect the negative battery cable.

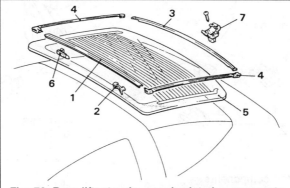

Fig. 70 Rear liftgate glass and related components

1. Liftgate upper molding
2. Liftgate molding clip
3. Lower liftgate molding
4. Side liftgate molding
5. Liftgate window glass
6. Liftgate molding fastener
7. Liftgate molding clip

2. Remove the mounting screw and the side rail anchor cover.

3. Remove the seat anchor covers.

4. Disconnect the electrical connector from the seat, if equipped.

5. Remove the seat mounting nuts and bolts.

6. Remove the seat from the vehicle.
To install:

7. Install the seat into the vehicle. Make sure the seat adjusters on both sides of the seat are locked in position.

8. Provisionally tighten the front mounting nuts first. After the front fasteners have been tightened, temporarily tighten the rear seat mounting bolts.

9. Tighten the front seat mounting nuts to 26 ft. lbs. (36 Nm) and the rear mounting bolts to 40 ft. lbs. (55 Nm).

10. Install the seat anchor covers.

11. Install the seat side rail covers. Secure the rear pawl of the slider rail anchor cover to the mounting bracket on the rear of the slider rail and position over rail.

12. Connect the electrical harness connector.

13. Connect the negative battery cable.

Rear Seat

REMOVAL & INSTALLATION

▶ See Figures 74, 75 and 76

1. Disconnect the negative battery cable.

2. To remove the rear seat cushion, pull the levers under both side seat cushions and lift upward.

3. Remove the seat cushions from the vehicle.

4. Remove the seat back cushion retainer bolts. Release the lock from the striker and remove the seat back cushions from the vehicle.
To install:

5. Install the seat back cushions into the vehicle. Align the mounting bolt holes and install bolts loosely.

6. Press the seat back cushion into the striker assembly. Tighten the mounting bolts to 12 ft. lbs. (17 Nm).

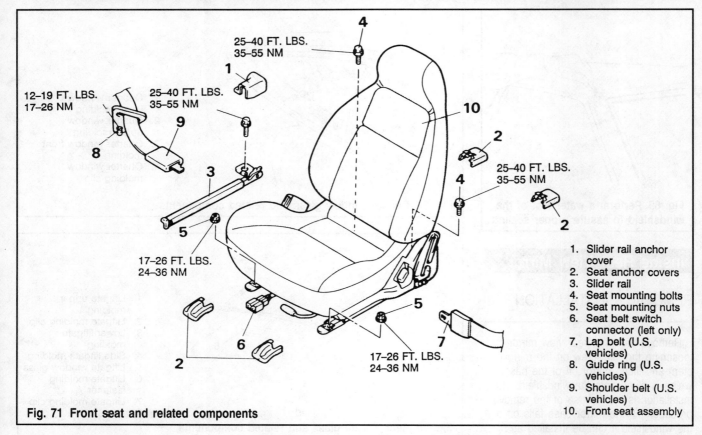

12–19 FT. LBS.
17–26 NM

25–40 FT. LBS.
35–55 NM

25–40 FT. LBS.
35–55 NM

25–40 FT. LBS.
35–55 NM

17–26 FT. LBS.
24–36 NM

17–26 FT. LBS.
24–36 NM

1. Slider rail anchor cover
2. Seat anchor covers
3. Slider rail
4. Seat mounting bolts
5. Seat mounting nuts
6. Seat belt switch connector (left only)
7. Lap belt (U.S. vehicles)
8. Guide ring (U.S. vehicles)
9. Shoulder belt (U.S. vehicles)
10. Front seat assembly

Fig. 71 Front seat and related components

7. Install the rear seat cushion as follows:

a. Securely insert the attachment wire of the rear seat cushion under the hinge bracket of the seat back.

b. Guide the seat belt buckles through the back seat cushions.

c. Securely insert the hook of the seat cushion into the seat support bracket in the floor of the vehicle. Make sure the hooks are locked in place.

8. Connect the negative battery cable.

Automatic Seat Belt System

▶ See Figure 77

REMOVAL & INSTALLATION

Lap Belt

1. Remove the retainer screws and the scuff plate from the inside rocker panel.

2. Remove the rear seat from the vehicle.

3. Remove the quarter trim mounting screws and the trim clip. Using plastic trim tool behind the trim panel, release the retainer clips and remove the quarter trim panel from the vehicle.

4. Remove the retainer and the bezel from the lap belt retractor assembly.

5. Remove the 2 mounting bolts and the lap belt retractor assembly from the vehicle.

6. Installation is the reverse of the removal procedure. Tighten the retractor assembly mounting bolts to 4 ft. lbs. (6 Nm).

Seat Belt Buckle

1. Disconnect the shoulder belt from the buckle.

2. Remove the front seat of the seat belt buckle to be removed.

3. If removing the left seat belt buckle, disconnect the seat belt switch connector.

4. Remove the seat belt buckle cover and the buckle retaining bolt. Remove the seat belt buckle.

5. Installation is the reverse of the removal procedure. Tighten the seat belt buckle retainer bolt to 40 ft. lbs. (55 Nm) torque.

Automatic Seat Belt Control Unit

1. Remove the retainer screws and the scuff plate from the inside rocker panel.

2. Remove the rear seat from the vehicle.

3. Remove the quarter trim mounting screws and the trim clip. Using plastic trim tool behind the trim panel, release the retainer clips and remove the quarter trim panel from the vehicle.

4. Disconnect the electrical connector at the control unit. Remove the retainer bolts and the seat belt control unit from the vehicle.

5. Installation is the reverse of the removal procedure.

Shoulder Belt

1. Disconnect the negative battery cable.

2. Remove the guide ring and retainer bolt.

3. Disconnect the shoulder belt.

4. Remove the floor console as follows:

a. Remove the screw plugs in the side covers. Remove the retainer screws and the side covers from the vehicle.

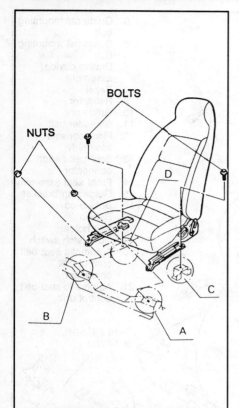

Fig. 72 Tighten front seat retainers temporarily in alphabetical sequence, then fully tighten to specified torque

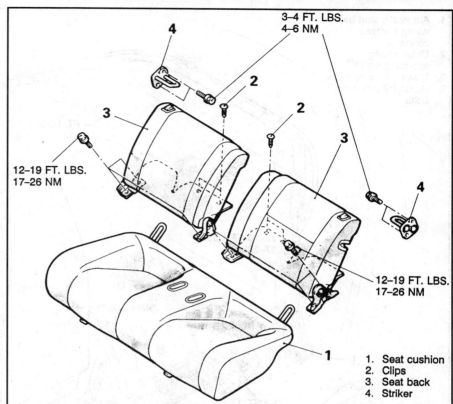

1. Seat cushion
2. Clips
3. Seat back
4. Striker

Fig. 74 Rear seat and related components. Some models may have a bench style back cushion.

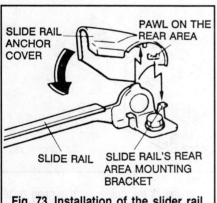

Fig. 73 Installation of the slider rail anchor covers

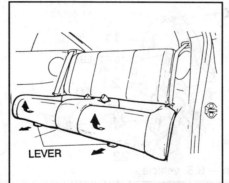

Fig. 75 Pull on the levers and raise the front of the seat cushion up to remove

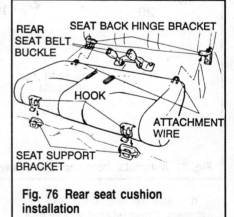

Fig. 76 Rear seat cushion installation

b. Remove the front mounting screw cover from the floor console. Remove the manual transaxle shift lever knob.

c. Remove the cup holder and the carpet inserts from the floor console assembly.

d. Label and disconnect the electrical wire harness connections for the floor console.

e. Remove the mounting bolts and the floor console from the vehicle.

5. Disconnect the outer switch connector at the retractor.

6. Remove the retainer bolts and the retractor assembly from the vehicle.

To install:

7. Install the retractor assembly into the vehicle. Install the retainer bolts and tighten to 40 ft. lbs. (55 Nm) torque.

8. Connect the outer switch connector at the retractor.

9. Install the floor console as follows:

a. Install the floor console in position in the vehicle. Position the seat shoulder belt in position. Install the console retainer bolts.

b. Reconnect the electrical harness connectors to the vehicle body harness.

c. Install the carpet and the cup holder to the console assembly.

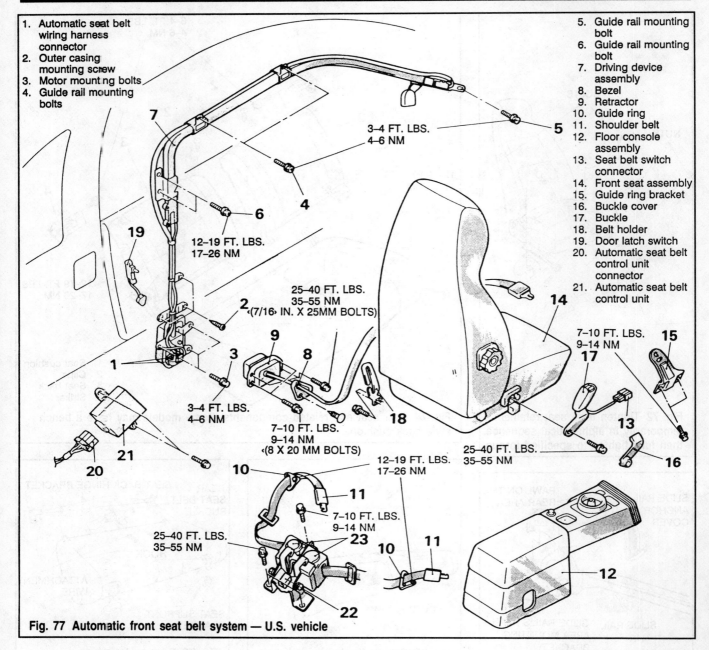

1. Automatic seat belt wiring harness connector
2. Outer casing mounting screw
3. Motor mounting bolts
4. Guide rail mounting bolts
5. Guide rail mounting bolt
6. Guide rail mounting bolt
7. Driving device assembly
8. Bezel
9. Retractor
10. Guide ring
11. Shoulder belt
12. Floor console assembly
13. Seat belt switch connector
14. Front seat assembly
15. Guide ring bracket
16. Buckle cover
17. Buckle
18. Belt holder
19. Door latch switch
20. Automatic seat belt control unit connector
21. Automatic seat belt control unit

3–4 FT. LBS.
4–6 NM

12–19 FT. LBS.
17–26 NM

25–40 FT. LBS.
35–55 NM
(7/16 IN. X 25MM BOLTS)

3–4 FT. LBS.
4–6 NM

7–10 FT. LBS.
9–14 NM
(8 X 20 MM BOLTS)

12–19 FT. LBS.
17–26 NM

7–10 FT. LBS.
9–14 NM

7–10 FT. LBS.
9–14 NM

25–40 FT. LBS.
35–55 NM

25–40 FT. LBS.
35–55 NM

7–10 FT. LBS.
9–14 NM

Fig. 77 Automatic front seat belt system — U.S. vehicle

d. Install the side covers and retainers. Cover retainer screws with plugs.

10. Install the shoulder belt guide ring and tighten the retainer to 19 ft. lbs. (26 Nm).

11. Connect the negative battery cable and check console electrical components for proper operation.

Driving Device Assembly

▶ **See Figure 77**

1. Disconnect the negative battery cable.

2. Remove the scuff plate, quarter trim, center pillar trim and front pillar trim.

3. Disconnect the automatic seat belt wiring harness connector at the base of the motor.

4. Remove the outer casing mounting screws. Remove the motor mounting bolts.

5. Remove the guide rail mounting bolts from the driving device assembly.

6. Remove the driving device from the vehicle.

To install:

7. Install the driving device into the vehicle.

8. Install the guide rail mounting bolts into the driving device assembly. Tighten mounting bolts No. 4 and No. 5 (as seen in the figure) to 4 ft. lbs. (6 Nm). Tighten mounting bolt No. 6, the mounting bolt on the vertical section of the driving device, to 12-19 ft. lbs. (17-26 Nm).

9. Install the outer casing mounting screws. Connect the harness connector to the seat belt wiring harness.

10. Install the trim removed during this operation.

11. Connect the negative battery cable and check for proper operation of the seat belt motor.

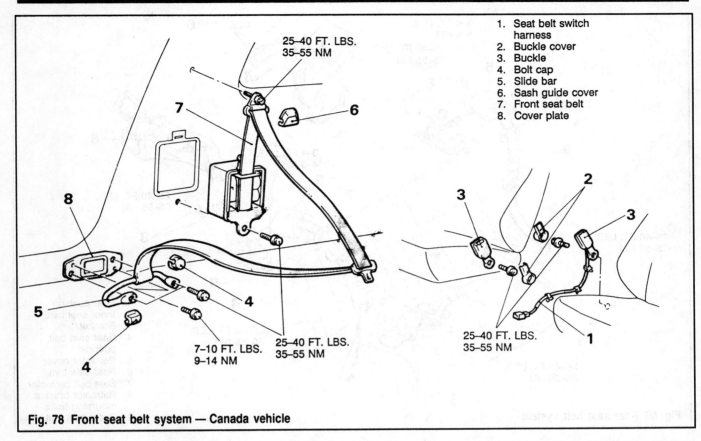

1. Seat belt switch harness
2. Buckle cover
3. Buckle
4. Bolt cap
5. Slide bar
6. Sash guide cover
7. Front seat belt
8. Cover plate

25–40 FT. LBS.
35–55 NM

7–10 FT. LBS.
9–14 NM

25–40 FT. LBS.
35–55 NM

25–40 FT. LBS.
35–55 NM

Fig. 78 Front seat belt system — Canada vehicle

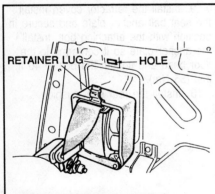

RETAINER LUG   HOLE

Fig. 79 Install front seat belt retractor retaining lug in the slot provided in the vehicle body

Manual Seat Belt System (Canada)

▶ See Figures 78 and 79

REMOVAL & INSTALLATION

Front Seat Belt Buckle

1. Remove the front seat on the side of the belt to be replaced.

2. Remove the quarter trim panel.

3. Disconnect the seat belt switch electrical harness. Remove the seat belt buckle cover.

4. Remove the buckle retainers and the buckle from the vehicle.

5. Installation is the reverse of the removal procedure. Tighten the buckle mounting bolt to 40 ft. lbs. (55 Nm).

Front Seat Belt Assembly

▶ See Figures 78 and 79

1. Remove the front seat on the side of the belt to be replaced.

2. Remove the quarter trim panel.

3. Remove the bolt cap. Remove the slide bar mounting bolts and slide bar.

4. Remove the sash guide cover. Insert a plastic trim tool from behind the lower part of the sash guide cover and pry off cover.

5. Remove the seat belt mounting bolts and the seat belt assembly from the vehicle.

To install:

6. Install the front seat belt. Fit the retractor retaining lug in the slot provided in the vehicle body. Secure the retractor with the retainer bolt tightened to 40 ft. lbs. (55 Nm).

7. Install the retainer bolt through the sash guide and into the body of the vehicle, tightening to 40 ft. lbs. (55 Nm).

8. Install the sash guide cover. Install the slide bar and secure in position. Tighten the forward most slide bar retaining bolt to 40 ft. lbs. (55 Nm) while tightening the rearward retainer bolt to 10 ft. lbs. (14 Nm).

9. Install the quarter trim panel and the front seat.

Rear Inner Seat Belt

▶ See Figure 80

1. Remove the rear seat cushion.

2. Remove the inner seat belt bracket mounting bolts and the bracket.

3. Remove the inner seat belt retainer bolts and the inner seat belt.

4. Installation is the reverse of the removal procedure. Tighten the bracket mounting bolts and the inner seat belt mounting bolts to 40 ft. lbs. (55 Nm).

Rear Outer Seat Belt

▶ See Figures 80 and 81

1. Remove the rear seat cushion.

2. Remove the seat belt anchor plate attaching bolt.

3. Remove the retractor cover.

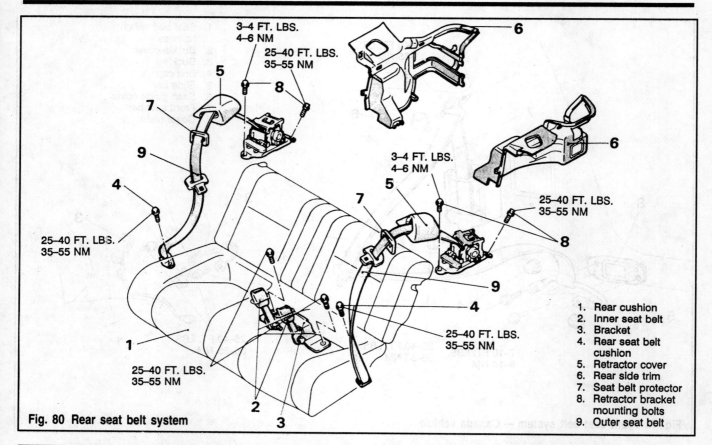

Fig. 80 Rear seat belt system

1. Rear cushion
2. Inner seat belt
3. Bracket
4. Rear seat belt cushion
5. Retractor cover
6. Rear side trim
7. Seat belt protector
8. Retractor bracket mounting bolts
9. Outer seat belt

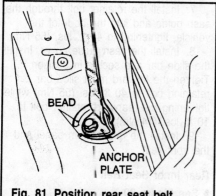

Fig. 81 Position rear seat belt anchor plate so it conforms to the floor bead

4. Remove the rear side trim panel.

5. Remove the seat belt protector.

6. Remove the seat belt retractor bracket mounting bolts and the outer seat belt assembly from the vehicle.

To install:

7. Install the outer seat belt assembly in position and install the mounting bolts. Tighten the rearward mounting bolt to 40 ft. lbs. (55 Nm) and the front bolt to 4 ft. lbs. (6 Nm).

8. Install the seat belt protector and the rear side trim.

9. Install the retractor cover. Install the seat belt anchor plate and secure in position with the attaching bolt. Install the anchor plate so it conforms to the floor bead and tighten the bolt to 40 ft. lbs. (55 Nm).

10. Install the rear seat cushion.

TORQUE SPECIFICATIONS

Component	US	Metric
Door hinge-to-door mounting bolts	19 ft. lbs.	26 Nm
Door hinge-to-body mounting bolts	19 ft. lbs.	26 Nm
Door striker bolts	10 ft. lbs.	14 Nm
Fender mounting bolts	4 ft. lbs.	6 Nm
Front bumper absorber mounting nuts	32 ft. lbs.	44 Nm
Front bumper absorber mounting bolts	12–19 ft. lbs.	17–26 Nm
Front door window regulator to door window glass bolts	3 ft. lbs.	4 Nm
Front seat mounting bolts	25–40 ft. lbs.	35–55 Nm
Front seat mounting nut	17–26 ft. lbs.	24–36 Nm
Hood latch mounting bolts/nut	5 ft. lbs.	7 Nm
Hood hinge-to-hood bolts	10 ft. lbs.	14 Nm
Hood hinge-to-body bolts	12 ft. lbs.	17 Nm
Liftgate hinge-to-liftgate bolts	10 ft. lbs.	14 Nm
Liftgate hinge-to-body bolts	10 ft. lbs.	14 Nm
Rear bumper absorber mounting nuts	25–39 ft. lbs.	35–55 Nm
Rear bumper absorber mounting bolts	12–19 ft. lbs.	17–26 Nm
Seat belt (automatic) guide ring attaching bolt	12–19 ft. lbs.	17–26 Nm
Seat belt (automatic) motor mounting bolts	4 ft. lbs.	6 Nm
Splash shield mounting screws	1.6 ft. lbs.	2.2 Nm

How to Remove Stains from Fabric Interior

For best results, spots and stains should be removed as soon as possible. Never use gasoline, lacquer thinner, acetone, nail polish remover or bleach. Use a 3' x 3" piece of cheesecloth. Squeeze most of the liquid from the fabric and wipe the stained fabric from the outside of the stain toward the center with a lifting motion. Turn the cheesecloth as soon as one side becomes soiled. When using water to remove a stain, be sure to wash the entire section after the spot has been removed to avoid water stains. Encrusted spots can be broken up with a dull knife and vacuumed before removing the stain.

Type of Stain	How to Remove It
Surface spots	Brush the spots out with a small hand brush or use a commercial preparation such as K2R to lift the stain.
Mildew	Clean around the mildew with warm suds. Rinse in cold water and soak the mildew area in a solution of 1 part table salt and 2 parts water. Wash with upholstery cleaner.
Water stains	Water stains in fabric materials can be removed with a solution made from 1 cup of table salt dissolved in 1 quart of water. Vigorously scrub the solution into the stain and rinse with clear water. Water stains in nylon or other synthetic fabrics should be removed with a commercial type spot remover.
Chewing gum, tar, crayons, shoe polish (greasy stains)	Do not use a cleaner that will soften gum or tar. Harden the deposit with an ice cube and scrape away as much as possible with a dull knife. Moisten the remainder with cleaning fluid and scrub clean.
Ice cream, candy	Most candy has a sugar base and can be removed with a cloth wrung out in warm water. Oily candy, after cleaning with warm water, should be cleaned with upholstery cleaner. Rinse with warm water and clean the remainder with cleaning fluid.
Wine, alcohol, egg, milk, soft drink (non-greasy stains)	Do not use soap. Scrub the stain with a cloth wrung out in warm water. Remove the remainder with cleaning fluid.
Grease, oil, lipstick, butter and related stains	Use a spot remover to avoid leaving a ring. Work from the outisde of the stain to the center and dry with a clean cloth when the spot is gone.
Headliners (cloth)	Mix a solution of warm water and foam upholstery cleaner to give thick suds. Use only foam—liquid may streak or spot. Clean the entire headliner in one operation using a circular motion with a natural sponge.
Headliner (vinyl)	Use a vinyl cleaner with a sponge and wipe clean with a dry cloth.
Seats and door panels	Mix 1 pint upholstery cleaner in 1 gallon of water. Do not soak the fabric around the buttons.
Leather or vinyl fabric	Use a multi-purpose cleaner full strength and a stiff brush. Let stand 2 minutes and scrub thoroughly. Wipe with a clean, soft rag.
Nylon or synthetic fabrics	For normal stains, use the same procedures you would for washing cloth upholstery. If the fabric is extremely dirty, use a multi-purpose cleaner full strength with a stiff scrub brush. Scrub thoroughly in all directions and wipe with a cotton towel or soft rag.

GLOSSARY

AIR/FUEL RATIO: The ratio of air to gasoline by weight in the fuel mixture drawn into the engine.

AIR INJECTION: One method of reducing harmful exhaust emissions by injecting air into each of the exhaust ports of an engine. The fresh air entering the hot exhaust manifold causes any remaining fuel to be burned before it can exit the tailpipe.

ALTERNATOR: A device used for converting mechanical energy into electrical energy.

AMMETER: An instrument, calibrated in amperes, used to measure the flow of an electrical current in a circuit. Ammeters are always connected in series with the circuit being tested.

AMPERE: The rate of flow of electrical current present when one volt of electrical pressure is applied against one ohm of electrical resistance.

ANALOG COMPUTER: Any microprocessor that uses similar (analogous) electrical signals to make its calculations.

ARMATURE: A laminated, soft iron core wrapped by a wire that converts electrical energy to mechanical energy as in a motor or relay. When rotated in a magnetic field, it changes mechanical energy into electrical energy as in a generator.

ATMOSPHERIC PRESSURE: The pressure on the Earth's surface caused by the weight of the air in the atmosphere. At sea level, this pressure is 14.7 psi at 32{248}F (101 kPa at 0{248}C).

ATOMIZATION: The breaking down of a liquid into a fine mist that can be suspended in air.

AXIAL PLAY: Movement parallel to a shaft or bearing bore.

BACKFIRE: The sudden combustion of gases in the intake or exhaust system that results in a loud explosion.

BACKLASH: The clearance or play between two parts, such as meshed gears.

BACKPRESSURE: Restrictions in the exhaust system that slow the exit of exhaust gases from the combustion chamber.

BAKELITE: A heat resistant, plastic insulator material commonly used in printed circuit boards and transistorized components.

BALL BEARING: A bearing made up of hardened inner and outer races between which hardened steel balls roll.

BALLAST RESISTOR: A resistor in the primary ignition circuit that lowers voltage after the engine is started to reduce wear on ignition components.

BEARING: A friction reducing, supportive device usually located between a stationary part and a moving part.

BIMETAL TEMPERATURE SENSOR: Any sensor or switch made of two dissimilar types of metal that bend when heated or cooled due to the different expansion rates of the alloys. These types of sensors usually function as an on/off switch.

BLOWBY: Combustion gases, composed of water vapor and unburned fuel, that leak past the piston rings into the crankcase during normal engine operation. These gases are removed by the PCV system to prevent the buildup of harmful acids in the crankcase.

BRAKE PAD: A brake shoe and lining assembly used with disc brakes.

BRAKE SHOE: The backing for the brake lining. The term is, however, usually applied to the assembly of the brake backing and lining.

BUSHING: A liner, usually removable, for a bearing; an anti-friction liner used in place of a bearing.

CALIPER: A hydraulically activated device in a disc brake system, which is mounted straddling the brake rotor (disc). The caliper contains at least one piston and two brake pads. Hydraulic pressure on the piston(s) forces the pads against the rotor.

CAMSHAFT: A shaft in the engine on which are the lobes (cams) which operate the valves. The camshaft is driven by the crankshaft, via a belt, chain or gears, at one half the crankshaft speed.

CAPACITOR: A device which stores an electrical charge.

CARBON MONOXIDE (CO): A colorless, odorless gas given off as a normal byproduct of combustion. It is poisonous and extremely dangerous in confined areas, building up slowly to toxic levels without warning if adequate ventilation is not available.

CARBURETOR: A device, usually mounted on the intake manifold of an engine, which mixes the air and fuel in the proper proportion to allow even combustion.

CATALYTIC CONVERTER: A device installed in the exhaust system, like a muffler, that converts harmful byproducts of combustion into carbon dioxide and water vapor by means of a heat-producing chemical reaction.

CENTRIFUGAL ADVANCE: A mechanical method of advancing the spark timing by using flyweights in the distributor that react to centrifugal force generated by the distributor shaft rotation.

CHECK VALVE: Any one-way valve installed to permit the flow of air, fuel or vacuum in one direction only.

CHOKE: A device, usually a moveable valve, placed in the intake path of a carburetor to restrict the flow of air.

CIRCUIT: Any unbroken path through which an electrical current can flow. Also used to describe fuel flow in some instances.

CIRCUIT BREAKER: A switch which protects an electrical circuit from overload by opening the circuit when the current flow exceeds a predetermined level. Some circuit breakers must be reset manually, while most reset automatically

COIL (IGNITION): A transformer in the ignition circuit which steps up the voltage provided to the spark plugs.

COMBINATION MANIFOLD: An assembly which includes both the intake and exhaust manifolds in one casting.

COMBINATION VALVE: A device used in some fuel systems that routes fuel vapors to a charcoal storage canister instead of venting them into the atmosphere. The valve relieves fuel tank pressure and allows fresh air into the tank as the fuel level drops to prevent a vapor lock situation.

COMPRESSION RATIO: The comparison of the total volume of the cylinder and combustion chamber with the piston at BDC and the piston at TDC.

CONDENSER: 1. An electrical device which acts to store an electrical charge, preventing voltage surges.
2. A radiator-like device in the air conditioning system in which refrigerant gas condenses into a liquid, giving off heat.

CONDUCTOR: Any material through which an electrical current can be transmitted easily

CONTINUITY: Continuous or complete circuit. Can be checked with an ohmmeter.

COUNTERSHAFT: An intermediate shaft which is rotated by a mainshaft and transmits, in turn, that rotation to a working part.

CRANKCASE: The lower part of an engine in which the crankshaft and related parts operate.

CRANKSHAFT: The main driving shaft of an engine which receives reciprocating motion from the pistons and converts it to rotary motion.

CYLINDER: In an engine, the round hole in the engine block in which the piston(s) ride.

CYLINDER BLOCK: The main structural member of an engine in which is found the cylinders, crankshaft and other principal parts.

CYLINDER HEAD: The detachable portion of the engine, fastened, usually, to the top of the cylinder block, containing all or most of the combustion chambers. On overhead valve engines, it contains the valves and their operating parts. On overhead cam engines, it contains the camshaft as well.

DEAD CENTER: The extreme top or bottom of the piston stroke.

DETONATION: An unwanted explosion of the air/fuel mixture in the combustion chamber caused by excess heat and compression, advanced timing, or an overly lean mixture. Also referred to as "ping".

DIAPHRAGM: A thin, flexible wall separating two cavities, such as in a vacuum advance unit.

DIESELING: A condition in which hot spots in the combustion chamber cause the engine to run on after the key is turned off.

DIFFERENTIAL: A geared assembly which allows the transmission of motion between drive axles, giving one axle the ability to turn faster than the other.

DIODE: An electrical device that will allow current to flow in one direction only.

DISC BRAKE: A hydraulic braking assembly consisting of a brake disc, or rotor, mounted on an axle, and a caliper assembly containing, usually two brake pads which are activated by hydraulic pressure. The pads are forced against the sides of the disc, creating friction which slows the vehicle.

DISTRIBUTOR: A mechanically driven device on an engine which is responsible for electrically firing the spark plug at a predetermined point of the piston stroke.

DOWEL PIN: A pin, inserted in mating holes in two different parts allowing those parts to maintain a fixed relationship.

DRUM BRAKE: A braking system which consists of two brake shoes and one or two wheel cylinders, mounted on a fixed backing plate, and a brake drum, mounted on an axle, which revolves around the assembly.

DWELL: The rate, measured in degrees of shaft rotation, at which an electrical circuit cycles on and off.

ELECTRONIC CONTROL UNIT (ECU): Ignition module, module, amplifier or igniter. See Module for definition.

ELECTRONIC IGNITION: A system in which the timing and firing of the spark plugs is controlled by an electronic control unit, usually called a module. These systems have no points or condenser.

ENDPLAY: The measured amount of axial movement in a shaft.

ENGINE: A device that converts heat into mechanical energy.

EXHAUST MANIFOLD: A set of cast passages or pipes which conduct exhaust gases from the engine.

FEELER GAUGE: A blade, usually metal, of precisely predetermined thickness, used to measure the clearance between two parts.

FIRING ORDER: The order in which combustion occurs in the cylinders of an engine. Also the order in which spark is distributed to the plugs by the distributor.

FLOODING: The presence of too much fuel in the intake manifold and combustion chamber which prevents the air/fuel mixture from firing, thereby causing a no-start situation.

FLYWHEEL: A disc shaped part bolted to the rear end of the crankshaft. Around the outer perimeter is affixed the ring gear. The starter drive engages the ring gear, turning the flywheel, which rotates the crankshaft, imparting the initial starting motion to the engine.

FOOT POUND (ft.lb. or sometimes, ft. lbs.): The amount of energy or work needed to raise an item weighing one pound, a distance of one foot.

FUSE: A protective device in a circuit which prevents circuit overload by breaking the circuit when a specific amperage is present. The device is constructed around a strip or wire of a lower amperage rating than the circuit it is designed to protect. When an amperage higher than that stamped on the fuse is present in the circuit, the strip or wire melts, opening the circuit.

GEAR RATIO: The ratio between the number of teeth on meshing gears.

GENERATOR: A device which converts mechanical energy into electrical energy.

HEAT RANGE: The measure of a spark plug's ability to dissipate heat from its firing end. The higher the heat range, the hotter the plug fires.

HUB: The center part of a wheel or gear.

HYDROCARBON (HC): Any chemical compound made up of hydrogen and carbon. A major pollutant formed by the engine as a byproduct of combustion.

HYDROMETER: An instrument used to measure the specific gravity of a solution.

INCH POUND (in.lb. or sometimes, in. lbs.): One twelfth of a foot pound.

INDUCTION: A means of transferring electrical energy in the form of a magnetic field. Principle used in the ignition coil to increase voltage.

INJECTOR: A device which receives metered fuel under relatively low pressure and is activated to inject the fuel into the engine under relatively high pressure at a predetermined time.

INPUT SHAFT: The shaft to which torque is applied, usually carrying the driving gear or gears.

INTAKE MANIFOLD: A casting of passages or pipes used to conduct air or a fuel/air mixture to the cylinders.

JOURNAL: The bearing surface within which a shaft operates.

KEY: A small block usually fitted in a notch between a shaft and a hub to prevent slippage of the two parts.

MANIFOLD: A casting of passages or set of pipes which connect the cylinders to an inlet or outlet source.

MANIFOLD VACUUM: Low pressure in an engine intake manifold formed just below the throttle plates. Manifold vacuum is highest at idle and drops under acceleration.

MASTER CYLINDER: The primary fluid pressurizing device in a hydraulic system. In automotive use, it is found in brake and hydraulic clutch systems and is pedal activated, either directly or, in a power brake system, through the power booster.

MODULE: Electronic control unit, amplifier or igniter of solid state or integrated design which controls the current flow in the ignition primary circuit based on input from the pick-up coil. When the module opens the primary circuit, the high secondary voltage is induced in the coil.

NEEDLE BEARING: A bearing which consists of a number (usually a large number) of long, thin rollers.

OHM:(Ω) The unit used to measure the resistance of conductor to electrical flow. One ohm is the amount of resistance that limits current flow to one ampere in a circuit with one volt of pressure.

OHMMETER: An instrument used for measuring the resistance, in ohms, in an electrical circuit.

OUTPUT SHAFT: The shaft which transmits torque from a device, such as a transmission.

OVERDRIVE: A gear assembly which produces more shaft revolutions than that transmitted to it.

OVERHEAD CAMSHAFT (OHC): An engine configuration in which the camshaft is mounted on top of the cylinder head and operates the valve either directly or by means of rocker arms.

OVERHEAD VALVE (OHV): An engine configuration in which all of the valves are located in the cylinder head and the camshaft is located in the cylinder block. The camshaft operates the valves via lifters and pushrods.

OXIDES OF NITROGEN (NOx): Chemical compounds of nitrogen produced as a byproduct of combustion. They combine with hydrocarbons to produce smog.

OXYGEN SENSOR: Used with the feedback system to sense the presence of oxygen in the exhaust gas and signal the computer which can reference the voltage signal to an air/fuel ratio.

PINION: The smaller of two meshing gears.

PISTON RING: An open ended ring which fits into a groove on the outer diameter of the piston. Its chief function is to form a seal between the piston and cylinder wall. Most automotive pistons have three rings: two for compression sealing; one for oil sealing.

PRELOAD: A predetermined load placed on a bearing during assembly or by adjustment.

PRIMARY CIRCUIT: Is the low voltage side of the ignition system which consists of the ignition switch, ballast resistor or resistance wire, bypass, coil, electronic control unit and pick-up coil as well as the connecting wires and harnesses.

PRESS FIT: The mating of two parts under pressure, due to the inner diameter of one being smaller than the outer diameter of the other, or vice versa; an interference fit.

RACE: The surface on the inner or outer ring of a bearing on which the balls, needles or rollers move.

REGULATOR: A device which maintains the amperage and/or voltage levels of a circuit at predetermined values.

RELAY: A switch which automatically opens and/or closes a circuit.

RESISTANCE: The opposition to the flow of current through a circuit or electrical device, and is measured in ohms. Resistance is equal to the voltage divided by the amperage.

RESISTOR: A device, usually made of wire, which offers a preset amount of resistance in an electrical circuit.

RING GEAR: The name given to a ring-shaped gear attached to a differential case, or affixed to a flywheel or as part a planetary gear set.

ROLLER BEARING: A bearing made up of hardened inner and outer races between which hardened steel rollers move.

ROTOR: 1. The disc-shaped part of a disc brake assembly, upon which the brake pads bear; also called, brake disc. 2. The device mounted atop the distributor shaft, which passes current to the distributor cap tower contacts.

SECONDARY CIRCUIT: The high voltage side of the ignition system, usually above 20,000 volts. The secondary includes the ignition coil, coil wire, distributor cap and rotor, spark plug wires and spark plugs.

SENDING UNIT: A mechanical, electrical, hydraulic or electromagnetic device which transmits information to a gauge.

SENSOR: Any device designed to measure engine operating conditions or ambient pressures and temperatures. Usually electronic in nature and designed to send a voltage signal to an on-board computer, some sensors may operate as a simple on/off switch or they may provide a variable voltage signal (like a potentiometer) as conditions or measured parameters change.

SHIM: Spacers of precise, predetermined thickness used between parts to establish a proper working relationship.

SLAVE CYLINDER: In automotive use, a device in the hydraulic clutch system which is activated by hydraulic force, disengaging the clutch.

SOLENOID: A coil used to produce a magnetic field, the effect of which is produce work.

SPARK PLUG: A device screwed into the combustion chamber of a spark ignition engine. The basic construction is a conductive core inside of a ceramic insulator, mounted in an outer conductive base. An electrical charge from the spark plug wire travels along the conductive core and jumps a preset air gap to a grounding point or points at the end of the conductive base. The resultant spark ignites the fuel/air mixture in the combustion chamber.

SPLINES: Ridges machined or cast onto the outer diameter of a shaft or inner diameter of a bore to enable parts to mate without rotation.

TACHOMETER: A device used to measure the rotary speed of an engine, shaft, gear, etc., usually in rotations per minute.

THERMOSTAT: A valve, located in the cooling system of an engine, which is closed when cold and opens gradually in response to engine heating, controlling the temperature of the coolant and rate of coolant flow.

TOP DEAD CENTER (TDC): The point at which the piston reaches the top of its travel on the compression stroke.

TORQUE: The twisting force applied to an object.

TORQUE CONVERTER: A turbine used to transmit power from a driving member to a driven member via hydraulic action, providing changes in drive ratio and torque. In automotive use, it links the driveplate at the rear of the engine to the automatic transmission.

TRANSDUCER: A device used to change a force into an electrical signal.

TRANSISTOR: A semi-conductor component which can be actuated by a small voltage to perform an electrical switching function.

TUNE-UP: A regular maintenance function, usually associated with the replacement and adjustment of parts and components in the electrical and fuel systems of a vehicle for the purpose of attaining optimum performance.

TURBOCHARGER: An exhaust driven pump which compresses intake air and forces it into the combustion chambers at higher than atmospheric pressures. The increased air pressure allows more fuel to be burned and results in increased horsepower being produced.

VACUUM ADVANCE: A device which advances the ignition timing in response to increased engine vacuum.

VACUUM GAUGE: An instrument used to measure the presence of vacuum in a chamber.

VALVE: A device which control the pressure, direction of flow or rate of flow of a liquid or gas.

VALVE CLEARANCE: The measured gap between the end of the valve stem and the rocker arm, cam lobe or follower that activates the valve.

VISCOSITY: The rating of a liquid's internal resistance to flow.

VOLTMETER: An instrument used for measuring electrical force in units called volts. Voltmeters are always connected parallel with the circuit being tested.

WHEEL CYLINDER: Found in the automotive drum brake assembly, it is a device, actuated by hydraulic pressure, which, through internal pistons, pushes the brake shoes outward against the drums.

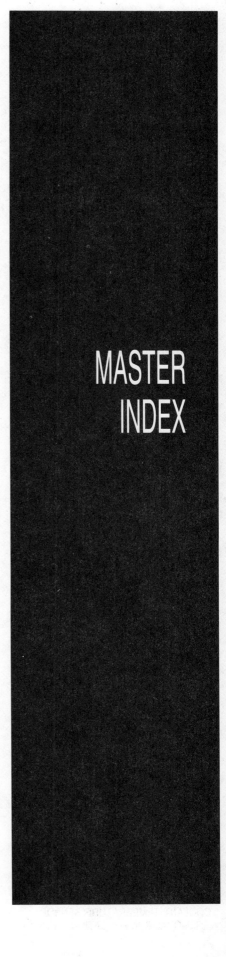

MASTER
INDEX